Progress in Nonlinear Differential Equations and Their Applications
Volume 19

Editor

Haim Brezis
Université Pierre et Marie Curie
Paris
and
Rutgers University
New Brunswick, N.J.

Nonlinear Dynamical Systems and Chaos

H. W. Broer
S. A. van Gils
I. Hoveijn
F. Takens
Editors

1996   Springer Basel AG

H. W. Broer, I. Hoveijn and F. Takens
Institute of Mathematics and Computer Science
University of Groningen
P.O. Box 800
9700 AV Groningen
The Netherlands

S. A. van Gils
Faculty of Mathematics and Computer Science
University of Twente
P.O. Box 217
7500 AE Enschede
The Netherlands

A CIP catalogue record for this book is available from the Library of Congress. Washington D.C., USA

Deutsche Bibliothek Cataloging-in-Publication Data
Nonlinear dynamical systems and chaos / H. W. Broer ... ed.
– Basel ; Boston ; Berlin : Birkhäuser, 1996
  (Progress in nonlinear differential equations and their applications ; Vol. 19)
  ISBN 978-3-0348-7520-2   ISBN 978-3-0348-7518-9 (eBook)
  DOI 10.1007/978-3-0348-7518-9
NE: Broer, H. W. [Hrsg.]; GT

© 1996 Springer Basel AG
Originally published by Birkhäuser Verlag Basel, Switzerland in 1996

Printed on acid-free paper produced from chlorine-free pulp. TCF ∞
9 8 7 6 5 4 3 2 1

Contents

Infinite dimensional systems

Time series analysis

Numerical continuation and bifurcation analysis

Preface

This volume contains the proceedings of the 'Dynamical Systems Conference' held at the University of Groningen in December 1995. in honour of Johann Bernoulli who was appointed at the university in the year 1695. This conference is also part of a rather young tradition of similar conferences in 1984 and 1989.

In the field of 'Nonlinear Dynamical Systems'. there is quite a rich variety of approaches. In this conference the main topics are:

- Symmetries in dynamical systems
- KAM theory and other perturbation theories
- Infinite dimensional systems
- Time series analysis
- Numerical continuation and bifurcation analysis

We would like to thank all contributors and referees for their cooperation. Furthermore we thank all attendants and lecturers of the conference. In particular, we thank those institutions which made this conference possible by their financial support: the Netherlands Organization for Scientific Research (NWO), the Royal Dutch Academy of Sciences (KNAW), the Mathematical Research Institute ($\mathcal{MRI}$, a joint Research School of the Universities of Groningen, Nijmegen, Twente and Utrecht) and the Institute of Mathematics and Computer Science of the University of Groningen.

Groningen, September 1995.

The editors:

H.W. Broer, I. Hoveijn, F. Takens

Institute of Mathematics and Computer Science.
University of Groningen
P.O. Box 800, 9700 AV Groningen
The Netherlands

S.A. van Gils

Faculty of Mathematics and Computer Science.
University of Twente
P.O. Box 217, 7500 AE Enschede
The Netherlands

Progress in Nonlinear Differential Equations
and Their Applications, Vol. 19
© 1996 Birkhäuser Verlag Basel/Switzerland

Symplecticity, reversibility and elliptic operators

Thomas J. Bridges *

Abstract

The concepts of symplecticity and reversibility are generalised and then used
as a framework for analyzing some aspects of gradient elliptic operators.
In this paper we focus on semi-linear elliptic equations of the form $\Delta\psi +
V'(|\dot\psi|^2)\psi = 0$ where ψ is a scalar-valued function, Δ is the Laplacian and
V is some smooth function.

1 Introduction

There has been recent and active interest in the literature on the interplay between
the theory of elliptic operators and symplecticity from various viewpoints (cf. Floer
[1988], Mielke [1991], McDuff [1990], Salamon [1990] and references therein).

The work of Floer and the related work in symplectic geometry (cf. the review
articles of McDuff [1990] and Salamon [1990]) use the theory of elliptic operators to
study the symplectic geometry of *finite-dimensional* time-dependent Hamiltonian
systems on the loop space. Ellipticity comes in when considering the gradient flow
associated with the finite-dimensional system,

$$\frac{\partial u}{\partial s} + \mathbf{J}\frac{\partial u}{\partial t} - \nabla H(u,t) = 0 \qquad (1.1)$$

where $u : \mathbb{R} \times \mathbb{S}^1 \to \mathbb{R}^{2n}$ and $\mathbf{J}$ is a symplectic operator (cf. Salamon [1990, equation
(4.4)]). Equation (1.1) is an elliptic partial differential equation and when $\nabla H \equiv 0$
it is the Cauchy-Riemann operator. Methods of elliptic operator theory are applied
to (1.1) in order to construct a relative Morse index on the space of bounded
solutions of (1.1).

The work of Mielke [1991] takes the opposite point of view: methods of
infinite-dimensional symplectic geometry are used as a framework for studying
gradient elliptic operators on an infinite cylinder with a preferred direction and a
symplectic structure is introduced for the preferred direction. The resulting evolu-

---
*Department of Mathematical and Computing Sciences, University of Surrey, Guildford, Surrey GU2 5XH, UK

tion equation is ill-posed and bounded solutions are obtained by symplectic centre manifold reduction.

In this paper we consider gradient elliptic operators on $\mathbb{R}^n$ (particularly $\mathbb{R}^2$). The main idea is that an independent symplectic structure and a reversible structure – with particular attention to the interaction between reversibility and symplecticity – may be introduced for each unbounded direction – a multi-symplectic structure (cf. Bridges [1993,1994,1995]). Moreover, even though the basic system is infinite dimensional – a partial differential equation – each of the reversible and symplectic structures are associated with a finite dimensional space; in this way much of the classic theory of differential forms and finite-dimensional symplectic geometry can be appealed to in the analysis. Such a structure is useful for analysing gradient elliptic operators without a preferred direction such as pattern formation systems.

For example consider the semilinear elliptic equation

$$\frac{\partial^2 u}{\partial x^2} + \frac{\partial^2 u}{\partial y^2} + V'(u) = 0 \qquad (x,y) \in \mathbb{R}^2 \tag{1.2}$$

for the scalar-valued function u where $V : \mathbb{R} \to \mathbb{R}$ is some smooth function. Letting $v = u_x$ an $w = u_y$ the equation is easily reformulated to

$$\begin{bmatrix} 0 & -1 & 0 \\ 1 & 0 & 0 \\ 0 & 0 & 0 \end{bmatrix} \begin{pmatrix} u \\ v \\ w \end{pmatrix}_x + \begin{bmatrix} 0 & 0 & -1 \\ 0 & 0 & 0 \\ 1 & 0 & 0 \end{bmatrix} \begin{pmatrix} u \\ v \\ w \end{pmatrix}_y = \begin{pmatrix} V'(u) \\ v \\ w \end{pmatrix}$$

or

$$\mathbf{J}_1 Z_x + \mathbf{J}_2 Z_y = \nabla S(Z) \qquad Z \in \mathcal{M} = \mathbb{R}^3 \tag{1.3}$$

where the definitions of Z, $\mathbf{J}_1$ and $\mathbf{J}_2$ are clear and

$$S(Z) = \tfrac{1}{2}(v^2 + w^2) + V(u) \tag{1.4}$$

where the gradient of S in (1.3) is with respect to the standard inner product on $\mathbb{R}^3$.

When $Z_x = 0$, $\mathbf{J}_2 Z_y = \nabla S(Z)$ is a classical Hamiltonian system on $\mathrm{Ker}(\mathbf{J}_2)^\perp$ and when $Z_y = 0$, $\mathbf{J}_1 Z_x = \nabla S(Z)$ is a classical Hamiltonian system on $\mathrm{Ker}(\mathbf{J}_1)^\perp$. In general we refer to the complete system (1.3) as a Hamiltonian system on a multi-symplectic structure: $(\mathcal{M}, \omega^{(1)}, \omega^{(2)})$ where

$$\omega^{(j)}(U_1, U_2) = \langle \mathbf{J}_j U_1, U_2 \rangle \qquad j = 1,2 \tag{1.5}$$

for any $U_1, U_2 \in \mathbb{R}^3$ where $\langle \cdot, \cdot \rangle$ is the standard inner product on $\mathbb{R}^3$. Note that both $\mathbf{J}_1$ and $\mathbf{J}_2$ are degenerate and so, in principle, correspond to pre-symplectic structures or indeed to a contact structure – since $\mathcal{M}$ in the above case is odd dimensional. However it is the concepts and results of symplectic geometry which lead to interesting results when generalised to the multi-component setting and therefore we refer to the structure $(\mathcal{M}, \omega^{(1)}, \omega^{(2)})$ as a multi-*symplectic* structure although the above qualifications should be kept in mind.

Note that there are other ways of formulating equation (1.2) (and other such equations) using the theory of differential forms. For example the system (1.2) can be formulated as an exterior differential system in the sense of Cartan (cf. Edelen [1985, Chapter 6]) and there is the Hodge theory for elliptic operators on manifolds (cf. Warner [1983]). However the the form (1.3) has additional structure and this is the main advantage of the formulation. The structures in the x an y directions are separated and identified precisely with particular closed exact two-forms and there is a special functional, here denoted $S(Z)$, that acts as an organising centre much like (but different from in essential ways – for example $S(Z)$ is *not* an invariant) a Hamiltonian functional in classical Hamiltonian mechanics. Moreover the "phase space" $\mathcal{M}$ is finite-dimensional and therefore many of the familiar operations from finite-dimensional Hamiltonian mechanics carry over.

The paper has six sections and each section treats a particular consequence of formulating a gradient elliptic operator as a Hamiltonian system on a multi-symplectic structure with a finite-dimensional phase space. Many of the familiar results from finite-dimensional Hamiltonian mechanics generalise to elliptic operators – without requiring infinite-dimensional techniques.

In §2 the formulation of gradient elliptic operators as a Hamiltonian system is introduced with particular attention to two examples along with a multi-symplectic formulation of symmetry and the "momentum map".

In §3 the concept of reversibility is generalised to include more than one dimension in which case reversors can be both symplectic and anti-symplectic depending on the direction of reversibility.

Restriction of an elliptic operator to the loop space – of the finite-dimensional phase space – leads to a geometric formulation of diagonal periodic patterns. In §4 such patterns are formulated as generalised relative equilibria and it is shown that families of such patterns have a naturally defined index. In §5 a model for pattern formation, the cubic Ginzburg-Landau equation on $\mathbb{R}^2$, is considered and it is shown that the loop space index, evaluated along a branch of diagonal periodic patterns changes by an integer precisely at the Eckhaus instability threshold.

Symplecticity and reversibility constrict the way eigenvalues can appear for linearised equations and this property is generalised to the multi-symplectic, multi-reversible setting in §6.

Radially symmetric states of elliptic equations, in the present framework, appear as bounded surfaces of revolution in the finite-dimensional phase space. Some geometrical properties of such states are considered in §7.

2   Multi-symplectic structure and elliptic operators

Although sufficient conditions for a general formulation can be given for gradient elliptic operators starting with a multiple integral formulation and the multi-dimensional Legendre transform we will restrict attention here to giving complete details for two simple examples. Generalisation to other semilinear and quasilinear elliptic operators on $\mathbb{R}^n$ will be clear in principle but will require substantially more technicalities.

First consider the semilinear elliptic equation (1.2). With coordinates $Z = (u, v, w) \in \mathcal{M} = \mathbb{R}^3$ this system was formulated in the introduction as a Hamiltonian system on a bi-symplectic structure $(\mathcal{M}, \omega^{(1)}, \omega^{(2)}; S)$ with Hamiltonian functional $S(Z)$ (cf. equations (1.3)-(1.4)). Here, several further observations will be presented about this structure.

The Hamiltonian functional is *not* an invariant as in the classical case. Indeed,

$$S_x = \langle \nabla S(Z), Z_x \rangle = \langle \mathbf{J}_1 Z_x + \mathbf{J}_2 Z_y, Z_x \rangle = \langle \mathbf{J}_2 Z_y, Z_x \rangle$$

and so

$$S_x = \omega^{(2)}(Z_y, Z_x) \qquad S_y = \omega^{(1)}(Z_x, Z_y)$$

which do not in general vanish. Note however that $\partial_x S_y = \partial_y S_x$ generates a conservation law

$$\frac{\partial}{\partial x}(\omega^{(1)}(Z_x, Z_y)) + \frac{\partial}{\partial y}(\omega^{(2)}(Z_x, Z_y)) = 0$$

which may be interpreted as a generalisation of the absolute invariance, under the flow, of the symplectic form (in the one dimensional case), to the multi-symplectic framework where a conservation law – conservation of symplecticity – appears.

The two operators $\mathbf{J}_1$ and $\mathbf{J}_2$ are a two-dimensional subspace (but not a Lie subalgebra) of the Lie algebra $so(3)$. It is interesting that the missing element of the standard basis for $so(3)$ appears as a symmetry. Let

$$\mathbf{J}_3 = \begin{pmatrix} 0 & 0 & 0 \\ 0 & 0 & -1 \\ 0 & 1 & 0 \end{pmatrix} \qquad \text{and} \quad \mathcal{G}_\theta = \exp(\mathbf{J}_3 \theta). \qquad (2.1)$$

Then $\{\mathbf{J}_1, \mathbf{J}_2, \mathbf{J}_3\}$ span the Lie algebra $so(3)$ and $\mathcal{G}_\theta$ is a rotation, an action for the group $SO(2)$ on $\mathcal{M}$. The fact that (1.2) is invariant under rotation in the (x, y)-plane appears in the formulation (1.3) in the following way. Let $\mathrm{Ad}_G \mathbf{J}$ denote the adjoint action of the Lie group $SO(3)$ on its Lie algebra $so(3)$: $\mathrm{Ad}_G \mathbf{J} = G \mathbf{J} G^{-1}$ for $G \in SO(3)$ and $\mathbf{J} \in so(3)$. Then for the action of $\mathcal{G}_\theta$ on $(\mathcal{M}, \omega^{(1)}, \omega^{(2)})$

$$\begin{array}{rcl} \mathrm{Ad}_{\mathcal{G}_\theta} \mathbf{J}_1 & = & \cos\theta\, \mathbf{J}_1 + \sin\theta\, \mathbf{J}_2 \\ \mathrm{Ad}_{\mathcal{G}_\theta} \mathbf{J}_2 & = & -\sin\theta\, \mathbf{J}_1 + \cos\theta\, \mathbf{J}_2 \,. \end{array} \qquad (2.2)$$

Define an action for $SO(2)$ as follows

$$\theta \cdot Z(x, y) = \mathcal{G}_\theta Z(x \cos \theta + y \sin \theta, -x \sin \theta + y \cos \theta).$$

Then acting on (1.3),

$$\mathcal{G}_\theta \mathbf{J}_1 Z_x + \mathcal{G}_\theta \mathbf{J}_2 Z_y = \mathcal{G}_\theta \nabla S(Z).$$

Noting that $S(\mathcal{G}_\theta Z) = S(Z)$ it follows that $\mathcal{G}_\theta \nabla S(Z) = \nabla S(\mathcal{G}_\theta Z)$ and so

$$(\mathrm{Ad}_{\mathcal{G}_\theta} \mathbf{J}_1) \, \mathcal{G}_\theta Z_x + (\mathrm{Ad}_{\mathcal{G}_\theta} \mathbf{J}_2) \, \mathcal{G}_\theta Z_y = \mathcal{G}_\theta \nabla S(Z) = \nabla S(\mathcal{G}_\theta Z)$$

or, using (2.2)-(2.3),

$$\mathbf{J}_1 (\theta \cdot Z)_x + \mathbf{J}_2 (\theta \cdot Z)_y = \nabla S(\theta \cdot Z);$$

that is, $\theta \cdot Z$ is a solution of (1.3) whenever Z is a solution.

The second example to be considered is

$$\frac{\partial^2 \Psi}{\partial x^2} + \frac{\partial^2 \Psi}{\partial y^2} + V'(|\Psi|^2)\Psi = 0 \qquad (x, y) \in \mathbb{R}^2 \qquad (2.4)$$

where $\Psi : \mathbb{R}^2 \to \mathbb{C}$ and $V : \mathbb{R} \to \mathbb{R}$ is again some, in general arbitrary, smooth function. The interesting feature of (2.4) is the presence of additional symmetry: $e^{i\theta} \Psi$ is a solution of (2.4) for any $\theta \in SO(2)$ when Ψ is a solution.

In order to reformulate (2.4) let

$$Z = \begin{pmatrix} u \\ v \end{pmatrix} \in \mathcal{M} = \mathbb{R}^6 \quad \text{with} \quad u + iv \overset{\mathrm{def}}{=} \begin{pmatrix} \Psi \\ \Psi_x \\ \Psi_y \end{pmatrix}. \qquad (2.5)$$

Then system (2.4) has the Hamiltonian formulation

$$\mathbf{K}_1 Z_x + \mathbf{K}_2 Z_y = \nabla S(Z) \qquad Z \in \mathcal{M} = \mathbb{R}^6 \qquad (2.6)$$

where

$$\mathbf{K}_1 = \mathbf{J}_1 \oplus \mathbf{J}_1, \quad \mathbf{K}_2 = \mathbf{J}_2 \ominus \mathbf{J}_2. \qquad (2.7)$$

The operators $\mathbf{J}_1$ and $\mathbf{J}_2$ are the skew-symmetric operators acting on $\mathbb{R}^3$ as defined in the introduction and

$$S(Z) = \tfrac{1}{2}\langle Z, Z \rangle + \tfrac{1}{2}V(u_1^2 + v_1^2) - \tfrac{1}{2}(u_1^2 + v_1^2) \qquad (2.8)$$

where $\langle \cdot, \cdot \rangle$ is the standard inner product on $\mathbb{R}^6$. Note that the gradient of S in (2.6) is with respect to the inner product on $\mathbb{R}^6$. The two symplectic forms in this case are

$$\omega^{(j)}(U_1, U_2) = \langle \mathbf{K}_j U_1, U_2 \rangle \quad j = 1, 2 \qquad \text{for any } U_1, U_2 \in \mathbb{R}^6. \qquad (2.9)$$

In terms of the multi-symplectic structure, the $SO(2)$-symmetry and generated conservation law for (2.4) can be given an interesting geometrical formulation that generalises the framework used in the finite-dimensional setting. First we introduce an action for the $SO(2)$ symmetry and its generator on $\mathcal{M}$. The action of $SO(2)$ on Ψ is $e^{i\theta}\Psi$. With the definition of Z in (2.5) an action on $\mathcal{M}$ is

$$\theta \cdot Z \stackrel{\text{def}}{=} \mathcal{R}_\theta Z \quad \text{with} \quad \mathcal{R}_\theta = \mathbf{r}_\theta \otimes \mathbf{I}_3 \quad \text{where} \quad \mathbf{r}_\theta = \begin{pmatrix} \cos\theta & -\sin\theta \\ \sin\theta & \cos\theta \end{pmatrix}$$

and $\mathbf{I}_3$ is the identity on $\mathbb{R}^3$. The infinitesimal generator is clearly

$$\xi \stackrel{\text{def}}{=} \frac{d}{d\theta}\mathcal{R}_\theta Z \Big|_{\theta=0} = \begin{pmatrix} \mathbf{0} & -\mathbf{I}_3 \\ \mathbf{I}_3 & \mathbf{0} \end{pmatrix} Z \tag{2.10}$$

and therefore

$$\mathcal{R}_\theta = \exp\left[\theta \begin{pmatrix} 0 & -1 \\ 1 & 0 \end{pmatrix} \otimes \mathbf{I}_3\right] = \mathbf{r}_\theta \otimes \mathbf{I}_3. \tag{2.11}$$

We now generalise the usual symplectic theory of symmetry and conservation laws in finite dimensions to the present case. The finite-dimensional theory is quite adequate here even though we are considering partial differential equations since the "phase" space $\mathcal{M}$ is finite dimensional and therefore the action of the group is on a finite dimensional space.

It is easily verified that $\mathcal{R}_\theta$ (in (2.11)) commutes with both $\mathbf{K}_1$ and $\mathbf{K}_2$ (noting that $\mathbf{K}_1$ and $\mathbf{K}_2$ have a tensor product representation: $\mathbf{K}_j = \mathbf{I}_2 \otimes \mathbf{J}_j$ for $j = 1, 2$ where $\mathbf{I}_2$ is the indentity on $\mathbb{R}^2$). Therefore we say that $\mathcal{R}_\theta$ in (2.11) is a multi-symplectic action of the Lie group $SO(2)$. We now have the following generalisation of the usual definition of a momentum mapping in the finite dimensional case (cf. Abraham & Marsden [1978, p. 276]).

Definition 2.1. *Let $(\mathcal{M}, \omega^{(1)}, \omega^{(2)})$ be a manifold on which there is a pair of closed two forms and suppose $\mathcal{R}_\theta$ is a one parameter Lie group with bi-symplectic action on $\mathcal{M}$ and generator ξ. We say that (P_1, P_2) is a basis for a conservation law if there exists functionals $P_j : \mathcal{M} \to \mathbb{R}$, $j = 1, 2$ such that*

$$\mathbf{i}_\xi \omega^{(j)} = \mathbf{d}P_j \qquad j = 1, 2. \tag{2.12}$$

Remark: note that the generalisation to $(\mathcal{M}, \omega^{(1)}, \dots, \omega^{(n)})$ is straightforward.

The above definition follows precisely the definition in the finite-dimensional case. The PDE properties of the elliptic equation appear here only by the presence of an additional two-form.

Bi-symplecticity of the action of $\mathcal{R}_\theta$ is equivalent to

$$\mathcal{L}_\xi \omega^{(j)} = 0 \qquad j = 1, 2$$

where $\mathcal{L}$ is the Lie derivative (Abraham & Marsden [1978, Chapter 2]) which satisfies (restricting attention to the present case)

$$\mathcal{L}_X \omega = \mathbf{d} \, \mathbf{i}_X \omega + \mathbf{i}_X \mathbf{d} \omega \qquad (2.13)$$

where ω is a two form and X a vectorfield and so. since $\omega^{(1)}$ and $\omega^{(2)}$ are closed,

$$0 = \mathcal{L}_\xi \omega^{(j)} = \mathbf{d} \, \mathbf{i}_\xi \omega^{(j)} + \mathbf{i}_\xi \mathbf{d} \omega^{(j)} = \mathbf{d} \, \mathbf{i}_\xi \omega^{(j)} \, .$$

Therefore, by the Poincaré Lemma it follows that $\mathbf{i}_\xi \omega^{(j)}$ for $j = 1, 2$ are exact one forms which is the basis for the form (2.12). For the present example (equations (2.4)-(2.11)) this result is trivial since $\mathcal{M} = \mathbb{R}^6$ and the two forms are exact and in fact more precise results are possible and are stated in the following Theorem along with the connection between (P_1, P_2) and conservation laws.

Theorem 2.2. *Let $\mathcal{R}_\theta$ be a bi-symplectic action of a one-parameter Lie group acting on $\mathcal{M}$ generated by ξ; that is, $\mathcal{L}_\xi \omega^{(j)} = 0$ for $j = 1. 2$. Suppose the two forms on $\mathcal{M}$ are exact: $\omega^{(j)} = \mathbf{d} \alpha^{(j)}$ for $j = 1, 2$ and that the one forms $\alpha^{(1)}$ and $\alpha^{(2)}$ are also $\mathcal{R}_\theta$ invariant*

$$\mathcal{L}_\xi \alpha^{(j)} = 0 \qquad for \quad j = 1. 2 \, . \qquad (2.14)$$

Let $Z(x, y)$ be a solution of the equations

$$\mathbf{i}_{Z_x} \omega^{(1)} + \mathbf{i}_{Z_y} \omega^{(2)} = \mathbf{d} S(Z) \, . \qquad (2.15)$$

There there exists a basis for a conservation law $P : \mathcal{M} \to \mathbb{R}^2$ defined by

$$P = (P_1, P_2) \quad with \quad P_j = -\mathbf{i}_\xi \alpha^{(j)} \qquad j = 1, 2 \qquad (2.16)$$

satisfying

$$\frac{\partial P_1}{\partial x} + \frac{\partial P_2}{\partial y} = -\mathcal{L}_\xi S \, . \qquad (2.17)$$

In particular. if $S(Z)$ is also $\mathcal{R}_\theta$ invariant then $(P_1. P_2)$ satisfy a conservation law.

Proof. The existence of the function $P : \mathcal{M} \to \mathbb{R}^2$ is proved using the $\mathcal{R}_\theta$ invariance of the one forms $\alpha^{(1)}$ and $\alpha^{(2)}$ as follows. Using the identity (2.13) and the hypothesis (2.14)

$$0 = \mathcal{L}_\xi \alpha^{(j)} = \mathbf{d} \, \mathbf{i}_\xi \alpha^{(j)} + \mathbf{i}_\xi \mathbf{d} \alpha^{(j)} \quad j = 1, 2$$

and so

$$\mathbf{i}_\xi \omega^{(j)} = \mathbf{i}_\xi \mathbf{d} \alpha^{(j)} = -\mathbf{d}(\mathbf{i}_\xi \alpha^{(j)}) = \mathbf{d} P_j \quad j = 1, 2$$

for some 0-forms (i.e. functions) $P_j : \mathcal{M} \to \mathbb{R}. \ j = 1. 2$. Therefore according to Definition 2.1, (P_1, P_2) form a basis for a conservation law. The connection with a conservation law is established as follows

$$
\begin{aligned}
\frac{\partial P_1}{\partial x} + \frac{\partial P_2}{\partial y} &= \mathbf{i}_{Z_x}\mathbf{d}P_1 + \mathbf{i}_{Z_y}\mathbf{d}P_2 && \text{(by definition)} \\
&= \mathbf{i}_{Z_x}\mathbf{i}_\xi\omega^{(1)} + \mathbf{i}_{Z_y}\mathbf{i}_\xi\omega^{(2)} && \text{(using (2.12) and (2.16))} \\
&= -\mathbf{i}_\xi\mathbf{i}_{Z_x}\omega^{(1)} - \mathbf{i}_\xi\mathbf{i}_{Z_y}\omega^{(2)} && \text{(skew-symmetry of } \omega^{(1)} \text{ and } \omega^{(2)}) \\
&= -\mathbf{i}_\xi(\mathbf{d}S(Z)) && \text{(using (2.15))} \\
&= -\mathcal{L}_\xi S && \text{(using (2.13))}
\end{aligned}
$$

proving (2.17). If S is in addition $\mathcal{R}_\theta$ invariant then $\mathcal{L}_\xi S = 0$ completing the proof. ∎

To apply Theorem 2.2 to the problem (2.4) note that $\omega^{(j)} = \mathbf{d}\alpha^{(j)}$ for $j = 1, 2$ with

$$
\alpha^{(j)} = \langle \alpha_j(Z), \mathbf{d}Z \rangle \quad \text{and} \quad \alpha_1(Z) = \begin{pmatrix} u_2 \\ 0 \\ 0 \\ v_2 \\ 0 \\ 0 \end{pmatrix}, \quad \alpha_2(Z) = \begin{pmatrix} u_3 \\ 0 \\ 0 \\ v_3 \\ 0 \\ 0 \end{pmatrix}.
$$

It is easily verified that $\mathcal{R}_\theta^*\alpha^{(j)} = \alpha^{(j)}$ for $j = 1, 2$ and therefore

$$
\begin{aligned}
P_1 &= -\mathbf{i}_\xi\alpha^{(1)} = v_1 u_2 - u_1 v_2 = \langle \mathbf{J}_1 v, u \rangle_{\mathbb{R}^3} \\
P_2 &= -\mathbf{i}_\xi\alpha^{(2)} = v_1 u_3 - u_1 v_3 = \langle \mathbf{J}_2 v, u \rangle_{\mathbb{R}^3}
\end{aligned} \tag{2.18}
$$

where $\mathbf{J}_1$ and $\mathbf{J}_2$ are the skew-symmetric operators on $\mathbb{R}^3$ defined in the introduction. According to Definition 2.1, (P_1, P_2) is a basis for a conservation law.

To prove that $\partial_x P_1 + \partial_y P_2 = 0$ it remains to verify that $S(Z)$, defined in (2.8), is $\mathcal{R}_\theta$ invariant but

$$
S(\mathcal{R}_\theta Z) = \tfrac{1}{2}\langle \mathcal{R}_\theta Z, \mathcal{R}_\theta Z \rangle + \tfrac{1}{2}V(u_1^2 + v_1^2) - \tfrac{1}{2}(u_1^2 + v_1^2) = S(Z)
$$

and the result follows.

3   On generalising reversibility

For finite-dimensional Hamiltonian systems there is an interesting interplay between reversibility and symplecticity. In finite dimensions, although reversors can be either symplectic or anti-symplectic, the system is rather degenerate when the reversor is not anti-symplectic (cf. Hoveijn [1995]) and therefore there is good reason to restrict attention to anti-symplectic involutions. In this section the effect of more than one symplectic operator and more than one reversor is considered in order to generalize the concept of reversibility. When there are two symplectic operators, say $\mathbf{J}_1$ and $\mathbf{J}_2$, a reversor and be both symplectic and anti-symplectic; for example, $\mathbf{J}_1$-symplectic and $\mathbf{J}_2$-anti-symplectic.

Recall the definition of reversibility when the involution is anti-symplectic for a classical Hamiltonian system written in the form

$$\mathbf{J} U_t = \nabla H(U) \qquad U \in \mathcal{M} = \mathbb{R}^{2n} \tag{3.1}$$

where $\mathcal{M}$ is the phase space, $\mathbf{J}$ is a skew-symmetric operator associated with a closed two-form and $\nabla H(U)$ is the gradient of the Hamiltonian functional. The system (3.1) is called a reversible-Hamiltonian system if there exists an anti-symplectic involution $\mathbf{R}$ (we will consider only linear reversors); that is, $\mathbf{R}$ satisfies

$$\mathbf{R} = \mathbf{R}^{-1} \qquad \text{and} \qquad \mathbf{R}^T \mathbf{J} \mathbf{R} = -\mathbf{J},$$

where $\mathbf{R}^T$ is the transpose of $\mathbf{R}$, such that H is $\mathbf{R}$-invariant: $H(\mathbf{R} \cdot U) = H(U)$. For such systems $\mathbf{R}U(-t)$ is a solution whenever $U(t)$ is a solution since, acting on (3.1) with $\mathbf{R}^T$ results in

$$\mathbf{R}^T \mathbf{J} U_t = \mathbf{R}^T \nabla H(U). \tag{3.2}$$

But $\mathbf{R}^T \mathbf{J} = -\mathbf{J} \mathbf{R}$, since $\mathbf{R}$ is anti-symplectic, and $\mathbf{R}$-invariance of H implies that $\mathbf{R}^T \nabla H(U) = \nabla H(\mathbf{R} \cdot U)$ and therefore

$$-\mathbf{J}(\mathbf{R}U)_t = \nabla H(\mathbf{R}U), \tag{3.3}$$

showing that $\mathbf{R}U(-t)$ is also a solution.

The concept of reversibility is now generalised to higher dimension with particular attention to the role of the multi-symplectic structure. Consider a Hamiltonian system on a bi-symplectic structure with governing equation

$$\mathbf{J}_1 Z_x + \mathbf{J}_2 Z_y = \nabla S(Z) \qquad Z \in \mathcal{M} = \mathbb{R}^N, \tag{3.4}$$

where $\mathcal{M}$ is the phase space and the operators $\mathbf{J}_1$ and $\mathbf{J}_2$ are skew-symmetric and associated with closed two-forms. There are two independent directions, x and y, in which the system can be reversible. We introduce the following definitions.

The system (3.4) is called x-reversible if there exists an involution $\mathbf{R}_1$ acting on $\mathcal{M}$, $\mathbf{R}_1 = \mathbf{R}_1^{-1}$, such that $S(\mathbf{R}_1 \cdot Z) = S(Z)$,

$$\mathbf{R}_1^T \mathbf{J}_1 \mathbf{R}_1 = -\mathbf{J}_1 \qquad \text{and} \qquad \mathbf{R}_1^T \mathbf{J}_2 \mathbf{R}_1 = \mathbf{J}_2.$$

In other words the involution is *anti-symplectic* with respect to the operator $\mathbf{J}_1$ but *symplectic* with respect to the operator $\mathbf{J}_2$.

Similarly, the system (3.4) is called y-reversible if there exists an involution $\mathbf{R}_2$ acting on $\mathcal{M}$, $\mathbf{R}_2 = \mathbf{R}_2^{-1}$, such that $S(\mathbf{R}_2 \cdot Z) = S(Z)$,

$$\mathbf{R}_2^T \mathbf{J}_1 \mathbf{R}_2 = \mathbf{J}_1 \qquad \text{and} \qquad \mathbf{R}_2^T \mathbf{J}_2 \mathbf{R}_2 = -\mathbf{J}_2.$$

In this case the involution is required to be *symplectic* with respect to the operator $\mathbf{J}_1$ but *anti-symplectic* with respect to the operator $\mathbf{J}_2$.

The above definition of reversibility, in the context of multi-symplectic structures, has natural further generalisation to elliptic systems with additional space dimensions.

An example of the utility of the above definitions is given by the semilinear elliptic equation

$$u_{xx} + u_{yy} + V'(u) = 0 \qquad (3.5)$$

introduced in §1-2. This system is clearly reversible in both x and y; that is, if $u(x, y)$ is a solution then so is $u(-x, y)$ and $u(x, -y)$. By letting $v = u_x$ and $w = u_y$ the system (3.5) can be reformulated as a Hamiltonian system on a multi-symplectic structure (cf. §1)

$$\mathbf{J}_1 Z_x + \mathbf{J}_2 Z_y = \nabla S(Z) \qquad Z \in \mathcal{M} = \mathbb{R}^3 \,.$$

Introduce the following reversors

$$\mathbf{R}_1 = \begin{pmatrix} 1 & 0 & 0 \\ 0 & -1 & 0 \\ 0 & 0 & 1 \end{pmatrix} \quad \text{and} \quad \mathbf{R}_2 = \begin{pmatrix} 1 & 0 & 0 \\ 0 & 1 & 0 \\ 0 & 0 & -1 \end{pmatrix} . \qquad (3.6)$$

It is clear that $\mathbf{R}_1 = \mathbf{R}_1^{-1}$ and $\mathbf{R}_2 = \mathbf{R}_2^{-1}$ and since $S(Z)$ is even in both v and w (cf. equation (1.4)) it follows that $S(\mathbf{R}_1 \cdot Z) = S(Z) = S(\mathbf{R}_2 \cdot Z)$. It remains to check the symplecticity of $\mathbf{J}_1$ and $\mathbf{J}_2$. Using the definitions of $\mathbf{J}_1$ and $\mathbf{J}_2$ we find

$$\begin{array}{llll} \mathbf{R}_1 \mathbf{J}_1 \mathbf{R}_1 & = & -\mathbf{J}_1 \\ \mathbf{R}_1 \mathbf{J}_2 \mathbf{R}_1 & = & \mathbf{J}_2 \end{array} \quad \text{and} \quad \begin{array}{llll} \mathbf{R}_2 \mathbf{J}_1 \mathbf{R}_2 & = \mathbf{J}_1 \\ \mathbf{R}_2 \mathbf{J}_2 \mathbf{R}_2 & = -\mathbf{J}_2 \,. \end{array}$$

Therefore $\mathbf{R}_1$ is symplectic (anti-symplectic) with respect to $\mathbf{J}_2$ (respectively $\mathbf{J}_1$) and $\mathbf{R}_2$ is symplectic (anti-symplectic) with respect to $\mathbf{J}_1$ (respectively $\mathbf{J}_2$); establishing a connection between the generalised reversibility and the generalised symplecticity for the elliptic operator (3.5).

4   Action, index and the loop space

Consider a gradient elliptic operator on $\mathbb{R}^2$, formulated as a Hamiltonian system on a bi-symplectic structure as in §1-2, with governing equation

$$\mathbf{i}_{Z_x} \omega^{(1)} + \mathbf{i}_{Z_y} \omega^{(2)} = \mathbf{d}S(Z) \qquad Z \in \mathcal{M} = \mathbb{R}^N \qquad (4.1)$$

where $\omega^{(j)} = \mathbf{d}\alpha^{(j)}$, $j = 1, 2$ are closed and exact two forms on $\mathcal{M}$.

A diagonal periodic pattern of (4.1) is a solution of the form

$$Z(x, y) = \gamma(\theta) \quad \text{where} \quad \theta = k_1 x + k_2 y \quad \text{and} \quad \gamma(\theta + 2\pi) = \gamma(\theta) \,. \qquad (4.2)$$

In this section a geometric framework for diagonal periodic patterns is presented along with an index for such solutions which, as will be shown, appear in 2-parameter (or in general n when the elliptic operator is posed on $\mathbb{R}^n$) families.

Diagonal periodic patterns correspond to the restriction of (4.1) to the loop space of $\mathcal{M}$. Recall that the loop space of a manifold is the space of parametrised closed curves $\gamma : \mathbb{S}^1 \to \mathcal{M}$ (cf. Weinstein [1978, §1.1]). We shall identify $\mathbb{S}^1 = \mathbb{R}/\mathbb{Z}$ and represent a loop in $\mathcal{M}$ by its cover $\gamma : \mathbb{R} \to \mathcal{M}$ satisfying $\gamma(\theta + 2\pi) = \gamma(\theta)$. By definition a diagonal periodic pattern satisfies

$$k_1 \, \mathbf{i}_{\dot\gamma} \omega^{(1)} + k_2 \, \mathbf{i}_{\dot\gamma} \omega^{(2)} = \mathbf{d}S(\gamma) \quad \text{for} \quad \gamma : \mathbb{S}^1 \to \mathcal{M}. \tag{4.3}$$

However, since $\omega^{(1)}$ and $\omega^{(2)}$ are exact, the governing equation (4.3) can be characterised, formally, as the Lagrange necessary condition for a constrained variational principle. Define the following family of *action* functionals

$$\mathcal{A}_j(\gamma) = \int_{\mathbb{S}^1} \mathbf{i}_{\dot\gamma} \alpha^{(j)} \qquad j = 1,2 \tag{4.4}$$

and the Hamiltonian functional on the loop space

$$\mathcal{S}(\gamma) = \int_{\mathbb{S}^1} S \circ \gamma. \tag{4.5}$$

In terms of the functionals $\mathcal{S}$, $\mathcal{A}_1$ and $\mathcal{A}_2$ diagonal periodic patterns therefore correspond, formally, to critical points of $\mathcal{S}$ on level sets of the action functionals

$$\operatorname{crit}\{\, \mathcal{S} \; : \; \mathcal{A}_1 = I_1, \quad \mathcal{A}_2 = I_2 \,\}. \tag{4.6}$$

for real numbers I_1 and I_2, with Lagrange necessary condition $\mathbf{d}\mathcal{F}(\gamma; k) = 0$ where

$$\mathcal{F}(\gamma; k) = \int_{\mathbb{S}^1} S \circ \gamma - k_1 \int_{\mathbb{S}^1} \mathbf{i}_{\dot\gamma} \alpha^{(1)} - k_2 \int_{\mathbb{S}^1} \mathbf{i}_{\dot\gamma} \alpha^{(2)}.$$

This follows since

$$\mathbf{d}\mathcal{F}(\gamma; k) \cdot \xi = \int_{\mathbb{S}^1} (\mathbf{d}S - k_1 \mathbf{i}_{\dot\gamma} \omega^{(1)} - k_2 \mathbf{i}_{\dot\gamma} \omega^{(2)}) \cdot \xi;$$

and when required to vanish recovers (4.3).

It follows from the theory of Lagrange multipliers (cf. Maddocks [1994]) that

$$k_j = \frac{\partial \mathcal{S}}{\partial I_j} \qquad j = 1,2 \tag{4.7}$$

and therefore, assuming sufficient differentiability,

$$\begin{pmatrix} \frac{\partial k_1}{\partial I_1} & \frac{\partial k_1}{\partial I_2} \\ \frac{\partial k_2}{\partial I_1} & \frac{\partial k_2}{\partial I_2} \end{pmatrix} = \begin{pmatrix} \frac{\partial^2 \mathcal{S}}{\partial I_1^2} & \frac{\partial^2 \mathcal{S}}{\partial I_1 \partial I_2} \\ \frac{\partial^2 \mathcal{S}}{\partial I_2 \partial I_1} & \frac{\partial^2 \mathcal{S}}{\partial I_2^2} \end{pmatrix} = \operatorname{Hess}_I(\mathcal{S}). \tag{4.8}$$

In the calculus of variations such matrices are called sensitivity matrices and provide information about critical point type (cf. Maddocks [1994]). We say that a diagonal periodic pattern is non-degenerate if

$$\Delta_K \stackrel{\text{def}}{=} \det[\text{Hess}_I(\mathcal{S})] \neq 0 \,.$$

This leads to the following naturally defined index for families of such patterns

$$\text{index}_k(\gamma) = \# \text{ negative eigenvalues of Hess}_I(\mathcal{S}). \qquad (4.9)$$

In the present case $\text{index}_k(\gamma)$ is either 0, 1 or 2. It should be noted that the above index can be naturally generalised to gradient elliptic operators on $\mathbb{R}^n$ in which case the index is either 0 or a natural number in $\{1, \ldots, n\}$. In the next section an example of the use of the above index is given; in particular it carries information about the linear stability of periodic patterns.

5   Example: loop space index and Eckhaus instability

Consider the system (2.4) as the stationary part of a parabolic partial differential equation

$$\Psi_t = \Delta\Psi + V'(|\Psi|^2)\Psi \qquad (5.1)$$

where, for simplicity, we will take, in this section, V to be of the form $V'(|\Psi|^2) = 1 - |\Psi|^2$. The equation (5.1) is generally called the Ginzburg-Landau equation with real coefficients and, with V taking the above special form, the cubic Ginzburg-Landau equation.

We will consider a family of stationary diagonal periodic patterns of (5.1), compute the loop space index of the branch and then show the relevance of the index for the Eckhaus instability.

There is a family of diagonal periodic patterns of (5.1) whose solution can be written down explicitly

$$\Psi(x,y) = A\,e^{i\theta} \qquad A \in \mathbb{C} \quad \theta = k_1 x + k_2 y \qquad (5.2)$$

which when substituted into (5.1) satisfies

$$\mathbf{k} \cdot \mathbf{k} + |A|^2 = 1 \qquad \text{where } \mathbf{k} = (k_1, k_2)\,. \qquad (5.3)$$

Now consider this family of solutions in the bi-symplectic framework. The formulation of the steady part of (5.1) as a Hamiltonian system on a bi-symplectic structure is given in equations (2.5)-(2.9). The solution (5.2) corresponds to a (x,y)-dependent flow along the $SO(2)$ group orbit and therefore has the explicit form

$$Z(x,y) = \gamma(\theta) = \mathcal{R}_\theta \widehat{Z} \qquad \theta = k_1 x + k_2 y\,. \qquad (5.4)$$

In fact the solution (5.4), more precisely the vector $\widehat{Z} \in \mathbb{R}^6$, is a generalised form of relative equilibrium in the sense that the solution (5.4) reduces to a constant vector $\widehat{Z} \in \mathbb{R}^6$ on the orbit space.

The governing equation for a solution of (2.6) can be written

$$\mathbf{i}_{\partial Z/\partial x}\omega^{(1)} + \mathbf{i}_{\partial Z/\partial y}\omega^{(2)} = \mathbf{d}S(Z) \tag{5.5}$$

where $\omega^{(1)}$ and $\omega^{(2)}$ are defined in (2.7) and (2.9) and $S(Z)$ is defined in (2.8) (with $V'(a) = 1 - a$). For the solution (5.4)

$$\left(\frac{\partial Z}{\partial x}, \frac{\partial Z}{\partial y}\right) = (\theta_x, \theta_y) \begin{pmatrix} \mathbf{0} & -\mathbf{I}_3 \\ \mathbf{I}_3 & \mathbf{0} \end{pmatrix} \mathcal{R}_\theta \widehat{Z} = k\mathcal{R}_\theta \widehat{\xi}$$

where

$$\widehat{\xi} = \frac{d}{d\theta}\mathcal{R}_\theta \widehat{Z}\Big|_{\theta=0} = \begin{pmatrix} \mathbf{0} & -\mathbf{I}_3 \\ \mathbf{I}_3 & \mathbf{0} \end{pmatrix} \widehat{Z}$$

and so, using the $\mathcal{R}_\theta$-invariance of $\omega^{(1)}$,

$$\mathbf{i}_{\partial Z/\partial x}\omega^{(1)} = \mathbf{i}_{k_1 \mathcal{R}_\theta \widehat{\xi}}\omega^{(1)} = k_1 \mathcal{R}_\theta^* \mathbf{i}_{\widehat{\xi}}\omega^{(1)}.$$

Similarly,

$$\mathbf{i}_{\partial Z/\partial y}\omega^{(2)} = \mathbf{i}_{k_2 \mathcal{R}_\theta \widehat{\xi}}\omega^{(2)} = k_2 \mathcal{R}_\theta^* \mathbf{i}_{\widehat{\xi}}\omega^{(2)}$$

and so (5.4) satisfies

$$k_1 \mathcal{R}_\theta^* \mathbf{i}_{\widehat{\xi}}\omega^{(1)} + k_2 \mathcal{R}_\theta^* \mathbf{i}_{\widehat{\xi}}\omega^{(2)} = \mathbf{d}S(\mathcal{R}_\theta \widehat{Z})$$

or, using (2.12) and (2.13),

$$k_1 \mathbf{d}P_1(\widehat{Z}) + k_2 \mathbf{d}P_2(\widehat{Z}) = \mathbf{d}S(\widehat{Z}) \tag{5.6}$$

which is the Lagrange necessary condition for a constrained variational principle on $\mathbb{R}^6$, with (k_1, k_2) as Lagrange multipliers, for the vector $\widehat{Z}$. Solution of (5.6) results in

$$\widehat{Z} = \begin{pmatrix} \widehat{u} \\ \widehat{v} \end{pmatrix} = \begin{pmatrix} 1 & 0 \\ 0 & k_1 \\ 0 & k_2 \\ 0 & 1 \\ -k_1 & 0 \\ -k_2 & 0 \end{pmatrix} \begin{pmatrix} \widehat{u}_1 \\ \widehat{v}_1 \end{pmatrix} \quad \text{with} \quad \widehat{u}_1^2 + \widehat{v}_1^2 + \mathbf{k} \cdot \mathbf{k} = 1$$

recovering (5.2)-(5.3).

To evaluate the index defined in §4, along the above branch of solutions, the action functionals are necessary and a straightforward calculation using the exact solution results in

$$\mathcal{A}_j = \int_{\mathbb{S}^1} \mathbf{i}_\gamma \alpha^{(j)} = k_j |A|^2 \qquad j = 1, 2$$

where $|A|^2 = \hat{u}_1^2 + \hat{v}_1^2$. Now, using (5.3),

$$\frac{\partial |A|^2}{\partial k_j} = -2k_j \qquad j = 1, 2$$

for the cubic Ginzburg-Landau equation and so

$$[\text{Hess}_I(\mathcal{S})]^{-1} = \begin{pmatrix} \frac{\partial \mathcal{A}_1}{\partial k_1} & \frac{\partial \mathcal{A}_1}{\partial k_2} \\ \frac{\partial \mathcal{A}_2}{\partial k_1} & \frac{\partial \mathcal{A}_2}{\partial k_2} \end{pmatrix} = \begin{pmatrix} 1 - 3k_1^2 - k_2^2 & -2k_1 k_2 \\ -2k_1 k_2 & 1 - 3k_2^2 - k_1^2 \end{pmatrix}.$$

Let τ, Δ denote respectively the trace and determinant of the inverse of $\text{Hess}_I(\mathcal{S})$. Then

$$\tau = 2(1 - 2\mathbf{k} \cdot \mathbf{k}) \qquad \Delta = |A|^2(1 - 3\mathbf{k} \cdot \mathbf{k})$$

and therefore the index changes from 0 to 1 when $\mathbf{k} \cdot \mathbf{k} = \frac{1}{3}$ which is precisely the Eckhaus threshold (cf. Eckhaus [1965]).

The explicit connection of the above index calculation and the Eckhaus instability of (5.1) is seen as follows. Let $\Psi(x, y, t) = \Psi_0(x, y) + U(x, y, t)$ where $\Psi_0(x, y) = Ae^{i\theta}$ is the basic state (5.2) (equivalently (5.4)). Then substitution into (5.1) and linearisation about Ψ_0 results in

$$U_t = \Delta U + [V'(|\Psi_0|^2) + |\Psi_0|^2 V''(|\Psi_0|^2)]U + V''(|\Psi_0|^2)\Psi_0^2 \overline{U}$$

which has a general solution

$$U(x, y, t) = e^{i\theta}[U_1 e^{i(\mu_1 x + \mu_2 y) + \lambda t} + U_2 e^{-i(\mu_1 x + \mu_2 y) + \overline{\lambda} t}]$$

for complex scalars U_1 and U_2, real numbers μ_1 and μ_2 and complex stability exponent λ. With $V''(|\Psi_0|^2) = -1$ the eigenvalue problem for $\lambda \in \mathbb{C}$ reduces to

$$\lambda^2 + 2(\mu_1^2 + \mu_2^2 + |A|^2)\lambda + (\mu_1^2 + \mu_2^2)^2 + 2(\mu_1^2 + \mu_2^2)|A|^2 - 4(k_1\mu_1 + k_2\mu_2)^2 = 0.$$

First, it is easily verified that if μ_1 and μ_2 are real the above quadratic equation has two real roots. Moreover one of the roots is always negative and hence stable. The sign of the other root can be either positive or negative and it is easily verified that both roots are negative if and only if $\mathbf{k} \cdot \mathbf{k} < \frac{1}{3}$ and when $\mathbf{k} \cdot \mathbf{k} > \frac{1}{3}$ there exists a real positive (unstable) linear stability exponent for some value of $\mu \in \mathbb{R}^2$.

6   Symplecticity, reversibility and eigenvalues

An elliptic operator, of the form considered in §1-2, on a multi-symplectic structure, linearised about a constant solution, has the following form

$$\mathbf{A}Z - \mathbf{J}_1 Z_x - \mathbf{J}_2 Z_y = 0 \qquad Z \in \mathbb{R}^N \qquad (6.1)$$

where $\mathbf{A} \in g\ell_N(\mathbb{R})$ and symmetric and $\mathbf{J}_i \in g\ell_N(\mathbb{R})$ i=1,2 and anti-symmetric. The generalisation of (6.1) to include additional symplectic operators is clear; for example for elliptic operators on $\mathbb{R}^n$, $n \geq 2$.

The eigenvalue problem corresponding to (6.1) is

$$(\mathbf{A} - \lambda_1 \mathbf{J}_1 - \lambda_1 \mathbf{J}_1)\xi = 0 \tag{6.2}$$

where $\lambda = (\lambda_1, \lambda_2) \in \mathbb{C}^2$ and $\xi \in \mathbb{C}^N$. We say that $\lambda \in \mathbb{C}^2$ is an eigenvalue of (6.2) if $(\lambda, \xi) \in \mathbb{C}^2 \times \mathbb{C}^N$ satisfies (6.2) with $\|\xi\| > 0$ or equivalently if

$$\Delta(\lambda_1, \lambda_2) = \det(\mathbf{A} - \lambda_1 \mathbf{J}_1 - \lambda_2 \mathbf{J}_2) = 0.$$

Proposition 6.1. *Under the above hypotheses on* $\mathbf{A}$, $\mathbf{J}_1$ *and* $\mathbf{J}_2$

$$\Delta(\lambda_1, \lambda_2) = \widehat{\Delta}(\lambda_1^2, \lambda_1 \lambda_2, \lambda_2^2).$$

Remark: This is a generalisation of the eigenvalue relation induced by a single symplectic structure for ordinary differential equations (cf. Abraham & Marsden [1978, p. 170; Theorem 3.1.17]).

Proof. The determinant is invariant under transposition; therefore

$$\begin{aligned}
\Delta(\lambda_1, \lambda_2) &\overset{\text{def}}{=} \det(\mathbf{A} - \lambda_1 \mathbf{J}_1 - \lambda_2 \mathbf{J}_2) \\
&= \det(\mathbf{A}^T - \lambda_1 \mathbf{J}_1^T - \lambda_2 \mathbf{J}_2^T) \\
&= \det(\mathbf{A} + \lambda_1 \mathbf{J}_1 + \lambda_2 \mathbf{J}_2) \\
&= \Delta(-\lambda_1, -\lambda_2).
\end{aligned}$$

The determinant $\Delta(\lambda_1, \lambda_2)$ is a polynomial and every such function can be expressed as a polynomial in terms of the $\mathbb{Z}_2$ invariants: λ_1^2, $\lambda_1 \lambda_2$ and λ_2^2 completing the proof. ■

Proposition 6.2. *In addition to the hypotheses of Proposition 6.1, suppose the system is also either x-reversible or y-reversible. Then*

$$\Delta(\lambda_1, \lambda_2) = \widehat{\Delta}(\lambda_1^2, \lambda_2^2). \tag{6.3}$$

Proof. Suppose the system is x-reversible. The proof is similar if it is y-reversible. x-reversibility implies

$$\mathbf{R}_1 \mathbf{A} \mathbf{R}_1 = \mathbf{A}, \ \mathbf{R}_1 \mathbf{J}_1 \mathbf{R}_1 = -\mathbf{J}_1, \mathbf{R}_1 \mathbf{J}_2 \mathbf{R}_1 = \mathbf{J}_2$$

with $\mathbf{R}_1^2 = \mathbf{I}$. Then, since $\det(\mathbf{R}_1)\det(\mathbf{R}_1) = 1$,

$$\begin{aligned}
\Delta(\lambda_1, \lambda_2) &\overset{\text{def}}{=} \det(\mathbf{A} - \lambda_1 \mathbf{J}_1 - \lambda_2 \mathbf{J}_2) \\
&= \det[\mathbf{R}_1(\mathbf{A} - \lambda_1 \mathbf{J}_1 - \lambda_2 \mathbf{J}_2)\mathbf{R}_1] \\
&= \det[\mathbf{R}_1 \mathbf{A} \mathbf{R}_1 - \lambda_1 \mathbf{R}_1 \mathbf{J}_1 \mathbf{R}_1 - \lambda_2 \mathbf{R}_1 \mathbf{J}_2 \mathbf{R}_1] \\
&= \det[\mathbf{A} + \lambda_1 \mathbf{J}_1 - \lambda_2 \mathbf{J}_2] \\
&= \Delta(-\lambda_1, \lambda_2).
\end{aligned}$$

By Proposition 6.1 $\Delta(\lambda_1, \lambda_2)$ is a polynomial function of λ_1^2, $\lambda_1 \lambda_2$ and λ_2^2. Evenness in λ_1 eliminates the dependence on the invariant $\lambda_1 \lambda_2$ resulting in the form (6.3) where $\widehat{\Delta}$ is also a polynomial. ■

We now consider the case where there is further symmetry: in particular
when the basic elliptic operator is invariant under rotation in the (x, y)-plane
which arises in the case of a semilinear elliptic equation. Such a symmetry leads
to further simplification of the eigenvalue structure.

For example consider the semilinear elliptic equation $\Delta u + V'(u) = 0$ which
has a formulation as a Hamiltonian system on a multi-symplectic structure (cf.
§1–2)

$$\mathbf{J}_1 Z_x + \mathbf{J}_2 Z_y = \nabla S(Z) \tag{6.4}$$

where the form of $S(Z)$ is given in §1. Let

$$\mathcal{G}_\theta = \exp(\mathbf{J}_3 \theta) \quad \text{where} \quad \mathbf{J}_3 = \begin{pmatrix} 0 & 0 & 0 \\ 0 & 0 & -1 \\ 0 & 1 & 0 \end{pmatrix}.$$

Then $\mathcal{G}_\theta$ is an action for $SO(2)$ and the rotation invariance in the (x, y)-plane leads
to the following symmetry properties for (6.4)

$$\begin{aligned}
\mathcal{G}_\theta^T \mathbf{J}_1 \mathcal{G}_\theta &= \cos\theta \mathbf{J}_1 - \sin\theta \mathbf{J}_2 \\
\mathcal{G}_\theta^T \mathbf{J}_2 \mathcal{G}_\theta &= \sin\theta \mathbf{J}_1 + \cos\theta \mathbf{J}_2 \\
\nabla S(\mathcal{G}_\theta Z) &= \mathcal{G}_\theta \nabla S(Z)
\end{aligned} \tag{6.5}$$

and in the linear case, when $\nabla S(Z) = \mathbf{A}Z$ we have $\mathcal{G}_\theta^T \mathbf{A} \mathcal{G}_\theta = \mathbf{A}$.

Proposition 6.3. *Suppose that (6.2) is posed on $\mathbb{C}^3$ and is $SO(2)$-equivariant in the
sense that the basic operators $\mathbf{A}$, $\mathbf{J}_1$ and $\mathbf{J}_2$ satisfy (6.5). Then*

$$\Delta(\lambda_1, \lambda_2) = \widehat{\Delta}(\lambda_1^2 + \lambda_2^2).$$

Remark. Note that for the semilinear elliptic equation (6.4) this result is trivial.
The interest in the formulation lies in its potential for generalisation; for example,
it is clear how to generalise this result to other group actions and to elliptic systems
on higher dimensional spaces.

Proof. Since $\det \mathcal{G}_\theta = 1$,

$$\begin{aligned}
\Delta(\lambda_1, \lambda_2) &\overset{\text{def}}{=} \det(\mathbf{A} - \lambda_1 \mathbf{J}_1 - \lambda_2 \mathbf{J}_2) \\
&= \det[\mathcal{G}_\theta^T (\mathbf{A} - \lambda_1 \mathbf{J}_1 - \lambda_2 \mathbf{J}_2)\mathcal{G}_\theta] \\
&= \det[\mathcal{G}_\theta^T \mathbf{A} \mathcal{G}_\theta - \lambda_1 \mathcal{G}_\theta^T \mathbf{J}_1 \mathcal{G}_\theta - \lambda_2 \mathcal{G}_\theta^T \mathbf{J}_2 \mathcal{G}_\theta] \\
&= \det[\mathbf{A} - (\lambda_1 \cos\theta + \lambda_2 \sin\theta)\mathbf{J}_1 - (-\lambda_1 \sin\theta + \lambda_2 \cos\theta)\mathbf{J}_2] \\
&= \Delta(\lambda_1 \cos\theta + \lambda_2 \sin\theta, -\lambda_1 \sin\theta + \lambda_2 \cos\theta)
\end{aligned}$$

and so $\Delta(\lambda_1, \lambda_2)$ is an $SO(2)$-invariant polynomial on $\mathbb{C}^2$ and every such function
can be expressed as a polynomial function of $\lambda_1^2 + \lambda_2^2$. ∎

7   Radially symmetric states and phase space geometry

An interesting and important class of solutions of. for example, semilinear elliptic equations of the form $\Delta u + V'(u) = 0$ is the bound states or "higher dimensional solitary waves" (cf. Berestyki et al. [1981] and references therein). Such states are radially symmetric and asymptotic to the trivial state as $x^2 + y^2 \to \infty$ (in $\mathbb{R}^2$ for example). In the multi-symplectic structures framework such states are characterised, geometrically, as follows.

Consider the elliptic equation $\Delta u + V'(u) = 0$ formulated as a Hamiltonian system (cf. equations (1.3)-(1.5))

$$\mathbf{J}_1 Z_x + \mathbf{J}_2 Z_y = \nabla S(Z). \tag{7.1}$$

Introducing polar coordinates $x = r \cos \theta$ and $y = r \cos \theta$ into (7.1) results in

$$(\cos \theta \mathbf{J}_1 + \sin \theta \mathbf{J}_2) Z_r + \tfrac{1}{r}(-\sin \theta \mathbf{J}_1 + \cos \theta \mathbf{J}_2) Z_\theta = \nabla S(Z) \tag{7.2}$$

which is again a Hamiltonian system on a multi-symplectic structure of the form $(\mathbb{R}^3, \Omega^{(1)}, \Omega^{(2)}, S)$ with

$$\begin{aligned}
\Omega^{(1)} &= \cos \theta \, \mathbf{d}v \wedge \mathbf{d}u + \sin \theta \, \mathbf{d}w \wedge \mathbf{d}u \\
\Omega^{(2)} &= -\tfrac{1}{r} \sin \theta \, \mathbf{d}v \wedge \mathbf{d}u + \tfrac{1}{r} \cos \theta \, \mathbf{d}w \wedge \mathbf{d}u
\end{aligned}$$

where, to avoid the singularity at $r = 0$, the limits $\mathbf{d}v/r$ and $\mathbf{d}w/r$ as $r \to 0$ should be finite.

In the phase space $\mathbb{R}^3$ radially symmetric states correspond precisely to parametrisations of solutions of (7.2) as a surface of revolution in $\mathbb{R}^3$, or (cf. O'Neill [1966, P. 234]),

$$Z(r, \theta) = \begin{pmatrix} g(r) \\ h(r) \cos \theta \\ h(r) \sin \theta \end{pmatrix}. \tag{7.3}$$

Substitution of (7.3) into (7.2) results in

$$[\cos\theta \mathbf{J}_1 + \sin\theta \mathbf{J}_2] \begin{pmatrix} \dot{g} \\ \dot{h}\cos\theta \\ \dot{h}\sin\theta \end{pmatrix} + \frac{1}{r}[-\sin\theta \mathbf{J}_1 + \cos\theta \mathbf{J}_2] \begin{pmatrix} 0 \\ -h\sin\theta \\ h\cos\theta \end{pmatrix} = \begin{pmatrix} V'(g) \\ h\cos\theta \\ h\sin\theta \end{pmatrix}$$

$$\text{or} \qquad \begin{aligned} -\dot{h} - \tfrac{1}{r}h &= V'(g) \\ \cos\theta \, \dot{g} &= h\cos\theta \\ \sin\theta \, \dot{g} &= h\sin\theta \end{aligned}$$

which together imply

$$\frac{d^2 g}{dr^2} + \frac{1}{r}\frac{dg}{dr} + V'(g) = 0: \quad h = \frac{dg}{dr} \tag{7.4}$$

recovering the usual equation for radially symmetric states of (1.2) on $\mathbb{R}^2$ (cf. Berestyki et al. [1981, equation (2)]).

Going back to the formulation (7.1), the radially symmetric solutions correspond to surfaces of revolution in $\mathbb{R}^3$ and therefore the differential geometry of such surfaces can be applied in order to analyze the geometry of solutions as well as the geometry of perturbed problems such as

$$\Delta u + V'(u) + \varepsilon F(x, y, u) = 0. \tag{7.5}$$

An analytic approach for perturbed problems of the type (7.5) is given in Angenet [1981, §D]) based on an extension of the Melnikov theory. However. the Melnikov theory for perturbation of homoclinic orbits can be formulated geometrically and by studying (7.5) in the phase space, with the basic state considered as a bounded surface of revolution in the phase space – a generalised homoclinic orbit, geometric methods can be used for analyzing the perturbation problem.

A simple example of a radially symmetric solution is as follows. Let $V'(u) = u^3 - u^2$ and so

$$u_{xx} + u_{yy} - u^2 + u^3 = 0. \tag{7.6}$$

It is easily verified that (7.6) has the exact solution

$$u(x, y) = \frac{4}{2 + x^2 + y^2} \tag{7.7}$$

which decays algebraically as $x^2 + y^2 \to \infty$. In the phase space the above solution has the form

$$Z(x, y) = \begin{pmatrix} u(x, y) \\ v(x, y) \\ w(x, y) \end{pmatrix} \overset{\text{def}}{=} \begin{pmatrix} u \\ u_x \\ u_y \end{pmatrix} = u \begin{pmatrix} 1 \\ -\frac{1}{2}xu \\ -\frac{1}{2}yu \end{pmatrix}$$

and the image of $Z : \mathbb{R}^2 \to \mathbb{R}^3$ is a "balloon" in the phase space with a cusp singularity at the origin and is shown plotted in Figure 1. The cusp point at the origin is due to the algebraic decay of u at infinity.

In general radially symmetric solutions of semi-linear elliptic equations (on $\mathbb{R}^2$ or in general $\mathbb{R}^n$) can be characterised geometrically as bounded surfaces of revolution in the phase space associated with the Hamiltonian formulation.

Secondly in this section a geometrical characterisation of diagonal patterns (cf. §4) for the system (1.2) will be given. For simplicity suppose that $V''(u) > 0$ for all $u \in \mathbb{R}$ in which case

$$S(Z) = \tfrac{1}{2}(v^2 + w^2) + V(u) \tag{7.8}$$

is diffeomorphic to a sphere (and henceforth $S(Z) = s$ will be referred to as a sphere).

Diagonal periodic patterns of (7.1) satisfy

$$(k_1\mathbf{J}_1 + k_2\mathbf{J}_2)Z_\theta = \nabla S(Z) \qquad \theta = k_1 x + k_2 y \tag{7.9}$$

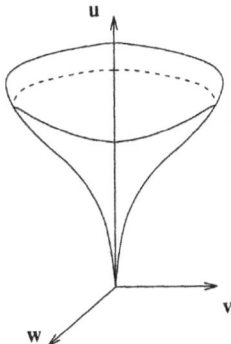

Figure 1. Image in the phase space of the radially
symmetric solution (7.7)

for some $\mathbf{k} = (k_1, k_2)$ with $\mathbf{k} \cdot \mathbf{k} \neq 0$. The system (7.9) is an ODE and in this case $S(Z)$ is an invariant surface in $\mathbb{R}^3$ for solutions of (7.9) since

$$\frac{d}{d\theta} S(Z) = \langle \nabla S(Z), Z_\theta \rangle = \langle (k_1 \mathbf{J}_1 + k_2 \mathbf{J}_2) Z_\theta, Z_\theta \rangle = 0$$

using the fact that $(k_1 \mathbf{J}_1 + k_2 \mathbf{J}_2)$ is skew-symmetric. Therefore all diagonal periodic patterns lie on invariant spheres in $\mathbb{R}^3$. However $k_1 \mathbf{J}_1 + k_2 \mathbf{J}_2$ is singular with

$$\mathrm{Ker}(k_1 \mathbf{J}_1 + k_2 \mathbf{J}_2) = \mathrm{span}\{\xi\} \quad \text{with} \quad \xi = \begin{pmatrix} 0 \\ k_2 \\ -k_1 \end{pmatrix}.$$

Therefore solutions of (7.9) are restricted to

$$\langle \nabla S(Z), \xi \rangle = k_2 v - k_1 w = 0$$

which is a plane passing through the origin. Geometrically diagonal periodic patterns of (7.1) are therefore closed curves in $\mathbb{R}^3$ given by the intersection of a sphere $S(Z) = s$ and the plane $\langle Z, \xi \rangle = 0$. Note that in the special case $V'(u) = u$ the diagonal periodic patterns correspond precisely to geodesics on $\mathbb{S}^2 = \{ Z \in \mathbb{R}^3 : \langle Z, Z \rangle = s \in \mathbb{R}_+ \}$.

As a third example, doubly periodic patterns; that is solutions of (7.1) of the form

$$Z(x + \ell_1, y) = Z(x, y + \ell_2) = Z(x.y) \qquad \forall (x, y) \in \mathbb{R}^2$$

with wavelengths ℓ_1 and ℓ_2, correspond geometrically to tori in the phase space.

References

[1] R. Abraham & J. E. Marsden [1978] *Foundations of Mechanics*, Second Edition, Benjamin-Cummings Publ. Co.: Reading, Massachusetts

[2] S. B. Angenet [1988] *The shadowing lemma for elliptic PDE*, In *Dynamics of Infinite-Dimensional Systems*, edited by S.-N. Chow & J. K. Hale, NATO ASI Series **F37**, Springer-Verlag, pp. 7–22

[3] H. Berestyki, P. L. Lions & L. A. Peletier [1981] *An ODE approach to the existence of positive solutions for semilinear problems in $\mathbb{R}^N$*, Indiana Univ. Math. J. **30**, pp. 141–57

[4] T. J. Bridges [1993] *Hamiltonian structure of plane-wave instabilities*, Field Institute Comm., published by AMS, in press.

[5] T. J. Bridges [1994] *Periodic patterns, linear instability, symplectic structure and mean-flow dynamics for 3D surface waves*, Phil. Trans. Royal Soc. Lond., in press.

[6] T. J. Bridges [1995] *Multi-symplectic structures and wave propagation*, Math. Proc. Camb. Phil. Soc. **121** (1996), to appear.

[7] W. Eckhaus [1965] *Studies in Nonlinear Stability Theory*, Springer-Verlag: Heidelberg

[8] D. G. B. Edelen [1985] *Applied Exterior Calculus*, John Wiley and Sons: New York

[9] A. Floer [1988] *The unregularized gradient flow of the symplectic action*, Comm. Pure & Appl. Math. **41**, pp. 775–813

[10] I. Hoveijn [1995] *Versal deformations and normal forms for reversible and Hamiltonian linear systems*, J. Diff. Eqns., in press.

[11] J. H. Maddocks [1994] *On second-order conditions in constrained variational principles*, J. Opt. & Applic., in press

[12] D. McDuff [1990] *Elliptic methods in symplectic geometry*, Bull. Amer. Math. Soc. **23**, pp. 311–58

[13] A. Mielke [1991] *Hamiltonian and Lagrangian flows on center manifolds with application to elliptic variational problems*, Lect. Notes in Math. **1489**, Springer-Verlag: Heidelberg

[14] B. O'Neill [1966] *Elementary Differential Geometry*, Academic Press: NY

[15] D. Salamon [1990] *Morse theory, the Conley index and Floer homology*, Bull. Lond. Math. Soc. **22**, pp. 113–40

[16] F. Warner [1983] *Foundations of differentiable manifolds and Lie groups*, Springer-Verlag: New York

[17] A. Weinstein [1978] *Bifurcations and Hamilton's principle*, Math. Z. **159**, pp. 235–48

Progress in Nonlinear Differential Equations
and Their Applications, Vol. 19
© 1996 Birkhäuser Verlag Basel/Switzerland

The Rolling Disc

R. Cushman   J. Hermans*   D. Kemppainen[†]

1 Introduction

In this paper we consider a homogeneous disc, which rolls without slipping on a horizontal plane under the influence of a vertical gravitational field. Due to the four dimensional symmetry group consisting of translations in the plane, rotations around a vertical axis and rotations around the axis perpendicular to the disc, one can reduce the dynamics to a four dimensional system, see section 2. This reduced system was obtained for the first time by Ferrers [6] in 1872, using Euler-angles. The first serious analysis of these equations was given by Vierkandt [18] in 1892. He showed that almost all orbits are periodic.

The easiest way to obtain this periodicity result is to use the reversibility of the system, which is typical for mechanical systems. A vector field V is *reversible* if there is an involution R such that $R_*V = -V$. Birkhoff [4] proved that orbits which intersect the fixed point set of R twice are periodic. The interesting thing here is that the fixed point set of the reverser R has only codimension one. For other applications of this idea to describe the motion of rolling bodies, see Hermans [7]. The motion of the disc is complicated by the existence of *hyperbolic* relative equilibria and by the *incompleteness* of the flow of the reduced vector field. The non-periodic orbits are of two types: those corresponding to the stable and unstable manifolds of hyperbolic equilibria, and those where the disc falls flat in finite time (incompleteness of the flow). (The 'complete analogue' of the rolling disc is the problem of a rolling body of revolution whose reduced dynamics was first analyzed by Chaplygin, see Moschuk [13].) The manifolds of initial conditions giving rise to non-periodic orbits are of codimension one in a given energy surface, which is remarkably low. The main achievement of this paper is to describe both the periodic and the non-periodic orbits qualitatively.

Another, more classical approach to solving the reduced equations of motions is to construct a second order linear differential equation. Surprisingly this equation is Legendre's equation, see Routh [17], §244. In 1899, Korteweg [10] first observed

---

*Mathematics Institute, University of Utrecht, 3508 TA Utrecht, The Netherlands

[†]Department of Mathematics and Statistics, University of Calgary, Calgary, Alberta, Canada

that hypergeometric functions appear in the solution of the reduced equations of motion. Appell [2] independently made the same observation in 1900. In the past one hundred years there has been little interest in the rolling disc. In this period, many authors were content only to repeat the derivation of the equations of motion and the proof of integrability by quadratures and leave it at that. Only recently has interest arisen in the geometric behavior of the solutions of the rolling disc in phase space. Kemppainen [9] was the first to construct a two parameter family of one degree of freedom Hamiltonian subsystems of the reduced vector field. These systems are our basic tool for analyzing the reduced vector field. In a recent preprint O'Reilly [15] has an interesting discussion of the motion of the rolling disc and numerically finds several bifurcations.

This paper is organized as follows. Section 2 consists of a (geometrical) reduction of the symmetries which results in a four dimensional dynamical system. In section 3 we consider the relative equilibria of this system. In section 4 we give a rigorous qualitative description of *all* the motions of the disc and in section 5 we reconstruct the motion from the reduced space to the eight dimensional constraint manifold. Unfortunately, we have not succeeded in giving a description of how the constraint manifold is singularly foliated by 3-tori.

2 Reduced equations of motion

In this section we derive the equations of motion for a rolling disc.

2.1 Equations of motion

Consider a reference disc of radius r, with uniformly distributed mass m, lying flat in a fixed reference frame with center of mass at the origin, (see figure 2.1). The position of the moving disc is given by applying an element (A, a) of the Euclidian group $\mathcal{E}(3) \subseteq SO(3) \times \mathbf{R}^3$ to a position of the reference disc. Using the left trivialization

$$\mathcal{E}(3) \times e(3) \rightarrow T\mathcal{E}(3) : (A, a, \Omega, b) = (A, a, A^{-1}\dot{A}, A^{-1}\dot{a}) \rightarrow (\dot{A}, \dot{a}),$$

where e(3) is the Lie algebra of $\mathcal{E}(3)$, we pull back the Lagrangian of the unconstrained disc to $\mathcal{E}(3) \times e(3)$ and obtain the Lagrangian

$$\mathcal{L} = T_{\text{rot}} + T_{\text{trans}} - V.$$

Here $T_{\text{rot}} = \frac{1}{2} \langle I\omega, \omega \rangle$ is the rotational kinetic energy of the disc with $I = \text{diag}(I_1, I_1, I_3)$ the moment of inertia tensor with respect to the principal axes and $\langle \, , \, \rangle$ is the Euclidian inner product on $\mathbf{R}^3$. We have identified $\Omega = \begin{pmatrix} 0 & -\omega_3 & \omega_2 \\ \omega_3 & 0 & -\omega_1 \\ -\omega_2 & \omega_1 & 0 \end{pmatrix} \in$

so(3) with the angular velocity vector $\omega = (\omega_1, \omega_2, \omega_3) \in \mathbf{R}^3$. This identification defines the mapping $i : so(3) \to \mathbf{R}^3$. The translational kinetic energy of the center of mass is $T_{\text{trans}} = \frac{1}{2} \langle b, b \rangle$ and the potential energy is $V = \mathrm{mg} \langle a, e_3 \rangle$. The Lagrangian derivative of $\mathcal{L}$ is

$$\delta\mathcal{L} = \left(\frac{d}{dt}(I\omega) - I\omega \times \omega, \; \mathrm{m}\frac{db}{dt} - \mathrm{m}(b \times \omega) - \mathrm{mg}\, A^{-1}e_3 \right), \tag{1}$$

which is a linear form on $e(3)$ depending on a base point in $\mathcal{E}(3)$. (For more details see Arnol'd [3, p.14].)

2.2 Constraints

The moving disc is subject to two kinds of constraint: a holonomic constraint of moving on a horizontal plane and a nonholonomic constraint of rolling. To treat these constraints, let

$$u = -A^{-1}e_3 \tag{2}$$

and suppose that $u \neq \pm e_3$, otherwise the disc is lying flat. Choose a vector

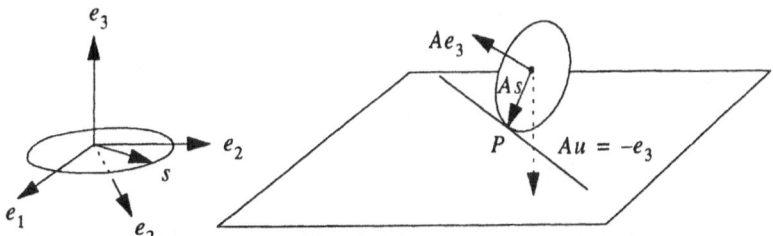

Figure 2.1. The rolling disc.

$s = s(A)$ in the e_1-e_2 plane of length r which lies in the plane spanned by u and e_3. Applying the element (A, a) of $\mathcal{E}(3)$ to the reference disc gives the vector $As + a$ which is the point of contact P of the moving disc with the horizontal plane. Therefore

$$0 = \langle As + a, e_3 \rangle = -\langle s, u \rangle + \langle a, e_3 \rangle. \tag{3}$$

Let

$$\widehat{u} = u - \langle u, e_3 \rangle e_3. \tag{4}$$

Then $\widehat{u}$ is the projection of u onto the plane of the reference disc and hence is colinear with s. Consequently,

$$s = s(A) = \frac{r}{\| \widehat{u} \|} \widehat{u}. \tag{5}$$

The condition that the disc rolls without slipping is the same as requiring that the velocity of the point of contact with the plane is zero: $0 = \dot{A}s + \dot{a}$, that is,

$$0 = A^{-1}\dot{A}s + A^{-1}\dot{a} = \omega \times s + b. \qquad (6)$$

Equations (3) and (6) define the constraint manifold $\mathcal{C}$ as a subset of $\mathcal{E}(3) \times T\mathbf{R}^3$:

$$\{(A,a,\omega,b)| \, b = -\omega \times s(A) \ \& \ \langle a, e_3 \rangle = \langle s(A), u(A) \rangle \}. \qquad (7)$$

2.3   Symmetry group

The rolling disc is invariant under translations in the horizontal plane and rotations about the vertical axis. Consider the 2-dimensional Euclidian group

$$\mathcal{E}(2) = \{ (R_\varphi, x) \in \mathcal{E}(3)| \, R_\varphi = \begin{pmatrix} \cos\varphi & -\sin\varphi & 0 \\ \sin\varphi & \cos\varphi & 0 \\ 0 & 0 & 1 \end{pmatrix} \ \& \ x = (x_1, x_2, 0) \}.$$

Viewed as a subgroup of $\mathcal{E}(3)$, $\mathcal{E}(2)$ acts on the constraint manifold $\mathcal{C}$ by

$$\Phi : \mathcal{E}(2) \times \mathcal{C} \to \mathcal{C} : (R_\varphi, x), \, (A, a, \omega, b) \to (R_\varphi A, R_\varphi a + x. \omega, b). \qquad (8)$$

Because the $\mathcal{E}(2)$ action Φ is free and proper, its orbit space $\mathcal{C}/\mathcal{E}(2)$ is a smooth manifold, which after a little thought is seen to be diffeomorphic to $S^2 \times \mathbf{R}^3$. Since

$$IR_\varphi = R_\varphi I \ \text{ and } \ \langle R_\varphi a + x, e_3 \rangle = \langle a, e_3 \rangle + \langle x, e_3 \rangle = \langle a. e_3 \rangle,$$

the action Φ preserves the Lagrangian $\mathcal{L}$.

2.4   Reduced equations of motion

To derive the $\mathcal{E}(2)$-reduced equations of motion on $S^2 \times \mathbf{R}^3$ we apply the d'Alembert principle, which states that $\delta\mathcal{L}$ is perpendicular to all vectors $(\widetilde{\omega}, \widetilde{b})$ satisfying (6). This gives

$$\begin{aligned} 0 &= \left\langle \frac{d}{dt}(I\omega) - I\omega \times \omega, \widetilde{\omega} \right\rangle + \left\langle m\frac{db}{dt} - m(b \times \omega) - mg\, u, \widetilde{b} \right\rangle \\ &= \left\langle \frac{d}{dt}(I\omega) - I\omega \times \omega - \left\{ m\frac{d}{dt}(\omega \times s) - m(\omega \times s) \times \omega + mg\, u \right\} \times s, \widetilde{\omega} \right\rangle \end{aligned}$$

for all $\widetilde{\omega} \in \mathbf{R}^3$. Therefore the terms in the angled brackets before the comma are zero. After expanding the vector product $((\omega \times s) \times \omega) \times s$ we obtain

$$\begin{aligned} \frac{d(I\omega)}{dt} &= I\omega \times \omega - mr^2\frac{d\omega}{dt} + m\left\langle \frac{d\omega}{dt}, s \right\rangle s + \\ &\quad + m\langle s, \omega \rangle \frac{ds}{dt} + m\langle \omega, s \rangle \, \omega \times s + mg\, u \times s \end{aligned} \qquad (9)$$

and

$$\frac{du}{dt} = -\omega \times u, \tag{10}$$

by differentiating (2). These are the desired equations of motion of the disc after the $\mathcal{E}(2)$ symmetry has been removed. Here s and u are related by (5). Equations (9) and (10) define the integral curves of a vector field V on $S^2 \times \mathbf{R}^3$.

A straightforward calculation shows that the total energy of the disc

$$E = \tfrac{1}{2} \langle I(\omega), \omega \rangle + \tfrac{1}{2} \langle \omega \times s, \omega \times s \rangle + \mathrm{mg} \langle s, u \rangle \tag{11}$$

is constant on the integral curves of V.

There is still another symmetry–that of rotation about the principal axis perpendicular to the plane of the disc. In more detail, consider the S^1 action

$$\Psi : \mathcal{C} \times S^1 \to \mathcal{C} : \Big((A, a, b, \omega), R_\psi \Big) \to (A R_\psi^{-1}; a, R_\psi b, R_\psi \omega), \tag{12}$$

where $R_\psi = \begin{pmatrix} \cos\psi & -\sin\psi & 0 \\ \sin\psi & \cos\psi & 0 \\ 0 & 0 & 1 \end{pmatrix}$. Since the $\mathcal{E}(2)$-action Φ (8) and the S^1-action Ψ (12) commute, the S^1-action Ψ induces an S^1 action $\widetilde{\Psi}$ on the $\mathcal{E}(2)$-orbit space $S^2 \times \mathbf{R}^3$ defined by

$$\widetilde{\Psi} : S^1 \times (S^2 \times \mathbf{R}^3) \to S^2 \times \mathbf{R}^3 :$$
$$R_\psi, (u, \omega) \to (R_\psi u, R_\psi \omega) = \left(\mathrm{e}^{\psi E_3} u, \mathrm{e}^{\psi E_3} \omega \right),$$

where E_3 is the matrix $\begin{pmatrix} 0 & -1 & 0 \\ 1 & 0 & 0 \\ 0 & 0 & 0 \end{pmatrix}$. A straightforward calculation shows that the vector field V is invariant under the S^1 action $\widetilde{\Psi}$. Thus V induces a vector field on the space of S^1 orbits on $S^2 \times \mathbf{R}^3$.

In order to determine this vector field we need an explicit description of this orbit space. We use invariant theory. The algebra of polynomials on $\mathbf{R}^3 \times \mathbf{R}^3$, which are invariant under the diagonal action of $\mathrm{e}^{\psi E_3}$, is generated by

$$\begin{aligned}
\sigma_1 &= u_3 = \langle u, e_3 \rangle & \sigma_4 &= \omega_3 = \langle \omega, e_3 \rangle \\
\sigma_2 &= -(u_1\omega_2 - u_2\omega_1) = -\langle \omega \times u, e_3 \rangle & \sigma_5 &= \omega_1^2 + \omega_2^2 \\
\sigma_3 &= u_1\omega_1 + u_2\omega_2 = \langle \omega, \widehat{u} \rangle & \sigma_6 &= u_1^2 + u_2^2
\end{aligned} \tag{13}$$

and satisfy the relation

$$\sigma_2^2 + \sigma_3^2 = \sigma_5 \sigma_6 \quad \sigma_5 \geq 0. \quad \sigma_6 \geq 0. \tag{14}$$

The orbit space of the S^1-action $\widetilde{\Phi}$ is defined by (14) and

$$1 = \sigma_6 + \sigma_1^2, \tag{15}$$

because $\widetilde{\Psi}$ acts on $S^2 \times \mathbf{R}^3$. Using (15) to eliminate σ_6 from (14) shows that this orbit space is the semialgebraic variety $\mathcal{M} \subseteq \mathbf{R}^5$ defined by

$$\sigma_2^2 + \sigma_3^2 = \left(1 - \sigma_1^2\right)\sigma_5 \quad \sigma_5 \geq 0 \quad \text{and} \quad |\sigma_1| \leq 1. \tag{16}$$

$\mathcal{M}$ is *not* smooth, because the S^1 action $\widetilde{\Psi}$ has fixed points. In fact it is creased along the half-planes

$$\{\pm(1, 0, 0, \sigma_4, \sigma_5) \in \mathbf{R}^5 \,|\, \sigma_4 \in \mathbf{R}, \ \sigma_5 \geq 0\}.$$

In order that $s = \frac{\mathbf{r}}{\|u\|}\,\widehat{u}$ (see (5)) makes sense, we must have $|u_3| < 1$ and therefore $|\sigma_1| < 1$. Equation (16) with the condition $|\sigma_1| \leq 1$ replaced by $|\sigma_1| < 1$ defines a smooth S^1-invariant open dense subset $\overline{M}$ of $\mathcal{M}$, namely

$$\sigma_5 = \frac{\sigma_2^2 + \sigma_3^2}{1 - \sigma_1^2}, \qquad |\sigma_1| < 1. \tag{17}$$

Note that $\overline{M}$ is diffeomorphic to $\mathbf{R}^4$.

To determine the induced vector field $\overline{V}$ on $\overline{M}$ we compute the Lie derivative of σ_i with respect to V. Using $I_1 = \frac{1}{4}\,\mathrm{mr}^2$ and $I_3 = \frac{1}{2}\,\mathrm{mr}^2$. after a lengthy but straightforward calculation we obtain

$$\begin{aligned}
\frac{d\sigma_1}{dt} &= \sigma_2 \\
\frac{d\sigma_2}{dt} &= \tfrac{6}{5}\,\sigma_3\sigma_4 - \tfrac{1}{5}\frac{\sigma_1}{1 - \sigma_1^2}\,\sigma_3^2 - \frac{\sigma_1}{1 - \sigma_1^2}\,\sigma_2^2 + \lambda\sigma_1\sqrt{1 - \sigma_1^2} \\
\frac{d\sigma_3}{dt} &= -2\,\sigma_2\sigma_4 \\
\frac{d\sigma_4}{dt} &= -\tfrac{2}{3}\frac{1}{1 - \sigma_1^2}\,\sigma_2\sigma_3.
\end{aligned} \tag{18}$$

Here $\lambda = \dfrac{4\mathrm{g}}{5\mathrm{r}}$, and $|\sigma_1| < 1$.

Since the $\frac{4}{5\mathrm{mr}^2}$ times the total energy E (11) is invariant under the S^1 action $\widetilde{\Psi}$, it induces a function

$$\overline{E} = \tfrac{1}{2}\frac{\sigma_2^2}{1 - \sigma_1^2} + \tfrac{1}{10}\frac{\sigma_3^2}{1 - \sigma_1^2} + \tfrac{3}{5}\,\sigma_4^2 + \lambda\sqrt{1 - \sigma_1^2} \tag{19}$$

on $\overline{M}$ called the *reduced energy*. A straightforward calculation shows that $\overline{E}$ is constant on the integral curves of $\overline{V}$.

3 Relative equilibria

In this section we give a physical interpretation of the equilibrium points of the reduced vector field $\overline{V}$ (18), which give rise to the $\mathcal{E}(2) \times S^1$ *relative* equilibria of the rolling disc. These equilbria were called steady motions by Routh [17] and can be found in Neimark and Fufaev [14, p.304].

Because the $\mathcal{E}(2) \times S^1$-action on the constraint manifold $\mathcal{C}$ is free and proper with orbit mapping

$$\pi : \mathcal{C} \to \mathcal{C}/(\mathcal{E}(2) \times S^1) = \overline{M}.$$

an equilibrium point σ^0 of $\overline{V}$ gives rise to a motion on the fiber $\pi^{-1}(\sigma^0)$ which is conjugate to a 1-parameter group $t \to \exp t\xi = c(t)$ of $\mathcal{E}(2) \times S^1$. (See Hermans [8].) From the definitions of the $\mathcal{E}(2)$-action Φ (8) and the S^1-action Ψ (12) it follows that

$$\begin{aligned}
c(t) &= (A(t), a(t), \omega(t), b(t)) \\
&= (R_{\varphi t} A^0 R_{\psi t}^{-1}, R_{\varphi t} a^0 + x(t), R_{\psi t} \omega^0, R_{\psi t} b^0),
\end{aligned} \tag{20}$$

where $c(0) = (A^0, a^0, \omega^0, b^0) \in \mathcal{C}$. To determine the element ξ in the Lie algebra $e(2) \times \mathbf{R}$ of $\mathcal{E}(2) \times S^1$ we need only compute $(\varphi, A^0 b^0, \psi)$ as a function of the components of σ^0. We need the following

Facts.

$$\begin{aligned}
\omega^0 &= -\varphi u^0 - \psi e_3 \tag{21} \\
u(t) &= R_{\psi t} u^0 \tag{22} \\
s(A(t)) &= R_{\psi t} s^0 \tag{23} \\
\langle a(t), e_3 \rangle &= \langle a^0, e_3 \rangle \tag{24} \\
a(t) &= a^0 + \frac{1}{\varphi} \begin{pmatrix} \sin \varphi t & \cos \varphi t - 1 & 0 \\ 1 - \cos \varphi t & \sin \varphi t & 0 \\ 0 & 0 & 0 \end{pmatrix} A^0 b^0. \tag{25}
\end{aligned}$$

(3.1) **Proof.** First we check that (21) holds. Since $A(t) = R_{\varphi t} A^0 R_{\psi t}^{-1}$ and $\omega(t) = R_{\psi t} \omega^0 = \exp(\psi t E_3) \omega^0$, using the definition of $\omega(t)$ and the mapping i we obtain

$$\begin{aligned}
\exp(\psi t E_3) \omega^0 &= \omega(t) = i\Big(A(t)^{-1} \dot{A}(t)\Big) \\
&= i\Big(\mathrm{Ad}_{\exp(\psi t E_3)(A^0)^{-1}}(\varphi E_3)\Big) - i(\psi E_3) \\
&= \exp(\psi t E_3)\Big((A^0)^{-1}(\varphi e_3) - \psi e_3\Big).
\end{aligned}$$

Using the definition of u^0 we obtain (21). Equations (22), (23) and (24) follow directly from the definition of the $\mathcal{E}(2) \times S^1$-action. To prove (25) we note that from the definition of $b(t)$ it follows that $\dot{a}(t) = A(t)b(t) = R_{\varphi t} A^0 b^0$.

On the other hand, differentiating the definition of $a(t)$ in (20) gives

$$\dot{a}(t) = R_{\varphi t}(\varphi E_3 a^0) + \dot{x}.$$

Therefore

$$\dot{x} = R_{\varphi t}(A^0 b^0 - \varphi E_3 a^0). \tag{26}$$

Equation (26) integrates to

$$x(t) = \frac{1}{\varphi} \begin{pmatrix} \sin \varphi t & \cos \varphi t - 1 & 0 \\ 1 - \cos \varphi t & \sin \varphi t & 0 \\ 0 & 0 & 0 \end{pmatrix} (A^0 b^0 - \varphi E_3 a^0),$$

since $x(0) = x^0 = 0$ using (20). From the definition of $a(t)$ we obtain (25). □

▷ The following argument shows that the Lie algebra element $\xi = (\varphi, A^0 b^0, \psi)$ corresponding to the equilibrium point σ^0 of the reduced vector field $\overline{V}$ is

$$\varphi = -\frac{\sigma_3^0}{1 - (\sigma_1^0)^2} \qquad \text{and} \qquad \psi = -\sigma_4^0 + \frac{\sigma_1^0 \sigma_4^0}{1 - (\sigma_1^0)^2}. \tag{27}$$

(3.2) **Proof.** From the definition of σ_3 (13) we obtain

$$\begin{aligned} \sigma_3^0 &= \langle u^0, \omega^0 \rangle - u_3^0 \omega_3^0 \\ &= -\varphi - \psi \sigma_1^0 - \sigma_1^0 \sigma_4^0, \text{ using (21) and the definition of } \sigma_1 \text{ and } \sigma_4. \end{aligned}$$

Using (21) and the definition of σ_4 we find that

$$\sigma_4^0 = \langle \omega^0, e_3 \rangle = -\varphi \sigma_1^0 - \psi. \tag{28}$$

Solving the above two equations gives (27). □

Note that the choice of c^0 in the constraint manifold $\mathcal{C}$ which lies in the fiber of the orbit map π over the equilibrium point σ^0 of the reduced vector field $\overline{V}$ depends on $\dim \mathcal{C} - \dim(\mathcal{E}(2) \times S^1) = 8 - 4 = 4$ parameters: two which fix the initial position a^0 of the center of mass of the disc, one which fixes the initial amount of rotation φ^0 of the disc about the vertical and another ψ_0 which fixes the initial amount of rotation of the disc about an axis perpendicular to the plane of the disc.

▷ We now are in position to give a physical interpretation of the motion of the disc corresponding to an equilibrium point σ^0. The disc rolls uniformly with a frequency $-\frac{\sigma_3^0}{1-(\sigma_1^0)^2}$ at a constant inclination $\cos^{-1}(\sigma_1^0)$ and the point of contact traces out a circle with center at $a^0 + \frac{1}{\varphi} E_3 A^0 b^0$ and of radius

$$R = r \sqrt{1 - (\sigma_1^0)^2 + \frac{(\sigma_4^0)^2}{(\sigma_3^0)^2}(1 - \sigma_1^0 - (\sigma_1^0)^2)^2}. \tag{29}$$

(3.3) **Proof.** Using (23) and (25) we see that the position of the point of contact of the disc is given by

$$
\begin{aligned}
A(t)s(t) + a(t) &= R_{\varphi t}A^0 R_{\psi t}^{-1}(R_{\psi t}s^0) + a^0 + \frac{1}{\varphi}E_3 A^0 b^0 - \frac{1}{\varphi}R_{\varphi t}E_3 A^0 b^0 \\
&= a^0 + \frac{1}{\varphi}E_3 A^0 b^0 + R_{\varphi t}A^0\Big(s^0 - \frac{1}{\varphi}(A^0)^{-1}e_3 \times b^0\Big).
\end{aligned}
$$

Therefore the point of contact traces out a circle with center $a^0 + \frac{1}{\varphi}E_3 A^0 b^0$ at a frequency $\varphi = -\frac{\sigma_3^0}{1-(\sigma_1^0)^2}$, using (27). The square of the radius is

$$
\begin{aligned}
R^2 &= \left\|A^0(s^0 + \frac{1}{\varphi}u^0 \times b^0)\right\|^2 = \|s^0\|^2 + \frac{2}{\varphi}\langle s^0, u^0 \times b^0\rangle + \frac{1}{\varphi^2}\|u^0 \times b^0\|^2 \\
&= r^2 + \frac{2}{\varphi}\langle u^0, e_3\rangle\langle s^0 \times e_3, s^0 \times \omega^0\rangle + \frac{1}{\varphi^2}\|s^0 \times \omega^0\|^2,
\end{aligned}
$$

since $u^0 = \widehat{u^0} + \langle u^0, e_3\rangle e_3$, $b^0 = s^0 \times \omega^0$ and $\langle u^0, b^0\rangle = 0$. The last equality here follows because $\langle u^0, s^0 \times \omega^0\rangle = -\langle u^0, s^0 \times (\varphi u^0 + \psi e_3)\rangle = -\psi\langle u^0 \times s^0, e_3\rangle = -\psi\langle u^0, e_3\rangle\langle e_3 \times s^0, e_3\rangle = 0$. Therefore

$$
\begin{aligned}
R^2 &= r^2 + \frac{2}{\varphi}\langle u^0, e_3\rangle\langle s^0 \times (s^0 \times e_3), \varphi u^0 + \psi e_3\rangle + \frac{1}{\varphi^2}\|s^0 \times (\varphi u^0 + \psi e_3)\|^2 \\
&= r^2 - \frac{2}{\varphi}r^2\langle u^0, e_3\rangle(\varphi\langle u^0, e_3\rangle + \psi) + \frac{r^2}{\varphi^2}(\varphi\langle u^0, e_3\rangle + \psi)^2 \\
&= r^2 - r^2\langle u^0, e_3\rangle^2 + r^2\frac{\psi^2}{\varphi^2},
\end{aligned}
$$

which using (27) gives the desired result. $\qquad\square$

To determine which way the disc tilts as it rolls in a circle, it is enough to determine the sign of $\langle A^0 s^0 \times A^0 b^0, e_3\rangle$, since the vector $A^0 s^0$, which points from the initial position of the center of mass to the initial point of contact, is perpendicular to the vector $A^0 b^0$, which is the initial velocity of the center of mass. Now

$$
\begin{aligned}
\langle A^0 s^0 \times A^0 b^0, e_3\rangle &= \langle s^0, u^0 \times b^0\rangle = -r^2\langle u^0, e^3\rangle(\varphi\langle u^0, e_3\rangle + \psi) \\
&= r^2 \sigma_1^0 \sigma_4^0, \quad \text{using (28)}.
\end{aligned}
$$

Thus when $\sigma_1^0 > 0$ the disc tilts inward when $\sigma_4^0 > 0$ and outward when $\sigma_4^0 < 0$. Both cases can occur.

4 Analysis of the reduced vector field

In this section we give a complete qualitative description of the integral curves of the reduced vector field $\overline{V}$ on $\overline{M}$. We begin by showing that $\overline{V}$ has a 2-parameter family of invariant 2-dimensional manifolds on which its orbits are level sets of the

2-parameter family of functions $(\overline{\sigma}_3, \overline{\sigma}_4) \to H_{\overline{\sigma}_3, \overline{\sigma}_4} = K + U_{\overline{\sigma}_3, \overline{\sigma}_4}$ (37). We analyze this family using ideas from singularity theory. The results are given in figure 3.1. In more detail, first we find the discriminant locus, that is the set $\widetilde{\mathcal{D}}$ of parameter values where $H_{\overline{\sigma}_3, \overline{\sigma}_4}$ is *not* a Morse function. Actually we show in section 4.1 that when $(\overline{\sigma}_3, \overline{\sigma}_4)$ lies on $\widetilde{\mathcal{D}}$, the function $H_{\overline{\sigma}_3, \overline{\sigma}_4}$ has exactly one degenerate critical point. Next, we show ((4.1)) that $H_{\overline{\sigma}_3, \overline{\sigma}_4}$ is a proper Morse functon if and only if $U_{\overline{\sigma}_3, \overline{\sigma}_4}$ is. This happens precisely when $(\overline{\sigma}_3, \overline{\sigma}_4)$ does not lie either on $\widetilde{\mathcal{D}}$ or on two lines ℓ_1 (43) and ℓ_2 (44). Consequently, when $(\overline{\sigma}_3, \overline{\sigma}_4)$ lies in one of the eight connected components of $\mathbf{R}^2 - (\widetilde{\mathcal{D}} \cup (\ell_1 \cup \ell_2))$, the number and type of the nondegenerate critical points of $U_{\overline{\sigma}_3, \overline{\sigma}_4}$ does *not* change.

Next we need a lengthy analysis to establish what the graph of $U_{\overline{\sigma}_3, \overline{\sigma}_4}$ looks like. The results are given in figure 4.2. In more detail, we begin showing (see ((4.3)) – ((4.4))) that at the cusp point $(\overline{\sigma}_3^0, \overline{\sigma}_4^0) = (0, \sqrt{\frac{5\lambda}{12}})$ of $\widetilde{\mathcal{D}}$, the $\mathbf{Z}_2$-versal normal form of the 1-parameter family $\overline{\sigma}_4 \to U_{0,\overline{\sigma}_4}$ of even functions is $\mu \to b + \mu \sigma_1^2 + a \sigma_1^4$ where a and b are smooth functions of μ and $a(0) \neq 0$. Thus near $\sigma_1 = 0$ we have found the number and type of the critical points of $U_{0,\overline{\sigma}_4}$ for all $\overline{\sigma}_4$ near $\bar{\sigma}_4^0$. A global argument (see ((4.5)) – ((4.7))) shows that there are no other critical points on $(-1, 1)$. A similar argument works for the other cusp points of the discriminant locus. Thus we know the graph of $U_{\overline{\sigma}_3, \overline{\sigma}_4}$ on each connected component of $\mathbf{R}^2 - (\widetilde{\mathcal{D}} \cup (\ell_1 \cup \ell_2))$. To complete our description of the graphs of $U_{\overline{\sigma}_3, \overline{\sigma}_4}$, we look at the parameter values on the segments of $\widetilde{\mathcal{D}}$ between the cusp points and the points where the lines ℓ_1 and ℓ_2 meet $\widetilde{\mathcal{D}}$. Another analysis (see ((4.8)) – ((4.10))) using singularity theory shows that $U_{\overline{\sigma}_3, \overline{\sigma}_4}$ undergoes a Morse cancellation of a critical point of index 2 or 0 with a critical point of index 1 as the parameters $(\overline{\sigma}_3, \overline{\sigma}_4)$ cross each segment. Special arguments (see ((4.11)) – ((4.13))) take care of graphs of $U_{\overline{\sigma}_3, \overline{\sigma}_4}$ for the remaining parameter values on $\widetilde{\mathcal{D}}$. We now know the graphs of $U_{\overline{\sigma}_3, \overline{\sigma}_4}$ and hence the level sets of $H_{\overline{\sigma}_3, \overline{\sigma}_4}$.

Using the integral map $J : \mathbf{R}^4 \to \mathbf{R}^3 : (\sigma_1, \sigma_2, \overline{\sigma}_3, \overline{\sigma}_4) \to (H_{\overline{\sigma}_3, \overline{\sigma}_4}, \overline{\sigma}_3, \overline{\sigma}_4)$ of the reduced vector field, we investigate how these level sets foliate a reduced energy level set. First we describe the range and critical values of J (see ((4.14)) – ((4.15))) and then we determine the topology of each fiber (see figure 4.4 and table 1). Finally we investigate how these fibers fit together. This determines the topology of a reduced energy surface (see table 1) and shows how it is foliated.

4.1   Equilibrium points and linearization

We begin by looking for equilibrium points of $\overline{V}$, which are the relative equilbria of section 3. The set $\mathcal{E}$ of equilibrium points is defined by

$$
\begin{aligned}
E_1(\sigma) &= \sigma_2 = 0 \\
E_2(\sigma) &= \tfrac{6}{5} \, \sigma_3 \sigma_4 - \tfrac{1}{5} \, \frac{\sigma_1}{1 - \sigma_1{}^2} \, \sigma_3^2 + \lambda \sigma_1 \sqrt{1 - \sigma_1^2} = 0.
\end{aligned}
\tag{30}
$$

Since 0 is a regular value of $E_2|\{\sigma_2 = 0\}$, $\mathcal{E}$ is a smooth 2-dimensional submanifold of $(-1, 1) \times \mathbf{R}^3$. An additional argument, which we will not give, shows that every connected component of $\mathcal{E}$ is diffeomorphic to $\mathbf{R}^2$.

Linearizing the reduced vector field $\overline{V}$ at a point $\sigma' \in \mathcal{E}$ gives the linear map $D\overline{V}(\sigma')$, whose characteristic polynomial χ is

$$\chi(\mu) = \mu^4 + E_3(\sigma')\mu^2 = 0,$$

where

$$E_3(\sigma') = \tfrac{12}{5}\, \sigma_4^2 - \tfrac{4}{5}\, \frac{\sigma_1}{1-\sigma_1^2}\, \sigma_3\sigma_4 + \tfrac{1}{5}\, \frac{5 - 3\sigma_1^2}{(1-\sigma_1^2)^2}\, \sigma_3^2 - \lambda\, \frac{1 - 2\sigma_1^2}{\sqrt{1-\sigma_1^2}}. \tag{31}$$

Because $\mathcal{E}$ is a smooth 2-dimensional submanifold of $\{\sigma_2 = 0\}$, the nonzero roots of χ change type when

$$E_3(\sigma') = 0. \tag{32}$$

Equations (30) and (32) define the *eigenvalue locus* $\mathcal{E}V$ where $D\overline{V}(\sigma')$ has an eigenvalue zero of multiplicity 4. $\mathcal{E}V$ has four branches $\mathcal{B}^{\eta_1,\eta_2}$. As σ_1 varies over $[-\varepsilon_1, \varepsilon_1]$ with $\varepsilon_1 = \sqrt{\frac{24-9\sqrt{5}}{38}}$, $\mathcal{B}^{\eta_1,\eta_2}$ is defined by

$$\begin{aligned}
\sigma_3^{\eta_1,+}(\sigma_1) &= \eta_1 \sqrt{\frac{\lambda}{2}}\, \frac{(1-\sigma_1^2)^{3/4}}{(1 - \tfrac{2}{3}\sigma_1^2)^{1/2}}\, S \\
\sigma_4^{\eta_1,+}(\sigma_1) &= -\eta_1 \frac{5\sqrt{2\lambda}}{6}\, \frac{(1 - \tfrac{2}{3}\sigma_1^2)^{1/2}}{(1-\sigma_1^2)^{1/4}}\, \frac{\sigma_1}{S} + \\
&\quad + \eta_1 \tfrac{1}{12}\, \sqrt{2\lambda}\, \frac{\sigma_1}{(1-\sigma_1^2)^{1/4}(1 - \tfrac{2}{3}\sigma_1^2)^{1/2}}\, S,
\end{aligned} \tag{33}$$

where

$$S = \begin{cases} \sqrt{1 - 2\sigma_1^2 + \sqrt{\delta(\sigma_1^2)}}. & \text{if } \eta_2 = + \\[2mm] \sqrt{\tfrac{10}{3} + \widetilde{\delta}(\sigma_1^2)}, & \text{if } \eta_2 = -. \end{cases}$$

and

$$\begin{aligned}
\delta(\sigma_1^2) &= 1 - \tfrac{32}{3}\, \sigma_1^2 + \tfrac{76}{9}\, \sigma_1^4, \\
\widetilde{\delta}(\sigma_1^2) &= \frac{1 - 2\sigma_1^2 - \sqrt{\delta(\sigma_1^2)}}{\sigma_1^2} - \tfrac{10}{3}.
\end{aligned}$$

Because

$$\begin{aligned}
(\sigma_3^{++}(\varepsilon_1), \sigma_4^{++}(\varepsilon_1)) &= (\sigma_3^{--}(-\varepsilon_1), \sigma_4^{--}(-\varepsilon_1)), \\
(\sigma_3^{--}(\varepsilon_1), \sigma_4^{--}(\varepsilon_1)) &= (\sigma_3^{-+}(-\varepsilon_1), \sigma_4^{-+}(-\varepsilon_1)), \\
(\sigma_3^{-+}(\varepsilon_1), \sigma_4^{-+}(\varepsilon_1)) &= (\sigma_3^{+-}(-\varepsilon_1), \sigma_4^{+-}(-\varepsilon_1)),
\end{aligned}$$

and

$$(\sigma_3^{+-}(\varepsilon_1), \sigma_4^{+-}(\varepsilon_1)) = (\sigma_3^{++}(-\varepsilon_1), \sigma_4^{++}(-\varepsilon_1)),$$

we see that the projection of the branches of the eigenvalue locus $\mathcal{EV}$ along the σ_1-axis onto the σ_3-σ_4 plane fit together to form a topological circle $\widetilde{\mathcal{EV}}$, which is actually smooth.

4.2   A family of Hamiltonian systems

Introduce σ_1 as a new time scale by setting $' = \frac{d}{d\sigma_1} = \frac{1}{\sigma_2}\frac{d}{dt}$. In the new time scale the integral curves of $\overline{V}$ satisfy

$$\frac{d\sigma_1}{d\sigma_1} = 1$$

$$\sigma_2\frac{d\sigma_2}{d\sigma_1} = \tfrac{6}{5}\,\sigma_3\sigma_4 - \tfrac{1}{5}\,\frac{\sigma_1}{1-\sigma_1^2}\,\sigma_3^2 - \frac{\sigma_1}{1-\sigma_1^2}\,\sigma_2^2 + \lambda\,\sigma_1\sqrt{1-\sigma_1^2}$$

$$\frac{d\sigma_3}{d\sigma_1} = -2\,\sigma_4$$

$$\frac{d\sigma_4}{d\sigma_1} = -\tfrac{2}{3}\,\frac{1}{1-\sigma_1^2}\,\sigma_3. \tag{34}$$

The third and fourth equations in (34) form an *invariant* subsystem $\overline{U}$. Let Φ_{σ_1} be the flow of $\overline{U}$, that is, for every initial condition $(\overline{\sigma}_3, \overline{\sigma}_4)$

$$\sigma_1 \to \Big(\sigma_3(\sigma_1; \overline{\sigma}_3, \overline{\sigma}_4), \sigma_4(\sigma_1; \overline{\sigma}_3, \overline{\sigma}_4)\Big) = \Phi_{\sigma_1}(\overline{\sigma}_3, \overline{\sigma}_4), \tag{35}$$

is an integral curve of the vector field $\overline{U}$, which is real analytic in $(-1, 1)$. Substituting (35) into the first two equations of (18) gives

$$\frac{d\sigma_1}{dt} = \sigma_2$$

$$\frac{d\sigma_2}{dt} = \tfrac{6}{5}\,\sigma_3(\sigma_1; \overline{\sigma}_3, \overline{\sigma}_4)\,\sigma_4(\sigma_1; \overline{\sigma}_3, \overline{\sigma}_4) - \tfrac{1}{5}\,\frac{\sigma_1}{1-\sigma_1^2}\,\sigma_3^2(\sigma_1; \overline{\sigma}_3, \overline{\sigma}_4) \tag{36}$$

$$- \frac{\sigma_1}{1-\sigma_1^2}\,\sigma_2^2 + \lambda\,\sigma_1\,\sqrt{1-\sigma_1^2}.$$

A calculation verifies that (36) is in Hamiltonian form on $\mathbf{R}^2$ with coordinates (σ_1, σ_2) and symplectic form $\frac{1}{1-\sigma_1^2}\,d\sigma_1 \wedge d\sigma_2$. The corresponding Hamiltonian function is

$$H_{\overline{\sigma}_3, \overline{\sigma}_4} = K + U_{\overline{\sigma}_3, \overline{\sigma}_4}$$

$$= \tfrac{1}{2}\,\frac{\sigma_2^2}{1-\sigma_1^2} + \tfrac{1}{10}\,\frac{1}{1-\sigma_1^2}\,\sigma_3^2(\sigma_1; \overline{\sigma}_3, \overline{\sigma}_4) + \tfrac{3}{5}\,\sigma_4^2(\sigma_1; \overline{\sigma}_3, \overline{\sigma}_4) + \lambda\sqrt{1-\sigma_1^2}. \tag{37}$$

Let us now interpret the Hamiltonian system $X_{H_{\overline{\sigma}_3, \overline{\sigma}_4}}$ geometrically. Let Ψ_{σ_1} be the flow of (34), which is not defined on $\sigma_2 = 0$. Because the vector field $\overline{U}$ is an invariant subsystem of (34), we find that

$$\Psi_{\sigma_1'}\left(\overline{\sigma}_1, \overline{\sigma}_2, \overline{\sigma}_3, \overline{\sigma}_4\right) = \left(\sigma_1' + \overline{\sigma}_1, \sigma_2(\overline{\sigma}_1, \overline{\sigma}_2. \overline{\sigma}_3, \overline{\sigma}_4). \Phi_{\sigma_1' + \overline{\sigma}_1}(\overline{\sigma}_3, \overline{\sigma}_4)\right).$$

Thus for every fixed initial condition $(\overline{\sigma}_3, \overline{\sigma}_4) \in \mathbf{R}^2$, the image of the 2-plane

$$\Pi_{\overline{\sigma}_3, \overline{\sigma}_4} = \{(\sigma_1, \sigma_2, \overline{\sigma}_3, \overline{\sigma}_4) \in \mathbf{R}^4 \,|\, (\sigma_1, \sigma_2) \in \mathbf{R}^2\}$$

under the diffeomorphism

$$\varphi : (-1, 1) \times \mathbf{R}^3 \to (-1, 1) \times \mathbf{R}^3 : (\sigma_1, \sigma_2, \overline{\sigma}_3, \overline{\sigma}_4) \to \left(\sigma_1, \sigma_2, \Phi_{\sigma_1}(\overline{\sigma}_3, \overline{\sigma}_4)\right)$$

is a smooth *invariant* 2-dimensional manifold of the reduced vector field $\overline{V}$. Moreover, the image under φ of an *orbit* of $X_{H_{\overline{\sigma}_3, \overline{\sigma}_4}}$ on $\Pi_{\overline{\sigma}_3, \overline{\sigma}_4}$ is an *orbit* of $\overline{V}$. As $(\overline{\sigma}_3, \overline{\sigma}_4)$ varies over $\mathbf{R}^2$, the invariant manifolds $\varphi(\Pi_{\overline{\sigma}_3, \overline{\sigma}_4})$ foliate $(-1, 1) \times \mathbf{R}^3$. Hence we can obtain a full description of the phase portrait of the reduced vector field on all of $(-1, 1) \times \mathbf{R}^3$ from the phase portraits of the Hamiltonian systems $X_{H_{\overline{\sigma}_3, \overline{\sigma}_4}}$.

We can also look at the Hamiltonian $H_{\overline{\sigma}_3, \overline{\sigma}_4}$ in an algebraic way. This will allow us to compute with coordinates $\sigma = (\sigma_1, \sigma_2, \sigma_3, \sigma_4)$ instead of with coordinates $\widetilde{\sigma} = (\sigma_1, \sigma_2, \overline{\sigma}_3, \overline{\sigma}_4)$. Consider the function

$$\widetilde{H} = \widetilde{K} + \widetilde{U} = \tfrac{1}{2} \frac{1}{1 - \sigma_1^2} \sigma_2^2 + \tfrac{1}{10} \frac{1}{1 - \sigma_1^2} \sigma_3^2 + \tfrac{3}{5} \sigma_4^2 + \lambda \sqrt{1 - \sigma_1^2}. \qquad (38)$$

Then $H_{\overline{\sigma}_3, \overline{\sigma}_4} = \varphi^* \widetilde{H}$. Let $\mathcal{S} = \mathbf{R}\{\sigma_1\} \mathbf{R}[\sigma_2, \sigma_3, \sigma_4]$ be the ring of polynomials in $\sigma_2, \sigma_3, \sigma_4$ with coefficients in the ring of real analytic functions on $|\sigma_1| < 1$. Think of $\mathcal{S}$ as a *differential algebra* with derivations $\frac{\partial}{\partial \sigma_1}$ and $\frac{\partial}{\partial \sigma_2}$ whose action on the generators is given by

$$\frac{\partial}{\partial \sigma_1} \sigma_1 = 1, \quad \frac{\partial}{\partial \sigma_1} \sigma_2 = 0, \quad \frac{\partial}{\partial \sigma_1} \sigma_3 = -2\,\sigma_4. \quad \frac{\partial}{\partial \sigma_1} \sigma_4 = -\tfrac{2}{3} \frac{1}{1 - \sigma_1^2} \sigma_3,$$

and

$$\frac{\partial}{\partial \sigma_2} \sigma_1 = 0, \quad \frac{\partial}{\partial \sigma_2} \sigma_2 = 1, \quad \frac{\partial}{\partial \sigma_2} \sigma_3 = 0, \quad \frac{\partial}{\partial \sigma_2} \sigma_4 = 0,$$

respectively. Then every computation involving taking derivatives of $H_{\overline{\sigma}_3, \overline{\sigma}_4}$ can be performed on $\widetilde{H}$ considered as an element of $\mathcal{S}$.

4.3 Qualitative properties

This section is concerned with describing the qualitative features of the level sets of the 2-parameter family of Hamiltonian functions $(\overline{\sigma}_3. \overline{\sigma}_4) \to H_{\overline{\sigma}_3, \overline{\sigma}_4}$. The results are summarized in figure 4.1. We begin by finding the locus $\widetilde{\mathcal{D}}$ of parameter values

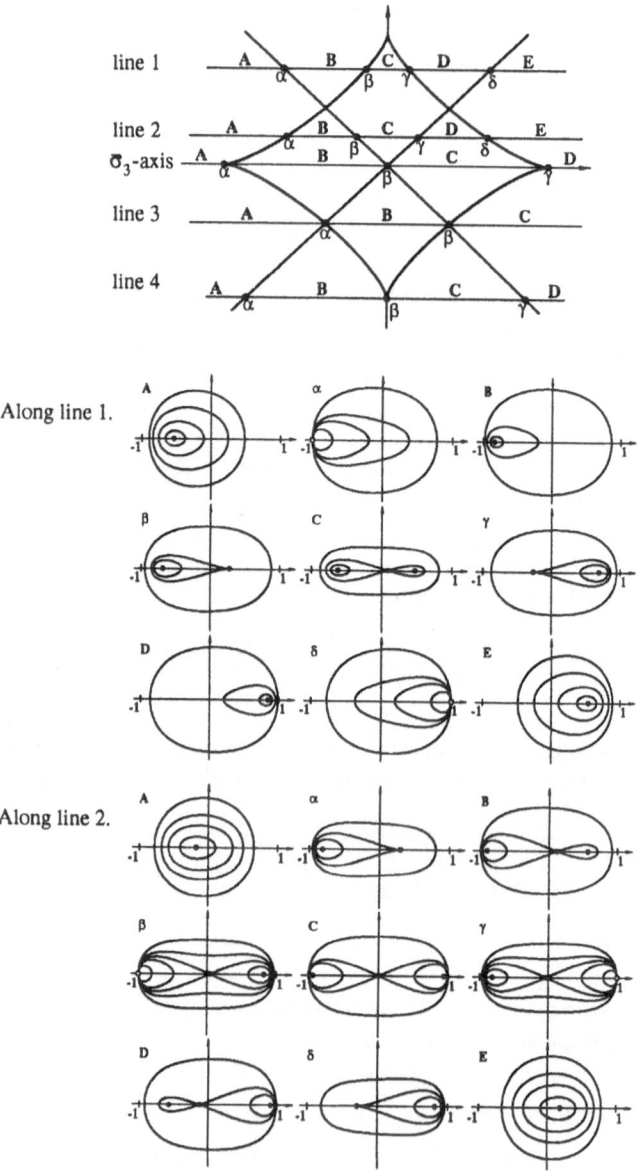

Figure 4.1. Level sets of $H_{\overline{\sigma}_3, \overline{\sigma}_4}$.

Along σ_3-axis.

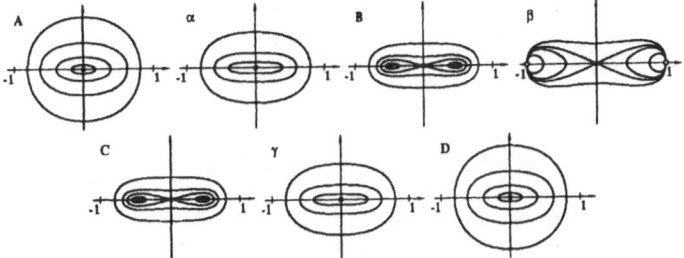

Along line 3.

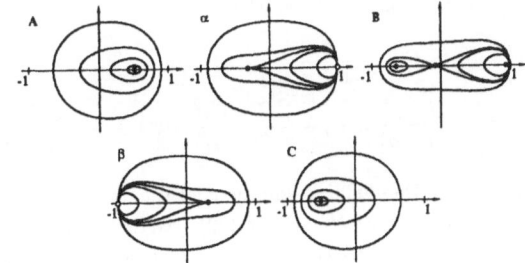

Along line 4.

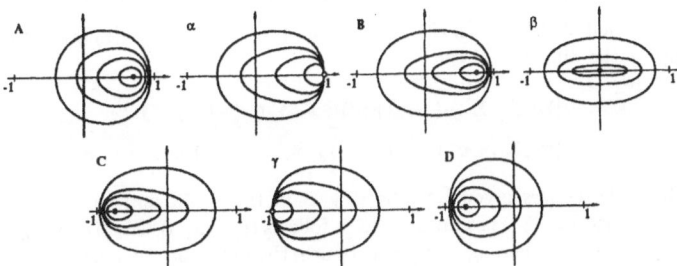

Figure 4.1 continued.

$(\overline{\sigma}_3, \overline{\sigma}_4)$ where $H_{\overline{\sigma}_3, \overline{\sigma}_4}$ has a *degenerate* critical point $\widehat{\sigma}$. A straightforward argument shows that

$$\frac{\partial \widetilde{H}}{\partial \sigma_1}(\widehat{\sigma}) = \frac{\partial \widetilde{H}}{\partial \sigma_2}(\widehat{\sigma}) = 0 \ \text{ and } \ \det\left(\frac{\partial^2 \widetilde{H}}{\partial \sigma_i \partial \sigma_j}(\widehat{\sigma})\right)_{1 \le i,j \le 2} = 0, \qquad (39)$$

if and only if

$$\widehat{\sigma}_2 = 0, \ \ \frac{\partial \widetilde{U}}{\partial \sigma_1}(\widehat{\sigma}) = 0, \ \text{ and } \ \frac{\partial^2 \widetilde{U}}{\partial \sigma_1{}^2}(\widehat{\sigma}) = 0. \qquad (40)$$

A calculation shows that the potential

$$\widetilde{U}(\sigma) = \tfrac{1}{10} \frac{1}{1 - \sigma_1^2} \sigma_3^2 + \tfrac{3}{5} \sigma_4^2 + \lambda \sqrt{1 - \sigma_1^2}$$

satisfies the last two equations in (40) if and only if

$$\tfrac{6}{5} \sigma_3 \sigma_4 - \tfrac{1}{5} \frac{\sigma_1}{1 - \sigma_1^2} \sigma_3^2 + \lambda \sigma_1 \sqrt{1 - \sigma_1^2} = 0$$

$$-\tfrac{12}{5} \sigma_4^2 + \tfrac{4}{5} \frac{\sigma_1}{1 - \sigma_1^2} \sigma_3 \sigma_4 - \tfrac{1}{5} \frac{5 - 3\sigma_1^2}{(1 - \sigma_1^2)^2} \sigma_3^2 + \lambda \frac{1 - 2\sigma_1^2}{\sqrt{1 - \sigma_1^2}} = 0. \qquad (41)$$

But $\sigma_2 = 0$ and (41) are the defining equations of the eigenvalue locus $\mathcal{EV}$; thus, $\mathcal{EV}$ is the locus of points where $\widetilde{H}$ satisfies (39). We determine the *discriminant locus* $\widetilde{\mathcal{D}}$ of the smooth family $(\overline{\sigma}_3, \overline{\sigma}_4) \rightarrow H_{\overline{\sigma}_3, \overline{\sigma}_4}$ as follows. First, consider the real analytic curve $\mathcal{B}$ in the first octant of σ_1-$\overline{\sigma}_3$-$\overline{\sigma}_4$ space made up of half of the branch $\mathcal{B}^{+-}$ of the eigenvalue locus $\mathcal{EV}$ going from $(0, 0, \sqrt{\tfrac{5\lambda}{12}})$ to $(\varepsilon_1, \sigma_3^{+-}(\varepsilon_1), \sigma_4^{+-}(\varepsilon_1))$, followed by half of the branch $\mathcal{B}^{++}$ going from $(\varepsilon_1, \sigma_3^{++}(\varepsilon_1), \sigma_4^{++}(\varepsilon_1))$ to $(0, 0, \sqrt{\lambda})$. Next apply the diffeomorphism

$$\widetilde{\psi} = (\varphi|\Sigma)^{-1} : \Sigma \subseteq (-1, 1) \times \mathbf{R}^3 \rightarrow \Sigma \subseteq (-1, 1) \times \mathbf{R}^3 :$$
$$(\sigma_1, 0, \sigma_3, \sigma_4) \rightarrow (\sigma_1, 0, \overline{\sigma}_3, \overline{\sigma}_4) = (\sigma_1, 0, \Phi_{-\sigma_1}(\sigma_3, \sigma_4)) \qquad (42)$$

to each of the curves given by reflecting $\mathcal{B}$ in the σ_1-$\overline{\sigma}_3$ and σ_1-$\overline{\sigma}_4$ coordinate planes. Because Φ_{σ_1} (35) is a linear map, it follows that $(\sigma_1, 0, \Phi_{-\sigma_1}(\eta_1 \sigma_3, \eta_2 \sigma_4)) = (\sigma_1, 0, \eta_1 \overline{\sigma}_3, \eta_2 \overline{\sigma}_4)$, where $\eta_1^2 = \eta_2^2 = 1$. Hence, the four reflected curves fit together to form a real analytic locus $\mathcal{D}$, which is singular at $(0, \pm\sqrt{\lambda}, 0)$ and $(0, 0, \pm\sqrt{\tfrac{5\lambda}{12}})$. Project $\mathcal{D}$ along the σ_1-axis onto the $\overline{\sigma}_3$-$\overline{\sigma}_4$ plane to obtain the discriminant locus $\widetilde{\mathcal{D}}$. Because this projection is a diffeomorphism when restricted to $\mathcal{B}$, we deduce that $\widetilde{\mathcal{D}}$ is a smooth circle with singular points at $(\pm\sqrt{\lambda}, 0)$ and $(0, \pm\sqrt{\tfrac{5\lambda}{12}})$. (See the diamond shaped curve in figure 4.1.)

▷ Next we show that the function $H_{\overline{\sigma}_3, \overline{\sigma}_4}$ on $(-1, 1) \times \mathbf{R}$ is proper except when $(\overline{\sigma}_3, \overline{\sigma}_4)$ lies on two lines ℓ_1 and ℓ_2 (see (43) and (44) below).

(4.1) **Proof.** From (37) we see that $H_{\overline{\sigma}_3,\overline{\sigma}_4}$ is the sum of two nonnegative functions K and $U_{\overline{\sigma}_3,\overline{\sigma}_4}$. Therefore $H_{\overline{\sigma}_3,\overline{\sigma}_4}$ is nonnegative and for every $h \geq 0$

$$H_{\overline{\sigma}_3,\overline{\sigma}_4}^{-1}([0,h]) \subseteq K^{-1}([0,h]) \cap \left(U_{\overline{\sigma}_3,\overline{\sigma}_4}^{-1}([0,h]) \times \mathbf{R} \right).$$

But

$$K^{-1}([0,h]) = \{(\sigma_1,\sigma_2) \in (-1,1) \times \mathbf{R} \,|\, \sigma_2^2 \leq h(1-\sigma_1^2)\}$$

is bounded in $(-1,1) \times \mathbf{R}$. Hence $H_{\overline{\sigma}_3,\overline{\sigma}_4}^{-1}([0,h])$ is compact if $U_{\overline{\sigma}_3,\overline{\sigma}_4}^{-1}([0,h])$ is a closed subset of $\mathbf{R}$ and hence is compact, that is. if $U_{\overline{\sigma}_3,\overline{\sigma}_4}$ is proper.

To prove that the potential $U_{\overline{\sigma}_3,\overline{\sigma}_4}$ is proper, it suffices to show that each of the nonnegative functions $\lambda\sqrt{1-\sigma_1^2}$, $\sigma_3^2(\sigma_1;\overline{\sigma}_3,\overline{\sigma}_4)$ and $\sigma_4^2(\sigma_1;\overline{\sigma}_3,\overline{\sigma}_4)$ (35) is proper on $(-1,1)$. Because $\sigma_1 \to \lambda\sqrt{1-\sigma_1^2}$ has a continuous extension to $[-1,1]$, it is proper. From the asymptotic properties of the function $\sigma_1 \to \sigma_3(\sigma_1;\overline{\sigma}_3,\overline{\sigma}_4)$ (see appendix 1), it follows that its square has a continuous extension to $[-1,1]$ and hence is proper. Now consider the lines ℓ_1 and ℓ_2 of initial conditions defined by

$$\ell_1: \qquad \mu \to \mu \left(\frac{\frac{4}{3}\Gamma(\frac{1}{2})}{|\Gamma(\frac{3}{4}+i\sqrt{\frac{13}{3}})|^2}, \frac{2\Gamma(\frac{3}{2})}{|\Gamma(\frac{3}{4}+i\sqrt{\frac{13}{3}})|^2} \right) \tag{43}$$

$$\ell_2: \qquad \mu \to \mu \left(\frac{\frac{4}{3}\Gamma(\frac{1}{2})}{|\Gamma(\frac{3}{4}+i\sqrt{\frac{13}{3}})|^2}, -\frac{2\Gamma(\frac{3}{2})}{|\Gamma(\frac{3}{4}+i\sqrt{\frac{13}{3}})|^2} \right), \tag{44}$$

respectively. From the results in appendix 1 we know that if $(\overline{\sigma}_3,\overline{\sigma}_4) \in \ell_1$ the function $\sigma_1 \to \sigma_4(\sigma_1;\overline{\sigma}_3,\overline{\sigma}_4)$ has a finite limit as $\sigma_1 \to 1^-$ and blows up like $\pm \ln\frac{1}{1+\sigma_1}$ as $\sigma_1 \to -1^+$. Also when $(\overline{\sigma}_3,\overline{\sigma}_4) \in \ell_2$, the function $\sigma_1 \to \sigma_4(\sigma_1;\overline{\sigma}_3,\overline{\sigma}_4)$ has a finite limit as $\sigma_1 \to -1^+$ and blows up like $\pm \ln\frac{1}{1-\sigma_1}$ as $\sigma_1 \to 1^-$. Therefore when $(\overline{\sigma}_3,\overline{\sigma}_4) \in \ell_1 \cup \ell_2$ the function $\sigma_1 \to \sigma_4^2(\sigma_1;\overline{\sigma}_3,\overline{\sigma}_4)$ is *not* proper; therefore neither is the potential $U_{\overline{\sigma}_3,\overline{\sigma}_4}$. For $(\overline{\sigma}_3,\overline{\sigma}_4) \notin \ell_1 \cup \ell_2$, the function $\sigma_1 \to \sigma_4^2(\sigma_1;\overline{\sigma}_3,\overline{\sigma}_4)$ blows up like $\left(\ln\frac{1}{1\mp\sigma_1}\right)^2$ as $\sigma_1 \to \pm1^\mp$. Therefore $U_{\overline{\sigma}_3,\overline{\sigma}_4}^{-1}([0,h])$ is a compact subset of $(-1,1)$ and hence $U_{\overline{\sigma}_3,\overline{\sigma}_4}$ is proper. $\quad\square$

Because $H_{\overline{\sigma}_3,\overline{\sigma}_4}$ is a proper Morse function when $(\overline{\sigma}_3,\overline{\sigma}_4) \notin \widetilde{\mathcal{D}} \cup \ell_1 \cup \ell_2$, it has a finite number of critical points all of which are nondegenerate. Since a nondegenerate critical point is isolated, their number and type remain the same as $(\overline{\sigma}_3,\overline{\sigma}_4)$ varies over each connected component of $\mathbf{R}^2 - (\widetilde{\mathcal{D}} \cup \ell_1 \cup \ell_2)$. (See figure 4.1).

To determine the topology of the level sets of the Hamiltonian $H_{\overline{\sigma}_3,\overline{\sigma}_4}$ as $(\overline{\sigma}_3,\overline{\sigma}_4)$ varies over $\mathbf{R}^2$, we need only determine the graph of the potential $U_{\overline{\sigma}_3,\overline{\sigma}_4}$. Our results are summarized in figure 4.2.

Let $\mathcal{R}$ be the compact region bounded by the discriminant locus $\widetilde{\mathcal{D}}$. First we determine what the graph of $U_{\overline{\sigma}_3,\overline{\sigma}_4}$ looks like when $(\overline{\sigma}_3,\overline{\sigma}_4) \in \Delta_1 = \mathcal{R} - \mathcal{R} \cap (\ell_1 \cup \ell_2)$. We will use singularity theory to show how $U_{\overline{\sigma}_3,\overline{\sigma}_4}$ changes along the $\overline{\sigma}_4$-axis

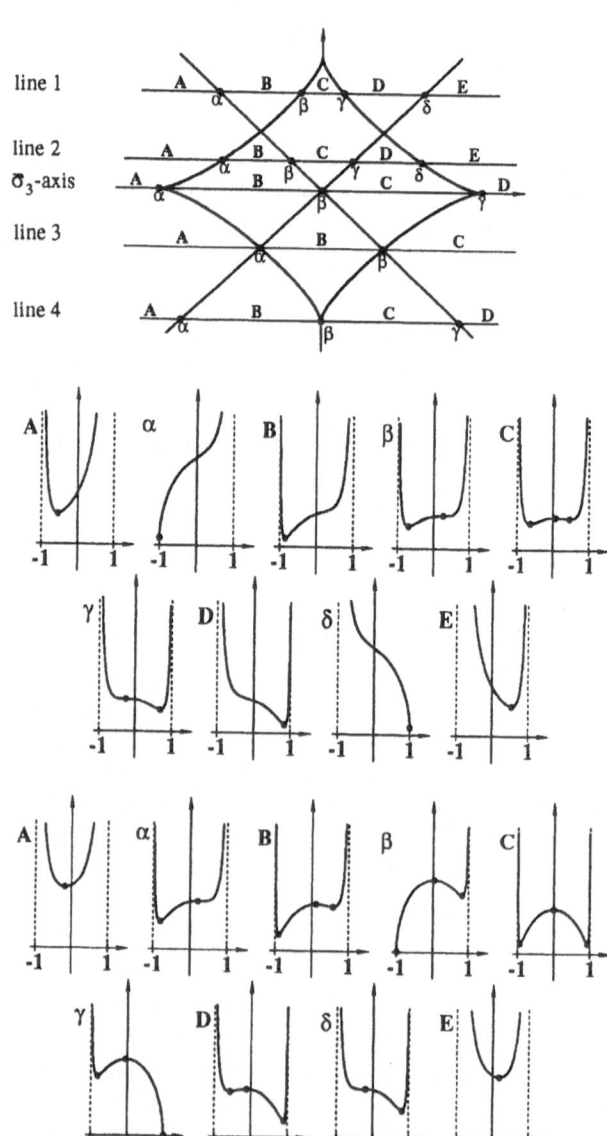

Figure 4.2. The potentials $U_{\overline{\sigma}_3, \overline{\sigma}_4}$.

Along $\overline{\sigma}_3$-axis.

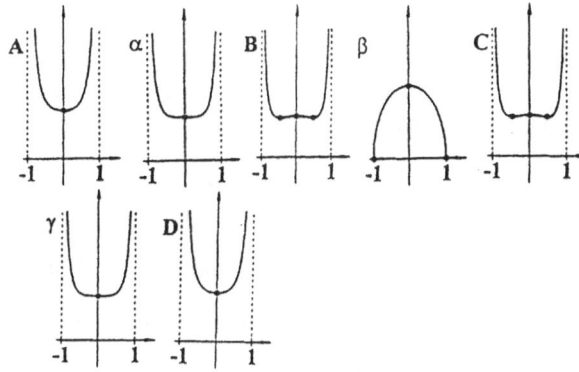

Along line 3.

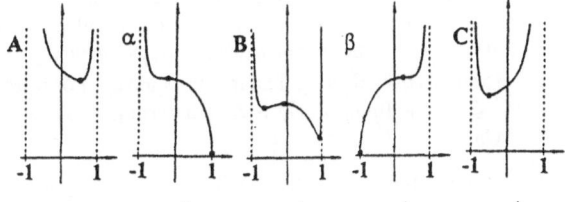

Along line 4.

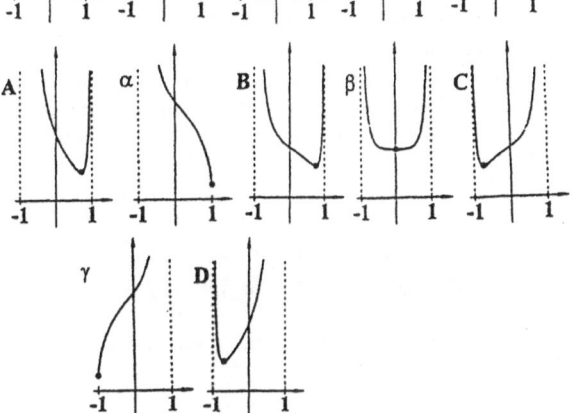

Figure 4.2 continued.

near $(0, \sqrt{\frac{5\lambda}{12}})$ where it intersects $\widetilde{\mathcal{D}}$. (See Poston and Stewart [16] for background on the singularity theory used here.) The argument for the other intersection point $(0, -\sqrt{\frac{5\lambda}{12}})$, and for the intersection points of the $\overline{\sigma}_3$-axis with $\widetilde{\mathcal{D}}$. is similar to the one given below and is omitted.

▷ We begin by proving the smooth 1-parameter family of smooth *even* functions $\mu \to G_\mu$, where

$$G_\mu(\sigma_1) = \tfrac{1}{2}\,\mu\,\sigma_1^2 + a(\mu)\,\sigma_1^4 \qquad \text{with } a(0) \neq 0 \tag{45}$$

is *versal* near 0 in the space of smooth even functions for all μ near 0.

(4.2) **Proof.** Consider the $\mathbf{Z}_2$-action on $\mathbf{R}$ generated by $\sigma_1 \to -\sigma_1$. It suffices to work in the space $\mathbf{R}[[\sigma_1^2]]$ of formal power series in the generator σ_1^2 of the algebra of $\mathbf{Z}_2$-invariant polynomials. The $\mathbf{Z}_2$-Jacobian ideal $\mathcal{J}$ of G_0 is generated by $\partial G_0 = 4a(0)\,\sigma_1^4$, since the vector field $\partial = \sigma_1 \frac{\partial}{\partial \sigma_1}$ generates the module of $\mathbf{Z}_2$-invariant vector fields over $\mathbf{R}[[\sigma_1^2]]$. The maximal ideal $\mathcal{M}$ of $\mathbf{R}[[\sigma_1^2]]$ is generated by σ_1^2. Since $\mathcal{M}^2 \subseteq \mathcal{J}$, the function G_0 is $\mathbf{Z}_2$ 1-determined. Thus G_μ is the $\mathbf{Z}_2$-versal unfolding of G_0. □

From versality of the family $\mu \to G_\mu$ it follows that for any smooth family $\nu \to L_\nu$ of smooth even functions, whose 4-jet at the origin is close to G_0, there is a local diffeomorphism $\vartheta : \mathbf{R} \to \mathbf{R}$ fixing the origin with $\vartheta(-x) = -\vartheta(x)$ and a smooth map $\rho : \mathbf{R} \to \mathbf{R}$ with $\rho(0) = 0$ such that in a neighborhood of 0 we have $\vartheta^* L_{\rho(\mu)} = G_\mu$ for every μ near 0. Consequently, the graphs of $L_{\rho(\mu)}$ and G_μ are the same near 0 for every sufficiently small μ.

▷ To show that $\mu \to G_\mu$ (45) is the versal normal form of the smooth family $\overline{\sigma}_4 \to U_{0,\overline{\sigma}_4} - U_{0,\overline{\sigma}_4^0}(0) = U_{\overline{\sigma}_4} - U_{\overline{\sigma}_4^0}(0)$ when $\overline{\sigma}_4$ is near $\overline{\sigma}_4^0 = \sqrt{\frac{5\lambda}{12}}$. we need only verify that the 4th derivative of $U_{\overline{\sigma}_4}$ at 0 is nonzero.

(4.3) **Proof.** By definition of the potential function $U_{\overline{\sigma}_3, \overline{\sigma}_4}$ (37),

$$U_{\overline{\sigma}_4}(\sigma_1) = \tfrac{1}{10}\,\frac{1}{1-\sigma_1^2}\,\sigma_3^2(\sigma_1; 0, \overline{\sigma}_4) + \tfrac{3}{5}\,\sigma_4^2(\sigma_1; 0, \overline{\sigma}_4) + \lambda\sqrt{1-\sigma_1^2}.$$

From equations (71) and (74) of appendix 1 we obtain

$$\sigma_3(\sigma_1) = \sigma_3(\sigma_1; 0, \overline{\sigma}_4) = -3\,\overline{\sigma}_4\,\sigma_1(1-\sigma_1^2)\,\frac{dF}{d\sigma_1^2}(\tfrac{\nu}{2}+\tfrac{1}{2}, -\tfrac{\nu}{2}, \tfrac{1}{2}; \sigma_1^2)$$

and

$$\sigma_4(\sigma_1) = \sigma_4(\sigma_1; 0, \overline{\sigma}_4) = \overline{\sigma}_4\,F(\tfrac{\nu}{2}+\tfrac{1}{2}, -\tfrac{\nu}{2}, \tfrac{1}{2}; \sigma_1^2);$$

hence $U_{\bar{\sigma}_4}$ is an even function. Therefore 0 is a critical point for every $\bar{\sigma}_4$. From the Taylor series of F (72) and the fact that $\frac{\nu}{2} + \frac{1}{2} = \frac{1}{4} + i\frac{1}{4}\sqrt{\frac{13}{3}}$ we obtain

$$\sigma_4(\sigma_1) = \bar{\sigma}_4 \left(1 + \frac{2}{3}\,\sigma_1^2 + \frac{11}{27}\,\sigma_1^4 + O(\sigma_1^6)\right)$$

and

$$\sigma_3(\sigma_1) = -3\,\bar{\sigma}_4\,\sigma_1(1 - \sigma_1^2)\left(\frac{2}{3} + \frac{22}{27}\,\sigma_1^2 + O(\sigma_1^4)\right).$$

Therefore the Taylor series of $U_{\bar{\sigma}_4}$ about 0 up through terms of order 4 is

$$U_{\bar{\sigma}_4}(\sigma_1) = U_{\bar{\sigma}_4}(0) + (\tfrac{6}{5}\,\bar{\sigma}_4^2 - \tfrac{\lambda}{2})\sigma_1^2 + (\tfrac{4}{3}\,\bar{\sigma}_4^2 - \tfrac{\lambda}{8})\sigma_1^4 + O(\sigma_1^6). \tag{46}$$

When $\bar{\sigma}_4 = \bar{\sigma}_4^0 = \sqrt{\frac{5\lambda}{12}}$, the coefficient of σ_1^2 in (46) vanishes and the coefficient of σ_1^4 is positive. $\qquad\square$

$\triangleright$ We now show that for every $(\bar{\sigma}_3, \bar{\sigma}_4)$ in the connected component of Δ_1 containing the portion of the $\bar{\sigma}_4$-axis where $0 < \bar{\sigma}_4 < \bar{\sigma}_4^0$, the function $U_{\bar{\sigma}_4}$ is a Morse function with three critical points: one of index 1 and two of index 0.

(4.4) **Proof.** Since the fourth derivative of $U_{\bar{\sigma}_4^0}$ at 0 is positive, from (46) and the versal normal form (45) it follows that for $\bar{\sigma}_4$ slightly less than $\bar{\sigma}_4^0$ the function $U_{\bar{\sigma}_4}$ is a Morse function with 0 a critical point of index 1 and two critical points $\pm\sigma_1^0$, $\sigma_1^0 > 0$ of index 1. The following argument shows that $U_{\bar{\sigma}_4}$ has no other critical points. Because $U_{\bar{\sigma}_4}$ is even, we need only look at $U_{\bar{\sigma}_4}$ on the interval $[0, 1)$. Suppose that $U_{\bar{\sigma}_4}$ has more than three critical points. Let $\hat{\sigma}_1$ be a fourth critical point. Since $(0, \bar{\sigma}_4) \notin \tilde{\mathcal{D}}$, $\hat{\sigma}_1$ is nondegenerate. Because $U_{\bar{\sigma}_4}$ is proper, it has only a finite number of nondegenerate critical points. Choose $\hat{\sigma}_1 \in (0, \sigma_1^0)$ as close to 0 as possible. Then $\hat{\sigma}_1$ is of index 0. For suppose that it has index 1. Then there must be a critical point between 0 and $\hat{\sigma}_1$ which is nondegenerate and of index 0. This contradicts the definition of $\hat{\sigma}_1$. Hence $\hat{\sigma}_1$ is of index 0. By the same argument there is another critical point $\tilde{\sigma}_1 \in (\hat{\sigma}_1, \sigma_1^0)$ which is nondegenerate and of index 1. Let $\bar{\sigma}_4$ increase toward $\bar{\sigma}_4^0$. Then neither $\hat{\sigma}_1$ nor $\tilde{\sigma}_1$ can run off to 1. Hence, either $\hat{\sigma}_1$ or $\tilde{\sigma}_1$ remain unequal and converge to 0 or σ_1^0 or for some $\bar{\sigma}_4'$ the critical points $\hat{\sigma}_1$ and $\tilde{\sigma}_1$ are equal. Neither possibility can occur: the first because 0 and σ_1^0 are nondegenerate critical points and hence are isolated; and the second because the resulting critical point is degenerate. But $(0, \bar{\sigma}_4') \in \tilde{\mathcal{D}}$. Thus neither $\hat{\sigma}_1$ nor $\tilde{\sigma}_1$ exists. Now suppose that $\hat{\sigma}_1 \in (\sigma_1^0, 1)$ and $\hat{\sigma}_1$ is as close to σ_1^0 as possible. Then $\hat{\sigma}_1$ has index 1. Since $(0, \bar{\sigma}_4) \notin \ell_1 \cup \ell_2$, the function $U_{\bar{\sigma}_4}$ blows up as $\sigma_1 \to 1^-$. Hence there is a critical point $\tilde{\sigma}_1 \in (\hat{\sigma}_1, 1)$ which is of index 0. Again let $\bar{\sigma}_4$ increase toward $\bar{\sigma}_4^0$. Then neither $\hat{\sigma}_1$ nor $\tilde{\sigma}_1$ can run off to 1. Repeating the argument above shows that they cannot fuse or converge to σ_1^0. Thus they do not exist. Hence $U_{\bar{\sigma}_4}$ is a Morse function with three critical points: one of index 1 and two of index 0 for $\bar{\sigma}_4$ slightly less than $\bar{\sigma}_4^0$.

Because $U_{\overline{\sigma}_3, \overline{\sigma}_4}$ is a *proper* Morse function for every $(\overline{\sigma}_3, \overline{\sigma}_4)$ in the connected component of Δ_1 containing the portion of the $\overline{\sigma}_4$-axis where $\overline{\sigma}_4 \in (0, \bar{\sigma}_4^0)$, the potential $U_{\overline{\sigma}_3, \overline{\sigma}_4}$ is a Morse function with *three* critical points: one of index 1 and two of index 0.   □

A similar argument holds for every connected component of Δ_1.

▷ Next we show that 0 is the only critical point of $U_{\bar{\sigma}_4^0}$ in $(-1, 1)$. It is degenerate and $U_{\bar{\sigma}_4^0} - U_{\bar{\sigma}_4^0}(0)$ has a nonzero 4-jet there.

(4.5) **Proof.** Suppose not. Then $U_{\bar{\sigma}_4^0}$ has a nonzero critical point $\hat{\sigma}_1$. Since $(\hat{\sigma}_1, 0, \bar{\sigma}_4^0)$ does not lie on the locus $\mathcal{D}$, the critical point $\hat{\sigma}_1$ is a nondegenerate. Choose $\hat{\sigma}_1$ to be the closest nonzero critical point to 0. Then $\hat{\sigma}_1$ has index 1. Hence for $\overline{\sigma}_4$ slightly smaller than $\bar{\sigma}_4^0$, the function $U_{\overline{\sigma}_4}$ has a nonzero critical point of index 1. Since $(0, \overline{\sigma}_4) \in \Delta_1$, by ((4.4)) $U_{\overline{\sigma}_4}$ has a unique critical point of index 1, which lies at the origin, because $U_{\overline{\sigma}_4}$ is even. This is a contradiction.   □

▷ We now show that in the connected component of $\Delta_2 = \mathbf{R}^2 - (\mathcal{R} \cup \ell_1 \cup \ell_2)$ containing that portion of the $\overline{\sigma}_4$-axis where $\overline{\sigma}_4 > \bar{\sigma}_4^0$ the potential $U_{\overline{\sigma}_4}$ is a Morse function with *one* critical point 0 of index 0.

(4.6) **Proof.** From the versal normal form (45) it follows that there is a neighborhood $\mathcal{N}$ of 0 such that for every $\overline{\sigma}_4$ slightly larger than $\bar{\sigma}_4^0$, the only critical point of $U_{\overline{\sigma}_4}$ in $\mathcal{N}$ is 0. Because the function $U_{\overline{\sigma}_4}$ is even, we may restrict our attention to the interval $[0, 1)$. Suppose that σ_1' is a nonzero critical point of $U_{\overline{\sigma}_4}$ for some $\overline{\sigma}_4$ slightly larger than $\bar{\sigma}_4^0$. As in ((4.5)) we may choose σ_1' to be the closest nonzero critical point to 0. Thus σ_1' has index 1. Since $U_{\overline{\sigma}_4}$ blows up to $+\infty$ as $\sigma_1 \to \pm 1^{\pm}$, there is another critical point $\widehat{\sigma_1}$ between σ_1' and the closer of -1 or 1. Let $\overline{\sigma}_4$ decrease to $\bar{\sigma}_4^0$. Because $U_{\overline{\sigma}_4}$ is proper when $\overline{\sigma}_4 > \bar{\sigma}_4^0$ and because 0 is the only critical point of $U_{\bar{\sigma}_4^0}$, one of the following two situations occur. Either for some $\bar{\sigma}_4'$ slightly larger than $\bar{\sigma}_4^0$ the function $U_{\bar{\sigma}_4'}$ has a degenerate critical point $\tilde{\sigma}_1$ (that is, the critical points σ_1' and $\hat{\sigma}_1$ fuse together at $\tilde{\sigma}_1$) or σ_1' and $\hat{\sigma}_1$ remain unequal and lie in $\mathcal{N}$ (that is, they converge to 0). Neither of these possibilities can occur: the first because $(\tilde{\sigma}_1, 0, \bar{\sigma}_4') \notin \mathcal{D}$ and the second by the construction of the neighborhood $\mathcal{N}$.

Since $U_{\overline{\sigma}_3, \overline{\sigma}_4}$ is proper when $(\overline{\sigma}_3, \overline{\sigma}_4)$ lies in the connected component of Δ_2 containing the segment $(\bar{\sigma}_4^0, \infty)$ of the $\overline{\sigma}_4$-axis, we deduce that $U_{\overline{\sigma}_3, \overline{\sigma}_4}$ has a unique nondegenerate minimum there.   □

A similar argument shows that the conclusion of ((4.6)) holds on every connected component of Δ_2. This completes the description of the graph of $U_{\overline{\sigma}_3, \overline{\sigma}_4}$ when $(\overline{\sigma}_3, \overline{\sigma}_4)$ lies in a connected component of $\mathbf{R}^2 - (\widetilde{\mathcal{D}} \cup \ell_1 \cup \ell_2)$.

We now determine what the graph of $U_{\overline{\sigma}_3, \overline{\sigma}_4}$ looks like when $(\overline{\sigma}_3, \overline{\sigma}_4)$ lies in $\widetilde{\mathcal{D}} \cup \ell_1 \cup \ell_2$.

▷ First we treat the case when $(\bar{\sigma}_3^0, \bar{\sigma}_4^0) \in \widetilde{\mathcal{D}}^* = \widetilde{\mathcal{D}} - \{(\pm\sqrt{\lambda}, 0), (0, \pm\sqrt{\frac{5\lambda}{12}})\}$ The 3-jet of $U_{\bar{\sigma}_3^0, \bar{\sigma}_4^0} - U_{\bar{\sigma}_3^0, \bar{\sigma}_4^0}(\sigma_1^0)$ at the degenerate critical point σ_1^0 is nonzero. It suffices to verify that $\frac{\partial^3 \widetilde{U}}{\partial \sigma_1^3}(\sigma)$ is nonzero at every point $\sigma \in \mathcal{EV}^* = \mathcal{EV} - \{(0, \pm\sqrt{\lambda}, 0), (0, 0, \pm\sqrt{\frac{5\lambda}{12}})\}$.

(4.7) **Proof.** Let $\widetilde{V} = -\frac{1}{1-\sigma_1^2} \frac{\partial \widetilde{U}}{\partial \sigma_1}$ and define $\widetilde{W}$ by $\frac{\partial \widetilde{V}}{\partial \sigma_1} = \frac{1}{1-\sigma_1^2} \widetilde{W}$. A calculation shows that

$$\widetilde{W} = -\frac{12}{5}(1-\sigma_1^2)\sigma_4^2 + \frac{4}{5}\sigma_1\sigma_3\sigma_4 - \frac{1}{5}\frac{5-3\sigma_1^2}{1-\sigma_1^2}\sigma_3^2 + \lambda(1-2\sigma_1^2)\sqrt{1-\sigma_1^2}. \quad (47)$$

Since $\frac{\partial \widetilde{U}}{\partial \sigma_1}(\sigma) = \frac{\partial^2 \widetilde{U}}{\partial \sigma_1^2}(\sigma) = 0$, it follows that $\frac{\partial^3 \widetilde{U}}{\partial \sigma_1^3}(\sigma) = -\frac{1}{(1-\sigma_1^2)^2}\frac{\partial \widetilde{W}}{\partial \sigma_1}(\sigma)$, where

$$\frac{\partial \widetilde{W}}{\partial \sigma_1}(\sigma) = \frac{16}{5}\sigma_1\sigma_4^2 - \frac{4}{15}\frac{\sigma_1(5-2\sigma_1^2)}{(1-\sigma_1^2)^2}\sigma_3^2 + \frac{8}{5}\frac{5-4\sigma_1^2}{1-\sigma_1^2}\sigma_3\sigma_4 - \lambda\frac{5-6\sigma_1^2}{\sqrt{1-\sigma_1^2}}. \quad (48)$$

Note that $\sigma \in \mathcal{EV}^*$ if an only if (41) holds and $\sigma \neq 0$. Using the second equation in (41) to eliminate σ_4^2 in (48) gives

$$\frac{\partial \widetilde{W}}{\partial \sigma_1}(\sigma) = \frac{8}{3}\frac{3-2\sigma_1^2}{1-\sigma_1^2}\sigma_3\sigma_4 + \frac{4}{3}\frac{\sigma_1(2-\sigma^2)}{(1-\sigma_1^2)^2}\sigma_3^2 - \frac{1}{3}\lambda\frac{\sigma_1(10\sigma_1^2-11)}{\sqrt{1-\sigma_1^2}}. \quad (49)$$

Using the first equation in (41) to eliminate $\sigma_3\sigma_4$ in (49) gives

$$\frac{\partial \widetilde{W}}{\partial \sigma_1}(\sigma) = \frac{4}{9}\frac{\sigma_1(\sigma^2-3)}{(1-\sigma_1^2)^2}\sigma_3^2 + \frac{1}{9}\lambda\frac{\sigma_1(70\sigma_1^2-93)}{\sqrt{1-\sigma_1^2}}. \quad (50)$$

Finally using the parametrization (33) of $\mathcal{EV}$ we obtain

$$\frac{\partial \widetilde{W}}{\partial \sigma_1}(\sigma) = -\frac{1}{9}\lambda\frac{\sigma_1}{(3-2\sigma_1^2)\sqrt{1-\sigma_1^2}}\left(T_1 + \eta_2 T_2\sqrt{\delta(\sigma_1^2)}\right), \quad (51)$$

where $\eta_2^2 = 1$,
$$T_1 = -297 + 438\sigma_1^2 - 152\sigma_1^4.$$

and
$$T_2 = 6(\sigma_1^2 - 3).$$

A short argument shows that $T_1^2 > T_2^2\,\delta(\sigma_1^2)$ when $|\sigma_1| \leq \varepsilon_1$ and that $\mathrm{sgn}\,T_1 = \mathrm{sgn}\,\sigma_1$. Therefore the third derivative of $\widetilde{U}$ at $\sigma \in \mathcal{EV}^*$ is nonzero. □

Using singularity theory one can show that the versal normal form of the function $F_{0,0}(\sigma_1) = a(0,0)\,\sigma_1^3$, $a(0,0) \neq 0$, which has a cubic degenerate critical point at 0, is the smooth 2-parameter family

$$(\nu, \lambda) \to F_{\nu,\lambda}(\sigma_1) = \nu\,\sigma_1 + \lambda\sigma_1^2 + a(\nu, \lambda)\,\sigma_1^3. \quad (52)$$

In figure 4.3 we have sketched the discriminant locus D and some graphs of (52). Let $\ell : \mu \to (\overline{\sigma}_3(\mu), \overline{\sigma}_4(\mu))$ be a line which intersects $\widetilde{\mathcal{D}}^*$ transversely at $(\overline{\sigma}_3(\mu_0),$ $\overline{\sigma}_4(\mu_0))$. From the versality of the family $(\nu, \lambda) \to F_{\nu, \lambda}$ it follows that there are smooth functions $a(\mu)$, $\nu(\mu)$, and $\lambda(\mu)$ such that the 1-parameter family

$$\mu \to F_{\nu(\mu), \lambda(\mu)} = \nu(\mu)\,\sigma_1 + \lambda(\mu)\,\sigma_1^2 + a(\nu(\mu), \lambda(\mu))\,\sigma_1^3 \tag{53}$$

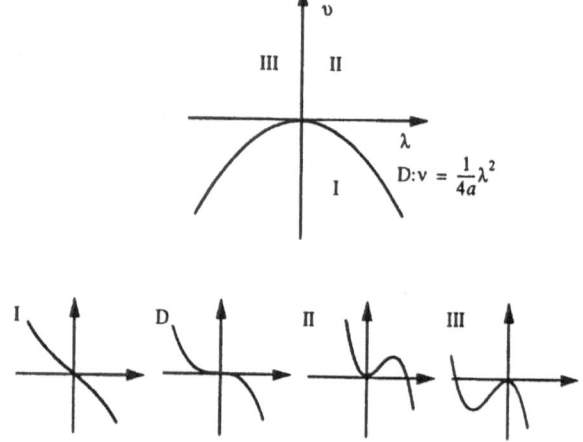

Figure 4.3. The family $F_{\nu, \lambda}$, when $a(0,0) < 0$.

with $a(\nu(\mu_0), \lambda(\mu_0)) \neq 0$ and $\nu(\mu_0) = \lambda(\mu_0) = 0$ is the versal normal form of the 1-parameter family

$$\mu \to U_\mu(\sigma_1) = U_{\overline{\sigma}_3(\mu), \overline{\sigma}_4(\mu)} - U_{\overline{\sigma}_3(\mu_0), \overline{\sigma}_4(\mu_0)}(\sigma_1^0) \tag{54}$$

near the critical point σ_1^0 for every μ near μ_0. From ((4.4)) and ((4.5)) we see that the smooth curve

$$\mu \to (\lambda(\mu), \nu(\mu)) \tag{55}$$

starts in the region in figure 4.3 above the discriminant locus D when $\mu < \mu_0$, passes through $(0,0)$ when $\mu = \mu_0$, and returns to the original region when $\mu > \mu_0$ *without* crossing D.

We now concentrate on the line $\ell_1 : \mu \to (\overline{\sigma}_3(\mu), \overline{\sigma}_4(\mu))$ (43) which intersects the discriminant locus $\widetilde{\mathcal{D}}^*$ transversely when $\mu = \mu_0$. From ((4.7)) it follows that the 3-jet of $U_{\mu_0} - U_{\mu_0}(\sigma_1^0)$ (54) at the degenerate critical point σ_1^0 is nonzero. Hence (53) is the versal normal form of the family $\mu \to U_\mu - U_{\mu_0}(\sigma_1^0)$.

▷ The following argument shows that the smooth curve (55) *crosses* the λ-axis transversely and thus goes from the region above the discriminant locus D in figure 4.3 to the region below as μ increases through μ_0.

(4.8) **Proof.** From the definition of the line ℓ_1 (43) and the potential $U_{\bar{\sigma}_3, \bar{\sigma}_4}$ (37) we obtain

$$U_\mu(\sigma_1) = \mu^2 \left[\tfrac{9}{40} (1 - \sigma_1^2)(P_\nu'(\sigma_1))^2 + \tfrac{3}{5} (P_\nu(\sigma_1))^2 \right] + \lambda \sqrt{1 - \sigma_1^2}.$$

Substituting $\sigma_3(\sigma_1) = -\tfrac{3}{2}\mu(1 - \sigma_1^2)P_\nu'(\sigma_1)$, $\sigma_4(\sigma_1) = \mu P_\nu(\sigma_1)$ into

$$\frac{d\tilde{U}_\mu}{d\sigma_1} = -\tfrac{6}{5} \frac{1}{1 - \sigma_1^2} \sigma_3\sigma_4 + \tfrac{1}{5} \frac{\sigma_1}{(1 - \sigma_1^2)^2} \sigma_3^2 - \lambda \frac{\sigma_1}{\sqrt{1 - \sigma_1^2}}$$

and combining terms gives

$$\frac{dU_\mu}{d\sigma_1}(\sigma_1) = \tfrac{9}{20} \mu^2 \left[4 P_\nu(\sigma_1)P_\nu'(\sigma_1) + \sigma_1(P_\nu'(\sigma_1))^2 \right] - \lambda \frac{\sigma_1}{\sqrt{1 - \sigma_1^2}}. \tag{56}$$

About the critical point σ_1^0 the Taylor expansion of U_μ is

$$\begin{aligned}
U_\mu(\sigma_1) &= U_\mu(\sigma_1^0) + \frac{\partial U_\mu}{\partial \sigma_1}(\sigma_1^0)(\sigma_1 - \sigma_1^0) + \mathrm{O}\!\left((\sigma_1 - \sigma_1^0)^2 \right) \\
&= U_\mu(\sigma_1^0) + \left(\frac{\partial}{\partial \mu}\!\left(\frac{\partial U_\mu}{\partial \sigma_1}(\sigma_1^0) \right)(\mu_0)(\mu - \mu_0) + \mathrm{O}((\mu - \mu_0)^2) \right)(\sigma_1 - \sigma_1^0) \\
&\quad + \mathrm{O}\!\left((\sigma_1 - \sigma_1^0)^2. \right),
\end{aligned} \tag{57}$$

The last equality follows because σ_1^0 is a critical point of U_{μ_0}. To show that the curve $\mu \to (\lambda(\mu), \nu(\mu))$ crosses the λ-axis transversely at 0 when $\mu = \mu_0$ it suffices to show that $\frac{d\nu}{d\mu}(\mu_0)$ is nonzero. From (57) and the versality of (53) we see that we need only show that $\mathcal{U} = \frac{\partial}{\partial \mu}\!\left(\frac{\partial U_\mu}{\partial \sigma_1}(\sigma_1^0) \right)(\mu_0)$ is nonzero. Partially differentiating (56) with respect to μ and evaluating the result at (σ_1^0, μ_0) gives $\mathcal{U} = \tfrac{9}{10} \mu_0 C(\sigma_1^0)$, where $C(\sigma_1^0) = 4 P_\nu(\sigma_1^0)P_\nu'(\sigma_1^0) + \sigma_1(P_\nu'(\sigma_1^0))^2$. Suppose that $C(\sigma_1^0) = 0$. Since σ_1^0 is a critical point of U_{μ_0}, evaluating (56) at (σ_1^0, μ_0) gives

$$0 = \tfrac{9}{20} \mu_0^2 C(\sigma_1^0) - \lambda \frac{\sigma_1^0}{\sqrt{1 - (\sigma_1^0)^2}}. \tag{58}$$

From the hypothesis and (58) it follows that $\sigma_1^0 = 0$. Using (82) of appendix 1 we obtain

$$C(0) = 4 P_\nu(0)P_\nu'(0) = -4 \frac{\Gamma(\tfrac{1}{2})\Gamma(\tfrac{3}{2})}{\left|\Gamma(\tfrac{1}{2} - \tfrac{\nu}{2})\right|^2 \left|\Gamma(\tfrac{\nu}{2} + 1)\right|^2} \neq 0,$$

which is a contradiction. Therefore $C(\sigma_1^0) \neq 0$ and thus $\frac{\partial}{\partial \mu}\!\left(\frac{\partial U_\mu}{\partial \sigma_1}(\sigma_1^0) \right)(\mu_0)$ is nonzero. $\square$

▷ Now we are in position to show that $U_{\bar{\sigma}_3^0, \bar{\sigma}_4^0}$ has only *one* critical point σ_1^0 when $(\bar{\sigma}_3^0, \bar{\sigma}_4^0) \in \widetilde{\mathcal{D}}^{**} = \widetilde{\mathcal{D}}^* - \widetilde{\mathcal{D}}^* \cap (\ell_1 \cup \ell_2)$. Moreover, $U_{\bar{\sigma}_3^0, \bar{\sigma}_4^0} - U_{\bar{\sigma}_3^0, \bar{\sigma}_4^0}(\sigma_1^0)$ has a nonzero 3-jet at σ_1^0.

(4.9) **Proof.** We begin by noting that the above discussion shows that as $(\bar{\sigma}_3, \bar{\sigma}_4) \in \mathcal{R} - \mathcal{R} \cap (\ell_1 \cup \ell_2)$ converges to a point $(\bar{\sigma}_3^0, \bar{\sigma}_4^0)$ on $\widetilde{\mathcal{D}}^{**}$, two nondegenerate critical points (one of index 0 and the other of index 1) of $U_{\bar{\sigma}_3, \bar{\sigma}_4}$ fuse together to form the cubic degenerate critical point σ_1^0 of $U_{\bar{\sigma}_3^0, \bar{\sigma}_4^0}$. Because $(\bar{\sigma}_3^0, \bar{\sigma}_4^0) \notin \ell_1 \cup \ell_2$ by hypothesis, it follows that $U_{\bar{\sigma}_3^0, \bar{\sigma}_4^0}$ blows up to $+\infty$ as $\sigma_1 \to \pm 1^{\mp}$. Since $U_{\bar{\sigma}_3^0, \bar{\sigma}_4^0}$ has a cubic degeneracy at σ_1^0, it is strictly decreasing near σ_1^0 in one of the intervals $(\sigma_1^0, 1)$ or $(-1, \sigma_1^0)$, depending on the sign of its third derivative at σ_1^0. Therefore in that interval $U_{\bar{\sigma}_3^0, \bar{\sigma}_4^0}$ has a nondegenerate critical point $\widehat{\sigma_1}$ of index 0. Suppose that $U_{\bar{\sigma}_3^0, \bar{\sigma}_4^0}$ has a nondegenerate critical point $\tilde{\sigma}_1$ in the interval $(\widehat{\sigma_1}, 1)$ or $(-1, \widehat{\sigma_1})$ which contains σ_1^0. Choose $\tilde{\sigma}_1$ as close to $\widehat{\sigma_1}$ as possible. Then $\tilde{\sigma}_1$ is of index 1. Therefore for $(\bar{\sigma}_3, \bar{\sigma}_4) \in \mathcal{R} - \mathcal{R} \cap (\ell_1 \cup \ell_2)$ near $(\bar{\sigma}_3^0, \bar{\sigma}_4^0)$, the function $U_{\bar{\sigma}_3, \bar{\sigma}_4}$ has two nondegenerate critical points $\bar{\sigma}_1$ of index 1. This is a contradicts $((4.4))$. Hence on the interval $(\widehat{\sigma_1}, 1)$ or $(-1, \widehat{\sigma_1})$ which contains σ_1^0, $U_{\bar{\sigma}_3^0, \bar{\sigma}_4^0}$ is strictly decreasing. A similar argument shows that $U_{\bar{\sigma}_3^0, \bar{\sigma}_4^0}$ is strictly increasing on the interval $(-1, \widehat{\sigma_1})$ or $(\widehat{\sigma_1}, 1)$, which does not contain σ_1^0. Thus $U_{\bar{\sigma}_3^0, \bar{\sigma}_4^0}$ has *one* critical point σ_1^0 which is degenerate and $U_{\bar{\sigma}_3^0, \bar{\sigma}_4^0} - U_{\bar{\sigma}_3^0, \bar{\sigma}_4^0}(\sigma_1^0)$ has a nonzero 3-jet at σ_1^0. □

We now concentrate on the case when $(\bar{\sigma}_3(\mu), \bar{\sigma}_4(\mu))$ lies on the line ℓ_1. The argument for the other line ℓ_2 is similar and is omitted.

▷ We first show that when $(\bar{\sigma}_3(\mu), \bar{\sigma}_4(\mu)) \in (\mathcal{R} - \{0\}) \cap \ell_1$ the potential U_μ is a Morse function with *two* critical points: one of index 0 and the other of index 1.

(4.10) **Proof.** Suppose that $(\bar{\sigma}_3(\mu)), \bar{\sigma}_4(\mu)) \in (\mathcal{R} - \{0\}) \cap \ell_1$ is close to $(\bar{\sigma}_3(\mu_0))$, $\bar{\sigma}_4(\mu_0))$. From $((4.9))$ we see that U_μ has two nondegenerate critical points: one σ_1^0 of index 0 and the other σ_1^1 of index 1. Moreover, between σ_1^0 and σ_1^1, U_μ has no critical points. Suppose that $\sigma_1^0 > \sigma_1^1$. Because $P_\nu(1) = 1$ and $P_\nu'(1) = \frac{4}{3}$, from (56) we deduce that $\lim_{\sigma_1 \to 1^{-1}} \frac{dU_\mu}{d\sigma_1} = -\infty$. Consequently, there is another critical point σ_1' which is nondegenerate and lies to the right of σ_1^0. Choose σ_1' as close to σ_1^0 as possible. Then σ_1' has index 1. Take $(\bar{\sigma}_3, \bar{\sigma}_4) \in \mathcal{R} - (\ell_1 \cup \ell_2)$ close to $(\bar{\sigma}_3(\mu)), \bar{\sigma}_4(\mu))$. From $((4.4))$ we know that $U_{\bar{\sigma}_3, \bar{\sigma}_4}$ has only one critical point of index 1. This is a contradiction. Hence $\sigma_1^0 < \sigma_1^1$. We now show that U_μ is strictly decreasing on $[\sigma_1', 1)$. Suppose not. Then there is a nondegenerate critical point $\tilde{\sigma}_1$ to the right of σ_1' which when chosen as close to σ_1' as possible has index 0. Because $\lim_{\sigma_1 \to 1^{-1}} \frac{dU_\mu}{d\sigma_1} = -\infty$ there is another critical point $\widehat{\sigma}_1$ to the right of $\tilde{\sigma}_1$ of index 1. As before this is a contradiction. Now we show that on $(-1, \sigma_1^0]$ the function U_μ is strictly decreasing. Suppose not. There there is a critical point $\tilde{\sigma}_1 \in (-1, \sigma_1^0)$ which is of index 1. Again this leads to a contradiction. Thus σ_1^0 and σ_1^1 are the only critical points of U_μ for μ slightly larger than μ_0.

We now show that U_μ is a Morse function with *two* critical points when $(\overline{\sigma}_3(\mu),$ $\overline{\sigma}_4(\mu))$ lies in $\mathcal{R} \cap (\ell_1 \cup \ell_2)$ and is *not* close to $(\overline{\sigma}_3(\mu_0), \overline{\sigma}(\mu_{04}))$. Suppose that $\tilde{\mu}$ is the infimum of the set of all $\mu \in (0, \mu_0)$ where U_μ is a Morse function with two critical points: σ_1^0 and σ_1^1. Suppose that $\tilde{\mu} > 0$. Because neither of these critical points runs off the interval $(-1, 1)$ as μ converges to $\tilde{\mu}$, they must fuse to form a degenerate critical point at $\hat{\sigma}_1$. But this cannot happen because $(\hat{\sigma}_1, \overline{\sigma}_3(\tilde{\mu}), \overline{\sigma}_4(\tilde{\mu})) \notin \mathcal{D}$. Therefore for every $\mu \in (\tilde{\mu}, \mu_0)$, U_μ is a Morse function with two critical points. For $\hat{\mu}$ slightly less than $\tilde{\mu}$ but larger than 0, the argument of the first paragraph with μ_0 replaced by $\tilde{\mu}$ shows that $U_{\hat{\mu}}$ is a Morse function with two critical points. This contradicts the definition of $\tilde{\mu}$. $\qquad\square$

At $\mu = \mu_0$ two of the critical points of U_μ with $\mu \in (0, \mu_0)$ fuse together to form a degenerate critical point σ_1^0 where $U_\mu - U_\mu(\sigma_1^0)$ has a nonvanishing 3-jet.

▷ We now show that U_{μ_0} has no other critical points.

(4.11) **Proof.** If it did, say at $\hat{\sigma}_1 \neq \sigma_1^0$, then $\hat{\sigma}_1$ is nondegenerate. For every μ slightly less than μ_0, the function U_μ has a nondegenerate critical point $\hat{\sigma}_1(\mu)$ which converges to $\hat{\sigma}_1$ as $\mu \to \mu_0$. But from $((4.10))$ we see that for $\mu < \mu_0$ every critical point of U_μ converges to the degenerate critical point σ_1^0 as $\mu \to \mu_0$. Since $\hat{\sigma}_1 \neq \sigma_1^0$, we have a contradiction. Therefore U_{μ_0} has only *one* critical point, which is degenerate. Therefore the 3-jet of $U_\mu - U_\mu(\sigma_1^0)$ at σ_1^0 is nonzero. $\qquad\square$

▷ We now show that if μ is larger than μ_0, then U_μ has *no* critical points.

(4.12) **Proof.** Suppose that U_μ has a critical point σ_1^0 when μ is slightly larger than μ_0. Then σ_1^0 is nondegenerate. It must be of index 0 because for every $(\overline{\sigma}_3, \overline{\sigma}_4)$ not on ℓ_1 but close to $(\overline{\sigma}_3(\mu), \overline{\sigma}_4(\mu))$ the function $U_{\overline{\sigma}_3, \overline{\sigma}_4}$ has a unique nondegenerate critical point of index 0. Since $\lim_{\sigma_1 \to 1^-} \frac{dU_\mu}{d\sigma_1} = -\infty$, the function U_μ has another nondegenerate critical point $\tilde{\sigma}_1$ which lies to the right of σ_1^0 and has index 1. Hence for every $(\overline{\sigma}_3, \overline{\sigma}_4)$ not on ℓ_1 but close to $(\overline{\sigma}_3(\mu), \overline{\sigma}_4(\mu))$ the function $U_{\overline{\sigma}_3, \overline{\sigma}_4}$ has two nondegenerate critical points. This contradicts $((4.6))$. Thus for μ slightly larger than μ_0, the function U_μ has *no* critical points.

Let $\tilde{\mu}$ be the supremum of the set W of numbers such that for every $\mu \in (\mu_0, \tilde{\mu})$, the function U_μ has no critical points. The same argument as in the preceding paragraph now applied to $U_{\tilde{\mu}}$ shows that $\tilde{\mu} \in W$. Again this argument applied to $\hat{\mu}$ slightly greater than $\tilde{\mu}$ shows that $\hat{\mu} \in W$. This contradicts the definition of $\tilde{\mu}$. Hence for every $\mu > \mu_0$, the function U_μ has no critical points. $\qquad\square$

This completes our treatment of the graph of $U_{\overline{\sigma}_3, \overline{\sigma}_4}$ when $(\overline{\sigma}_3, \overline{\sigma}_4)$ lies on ℓ_1.

Thus we have determined the graph of $U_{\overline{\sigma}_3, \overline{\sigma}_4}$ when $(\overline{\sigma}_3, \overline{\sigma}_4)$ lies in $\mathbf{R}^2$.

From the graph of $U_{\overline{\sigma}_3, \overline{\sigma}_4}$ it is straightforward to read off the level sets of $H_{\overline{\sigma}_3, \overline{\sigma}_4}$. Thus we obtain figure 4.1.

4.4   Foliation of reduced energy surfaces

In this section we will describe how the level set of the reduced energy function

$$H : \mathbf{R}^4 \to \mathbf{R} : (\sigma_1, \sigma_2, \overline{\sigma}_3, \overline{\sigma}_4) \to H_{\overline{\sigma}_3, \overline{\sigma}_4}(\sigma_1, \sigma_2)$$

are foliated by the integral curves of the 2-parameter family of Hamiltonian vector fields $X_{H_{\overline{\sigma}_3, \overline{\sigma}_4}}$. Because the flow of the vector field $X_{H_{\overline{\sigma}_3, \overline{\sigma}_4}}$ is not always complete, we must regularize it and also the reduced energy $\tilde{H}$. For the details of this regularization, which amounts to changing the time scale and compactifying the energy surfaces, see appendix 2. The results of our analysis is given in figure 4.4. From now on we will assume that the vector field $X_{H_{\overline{\sigma}_3, \overline{\sigma}_4}}$ and the energy function H have been regularized. From the construction of the 2-parameter family $(\overline{\sigma}_3, \overline{\sigma}_4) \to H_{\overline{\sigma}_3, \overline{\sigma}_4}(\sigma_1, \sigma_2)$ of Hamiltonians in section 4.2, it follows that the functions $H_{\overline{\sigma}_3, \overline{\sigma}_4}$, $\overline{\sigma}_3$ and $\overline{\sigma}_4$ on $\mathbf{R}^2$ with coordinates (σ_1, σ_2) are *constant* on the integral curves of $X_{H_{\overline{\sigma}_3, \overline{\sigma}_4}}$. Therefore we define the *integral map*

$$J : \mathbf{R}^4 \to \mathbf{R}^3 : (\sigma_1, \sigma_2, \overline{\sigma}_3, \overline{\sigma}_4) \to \left(H_{\overline{\sigma}_3, \overline{\sigma}_4}(\sigma_1, \sigma_2), \overline{\sigma}_3, \overline{\sigma}_4 \right). \tag{59}$$

This map encodes all the qualitative behavior of the 2-parameter family of vector fields $X_{H_{\overline{\sigma}_3, \overline{\sigma}_4}}$. Our task is to find the image of J, describe the topology of *all* of ▷ its fibers, and show how these fibers foliate a level set of H. We begin by observing that the point $(\sigma_1^0, \sigma_2^0, \overline{\sigma}_3^0, \overline{\sigma}_4^0)$ is a critical point of the integral mapping J if and only if (σ_1^0, σ_2^0) is a critical point of the Hamiltonian $H_{\overline{\sigma}_3^0, \overline{\sigma}_4^0}$.

(4.13) **Proof.** From

$$\operatorname{rank} DJ(\sigma) = \operatorname{rank} \begin{pmatrix} \frac{\partial H}{\partial \sigma_1}(\sigma) & \frac{\partial H}{\partial \sigma_2}(\sigma) & \frac{\partial H}{\partial \sigma_3}(\sigma) & \frac{\partial H}{\partial \sigma_4}(\sigma) \\ 0 & 0 & 1 & 0 \\ 0 & 0 & 0 & 1 \end{pmatrix}$$

$$= \operatorname{rank} \begin{pmatrix} \frac{\partial H_{\overline{\sigma}_3, \overline{\sigma}_4}}{\partial \sigma_1}(\sigma_1, \sigma_2) & \frac{\partial H_{\overline{\sigma}_3, \overline{\sigma}_4}}{\partial \sigma_2}(\sigma_1, \sigma_2) & 0 & 0 \\ 0 & 0 & 1 & 0 \\ 0 & 0 & 0 & 1 \end{pmatrix}$$

the assertion follows. □

Therefore the set $\mathcal{CV}$ of critical values of J is the image under J of the set of critical points of $H_{\overline{\sigma}_3, \overline{\sigma}_4}$ as $(\overline{\sigma}_3, \overline{\sigma}_4)$ varies over $\mathbf{R}^2$. Let $\mathcal{CV}_h$ be the h-slice of $\mathcal{CV}$ formed by setting the first coordinate in $\mathbf{R}^3$ equal to h. ($\mathcal{CV}_h$ are the darkened curves in figure 4.4). To obtain the critical locus $\mathcal{CV}_h$ we first find a parametrization of the critical points of the regularized energy function $F^h = (1 - \sigma_1^2)(\tilde{H} - h)$ using (38) and the first two equations of (40). On the piece of this locus in the first octant of σ_1-σ_3-σ_4 space we apply the diffeomorphism $\tilde{\psi}$ (42), reflect the result in the coordinate planes of σ_1-$\overline{\sigma}_3$-$\overline{\sigma}_4$ space, and then project onto the $\overline{\sigma}_3$-$\overline{\sigma}_4$ plane. From this procedure (which we carried out numerically to produce figure 4.4) we see that $\mathcal{CV}_h$ is a union of finitely many piecewise real analytic curves. Let $\mathcal{D}_h$ be the

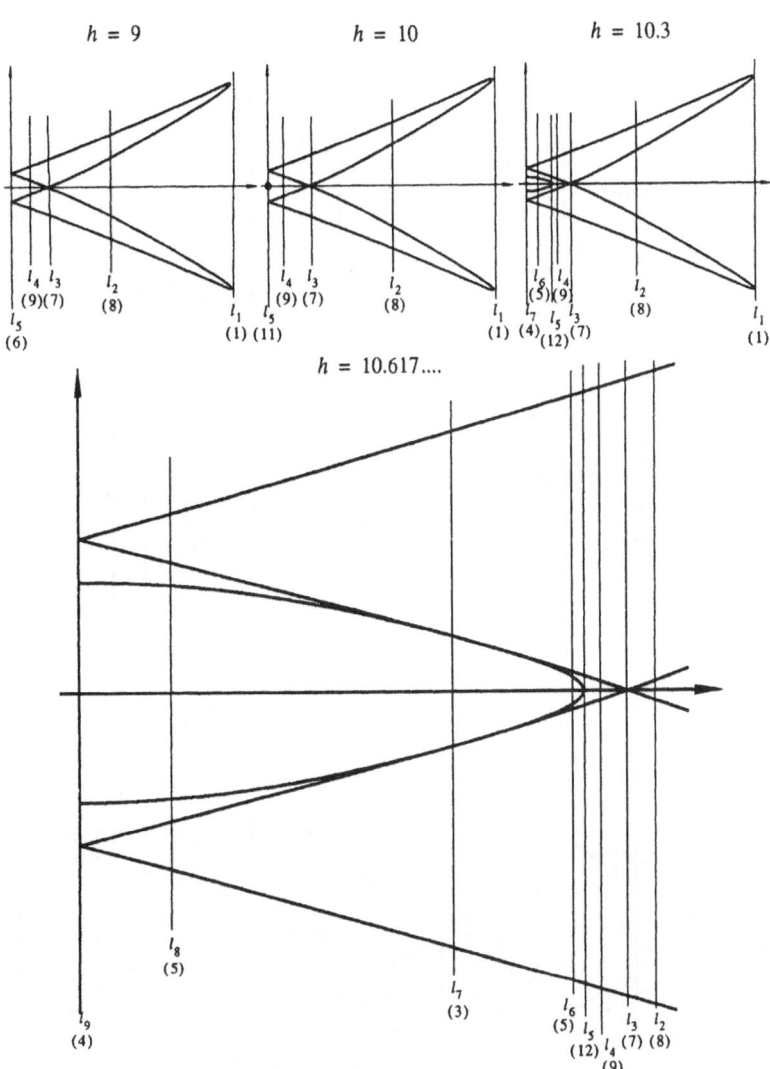

Figure 4.4. The bifurcation diagram for half of a h-slice of the integral map J of the rolling disc. Here λ is equal to 10. The darkened curves are the critical values of J in the h-slice and the numbers in parentheses under the vertical lines correspond to the row of table 1.

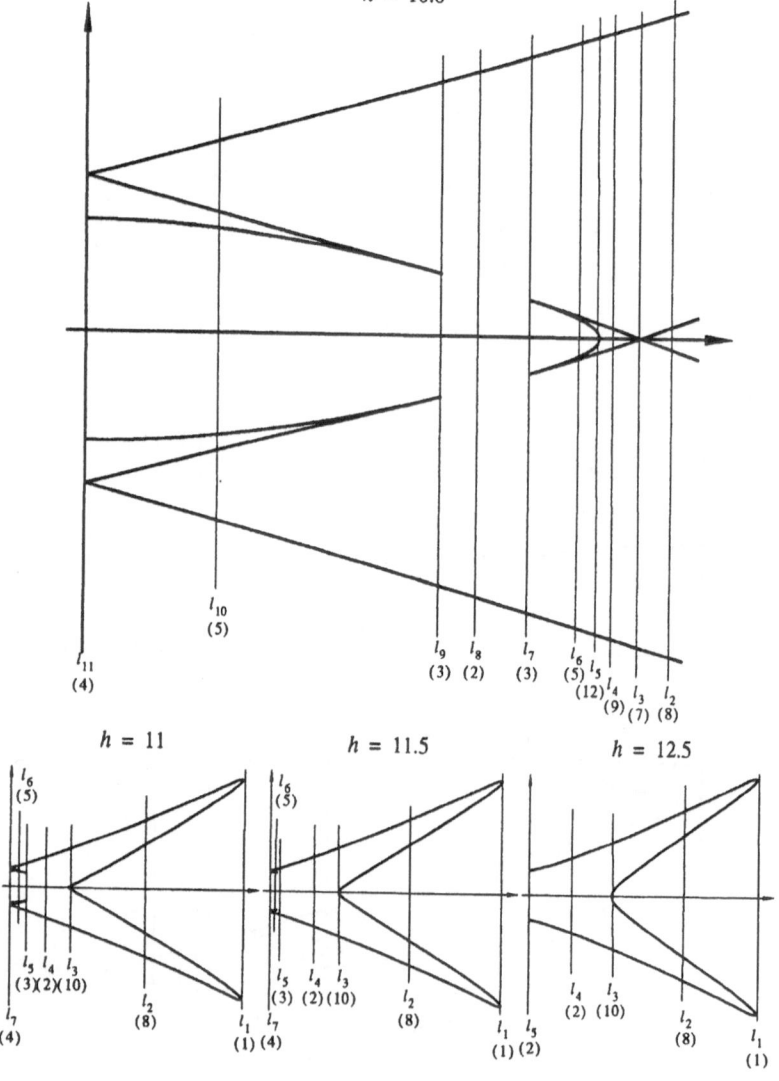

Figure 4.4 continued.

largest connected region containing the origin which is bounded by a subset of the real analytic curves making make up $\mathcal{CV}_h$. Then $\mathcal{D}_h$ is an h-slice of the image of J. Since F^h is a proper map, the preimage of 0 is a compact subset K_h of $\mathbf{R}^4$. Hence $\widetilde{\psi}(K_h)$ is compact. Since $J(\widetilde{\psi}(K_h)) = \mathcal{D}_h$, we see that $\mathcal{D}_h$ is compact. From the definition of the integral map J it follows that for every $h \in \mathbf{R}$ and every $(\overline{\sigma}_3, \overline{\sigma}_4) \in \mathcal{D}$ the fiber of J over the point $(h, \overline{\sigma}_3, \overline{\sigma}_4)$ is the 0-level set of F^h, which is the regularized h-level set of $H_{\overline{\sigma}_3, \overline{\sigma}_4}$. From figure 3.1 we see that every such level set is a *finite* disjoint union of one or more of the following sets: a point (pt), a circle (S^1) or two circles joined at a point $(S^1 \vee S^1)$. Having determined the fibers of J over every point of $\mathcal{D}_h$ (again we used a computer here to help to draw figure 4.4), we must find out how they fit together. First, we look at the fibers over a $\overline{\sigma}_3$-slice of $\mathcal{D}_h$. We draw the following diagram. (See table 2). Whenever the $\overline{\sigma}_3$-slice intersects the critical locus $\mathcal{CV}_h$ we draw a point in the diagram. Under each point we write the topological type of the fiber of the map J over that point. Next we join two consecutive points of the diagram with a line segment if the portion P of the $\overline{\sigma}_3$-slice between these two points lies in $\mathcal{D}_h$. Above this line segment we write the topological type of a fiber of J over a point of P. This type does *not* depend of the point chosen on P, because P lies in the set regular values of J. An inspection of figure 4.4 shows that only the diagrams appearing in the first column of table 1 occur.

To obtain the topology of the 2-manifolds defined by the preimage of the $\overline{\sigma}_3$-slice of $\mathcal{D}_h$, one can draw pictures or more rigorously use a formula of Viro [19] to compute the Euler characteristic.

We now have to piece together the topology of the fibers over the $\overline{\sigma}_3$-slices of $\mathcal{D}_h$ to obtain the topology of the regularized h-level set of the energy. Let $\ell^+_{\overline{\sigma}_3,h}$ be the set of points in $\mathcal{D}_h$ which lie on or to the *right* of the vertical line $\ell_{\overline{\sigma}_3}$ in the $\overline{\sigma}_3$-$\overline{\sigma}_4$ plane. In figure 4.4 as $\overline{\sigma}_3$ decreases to 0 we obtain the following sequences for the topology of $J^{-1}(\ell^+_{\overline{\sigma}_3,h})$. (See table 1.) From the sequences in the first column of table 1 and using the axial symmetry of $\mathcal{D}_h$ in the $\overline{\sigma}_4$-axis, we see that $(F^h)^{-1}(0)$ is the union of *two* copies of $J^{-1}(\ell^+_{0,h})$ glued along their boundary. This explains the entries in rows 1 and 2 in table 1. To explain the entry in row 3, think of a height function on S^3 with 4 nondegenerate critical points: two of index 3, one of index 2 and one of index 0. (The situation on S^2 is easier to visualize.) The points on S^3 above a fixed height (which is greater than or equal to the height of the equator) gives the sequence in row 3. This is an example of the well known phenomenon of cancellation of critical points in Morse theory (see Milnor [12]). Even though the regularized energy level set $(F^h)^{-1}(0)$ is topologically a 3-sphere for every $h > \lambda$, the fibration of $(F^h)^{-1}(0)$ by regularized integral curves of $X_{H_{\overline{\sigma}_3,\overline{\sigma}_4}}$ with $(\overline{\sigma}_3, \overline{\sigma}_4) \in \mathcal{D}_h$ *changes* as the topology of the set of critical values $\mathcal{CV}_h$ changes. We now explain the origin of three bifurcations in the geometry of $\mathcal{CV}_h$ which are clearly visible in figure 4.4.

▷ The first bifurcation is is the appearance of sections of *two* symmetrical swallowtail surfaces as h decreases through $\frac{11}{10}\lambda$ and $\frac{5}{4}\lambda$. Physically when h decreases

R. Cushman, J. Hermans, D. Kemppainen

	Sequence	Topology
1.	pt $\coprod$ pt, $\bar{D}^3 \coprod \bar{D}^3$	$S^3 \coprod S^3$
2.	pt $\coprod$ pt, $\bar{D}^3 \coprod \bar{D}^3$, $\bar{D}^3 \vee \bar{D}^3$	$S^3 \vee S^3$
3.	pt $\coprod$ pt, $\bar{D}^3 \coprod \bar{D}^3$, $\bar{D}^3 \vee \bar{D}^3$, $\bar{D}^3$	S^3

Table 1. Topology of regularized energy level set $(F^h)^{-1}(0)$.

	Diagram	Topology
1.	• • pt pt	pt $\coprod$ pt
2.	•—S^1—• pt pt	S^2
3.	•—S^1—•—S^1—•—S^1—• pt S^1 S^1 pt	S^2
4.	•—$S^1 \coprod S^1$—•—S^1—•—$S^1 \coprod S^1$—• pt $\coprod$ pt $S^1 \vee S^1$ $S^1 \vee S^1$ pt $\coprod$ pt	S^2
5.	•—S^1—•—$S^1 \coprod S^1$—•—S^1—•—$S^1 \coprod S^1$—•—S^1—• pt pt $\coprod S^1$ $S^1 \vee S^1$ $S^1 \vee S^1$ pt $\coprod S^1$ pt	S^2
6.	•—$S^1 \coprod S^1$—• pt $\coprod$ pt pt $\coprod$ pt	$S^2 \coprod S^2$
7.	•—S^1—•—S^1—• pt pt $\coprod$ pt pt	$S^2 \coprod S^2$
8.	•—S^1—• •—S^1—• pt pt pt pt	$S^2 \coprod S^2$
9.	•—S^1—•—$S^1 \coprod S^1$—•—S^1—• pt pt $\coprod S^1$ pt $\coprod S^1$ pt	$S^2 \coprod S^2$
10.	•—S^1—•—S^1—• pt pt pt	$S^2 \vee S^2$
11.	•—$S^1 \coprod S^1$—$S^1 \coprod S^1$—• pt $\coprod$ pt $S^1 \vee S^1$ pt $\coprod$ pt	$S^2 \vee S^2$
12.	•—S^1—•—$S^1 \coprod S^1$—•—$S^1 \coprod S^1$—•—S^1—• pt pt $\coprod S^1$ $S^1 \vee S^1$ pt $\coprod S^1$ pt	$S^2 \vee S^2$

Table 2. Diagram of a $\bar{\sigma}_3$-slice of $\mathcal{D}_h$ and topology of preimage under the map J of this slice. The symbol $\coprod$ means disjoint union and the symbol $\vee$ means union with one point in common.

through $\frac{11}{10} \lambda$ the vertically spinning disc losses stability: whereas when h decreases through $\frac{5}{4} \lambda$ the vertically rolling disc losses stability. The second bifurcation is the appearance of a topological circle of critical points which springs from the origin as h increases through the value λ. The third bifurcation is the pairwise fusion of two swallowtail cusps at $h = 1.0617\dots\lambda$. Here the tilted rolling disc gains stability as h increases through $1.0617\dots\lambda$.

(4.14) **Proof.** The proof of the first two bifurcations begins with finding the Taylor expansion of the regularized Hamiltonian

$$F^h(\sigma_1, \sigma_2, \overline{\sigma}_3, \overline{\sigma}_4) = (1 - \sigma_1^2)(H - h) = \tfrac{1}{2} \sigma_1^2 + \tfrac{1}{10} \sigma_3^2(\sigma_1, \overline{\sigma}_3, \overline{\sigma}_4)$$
$$+ \tfrac{3}{5} \sigma_4^2(\sigma_1, \overline{\sigma}_3, \overline{\sigma}_4) + \lambda(1 - \sigma_1^2)^{3/2} - h(1 - \sigma_1^2) \quad (60)$$

about $(0, 0, \overline{\sigma}_3, \overline{\sigma}_4)$ up through terms of order four. Using the results of appendix 1, a calculation gives

$$F^h(\sigma_1, \sigma_2, \overline{\sigma}_3, \overline{\sigma}_4) = \tfrac{1}{2} \sigma_2^2 + (\tfrac{1}{10} \overline{\sigma}_3^2 + \tfrac{3}{5} \overline{\sigma}_4^2 + \lambda - h) + (-\tfrac{6}{5} \overline{\sigma}_3 \overline{\sigma}_4) \sigma_1$$
$$+ (\tfrac{2}{5} \overline{\sigma}_3^2 + \tfrac{3}{5} \overline{\sigma}_4^2 - \tfrac{3}{2} \lambda + h) \sigma_1^2 + (-\tfrac{8}{15} \overline{\sigma}_3 \overline{\sigma}_4) \sigma_1^3$$
$$+ (-\tfrac{1}{18} \overline{\sigma}_3^2 + \tfrac{2}{15} \overline{\sigma}_4^2 - \tfrac{3}{8} \lambda) \sigma_1^4 + O(\sigma_1^5). \quad (61)$$

When $h = \frac{11}{10} \lambda$ and $(\overline{\sigma}_3^0, \overline{\sigma}_4^0) = (\pm\sqrt{\lambda}, 0)$ or when $h = \frac{5}{4} \lambda$ and $(\overline{\sigma}_3^0, \overline{\sigma}_4^0) = (0, \pm\sqrt{\frac{5\lambda}{12}})$, using (61) it is straightforward to check that the 3-jet of F^h at $(0, 0, \overline{\sigma}_3^0, \overline{\sigma}_4^0)$ vanishes and the 4-jet does not. From singularity theory we know that the versal unfolding of $g(\sigma_1, \sigma_2) = \tfrac{1}{2} \sigma_2^2 + a \sigma_1^4$ with $a \neq 0$ is

$$f(\sigma_1, \sigma_2) = \tfrac{1}{2} \sigma_2^2 + e + d\sigma_1 + c\sigma_1^2 + b\sigma_1^3 + a\sigma_1^4. \quad (62)$$

After rescaling and translating σ_1, (62) becomes

$$F(\sigma_1, \sigma_2) = \tfrac{1}{2} \sigma_2^2 + C + B\sigma_1 + A\sigma_1^2 + \sigma_1^4. \quad (63)$$

Thus the versal normal form of F^h is given by (63). Using (61) the parameters A, B, C are explicitly known functions of $h, \overline{\sigma}_3, \overline{\sigma}_4$. As a function of A, B, C the critical values of F corresponding to critical points in $F^{-1}(0)$ form a swallowtail surface. This proves the existence of the swallowtail surfaces in figure 4.4.

To explain the appearance of the circle of critical points coming from the origin, we know that $F^h_{\overline{\sigma}_3, \overline{\sigma}_4}$ has a critical point when

$$0 = \frac{\partial F^h_{\overline{\sigma}_3, \overline{\sigma}_4}}{\partial \sigma_1} = (-\tfrac{6}{5} \overline{\sigma}_3 \overline{\sigma}_4) \sigma_1 + 2(\tfrac{2}{5} \overline{\sigma}_3^2 + \tfrac{3}{5} \overline{\sigma}_4^2 - \tfrac{3}{2} \lambda + h) \sigma_1$$
$$+ 3(-\tfrac{8}{15} \overline{\sigma}_3 \overline{\sigma}_4) \sigma_1^2 + 4(-\tfrac{1}{18} \overline{\sigma}_3^2 + \tfrac{2}{15} \overline{\sigma}_4^2 - \tfrac{3}{8} \lambda) \sigma_1^3 + O(\sigma_1^4). \quad (64)$$

$$0 = \frac{\partial F^h_{\overline{\sigma}_3, \overline{\sigma}_4}}{\partial \sigma_2} = \sigma_2.$$

Since $F^\lambda_{(0,0)} = 0$ and

$$\frac{\partial^2 F^\lambda_{(0,0)}}{\partial \sigma_1{}^2}(0,0) = 2(\tfrac{2}{5}\,\bar\sigma_3^2 + \tfrac{3}{5}\,\bar\sigma_4^2 - \tfrac{3}{2}\,\lambda + h)\Big|_{(\bar\sigma_3,\bar\sigma_4)=(0.0)} = -\lambda < 0, \qquad (65)$$

the implicit function theorem gives a smooth function $\sigma_1 = s(h,\bar\sigma_3.\bar\sigma_4)$ with $0 = s(\lambda,0,0)$ such that for every $(h,\bar\sigma_3,\bar\sigma_4)$ near $(0,0,0)$, the point $p = (s(h,\bar\sigma_3,\bar\sigma_4),0)$ is a critical point of $F^\lambda_{(0,0)}$ near $(0,0)$. In order that p lie in $(F^h)^{-1}_{\bar\sigma_3.\bar\sigma_4}(0)$

$$\begin{aligned}
0 &= G_h(\bar\sigma_3.\bar\sigma_4) = F^h(s(h,\bar\sigma_3,\bar\sigma_4),0,\bar\sigma_3,\bar\sigma_4)\\
&= (\tfrac{1}{10}\,\bar\sigma_3^2 + \tfrac{3}{5}\,\bar\sigma_4^2 + \lambda - h) + (-\tfrac{6}{5}\,\bar\sigma_3\bar\sigma_4)s(h,\bar\sigma_3,\bar\sigma_4) + \mathrm{O}(\bar\sigma_3^2 + \bar\sigma_4^2),\\
&\qquad\qquad \text{using (61).}\\
&= (\tfrac{1}{10}\,\bar\sigma_3^2 + \tfrac{3}{5}\,\bar\sigma_4^2 + \lambda - h) + \mathrm{O}((\bar\sigma_3^2 + \bar\sigma_4^2)^{3/2}),\\
&\qquad\qquad \text{since } s(\lambda,0,0) = 0.
\end{aligned}$$

From (65) it is obvious that $(0,0)$ is a nondegenerate critical point of G_h of index 0. Therefore, using the Morse lemma, we find that for every h slightly larger than λ, the level set $G_h^{-1}(0)$ is diffeomorphic to a circle $\mathcal{C}$ in the $\bar\sigma_3$-$\bar\sigma_4$ plane. From (65), $\frac{\partial^2 F^\lambda_{0.0}}{\partial\sigma_1\partial\sigma_2}(0,0) = 0$. and $\frac{\partial^2 F^\lambda_{0.0}}{\partial\sigma_2{}^2}(0,0) = 1$ it follows that every critical point of F^h corresponding to $(\bar\sigma_3,\bar\sigma_4)$ of $\mathcal{C}$ is nondegenerate and of index 1. This proves the existence of a circle of critical points of F^h in figure 4.4.

To explain the pairwise fusion of two swallowtail cusps, we look at the locus $\mathcal{E}_h$ of equilibrium points of the reduced vector field $\overline{V}$ of reduced energy h which is defined by equations (30) of section 3.1 and (19) of section 2.4. These equations can be solved to give

$$\sigma_3^\pm(h,\sigma_1) = 5h\,\frac{(1-\sigma_1^2)^{3/4}}{(6-5\sigma_1^2)^{1/2}}\big[6h\sqrt{1-\sigma_1^2} + \lambda(7\sigma_1^2 - 6) \pm \sqrt{\Delta(h,\sigma_1)}\,\big]$$

and

$$\sigma_4^\pm(h,\sigma_1) = \tfrac{1}{6}\,\frac{\sigma_1}{1-\sigma_1^2}\,\sigma_3^\pm(h,\sigma_1) - \tfrac{5}{6}\,\lambda\,\frac{\sigma_1}{\sigma_3^\pm(h,\sigma_1)}\sqrt{1-\sigma_1^2}$$

where $|\sigma_1| < 1$ and

$$\Delta(h,\sigma_1) = 12\lambda h\,(7\sigma_1^2 - 6)\sqrt{1-\sigma_1^2} + (1-\sigma_1^2)[36h^2 + \lambda^2(36 - 54\sigma_1^2)] \geq 0,$$

which are a parametrization of the branches of $\mathcal{E}_h$. The image of the piece of $\mathcal{E}_h$ in the first octant of σ_1-σ_3-σ_4 space under the diffeomorphism $\widetilde{\psi}$ (see (42) of section 4.3) gives a branch of the critical value locus $\mathcal{CV}$ of the integral map J. The other branches are obtained by reflection in the h-$\bar\sigma_3$ and h-$\bar\sigma_4$ planes. For a fixed value of h, the cusp points of the h-slice $\mathcal{CV}_h$ of $\mathcal{CV}$ occur when two branches of $\mathcal{CV}$ coincide, that is, when

$$\Delta(h,\sigma_1) = 0. \qquad (66)$$

Furthermore two cusps coincide when these two branches coincide with multiplicity greater than 1, that is, when $\frac{\partial \Delta}{\partial \sigma_1} = 0$. This is equivalent to

$$0 = -12h\lambda \frac{(7\sigma_1^2 - 6)}{\sqrt{1 - \sigma_1^2}} + 168h\lambda \sigma_1 \sqrt{1 - \sigma_1^2} + 36\sigma_1 [6\lambda^2 \sigma_1^2 - 5\lambda^2 - 6h^2]. \qquad (67)$$

Solving (66) and (67) gives $(h, \sigma_1) = (\lambda, 0)$ or $(\frac{\lambda}{18} \sqrt{552 - 2238\sigma_1^2 + 1260\sigma_1^4}, \sigma_1)$ where $x = \sigma_1^2$ is the unique positive root of $15x^3 - 38x^2 + 26x - 2 = 0$, which is approximately, $(1.0617\lambda, \pm.296307)$.

This completes the proof of the bifurcations in figure 4.4.   □

5   Reconstruction

In this section we reconstruct the physical motion of the disc corresponding to a relative equilibria. It turns out that the motion can be described by two angles. This is related to general theorem about reconstruction of the motion of relative equilibria. In our case this theorem says that the motion through a relative equilibrium is quasi-periodic on a two dimensional torus (see Hermans [8, Theorem 2.2.1]).

▷ Consider a principal G-bundle with total space C, base space $\overline{M}$, and bundle projection map $\pi : C \to \overline{M}$. Suppose that G is a compact connected Lie group of rank r and suppose that the G-action Φ on C is free and proper. Let V be a G-invariant vector field on C with flow φ^t. Hence V induces a vector field $\overline{V}$ on $\overline{M}$. Assuming that the monodromy element associated to a relative equilibrium is semisimple, we have

(i) the motion through a relative equilibrium is quasi-periodic on a torus of rank r.

(ii) if the reduced flow is periodic then the motion in C is quasi-periodic on tori of dimension $r + 1$. The number of parameters the vector field on these tori depends on is equal to the dimension of M minus 1.

(5.1) **Proof.** See [7].   □

We can weaken the compactness condition on G in ((5.1)). The essence of the proof of ((5.1)) is to consider the monodromy element: namely, that element $\mu \in G$ for which $\varphi^\tau(c) = \Phi_\mu(c)$, where τ is the period of the orbit through $\pi(c)$ of the reduced vector field. To carry through the argument in ((5.1)) it is sufficient that the centralizer of a generic monodromy element in G is compact. In the case of the disc $G = \mathcal{E}(2) \times S^1$. Hence we need to verify whether the centralizer of an element $(A, a) \in \mathcal{E}(2)$ is compact. If $A \neq I$ the centralizer of (A, a) is

$$\{(b, B) \in \mathcal{E}(2) \mid b = (A - I)^{-1}(B - I)a \}, \qquad (68)$$

which is diffeomorphic to a circle. Looking at the action of $\mathcal{E}(2)$ on $\mathbf{R}^3$, we see that the group orbit $(B, b) \cdot x$ is a rotation about the point $(I - A)^{-1}a$. If $A = I$ and $a \neq 0$, then the orbit is translation by any vector in $\mathbf{R}^2 \times \{0\}$.

Consequently we see that all reduced periodic orbits correspond to quasi-periodic motion on three dimensional tori inside the constraint manifold. The tori are easily interpreted in terms of the physical motion. One angle ψ measures the amount the disc has rotated about its vertical principal axis. Another angle φ measures the position of the point of contact, and the third angle θ measures the wobbling of the disc with respect to the vertical.

The three dimensional tori lie around the two dimensional tori comming from the relative equilibria. Since a 3-torus is determined by a single periodic orbit in the reduced space we have a three parameter family of tori. The three parameters are the energy and the two integrals coming from Legendre's equation. The eight dimensional constraint manifold $\mathcal{C}$ is filled up by the three parameter family of 3-tori. The remaining two dimensional gap is explained by the action of the remaining of the symmetry group $(\mathcal{E}(2) \times S^1)/T^2$, where T^2 is the 2-torus coming from the two dimensional centralizer of monodromy element μ.

Appendix 1 Asymptotic behavior

In this appendix we study the asymptotic behavior as $\sigma_1 \to \pm 1^{\mp}$ of solutions of the time dependent linear system

$$\frac{d}{d\sigma_1} \begin{pmatrix} \sigma_3 \\ \sigma_4 \end{pmatrix} = \frac{1}{1 - \sigma_1^2} \begin{pmatrix} 0 & -2(1 - \sigma_1^2) \\ -\frac{2}{3} & 0 \end{pmatrix} \begin{pmatrix} \sigma_3 \\ \sigma_4 \end{pmatrix}, \tag{69}$$

which has regular singular points at $\sigma_1 = \pm 1$. The main result is that on two lines of initial conditions σ_4 is *bounded* at one endpoint, whereas in all other cases it has a logarithmic singularity at both endpoints.

Since

$$0 = \frac{d}{d\sigma_1}\left((1 - \sigma_1^2)\frac{d\sigma_4}{d\sigma_1} \right) - \frac{4}{3}\,\sigma_4 = (1 - \sigma_1^2)\frac{d^2\sigma_4}{d\sigma_1^2} - 2\,\sigma_1\,\frac{d\sigma_4}{d\sigma_1} - \frac{4}{3}\,\sigma_4, \tag{70}$$

σ_4 satisfies Legendre's equation with parameter $\nu = -\frac{1}{2} + i\frac{1}{2}\sqrt{\frac{13}{3}}$, because $\nu(\nu + 1) = -\frac{4}{3}$. A general solution of (70) is given by

$$\sigma_4(\sigma_1) = A\,F(\tfrac{\nu}{2} + \tfrac{1}{2}, -\tfrac{\nu}{2}, \tfrac{1}{2}; \sigma_1^2) + B\,\sigma_1\,F(\tfrac{1}{2} - \tfrac{\nu}{2}, \tfrac{\nu}{2} + 1, \tfrac{3}{2}; \sigma_1^2), \tag{71}$$

where $A, B \in \mathbf{R}$ and

$$F(a, b, c; z) = \sum_{k=0}^{\infty} \frac{(a)_k\,(b)_k}{(c)_k\,k!}\,z^k \qquad a, b, c \in \mathbf{C} \tag{72}$$

with $(a)_k = a(a+1)\cdots(a+k-1)$ is the hypergeometric function. The function F is holomorphic in $|z| < 1$ and satisfies the differential equation

$$z(1-a)\frac{d^2F}{dz^2} + [c - (a+b+1)z]\frac{dF}{dz} - ab\,z = 0. \tag{73}$$

see [11]. When $a = \bar{b}$ and c is real, then $\dfrac{(a)_k(b)_k}{(c)_k} = \dfrac{|(a)_k|^2}{(c)_k}$ is real and hence $F(a,b,c;z)$ is real valued on $(-1,1)$. Note that $\frac{\nu}{2} + \frac{1}{2} = -\frac{\nu}{2}$ and $\frac{1}{2} - \frac{\nu}{2} = \overline{\frac{\nu}{2} + 1}$. Thus (71) is real valued when $|\sigma_1| < 1$. Substituting (71) into the second component of (44) gives

$$\begin{aligned}
\sigma_3(\sigma_1) = {} & -3A\,\sigma_1(1-\sigma_1^2)\frac{dF}{d\sigma_1^2}(\tfrac{\nu}{2}+\tfrac{1}{2}, -\tfrac{\nu}{2}\cdot\tfrac{1}{2};\sigma_1^2) \\
& -\tfrac{3}{2}B\,(1-\sigma_1^2)F(\tfrac{1}{2}-\tfrac{\nu}{2}, \tfrac{\nu}{2}+1, \tfrac{3}{2};\sigma_1^2) \\
& -3B\,\sigma_1^2(1-\sigma_1^2)\frac{dF}{d\sigma_1^2}(\tfrac{1}{2}-\tfrac{\nu}{2}\cdot\tfrac{\nu}{2}+1.\tfrac{3}{2};\sigma_1^2).
\end{aligned} \tag{74}$$

From (71) and (74) we obtain the initial conditions

$$\begin{aligned}
\bar{\sigma}_3 = \sigma_3(0) = -\tfrac{3}{2}B\,F(\tfrac{1}{2}-\tfrac{\nu}{2}, \tfrac{\nu}{2}+1, \tfrac{3}{2}:0) = -\tfrac{3}{2}B \\
\bar{\sigma}_4 = \sigma_4(0) = A\,F(\tfrac{\nu}{2}+\tfrac{1}{2}, -\tfrac{\nu}{2}, \tfrac{1}{2};0) = A.
\end{aligned} \tag{75}$$

Let $\psi(z) = \Gamma'(z)/\Gamma(z)$. Using the series [1, formula 15.3.10 on p.559]

$$\begin{aligned}
F(a,b,a+b;z) = {} & \\
= {} & \frac{\Gamma(a+b)}{\Gamma(a)\Gamma(b)}\sum_{k=0}^{\infty}\frac{(a)_k(b)_k}{(k!)^2}\left[\psi\Big(\frac{\Gamma(k+1)^2}{\Gamma(a)\Gamma(b)}\Big) - \ln(1-z)\right](1-z)^k,
\end{aligned} \tag{76}$$

which converges when $|1-z| < 1$ and $|\arg z| < \pi$. We obtain the formulæ

$$\begin{aligned}
F(a,b,a+b;z) = {} & \frac{\Gamma(a+b)}{\Gamma(a)\Gamma(b)}\left[\psi\Big(\frac{\Gamma(1)^2}{\Gamma(a)\Gamma(b)}\Big) - \ln(1-z)\right] + \\
& + O\Big((1-z)\ln(1-z)\Big)
\end{aligned} \tag{77}$$

and

$$\begin{aligned}
\frac{dF}{dz}(a,b,a+b;z) = {} & \frac{\Gamma(a+b)}{\Gamma(a)\Gamma(b)}\left[\frac{1}{1-z} + \ln(1-z) + ab\,\psi\Big(\frac{\Gamma(2)^2}{\Gamma(a)\Gamma(b)}\Big)\right] + \\
& + O\Big((1-z)\ln(1-z)\Big).
\end{aligned} \tag{78}$$

Substituting (77) and (78) into (74) gives

$$\sigma_3(\sigma_1) \sim \mp 3A\,\frac{\Gamma(\tfrac{1}{2})}{\Gamma(\tfrac{\nu}{2}+1)\Gamma(-\tfrac{\nu}{2})} - 3B\,\frac{\Gamma(\tfrac{3}{2})}{\Gamma(\tfrac{1}{2}-\tfrac{\nu}{2})\Gamma(\tfrac{\nu}{2}+1)} \tag{79}$$

as $\sigma_1 \to \pm 1^{\mp}$. Thus for every initial condition $(\bar{\sigma}_3, \bar{\sigma}_4)$ the function $\sigma_1 \to \sigma_3(\sigma_1; \bar{\sigma}_3, \bar{\sigma}_4)$ remains bounded on $[-1, 1]$. Substituting (77) and (78) into (71) gives

$$\sigma_4(\sigma_1) \sim C_0^{\pm} + C_1^{\pm} \ln \frac{1}{1 - \sigma_1^2} \tag{80}$$

as $\sigma_1 \to \pm 1^{\mp}$. Here

$$C_1^{\pm} = A \frac{\Gamma(\frac{1}{2})}{\Gamma(\frac{\nu}{2} + 1)\Gamma(-\frac{\nu}{2})} \pm 3B \frac{\Gamma(\frac{3}{2})}{\Gamma(\frac{1}{2} - \frac{\nu}{2})\Gamma(\frac{\nu}{2} + 1)}. \tag{81}$$

Suppose that $\sigma_1 \to 1^-$. From the fact that

$$\tfrac{1}{2} P_\nu(0) = \frac{\Gamma(\frac{3}{2})}{\Gamma(\frac{1}{2} - \frac{\nu}{2})\Gamma(\frac{\nu}{2} + 1)} \quad \text{and} \quad \tfrac{1}{2} P_\nu'(0) = -\frac{\Gamma(\frac{1}{2})}{\Gamma(\frac{\nu}{2} + 1)\Gamma(-\frac{\nu}{2})}, \tag{82}$$

where

$$P_\nu(z) = F(-\nu, \nu + 1, 1; \tfrac{1}{2}(1 - z))$$

on $|1 - z| < 2$, see [11, formulæ (7.66) and (7.68) on p. 178], it follows that $C_1^+ = 0$ if and only if $A = \mu P_\nu(0)$ and $B = \mu P_\nu'(0)$ for every $\mu \in \mathbf{R}$. In view of (75) and the definition of the line ℓ_1 (43), we see that the function $\sigma_1 \to \sigma_4(\sigma_1; \bar{\sigma}_3, \bar{\sigma}_4)$ remains bounded as $\sigma_1 \to 1^-$ if $(\bar{\sigma}_3, \bar{\sigma}_4)$ lies on the line ℓ_1. If $C^+ \neq 0$, then it follows from (81) that $\sigma_4(\sigma_1)$ has a logrithmic singularity at 1. More precisely,

$$\sigma_4(\sigma_1) \sim C_0^+ + C^+ \ln \frac{1}{1 - \sigma_1} \qquad \text{as } \sigma_1 \to 1^-.$$

Therefore the function $\sigma_1 \to \sigma_4(\sigma_1; \bar{\sigma}_3, \bar{\sigma}_4)$ remains bounded as $\sigma_1 \to 1^-$ if and only if $(\bar{\sigma}_3, \bar{\sigma}_4)$ lies on the line ℓ_1. A similar argument shows that the function $\sigma_1 \to \sigma_4(\sigma_1; \bar{\sigma}_3, \bar{\sigma}_4)$ remains bounded as $\sigma_1 \to -1^+$ if and only if $(\bar{\sigma}_3, \bar{\sigma}_4)$ lies on the line ℓ_2 (see (44)). If $(\bar{\sigma}_3, \bar{\sigma}_4)$ does not lie on the line ℓ_2 then

$$\sigma_4(\sigma_1) \sim C_0^- + C^- \ln \frac{1}{1 + \sigma_1} \qquad \text{as } \sigma_1 \to -1^+.$$

Appendix 2 Regularization

In this appendix we show how to regularize the family of Hamiltonian vector fields $X_{H_{\bar{\sigma}_3, \bar{\sigma}_4}}$ on $(-1, 1) \times \mathbf{R}$ with symplectic form $\frac{1}{1 - \sigma_1^2} d\sigma_1 \wedge d\sigma_2$ corresponding to the 2-parameter family of Hamiltonian functions

$$\begin{aligned} H_{\bar{\sigma}_3, \bar{\sigma}_4}(\sigma_1, \sigma_2) &= \tfrac{1}{2} \frac{\sigma_2^2}{1 - \sigma_1^2} + U_{\bar{\sigma}_3, \bar{\sigma}_4}(\sigma_1) \\ &= \tfrac{1}{2} \frac{\sigma_2^2}{1 - \sigma_1^2} + \tfrac{1}{10} \frac{1}{1 - \sigma_1^2} \sigma_3^2(\sigma_1, \bar{\sigma}_3, \bar{\sigma}_4) + \tfrac{3}{5} \sigma_4^2(\sigma_1, \bar{\sigma}_3, \bar{\sigma}_4) + \lambda \sqrt{1 - \sigma_1^2}. \end{aligned}$$

The integral curves of $X_{H_{\bar{\sigma}_3, \bar{\sigma}_4}}$ satisfy

$$\frac{d\sigma_1}{dt} = (1 - \sigma_1^2) \frac{\partial H_{\bar{\sigma}_3, \bar{\sigma}_4}}{\partial \sigma_2} \qquad \frac{d\sigma_2}{dt} = -(1 - \sigma_1^2) \frac{\partial H_{\bar{\sigma}_3, \bar{\sigma}_4}}{\partial \sigma_1}. \tag{83}$$

The flow of $X_{H_{\bar{\sigma}_3, \bar{\sigma}_4}}$ is *not* defined for all time, because for some initial conditions the disc falls flat in finite time. After regularization this does *not* occur.

To begin the regularization process we restrict our attention to the h-level set of $H_{\bar{\sigma}_3, \bar{\sigma}_4}$ when $h \geq 0$. Look at the function

$$F^h_{\bar{\sigma}_3, \bar{\sigma}_4} = (1 - \sigma_1^2)(H_{\bar{\sigma}_3, \bar{\sigma}_4} - h)$$
$$= \tfrac{1}{2}\,\sigma_2^2 + \tfrac{1}{10}\,\sigma_3^2(\sigma_1, \bar{\sigma}_3, \bar{\sigma}_4) + \tfrac{3}{5}\,(1 - \sigma_1^2)\sigma_4^2(\sigma_1, \bar{\sigma}_3, \bar{\sigma}_4) + \lambda\,(1 - \sigma_1^2)^{3/2}.$$

On $H^{-1}_{\bar{\sigma}_3, \bar{\sigma}_4}(h)$, considered as a subset of $\mathbf{R}^2$ with symplectic form $d\sigma_1 \wedge d\sigma_2$, Hamilton's equations defining the integral curves of $X_{F^h_{\bar{\sigma}_3, \bar{\sigma}_4}}$ are

$$\frac{d\sigma_1}{dt} = \frac{\partial F^h_{\bar{\sigma}_3, \bar{\sigma}_4}}{\partial \sigma_2} = (1 - \sigma_1^2)\frac{\partial H_{\bar{\sigma}_3, \bar{\sigma}_4}}{\partial \sigma_2}$$

$$\frac{d\sigma_2}{dt} = -\frac{\partial F^h_{\bar{\sigma}_3, \bar{\sigma}_4}}{\partial \sigma_1} = -\frac{\partial(1 - \sigma_1^2)}{\partial \sigma_1}(H_{\bar{\sigma}_3, \bar{\sigma}_4} - h) - (1 - \sigma_1^2)\frac{\partial H_{\bar{\sigma}_3, \bar{\sigma}_4}}{\partial \sigma_1}$$

$$= -(1 - \sigma_1^2)\frac{\partial H_{\bar{\sigma}_3, \bar{\sigma}_4}}{\partial \sigma_1}, \tag{84}$$

since we have restricted the equations of motion to the level set $H^{-1}_{\bar{\sigma}_3, \bar{\sigma}_4}(h)$. These are the same equations as (83). Because $F^h_{\bar{\sigma}_3, \bar{\sigma}_4}$ is a once differentiable at $\sigma_1 = \pm 1$, $F^h_{\bar{\sigma}_3, \bar{\sigma}_4}$ is the regularization of $H^{-1}_{\bar{\sigma}_3, \bar{\sigma}_4}$ on $H^{-1}_{\bar{\sigma}_3, \bar{\sigma}_4}(h)$.

▷ Note that the 0-level set of $F^h_{\bar{\sigma}_3, \bar{\sigma}_4}$ is compact, having been obtained by adding the points $(\pm 1, 0)$ to $H^{-1}_{\bar{\sigma}_3, \bar{\sigma}_4}(h)$.

(A2.1) **Proof.** Since $0 = (1 - \sigma_1^2)(H_{\bar{\sigma}_3, \bar{\sigma}_4} - h)$ if and only if $\sigma_1 = \pm 1$ or $H_{\bar{\sigma}_3, \bar{\sigma}_4} = h$, we see that $\{F^h_{\bar{\sigma}_3, \bar{\sigma}_4} = 0\} = \{\sigma_1 = \pm 1\} \cup \{H_{\bar{\sigma}_3, \bar{\sigma}_4} = h\}$.

Because $H_{\bar{\sigma}_3, \bar{\sigma}_4}$ is nonnegative,

$$H_{\bar{\sigma}_3, \bar{\sigma}_4} = h\} \subseteq \{\sigma_2^2 \leq 2h(1 - \sigma_1^2)\} \cup \left(\{U_{\bar{\sigma}_3, \bar{\sigma}_4} \leq h\} \times \mathbf{R}\right)$$

$$\subseteq \{\sigma_2^2 \leq 2h(1 - \sigma_1^2)\} \cup \left((-1.1) \times \mathbf{R}\right).$$

Consequently, $F^h_{\bar{\sigma}_3, \bar{\sigma}_4} = 0\} \subseteq \left([-1, 1] \times \mathbf{R}\right) \cup \{\sigma_2^2 \leq 2h(1 - \sigma_1^2)\}$, which is compact. □

Therefore the flow of $X_{F^h_{\bar{\sigma}_3, \bar{\sigma}_4}}$ on the 0-level set of $F^h_{\bar{\sigma}_3, \bar{\sigma}_4}$ is defined for *all* time, that is, the vector field $X_{F^h_{\bar{\sigma}_3, \bar{\sigma}_4}}$ is complete.

Acknowledgment. We would like to thank Prof. J.J. Duistermaat of the University of Utrecht for giving us a copy of his notes on the rolling disc. This was the starting point of our research. The Lie group formulation of the problem given in section 2 is due to him.

References

[1] Abromowitz, M. and Stegun, I., "The Handbook of Mathematical Functions", Dover, New York, 1972.

[2] Appell, P. 1900: *Sur l'intégration des équations du mouvement d'un corps pesant de révolution roulant par une arête circulaire sur un plan horizontal; cas particulier du cerceau*, Rend. Palermo vol **14**, 1–6.

[3] Arnol'd, V.I., "Encyclopaedia of Mathematical Sciences", volume 3. *Dynamical Systems, III*, Springer Verlag, New York, 1988.

[4] Birkhoff, G. D. 1915: *The restricted problem of three bodies*, Rend. Circ. Mat. Palermo, vol **39**. 265–334. ≡ *Collected Mathematical Papers*, American Mathematical Society, New York, 1950; vol **1**, 682–751.

[5] Chaplygin, S. A. 1897: *On the motion of a heavy body on a horizontal plane*, Physics Section of the Imperial Friends of Physics, Anthropology and Ethnographics, Moscow **9**. Reproduced in *Selected Works on Mechanics and Mathematics*, State Publ. House, Technical Theoretical Literature, Moscow 1954, 413–425, (both in Russian).

[6] Ferrers, N. M. 1872: *Extension of Lagrange's equations*, Quartely Journal of Mathematics, vol **12**. 1–5.

[7] Hermans, J. 1995: *A symmetric sphere rolling on a surface*, Nonlinearity, **8** (4), 493–515.

[8] Hermans, J. 1995: "Rolling Rigid Bodies with and without Symmetries", Ph. D. thesis, University of Utrecht, ISBN 90-393-0680-X.

[9] Kemppainen, D. 1993: Master thesis, University of Calgary.

[10] Korteweg, D. J. 1899: *Über eine ziemlich verbreitete unrichtige Behandlungsweise eines Prcblems der rollenden Bewegung*, Nieuw Archief voor Wiskunde, **4** (2), 130–155.

[11] Lebedev, N. N., "Special Functions and their Applications", Dover, New York, 1972.

[12] Milnor, J., "Lectures on the h-cobordism theorem", Princeton University Press, Princeton, N.J., 1965.

[13] Moshchuk, N. K. 1988: *A qualitative analysis of the motion of a heavy solid of revolution on an absolutely rough plane*, J. Appl. Math. Mech. (2) **52**, 159–165.

[14] Neimark, J. I. and Fufaev, N. A., "Dynamics of Nonholonomic Systems", Translations of Mathematical Monographs, volume **33**, American Mathematical Society, Providence, R.I., 1972.

[15] O'Reilly, O. 1995: *The dynamics of rolling disks and sliding disks.* preprint, Department of Mechanical Engineering, University of California, Berkeley.

[16] Poston, T. and Stewart, I., "Taylor Expansions and Catastrophe Theory", Pitman, London, 1976.

[17] Routh, E. J.: "Advanced Dynamics of a System of Rigid Bodies", Sixth edition, Dover, New York, 1960.

[18] Vierkandt, A. 1892: *Über gleitende und rollende Bewegung*, Monatsh. f. Math. u. Physik, vol **3**, 31–54, 97–134.

[19] Viro, Ya., *Some integral calculus based on Euler characteristic*, in: "Topology and Geometry – Rohlin Seminar", Lecture Notes in Mathematics, vol. **1346**, 1988, 127–138.

Progress in Nonlinear Differential Equations
and Their Applications, Vol. 19
© 1996 Birkhäuser Verlag Basel/Switzerland

Testing for S_n-Symmetry with a Recursive Detective

Karin Gatermann[*]

Abstract

The theory of the symmetry increasing or decreasing bifurcations of an attractor in symmetric dynamical systems has been studied intensively. Special functions called detectives enables one to determine the symmetry properties of a certain attractor. But it can be difficult to find a detective and moreover the numerical evaluation can be costly. Representation theory and invariant theory are used to derive efficient methods. A method is proposed to evaluate a detective recursively for the symmetric group S_n. Comparision shows that this is very efficient both in CPU time and in storage.

1 Introduction

The theory of the symmetry increasing or decreasing bifurcations of an attractor in symmetric dynamical systems is largely complete, see [12], [6], [9], [10], [5], [20], [3], [13], [4]. Given a symmetric dynamical system, then there exists a set of functions, called *detectives* that enables one to determine the symmetry properties of a certain attractor. However, in a practical situation it can be difficult to find this set of detectives within a reasonable amount of time. Furthermore it can be quite costly to evaluate them on an attractor in order to determine its symmetry.

The aim of the article is to present an efficient numerical method to obtain a set of detectives for a given symmetry group. The methods are from representation theory and invariant theory. In connection with the latter, the recently developed algorithms for polynomial rings are very useful. Applying the results to the group S_n shows that the presented method can be very efficient indeed.

Section 2 introduces the concept of detectives and gives a precise definition. Then some aspects of using them to determine the symmetry properties of a given set (attractor) are discussed. As an example and for later use two possibilities of detectives for S_n are given.

*Konrad-Zuse-Zentrum für Informationstechnik Berlin, Heilbronner Str. 10, D-10711 Berlin-Wilmersdorf, and Freie Universität Berlin, Institut für Mathematik I, Fachbereich Mathematik und Informatik, Arnimalle 2-6, D-14195 Berlin

In section 3 the structure of the detectives themselves gives rise to decomposition properties that can be used to express the detectives in a smaller set of functions. This property turns out to be very powerful for the group S_n since it allows one to recursively define the detectives. This is the subject of section 4. First I present the recursion for the detectives, which is then applied to an array of Josephson junctions with S_{10} symmetry. Compared to previous results in [25] for S_4 and S_5 the recursive method turns out to be very efficient both in CPU time as in storage.

2 The concept of detectives

In this section the detection of the symmetry of attractors is recalled.

We are interested in dynamical systems

$$\dot{x} = f(x, \lambda), \quad x \in R^n, \lambda \in R \tag{1}$$

or discrete dynamical systems

$$x^k = f(x^{k-1}, \lambda), \quad x \in R^n, \lambda \in R \tag{2}$$

which are equivariant with respect to a faithful, orthogonal representation $\vartheta : G \to Gl(R^n)$ of a finite group G. In the following ϑ always refers to this representation.

To make the notion of equivariance more precise we give the following definition.

Definition 2.1 *([16]) Let ϑ, ρ be two representations of G. A C^∞ function h is called ϑ-ρ-equivariant, if*

$$h(\vartheta(t)x) = \rho(t)h(x) \quad \forall\, t \in G, \forall\, x \in R^n.$$

The mapping f in (1) or (2) is assumed to be ϑ-ϑ-equivariant which usually is called G-equivariant.

In [6] attractors which do not lie completely in a fixed point space are thickened to open sets. So let $\mathcal{A}$ be the class of all open subsets A of R^n with piecewise smooth boundary that satisfy the dichotomy

$$\vartheta(t)A = A \text{ or } \vartheta(t)A \cap A = \emptyset \quad \forall\, t \in G,$$

where ϑ is the faithful representation in (1).

$$H(A) = \{t \in G | \vartheta(t)A = A\}$$

denotes the isotropy group of an attractor. Observables transform the symmetry of attractors into a *physical space* W.

Definition 2.2 *([6]) Let $\rho : G \to Gl(W)$ be a linear representation. A ϑ-ρ-equivariant C^∞ function $\phi : R^n \to W$ is called an observable. The vector*

$$K_\phi(A) := \int_A \phi d\mu$$

is an observation, where μ is assumed to be the Lebesgue measure.

The determination of $H(A)$ is thus shifted to determining the isotropy group of $K_\phi(A) \in W$ denoted by $H_\phi(A)$.

Checking isotropy may be done with *distances*:
Let $\text{Fix}(H, W)$ be the fixed point space of a subgroup H of G within W and

$$P^{\rho, H}(y) = \frac{1}{|H|} \sum_{t \in H} \rho(t)(y). \tag{3}$$

the projection onto $\text{Fix}(H, W)$. Then

$$d^H(y) = \|y - P^{\rho, H} y\|_2^2 = \|(Id - P^{\rho, H}) y\|_2^2, \tag{4}$$

gives the distance to the fixed point space. Clearly, the isotropy of y is the maximal subgroup H with distance zero.

For the detection of symmetry of attractors it becomes important that ρ distinguishes all subgroups, i.e. all subgroups $H = G_y$ of G appear to be isotropy groups in W for one $y \in W$.

Definition 2.3 *([6] Def. 4.2): Two representations $\rho_1 : G \to Gl(W_1)$ and $\rho_2 : G \to Gl(W_2)$ are lattice equivalent if there exists a linear isomorphism $L : W_1 \to W_2$ such that*

$$L(Fix(H, W_1)) = Fix(H, W_2).$$

for every subgroup H of G.

Let $\vartheta^i, i = 1, \ldots, h$ denote the inequivalent irreducible representations of G. ϑ^1 denotes the unit representation. For a linear representation $\rho : G \to Gl(R^n)$ let $m_i(\rho)$ be the multiplicity in the canonical decomposition $\rho = \sum_{i=1}^h m_i(\rho) \vartheta^i$. Let P_i denote the projection onto the isotypic component with respect to ϑ^i.

Lemma 2.4 *([6] Thm. 4.3): $\rho = \sum_{i=1}^h \vartheta^i$ distinguishes all subgroups of G, where ϑ^i denote the irreducible representations of G.*

A *detective* is an observable which generically determines all symmetries of sets in $\mathcal{A}$.

Definition 2.5 *([6] Def. 5.1) The observable ϕ is a detective for G if for each subset $A \in \mathcal{A}$ almost all near identity ϑ-ϑ-equivariant diffeomorphism ψ satisfy*

$$H_\phi(\psi(A)) = H(A).$$

Theorem 2.6 ([6] Thm. 5.2): Let $\phi^i, i = 1, \ldots, h$ be ϑ-ϑ^i-equivariant observables which are polynomial and $\phi^i \not\equiv 0$. Then $\phi = (\phi^1, \ldots, \phi^h)$ is a detective for G.

It is clear that in Thm. 2.6 it is sufficient to consider all lattice inequivalent irreducible representations.

It turned out in case where the attractor is contained within a fixed point space of K one has to be more careful, see [15]. Then the symmetry of A may be one of the subgroups of the normalizer $N_G(K)$ of K. So the requirement is that $\phi_{|Fix(K)}$ is a detective for the group $N_G(K)$.

Before we discuss special detectives we shortly discuss the practical evaluation of the observation. Precise descriptions can be found in the literature. For discrete dynamical systems one uses

$$\lim_{N \to \infty} \frac{1}{N} \sum_{k=0}^{N} \phi(f^k(x_0)),$$

provided the ergodic theorem is valid. For ordinary differential equations the attractor is $\{x(t)|t \geq 0\}$ and the observation becomes

$$\lim_{T \to \infty} \frac{1}{T} \int_0^T \phi(x(t))dt.$$

There are two suggestions for detectives:

1.) In [6],[9] the observable $\phi(x) = xx^t$ was chosen as detective. ϕ is isomorphic to a ρ-ϑ-equivariant mapping where ρ is a subrepresentation of the *tensor product* $\vartheta \otimes \vartheta^*$ where $\vartheta^* : G \to Gl(R^n), \vartheta^*(t) = \vartheta(t^{-1})^t$ is the contragredient representation. In our case when ϑ is orthogonal then $\vartheta^* = \vartheta$. Obviously, range $\phi \neq R^{n,n}$. If $\vartheta = \sum_{i=1}^h m_i(\vartheta)\vartheta^i$ is the canonical decomposition and

$$\vartheta^i \otimes \vartheta^j = \sum_{k=1}^h c_k^{ij}\vartheta^k$$

with the multiplicities c_k^{ij} being the Clebsch-Gordan coefficients, then

$$\vartheta \otimes \vartheta = \sum_{k=1}^h \left(\sum_{i,j=1}^h m_i(\vartheta) \cdot m_j(\vartheta) \cdot c_k^{ij} \right) \vartheta^k = \sum_{k=1}^h m_k(\vartheta \otimes \vartheta) \cdot \vartheta^k.$$

Assume ϑ has the following property:
For each $k = 1, \ldots, h$ there exists i and j such that

$$m_i(\vartheta) \neq 0, m_j(\vartheta) \neq 0 \text{ and } c_k^{ij} \neq 0.$$

Assuming this property we have $m_k(\vartheta \otimes \vartheta) > 0, k = 1, \ldots, h$. To prove that ϕ is a detective for G it remains to show that $P_k\phi \not\equiv 0, k = 1, \ldots, h$. Using the decomposition one needs to find for each k irreducible representations ϑ^i, ϑ^j with $c_k^{ij} > 0$ and $P_k x_i x_j^t \not\equiv 0$, where x_i, x_j behave like ϑ^i and ϑ^j, respectively. This can easily be implemented and checked in a Computer Algebra environment.

In [6] it is shown that $\vartheta \otimes \vartheta$ distinguishes all subgroups in the case of rings of coupled cells with $G = D_p$ symmetry. For this $p \geq 3$ and the number of equations per cell $m \geq 2$ is essential.

2.) A second detective was given by the left regular representation $L : G \to Gl(R^{|G|})$, see [9]. Its well-known decomposition is

$$L = \sum_{i=1}^{h} m_i(L)\vartheta^i = \sum_{i=1}^{h} dim(\vartheta^i)\vartheta^i.$$

So any ϑ-L-equivariant mapping ϕ with $P_i^L \phi \not\equiv 0, i = 1, \ldots, h$ is a detective. L and ϑ define a group action on the space of polynomial mappings

$$\begin{aligned} \delta : \quad & G \to \quad Gl(R[x]^{|G|}) \\ & t \hookrightarrow \quad \delta(t) \\ \delta(t) : \quad & R[x]^{|G|} \to \quad R[x]^{|G|} \\ & \delta(t)(\phi(x)) = L(t)\phi(\vartheta(t^{-1})x). \end{aligned}$$

This is a linear representation.

$$P^{\delta,G} = \frac{1}{|G|} \sum_{t \in G} \delta(t). \tag{5}$$

is a projection onto the trivial component with respect to δ which consists of ϑ-L-equivariant mappings. In [25] it is suggested to choose a polynomial $p(x)$ and $q(x) = (p(x), 0, \ldots, 0)^t$. An observable is defined by

$$\phi(x) := P^{\delta,G}(q) = \frac{1}{|G|} \sum_{t \in G} \delta(t)(q) = \frac{1}{|G|} \sum_{t \in G} L(t^{-1})((p(\vartheta(t)x), 0, \ldots, 0)^t), \tag{6}$$

For the special choice $p(x) = x_1 x_2^2 \cdots x_{n-1}^{n-1}$ the mapping ϕ is a detective for S_n, see [25].

3 Poincaré series and irreducible observables

The aim of this section is to show the richness of polynomial detectives. Applying some theory we derive results for the degree of detectives and its decomposition into smaller functions.

The polynomial, ρ-ϑ-equivariant mappings $\phi(x)$ form a module over the ring of ϑ-invariant polynomials. Then each component $\phi_i(x), i = 1, \ldots, dim(\rho)$ is a polynomial. ϕ_i is said to be *homogeneous* of degree k if in the representation with monomials $\phi_i(x) = \sum_{j \in J} a_j x^j$ only monomials of degree k appear. The vector of polynomials $\phi(x)$ is said to be *homogeneous* of degree k if for each component either $\phi_i(x) \equiv 0$ or $o_i(x)$ is homogeneous of degree k. Let m_k denote the dimension of the vector space of ϑ-ρ-equivariant mappings which are homogeneous of degree k.

Theorem 3.1 *([26],[16]) Let ρ and ϑ be linear representations of G. Let m_k be the number of linear independent ρ-ϑ-equivariant polynomial mappings homogeneous of degree k. Then the Hilbert-Poincaré series is*

$$\sum_{k=0}^{\infty} m_k z^k = \frac{1}{|G|} \sum_{t \in G} \frac{tr(\rho(t^{-1}))}{det(Id - z \cdot \vartheta(t))}.$$

For $\rho = \vartheta$ this is exactly the series given by Sattinger [21]. For ρ being the trivial irreducible representation the series is the well-known Molien series for invariant polynomials which was proved at the end of the last century.

The righthand side is easily evaluated and thus the dimensions m_k for small k are cheaply determined with a Taylor expansion. This has been implemented in a Computer Algebra System.

The following theorem is a generalization of the results for the invariants, see e.g. [24] and for the isotypic components of $R[x]$.

Theorem 3.2 *([26],[16]) There exist n homogeneous invariants $\sigma_i, i = 1, \ldots, n$ such that the ϑ-ρ-equivariants form a free module over the subring $R[\sigma]$. (The module is Cohen-Macaulay.)*

Thm. 3.2 states that each equivariant has a unique representation $\sum A_i(\sigma_1, \ldots, \sigma_n) b_i$ with polynomials A_i. Note that if one considers the invariant ring instead of the subring $R[\sigma]$ such representations are non-unique in general.

The free basis is determined with the help of projections and the series, see [16]. From Computer Algebra graded Gröbner bases with respect to weighted orderings are used for this purpose.

Lemma 3.3 *The minimal degree of a detective for a given group action ϑ is*

$$d = max_{i=1,\ldots,h}(kmin(\vartheta^i, \vartheta)),$$

where $\vartheta^i, i = 1, \ldots, h$ are the pairwise lattice inequivalent irreducible representations which are necessary to distinguish all subgroups of G. $kmin(\vartheta^i, \vartheta)$ is the minimal degree of a non-zero, homogeneous ϑ^i-ϑ-equivariant.

Proof: Thm. 2.6 means that for each detective $\phi(x)$ the restrictions $P_i^\rho \phi(x) \not\equiv 0, i = 1, \ldots, h$ hold. The mapping $P_i^\rho \phi(x)$ contains at least one ϑ^i-ϑ-equivariant mapping unequal zero which has minimal degree $kmin(\vartheta^i, \vartheta)$. $\square$

The values $kmin(\vartheta^i, \vartheta)$ can easily be read off from the series in Theorem 3.1 with $\rho = \vartheta^i$.

Detectives may be build from smaller functions. In contrast to above where we used a decomposition on the image we will now study the consequences of a decomposition in the domain.

Definition 3.4 *A ϑ^i-ϑ^j-equivariant mapping $\phi^i_j(x) \not\equiv 0$ is called ϑ^i-ϑ^j-observable (irreducible observable).*

Remark: There are combinations ϑ^i, ϑ^j such that no ϑ^i-ϑ^j-observables exist.

Example: $D_6 = \{id, r, r^2, r^3, r^4, r^5, s, sr, sr^2, sr^3, sr^4, sr^5\}$,
$\vartheta^2(r) = 1, \vartheta^2(s) = -1$ and ϑ^5 the 2-dim. faithful representation. Then there exists no ϑ^2-ϑ^5-observable, since the series equals zero. But there are ϑ^5-ϑ^2-observables. $kmin(\vartheta^5, \vartheta^2) = 6$ can be read off from the series.

Tensor products of representations can be used to construct such small observables from known observables. Let ϕ be a η-δ-observable and ψ a η-ρ-observable where δ, η, ρ are irreducible representations. Then a η-$(\delta \otimes \rho)$-equivariant observable is defined by χ

$$\chi_{i+j}(x) = \phi_i(x) \cdot \psi_j(x) \quad i = 1, \ldots, dim(\delta), j = 1, \ldots, dim(\rho). \tag{7}$$

If $m_k(\delta \otimes \rho) = 1$ then $P_k\chi$ is a η-ϑ^k-observable.

Example: Let ϑ^2 be the non-trivial, 1-dim representation of S_4, ϑ^3 the 2-dim irreducible representation, and ϑ^4, ϑ^5 the 3-dim. irreducible representations of S_4. Let ϑ^4 be faithful. The 3-dim. irreducible representation ϑ^5 is equivalent to $\vartheta^2 \otimes \vartheta^4$. Let $\phi(x)$ be a ϑ^3-ϑ^2-observable and ψ a ϑ^3-ϑ^4-observable. Then $\chi(x) = \phi(x) \cdot \psi(x)$ is a ϑ^3-ϑ^5-observable.

Since every representation ϑ has a decomposition $\vartheta = \sum_{i=1}^h m_i(\vartheta)\vartheta^i$ the small ϑ^i-ϑ^j-observables can help to build a detective, but in general more complicated functions are necessary.

Another composition of observables is the following. Assume the representation ϑ in the domain decomposes into $\vartheta = \delta_1 + \delta_2$ and $R^n = V_1 + V_2$, respectively. Let $\phi_i : V_i \to R^m$ be δ_i-ρ-observables, $i = 1, 2$. Then $(o_1 + \phi_2) : R^n \to R^m$ is a ϑ-ρ-observable.

There may be detectives for G which do not depend on the full domain R^n. But the contrary seems to be appropriate. The following seems to be a reasonable demand for a detective $\phi : R^n \to R^m$

For all ϑ-invariant subspaces W of R^n the function $\Psi : W \to R^n, \Psi(w) = \psi(w_1 + w)$ is not a constant function, where $w_1 \in W^\perp$ is a fixed value in the direct complement $W^\perp$ of W in R^n.

An even sharper demand is that $P_i^\rho \circ \Psi$ is not a constant function for $i = 2, \ldots, h$.

4   Detectives with recursive evaluation

Assume that H is a proper subgroup of G and that we already have a detective for H. Is it possible to use this information in order to construct a detective for G? We looked at theoretical results on induced representations such as the Frobenius reciprocity and the Theorem by Mackey (see [23] or [1]), but they are not suitable for practical considerations since no explicit formulas are obtained. However, for the symmetric group S_n explicit formulas for the relation between the irreducible representations σ^α of S_n and S_{n-1} are known, see [7]. The main advantage is that $Res_{S_{n-1}}\sigma^\alpha$ is already in block diagonal form. So no innerconnectivity matrices as in [14] are needed. Based on the Young tableaux we give a detective for S_n which is evaluated recursively. In order to clarify the group theoretic structure we recall a simple, but useful lemma.

Lemma 4.1 Let $\eta : G \to Gl(V)$ be a linear represenation of G and H a proper subgroup of G. Furthermore let $t_i, i = 1,\ldots,[G : H]$ be representatives of left cosets and $w \in V$ a H-invariant vector with respect to $Res_H(\vartheta)$. Then

$$v := \frac{1}{[G : H]} \sum_{i=1}^{[G:H]} \eta(t_i)w$$

is G-invariant.

Proof: The left cosets form a partition of G and each $g \in G$ corresponds to a permutation of the left cosets.   □

The symmetric group S_n has irreducible representations σ^α, where $\alpha = (\alpha_1, \alpha_2, \ldots), \alpha_i \geq \alpha_{i+1}$ are the partitions of n, denoted by $\alpha \vdash n$. These representations can nicely be described with the Young diagrams. This is presented in a way suitable for applications in [7] from where the following recursive scheme for σ^α was taken. Also [18] and [19] are interesting references for the irreducible representations of the symmetric group.

Each partition α corresponds to an ordered collection of boxes, the Young diagrams. The numbers $1,\ldots,n$ are put into the diagram such that in each row and each column the numbers increase. The number of these so-called standard α-tableau equals the dimension of σ^α. The matrices $\sigma^\alpha(t)$ are given in a basis indexed by the standard α-tableau, which are ordered in the last letter sequence which successively compares the last entries in each row.

Now two facts are important:

a.) The restricted representation decomposes as $Res_{S_{n-1}}(\sigma^\alpha) = \sum_{\beta \vdash n-1, \beta \subset \alpha} \sigma^\beta$. Moreover one can work with the same coordinates. Thus $\sigma^\alpha(t) = \text{diag}(\sigma^\beta(t))$, $t \in S_{n-1}$.

b.) It is sufficient to give $\sigma^\alpha(t)$ for generators of the group, e.g. the neighboring transpositions $(i, i+1), i = 1,\ldots,n-1$.

The consequence is that the matrices $\sigma^\alpha(i, i + 1)$ are known from the matrices $\sigma^\beta(i, i + 1), \beta \vdash n - 1, i = 1, \ldots, n - 2$. For $i = n - 1$ a precise description of the sparse matrix $\sigma^\alpha(n - 1, n)$ is given in [7]. The sparse matrix $\sigma^\alpha(i, i + 1)$ can be stored in a vector of length $2 \cdot \dim(\sigma^\alpha)$, see [7].

Based on the two facts above we now develop a recursive detective for the representation ϑ which describes the permutation of variables. Let for all partitions $\beta \vdash (n-1)$ the mappings $f^\beta : R^{n-1} \to R^{\dim(\sigma^\beta)}$ be $Res_{S_{n-1}}(\vartheta)$-σ^β-equivariant. For each partition $\alpha \vdash n$ we have by condition a.) a mapping $f : R^{n-1} \to R^{\dim(\sigma^\alpha)}$, $f = (f^{\beta_1}, \ldots, f^{\beta_r})$. All mappings f^β with $m_\beta(Res_{S_{n-1}}(\sigma^\alpha)) = 1$ or equivalently $\beta \subset \alpha$ are involved. The ordering of the f^β's in f is given by the last letter ordering of the standard α-tableaux. Let P be the projection $R^n \to R^{n-1}$ $(x_1, \ldots, x_n) \to (x_1, \ldots, x_{n-1})$. $f \circ P$ is $Res_{S_{n-1}}(\vartheta)$-η-equivariant with $\eta = \sum_{\beta \vdash n-1, \beta \subset \alpha} \sigma^\beta$.

As representatives of left cosets of S_n/S_{n-1} the cyclic permutations

$$t_i = (i, i + 1, \ldots, n) = (i, i + 1)(i + 1, i + 2) \cdots (n - 1, n). i = 1, \ldots, n - 1, t_n = id$$

are chosen.

Lemma 4.2 *Assume the above notations. The mappings* $F^\alpha : R^n \to R^{\dim(\sigma^\alpha)}$

$$F^\alpha(x) = \sum_{i=1}^n \sigma^\alpha(t_i)[f \circ P](\vartheta(t_i^{-1})x) \tag{8}$$

are ϑ-σ^α-equivariant for all partitions $\alpha \vdash n$.

Proof: Apply Lemma 4.1 with $w = f \circ P$ and the representation $\eta(t)w = \sigma^\alpha(t)[f \circ P](\vartheta(t^{-1})x)$.

Remark 4.3   i.) *Evaluation of $F^\alpha(x)$ only needs evaluations at*

$$(x_1, \ldots, x_{n-1}, \hat{x}_n), (x_1, \ldots, \hat{x}_{n-1}, x_n), \ldots, (\hat{x}_1, x_2, \ldots, x_n),$$

where the symbol $\hat{x}_i$ means that the variable x_i is dropped.

ii.) *Since the matrices $\sigma^\alpha(i, i + 1)$ are sparse the necessary matrix-vector operations can be performed cheaply, see [7, p. 131].*

Example 4.4 *For $S_2 = Z_2 = \{id, (1, 2)\}$ the Young diagrams are*

and .

The corresponding irreducible representations are given in the notation above by $\sigma^\beta(1, 2) = 1$ for $\beta = 2$ and $\sigma^\beta(1, 2) = -1$ for $3 = (1, 1)$. For $n = 3$ and the partition $\alpha = (1, 2)$ we have 2 standard Young tableaux

$<$.

By the branching theorem we have

$$\sigma^\alpha(1.2) = \left(\begin{array}{cc} \sigma^{(1,1)}(1,2) & 0 \\ 0 & \sigma^2(1,2) \end{array} \right) = \left(\begin{array}{cc} -1 & 0 \\ 0 & 1 \end{array} \right).$$

The last matrix is

$$\sigma^\alpha(2,3) = \left(\begin{array}{cc} \frac{1}{2} & \frac{3}{4} \\ 1 & -\frac{1}{2} \end{array} \right) = \left(\begin{array}{cc} d^{-1} & 1-d^{-2} \\ 1 & -d^{-1} \end{array} \right).$$

where $d = |x-u| + |y-v| = 2$ is the distance between the positions (x,y) and (u,v) for $n=3$ in the standard α-tableaux. Representatives of left cosets are id, the transposition $(1,2)$, and the 3-cycle $(1,2)\cdot(2,3)$. Let f^2 and $f^{(1.1)}$ be invariant and equivariant functions, respectively,

$$f^2((1,2)(x_1,x_2)) = f^2(x_2,x_1) = \sigma^2(1,2)f^2(x_1,x_2) = f^2(x_1,x_2)$$
$$f^{(1,1)}((1,2)(x_1,x_2)) = f^{(1,1)}(x_2,x_1) = \sigma^{(1,1)}(1,2)f^{(1,1)}(x_1,x_2) = -f^{(1,1)}(x_1,x_2).$$

Define $f(x_1,x_2) = (f^{(1,1)}(x_1,x_2), f^2(x_1,x_2))$ and

$$F^{(2,1)}(x_1,x_2,x_3) = f(x_1,x_2) + \sigma^{(2,1)}(2,3)f(x_1,x_3) + \sigma^{(2,1)}(1,2)\sigma^{(2,1)}(2,3)f(x_2,x_3).$$

The mapping $F^{(2,1)}$ is ϑ-$\sigma^{(2,1)}$-equivariant, where ϑ is the permutation of variables.

In order to construct a detective we need that F^α is not the zero-mapping. This is true if there exists $x_0 \in R^n$ with

$$F^\alpha(x_0) = \sum_{i=1}^n \sigma^\alpha(t_i)[f \circ P](\vartheta(t_i^{-1})x_0) \neq 0.$$

Lemma 4.5 *Let $f^\beta \not\equiv 0, \beta \vdash n-1$ be given $Res_{S_{n-1}}\vartheta$-σ^β-equivariant functions. Then for each partition $\alpha \vdash n$ there exists a S_{n-1}-invariant function $g : R^{n-1} \to R$ such that F^α as defined in (8) with $f = g \cdot (f^{\beta_1}, \ldots, f^{\beta_r}), \beta_i \subset \alpha$ is ϑ-σ^α-equivariant and not the zero mapping. $F = (F^{\alpha_1}, \ldots, F^{\alpha_s})$ is a detective for S_n where $\alpha_1, \ldots, \alpha_s$ denote the partitions of n.*

Proof: Choose $x_0 \in R^n$ such that it is not S_n-invariant and such that $f^\beta(Px_0) \neq 0$. Either we already have $F^\alpha(x_0) \neq 0$ with $f = (f^{\beta_1}, \ldots, f^{\beta_r})$ or we choose an S_{n-1}-invariant g. Since $f^\beta(x_0) \neq 0$ and x_0 is not invariant it is possible that the values $g(P\vartheta(t_i^{-1})x_0)$ are such that the vectors $g(P\vartheta(t_i^{-1})x_0)\sigma^\alpha(t_i)f(P(\vartheta(t_i^{-1})x_0))$ do not sum to zero. $\square$

Of course the step of Lemma 4.2 can be repeated. Let s_j be representatives of the left cosets of S_{n-2} in S_{n-1}. For example choose $s_{n-1} = id$.

$$s_j = (j, j+1, \ldots, n-1) = (j, j+1) \cdots (n-2, n-1), j = 1, \ldots, n-2.$$

$k = 1$		$f(x_1)$		$f(x_2)$	$f(x_3)$	$f(x_4)$

$$
\begin{array}{ccccccc}
k=2 & f^2(x_1,x_2) & f^2(x_1,x_3) & f^2(x_1,x_4) & f^2(x_2,x_3) & f^2(x_2,x_4) & f^2(x_3,x_4) \\
& f^{(1,1)}(x_1,x_2) & f^{(1,1)}(x_1,x_3) & f^{(1,1)}(x_1,x_4) & f^{(1,1)}(x_2,x_3) & f^{(1,1)}(x_2,x_4) & f^{(1,1)}(x_3,x_4)
\end{array}
$$

$$
\begin{array}{cccc}
& f^3(x_1,x_2,x_3) & f^3(x_1,x_2,x_4) & f^3(x_1,x_3,x_4) & f^3(x_2,x_3,x_4) \\
k=3 & f^{(2,1)}(x_1,x_2,x_3) & f^{(2,1)}(x_1,x_2,x_4) & f^{(2,1)}(x_1,x_3,x_4) & f^{(2,1)}(x_2,x_3,x_4) \\
& f^{(1,1,1)}(x_1,x_2,x_3) & f^{(1,1,1)}(x_1,x_2,x_4) & f^{(1,1,1)}(x_1,x_3,x_4) & f^{(1,1,1)}(x_2,x_3,x_4)
\end{array}
$$

$$
\begin{array}{ll}
& f^4(x_1,x_2,x_3,x_4) \quad f^{(2,1,1)}(x_1,x_2,x_3,x_4) \\
k=4 & f^{(3,1)}(x_1,x_2,x_3,x_4) \quad f^{(1,1,1,1)}(x_1,x_2,x_3,x_4) \\
& f^{(2,2)}(x_1,x_2,x_3,x_4)
\end{array}
$$

Figure 1: The quantities which are computed in Algorithm 4.6 for $n = 4$.

Then

$$
F^\alpha(x) = \sum_{i=1}^n \sigma^\alpha(t_i) \bigoplus_{\beta \vdash n-1, \beta \subset \alpha} f^\beta(P_{n-1}(\vartheta(t_i^{-1})x) \tag{9}
$$

$$
= \sum_{i=1}^n \sigma^\alpha(t_i) \bigoplus_{\beta \vdash n-1, \beta \subset \alpha} \sum_{j=1}^{n-1} \sigma^\beta(s_j) \bigoplus_{\gamma \vdash n-2, \gamma \subset \beta} f^\gamma(P_{n-2}(\vartheta(s_j^{-1})P_{n-1}\vartheta(t_i^{-1})x).
$$

We use P_k for the restriction to the first k coordinates. Secondly, let us introduce the notation $t_j^k = (j, j+1) \cdots (k-1, k)$ for cyclic permutations. For convenience $t_k^k = id$.

In these notations it is clear that (9) needs intermediate evaluations of type

$$
f^\beta(P_{n-1}\vartheta(t_i^n)^{-1}x) = \sum_{j=1}^{n-1} \sigma^\beta(t_j^{n-1}) \bigoplus_{\gamma \subset n-2, \gamma \subset \beta} f^\gamma(P_{n-2}\vartheta(t_j^{n-1})^{-1}\vartheta(t_i^n)^{-1}x).
$$

Observe that the arguments $P_{n-2}\vartheta(t_j^{n-1})^{-1}\vartheta(t_i^n)^{-1}x$ may be equal although i and j are different. We have $P_{n-2}\vartheta(t_j^{n-1})^{-1}\vartheta(t_i^n)^{-1}x = (x_{\nu_1}, \ldots, x_{\nu_{n-2}}), \nu_i < \nu_{i+1}, i = 1, \ldots, n-2$, but $x_{\nu_{n-1}}, x_{\nu_n}$ with $\nu_{n-1}, \nu_n \in \{1, \ldots, n\} \setminus \{\nu_1, \ldots, \nu_{n-2}\}$ have been deleted. Since the order of $x_{\nu_{n-1}}$ and x_{ν_n} does not matter this can be achieved in several ways.

Repeating the division process we have the following algorithm

Algorithm 4.6 *(Recursive evaluation of a detective for S_n)*

Given: $f : R \to R.\, f \not\equiv c, c \in R$.

Stored: $\sigma^\beta(k - 1.\,k), \beta \vdash k, k = 2, \ldots, n$

Input: $x \in R^n$

Output: $F(x) \in R^m$, $m = \sum_{\alpha n} \dim \sigma^\alpha$　　　　where F is ϑ-$(\sum_{\alpha \vdash n} \sigma^\alpha)$-equivariant
and ϑ describes the permutation of variables.

Initialization: $k = 1$

Evaluate $f(x_1), \ldots. f(x_n)$.

for $k = 2$ to n do

　　for each $\beta \vdash k$ do

　　　for all possible values of

　　　　$y = P_k \vartheta(t_{j_{k+1}}^{k+1})^{-1} \vartheta(t_{j_{k+2}}^{k+2})^{-1} \cdots \vartheta(t_{j_n}^n)^{-1} x, 1 \le j_i \le i, i = k+1, \ldots, n$

　　　　(equivalently $y = (x_{\nu_1}, \ldots, x_{\nu_k}), \nu_i < \nu_{i+1}, i = 1, \ldots, k-1$)

　　　evaluate

　　　　$f^\beta(y) = \frac{1}{k} \sum_{i=1}^k \sigma^\beta(t_i^k) \bigoplus_{\gamma \vdash k-1, \gamma \subset \beta} f^\gamma(P_{k-1} \vartheta((t_i^k)^{-1})y)$

　　　　if $\beta \ne k$ and $k \le n-1$ then $f^\beta(y) := g \cdot f^\beta(y)$ where $g = \sum_{j=0}^k (-f^k(y))^j$

Figure 1 shows the numbers which are computed in layers $k = 1.2.3,4$ for $n = 4$.
The multiplication with g assures that no components are the zero mapping and
gives a numerical balancing in the components.

Lemma 4.7 *Algorithm 4.6 evaluates a detective for S_n at $x \in R^n$. More precisely:
Let the representation ϑ denote the permutation of variables and σ^α the irreducible
representations of S_n corresponding to the partitions $\alpha \vdash n$ (Young's seminormal
form). The values $f^\alpha(x)$ are computed for ϑ-σ^α-equivariant functions $f^\alpha \not\equiv 0$.*

Proof: From Lemma 4.2 and the multiple division it is clear that f^α is ϑ-σ^α-
equivariant. It remains to show that $f^\alpha \not\equiv 0$. Using Lemma 4.5 we need to show
that the S_k-invariant functions g have been chosen sufficiently generic. The main
point for a polynomial f for this to happen is that the polynomial degree of f^β is
sufficiently large.

The degree of a ϑ-$\sigma^{(1,\ldots,1)}$-equivariant polynomial $f^{(1,\ldots,1)} \not\equiv 0$ is $\ge \frac{n(n-1)}{2}$. (These
equivariants form a module over the invariant ring generated by $\prod_{i=1, j=i+1}^n (x_i - x_j)$). Since $(1, \ldots. 1)$ is the most sensitive case and the functions g are chosen
sufficiently generic the statement follows by Lemma 4.5.　　　　　　　　　$\square$

The proof includes already the proof of the following lemma.

Lemma 4.8 *Let ϑ denote the representation of S_n which permutes n variables.
Then the lowest degree $\max_{\alpha \vdash n} k\min(\sigma^\alpha, \vartheta)$ of a polynomial detective is $\binom{n}{2}$.*

n	$n!$	regular rep.	stored perms.	Young's seminormal form
5	120	0.01 sec	0.01 sec	0.01 sec
6	720	0.01 sec	0.02 sec	0.02 sec
7	5040	0.08 sec	0.04 sec	0.10 sec
8	40320	0.67 sec	0.29 sec	0.42 sec
9	362880	6.54 sec	6.60 sec	1.98 sec
10	3628800	77.55 sec	− sec	9.47 sec
11	39916800	− sec	− sec	48.66 sec

Table 1: Comparison between a detective based on regular representation (implementations with and without storage of permutations) and a detective based on Young's seminormal form (recursion): Computing times for one function evaluation on a Sun 4 implemented in C.

Recall the formula $\frac{1}{|G|} \sum \rho(t) q(\vartheta(t^{-1})x)$ for a ϑ-ρ-equivariant mapping. A simple choice is $\rho = L$ the regular representation and $q(x) = (p(x), 0, \ldots, 0)$ such that p is a monomial. Lemma 4.8 shows that $p(x) = x_1 x_2^2 \cdots x_{n-1}^{n-1}$ is a monomial of lowest possible degree in order to give a detective. In [25] it is shown that this is indeed a detective.

The numerical properties of Algorithm 4.6 are the following. In each step k evaluation at $\binom{n}{k}$ values y are needed. The tupel $(f^k, \ldots, f^{(1,\ldots,1)})$ has dimension $\sum_{\alpha \vdash k} \dim \sigma^\alpha$. The determination of these $\binom{n}{k} \sum_{\alpha \vdash k} \dim \sigma^\alpha$ values necessitates $\#(\alpha \vdash k) \frac{k(k-1)}{2}$ matrix-vector multiplications which are performed cheaply. Here $\#(\alpha \vdash k)$ denotes the number of partitions. The number of additions at layer k is $\binom{n}{k} (k-1) \sum_{\alpha \vdash k} \dim \sigma^\alpha$.

If we compare this detective $F = \oplus_{\alpha \vdash n} f^\alpha$ with the detective above based on the left regular representation we notice that the dimension is much smaller. We have $\sum_{\alpha \vdash n} \dim \sigma^\alpha$ with $\sum_{\alpha \vdash n} (\dim \sigma^\alpha)^2 = |S_n| = n!$ (the general formula for the dimensions of irreducible representations) in comparison to $n!$ itself. Table 1 shows the performance of one function evaluation as n increases. Various implementations for the detective with the left regular representation have been tested. The powers x_i^j are computed once and stored. The first alternative is to generate all permutations while the detective is evaluated, but this requires a lot of trivial computations. We have implemented the algorithm in [22] for this purpose. The second alternative is to generate the permutations in advance, which is necessary for the determination of the symmetry of the vector. For this one needs to know at which position which permutation is placed. But the storage needs $n!(n-1)$ integers for pointers. This leads to the effect that for n large a lot of system cpu is spent on administration of this storage. For $n = 10$ it even fails to allocate the storage.

The detective based on Young's seminormal form also necessitates some amount of storage shown in Table 2. Although the matrices $\sigma^\alpha(k-1,k), \alpha \vdash k$ and some pointers for $\sigma^\alpha(i, i+1), i < k-1$ need to be stored the computing time is smaller if all $\sigma^\alpha(i, i+1), i = 1, \ldots, k-1$ are stored. For $n = 10$ the representation matrices need space for 227376 integers and the intermediate function values are stored in 123108 reals. The experience shows that this detective is preferable for larger n due to lower dimension and smaller storage requirements.

n	$n!$	$\#(\alpha \vdash n)$	$\sum_{\alpha \vdash n} \dim \sigma^\alpha$	rep.matrices	f's
5	120	7	26	288 int	141 real
6	720	11	76	1048 int	498 real
7	5040	15	232	3832 int	1849 real
8	40320	22	764	14528 int	7192 real
9	362880	30	2620	56448 int	29185 real
10	3628800	42	9496	227376 int	123108 real
11	39916800	56	35696	941296 int	538077 real

Table 2: Comparison of storage requirements between detectives based on regular representation and based on Young's seminormal form.

Finally, we like to mention that Algorithm 4.6 uses a principle known as Divide and Conquer. For other divide and conquer algorithms see [22].

Due to [6], [9] [8] detecting the symmetry of an attractor means computing $w := \frac{1}{N} \sum_{i=1}^N F(y_i)$, and distances $||(P^H - Id)w||_2$ for all subgroups H. Here we denote by P^H the projection on the fixed point space of H. The maximal subgroup H with $||(P^H - Id)w||_2 = 0$ is the symmetry group of the attractor. One still needs to think about how to perform $P^H - Id$.

For completeness we state:

Lemma 4.9 If $\eta : H \to Gl(V)$ with $\eta(t) = diag(\vartheta^i(t)), \vartheta^i, i = 1 \ldots, h$ being the irreducible representations of H, ϑ^1 being the trivial irreducible representation then $(P^H - Id)w = (0, w_2, \ldots, w_m), m = \sum_{i=1}^h \dim(\vartheta^i)$.

This suggests to collect within algorithm 4.6 all tupel $\oplus_{\beta \vdash k, \beta \neq k} f^\beta(y)$ for all $k = 2, \ldots, n$ and all possible y. But this needs too much storage.

Lemma 4.10 Let S_n, generated by $(1,2), \ldots, (n-1, n)$ be represented by η acting as $diag(\sigma^\alpha)$, $\alpha \vdash n$ where σ^α are the irreducible representations given by the Young tableaux. Then for S_k generated by $(1,2), \ldots, (k-1, k), k = 2, \ldots, n$ there exists a set of indices I_k such that

$$((P^{S_k} - Id)w)_i = \begin{cases} w_i & i \in I_k \\ 0 & i \notin I_k \end{cases} \tag{10}$$

For conjugate subgroups sS_ks^{-1} we have the formula

$$P^{sS_ks^{-1}} - Id = \eta(s)(P^{S_k} - Id)\eta(s^{-1}).$$

Proof: The vector w decomposes as $w = (w^n, \ldots, w^\alpha, \ldots, w^{(1,\ldots,1)})$ into subvectors w^α, α a partition of n. Since $\sigma^\alpha = \sum_{\beta \vdash n-1, \beta \subset \alpha} \sigma^\beta$ each w^α decomposes into subvectors $w^{\alpha,\beta}, \beta \vdash n-1$. The ordering of $w^{\alpha,\beta}$ depends on the last letter ordering. Repeating this step we obtain $w^{\alpha_1, \ldots, \alpha_k}$ where α_n is a partition of n and α_i is a partition of i with $\alpha_i \subset \alpha_{i+1}, i = n-1, \ldots, k$. The trivial irreducible representation of S_k is denoted by σ^k. This yields

$$(P^{S_k} - Id)w^{\alpha_n, \ldots, \alpha_k} = \begin{cases} w^{\alpha_n, \ldots, \alpha_k}, & \text{if } \alpha_k \neq k \\ 0, & \text{if } \alpha_k = k \end{cases}$$

$\square$

Remark: For proper subgroups H of S_k it is more difficult to evaluate the distance $\|(P^H - Id)w\|$. But once the projections $P^{H,\beta}, \beta \vdash k$ in the coordinates of an irreducible representation σ^β of S_k are known, the distance can be evaluated using $P^{S_k} - Id$.

5 Example

Coupled arrays of Josephson junctions are a typical example of a dynamical system with S_n-symmetry given by permutations. These arrays have been discussed in various papers, e.g. [2]. In [25] a lot of numerical simulations of symmetric chaos are presented for the Josephson junctions for $n = 4$ and $n = 5$. Both, the pure capacitive and the pure resistive cases are treated in that article.

In contrast to [25] our aim is to perform calculations for larger n.

The equations for the pure capacitive load read

$$\begin{aligned} \dot{\xi}_k &= \psi_k \\ \dot{\psi}_k &= \frac{1}{3+\beta}I - \frac{1}{\beta}(\psi_k + sin(\xi_k) - \frac{3}{n(3+\beta)}\sum_{j=1}^n(\psi_j + sin(\xi_j))) \end{aligned} \quad k = 1, \ldots, n.$$

We have done computations for $n = 10$ using the program Code++ [17]. Figure 2 shows an S_{10}-invariant attractor, where the parameter values have been chosen to be $\beta = 0.2$ and $I = 1.05$. The solution seems to converge against a periodic orbit with S_{10}-symmetry. The triangular shape in the right picture is explained by the fact that $\{\xi_1 = \xi_2, \psi_1 = \psi_2\}$, $\{\xi_1 = \xi_3, \psi_1 = \psi_3\}$, and $\{\xi_2 = \xi_3, \psi_2 = \psi_3\}$ are fixed point spaces which are flow invariant.

The value of the distance $\|Id - P^{S_{10}}v\|$ to the fixed point space of S_{10} is $1.26185e - 07$ where the approximate observation $v = \sum_{i=1}^N F(\xi^i)$ was used and the recursive detective F was evaluated at $N = 3000$ points. This small value

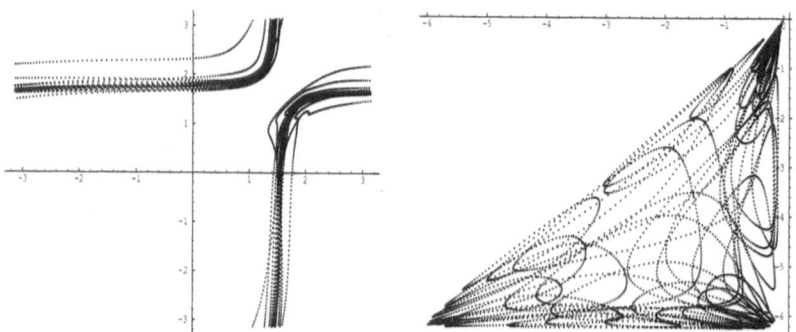

Figure 2: S_{10}-symmetric attractor in the coupled array of Josephson junctions (20 equations) for the parameter values $\beta = 0.2, I = 1.05$. The left picture shows ξ_1 versus ξ_2 plotted modulo 2π and the right picture shows $\xi_1 - \xi_3$ versus $\xi_2 - \xi_3$.

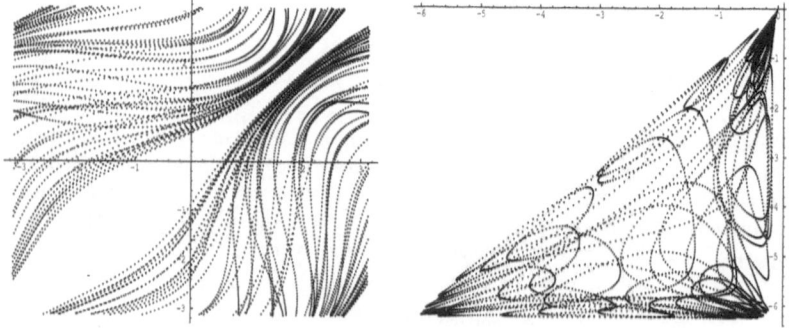

Figure 3: Symmetric attractor in the coupled array of Josephson junctions for the parameter values $\beta = 0.23, I = 1.13$. The left picture shows ξ_7 versus ξ_8 plotted modulo 2π and the right picture shows $\xi_1 - \xi_3$ versus $\xi_2 - \xi_3$.

clearly indicates that the type of symmetry is S_{10}. It is remarkably small since usually already a value of 0.05 is accepted to indicate a symmetry type.

In Figure 3 a more complicated attractor is presented. The parameter values are $\beta = 0.23$ and $I = 1.13$. The distances have been computed for $sS_k s^{-1}, k = 2, \ldots, n$ yielding a symmetry different from S_{10}.

In Algorithm 4.6, the recursive detective, the function f was chosen as $f(x_1) = \frac{x_1}{4} + \frac{1}{4x_1} + 1$.

These computations for S_{10} clearly demonstrated that one needs a sophisticated function for the detection of symmetry. The computing time depends on the detective since it is evaluated many times and secondly the distances are computed for a lot of subgroups of S_n. The recursive detective in Algorithm 4.6 was used successfully and is a typical example of modern algorithm technique.

Acknowledgments: Special thank are due to Michael Dellnitz for helpful discussions and to A. Kerber for the hint to the reference [7].

References

[1] W.A. Adkins and St.H. Weintraub. *Algebra An Approach via Module Theory*, volume 136 of *Graduate Texts in Mathematics*. Springer, New York, 1992.

[2] D.G. Aronson, M. Golubitsky, and M. Krupa. Coupled arrays of Josephson junctions and bifurcation of maps with S_N symmetry. *Nonlinearity*, 4:861–902, 1991.

[3] P. Ashwin and I. Melbourne. Symmetry groups of attractors. *Arch. Rat. Mech. Anal.*, 126:59–78, 1994.

[4] P. Ashwin and M. Nicol. Detection of symmetry of attractors from observations. Part I: Theory. Preprint, University of Warwick, 1995.

[5] Ph.J. Aston and M. Dellnitz. Symmetry breaking bifurcations and chaotic attractors. SC 94-27, Konrad-Zuse-Zentrum für Informationstechnik, Berlin, 1994.

[6] E. Barany, M. Dellnitz, and M. Golubitsky. Detecting the symmetry of attractors. *Physica D 1993*, 67:66–87, 1993.

[7] M. Clausen and U. Baum. *Fast Fourier Transforms*. BI Wissenschaftsverlag, Mannheim, Leipzig, Wien, Zürich, 1993.

[8] M. Dellnitz, M. Golubitsky, and I. Melbourne. Mechanisms of symmetry creation. In E. Allgower, K. Böhmer, and M. Golubitsky, editors. *Bifurcation and Symmetry*, volume ISNM 104, pages 99–109, Basel, Boston, Berlin, 1992. Birkhäuser.

[9] M. Dellnitz, M. Golubitsky, and I. Nicol. Symmetry of attractors and the Karhunen-Loève decomposition. In L. Sirovich, editor, *Trends and Perspectives in Applied Mathematics*, New York, 1994. Springer.

[10] M. Dellnitz and C. Heinrich. Admissible symmetry increasing bifurcations. Research Report UH/MD 187, University of Houston, Department of Mathematics, Houston, 1994.

[11] A. Fässler and E. Stiefel. *Group Theoretical Methods and Their Applications*. Birkhäuser, Boston, 1992.

[12] M. Field and M. Golubitsky. *Symmetry in Chaos*. Oxford University Press, Oxford, 1992.

[13] M. Field, I. Melbourne, and M. Nicol. Symmetric attractors for diffeomorphisms and flows. *Proc. London Math. Soc.*, 1995. To appear.

[14] K. Gatermann. Computation of bifurcation graphs. In E. Allgower, K. Georg, and R. Miranda, editors, *Exploiting Symmetry in Applied and Numerical Analysis*, volume 29 of *AMS Lectures in Applied Mathematics*, pages 187–201, Providence, Rhode Island, 1993. AMS.

[15] K. Gatermann. A remark on the detection of symmetry of attractors. In P. Chossat, editor, *Dynamics, Bifurcation and Symmetry New Trends and New Tools*, volume 437 of *NATO ASI Series C: Mathematical and Physical Sciences*. pages 123–125, Dordrecht, Boston, London, 1994. Kluwer.

[16] K. Gatermann. Semi-invariants, equivariants and algorithms. SC 94-11, Konrad-Zuse-Zentrum für Informationstechnik, Berlin, 1994. To appear in AAECC.

[17] A. Hohmann, 1994. Code++, a program written in C++.

[18] G. James and A. Kerber. *Representation Theory of the Symmetric Group*. Addison-Wesley Publ. Comp., Reading, Massachusetts, 1981.

[19] W. Ludwig and C. Falter. *Symmetries in Physics Group Theory Applied to Physical Problems*, volume 64 of *Springer Series in Solid-State Sciences*. Springer, Berlin, 1988.

[20] I. Melbourne. M. Dellnitz, and M. Golubitsky. Structure of symmetric attractors. *Arch. Rat. Mech. Anal.*, 123:75–98. 1993.

[21] D.H. Sattinger. *Group Theoretic Methods in Bifurcation Theory*. volume 762 of *Lecture Notes in Mathematics*. Springer Verlag, 1978.

[22] R. Sedgewick. *Algorithms*. Addison-Wesley Publ. Comp., Reading. Massachusetts, 1983.

[23] J.P. Serre. *Linear Representations of Finite Groups*. Springer, New York, 1977.

[24] B. Sturmfels. *Algorithms in Invariant Theory*, volume 1 of *Texts and Monographs in Symbolic Computation*. Springer, Wien, 1993.

[25] V. Tchistiakov. Detecting symmetry breaking bifurcations in the system describing the dynamics of coupled arrays of Josephson junctions. Preprint, University of Twente, Enschede, 1995.

[26] P. Worfolk. Zeros of equivariant vector fields: Algorithms for an invariant approach. *J. Symb. Comp.*, 17:487–511, 1994.

Progress in Nonlinear Differential Equations
and Their Applications, Vol. 19
© 1996 Birkhäuser Verlag Basel/Switzerland

Normal forms of vector fields satisfying certain geometric conditions

I.U. Bronstein   A.Ya. Kopanskii*

Abstract

The paper describes the simplified polynomial resonant normal forms of vector fields satisfying some geometric conditions (namely, preserving volume or symplectic forms or contact structures) with respect to C^k smooth changes of variables preserving the same structure.

Introduction

In the last years much attention has been paid to the theory of finitely smooth normal forms of vector fields in the neighbourhood of a singular point (see, for example, [1], [2]). The problem of reducing to normal form can be divided into three steps. At the first stage, one brings finite jets of the vector field to the *resonant* normal form by the aid of polynomial changes of variables. The second step deals with "cutting the tail", *i.e.*, reducing to the *polynomial* (resonant) normal form. Finally, one uses finitely smooth transformations to *simplify* the resonant normal form, *i.e.*, to eliminate as many monomial terms as possible. Many results obtained in these research directions are summed up in the book [2].

The recent work [3] is devoted to (the second step of) normalization of vector fields satisfying some geometric conditions (namely, preserving volume or symplectic forms or contact structures) by a smooth change of variables preserving the same structure. The method of proof used in [3] is based on the *deformation method* of singularity theory (R.Tom, J.Mather). As the authors emphasize, this method is ideally suited to discussing conjugacy problems in which a geometric structure is preserved. The deformation method reduces non-local non-linear problems to linear ones, namely, to solving *cohomology equations* (for the corresponding *hamiltonians*).

This paper is concerned with finitely smooth normal forms of volume preserving, symplectic and contact vector fields in the vicinity of a *hyperbolic* singular point. We make essential use of the techniques developed in [3] as well as of the results presented in [2]. Similar results hold for diffeomorphisms.

---

*Institute of Mathematics, Academy of Sciences of Moldova, Kishinev 277028 Moldova, comp@revel.moldova.su

We would like to warn the reader that in this paper (as well as in the book [2]) the term *resonance* refers to the *real parts* of the eigenvalues. This enables us to obtain normal form which smoothly depend on the vector fields. That is why one may hope that these normal forms are of interest to bifurcation theory.

The main results are given in Theorem 9.1. The proof makes use of a new version of the smooth invariant section theorem presented in the Appendix.

1 Geometric structures and differential forms

Let $\mathcal{M}$ be a contractible manifold of dimension $\dim \mathcal{M} = d$ (for example, a convex neighbourhood of the origin in $\mathbb{R}^n$). Let p be an integer, $0 \leq p \leq d$, and γ be a differential p-form. The form γ is said to be *non-degenerate* if the mapping $\xi \to i_\xi(\gamma)$ which to every vector field $\xi : \mathcal{M} \to T\mathcal{M}$ puts in correspondence the $(p-1)$-form $i_\xi(\gamma)$ is an isomorphism; that is, for each $x \in \mathcal{M}$ the mapping $T_x\mathcal{M} \to \Lambda^{p-1}(T_x^*\mathcal{M})$ defined by $v \mapsto i_v(\gamma(x))$ is an isomorphism of the tangent space $T_x\mathcal{M}$ onto the space of all skew-symmetric $(p-1)$-linear forms on $T_x\mathcal{M}$. This is possible only when

$$d = \dim(T_x\mathcal{M}) = \dim(\Lambda^{p-1}(T_x^*\mathcal{M})) = \binom{d}{p-1},$$

i.e., whenever $p = 2$ or $p = d$.

If γ is a non-degenerate p-form, then each differential $(p-1)$-form ω determines a unique vector field ξ such that $i_\xi(\gamma) = \omega$ (and *vice versa*).

If ω is a *closed* $(p-1)$-form (*i.e.*, $d\omega = 0$) then by Poincaré's lemma there exists a $(p-2)$-form F such that

$$i_\xi(\gamma) \equiv \omega = dF.$$

In this case the form F is called the *hamiltonian* of the vector field ξ. If, in addition, γ is a closed p-form, then the form $\omega = i_\xi(\gamma)$ is closed iff $L_\xi\gamma = 0$. The last equality means that the phase flow of ξ *respects* the p-form γ.

We shall consider three particular cases of geometric structures defined by differential forms.

Let μ be a *non-degenerate d-form* (*volume form*) on $\mathcal{M}$ (note that each volume form is closed). A vector field is called *volume preserving* (more exactly, preserving the volume form μ) if its (local) phase flow respects the form μ, *i.e.*, $L_\xi\mu = 0$. In this case, there exists a $(d-2)$-form F such that $i_\xi\mu = dF$. The hamiltonian F is determined up to adding a closed form. Each $(d-2)$-form F determines a μ-preserving vector field ξ.

Let $\mathcal{M}$ be a contractible manifold of dimension $2d$ provided with the *symplectic structure*, *i.e.*, with a *closed non-degenerate 2-form* ω. The vector field ξ is said to be *symplectic* if $L_\xi\omega = 0$. The *hamiltonian* F, $i_\xi(\omega) = dF$, is a function determined up to a constant. Each function may serve as a hamiltonian of some symplectic vector field.

Now let $\mathcal{M}$ be a contractible manifold of dimension $2d + 1$ and θ be a differential 1-form on $\mathcal{M}$ such that $\theta \wedge (d\theta)^d$ is a volume form. The *contact structure* on $\mathcal{M}$ is defined to be the codimension 1 subbundle of the tangent bundle $T\mathcal{M}$ determined by $\theta = 0$. Note that if $\lambda : \mathcal{M} \to \mathbb{R}$ is a non-vanishing function then the form $\lambda\theta$ gives the same contact structure. The vector field ξ is said to be a *contact vector field* if its flow preserves the contact structure, *i.e.*, there exists a function $\mu_\xi : \mathcal{M} \to \mathbb{R}$ such that

$$L_\xi\theta = \mu_\xi\theta.$$

If ξ is a contact vector field, then the function $F = i_\xi\theta \equiv \theta(\xi)$ is called the *contact hamiltonian*. Every function $F : \mathcal{M} \to \mathbb{R}$ determines a contact vector field ξ. The correspondence $\xi \to F$, $F = \theta(\xi)$, is one-to-one.

Let μ be a volume form on $\mathbb{R}^d$, ω be a symplectic form on $\mathbb{R}^{2d}$ and θ be a contact form on $\mathbb{R}^{2d+1}$. According to a theorem due to Darboux, the following statements are true:

1. There exists a system of coordinates $u = (u_1, \ldots, u_d)$ on $\mathbb{R}^d$ such that

$$\mu(u) = du_1 \wedge \ldots \wedge du_d. \tag{1.1}$$

2. There exist coordinates $(u, v) = (u_1, \ldots, u_d, v_1, \ldots, v_d)$ on $\mathbb{R}^{2d}$ such that

$$\omega(u, v) = du_1 \wedge dv_1 + \ldots + du_d \wedge dv_d. \tag{1.2}$$

3. One can find coordinates $(u, v, w) = (u_1, \ldots, u_d, v_1, \ldots, v_d, w)$ on $\mathbb{R}^{2d+1}$ so that

$$\theta(u, v, w) = u_1 dv_1 + \ldots + u_d dv_d + dw. \tag{1.3}$$

In what follows, the forms (1.1)–(1.3) will be called *standard*. The changes of variables which respect these forms are called *canonical*.

2 Transition formulas

Let $u \in \mathbb{R}^d$, $1 \leq i < j \leq d$. In order to simplify notation, we shall write

$$\widehat{du_i} = du_1 \wedge \ldots \wedge du_{i-1} \wedge du_{i+1} \wedge \ldots \wedge du_d.$$
$$\widehat{du_i} \wedge \widehat{du_j} = du_1 \wedge \ldots \wedge du_{i-1} \wedge du_{i+1} \wedge \ldots \wedge du_{j-1} \wedge du_{j+1} \wedge \ldots \wedge du_d.$$

For example, each $(d-1)$-form γ can be written as

$$\gamma(u) = \sum_{s=1}^{d} \gamma^s(u)\widehat{du_s}.$$

Similarly, if F is a $(d-2)$-form, then

$$F(u) = \sum_{p=1}^{d-1} \sum_{q=p+1}^{d} F^{pq}(u)\widehat{du_p} \wedge \widehat{du_q}$$

Let ξ be a vector field which preserves the volume form μ. With respect to canonical coordinates, we can choose a hamiltonian F so that

$$F^{pq}(u) = (-1)^{p+q} \int_0^1 [\xi^q(ut)u_p - \xi^p(ut)u_q]t^{d-2}dt \qquad (2.1)$$

$$(p = 1, \ldots, d-1; \quad q = p+1, \ldots, d).$$

Conversely, every $(d-2)$-form F serves as a hamiltonian of the following volume preserving vector field:

$$\xi^p(u) = \sum_{q=1}^{p-1}(-1)^{p+q}\frac{\partial F^{qp}(u)}{\partial u_q} - \sum_{q=p+1}^{d}(-1)^{p+q}\frac{\partial F^{pq}(u)}{\partial u_q} \quad (p = 1 \ldots, d). \qquad (2.2)$$

The relationship between symplectic vector fields $(\xi, \eta) = (\xi^1, \ldots, \xi^d, \eta^1, \ldots, \eta^d)$ on $\mathbb{R}^{2d}$ and the corresponding hamiltonians F can be expressed as follows:

$$F(u, v) = \sum_{p=1}^{d} \int_0^1 [\xi^p(ut, vt)v_p - \eta^p(ut, vt)v_p]dt; \qquad (2.3)$$

$$\xi^p(u, v) = \frac{\partial F(u, v)}{\partial v_p}, \quad \eta^p(u, v) = -\frac{\partial F(u, v)}{\partial u_p} \quad (p = 1, \ldots, d). \qquad (2.4)$$

Similarly, for contact vector fields (ξ, η, ζ) on $\mathbb{R}^{2d+1}$ and contact hamiltonians $F(u, v, w)$ one has

$$F(u, v, w) = \sum_{p=1}^{d} u_p\eta^p(u, v, w) + \zeta(u, v, w); \qquad (2.5)$$

$$\xi^p(u, v, w) = -\partial F/\partial v_p + u_p\partial F/\partial w,$$

$$\eta^p(u, v, w) = \partial F/\partial u_p \quad (p = 1, \ldots, d), \qquad (2.6)$$

$$\zeta(u, v, w) = F(u, v, w) - \sum_{p=1}^{d} u_p\partial F/\partial u_p$$

Moreover,

$$\mu_{(\xi, \eta, \zeta)}(u, v, w) = \partial F(u, v, w)/\partial w.$$

3   Normalization of linear parts of vector fields

We shall say that the hyperbolic linear vector field $\dot{x} = Ax$ ($x \in \mathbb{R}^d$) is reduced to the *block-diagonal normal form* if:

(1) $A = \mathrm{diag}\{A_1, \ldots, A_n\}$;

(2) A_i is a linear operator all of whose eigenvalues have one and the same real part θ_i ($i = 1, \ldots, n$);

(3) all the diagonal elements of the matrix A_i are equal to θ_i;

(4) if $i \neq j$, then $\theta_i \neq \theta_j$.

Let ξ be a vector field on $\mathbb{R}^d$ and the origin $0 \in \mathbb{R}^d$ be a hyperbolic singular point. Assume that ξ preserves the volume form μ. There exists a coordinate system $u = (u_1, \ldots, u_d)$ on $\mathbb{R}^d$ which reduces μ to the standard form (1.1). Denote $B = D\xi(0)$. It is well-known that one can find a real linear change of variables $y = Su$, $\det S = 1$, which brings the operator B to the Jordan real normal form. Clearly, the vector field $\dot{y} = Ay$, where $A = SBS^{-1}$, takes the block-diagonal normal form. Because $\det S = 1$ the volume form is still the standard one. Observe that $\mathrm{div} A = \sum_{i=1}^{n} \theta_i \cdot \dim A_i = 0$.

Now let ξ be a symplectic vector field on $\mathbb{R}^{2d}$, with the origin being a hyperbolic singular point. Choose coordinates $(u_1, \ldots, u_d, v_1, \ldots, v_d)$ so that the symplectic form ω takes the form (1.2). Denote $B = D\xi(0)$. It is known that $B = IC$, where C is a symplectic linear operator and

$$I = \begin{pmatrix} 0 & -E \\ E & 0 \end{pmatrix}.$$

According to Williamson's theorem (see [4]), there exists a canonical linear transformation which brings $B = IC$ to the form

$$\begin{pmatrix} A & 0 \\ 0 & -A \end{pmatrix}$$

where A is expressed in the block-diagonal normal form and all its eigenvalues have positive real parts.

Let ξ be a contact vector field on $\mathbb{R}^{2d+1}$ and the origin be a hyperbolic singular point of ξ. Suppose (u, v, w) are canonical coordinates, *i.e.*, the contact form θ is given by (1.3). Denote $B = D\xi(0)$, then we have

$$B(u, v, w) = \begin{pmatrix} Au \\ -Av + \gamma v \\ \gamma w \end{pmatrix}.$$

where A is a hyperbolic $d \times d$-matrix and γ is a non-zero number. Now "symplectize" the linear part of the vector field ξ by introducing coordinates (z, v, w, λ) on $\mathbb{R}^{2d+2}$ so that $z = u\lambda$. $\lambda \in \mathbb{R}\backslash\{0\}$, and letting

$$\dot{z} = Az - \gamma z, \quad \dot{v} = -Av + \gamma v, \quad \dot{w} = \gamma w, \quad \dot{\lambda} = -\gamma\lambda.$$

This linear symplectic vector field can be reduced to the block-diagonal normal form by some canonical linear transformation which leaves invariant the subspace $w = 0$, $\lambda = 0$. Because this transformation commutes with the natural action of the multiplicative group of non-zero real numbers, it induces a canonical contact transformation on $\mathbb{R}^{2d+1}$ which brings the linear part of the contact vector field ξ to the block-diagonal normal form.

4   Deformation method

The deformation method is a powerful tool for solving various conjugacy problems [1,3]. Let us outline this very useful approach (in the context of vector fields).

Let ξ and η be vector fields defined on a manifold $\mathcal{M}$. For $\varepsilon \in \mathbb{R}$ set $\xi_\varepsilon = \varepsilon\eta + (1 - \varepsilon)\xi$. Define the vector field Ξ on $\mathcal{M} \times \mathbb{R}$ by $\dot{x} = \xi_\varepsilon(x)$, $\dot{\varepsilon} = 0$. Suppose we have found a vector field Φ on $\mathcal{M} \times \mathbb{R}$ of the form $\dot{x} = \varphi_\varepsilon(x)$. $\dot{\varepsilon} = 1$, which commutes with Φ, i.e.

$$[\Phi, \Xi] = 0. \tag{4.1}$$

Let $F(x, \varepsilon, t)$ $(x \in \mathcal{M}$. $\varepsilon \in \mathbb{R}$, $t \in \mathbb{R})$ denote the phase flow of Φ. Put $h(x) = \mathrm{pr}_1 \circ F(x, 0, 1)$. It is easy to show that $(h)_*(\xi) \equiv Th \circ \xi \circ h^{-1} = \eta$. i.e., ξ and η are conjugate by h. Observe that (4.1) is equivalent to

$$D\varphi_\varepsilon \cdot \xi_\varepsilon(x) = D\xi_\varepsilon(x) \cdot \varphi_\varepsilon + \eta(x) - \xi(x) \quad (x \in \mathcal{M}). \tag{4.2}$$

which is called the *cohomology equation*. The corresponding characteristic system is

$$\dot{y} = D\xi_\varepsilon(x) \cdot y + \eta(x) - \xi(x), \quad \dot{x} = \xi_\varepsilon(x) \quad (x, y \in \mathcal{M};\ \varepsilon \in \mathbb{R}). \tag{4.3}$$

The solutions of (4.2) are nothing else than invariant sections $y = \varphi_\varepsilon(x)$ of (4.3). Thus, the deformation method reduces the conjugacy problem to finding a smooth invariant section $y = \varphi_\varepsilon(x)$ of the affine extension (4.3) and then to integrating the system $\dot{x} = \varphi_\varepsilon(x)$. $\dot{\varepsilon} = 1$.

Now let us turn to the case where the vector fields ξ and η satisfy a certain geometric condition (for example, they are symplectic). It is natural to ask whether one can choose the conjugation map to be canonical (in the corresponding sense). In order to solve this more delicate conjugacy problem, we need to use the relationships between vector fields and differential forms (see section 2). as suggested in [3].

Let γ be a closed non-degenerate k-form and ξ and η be vector fields which preserve γ. It is not hard to verify that

$$i_{[\xi_\varepsilon, \varphi_\varepsilon]}(\gamma) = d(i_{\xi_\varepsilon}(i_{\varphi_\varepsilon}(\gamma))),$$

therefore (4.2) implies that

$$d(i_{\xi_\varepsilon}(i_{\varphi_\varepsilon}(\gamma))) = i_{\eta-\xi}(\gamma).$$

Let F and G be the hamiltonians of the vector fields ξ and η. Denote by H_ε the hamiltonian of the vector field φ_ε we are seeking. Then

$$d(i_{\xi_\varepsilon}(dH_\varepsilon)) = dG - dF.$$

Taking into account the equality

$$L_{\xi_\varepsilon} H_\varepsilon = i_\varepsilon(dH_\varepsilon) + di_{\xi_\varepsilon}(H_\varepsilon)$$

we get

$$dL_{\xi_\varepsilon} H_\varepsilon = di_{\xi_\varepsilon}(dH_\varepsilon),$$

whence

$$d(L_{\xi_\varepsilon} H_\varepsilon - G + F) = 0.$$

It suffices to find the hamiltonian H_ε from the equation

$$L_{\xi_\varepsilon} H_\varepsilon = G - F, \tag{4.3}$$

which is called the *cohomology equation for hamiltonians*.

Let θ be a contact differential 1-form. Because

$$i_{[\xi_\varepsilon, \varphi_\varepsilon]}(\theta) = L_\xi \theta(\varphi_\varepsilon) - \mu_{\xi_\varepsilon} \theta(\varphi_\varepsilon), \quad L_{\xi_\varepsilon} \theta = \mu_{\xi_\varepsilon} \theta,$$

the equation (4.2) writes

$$L_{\xi_\varepsilon} H_\varepsilon - \mu_{\xi_\varepsilon} H_\varepsilon = G - F, \tag{4.4}$$

where F and G are the contact hamiltonians of the given contact vector fields ξ and η, and H_ε is the contact hamiltonian of the unknown vector field φ_ε.

Thus, in order to prove that two vector fields which preserve the volume form (the symplectic form, the contact structure) are conjugate with one another by a diffeomorphism respecting the corresponding structure, we have to solve the cohomology equation (4.3) or (4.4), then find the corresponding vector field φ_ε and integrate the system $\dot{x} = \varphi_\varepsilon(x)$, $\dot{\varepsilon} = 1$.

5 Cohomology equations for hamiltonians

Let us write out the cohomology equations for hamiltonians expressed in the canonical coordinates (see (1.1)–(1.3)) as well as the corresponding characteristic systems.

a) The volume preserving case

$$\sum_{t=1}^{d} \xi_\varepsilon^t \frac{\partial H_\varepsilon^{pq}}{\partial u_t} + \sum_{t=1}^{d} \frac{\partial \xi_\varepsilon^t}{\partial u_t} H_\varepsilon^{pq} - \left(\frac{\partial \xi_\varepsilon^p}{\partial u_p} + \frac{\partial \xi_\varepsilon^q}{\partial u_q}\right) H_\varepsilon^{pq}$$

$$+ \sum_{t=1}^{p-1} [(-1)^{q+t} \frac{\partial \xi_\varepsilon^q}{\partial u_t} H_\varepsilon^{tp} + (-1)^{p+t-1} \frac{\partial \xi_\varepsilon^p}{\partial u_t} H_\varepsilon^{tq}]$$

$$+ \sum_{t=p+1}^{q-1} [(-1)^{q+t-1} \frac{\partial \xi_\varepsilon^q}{\partial u_t} H_\varepsilon^{pt} + (-1)^{p+t-1} \frac{\partial \xi_\varepsilon^p}{\partial u_t} H_\varepsilon^{tq}]$$

$$+ \sum_{t=q+1}^{d} [(-1)^{q+t-1} \frac{\partial \xi_\varepsilon^q}{\partial u_t} H_\varepsilon^{pt} + (-1)^{p+t} \frac{\partial \xi_\varepsilon^p}{\partial u_t} H_\varepsilon^{qt}] \tag{5.1}$$

$$\equiv \sum_{t=1}^{d} \xi_\varepsilon^t \frac{\partial H_\varepsilon^{pq}}{\partial u_t} + \mathcal{A}_{pq} \cdot H_\varepsilon = G_{pq} - F_{pq}$$

$$(p = 1, \ldots, d-1; \quad q = p+1, \ldots, d):$$

$$\dot{H}_\varepsilon^{pq} = -\mathcal{A}_{pq}(u) \cdot H_\varepsilon + G^{pq}(u) - F^{pq}(u), \; \dot{u} = \xi_\varepsilon(u)$$

$$(1 \le p < q \le d; \; u \in \mathbb{R}^d, \; H_\varepsilon^{pq} \in \mathbb{R}). \tag{5.2}$$

Here $H_\varepsilon = \{H_\varepsilon^{pq} : 1 \le p < q \le d\}$ and, for each $u \in \mathbb{R}^d$, $\mathcal{A}_{pq}(u) \in L(\mathbb{R}^{d(d-1)/2}, \mathbb{R})$.

b) The symplectic case

$$\sum_{t=1}^{d} \left(\xi_\varepsilon^t \frac{\partial H_\varepsilon}{\partial u_t} + \eta_\varepsilon^t \frac{\partial H_\varepsilon}{\partial v_t}\right) = G - F: \tag{5.3}$$

$$\dot{H}_\varepsilon = G(u,v) - F(u,v), \quad \dot{u} = \xi_\varepsilon(u,v), \quad \dot{v} = \eta_\varepsilon(u,v). \tag{5.4}$$

c) The contact case

$$\sum_{t=1}^{d} \left(\xi_\varepsilon^t \frac{\partial H_\varepsilon}{\partial u_t} + \eta_\varepsilon^t \frac{\partial H_\varepsilon}{\partial v_t} \right) + \xi_\varepsilon \frac{\partial H_\varepsilon}{\partial w} - \mu_{\xi_\varepsilon} H_\varepsilon = G - F; \qquad (5.5)$$

$$\dot{H}_\varepsilon = \mu_{\xi_\varepsilon} \cdot H_\varepsilon + G(u, v, w) - F(u, v, w),$$

$$\dot{u} = \xi_\varepsilon(u, v, w), \quad \dot{v} = \eta_\varepsilon(u, v, w). \quad \dot{w} = \zeta_\varepsilon(u, v, w). \qquad (5.6)$$

6 Introducing integrated variables

Before proceeding to the normalization of jets of vector fields and their hamiltonians we have to introduce integrated canonical "coordinates". As it was shown in section 3, without loss of generality we may assume that the linear parts of vector fields are reduced to the block-diagonal normal form (with respect to the canonical coordinates).

Denote $d_i = \dim A_i$ $(i = 1, \ldots, n)$. Then $d_1 + \ldots + d_n = d$. Given a point $u = (u_1, \ldots, u_d) \in \mathbb{R}^d$, denote $x_i = (u_{j_{i-1}+1}, \ldots, u_{j_i})$, where $j_i = \sum_{p=1}^{i} d_p$ $(i = 1, \ldots, n)$, $j_0 = 0$. We shall also use the notation $J_i = \{j_{i-1} + 1, \ldots, j_i\}$.

Let $\omega \in \mathbb{Z}_+^d$ be some multi-index. Denote $\tau^i = \sum_{j \in J_i} \omega^j$ $(i = 1, \ldots, n)$. Clearly, $\tau = (\tau^1, \ldots, \tau^n) \in \mathbb{Z}_+^n$. Thus, we have defined an operator $S : \mathbb{Z}_+^d \to \mathbb{Z}_+^n$, $\tau = S(\omega)$. Every monomial $u^\omega \equiv u_1^{\omega^1} \cdot \ldots \cdot u_d^{\omega^d}$ can be written as $x^\tau = x_1^{\tau^1} \cdot \ldots \cdot x_n^{\tau^n}$, where $\tau = S(\omega)$.

Let $\bar{\xi}$ be a polynomial vector field of degree Q, i.e..

$$\bar{\xi}_j(u) = A_j u + \sum_{|\omega|=2}^{Q} \bar{a}_\omega^j u^\omega \quad (j = 1, \ldots, d). \qquad (6.1)$$

In terms of the integrated variables, $\bar{\xi}$ can be rewritten as

$$\xi_i(x) = A_i x_i + \sum_{|\tau|=2}^{Q} a_\tau^i x^\tau \quad (i = 1, \ldots, n), \qquad (6.2)$$

where $\xi_i = \{\bar{\xi}_j : j \in J_i\}$, $a_\tau^i = \{\bar{a}_\omega^j : j \in J_i, \tau = S(\omega)\}$.

Assume that the vector field (6.1) preserves the standard volume from (1.1), then its hamiltonian F defined by (2.1) is a polynomial of degree $Q + 1$, i.e.,

$$\bar{F}^{pq}(u) = \sum_{|\omega|=2}^{Q+1} \bar{b}_\omega^{pq} u^\omega \quad (p = 1, \ldots, d-1; \; q = p+1, \ldots, d). \qquad (6.3)$$

By using the integrated variables $x_1, \ldots, x_n$, we get

$$F^{ij}(x) = \sum_{|\tau|=2}^{Q+1} b_\tau^{ij} x^\tau \quad (1 \le i \le j \le n). \tag{6.4}$$

where

$$\begin{aligned}
F^{ij} &= (\bar{F}^{pq} : p \in J_i, q \in J_j), & b_\tau^{ij} &= (\bar{b}_\omega^{pq} : p \in J_i, q \in J_j. \tau = S(\omega)), \\
& & & (1 \le i < j \le n).
\end{aligned} \tag{6.5}$$

Here it is convenient to assume that $\bar{F}^{pq} = -\bar{F}^{qp}$ for $q, p \in J_i$, $q < p$.

If ξ is a symplectic polynomial vector field on $\mathbb{R}^{2d}$, then its hamiltonian expressed in the integrated variables becomes

$$F(x, y) = \sum_{|\alpha|+|\beta|=2}^{Q+1} b_{\alpha\beta} x^\alpha y^\beta, \tag{6.6}$$

where $\alpha, \beta \in \mathbb{Z}_+^n$. Similarly, for the contact hamiltonian we get

$$F(x, y, z) = \sum_{|\alpha+\beta+\gamma|=2}^{Q+1} b_{\alpha\beta\gamma} x^\alpha y^\beta z^\gamma + bz. \tag{6.7}$$

7   Normalization of jets of vector fields and hamiltonians

The polynomial vector field (6.2) is said to be *resonant* if the condition $\theta_i \ne \langle \tau, \theta \rangle \equiv \tau^1 \theta_1 + \ldots + \tau^n \theta_n$ implies $a_\tau^i = 0$.

Assume that the vector field (6.2) preserves the standard volume form (1.1) and F is its hamiltonian defined by (2.1). It is easy to verify that (6.2) is resonant iff the condition $\theta_i + \theta_j \ne \langle \tau, \theta \rangle$, $i < j$, implies that $b_\tau^{ij} = 0$ (see (6.4)). Thus, it is natural to say that the hamiltonian (6.4) is *resonant* if it contains only terms $b_\tau^{ij} x^\tau$ with $\theta_i + \theta_j = \langle \tau, \theta \rangle$.

Similarly, the symplectic hamiltonian (6.6) is called *resonant* if $\langle \alpha, \theta \rangle \ne \langle \beta, \theta \rangle$ implies $b_{\alpha\beta} = 0$. Recall that in the symplectic case the real parts of the eigenvalues of A are equal to $\theta_1, \ldots, \theta_n, -\theta_1, \ldots, -\theta_n$.

In the contact case, the real parts of the eigenvalues are equal to $\theta_1, \ldots, \theta_n$, $-\theta_1 + \nu, \ldots, -\theta_n + \nu, \nu$. Therefore, the contact hamiltonian (6.7) is said to be *resonant* if the condition $\langle \alpha, \theta \rangle - \langle \beta, \theta \rangle + (\gamma + \beta_1 + \ldots + \beta_n)\nu \ne \nu$ implies $b_{\alpha\beta\gamma} = 0$.

Let us consider the vector field ξ,

$$\xi_i(x) = A_i x_i + \sum_{|\tau|=2}^{Q} a_\tau^i x^\tau + \xi_Q(x), \tag{7.1}$$

where $\xi_Q \in C^K$, $K \geq Q + 1$, $D^p \xi_Q(0) = 0$ $(p = 0, 1, \ldots, Q)$. Assume that (7.1) preserves the standard volume form. Clearly, the corresponding hamiltonian $F = (F^{ij})$ is of the form

$$F^{ij}(x) = \sum_{|\tau|=2}^{Q+1} b_\tau^{ij} x^\tau + F_{Q+1}^{ij}(x). \tag{7.2}$$

where $F_{Q+1} \in C^K$, $D^p F_{Q+1}(0) = 0$ $(p = 0, 1, \ldots, Q + 1)$.

Lemma 7.1. Let $s \in \{1, \ldots, n\}$, $t \in \{1, \ldots, n\}$, $s \leq t$. Let $\sigma \in \mathbb{Z}_+^n$ be a multi-index such that $|\sigma| = Q + 1$ and $\theta_s + \theta_t \neq \langle \sigma, \theta \rangle$. There exists a canonical change of variables $y = g(x)$ such that $D^p(g - \mathrm{id})(0) = 0$ $(p = 0, 1, \ldots, Q - 1)$ and the hamiltonian $\tilde{F}$ of the vector field $\tilde{\xi} = g_*\xi$ takes the form

$$\tilde{F}^{ij}(y) = \sum_{|\tau|=2}^{Q+1} \tilde{b}_\tau^{ij} y^\tau + \tilde{F}_{Q+1}^{ij}(y).$$

where $\tilde{b}_\tau^{ij} = b_\tau^{ij}$ whenever $(i, j, \tau) \neq (s, t, \sigma)$; $\tilde{b}_\sigma^{st} = 0$. $D^p \tilde{F}_{Q+1}^{ij}(0) = 0$ for $p = 0, 1, \ldots, Q + 1$.

Proof. Let $\dot{x} = Ax$ be the linear part of ξ at the point $x = 0$. Let $\hat{f} = (\hat{f}^{ij})$ be the $(d - 2)$-form defined by $\hat{f}^{ij} = 0$ if $(i, j) \neq (s, t)$. $\hat{f}^{st}(x) = b_\sigma^{st} x^\sigma$. Consider the equation

$$L_{Ax} G = \hat{f}. \tag{7.3}$$

Try to find the unknown $(d - 2)$-form $G = (G^{ij})$ so that $G^{ij} = 0$ for $(i, j) \neq (s, t)$ and $G^{st}(x) = h_\sigma^{st} x^\sigma$. Note that (7.3) is a special case of (5.1). Hence, (7.3) reduces to

$$\sum_{l=1}^n \sum_{r \in J_l} \sum_{m \in J_l} a_{rm} u_m \frac{\partial G^{pq}}{\partial u_r} + \sum_{r \in J_s} (-1)^{p+r-1} a_{pr} G^{rq} \tag{7.4}$$

$$+ \sum_{r \in J_t} (-1)^{q+r-1} a_{qr} G^{pr} = \hat{f}^{pq} \quad (p \in J_s, \quad q \in J_t).$$

Let us show that the real parts of all the eigenvalues of the linear operator $L_{A,s,t} : G^{st} \mapsto L_{Ax} G^{st}$ are equal to the number $\langle \sigma, \theta \rangle - \theta_s - \theta_t$. To this end, approximate the (complexified) matrix A by a semi-simple matrix B. Then

$$L_{B,p,q}(h_\omega^{pq} u^\omega) = \langle \omega, \nu \rangle h_\omega^{pq} u^\omega - (\nu_p + \nu_q) h_\omega^{pq} u^\omega,$$

where $\nu_1, \ldots, \nu_d$ are the eigenvalues of $B, p \in J_s, q \in J_t, \sigma = S(\omega)$. Taking into account that the spectrum of a matrix depends continuously on its entries and that A is a block-diagonal matrix we easily arrive to the conclusion that the spectrum of $L_{A,s,t}$ lies on the line $\mathrm{Re}\lambda = \langle \sigma, \theta \rangle - \theta_s - \theta_t$, as asserted above.

Because $\langle \sigma, \theta \rangle - \theta_s - \theta_t \neq 0$, we see that system (7.4) (as well as equation (7.3)) has a solution of the desired form.

Identify the jet $j_0^Q(\xi)$ with the Taylor's polynomial of ξ of degree Q. Because ξ preserves the standard volume form, the polynomial vector field $j_0^Q(\xi)$ is also volume preserving. Its hamiltonian, f, coincides with $j_0^{Q+1}(F)$. Let

$$\tilde{f}^{ij}(x) = \sum_{|\tau|=2}^{Q+1} \tilde{b}_\tau^{ij} x^\tau, \quad \tilde{f} = (\tilde{f}^{ij}), \quad \hat{f} = \tilde{f} - f.$$

Clearly, $\hat{f}^{ij}(x) = 0$ if $(i,j) \neq (s,t)$; $\hat{f}^{st} = b_\sigma^{st} x^\sigma$. Let G denote the solution of (7.3) and η be the corresponding vector field. Let $g^t : \mathbb{R}^d \to \mathbb{R}^d$ be the phase flow of η. Define $\xi_\varepsilon = g_*^\varepsilon \xi$. Let F_ε be the hamiltonian of ξ_ε (see (2.1)). Then G is nothing else than the solution of the cohomology equation

$$L_{\xi_\varepsilon} G = \partial F_\varepsilon / \partial \varepsilon. \tag{7.5}$$

Observe that the components of G are $O(\|x\|^{Q+1})$ as $\|x\| \to 0$.

Therefore $j_0^{Q+1}(L_{\xi_\varepsilon} G) = L_{Ax} G$, whence $j_0^{Q+1}(\partial F_\varepsilon / \partial \varepsilon) = \hat{f}$.

Thus, $j_0^{Q+1}(F_\varepsilon) = f + \varepsilon \hat{f}$, hence $j_0^{Q+1}(F_1) = \tilde{f}$. We conclude that $g^1 : \mathbb{R}^d \to \mathbb{R}^d$ is the desired canonical C^∞ smooth change of variables and $\tilde{F} = F_1$. The proof is complete.

Analogous results hold for symplectic and contact vector fields.

Lemma 7.2. *Let F be a symplectic hamiltonian of the form*

$$F(x,y) = \sum_{|\alpha+\beta|=2}^{Q+1} b_{\alpha\beta} x^\alpha y^\beta + F_{Q+1}(x,y).$$

where $F_{Q+1} \in C^K$, $K \geq Q+1$ and $D^p F_{Q+1}(0) = 0$ $(p = 0.1.\ldots.Q+1)$. Let the pair $(\bar{\alpha}, \bar{\beta})$ be such that $|\bar{\alpha}| + |\bar{\beta}| = Q+1$, $\langle \bar{\alpha}, \theta \rangle \neq \langle \bar{\beta}, \theta \rangle$. Then there exists a C^∞ smooth canonical change of variables which brings the hamiltonian F to the form

$$\tilde{F}(x,y) = \sum_{|\alpha+\beta|=2}^{Q+1} \tilde{b}_{\alpha\beta} x^\alpha y^\beta + \tilde{F}_{Q+1}(x,y).$$

where $\tilde{b}_{\alpha\beta} = b_{\alpha\beta}$ whenever $(\alpha,\beta) \neq (\bar{\alpha},\bar{\beta})$, and $\tilde{b}_{\bar{\alpha}\bar{\beta}} = 0$.

Lemma 7.3. *Let $F(x,y,z)$ be a contact hamiltonian of the form*

$$F(x.y.z) = \sum_{|\alpha+\beta+\gamma|=2}^{Q+1} b_{\alpha\beta\gamma} x^\alpha y^\beta z^\gamma + bz + F_{Q+1}(x,y.z),$$

where $F_{Q+1} \in C^K$. $K \geq Q+1$, $D^p F_{Q+1}(0) = 0$ $(p = 0,1,\ldots.Q+1)$. Let $(\bar{\alpha}, \bar{\beta}, \bar{\gamma})$ be a triple of multi-indices such that $|\bar{\alpha}| + |\bar{\beta}| + |\bar{\gamma}| = Q+1$, and $\langle \bar{\alpha}.\theta \rangle + (\bar{\gamma} + |\bar{\beta}|)\nu \neq$

$\langle \bar{\beta}, \theta \rangle + \nu$. Then there exists a C^∞ smooth canonical change of variables which reduces F to the form

$$\tilde{F}(x, y, z) = \sum_{|\alpha+\beta+\gamma|=2}^{Q+1} \tilde{b}_{\alpha\beta\gamma} x^\alpha y^\beta z^\gamma + bz + \tilde{F}_{Q+1}(x, y, z),$$

where $\tilde{b}_{\alpha\beta\gamma} = b_{\alpha\beta\gamma}$ if $(\alpha, \beta, \gamma) \neq (\bar\alpha, \bar\beta, \bar\gamma)$, and $\tilde{b}_{\bar\alpha\bar\beta\bar\gamma} = 0$.

8   Resonant polynomial normal forms of hamiltonians

Let ξ be a C^K smooth vector field, $\xi(0) = 0$, $A = D\xi(0)$ and $\theta_1, \ldots, \theta_n$ denote the real parts of the eigenvalues of A. Assume that $\theta_i \neq 0$ ($i = 1, \ldots, n$), i.e., the origin is a *hyperbolic* singular point of ξ. Denote

$$\mathrm{M} = \max\{\theta_i : i = 1, \ldots, n\}, \quad \mu = \min\{\theta_i : \theta_i > 0\},$$

$$\Lambda = \max\{-\theta_i : \theta_i < 0\}, \quad \lambda = \min\{-\theta_i : \theta_i < 0\}.$$

Lemma 8.1. *Let ξ preserve the standard volume form and F be its hamiltonian defined by (2.1). Let k be an integer satisfying the inequality*

$$K > \frac{2\Lambda}{\lambda} + \frac{2\mathrm{M}}{\mu} + (k+1)\left(\frac{\mathrm{M}}{\lambda} + \frac{\Lambda}{\mu}\right) + 2. \tag{8.1}$$

Then there exists a canonical transformation of class C^k defined in the vicinity of the origin which conjugates F and $j_0^K(F)$.

Proof. Identify the jet $j_0^{K-1}(\xi)$ with the corresponding Taylor's polynomial. Clearly, the polynomial vector field $j_0^{K-1}(\xi)$ respects the standard volume form. Recall that the hamiltonian F is a $(d-2)$-form of class C^K. It easily follows that $j_0^K(F)$ is the hamiltonian for $j_0^{K-1}(\xi)$. Consider the cohomology equation (5.1) and the related affine extension (5.2), where $\xi_\varepsilon = \varepsilon j_0^{K-1}(\xi) + (1-\varepsilon)\xi$, $G = j_0^K(F)$. Then $\mathcal{A} = (\mathcal{A}_{pq}(u))$ is a family of linear morphisms of class $C^{K-1} \subset C^{k+1}$ and $j_0^K(G - F) = 0$. It is easily seen from (5.1) that the spectrum of the matrix $\mathcal{A}(0)$ lies between the lines $\mathrm{Re}\nu = -2\Lambda$ and $\mathrm{Re}\nu = 2\mathrm{M}$. Now apply Theorem 10.3 from the Appendix. The proof is complete.

Lemma 8.2. *Let ξ be a symplectic vector field on $\mathbb{R}^{2d}$, F be its hamiltonian, k be an integer satisfying*

$$K > \frac{2\Lambda(k+1)}{\lambda} + 2. \tag{8.2}$$

Then there exists a (locally defined) canonical change of variables of class C^k which conjugates F and $j_0^K(F)$.

The proof is similar to that of Lemma 8.1. Observe that the spectrum of the linear part of ξ is symmetric relatively the imaginary axis and the first eigenvalue of (5.4) equals to 0.

Now let ξ be a contact vector field on $\mathbb{R}^{2d+1}$, F be its contact hamiltonian and η be the symplectic vector field on $\mathbb{R}^{2d+2}$ that corresponds to ξ. Since $\xi(0) = 0$, we have $\eta(0) = 0$ and the origin is a hyperbolic singular point of η. Denote $A = D\eta(0)$. Define the numbers λ and Λ for A, as before.

Lemma 8.3. *If k satisfies the inequality* (8.2) *then there exists a canonical transformation of class C^k which conjugates F and $j_0^K(F)$ near the origin.*

Combining these results with the results presented in the preceding section, we obtain

Theorem 8.1. *Let ξ be a vector field of class C^K which preserves the (standard) volume form (symplectic form, contact structure) and the origin be a hyperbolic singular point of ξ. Let F be the corresponding local hamiltonian. If k satisfies the inequality* (8.1) *or, respectively,* (8.2), *then there exists a local canonical transformation of class C^k which reduces the hamiltonian F (and, consequently, the vector field ξ) to the polynomial resonant normal form.*

9   Simplified resonant normal forms

Basically, a resonant vector field cannot be linearized via real-analytic or infinitely smooth transformations. Nevertheless, it turns out that some resonant monomials x^τ can be eliminated by using C^k ($k < \infty$) changes of variables. The first condition on a multi-index τ providing this procedure was proposed by Samovol [5]. His main idea was further developed by the authors in the book [2]. In particular, we have introduced the following sufficient condition that comprises all the known ones.

Condition $\mathfrak{A}(k)$. The multi-index $\tau = (\tau^1, \ldots, \tau^n) \in \mathbb{Z}_+^n$ is said to satisfy the condition $\mathfrak{A}(k)$, $k \geq 1$, if there is a collection $\sigma = (\sigma_1, \ldots, \sigma_p)$ of n-vectors $\sigma_i = (\sigma_i^1, \ldots, \sigma_i^n)$ $(i = 1, \ldots, p)$ with non-negative components such that either

$$\langle \sigma_i, \theta \rangle \equiv \sum_{j=1}^n \sigma_i^j \theta_j = 1 \quad (i = 1, \ldots, p) \quad \text{or} \quad \langle \sigma_i, \theta \rangle = -1 \quad (i = 1, \ldots, p)$$

and the inequality

$$\langle \tau, \bar{u} \rangle > k \max \{\bar{u}_j : j = 1, \ldots, n\}$$

holds for every vertex $\bar{u} = (\bar{u}_1, \ldots, \bar{u}_n)$ of the convex polyhedral domain $D = D(\sigma)$ determined by the inequalities

$$u_j \geq 0 \quad (j = 1, \ldots, n), \quad \langle \sigma_i, u \rangle \geq 1 \quad (i = 1, \ldots, p).$$

In order to clarify the essence of the above condition, let us note that the mapping $\Phi : \mathbb{R}^n \to \mathbb{R}$, $\Phi(x) = C \ln(|x|^{\sigma_1} + \ldots + |x|^{\sigma_p})$ (here $|x|^{\sigma_i} \equiv |x_1|^{\sigma_i^1} \cdots |x_n|^{\sigma_i^n}$), is C^k smooth iff the multi-index τ satisfies the condition $\mathfrak{A}(k)$ with respect to the collection $\sigma = (\sigma_1, \ldots, \sigma_p)$. We point out that the changes of variables of the form $z = y + \Phi(x)$ arise naturally when linearizing the model vector field

$$\dot{y} = \langle \tau, \theta \rangle y + x^\tau, \quad \dot{x}_j = \theta_j x_j \quad (j = 1, \ldots, n).$$

Let Q be an integer, $Q \geq 2$. The polynomial

$$P(x) = \Big(\sum_{|\omega|=2}^{Q} P_\omega^1 x^\omega, \ldots, \sum_{|\omega|=2}^{Q} P_\omega^m x^\omega \Big)$$

is called τ-divisible if $P_\omega^j \neq 0$ implies $\tau^1 \leq \omega^1, \ldots, \tau^n \leq \omega^n$.

Lemma 9.1. *Let E and F be finite-dimensional real vector spaces, $dim\, E = d$; $\dot{x} = Ax$ be a hyperbolic linear vector field on E and $\dot{y} = By$ be a linear vector field on F. Let $\tau \in \mathbb{Z}_+^n$ satisfy condition $\mathfrak{A}(k)$ with respect to the vector field $\dot{x} = Ax$. Let $P : E \to F$ be a τ-divisible resonant polynomial. Let $Q_\varepsilon : E \times F \to F$ be a C^∞ smooth family of resonant polynomials, $Q_\varepsilon(x, 0) = 0$ $(x \in E)$, and $q_\varepsilon : E \to E$ be a C^∞ smooth family of resonant polynomials. Then the extension*

$$\dot{y} = By + Q_\varepsilon(x, y) + P(x), \quad \dot{x} = Ax + q_\varepsilon(x) \tag{9.1}$$

admits a C^k smooth invariant section $y = \varphi_\varepsilon(x)$ which C^∞ smoothly depends on the parameter ε.

The proof of this lemma is based on the techniques elaborated in [2], Chapter II. It uses the iterative method of introducing additional variables (see Lemma II.8.1) as well as Theorem II.5.23 adapted to the case of vector fields. Because of the size limitations, the details are left to the interested reader.

Let us say that the hamiltonian F of the vector field ξ is reduced to the *simplified resonant polynomial k-normal form* if F is a polynomial which contains only such resonant terms x^τ that τ does not satisfy condition $\mathfrak{A}(k+1)$.

Theorem 9.1. *Let the vector field ξ and its hamiltonian F satisfy the hypotheses of Theorem 8.1, then F can be reduced to the simplified resonant polynomial k-normal form by a local canonical transformation of class C^k.*

Proof. Let F be the resonant polynomial hamiltonian of the vector field ξ. Represent F in the form $F = P + Q$, where P is a τ-divisible resonant polynomial and τ satisfies condition $\mathfrak{A}(k+1)$. Let $A = D\xi(0)$ and p be the polynomial vector field which corresponds to the hamiltonian P. Denote $F_\varepsilon = Q + \varepsilon P$, then $\xi_\varepsilon(x) = Ax + q(x) + \varepsilon p(x)$.

Let us prove that the cohomology equations (5.1), (5.3), (5.5) with $G - F = P$ have C^{k+1} smooth solutions. Consider the corresponding characteristic systems (5.2), (5.4) and (5.6). In each of the three cases, the characteristic system is of the form (9.1). According to Lemma 9.1, there exists a family of invariant sections

$y = H_\varepsilon(x)$ of class C^{k+1}. The hamiltonian H_ε determines a change of variables of class C^k (see Section 4).

Applying successively these arguments to all multi-indices τ satisfying condition $\mathfrak{A}(k+1)$, we get the desired result.

10 Appendix

At first, let us recall some definitions and state the standing assumptions. We keep the terminology and notation adopted in [2].

Let $p : (E, \mathbb{R}, \pi) \to (B, \mathbb{R}, \rho)$ be a linear extension. The numbers

$$\Omega(\pi, b) = \limsup_{t \to +\infty} \tfrac{1}{t} \ln \|\pi^t(b)\|,$$
$$\omega(\pi, b) = -\limsup_{t \to +\infty} \tfrac{1}{t} \ln \|\pi^{-t}(\rho^t(b))\| \quad (b \in B)$$

are called *the upper and the lower Lyapunov exponents*. They do not depend on the choice of the Riemannian metric on (E, ρ, B).

Let $\mathcal{M}$ be a compact smooth manifold, $\xi : \mathcal{M} \to T\mathcal{M}$ be a C^{k+1} vector field and $(\mathcal{M}, \mathbb{R}, f)$ be the phase flow determined by ξ. Let Δ be a submanifold of $\mathcal{M}$ of class C^1 invariant under the flow f and $T_\Delta \mathcal{M}$ denote the restriction of the tangent bundle $T\mathcal{M}$ on Δ.

In what follows, we shall assume that there exist Tf-invariant vector subbundles X and Y of $T_\Delta \mathcal{M}$ such that

(a) $T_\Delta \mathcal{M} = T\Delta \oplus X \oplus Y$;
(b) $\Omega(Tf|Y, b) < 0, \quad -\omega(Tf|X, b) < 0,$
     $\Omega(Tf|Y, b) - k\omega(Tf|T\Delta, b) < 0,$
     $-\omega(Tf|X, b) + k\Omega(Tf|T\Delta, b) < 0 \quad (b \in B).$

These conditions mean that the submanifold Δ is *normally k-hyperbolic*. According to the well-known results (see, for example, [2], Theorem V.2.2), there exist the stable manifold $W^s(\Delta)$ and the unstable manifold $W^u(\Delta)$, both of class C^k, such that

$$W^s(\Delta) \cap W^u(\Delta) = \Delta, \quad T_b W^s(\Delta) = T_b \Delta \oplus Y_b, \quad T_b W^u(\Delta) = T_b \Delta \oplus X_b \quad (b \in B).$$

Consequently, Δ is a C^k submanifold.

Assume, in addition, that

(c) $\Omega(Tf|Y, b) - \omega(Tf|T\Delta, b) + k \cdot \Omega(Tf|T\Delta, b) < 0,$
     $-\omega(Tf|X, b) + \Omega(Tf|T\Delta, b) - k \cdot \omega(Tf|T\Delta, b) < 0 \quad (b \in B).$

By [2], Theorem V.4.6, there exist f-invariant foliations of class $C^{k,k+1}$

$$W^s(\Delta) = \cup_{b \in \Delta} W^{ss}(b), \quad W^u(\Delta) = \cup_{b \in \Delta} W^{uu}(b)$$

such that $T_b W^{ss}(b) = Y_b$, $T_b W^{uu}(b) = X_b$ $(b \in \Delta)$. Smoothness of class $C^{k,k+1}$ means that the leaves $W^{ss}(b)$ and $W^{uu}(b)$ are C^{k+1} submanifolds which C^k-smoothly depend on $b \in \Delta$. Hence it follows that X and Y are C^k vector subbundles of $T_\Delta \mathcal{M}$.

Let $U = U(\Delta)$ denote a sufficiently small neighbourhood of Δ in $\mathcal{M}$ and $H : X \oplus Y \to U(\Delta)$ be some tubular neighbourhood of class $C^{k,k+1}$. By using the map H^{-1} transfer the local flow $f|U$ to the neighbourhood of the zero section $Z(\Delta)$ of the vector bundle $X \oplus Y$. In other words, henceforth we shall assume that in the vicinity U of $Z(\Delta) \approx \Delta$ there is defined a local flow of class $C^{k,k+1}$ which preserves $Z(\Delta)$. Taking into account what has been said about the foliations $\{W^{ss}(b)\}$ and $\{W^{uu}(b)\}$. we can straighten out these foliations by using an appropriate $C^{k,k+1}$ local diffeomorphism. So we shall assume that $W^{ss}(b) = Y_b$, $W^{uu}(b) = X_b$ $(b \in B)$.

Let $p : X \oplus Y \to \Delta$, $p_X : X \oplus Y \to X$ and $p_Y : X \oplus Y \to Y$ be the natural projections. Given a point $z \in X_b \oplus Y_b$, we shall write $z = (b, x, y)$ $(b \in \Delta, x \in X_b, y \in Y_b)$ and

$$f^t(z) = (f_1^t(b, x, y), f_2^t(b, x, y), f_3^t(b, x, y)).$$

Recall that f_i^t $(i = 1, 2, 3)$ are $C^{k,k+1}$ smooth and

$$f_1^t(b, 0, 0) = f^t(b), \quad f_2^t(b, 0, y) = 0. \quad f_3^t(b, x, 0) = 0.$$

Now let us state the last but one standing assumption. Let $(E, \pi, X \oplus Y)$ be a vector bundle of class C^k and $F^t : E|U(\Delta) \to E|U(\Delta)$ be an affine extension of the local flow $f^t|U(\Delta)$, i.e.,

$$F^t(w) = A^t(b, x, y)w + \Phi^t(b, x, y) \quad (w \in E_z, \ z = (b, x, y)), \tag{10.1}$$

where A^t is a linear f^t-morphism. Moreover, assume that

$$\text{(d)} \quad \Phi^t(b, x, y) = \varphi^t(b, x, y)y^Q + \psi^t(b, x, y)x^S,$$

where $\varphi^t : X \oplus Y \to P_Q(Y, E)$ is a C^k section of the bundle $P_Q(Y, E)$ (the space $P_Q(Y_b, E_z)$ of homogeneous polynomials of degree Q serves as a fiber of $P_Q(Y, E)$ over $z = (b, x, y) \in X \oplus Y$) and $\psi^t : X \oplus Y \to P_S(X, E)$ is a C^k section of the bundle $P_S(X, E)$.

Let E_Δ be the restriction of the vector bundle E to Δ and $A_0^t = A^t|E_\Delta$.

Theorem 10.1. *If the standing assumption* (a). (b). (c), (d) *and*

$$\text{(e)} \quad \begin{aligned} &-\omega(A_0, b) + Q \cdot \Omega(Tf|Y, b) + k \cdot \Omega(Tf|X. b) < 0, \\ &\Omega(A_0, b) - S \cdot \omega(Tf|X, b) - k \cdot \omega(Tf|Y. b) < 0 \quad (b \in \Delta) \end{aligned}$$

are fulfilled. then there exists a C^k smooth local section $\sigma : U(\Delta) \to E$ invariant under the local flow F^t.

Proof. Because A^t is a linear morphism, it suffices to prove the statement in the case where $\psi^t \equiv 0$.

According to [2], Lemma III.2.3, the conditions

$$\sup_{b\in\Delta}\|Tf^t|Y_b\| < 1, \quad \sup_{b\in\Delta}\|Tf^{-t}|X_{f^t(b)}\| < 1,$$

$$\sup_{b\in\Delta}\|Tf^t|Y_b\|\|Tf^{-t}|T_{f^t(b)}\Delta\|^k < 1,$$

$$\sup_{b\in\Delta}\|Tf^{-t}|X_{f^t(b)}\|\|Tf^t|T_b\Delta\|^k < 1,$$

$$\sup_{b\in\Delta}\|A^{-t}|E_{(f^t(b),0,0)}\|\|Tf^t|Y_b\|^Q\|Tf^t|X_b\|^k < 1 \qquad (10.2)$$

hold for all sufficiently large numbers $t > 0$. Fix such a number t. Let us prove the existence of a C^k smooth section $\sigma : U \to E$ invariant under the local diffeomorphism F^t. For simplicity of notation we shall drop the upper index t.

At first, extend the map $f|U$ to $X \oplus Y$ and the map F to E in the following way. Denote $L|X_b \oplus Y_b = Df(b,0,0)$ ($b \in \Delta$). Clearly, $L : X \oplus Y \to X \oplus Y$ is a linear morphism of class C^k and covers the diffeomorphism $f|\Delta$. By shrinking the neighbourhood U of the zero section $Z(\Delta) \subset X \oplus Y$ we can make the map $L^{-1} \circ f|U$ as near to id as we like. Therefore there exists a vector field $v : U \to TU$ of class $C^{k,k+1}$ such that $L^{-1} \circ f = \exp \circ v$ on U. Let $\tilde{v} : X \oplus Y \to T(X \oplus Y)$ be a vector field of class $C^{k,k+1}$ which agrees with v on U and vanishes out of some larger neighbourhood U_0. Define $\tilde{f} = L \circ \exp \circ \tilde{v}$. Clearly, $\tilde{f}|U = f|U$ and $\tilde{f} = L$ on $(X \oplus Y)\backslash U_0$. Henceforth, we shall omit the tilda over f. Analogously, extend the linear morphism A to E so that $A(b,x,y) = A(b,0,0)$ outside of some neighbourhood U_0. At last, extend Φ to E by the cut-off procedure.

Because $f_3(b,x,0) = 0$ and $f_3 \in C^{k,k+1}$, there exists a mapping $\delta : X \oplus Y \to L(Y,Y)$ of class C^k such that $f_3(b,x,y) = \delta(b,x,y)y$. Moreover, $\delta = L|Y$ outside of some neighbourhood of the zero section.

That the section $\sigma : X \oplus Y \to E$ is F–invariant can be written in the form

$$A(b,x,y)\sigma(b,x,y) + \Phi(f(b,x,y)) = \sigma(f(b,x,y)),$$

or

$$\sigma(b,x,y) = [A(b,x,y)]^{-1}\sigma(f(b,x,y)) - [A(b,x,y)]^{-1}\Phi(f(b,x,y)),$$

which is equivalent to

$$\sigma(b,x,y) = A^{-1}(f(b,x,y))\sigma(f(b,x,y)) - A^{-1}(f(b,x,y))\Phi(f(b,x,y)). \qquad (10.3)$$

At first, let us prove that equation (10.3) has a continuous solution. Denote by $\Gamma^0(E)$ the set of all continuous sections of the vector bundle E and by $\Gamma^0_Q(E)$ the set of all $\sigma \in \Gamma^0(E)$ such that

$$|||\sigma|||_0 \equiv \sup\left\{\frac{\|\sigma(b,x,y)\|}{\|y\|^Q} : b \in \Delta, x \in X_b, y \in Y_b, \|y\| \neq 0\right\} < \infty.$$

Clearly, $\Gamma_Q^0(E)$ provided with the norm $||| \cdot |||_0$ is a Banach space. Consider the affine operator $F_\# : \Gamma^0(E) \to \Gamma^0(E)$ defined by

$$F_\# \sigma(b, x, y) = A^{-1}(f(b, x, y))\sigma(f(b, x, y)) - A^{-1}(f(b, x, y))\Phi(f(b, x, y)).$$

Let us show that $\Gamma_Q^0(E)$ is invariant under $F_\#$. In fact, let $\sigma \in \Gamma_Q^0(E)$. Note that $\Phi \in \Gamma_Q^0(E)$ according to hypothesis (d). Therefore

$$|||F_\# \sigma|||_0 = \sup \left\{ \frac{\|A^{-1}(f(b,x,y))\sigma(f(b,x,y)) - A^{-1}(f(b,x,y))\Phi(f(b,x,y))\|}{\|y\|^Q} \right\}$$

$$\leq \sup \frac{\|A^{-1}(f(b,x,y))\sigma(f(b,x,y))\|}{\|y\|^Q} + \sup \frac{\|A^{-1}(f(b,x,y))\Phi(f(b,x,y))\|}{\|y\|^Q}$$

$$\leq \sup_{b \in \Delta}(\|A^{-1}|E_{f(b,0,0)}\| + \varepsilon) \cdot \sup \frac{\|\Phi(f(b,x,y))\|}{\|f_3(b,x,y)\|^Q} \cdot \sup \frac{\|\delta(b,x,y)\|^Q \|y\|^Q}{\|y\|^Q}$$

$$+ \sup_{b \in \Delta}(\|A^{-1}|E_{f(b,0,0)}\| + \varepsilon) \cdot \sup \frac{\|\sigma(f(b,x,y))\|}{\|f_3(b,x,y)\|^Q} \cdot \sup \frac{\|\delta(b,x,y)\|^Q \|y\|^Q}{\|y\|^Q}$$

$$\leq \sup_{b \in \Delta}(\|A^{-1}|E_{f(b,0,0)}\| + \varepsilon) \cdot \sup_{b \in \Delta}(\|Tf|Y_b\| + \varepsilon)^Q \{|||\sigma|||_0 + |||\Phi|||_0\},$$

where ε is a positive number which depends on the neighbourhood U of $Z(\Delta)$ and can be made arbitrarily small by shrinking U. Thus, $F_\# \sigma \in \Gamma_Q^0(E)$.

Next let us show that $F_\# : \Gamma_Q^0(E) \to \Gamma_Q^0(E)$ is a contracting operator. It is easily seen from the above estimates that for any $\sigma_1, \sigma_2 \in \Gamma_Q^0(E)$ we have

$$|||F_\# \sigma_1 - F_\# \sigma_2|||_0 \leq (\sup_{b \in \Delta} \|A^{-1}|E_{f(b,0,0)}\| + \varepsilon) \cdot (\sup_{b \in \Delta} \|Tf|Y_b\| + \varepsilon)^Q |||\sigma_1 - \sigma_2|||_0.$$

Recalling (10.2) and assuming that $\varepsilon > 0$ is sufficiently small we conclude that $F_\#|\Gamma_Q^0(E)$ is a contraction. Let $\bar{\sigma} \in \Gamma_Q^0(E)$ be its fixed point. Then $\bar{\sigma}$ is a continuous F-invariant section of the vector bundle E.

Now let us pass to the proof that $\bar{\sigma}$ is C^k smooth. Denote $B = X \oplus Y$. Recall that $(E, \pi, X \oplus Y) \equiv (E, \pi, B)$ is a C^k vector bundle. Let us briefly present some notions and facts from [7] which we shall use later on. For $m = 0, 1, \ldots, k$, denote by $P^m(E)$ the set of all m-jets of sections of the bundle (E, π, B) endowed with the natural structure of a C^{k-m} vector bundle over B. There exists a canonical filtration

$$E \xleftarrow{\rho^1} P^1(E) \xleftarrow{\rho^2} \ldots \xleftarrow{\rho^m} P^m(E).$$

where ρ^i are surjective linear morphisms of class C^{k-i} ($i = 1, \ldots, m$). Let $N^m(E)$ denote the kernel of ρ^m. Then $P^m(E) = P^{m-1}(E) \oplus N^m(E)$ is a C^{k-m} vector bundle over $P^{m-1}(E)$. The kernel $N^m(E)$ is naturally isomorphic to $L_m(TB, E)$. There exists an isomorphism

$$\mu_m : P^m(E) \to E \oplus L(TB, E) \oplus \ldots \oplus L_m(TB, E) \equiv L^m(E),$$

but μ_m is not natural. Every C^k automorphism (F, f) of the bundle (E, π, B) induces a morphism $P^m(F) : P^m(E) \to P^m(E)$. By using the isomorphism μ_m, $P^m(F)$ can be written as follows:

$$P^m(F)(\eta_{m-1}, \xi^m) = (P^{m-1}(F)(\eta_{m-1}), D_V F(b) \circ \xi^m \circ (Tf^{-1})^m + \upsilon_m(\eta_{m-1})),$$

where $\eta_m \equiv (\eta_{m-1}, \xi^m) = (v, \xi^1, \ldots, \xi^{m-1}, \xi^m)$, $v \in E$, $\xi^i \in L_i(TB, E)$ $(i = 1, \ldots, m)$, $\psi_m \in L(L^{m-1}(E), L_m(TB, E))$ and D_V denotes the derivative along the fiber.

Let $\Gamma^0(P^m(E))$ denote the space of all continuous bounded sections of the vector bundle $P^m(E)$. Clearly,

$$\Gamma^0(P^m(E)) = \Gamma^0(P^{m-1}(E)) \oplus \Gamma^0(L_m(TB, E)).$$

Let $\rho_*^m : \Gamma^0(P^m(E)) \to \Gamma^0(P^{m-1}(E))$ denote the natural projection. The automorphism (F, f) induces $F_\# : \Gamma^0(E) \to \Gamma^0(E)$, where

$$F_\#(\sigma)(d) = F(\sigma(f^{-1}(d))) \quad (d \in B),$$

and $P_\#^m(F) : \Gamma^0(P^m(E)) \to \Gamma^0(P^m(E))$, which can be written as

$$P_\#^m(F)(\lambda, \mu)(d)$$
$$= \left(P_\#^{m-1}(F)(\lambda)(d), D_V F(\sigma(f^{-1}(d))) \circ \mu \circ (Tf^{-1}(d))^m + \upsilon_m^*(\lambda)(d) \right),$$
$$(10.4)$$

where $\lambda \in \Gamma^0(P^{m-1}(E))$, $\mu \in \Gamma^0(L_m(TB, E))$, $\sigma = \rho_*^1 \circ \ldots \circ \rho_*^{m-1}(\lambda)$, and $\psi_m^*(\lambda) \in \Gamma^0(L_m(TB, E))$ is a section which depends only on λ.

Let us return to the proof that $\bar{\sigma}$ is C^k smooth. Apply formula (10.4) to the morphism (F^{-1}, f^{-1}), where F is defined by (10.1). Because $N^m(E)$ can be decomposed as a direct sum of $L_{q,r,s}(T\Delta, X, Y; E)$ for all non-negative integers q, r, s such that $q + r + s = m$, we get

$$P_\#^m(F^{-1})(\lambda, \mu)(b, x, y)$$
$$= (P_\#^{m-1}(F^{-1})(\lambda)(b, x, y), \psi_m^*(\lambda)(b, x, y)$$
$$+ \sum_{q+r+s=m} A^{-1}(f(b, x, y)) \circ \mu_{q,r,s} \circ (Tf|T_b\Delta)^q (Tf|X_b)^r (Tf|Y_b)^s$$
$$+ \sum_{q+r+s=m} A^{-1}(f(b, x, y)) \circ \Phi_{q,r,s} \circ (Tf|T_b\Delta)^q (Tf|X_b)^r (Tf|Y_b)^s),$$

where $\mu_{q,r,s}$ denotes the projection of μ onto $L_{q,r,s}(T\Delta, X, Y; E)$ and $\Phi_{q,r,s}$ is the same projection of the main part of $j^m(\Phi)$.

Let $\Gamma_Q^0(P^m(E))$ denote the subspace of $\Gamma^0(P^m(E))$ which consists of sections $\xi \in \Gamma^0(P^m(E))$ such that the projection $\xi_{q,r,s}$ of ξ onto $\Gamma^0(L_{q,r,s}(T\Delta, X, Y : E))$ satisfies the condition

$$|||\xi|||_m \equiv \sup\left\{ \frac{\|\xi_{q,r,s}(b,x,y)\|}{\|y\|^{Q-s}} : (b,x,y) \in X \oplus Y, \right.$$
$$\left. \|y\| \neq 0, \ q + r + s \leq m \right\} < \infty.$$

Provide $\Gamma_Q^0(P^m(E))$ with the norm $||| \cdot |||_m$.

To prove that $\bar{\sigma}$ is C^k smooth, we proceed by induction.

Induction hypothesis. For $1 \leq m \leq k$ and all l, $l = 0.1, \ldots, m-1$, $\bar{\sigma}$ is a C^l smooth section and, moreover, $j^l(\bar{\sigma})$ is the attracting fixed point of the operator $P_\#^l(F^{-1}) : \Gamma_Q^0(P^l(E)) \to \Gamma_Q^0(P^l(E))$.

Consider the morphism $P_\#^m(F^{-1}) : \Gamma_Q^0(P^m(E)) \to \Gamma_Q^0(P^m(E))$. Let us show that $P_\#^m(F^{-1})$ transforms $\Gamma_Q^0(P^m(E))$ into itself. Let $\eta_{m-1} \in \Gamma_Q^0(P^{m-1}(E))$ and $(\eta_{m-1}, \xi^m) \in \Gamma_Q^0(P^m(E))$. Since $\Phi(b,x,y) = \varphi(b,x,y)y^Q$ where φ is bounded in the C^k norm on $X \oplus Y$, it is not hard to show that $j^l(\Phi) \in \Gamma_Q^0(P^l(E))$ for all $l = 0, 1, \ldots, k$ (see [2], section A.8). Hence

$$|||P_\#^m(F^{-1})(\eta_{m-1}, \xi^m)|||_m$$
$$= \sup_{q+r+s \leq m} \frac{\|\mathrm{pr}_{q,r,s} \circ P_\#^m(F^{-1})(\eta_{m-1}, \xi^m)(b,x,y)\|}{\|y\|^{Q-s}}$$
$$\leq \sup_{q+r+s \leq m} \frac{\|\mathrm{pr}_{q,r,s} \circ \psi_m^*(\eta_{m-1})(b,x,y)\|}{\|y\|^{Q-s}}$$
$$+ \sum_{q+r+s = m} \sup \frac{\|A^{-1}(f(b,x,y)) \circ \xi_{q,r,s}(f(b,x,y)) \circ (Tf|T_b\Lambda)^q (Tf|X_b)^r (Tf|Y_b)^s\|}{\|y\|^{Q-s}}$$
$$+ \sum_{q+r+s = m} \sup \frac{\|A^{-1}(f(b,x,y)) \circ \Phi_{q,r,s}(f(b,x,y)) \circ (Tf|T_b\Lambda)^q (Tf|X_b)^r (Tf|Y_b)^s\|}{\|y\|^{Q-s}}$$
$$< \infty.$$

In fact, the first term is bounded because $\eta_{m-1} \in \Gamma_Q^0(P^{m-1}(E))$, the second and the third terms are bounded since $j^m(\Phi) \in \Gamma_Q^0(P^m(E))$, $(\eta_{m-1}, \xi^m) \in \Gamma_Q^0(P^m(E))$ and $f_3(b,x,y) = \delta(b,x,y)y$.

Next let us show that $P_\#^m(F^{-1})$ is fiberwise contracting on $\Gamma_Q^0(P^m(E))$ over $\Gamma_Q^0(P^{m-1}(E))$. If (η_{m-1}, ξ^m) and $(\eta_{m-1}, \bar{\xi}^m)$ belong to $\Gamma_Q^0(P^m(E))$, then

$$|||P_\#^m(F^{-1})(\eta_{m-1}, \xi^m) - P_\#^m(F^{-1})(\eta_{m-1}, \bar{\xi}^m)|||_m$$

$$\leq \sum_{q+r+s = m} \sup \frac{\|A^{-1}(f(b,x,y)) \circ (\xi^m - \bar{\xi}^m)(f(b,x,y)) \circ (Tf|T_b\Delta)^q (Tf|X_b)^r (Tf|Y_b)^s\|}{\|y\|^{Q-s}}$$

$$\leq \sum_{q+r+s=m} \sup_{b\in\Delta}(\|A^{-1}|E_{f(b,0,0)}\|+\varepsilon)(\|Tf|T_b\Delta\|+\varepsilon)^q(\|Tf|X_b\|+\varepsilon)^r \times$$

$$\times(\|Tf|Y_b\|+\varepsilon)^s \cdot \sup\frac{\|(\xi^m-\bar\xi^m)(f(b,x,y))\|}{\|f_3(b,x,y)\|^{Q-s}}\cdot\sup\frac{\|\delta(b,x,y)y\|^{Q-s}}{\|y\|^{Q-s}}$$

$$\leq \sum_{q+r+s=m}\sup_{b\in\Delta}(\|A^{-1}|E_{f(b,0,0)}\|+\varepsilon)(\|Tf|Y_b+\varepsilon)^Q(\|Tf|T_b\Delta\|+\varepsilon)^q\times$$

$$\times(\|Tf|X_b\|+\varepsilon)^r\cdot\sup\frac{\|(\xi^m-\bar\xi^m)(f(b,x,y))\|}{\|f_3(b,x,y)\|^{Q-s}}$$

$$\leq\sup_{b\in\Delta}(\|A^{-1}|E_{f(b,0,0)}\|+\varepsilon)(\|Tf|Y_b\|+\varepsilon)^Q(\|Tf|X_b\|+\varepsilon)^k\cdot|\|\xi^m-\bar\xi^m|\|_m,$$

because $\Omega(Tf|T\Delta,b) < \omega(Tf|X,b) \leq \Omega(Tf|X,b)$. Taking into account (10.2) and assuming $\varepsilon > 0$ to be small enough, we derive that $P_\#^m(F^{-1})$ is fiberwise contracting on $\Gamma_Q^0(P^m(E))$ over $\Gamma_Q^0(P^{m-1}(E))$. According to the induction hypothesis, $\bar\sigma$ is C^{m-1} and $j^{m-1}(\bar\sigma)$ is the attracting fixed point of $P_\#^{m-1}(F^{-1}): \Gamma_Q^0(P^{m-1}(E)) \to \Gamma_Q^0(P^{m-1}(E))$. By Theorem A.25 from [2], there exists a section $\eta_m \in \Gamma_Q^0(N^m(E))$ such that $\eta_m \equiv (j^{m-1}(\bar\sigma).\xi^m)$ is the attracting fixed point of $P_\#^m(F^{-1})$.

Let us prove that $\bar\sigma \in C^m$ and $\eta_m = j^m(\bar\sigma)$. To this end, consider the sequence $\{j^m(F_\#^{-n}(\Phi))\}_n = \{[P_\#^m(F^{-1})]^n(j^m(\Phi))\}$ $(n = 1, 2, \ldots)$. Because η_m is the attracting fixed point of $P_\#^m(F^{-1})$, the sequence $\{j^m(F_\#^{-n}(\Phi))\}_{n=1,2,\ldots}$ uniformly converges to η_m and, moreover, $\{F_\#^{-n}(\Phi)\} \to \bar\sigma$ in the C^0 topology. Hence, $\bar\sigma$ is C^m smooth and $j^m(\bar\sigma) = \eta_m$. This means that $j^m(\bar\sigma)$ is the attracting fixed point of the morphism $P_\#^m(F^{-1}): \Gamma_Q^0(P^m(E)) \to \Gamma_Q^0(P^m(E))$. Because $\bar\sigma$ is the unique F-invariant section of $\Gamma_Q^0(E)$, it is necessarily invariant under the flow $\{F^t\}$. The proof is complete.

Let $\dot z = \xi(z)$ $(z \in \mathbb{R}^d)$ be a C^k smooth vector field. Assume that the origin $z = 0$ is a singular point of ξ. Denote $L = D\xi(0)$. Decompose $\mathbb{R}^d$ into a direct sum of L-invariant linear subspaces W, X and Y so that the spectrum of $L|W$ is pure imaginary and the real parts of the eigenvalues of $L|X$ $(L|Y)$ are positive (negative, respectively). Let $\{f^t\}$ denote the phase flow of ξ. Without loss of generality, we may suppose that W, X and Y are f^t-invariant. Besides, consider the vector field

$$\dot u = A(z)u + \Phi(z), \quad \dot z = \xi(z) \quad (z \in \mathbb{R}^d, u \in \mathbb{R}^c), \tag{10.5}$$

where $A: \mathbb{R}^d \to L(\mathbb{R}^c, \mathbb{R}^c)$ and $\Phi: \mathbb{R}^d \to \mathbb{R}^c$ are C^k smooth functions.

Let us introduce some more notation. Given a linear operator $\mathfrak{L}$, denote by $\sigma(\mathfrak{L})$ its spectrum and define

$$\omega(\mathfrak{L}) = \min\{\mathrm{Re}\lambda : \lambda \in \sigma(\mathfrak{L})\}, \quad \Omega(\mathfrak{L}) = \max\{\mathrm{Re}\lambda : \lambda \in \sigma(\mathfrak{L})\}.$$

Put $\Theta = \Omega(A(0)), \quad \theta = -\omega(A(0)), \quad M = \Omega(L|X), \quad \mu = \omega(L|X),$
$\Lambda = -\omega(L|Y), \quad \lambda = -\Omega(L|Y).$

Observe that M, μ, Λ and λ are positive numbers.

Theorem 10.2. *Assume that* $\Phi(b, x, y) = \varphi(b, x, y) \cdot y^Q + \psi(b, x, y) \cdot x^S$, *where* $b \in W, x \in X, y \in Y, \varphi, \psi \in C^k, \theta - Q\lambda + k\mathrm{M} < 0, \Theta - S\mu + k\Lambda < 0$. *Then there exists a* C^k *smooth function* $u = \sigma(z)$ *defined in the neighbourhood of* $z = 0$ *such that its graph is invariant under the vector field* (10.5).

The proof of this statement is similar to that of Theorem 10.1 and will be omitted. We only note that there is no need in assuming that the vector field (10.5) is of class C^{k+1} (as we did in Theorem 10.1).

Theorem 10.3. *Assume that the fuction* $\Phi : \mathbb{R}^d \to \mathbb{R}^c$ *is of class* C^K *and* $\|\Phi(b, x, y)\| \leq C\|(x, y)\|^Q$, *where*

$$K > \theta/\lambda + \Theta/\mu + k(\mathrm{M}/\lambda + \Lambda/\mu + 1) + 2,$$
$$Q > \theta/\lambda + \Theta/\mu + k(\mathrm{M}/\lambda + \Lambda/\mu) + 2.$$

Then there exists a (local) invariant C^k *smooth section of the affine extension determined by* (10.5).

Theorem 10.3 follows immediately from Theorem 10.2 by virtue of Taylor's formula.

References

[1] Yu.S. Il'yashenko and S.Yu. Yakovenko. Finitely smooth normal forms of local families of diffeomorphisms and vector fields. Uspekhi Mat. Nauk, 46, No.1 (1991), 3–39 (in Russian).

[2] I.U. Bronstein and A.Ya. Kopanskii. Smooth Invariant Manifolds and Normal Forms. World Scientific. An International Publisher, Singapore. 1994.

[3] A. Banyaga, R. de la Llave and C.E. Wayne. Cohomology equations near hyperbolic points and geometric versions of Sternberg linearization theorem. MP ARC Preprint 94–135.

[4] V.I. Arnol'd. Mathematical Methods in Classical Mechanics. Springer-Verlag, New York, 1978.

[5] V.S. Samovol. Linearization of systems of differential equations in the neighbourhood of a toroidal manifold. Trudy Moscow Math. Ob., 38 (1979), 187–219 (in Russian)

[6] N. Bourbaki. Differentiable and Analytic Manifolds (summary of results). Mir Publishers. Moscow, 1975 (in Russian).

Progress in Nonlinear Differential Equations
and Their Applications, Vol. 19
© 1996 Birkhäuser Verlag Basel/Switzerland

On symmetric ω-limit sets in reversible flows

Jeroen S.W. Lamb   Matthew Nicol[*]

Abstract

Let $\Gamma \subset O(n)$ be a finite group acting orthogonally on $\mathbb{R}^n$. We say that Γ is a reversing symmetry group of the flow f^t if Γ has an index two subgroup $\tilde{\Gamma}$ whose elements commute with f^t and for all elements $\rho \in \Gamma - \tilde{\Gamma}$ and all t, $f^t \circ \rho(x) = \rho \circ f^{-t}(x)$. In dimensions $n = 1, 2$ we describe all symmetry groups of ω-limit sets for such reversible flows. In case $n \geq 1$ we give group and representation theoretic restrictions on possible symmetry groups and show that for subgroups of $\tilde{\Gamma}$ our conditions are necessary and sufficient. We also describe in detail the possible symmetries of periodic orbits.

Finally, we show that if a Liapunov stable ω-limit set is fixed setwise by a reversing symmetry then it is transitive.

1 Introduction and outline

In this paper we will be concerned with the symmetry properties of ω-limit sets of reversible flows. For completeness, in this introductory discussion we will mention earlier results pertaining to the symmetry properties of ω-limit sets of mappings as well. In recent years, a number of studies [2, 5, 16, 17] have resulted in necessary and sufficient conditions for stable ω-limit sets to have symmetries in equivariant dynamical systems. In this paper we will discuss extensions of these results in the flow context by taking into account the presence of reversing symmetries.

We will write invertible dynamical systems in the form $f^t : \mathbb{R}^n \mapsto \mathbb{R}^n$, with the time variable $t \in \mathbb{Z}$ (discrete time) or $t \in \mathbb{R}$ (continuous time), such that f^t is continuous on $\mathbb{R}^n$, $f^{t_1} \circ f^{t_2} = f^{t_1+t_2}$ for all t_1, t_2, and $f^0 = \mathrm{id}$. In the case of discrete time we thus have the dynamics generated by a homeomorphism $f = f^1$ (with $f^n = f \circ \ldots \circ f$ (n times)). In case f is moreover continuously differentiable it is a diffeomorphism. In the case of continuous time f^t is the flow of a vector field in $\mathbb{R}^n$.

We will distinguish two types of symmetry properties. The homeomorphism $\sigma : \mathbb{R}^n \mapsto \mathbb{R}^n$ is a *symmetry* of f^t if for all t

$$f^t \circ \sigma = \sigma \circ f^t, \tag{1}$$

---

[*]Nonlinear Systems Laboratory, Mathematics Institute, University of Warwick, Coventry CV4 7AL, United Kingdom

where ∘ denotes composition. When f^t has a symmetry σ it is also called σ-*equivariant*. The collection of all symmetries of f^t forms a group under composition. This group is usually called the *symmetry group* of f^t.

In addition to symmetries, f^t may also possess *reversing symmetries*: a homeomorphism $\rho : \mathbb{R}^n \mapsto \mathbb{R}^n$ is a reversing symmetry of f^t if for all t

$$f^t \circ \rho = \rho \circ f^{-t}. \tag{2}$$

As the composition of two reversing symmetries yields a symmetry and the composition of a reversing symmetry and a symmetry yields a reversing symmetry, the union of the sets of symmetries and reversing symmetries of f^t form a group, which we will call the *reversing symmetry group* of f^t [9]. It is easily checked that any reversing symmetry group contains a normal subgroup of index two which contains only symmetries. A dynamical system which possesses a reversing symmetry will be called *reversible*. It should be noted that we do not require reversing symmetries to be involutions. Recall that a map ρ is called an involution if $\rho^2 = \mathrm{id}$.

The ω-limit set of a point $\mathbf{x} \in \mathbb{R}^n$ consists of all $\mathbf{y} \in \mathbb{R}^n$ for which there exists a strictly increasing set of positive times $\{t_k\}$ (with either $t_k \in \mathbb{Z}$ or $t_k \in \mathbb{R}$ and $\lim_{k \to \infty} t_k = \infty$) such that $f^{t_k}(\mathbf{x})$ converges to $\mathbf{y}$. The ω-limit set of a flow is always connected. We will be concerned with sets A which are ω-limit sets of a reversible flow $f^t : \mathbb{R}^n \mapsto \mathbb{R}^n$, i.e. $A = \omega(\mathbf{x})$ for some $\mathbf{x} \in \mathbb{R}^n$.

An ω-limit set A is called *Liapunov stable* if for any open neighbourhood V of A there exists an open neighbourhood $U \subset V$ of A such that $f^t(U) \subset V$ for all $t \geq 0$. Liapunov stability should be clearly distinguished from the stronger notion of asymptotic stability. An ω-limit set A is called *asymptotically stable* if there exists an open neighbourhood U of A such that $\omega(x) \subseteq A$ for all $\mathbf{x} \in U$.

Our interest in this paper will be with Liapunov stable ω-limit sets of flows on $\mathbb{R}^n$ having a reversing symmetry group Γ that is a finite subgroup of $O(n)$. Throughout this paper we will assume Γ to be acting as a linear representation on $\mathbb{R}^n$ (and orthogonal with respect to the natural inner product on $\mathbb{R}^n$).

We define the *symmetry group* $\Sigma \leq \Gamma$ of a subset $A \subset \mathbb{R}^n$ as

$$\Sigma = \{\gamma \in \Gamma \mid \gamma(A) = A\}. \tag{3}$$

Let $T \leq \Sigma$ denote the subgroup of instantaneous symmetries, i.e. the symmetries which fix A pointwise. Since in describing the symmetry of the set A it is relevant to know both Σ and T, we indicate the symmetry of A by the pair (Σ, T). Note that for all such pairs, T must be a normal subgroup of Σ.

In this context we have the following notion of *admissibility*:

Definition 1.1 Suppose that Σ is a subgroup of Γ. We say that Σ is *admissible* for flows (homeomorphisms, diffeomorphisms) if there exists a flow (homeomorphism, diffeomorphism) with reversing symmetry group Γ and a Liapunov stable ω-limit set with symmetry $(\Sigma, 1)$.

The central question we would like to address is:[1]

What subgroups of reversing symmetry groups are admissible as symmetry groups of Liapunov stable ω-limit sets for reversible dynamical systems ?

In the present paper we will address this question in the context of flows. The context of reversible mappings will be treated in a forthcoming paper [11].

The above question has been answered in a variety of equivariant contexts. An important observation is that the symmetry group of a periodic orbit must be a cyclic extension of an isotropy subgroup (see [17]). However, if the asymptotic dynamics are more complicated than this then the phenomenon of *symmetry on average* may occur in which an ω-limit set displays greater symmetry than that obtainable by periodic orbits.

In the continuous category necessary and sufficient conditions have been given for admissibility in the work of Melbourne *et al.* [17] and Ashwin and Melbourne [2]. Group elements which act as reflections play the crucial role in determining admissibility. In this context we call $\tau : \mathbb{R}^n \mapsto \mathbb{R}^n$ a reflection if $\tau^2 = $ id and $\dim \mathrm{Fix}(\tau) = n - 1$. The fixed point subspace of τ is called a *reflection hyperplane* and divides $\mathbb{R}^n$ in two.

If Δ is a subgroup of Γ we let L_Δ denote the union of reflection hyperplanes corresponding to group elements not contained in Δ. A key result of [17] was that if Σ is admissible then necessarily there is a subgroup Δ such that

a) Δ is a normal subgroup of Σ,

b) Σ/Δ is cyclic,

c) Δ fixes a connected component of $\mathbb{R}^n - L_\Delta$.

Ashwin and Melbourne [2] showed that this representation-theoretic condition is also sufficient by using a construction involving Cayley graphs. In particular they showed that all cyclic subgroups of Γ were admissible, recapturing a result of King and Stewart [8]. They also showed that if a Σ-symmetric attractor is connected then Σ must fix a connected component of $\mathbb{R}^n - L_\Sigma$. Furthermore the attractors constructed in the proof of this theorem were topologically transitive, possessed open basins of attraction and the connected attractors were topologically mixing.

When equivariant homeomorphisms or flows are considered then the restrictions on admissibility are more severe. Let L denote the union of the reflection hyperplanes corresponding to reflections in Γ. Melbourne [16] showed that if A is a Σ-symmetric ω-limit set for a Γ-equivariant homeomorphism then either

---

[1]We would like to note that the context of Liapunov stable ω-limit sets is chosen in connection with the setting in previous work on symmetries of attractors in [2, 5, 16, 17]. However, in particular in the reversible case, other types of ω-limit sets may be equally (or more) relevant. In this respect it is important to note that most of our results apply to ω-limit sets in general, rather than only to Liapunov stable ones. For further remarks we refer the reader to the discussion in Section 7.

a) Σ contains no reflections and fixes a connected component of $\mathbb{R}^n - L$, or

b) Σ has an index two subgroup that fixes a connected component of $\mathbb{R}^n - L$

If Σ is to be admissible for flows then Σ must fix a connected component of $\mathbb{R}^n - L$.

Precise necessary and sufficient conditions for admissibility of flows and homeomorphisms are given in [5]. In fact if $n \geq 4$ then the ω-limit sets that realise admissibility can be taken to be Axiom A attractors and in any case they can be taken to be asymptotically stable. The sufficiency conditions for flows is complicated somewhat by a topological obstruction in the case $n = 2$, but otherwise if Σ fixes a connected component of $\mathbb{R}^n - L$ then it is admissible and the admissibility can be realised by asymptotically stable ω-limit sets (which can be taken as Axiom A if $n \geq 5$). In the case of dimension 2 then the only subgroup of $\mathbb{D}_m$ that is admissible for flows is 1 and if $\Gamma = \mathbb{Z}_n$ then the only admissible subgroups are 1 and $\mathbb{Z}_m$.

In order to establish admissibility we use two ingredients: a collection of necessary conditions for admissibility which rule out certain subgroups and for the remaining subgroups a construction of ω-limit sets with the required symmetry.

We will now summarize our main results. We give necessary conditions on finite subgroups of $O(n)$ for admissibility for flows and necessary and sufficient conditions in the case $n \leq 2$. Furthermore, we extend the results of [5] on the admissibility of symmetry groups of ω-limit sets that do not contain reversing symmetries to the reversible flow context. We first introduce some notation and terminology.

We let $\hat{\Gamma}$ denote the symmetry group *associated* to Γ, i.e. $\hat{\Gamma}$ is the largest subgroup of Γ containing no reversing symmetries. Then, $\hat{\Gamma}$ is a normal subgroup of Γ and when $\Gamma \neq \hat{\Gamma}$ it is a subgroup of index two.

In equivariant dynamical systems restrictions on symmetries for ω-limit sets occur due to the presence of symmetries which are reflections. As in the equivariant case we let L denote the union of all the reflection hyperplanes of Γ. i.e.

$$L := \bigcup_{\text{refl } \tau \in \Gamma} \text{Fix}(\tau). \tag{4}$$

We define $\tilde{L}$ to be the union of all the reflection hyperplanes of $\tilde{\Gamma}$. i.e.

$$\tilde{L} := \bigcup_{\text{refl } \tau \in \tilde{\Gamma}} \text{Fix}(\tau). \tag{5}$$

If Γ does not contain reversing symmetries, the definitions of L and $\tilde{L}$ coincide.

We first note that, as a consequence of the results of [5]. if Γ is a finite subgroup of $O(n)$ ($n \geq 1$), $\Sigma \leq \Gamma$ and A is a Σ-symmetric ω-limit set for a flow $f^t : \mathbb{R}^n \mapsto \mathbb{R}^n$ with reversing symmetry group Γ then Σ must fix a connected component of $\mathbb{R}^n - \tilde{L}$.

In case Γ does not contain reversing symmetries it has been shown in [5] that subgroups Σ are admissible for flows only if they fix a connected component of $\mathbb{R}^n - L$ and that all such subgroups are admissible whenever $n \geq 3$. For subgroups Σ that do not contain reversing symmetries, this result extends to the reversible case:

Theorem 1.2 *Suppose Γ is a finite subgroup of $O(n)$, $n \geq 3$ and $\Sigma \leq \Gamma$. Then, a subgroup Σ that does not contain a reversing symmetry is admissible for flows $f^t : \mathbb{R}^n \mapsto \mathbb{R}^n$ with reversing symmetry group Γ if and only if Σ fixes a connected component of $\mathbb{R}^n - L$. The ω-limit sets that realize admissibility can be taken to be asymptotically stable. Moreover, if $n \geq 5$ the ω-limit sets that realize admissibility can be taken to be Axiom A attractors.*

We now give necessary and sufficient conditions for admissibility of subgroups that do not fix a connected component of $\mathbb{R}^n - L$. For an explanation of our notation of reversing symmetry groups we refer the reader to the Appendix.

Theorem 1.3 *Suppose Γ is a finite subgroup of $O(n)$ $n \geq 2$, and $\Sigma \leq \Gamma$. Then, a subgroup Σ that does not fix a connected component of $\mathbb{R}^n - L$ is admissible for flows $f^t : \mathbb{R}^n \mapsto \mathbb{R}^n$ with reversing symmetry group Γ if and only if*

(i) Σ fixes a connected component of $\mathbb{R}^n - \tilde{L}$

(ii) $\Sigma \simeq \mathbb{Z}_k \wedge \mathbb{D}'_1 \ (km')$ where the subgroup isomorphic to $\mathbb{D}'_1 \ (m')$ is generated by a reversing reflection ρ and the cyclic subgroup isomorphic to $\mathbb{Z}_k \ (k)$ is a maximal cyclic subgroup of $\tilde{\Gamma}.$[2]

(iii) when $k = 2$ in (ii), the subgroup of $\tilde{\Gamma}$ isomorphic to $\mathbb{Z}_2$ acts freely on the connected components of $\mathbb{R}^n - \text{Fix}(\rho)$.

The ω-limit sets that realise admissibility must be periodic orbits.

Further discussion on those subgroups for which admissibility may be realised by periodic orbits is given in Section 3, Remark 3.1.

In $\mathbb{R}$. the situation is very special due to topological restrictions.

Remark 1.4 Suppose Γ is a finite subgroup of $O(1)$, and $\Sigma \leq \Gamma$. Then the only admissible subgroup Σ for flows $f^t : \mathbb{R} \mapsto \mathbb{R}$ with reversing symmetry group Γ is 1.

Also in $\mathbb{R}^2$, extra restrictions arise. In summary, they lead to the following result, with reference to the tables in Section 5:

Theorem 1.5 *Suppose Γ is a finite subgroup of $O(2)$, and $\Sigma \leq \Gamma$. Then all subgroups Σ which are admissible for flows $f^t : \mathbb{R}^2 \mapsto \mathbb{R}^2$ with reversing symmetry group Γ are listed in Table 1 and Table 2. The ω-limit sets that realize admissibility can be taken to be Liapunov stable periodic orbits.*

---

[2] A maximal cyclic subgroup of $\tilde{\Gamma}$ is a cyclic subgroup that is not contained strictly in another cyclic subgroup of $\tilde{\Gamma}$.

This paper is organized as follows. In Section 2 we are concerned with symmetry properties of periodic orbits. In Section 3 we prove Theorem 1.2 which concerns the admissibility of subgroups that do not contain reversing symmetries by showing that the constructions of [5] extend to the reversible case. We also prove Theorem 1.3 concerning the admissibility of subgroups for ω-limit sets that intersect more than one connected component of $\mathbb{R}^n - L$, exploiting the fact that these ω-limit sets must be periodic orbits. In the same section, in Remark 3.1 we address the general question of for which symmetry groups admissibility is realisable by periodic orbits. In Section 4 we show that Liapunov stable ω-limit sets that are symmetric with respect to a reversing symmetry must be transitive. We also show that if A is a Liapunov stable ω-limit set and $\gamma(A) \cap A \neq \emptyset$ (where γ is a symmetry or a reversing symmetry) then $\gamma(A) = A$. We then focus on the construction of symmetric ω-limit sets that are periodic orbits in the context of $\mathbb{R}^2$ and establish in Section 5 the admissibility of subgroups of $O(1)$ and $O(2)$. In Section 6 we explain how our results extend to the case of symmetric ω-limit sets possessing nontrivial instantaneous symmetries. The paper is concluded with a discussion.

2   Symmetric periodic orbits

In this section we will focus on symmetric periodic orbits. They are the simplest examples of transitive ω-limit sets. We will include, for completeness, results concerning periodic orbits of reversible mappings as well.

An orbit of f^t is defined as

$$o(\mathbf{x}) := \{f^t(\mathbf{x})\}_{t \in \mathbb{R} \text{ or } t \in \mathbb{Z}}. \tag{6}$$

An orbit $o(\mathbf{x})$ is called periodic with period p if p is the smallest positive real (or integer) for which $f^p(\mathbf{x}) = \mathbf{x}$.

Our main result on symmetric periodic orbits is contained in the following proposition:[3]

Proposition 2.1 *Let $f^t : \mathbb{R}^n \to \mathbb{R}^n$ be a dynamical system with reversing symmetry group Γ (which is finite subgroup of $O(n)$) and associated symmetry group $\tilde{\Gamma}$. Suppose f^t has a periodic orbit and symmetry (Σ, T). Let $\tilde{\Sigma} := \Sigma \cap \tilde{\Gamma}$ and $\tilde{T} := T \cap \tilde{\Gamma}$. Then we have one of the following two situations*

(a) $T = \tilde{T}$ and either $\Sigma/T \simeq \mathbb{Z}_k$ (k), if $\Sigma = \tilde{\Sigma}$, or $\Sigma/T \simeq \mathbb{Z}_k \wedge \mathbb{D}'_1$ (km'), if $\Sigma \neq \tilde{\Sigma}$, and in the discrete time case k divides the period p of the periodic orbit.

(b) $T \neq \tilde{T}$ and in the discrete time case the period p is either 1 or 2 and $\Sigma/T \simeq \mathbb{Z}_k$ (k dividing p), while in the flow case $\Sigma/T \simeq 1$ and the periodic orbit is a fixed point.

---

[3]For an explanation of our notation of reversing symmetry groups the reader is referred to the Appendix.

A further discussion on for which groups Σ admissibility may be achieved by periodic orbits is presented in Section 3, Remark 3.1.

The proof of Proposition 2.1 uses the following Lemma.

Lemma 2.2 ([12]) *If an orbit $o(\mathbf{x})$ is symmetric with respect to a reversing symmetry ρ, then $o(\mathbf{x}) \subseteq \mathrm{Fix}(\rho^2)$.*

Proof. The proof is elementary. If $o(\mathbf{x})$ is symmetric with respect to ρ then for all $\mathbf{y} \in o(\mathbf{x})$, $\rho(\mathbf{y}) = f^t(\mathbf{y})$ for some t. Hence $\rho^2(\mathbf{y}) = \rho \circ f^t(\mathbf{y}) = f^{-t} \circ \rho(\mathbf{y}) = \mathbf{y}$ for all $\mathbf{y} \in o(\mathbf{x})$. $\qquad\square$

Proof of Proposition 2.1 We consider case (b) first. Suppose T contains a reversing symmetry ρ, then $f^{-t} \circ \rho(x) = f^{-t}(x) = \rho \circ f^t(x) = f^t(x)$. Hence in the discrete time case the periodic orbit has period 1 or 2 (and Σ/T is isomorphic to 1 or $\mathbb{Z}_2$) and in the flow case the periodic orbit is a fixed point.

Now we consider case (a) where $\tilde{T} = T$. Suppose that $\Sigma = \tilde{\Sigma}$. Then by [17] we have $\Sigma/T \simeq \mathbb{Z}_k$ (k), where k divides the period p in the discrete time case. Suppose now that $\Sigma \neq \tilde{\Sigma}$ then $\Sigma = \tilde{\Sigma} \cup \rho\tilde{\Sigma}$ for some reversing symmetry ρ. Now $\tilde{\Sigma}/T$ is a normal subgroup of Σ/T and the group generated by $\rho T/T$ is of order 2 (since $\rho^2 \in \tilde{T}$) and together with $\tilde{\Sigma}/T$ generates Σ/T. Thus Σ/T is the semi-direct product of $\tilde{\Sigma}/T$ by $\langle \rho T/T \rangle$. As before $\tilde{\Sigma}/T \simeq \mathbb{Z}_k$ (k) for some integer k (which divides the period p in the discrete time case). The order 2 subgroup generated by $\rho T/T$ is isomorphic to $\mathbb{D}'_1$ (m'). Note that the group $\mathbb{Z}_k \wedge \mathbb{D}'_1$ (km') is isomorphic to the dihedral group $\mathbb{D}_k$ if one ignores the difference between symmetries and reversing symmetries. $\qquad\square$

Finally, we recall a well-known characterization of orbits of flows that are symmetric with respect to reversing symmetries, cf. e.g. [4, 10], and a useful Corollary.

Theorem 2.3 *If f^t is a flow with reversing symmetry ρ and orbit $o(\mathbf{x})$, then $o(\mathbf{x})$ is symmetric with respect to ρ if and only if $o(\mathbf{x})$ intersects $\mathrm{Fix}(\rho)$. The orbit is nonperiodic if and only if the intersection consists of only one point and $o(\mathbf{x}) \nsubseteq \mathrm{Fix}(\rho)$. The orbit is periodic (and not stationary) if and only if $o(\mathbf{x})$ intersects $\mathrm{Fix}(\rho)$ in precisely two points.*

Corollary 2.4 *Suppose ρ is a reflection and a reversing symmetry of a flow f^t : $\mathbb{R}^n \mapsto \mathbb{R}^n$ possessing an ω-limit set that is symmetric with respect to ρ. Then this ω-limit set is either a fixed point (contained in $\mathrm{Fix}(\rho)$) or a periodic orbit (transversally intersecting $\mathrm{Fix}(\rho)$ in two points).*

3 Stable symmetric ω-limit sets

In this section we establish Theorem 1.2 and Theorem 1.3. We will begin with the proof of Theorem 1.3, which is based upon the ideas of the preceding section and [5].

Proof of Theorem 1.3 Suppose Σ is admissible for flows. Then Σ must fix a connected component of $\mathbb{R}^n - \tilde{L}$ by the results of [5]. Suppose Σ does not fix a connected component of $\mathbb{R}^n - L$. Then any flow which realises admissibility of Σ must intersect $\mathrm{Fix}(\rho)$ for some reversing reflection ρ. Thus by Theorem 2.3 the ω-limit of the flow is ρ-symmetric and so by Corollary 2.4 the ω-limit set must be a periodic orbit (as it is assumed to have trivial instantaneous symmetry). Proposition 2.1 shows, as Σ contains a reversing symmetry and $T = \tilde{T} = 1$, that $\Sigma \simeq \mathbb{Z}_k \wedge \mathbb{D}_1'$ (km') for some $k \geq 1$, where the subgroup isomorphic to $\mathbb{D}_1'$ (m') is generated by the reversing reflection ρ.

Moreover, if the ω-limit set intersects the fixed sets of two reversing reflections then the symmetry group of the ω-limit set is precisely the group generated by these two reflections. Therefore, if $k \geq 2$, the ω-limit set (which is a periodic orbit) must connect two different faces of a connected component of $\mathbb{R}^n - L$ originating from two different reversing reflection hyperplanes. Two of such reversing reflections generate a dihedral group whose nonreversing cyclic subgroup is maximal.

Conversely, if Γ contains a subgroup $\Sigma \simeq \mathbb{Z}_k \wedge D_1'$ (km') $k \geq 3$ that fixes a connected component of $\mathbb{R}^n - \tilde{L}$, but not of $\mathbb{R}^n - L$, with the group isomorphic to $\mathbb{D}_1'$ (m') being generated by a reversing reflection ρ and the subgroup of $\tilde{\Gamma}$ isomorphic to $\mathbb{Z}_k$ (k) being a maximal cyclic subgroup, one connected component of $\mathbb{R}^n - L$ will possess two faces made up of the fixed point subspaces of reversing reflections ρ_1, ρ_2 generating Σ. A flow with a Σ-symmetric (Liapunov stable) periodic orbit is constructed as follows: inside the connected component of $\mathbb{R}^n - L$ with faces consisting of parts of $\mathrm{Fix}(\rho_1)$ and $\mathrm{Fix}(\rho_2)$, these two faces are connected by a flow-line (satisfying the boundary condition of orthogonal impact on $\mathrm{Fix}(\rho_1)$ and $\mathrm{Fix}(\rho_2)$) avoiding intersections with fixed point subspaces of other (reversing)) symmetries. This flow-line can be smoothly embedded in a flow in the connected component of $\mathbb{R}^n - L$ satisfying appropriate boundary conditions at the fixed point subspaces. From here, the flow in all other connected components of $\mathbb{R}^n - L$ follows by symmetry, such that the flow-line intersecting ρ_1 and ρ_2 extends to a Σ-symmetric periodic orbit in $\mathbb{R}^n$. It is obvious that the embedding can be constructed in such a way that the orbit will be Liapunov stable. Note that the constructed periodic orbit cannot have more symmetry than Σ.

In case $k = 2$, one must add the requirement that the subgroup of $\tilde{\Gamma}$ isomorphic to $\mathbb{Z}_2$ (2) acts freely on the connected components of $\mathbb{R}^n - \mathrm{Fix}(\rho)$. Namely, if it does not, then Σ cannot be the symmetry of a periodic orbit. This follows from the fact that $\mathrm{Fix}(\rho)$ divides any ρ-symmetric orbit in two halves. Let σ denote generator of the cyclic subgroup of $\tilde{\Gamma}$ isomorphic to $\mathbb{Z}_2$ (2), and suppose the period of the symmetric periodic orbit is p, then for any point $\mathbf{x}(t)$ on this periodic orbit $\sigma(\mathbf{x}(t)) = \mathbf{x}(t + p/2)$. However, $\rho(\mathbf{x}(t)) = \mathbf{x}(a - t)$ for some $a \in \mathbb{R}$. Hence, if $\mathbf{x}(t)$ is in one connected component of $\mathbb{R}^n - \mathrm{Fix}(\rho)$ then $\mathbf{x}(t + p/2)$ must be in the other connected component. Moreover, σ fixes $\mathrm{Fix}(\rho)$ and thus should permute the connected components of $\mathbb{R}^n - \mathrm{Fix}(\rho)$.

In case $k \geq 3$, however, this requirement is redundant, as it follows directly from the (dihedral) group structure of Σ that the $\mathbb{Z}_k$ action does not fix Fix(ρ).

Finally, in case $k = 1$ the construction of a symmetric periodic orbit consists of constructing and embedding a flow line connecting two points on Fix(ρ) inside one connected component of $\mathbb{R}^n - L$.   □

Remark 3.1 In the case of flows, it is not difficult to obtain from Proposition 2.1 and Theorem 2.3 necessary and sufficient conditions for subgroups Σ of Γ containing a reversing symmetry for which admissibility can be realized by periodic orbits. As discussed in the proof of Theorem 1.3, for the construction of symmetric periodic orbits it suffices to construct a flow containing one flow-line connecting two points of the fixed sets of reversing symmetries generating Σ. As the periodic orbit must intersect each fixed set twice it is necessary that the fixed point subspaces of the reversing symmetries have dimensions greater or equal to 1. The case that Σ does not fix a connected component of $\mathbb{R}^n - L$ has been treated in Theorem 1.3. In the remaining case, in which Σ fixes a connected component of $\mathbb{R}^n - L$ and $\Sigma \simeq \mathbb{Z}_k \wedge \mathbb{D}'_1$ (km') one needs to consider the cases $k = 1$, $k = 2$ and $k \geq 3$ separately.

If $k = 1$ (and $\Sigma \simeq \mathbb{D}'_1$) a necessary and sufficient condition for the possibility of constructing a symmetric periodic orbit is that the fixed set of the reversing symmetry generating Σ has dimension greater or equal to 1. Within one connected component of $\mathbb{R}^n - L$ it is possible to locally construct a flow-line connecting two points of Fix(ρ) without intersecting any other fixed point subspace and this can be embedded in a flow which has a reversing symmetry group Γ.

If $k \geq 3$ it is easily verified from the (dihedral) group structure of Σ that indeed all reversing symmetries in Σ must have fixed point subspaces of dimension greater or equal to 1. Namely, the only linear involution in $O(n)$ that has a zero-dimensional fixed-point subspace is $-\mathrm{id}$, and $-\mathrm{id}$ does commute with all k-fold rotations in $O(n)$. Admissibility for these subgroups can always be realized by symmetric periodic orbits, in a similar way as described above.

In case $k = 2$ one has to take more care and check explicitly that the two reversing symmetries generating Σ indeed both have fixed point subspaces of dimension greater or equal to 1.

In a similar way, it can also be shown that for every cyclic subgroup of $\tilde{\Gamma}$ that fixes a connected component of $\mathbb{R}^n - L$, admissibility can be realized by a periodic orbit. Let us say that this cyclic group is isomorphic to $\mathbb{Z}_k$ and generated by σ. Then, inside one component one constructs a flow with a flow-line connecting two $\mathbb{Z}_k$-equivalent points (i.e. a point $\mathbf{x}$ and a point $\sigma(\mathbf{x})$), that does not intersect any fixed point subspace. This flow-line can be embedded in a Γ-symmetric flow by first embedding the flow line in a Γ-symmetric flow inside the connected component of $\mathbb{R}^n - L$ and then extending the flow by symmetry extends to a flow in $\mathbb{R}^n$. Taking appropriate boundary conditions for the initial flow-line (to avoid the occurrence of a singularity at $\mathbf{x}$) will then lead to a flow with a $\mathbb{Z}_k$-symmetric periodic orbit. Liapunov stability can also be realized in the embedding.

Proof of Theorem 1.2 We will first establish the necessity of the condition. In the case that $L = \tilde{L}$ then the results of [5] show that Σ must fix a connected component of $\mathbb{R}^n - L$. So we suppose that $L \neq \tilde{L}$. Note that, in this case, if Σ does not fix a connected component of $\mathbb{R}^n - L$ then the flow which realises the admissibilty of Σ must intersect $\text{Fix}(\rho)$ for some reversing reflection ρ and hence Σ contains ρ (by Theorem 2.3). This contradicts the assumption that Σ contains no reversing symmetries. So the condition is necessary.

The proof of sufficiency involves an obvious modification to the corresponding proofs of [5] and we present in detail only the construction of Axiom A attractors for flows in dimension $n \geq 5$. The case $n = 3, 4$ (in which we may construct asymptotically stable attractors) is handled with the same (mutatis mutandi) modification to the constructions given in [5].

Let $\tilde{\Gamma}$ be the symmetry group associated to Γ and $\tilde{\Sigma}$ be a subgroup of $\tilde{\Gamma}$ which fixes a connected component of $\mathbb{R}^n - L$. Let ρ be a reversing symmetry (recall that $\Gamma = \rho\tilde{\Gamma} \cup \tilde{\Gamma}$). Note that , by the results of [5], there exists a Γ- equivariant flow which has an attractor with symmetry group $\tilde{\Sigma}$. Hence there exists a $\tilde{\Sigma}$ invariant connected open set O such that $\gamma(O) \cap O = \emptyset$ for all $\gamma \in \Gamma - \tilde{\Sigma}$.

The existence of such an open set O shows that we may as in [5] construct a $\tilde{\Gamma}$-equivariant flow f^t with a $\tilde{\Sigma}$-symmetric attractor $A \subset O$ (note that $\tilde{\Sigma}$ is Class I in the terminology of [5]). This attractor has $|\tilde{\Gamma} : \tilde{\Sigma}|$ conjugate attractors. In the construction of f^t and A we may require that A has a $\tilde{\Sigma}$ invariant neighbourhood $U \subset O$ such that $f^t(U) = U$ and $f^t|_{\tilde{\Gamma}U^c} = \text{id}|_{\tilde{\Gamma}U^c}$ (this follows from the construction of [5]).

We may smoothly isotope $f^t|_U$ to a flow g^t which is the identity map on O^c (since by construction in [5] we may take f^t Γ-equivariantly isotopic to the identity and $f^t(U) = U$ for all t) and which has the properties

a) $g^t_{|U} = f^t_{|U}$ for all t

b) $g^t(O) = O$ for all t.

Now we extend g^t Γ-equivariantly to a flow h^t. Finally define a flow F^t by $F^t(x) = h^t(x)$ if $x \in \tilde{\Gamma}(O)$ and $F^t(x) = h^{-t}(x)$ if $x \in \rho\tilde{\Gamma}(O)$. Note that F^t has reversing symmetry group Γ and F^t has a $\tilde{\Sigma}$ symmetric Axiom A attractor contained in O.   □

4   Marginal stability and transitivity

In this section we will show that Liapunov stable ω-limit sets that are symmetric with respect to a reversing symmetry must be transitive. Recall that a set A is called transitive if $A = \omega(\mathbf{x})$ for some $\mathbf{x} \in A$.

Theorem 4.1 *Suppose that A is a Liapunov stable ω-limit set of a dynamical system f^t with reversing symmetry ρ. Then,*

$$\rho(A) = A \quad \Rightarrow \quad A \text{ is transitive.}$$

The proof of Theorem 4.1 will follow directly from Lemma 4.3 and Lemma 4.4 below.

A common notion of attractor in the literature is that of an asymptotically stable ω-limit set (for a definition see Section 1). In the context of symmetric (equivariant) dynamical systems, in order to establish admissibility of symmetry groups for attractors in [5] asymptotically stable ω-limit sets were constructed, although the main results were obtained under the assumption of only Liapunov stability.

However, in reversible systems an ω-limit set that is symmetric with respect to a reversing symmetry cannot be asymptotically stable, it can at most be *marginally stable*.

Definition 4.2 An ω-limit set A of a dynamical system f^t is *marginally stable* if for all open neighbourhoods V of A there exists an open neighbourhood U of A such that $f^t(U) \subset V$ for all $t \in \mathbb{R}$ (in the case of continuous time) or $t \in \mathbb{Z}$ (in the case of discrete time).

Lemma 4.3 *Suppose we have the setting of Theorem 4.1. Then,*

$$\rho(A) = A \quad \Rightarrow \quad A \text{ is marginally stable.}$$

Proof. Because A is Liapunov stable, by definition we have that for all neighbourhoods V there exists a neighbourhood $U \subset V$ such that $f^t(U) \subset V$ for all $t \geq 0$. We are thus left to prove the assertion for $t < 0$.

Suppose the assertion does not hold, then $\rho(A) = A$ and there exists a neighbourhood V such that for all $U \subset V$ there exists a $t < 0$ such that $f^t(U) \not\subset V$ (note that without loss of generality we may assume that all neighbourhoods V, U in the following discussion are ρ-symmetric). Thus, given $U \subset V$, there exists a point $\mathbf{x} \notin V$ for which $f^t(\mathbf{x}) = \tilde{\mathbf{x}} \in U \subset V$ for some $t > 0$. But then also $\rho(\tilde{\mathbf{x}}) = \rho(f^t(\mathbf{x})) = f^{-t}(\rho(\mathbf{x}))$, which implies that $f^t(\rho(\tilde{\mathbf{x}}))$ is not in V and hence A is not Liapunov stable since $\rho(\tilde{\mathbf{x}}) \in U$ and $U \subset V$ was arbitrary. $\qquad\square$

Lemma 4.4 *If an ω-limit set is marginally stable then it is also transitive.*

Proof. Suppose $A = \omega(\mathbf{x})$ and A is marginally stable. We show that $\mathbf{x} \notin A$ leads to a contradiction. Every neighbourhood $V \supset A$ contains a point of the f^t-orbit of $\mathbf{x}$. In turn this implies that there exists a neighbourhood V' for which there does not exist a smaller neighbourhood U such that $f^t(U) \subset V'$ for all $t < 0$. This implies that A cannot be marginally stable. $\qquad\square$

We now prove a theorem which is the reversible analogue to a result of [3, 17].

Theorem 4.5 *Suppose* $f^t : \mathbb{R}^n \to \mathbb{R}^n$ *has a Liapunov stable ω-limit set A. Suppose γ is either a symmetry or a reversing symmetry of f^t. Then if $\gamma(A) \cap A \neq \emptyset$ then $\gamma(A) = A$.*

Proof. The case where γ is a reversing symmetry of f^t has been covered in [17]. Suppose that γ is a reversing symmetry of f^t. For convenience we let $\omega_g(\mathbf{x})$ denote the ω-limit set of $\mathbf{x}$ under g^t (either a map or flow). Let $\mathbf{x}$ be such that $\omega_f(\mathbf{x}) = A$. Then, as γ is a reversing symmetry of f^t, $\omega_{f^{-1}}(\gamma(\mathbf{x})) = \gamma(A)$. Suppose that $\gamma(A)$ and A are not disjoint and $\gamma(A) \neq A$. We will derive a contradiction.

In this case there exists a neighbourhood V of A such that $\gamma(A) \cap V^c \neq \emptyset$, where V^c denotes the complement of V in $\mathbb{R}^n$. Then there exists a neighbourhood $U \subset V$ such that $f^t(U) \subset V$ for all $t \geq 0$. There exists an increasing sequence $\{t_k\}$ such that $f^{-t_k}(\gamma(\mathbf{x})) \in V^c$ (since $\omega_{f^{-1}}(\gamma(\mathbf{x})) = \gamma(A)$). But there exists a time $t' > t_1$ such that $f^{-t'}(\gamma(\mathbf{x})) \in U$ and hence $f^t(f^{-t'}(\gamma(\mathbf{x}))) \in V$ for all positive t. However if $t = t' - t_1 > 0$ then $f^t(f^{-t'}(\gamma(\mathbf{x}))) \in V^c$, which is a contradiction. Thus $\gamma(A) = A$ if $\gamma(A) \cap A \neq \emptyset$. □

5   Constructions in $\mathbb{R}$ and $\mathbb{R}^2$

In this section we outline the construction of symmetric periodic orbits in $\mathbb{R}$ and $\mathbb{R}^2$.

Before treating the constructions in $\mathbb{R}^2$, we first prove the results on admissibility in the one-dimensional case.

Proof of Remark 1.4, In the case that $\Gamma = \mathbb{Z}_2$ and does not contain a reversing symmetry, it is easily checked that its only admissible subgroup is 1, cf. [5]. We are left to consider only one group Γ containing a reversing symmetry, namely $\mathbb{Z}'_2 := \langle \rho \rangle$, with $\rho = -\mathrm{id}$. In the case of a flow it follows immediately that the only possible $\mathbb{Z}'_2$-symmetric ω-limit set is a stationary point at $\mathrm{Fix}(\rho)$. □

The rest of this section will be devoted to constructions of symmetric periodic orbits in $\mathbb{R}^2$.

Theorem 5.1 *Suppose A is a Liapunov stable ω-limit set of a flow f^t in $\mathbb{R}^2$ with reversing symmetry ρ. Then, if $\rho(A) = A$, A is a periodic orbit.*

Proof. By Theorem 4.1, A is transitive. Hence A is an ω-limit set for f^t as well as for f^{-t}. For flows in $\mathbb{R}^2$ it is well known, cf. [13], that this implies that A must be a periodic orbit. □

We now discuss the question of the admissibility of subgroups of $O(2)$.

From Theorem 5.1 and Theorem 2.3 it follows that $\mathbb{Z}'_n$ is not admissible for flows for any $n \geq 2$. In more detail, Theorem 5.1 shows that a Liapunov stable ω-limit set must be a periodic orbit and Theorem 2.3 shows that this periodic orbit must intersect $\mathrm{Fix}(\rho)$ (the origin) in precisely two points unless it is a fixed point at the origin (and hence possesses nontrivial instantaneous symmetry).

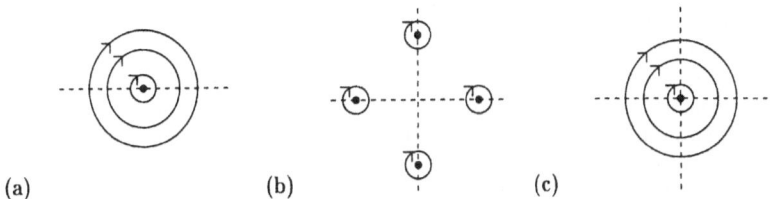

Figure 1: Schematic illustration of the construction of flows f^t with a periodic orbit that is symmetric with respect to a reversing symmetry ρ. The strategy is to construct flows constituting a local rotation in a small domain isotoped to the identity outside this domain. Some examples: **(a)** Local rotation constituting $\mathbb{D}'_1 (m')$ symmetric marginally stable ω-limit sets. **(b)** Construction with $\mathbb{D}'_1 (m')$ symmetry when $\Gamma \simeq \mathbb{Z}_2 \times \mathbb{D}'_1 (2m')$. **(c)** Construction with $\mathbb{Z}_2 \times \mathbb{D}'_1 (2m')$ symmetry when $\Gamma \simeq \mathbb{Z}_2 \times \mathbb{D}'_1 (2m')$.

Γ	Σ
$\mathbb{Z}'_n (n'), \ n \geq 1$	1
$\mathbb{D}'_1 (m')$	$\mathbb{D}'_1 (m')$
$\mathbb{Z}_n \wedge \mathbb{D}'_1 (nm'), \ n \geq 1$	$\mathbb{Z}_n \wedge \mathbb{D}'_1 (nm'). \ \mathbb{D}'_1 (m'), \ 1$
$\mathbb{D}_n \wedge \mathbb{D}'_1 ((2n)'m), \ n \geq 1$	$\mathbb{D}'_1 (m'), \ 1$

Table 1: Subgroups Σ of reversing symmetry groups Γ that are admissible for flows in the case that Γ is a finite subgroup of $O(2)$. Groups labeled with a prime indicate groups generated by a reversing symmetry. The notation between brackets follows conventions in crystallography (see the Appendix or [14] for more details).

Γ	Σ
$\mathbb{Z}_n (n), \ n \geq 1$	$\mathbb{Z}_n (n)$
$\mathbb{D}_n \ n \geq 1$	1

Table 2: Subgroups Σ of symmetry group Γ which are admissible for flows in the case that Γ is a finite subgroup of $O(2)$ (taken from [5]).

Now we look at the subgroups which fix a connected component of $\mathbb{R}^n - \tilde{L}$. The subgroup $\mathbb{D}'_1$ plays a distinguished role. The local construction of a $\mathbb{D}'_1$ symmetric periodic orbit is given in Fig. 1(a). The periodic orbit may be taken to be marginally stable. The constructions in Fig. 1 show that $\mathbb{D}'_1$ is admissible in case Γ is $\mathbb{D}'_1 (m')$ or $\mathbb{Z}_n \wedge \mathbb{D}'_1 (nm')$. In a similar way, one finds that $\mathbb{D}'_1$ is admissible in the case that $\Gamma = \mathbb{D}_n \wedge \mathbb{D}'_1 ((2n)'m)$.

In the case of $\mathbb{Z}_n \wedge \mathbb{D}'_1$ it follows from Theorem 1.3 that the subgroups $\mathbb{Z}_k$ ($k|n$, $k \neq 1$) and $\mathbb{Z}_k \wedge \mathbb{D}'_1$ ($k|n, k \neq 1, n$) are not admissible, while $\mathbb{Z}_n \wedge \mathbb{D}'_1$ is admissible. The construction of a $\mathbb{Z}_n \wedge \mathbb{D}'_1$ symmetric Liapunov stable attractor consists of a $\mathbb{D}'_1$ symmetric periodic orbit with $\mathbb{Z}_n$ symmetry, cf. Fig. 1(c).

The results are summarized in Table 1. Together with the results of [5] – which are summarized in Table 2 – they cover all admissible finite subgroups of $O(2)$ for flows in $\mathbb{R}^2$.

6   Symmetric ω-limit sets in fixed point subspaces

In this paper we have concentrated mainly on symmetric ω-limit sets with trivial instantaneous symmetry, i.e. in the notation of Section 1 we have studied the admissibility of subgroup (Σ, T) with $T = 1$. In this section we discuss how the results obtained in the previous sections extend to the case of symmetry groups (Σ, T) with $T \neq 1$, giving a reversible extension to the exposition in [5].

We first note T must be a normal subgroup of Σ and that T must be an isotropy subgroup (cf. [16]).[4] Moreover, if T contains a reversing symmetry then the ω-limit set must be a fixed point (and hence Σ must be an isotropy subgroup of Γ). The situation may become more interesting when T does not contain a reversing symmetry, and from now on we will assume that this is the case. Note that then $\bar{\mathrm{Fix}}(T)$ is an invariant subspace for f^t.

One obtains necessary conditions for admissibility of a pair of subgroups (Σ, T) by restricting to $\mathrm{Fix}(T)$, which is isomorphic to $\mathbb{R}^{n_T}$ for some $n_T \leq n$.

Let $N(T)$ denote the normalizer of T in Γ, i.e.

$$N(T) := \{\gamma \in \Gamma \mid \gamma \circ \tau \circ \gamma^{-1} \in T \text{ for all } \tau \in T\}. \tag{7}$$

Then a flow f^t on $\mathbb{R}^n$ with reversing symmetry group Γ restricts to a flow with reversing symmetry group $\Gamma_T := N(T)/T$ on $\mathrm{Fix}(T)$. Then it follows that if (Σ, T) is admissible for flows in $\mathbb{R}^n$ with reversing symmetry group Γ, then also $(\Sigma_T, 1)$ is admissible for flows in $\mathbb{R}^{n_T}$ with reversing symmetry group Γ_T, where $\Sigma_T := \Sigma/T$. However, this observation does not lead to an optimal result due to the presence of so-called *hidden symmetries* [7] which are elements γ of $\Gamma - N(T)$ with the property that $\gamma(\mathrm{Fix}(T)) \cap \mathrm{Fix}(T) \neq \emptyset$.

In order to determine the admissibility of a pair of subgroups (Σ, T) it suffices to take account of *hidden reflections*, which are elements $\gamma \in \Gamma - N(T)$ such that $\mathrm{Fix}(\gamma)$ intersects $\mathrm{Fix}(T)$ in a codimension one subspace. Following [5] we denote the set of $\gamma \in \Gamma$ for which $\mathrm{Fix}(\gamma)$ intersects $\mathrm{Fix}(T)$ in a codimension one subspace as K_T.

Hidden reflections in Γ give rise to invariant subspaces within $\mathrm{Fix}(T)$ separating $\mathrm{Fix}(T)$ in two connected components. In particular, hidden reflections due

---

[4]A group $T \leq \Gamma$ is called an isotropy subgroup of Γ if there exists a point $\mathbf{x} \in \mathbb{R}^n$ such that $T = \{\gamma \in \Gamma \mid \gamma(\mathbf{x}) = \mathbf{x}\}$.

to reversing symmetries give rise to invariant subspaces that are fixed pointwise by the flow f^t.

Lemma 6.1 *Let* $f^t : \mathbb{R}^n \mapsto \mathbb{R}^n$ *be a flow with reversing symmetry group* Γ *that is a finite subgroup of* $O(n)$. *Furthermore, let* $T \leq \tilde{\Gamma}$ *be an isotropy subgroup and* $\rho \in \Gamma - \tilde{\Gamma}$ *be a reversing symmetry that is not in* $N(T)$. *Suppose that* $\mathrm{Fix}(\rho)$ *intersects* $\mathrm{Fix}(T)$ *in a codimension one subspace, then for all* $\mathbf{x} \in \mathrm{Fix}(\rho) \cap \mathrm{Fix}(T)$ *we have* $f^t(\mathbf{x}) = \mathbf{x}$ *for all* t.

Proof. For all $\mathbf{x} \in \mathrm{Fix}(T)$ it follows from the fact that $T \leq \tilde{\Gamma}$, that $f^t(\mathbf{x}) \in \mathrm{Fix}(T)$ for all t. Using the fact that Γ is a finite subgroup of $O(n)$ we furthermore obtain that if $\mathrm{Fix}(\rho)$ intersects $\mathrm{Fix}(T)$ in a codimension one subspace and $\rho \notin N(T)$, then for all $\mathbf{x} \in \mathrm{Fix}(T) - \mathrm{Fix}(T) \cap \mathrm{Fix}(\rho)$, $\rho(\mathbf{x}) \notin \mathrm{Fix}(T)$.

Now consider some point $\mathbf{x} \in \mathrm{Fix}(\rho) \cap \mathrm{Fix}(T)$. For this point it follows that $f^t(\mathbf{x}) \in \mathrm{Fix}(T)$ and $f^t(\mathbf{x}) = \rho(f^{-t}(\mathbf{x}))$ for all t. As for all $\mathbf{x} \in \mathrm{Fix}(T) - \mathrm{Fix}(T) \cap \mathrm{Fix}(\rho)$, $\rho(\mathbf{x}) \notin \mathrm{Fix}(T)$, this implies that $f^t(\mathbf{x}) \in \mathrm{Fix}(\rho)$ for all t. In turn, this implies that $f^t(\mathbf{x}) = f^{-t}(\mathbf{x})$ for all t, and hence that $f^t(\mathbf{x}) = \mathbf{x}$ for all t. $\square$

Accordingly we define $\tilde{K}_T := K_T \cap (\Gamma - N(T) + N(T) \cap \tilde{\Gamma})$ which is precisely the subgroup of K_T giving rise to invariant codimension one subspaces within $\mathrm{Fix}(T)$.

In analogy to (4)-(5) one now defines

$$L_T := \bigcup_{\tau \in K_T} \mathrm{Fix}(\tau), \tag{8}$$

$$\tilde{L}_T := \bigcup_{\tau \in \tilde{K}_T} \mathrm{Fix}(\tau). \tag{9}$$

From here it is straighforward to show that the results of Theorem 1.2 and Theorem 1.3, as well as the results on symmetric periodic orbits in Section 2 and Section 3, Remark 3.1, translate directly to the context of ω-limit sets in fixed point subspaces by simply replacing $\mathbb{R}^n$ by $\mathrm{Fix}(T)$, L by L_T, and $\tilde{L}$ by $\tilde{L}_T$.

7 Discussion

We have given necessary and sufficient conditions for a subgroup to be admissible for flows in dimensions $n = 1, 2$.

We have also given restrictions on admissibility in all dimensions and shown ($n > 1$) that a subgroup containing no reversing symmetries is admissible for flows if and only if it fixes a connected component of $\mathbb{R}^n - L$. Moreover, we have given a full description of the symmetry groups for which admissibility can be realized by periodic orbits. It should be noted that in the case of subgroups Σ that do fix a connected componentof $\mathbb{R}^n - L$ and contain a reversing symmetry, our conditions are only shown to be necessary and not yet to be sufficient if Σ

is such that admissibility cannot be realized by periodic orbits. Further insight in the structure of marginally stable ω-limit sets other than periodic orbits will be needed to fully treat these cases.

We have concentrated on the symmetry properties of reversible flows. We address the issue of the symmetry properties of ω-limit sets of reversible maps in a forthcoming paper [11]. Similar questions could be asked about the symmetry properties of the ω-limit sets of symplectic maps and Hamiltonian flows with reversing symmetry groups. In this context it is worthwhile to note that the results on admissibility of Theorem 1.5 also apply in the case of Hamiltonian flows. Indeed, the constructions of periodic orbits in Section 5 can be done using solely Hamiltonian flows. For related work on the symmetries of periodic orbits for Hamiltonian flows see e.g. [6, 18].

In the Hamiltonian context (and indeed also in the reversible context) more consideration should be given to the notion of stability that is used. One could ask simply for the possible symmetries of ω-limit sets with no additional structure but to reflect the behavior of physical systems some form of stability is desirable. For Hamiltonian systems with no more than two degrees of freedom the concept of Liapunov stability is reasonable, e.g. in connection with the stability results implied by KAM-theory, cf. Moser [19]. However, especially in higher dimensional systems the phenomenon of Arnol'd diffusion makes topological notions less adequate. Thom [20] has proposed the measure theoretic notion of a *vague attractor* as a Hamiltonian analog of the attractors of differentiable dynamics (cf. also [1, p.583]). Similar considerations exist in the work of Melbourne [15] concerning the notion of *essential asymptotic* stability.

Another interesting line of investigation is the relation between symmetry and mixing properties in reversible (symplectic, Hamiltonian) systems. An elementary result in this direction is Theorem 4.1. Earlier work on equivariant systems has shown that topological mixing is incompatible with attractors having certain symmetries [17, 5]. As Hamiltonian systems have a natural invariant measure, relations between ergodic properties of the system and symmetry properties would be expected.

Acknowledgements

We thank Ian Melbourne for a helpful discussion and pointing out an error in an earlier version of this paper. We acknowledge the support of the University of Warwick and the University of Amsterdam for mutual visits in the early stages of this work. The research of JSWL is supported by the European Union through its Human Capital & Mobility research fellowship scheme (ERBCHBICT941533).

Appendix: notation

In this appendix we briefly explain our notation of reversing symmetry groups.

In the notation of reversing symmetry groups one needs to take account of the difference between symmetries and reversing symmetries of Γ. One way to do this is to adopt the notation used in crystallography in the description of magnetic groups, cf. [14]. In this notation, the finite subgroups of $O(2)$ are denoted in the following way. A single integer 'n' denotes a n-fold rotation, and the letter 'm' denotes a reflection (mirror). As all finite subgroups of $O(2)$ are generated by rotations and reflections they can just be denoted by their generators. Adopting the convention of denoting the cyclic subgroups of $O(2)$ that are generated by an n-fold rotation by $\mathbb{Z}_n$ and denoting the diheral subgroups of $O(2)$ generated by a reflection and an n-fold rotation by $\mathbb{D}_n$ we then have $n = \mathbb{Z}_n$ and $nm = \mathbb{D}_n$. In the case of reversing symmetry groups, in the crystallographic notation one labels reversing generators with a prime, e.g. m' or $2'm$.

Alternatively, we may write Γ as a group product $\Gamma = \mathcal{G} \cdot \mathcal{H}'$, with $\mathcal{G} \leq \tilde{\Gamma}$, $\mathcal{H}'$ is a cyclic group generated by a reversing symmetry (as indicated by the prime), and the product '$\cdot$' being either semi-direct '$\wedge$' or direct '$\times$'. Note that in case Γ contains a reversing symmetry that is an involution, we may choose $\mathcal{G} = \tilde{\Gamma}$ and $\mathcal{H} \simeq \mathbb{Z}_2$. This notation has the advantage that it shows how reversing symmetry groups can be viewed as cyclic extensions of symmetry groups. In the present discussion we will use primarily the latter notation, followed by the crystallographic notation between brackets, e.g. $\mathbb{D}_1 \times \mathbb{D}_1'$ $(2'm)$.

In denoting the symmetry of an ω-limit set it is in principle not relevant to distinguish between symmetries and reversing symmetries. It is however instructive to use a notation which shows the origin of the symmetry. One easily verifies the symmetry of the set by ignoring the primes in the notation.

References

[1] R. Abraham and J.E. Marsden. *Foundations of mechanics*. Addison-Wesley, Redwood City CA, 2nd edition, 1978. (updated 1985 printing).

[2] P. Ashwin and I. Melbourne. Symmetry groups of attractors. *Arch. Rat. Mech. Anal.*, 126:59–78, 1994.

[3] P. Chossat and M. Golubitsky. Symmetry-increasing bifurcation of chaotic attrcators. *Physica D*, 32:423–436, 1988.

[4] R.L. Devaney. Reversible diffeomorphisms and flows. *Trans. Am. Math. Soc.*, 218:89–113, 1976.

[5] M. Field. I. Melbourne, and M. Nicol. Symmetric attractors for diffeomorphisms and flows. Proceedings of the LMS, to appear, 1994.

[6] M. Golubitsky, J.-M. Mao, and M. Nicol. Symmetries of periodic solutions for planar potential systems. Proceedings of the AMS, to appear. 1994.

[7] M. Golubitsky, J.E. Marsden, and D.G. Schaeffer. Bifurcation problems with hidden symmetries. In W.E. Fitzgibbon III, editor, *Partial differential equations and dynamical systems*, volume 101 of *Research notes in mathematics*, pages 181–210. Pitman, San Fransisco, 1984.

[8] G.P. King and I.N. Stewart. Symmetric chaos. In W.F. Ames and C.F. Rogers, editors, *Nonlinear equations in the applied sciences*, pages 257–315. Academic Press, 1991.

[9] J.S.W. Lamb. Reversing symmetries in dynamical systems. *J. Phys. A: math. gen.*, 25:925–937, 1992.

[10] J.S.W. Lamb. *Reversing symmetries in dynamical systems*. PhD thesis, University of Amsterdam, 1994.

[11] J.S.W. Lamb and M. Nicol. On symmetric ω-limit sets in reversible dynamical systems. in preparation, 1995.

[12] J.S.W. Lamb and G.R.W. Quispel. Reversing k-symmetries in dynamical systems. *Physica D*, 73:277–304, 1994.

[13] S. Lefschetz. *Differential equations: geometric theory*, volume VI of *Pure and Applied Mathematics*. Wiley Interscience, New York, 2nd edition, 1963.

[14] A.L. Loeb. *Color and symmetry*. Wiley & Sons, New York, 1971.

[15] I. Melbourne. An example of a non-asymptotically stable attractor. *Nonlinearity*, 4:835–844, 1991.

[16] I. Melbourne. Generalizations of a result on symmetry groups of attractors. In J. Chadam and W. Langford, editors, *Pattern formation: symmetry methods and applications*, Fields Institute Communications. Am. Math. Soc., Providence RI, 1994. to appear.

[17] I. Melbourne, M. Dellnitz, and M. Golubitsky. The structure of symmetric attractors. *Arch. Rat. Mech. Anal.*, 123:75–98, 1993.

[18] J.A. Montaldi, R.M. Roberts, and I.N. Stewart. Existence of nonlinear normal modes of symmetric hamiltonian systems. *Nonlinearity*, 3:695–730, 1990.

[19] J. Moser. *Stable and random motions in dynamical systems*, volume 77 of *Annals of Mathematics studies*. Princeton University Press, Princeton NJ, 1973.

[20] R. Thom. *Structural stability and morphogenesis: an outline of a general theory of models*. Benjamin-Cummings, Reading MA, 1975.

Progress in Nonlinear Differential Equations
and Their Applications, Vol. 19
© 1996 Birkhäuser Verlag Basel/Switzerland

Symmetry Breaking in Dynamical Systems

Reiner Lauterbach*

Abstract

Symmetry breaking bifurcations and dynamical systems have obtained a lot
of attention over the last years. This has several reasons: real world applications give rise to systems with symmetry, steady state solutions and
periodic orbits may have interesting patterns, symmetry changes the notion
of structural stability and introduces degeneracies into the systems as well
as geometric simplifications. Therefore symmetric systems are attractive to
those who study specific applications as well as to those who are interested
in a the abstract theory of dynamical systems. Dynamical systems fall into
two classes, those with continuous time and those with discrete time. In this
paper we study only the continuous case, although the discrete case is as
interesting as the continuous one. Many global results were obtained for the
discrete case. Our emphasis are heteroclinic cycles and some mechanisms
to create them. We do not pursue the question of stability. Of course many
studies have been made to give conditions which imply the existence and stability of such cycles. In contrast to systems without symmetry heteroclinic
cycles can be structurally stable in the symmetric case. Sometimes the various solutions on the cycle get mapped onto each other by group elements.
Then this cycle will reduce to a homoclinic orbit if we project the equation
onto the orbit space. Therefore techniques to study homoclinic bifurcations
become available. In recent years some efforts have been made to understand
the behaviour of dynamical systems near points where the symmetry of the
system was perturbed by outside influences. This can lead to very complicated dynamical behaviour, as was pointed out by several authors. We will
discuss some of the technical difficulties which arise in these problems. Then
we will review some recent results on a geometric approach to this problem
near steady state bifurcation points.

1 Introduction

In this paper we would like to investigate the effects of symmetry breaking in
dynamical systems. One theme which seems to b closely related to it is the occurrence of structurally stable heteroclinic cycles in equivariant systems. There are
several well known examples, see for example GUCKENHEIMER & HOLMES [23]

---

*WIAS Berlin, Mohrenstr. 39, D-10117 Berlin, Germany

and the work of KRUPA and MELBOURNE [29, 30, 31, 40] which is directly related to heteroclinic cycles, and the papers by LAUTERBACH & ROBERTS [36] and also [35, 39]. Heteroclinic cycles may be generated in various ways, we distinguish

1. the "invariant plane case",

2. problems with higher codimension,

3. forced symmetry breaking.

Of course there is some relation with the topics of the very nice survey on heteroclinic cycles by KRUPA [29]. But our perspectives are somewhat different, our main emphasis are methods in equivariant systems, heteroclinic cycles are to be considered as a spin off. Let us briefly discuss the items mentioned before:

1. With the invariant plane case we mean a scenario for the occurrence of heteroclinic cycles in fixed point spaces of subgroups which was first described by Melbourne, Chossat & Golubitsky [41]. We shall see later the example by GUCKENHEIMER & HOLMES [23], which fits very nicely into this framework, where the group $\mathbf{T} \oplus \mathbf{Z}_2$ acts irreducibly on $\mathbf{R}^3$. However in many cases such a situation occurs when the group action is reducible, compare ARMBRUSTER & CHOSSAT [2, 8], CHOSSAT & GUYARD [10, 24].

2. In systems without symmetry it is well known that complicated dynamical behaviour can occur if the system under consideration has higher codimension. Of course the same is true for equivariant systems, however, since symmetric systems automatically have some degeneracies, it becomes increasingly difficult to study problems with higher codimension. We present a example due to LAUTERBACH & SANDERS [38], where invariant theory has been used to study a problem with topological codimension 3. Again, a heteroclinic cycle occurs. This cycle is constructed for the equation on the orbit space. Therefore the issue of lifting it back to the full space becomes important. With respect to this problem finite and infinite groups show a different behaviour. In the finite case it is clear that a heteroclinic cycle lifts to a heteroclinic cycle, which might involve more equilibria and more heteroclinic connections than the one on the orbit space, but in principle we find the same object. However, in the case of a continuous group this changes. Even an equilibrium does not necessarily lift to an equilibrium but to a so called relative equilibrium. The heteroclinic connections just connects two group orbits. There is another important difference between the discrete and the continuous case: the behaviour with respect to perturbations which do not respect the (full) symmetry. This leads to our last topic.

3. We speak of forced symmetry breaking when we perturb the system with terms having less symmetry than the original problem. We shall see that this is a natural problem from the application point of view. It leads to

interesting dynamical effects and again heteroclinic cycles come up. Our techniques evolved from the work of LAUTERBACH & ROBERTS [36, 37]. We shall look at some questions concerning group theoretic conditions for the existence of heteroclinic cycles and moreover how to prove the existence of them for PDE's when the symmetry has been slightly perturbed.

The example 3.3 was found in a discussion with Karin Gatermann, Frédéric Guyard and Matthias Rumberger.

2 Symmetric dynamical systems – why?

It is well known that many physical systems can be modeled in terms of dynamical systems. just consider the classical problems of mechanics. In the course of the last decades the applicability of dynamical system theory has widely expanded. Systems in chemistry, biology, economy and other sciences were translated into mathematical language and can be written as dynamical systems. In the course of this translation process many simplifying and abstracting assumptions are being made. In many cases these abstractions and simplifications lead to additional structures in the equations, which were not present in the original problem. One of those structures could be the occurrence of symmetries. However, symmetries do not only come into the game by the process of mathematical idealization. but can also be a very natural ingredient of the problem under consideration. Experiments can take place in a symmetric surrounding. the nature often finds beautifully symmetric forms or patterns. From this we see that symmetries can be a natural context for the study of real world phenomena. By now it is well known that the steady state or periodic solutions of a symmetric system can reflect less symmetry than we find originally in the system. This has been observed a long time ago, we usually refer to this as *spontaneous symmetry breaking*. see for example SATTINGER [45]. In contrast to this we can also imagine situations where a system, on the first glance, has a certain symmetry, but a closer look reveals that in fact some of these symmetries are present only approximately. Therefore. the full problem has less or no symmetry whatsoever. A typical example would be a problem in geophysics where the earth, in the first approximation, has the symmetry of a ball, if we look more closely we observe the flattening of the poles. reducing the symmetry of the ball to the symmetry of a circular disc, in the group theory language, which we will adopt. it has symmetry $O(2)$. Taking the rotation into account reduces the symmetry to the group $SO(2)$ and finally looking from very a close perspective, the typical human approach, we see no symmetry at all. Nevertheless we expect that a decent theory takes into account that we are close to a symmetric problem. We call this type of problem *forced symmetry breaking*. In fact in the example of our planet we described a *hierarchy* of forced symmetry breakings. As we shall point out the problem of forced symmetry breaking leads to very severe conceptual and computational difficulties. A global understanding of these problems is not in reach. However it has been noted and we shall see that this can lead to extremely rich and difficult dynamical behaviour.

Symmetries can be described in the language of group theory. The most obvious way of doing it is to consider a domain (or a compact embedded manifold without boundary) $\Omega \subset \mathbf{R}^n$ and its symmetry group G_Ω defined by

$$G_\Omega = \{A \in \mathbf{O}(n) \,|\, A\omega \in \Omega \,\forall \omega \in \Omega\}. \tag{1}$$

If the mathematical formulation of the problem leads to a partial differential equation of the type

$$\frac{\partial u}{\partial t} + Lu = f(u, \lambda) \tag{2}$$

with a sectorial operator L and f sufficiently smooth and "reasonable" boundary conditions, then (2) defines a semidynamical system on $L^2(\Omega)$ (or $H^{1,2}(\Omega)$). The group G_Ω acts on function spaces X (for example $X = L^2(\Omega)$ or a Sobolev space $W^{k,p}(\Omega)$ over Ω) simply by

$$G_\Omega \times X : (\gamma, u) \mapsto \gamma u, \ \gamma u(\omega) = u\left(\gamma^{-1}\omega\right). \tag{3}$$

We assume that the linear operator is *equivariant* with respect to this action, i.e.

$$L(\gamma u) = \gamma(Lu), \text{ for all } u \in X, \ \gamma \in G_\Omega \tag{4}$$

This assumption is always satisfied if L is the Laplace operator. Smaller groups allow some more general partial differential operators. In fact a reasonable modeling should lead to G_Ω equivariant operators.

We say that equation (2) is equivariant with respect to G_Ω if

$$-L(\gamma u) + f(\gamma u, \lambda) = \gamma(-Lu + f(u, \lambda)) \tag{5}$$

for all $u \in X$ and all $\gamma \in G_\Omega$ and if the boundary conditions are invariant under the action of G_Ω on u. Typical examples are the buckling of spheres, where the space X is a function space over the 2-sphere and no boundary condition are present or the spherical Bénard problem, where Ω is a spherical annulus and we have boundary conditions on the inner and the outer sphere. Let us just recall that the Bénard problem is to describe a fluid flow between two infinite plates, where the temperature on the plates is spatially constant and different. let T_l denote the temperature at the lower plate and T_u denote the temperature at the upper plate we requite $T_l > T_u$. It is known that if the difference exceeds a certain value some interesting states occur. In the spherical Bénard problem we consider a fluid confined between two concentric spherical shells with inner and outer temperatures T_i and T_o, respectively. If $T_i - T_o$ is sufficiently large. we again observe new and interesting states. The Navier equations, describing these problems. are equivariant with respect to the Euclidean group in the planar case and with respect to $\mathbf{O}(3)$ in the spherical case.

Observe that the equivariance was assumed for the operator L, the nonlinearity is automatically equivariant if it does not explicitly depend on the spatial variable ω. Therefore adding in small terms which are spatially non constant leads to forced symmetry breaking. A typical scenario for the Bénard problem is to assume a small deviation from spatially homogeneous temperatures on the boundary, which can be rewritten in terms of small perturbations in the interior with explicit space dependence. We will come back later to these issues. Before we go on, we collect some simple properties of dynamical systems with symmetry, which are easily verified.

Some simple facts

1. If $u(t)$ is a solution, then $\gamma u(t)$ is again a solution.

2. For $u \in X$ let H_u denote its isotropy subgroup, i.e. $H_u = \{\gamma \in G_\Omega \,\|\, \gamma u = u\}$. Then $H_{\gamma u} = \gamma H_u \gamma^{-1}$.

3. Along trajectories the isotropy is not decreasing. i.e. if $0 < s < t$ then $H_{u(s)} \subset H_{u(t)}$. If backward uniqueness holds, then we have equality.

The main issues to be studied are to characterize the symmetry type of bifurcating solutions, structural stability in equivariant systems and global behaviour. For the local questions singularity theory proved to be very successful, compare GOLUBITSKY, STEWART & SCHAEFFER [22]. The main ingredient in a local theory are the center manifold theory or Lyapunov–Schmidt reduction. It is important to note that these tools carry over directly to the equivariant context. We just recall these results. For this we need some group theory language. A *representation* of a group G is a homomorphism ρ into $\mathbf{Gl}(n)$ for some n. We also say that G acts on $\mathbf{R}^n$. If there is a continuous homomorphism into the bounded linear operators on a Banach space we speak about an *infinite dimensional* representation. Actions on the function space as described before are such infinite dimensional representations. A subspace U of $\mathbf{R}^n$ (or X) is called invariant if for all $\gamma \in G$ $\rho(\gamma)u \in U$ for all $u \in U$. A representation is called reducible if it has a nontrivial invariant subspace, otherwise it is *irreducible*. A representation is absolutely irreducible if the only equivariant linear mappings are scalar multiples of the identity. Two representations ρ_1, ρ_2 on spaces V_1, V_2 of a group G are called *equivalent* if there is an isomorphism $\tau : V_1 \to V_2$ such that for all $\gamma \in G$ we have $\tau \circ \rho_1(\gamma) = \rho_2(\gamma) \circ \tau$ as mappings $V_1 \to V_2$. For any finite group there are up to equivalence only finitely many irreducible representations, any representation can be written as a sum of irreducible ones. A similar statement is true for infinite dimensional representations of compact Lie groups. An important tool is the *character* of a representation. It is a function $\chi : G \to \mathbf{C}^* = \{z \in \mathbf{C} \mid |z| = 1\}$, defined by

$$\chi(g) = \mathrm{tr}(\rho(g)). \tag{6}$$

For an introduction into character theory, see for example SERRE [47]. A very nice tool for doing actual computations with characters is the program GAP [18].

Some more facts

1. Any (closed) invariant subspace has a (closed) invariant complement.

2. Any absolutely irreducible representation is irreducible. Over **C** the reversed statement is also true. The group **SO**(2) acts (by rotations) irreducibly on **R**^{2}, but not absolutely irreducibly.

3. The kernel of an equivariant linear mapping is invariant.

4. Consider $L^2(G)$ to be space of square integrable (with respect to Haar measure, see HEWITT & ROSS [25]) complex valued functions. The characters of all irreducible representations form a complete orthonormal system.

GOLUBITSKY, STEWART & SCHAEFFER [22] show that in generic one parameter families of linear equivariant mappings on **R**^{n} the kernel is either trivial or absolutely irreducible. From this it follows that in one parameter families of equivariant, finite dimensional bifurcation problems the kernels are generically absolutely irreducible representations of G_Ω. A similar theorem is true for one parameter families of sectorial operators with compact resolvent, see LAUTERBACH [34]. However this does not imply that it is sufficient to study only one parameter bifurcations with absolutely irreducible kernels. For an example, see the section on local bifurcations.

Theorem 2.1 *Let X be a Banach space, $F : X \times \mathbf{R} \to X$ be G-equivariant and sufficiently smooth. Assume $F(0, \lambda) = 0$ for all $\lambda \in \mathbf{R}$ and $D_x F(0,0)$ has a nontrivial kernel K. Let $f : K \times \mathbf{R} \to K$ denote the mapping obtained via a Lyapunov–Schmidt reduction, then f is G-equivariant.*

For the center manifold reduction we note, that if all choices are made in a reasonable way, then the equation on the center manifold is G–equivariant. Just choose a cut–off which makes the center manifold unique and apply the group elements to get another center manifold, which by uniqueness coincides with the first one. This sets the stage for a local theory, which of course if well known. For further reference and for setting up the notation we recall some of the fundamental results in the next section. First, however, we need some more notation. If G acts on the space V, then for $v \in V$ the *orbit* is given by

$$\mathcal{O}(v) = \{\gamma v \mid \gamma \in G\}. \tag{7}$$

The set of all orbits is denoted by V/G. To each point we associate the *orbit type* as the class of subgroups $[H]$ conjugate to the isotropy subgroup H of v, i.e.

$$[H] = \{\gamma H \gamma^{-1} \mid \gamma \in G\}, \tag{8}$$

where H is the isotropy subgroup of v. From our previous observation that the isotropy subgroup of γv is given by $\gamma H \gamma^{-1}$ we conclude that the orbit type is constant along orbits, which justifies this nomenclature.

3 Some Aspects of Local Bifurcations

The last section prepared to consider finite dimensional equivariant dynamical systems, which we consider to be the reduced equations obtained via the equivariant center manifold reduction. There are several methods to study and classify these problems. In [22] an equivariant singularity theory was developed, however in practical computations it is often difficult to get to satisfactory answers. In BUZANO, GEYMONAT & POSTON [6] this theory is applied to the low dimensional representations of the dihedral group D_n, which are used to study the buckling of thin rods with a cross–section of a regular n-gon. An attempt to give a classification of G–equivariant problems by their codimension (in the sense of contact equivalence, see [22]) is made in GATERMANN & LAUTERBACH [19]. Here computer algebra is used to construct G–equivariant bifurcation problems, ordered by codimension. The calculations use Poincaré–series and give lists of generating elements for the ring of invariant functions and the module of equivariant mappings. However, even here one cannot expect to treat large groups or high dimensional representations. A second approach is based on isotropy subgroups and the geometry of fixed point subspaces. Let V be a finite dimensional vector space and $\rho : G \to \mathrm{Gl}(V)$ be a representation of a group G. Let $f : V \times \mathbf{R} \to V$ be G–equivariant, i.e. $f(\rho(\gamma)v, \lambda) = \rho(\gamma)f(v, \lambda)$ and consider the differential equation

$$\dot{v} = f(v, \lambda). \tag{9}$$

A subgroup $H \subset G$ is called an isotropy subgroup, if there exists some $v \in V$, $v \neq 0$, such that H is the isotropy subgroup of v, i.e.

$$H = \{\gamma \in G \,|\, \rho(\gamma)v = v\}. \tag{10}$$

An isotropy subgroup H is called *maximal* if it is a maximal element in the partial ordered set of all isotropy subgroup. It is easy to see that subgroups with one dimensional fixed point space are always maximal. Due to the so called equivariant branching lemma these groups play an important rôle, sometimes they are called *axial* subgroups. The following result goes back to VANDERBAUWHEDE [48] and CICOGNA [12]. The present formulation is due to IHRIG & GOLUBITSKY [28].

Theorem 3.1 *Let v be an absolutely irreducible representation of G, then we have*

1. *$f(0, \lambda) = 0$,*

2. *$D_v f(0, \lambda) = c(\lambda)\mathbf{1}_V$ for some scalar $c(\lambda)$.*

Suppose $c(0) = 0$, $c'(0) \neq 0$ and H is a maximal isotropy subgroup of G, then there exists a bifurcating branch of solutions having isotropy H.

The proof of this important result is very simple. A similar theorem holds in the context of Hopf bifurcation, however it is much less trivial. Again we just recall the statement. It is due to GOLUBITSKY & STEWART [21]. It describes the bifurcation of periodic solutions with spatial–temporal symmetry. Bifurcation of periodic solutions can be studied using a Lyapunov–Schmidt reduction in a space of 2π periodic functions. The equations obtained by this reduction do not only allow G–equivariance, but due to the fact that the problem is invariant under time shifts, the group $G \times S^1$ acts on the kernel and the bifurcation equation is equivariant with respect to this action. Note that this action is not absolutely irreducible, in general it is an irreducible sum of two absolutely irreducible representations. The set of commuting matrices is isomorphic to **C**. Such representations are called *simple* ([22]). In fact a genericity result similar to the one that kernels at an eigenvalues zero are generically absolutely irreducible one shows that invariant with pure imaginary eigenvalues lead to generically to simple representations, see [22]. A subgroup H of $\Gamma = G \times S^1$ is called **C**-*axial* if it has a two dimensional fixed point space.

Theorem 3.2 *Let the trivial solution of equation(17) loose its stability through a pair of conjugate complex eigenvalues which cross the imaginary axis with nontrivial speed, and suppose that the representation of Γ on the kernel is simple. Then, if $H \subset \Gamma$ is a subgroup with $\dim\mathrm{Fix}(H) = 2$, then a branch of periodic solutions with isotropy H bifurcates.*

These results have been used to classify bifurcating solutions in many applications. Especially the equivariant Hopf theorem leads to very interesting patterns in coupled oscillators. One assumes that there are n identical oscillators and there are various possibilities of coupling. In general one has the internal symmetry of the single oscillator Σ and a group S of permutations acting on the set of oscillators. Depending on the coupling one can have a direct product structure $\Sigma \times S$ or a wreath-product $G \wr S$. For both types of coupling DIONNE, GOLUBITSKY & STEWART [14, 15] give a characterization of the **C**–axial subgroups. Although further solutions could occur, this gives a broad class of solutions leading to interesting patterns.

Example 3.3 *Now we would like to discuss an example which shows that although generically the kernels at bifurcation points are absolutely irreducible representations, it might be important to understand also bifurcations with non absolutely irreducible group actions. Look at the five dimensional absolutely irreducible representation of A_5, the group of even permutations of five letters, also known as the symmetry group of the icosahedron. The isomorphism is given by viewing the symmetry group of the icosahedron as a permutation group on its subgroups of order 12, which are isomorphic to the symmetry group of a tetrahedron. In this representation the 2-Sylow subgroup D_2 of A_5 (see LANG [32] for the notion of Sylow groups) of A_5 is a maximal isotropy subgroup and has a two dimensional fixed point space. The normalizer of D_2 is the group A_4 (isomorphic to the symmetry group of the tetrahedron) and it acts on $\mathrm{Fix}(D_2)$. This is a general fact that the normalizer of a subgroup H acts on the fixed point space of H. Since D_2 acts* *trivially on its fixed point space, the effective action is the action of* $\mathbf{Z}_3$. *This action is irreducible but not absolutely irreducible. The bifurcation within this fixed point subspace is not the generic* $\mathbf{Z}_3$ *bifurcation in* $\mathbf{R}^2$. *The later one would allow for a Hopf bifurcation which is not possible in the* A_5-*context. To compute the actual bifurcations in* Fix(D_2) *requires an understanding of the module of equivariant mappings. These computations touch the limitations of todays workstations.*

A different method of studying equivariant dynamical systems is to project these equations onto the orbit space. This can be combined with various other techniques. For continuous groups it leads to a reduction of dimension. Let us briefly describe the essential features.

Definition 3.4 *Let* G *act on a space* V *(or on a manifold, topological space etc.). The orbit space is given by* $V/\sim$, *where* $\sim$ *is the equivalence relation,* $v_1 \sim v_2$ *iff* v_1, v_2 *are on the same group orbit. This definition is equivalent to the previous definition of* V/G.

If V is a Hausdorff space and if G is compact, then V/G is again Hausdorff. In general V/G is not a manifold, but stratified, where each stratum is a smooth manifold. Since the orbit type is constant along orbits, we associate to each element in V/G an orbit type. The strata for the smooth stratification are given by orbits of the same orbit type. That this stratification is finite follows from the finiteness of orbit types which is part of the following theorem.

Theorem 3.5 *Let* G *be a compact Lie group,* ρ *a representation on a vector space* V. *Then we have:*

1. *There are only finitely many orbit types.*

2. *There is a uniquely determined minimal obit type, called the* principal orbit type.

3. *The set* $P_G(V)$ *of points of minimal orbit type is open and dense in* V.

4. *If* G *is connected, then* $P_G(V)$ *is connected.*

Proof: See BREDON [5]. □

SCHWARZ [46] shows that there is a C^∞-structure on the orbit space, that each smooth, stratum preserving vector field on V/G lifts to a G-equivariant vector field on V. This will be useful later. At the moment we are interested in projecting vector fields onto the orbit space. This was used in LAUTERBACH & SANDERS [38] to study a certain degenerate case of $\mathbf{O}(3)$-equivariant bifurcations. As we shall see later the orbit space is also an important tool for the study of forced symmetry breakings. In order to get differential equations on the orbit space, we recall some facts from invariant theory. As we have seen, to each action of a group G on a space V we have a natural action of G on function spaces over V. This defines an action on polynomials on V. A polynomial is called *invariant* if it is fixed under this action. Obviously the invariant polynomials form a ring.

Definition 3.6 *Let $\mathcal{R}_V$ denote the ring of all G-invariant polynomials. $\mathcal{M}_V$ is the module of G-equivariant polynomial mappings from V into itself.*

It is a fundamental result of HILBERT that $\mathcal{R}_V$ is a finitely generated algebra, as well as $\mathcal{M}_V$ is finitely generated over this ring. From here it follows that equivariant equations can be rewritten as

$$\dot{v} = f(v,\lambda) = \sum_{j=1}^{t_0} f_j(\pi_1(v), \pi_2(v), \ldots, \pi_s(v), \lambda) e_j(v). \tag{11}$$

The finiteness result together with this rewriting of the equation and the following lemma constitute the basis for the reduction to the orbit space.

Lemma 3.7 $\mathcal{R}_V$ *separates orbits, i.e. for two different orbits τ_1, τ_2 there exists an invariant polynomial p such p is 0 on τ_1 and 1 on τ_2.*

Proof: It is easy to construct a continuous function which is 0 on τ_1, and 1 on τ_2. From Weierstraß approximation theorem we know that for each $\varepsilon > 0$ there is a polynomial q, such that $q(v) > 1-\varepsilon$ for all $v \in \tau_2$ and $q(v) < \varepsilon$ for $v \in \tau_1$. Averaging this polynomial over the group gives p (up to a multiplicative constant).   □

Together with Hilbert's finiteness theorem this tells us, that there are finitely many invariant polynomials $\pi_1, \ldots, \pi_s$ which generate the algebra $\mathcal{R}_V$ and which separate orbits. Therefore the map

$$\Pi : V \to \mathbf{R}^s : v \mapsto (\pi_1(s), \ldots, \pi_s(v)) \tag{12}$$

gives rise to a continuous and injective mapping from $V/G \to \mathbf{R}^s$. Therefore the range of Π is a homeomorphic image of V/G and we can use the map Π to derive a differential equation on the orbit space. This can be done as follows: choose coordinates on $\mathbf{R}^s$ using $\pi_1, \ldots, \pi_s$ and compute for $i = 1, \ldots, s$

$$\frac{\partial \pi_i}{\partial t} = <\nabla \pi_i, \dot{v}> = <\nabla \pi_i, f(v,\lambda) = \sum_{j=1}^{t_0} f_j(\pi_1, \ldots, \pi_s, \lambda) <\nabla \pi_i, e_j> . \tag{13}$$

Now, we have reduced the computation of the equation on the orbit space to a computation of the scalar products $<\nabla \pi_i, e_j>$. Sometimes it is possible to compute these scalar products without knowing explicitly the functions π_i, e_j. In any case it is possible to derive these equations automatically from the invariants and equivariants. The explicit expressions for the invariants and equivariants are often very cumbersome, while the reduced equation has a reasonable form. However, it might be difficult to give a precise interpretation of the results on the orbit space to the full equation.

A *relative equilibrium* is a solution $v(t)$ which is part of a group orbit. Steady states on the orbit space are relative equilibria for the original equation, but need not be equilibria. In a similar fashion periodic orbits on the orbit space are *relative periodic orbits* for the full equation.

There are a few points to be observed. It is in general not true, that $s < \dim(V)$. This seems to indicate a gain in dimension rather than a loss. However it can be shown that the maximal number of algebraically independent generators is $r = \dim(V) - \dim G \leq \dim(V)$. In the reduction process described before one gets r differential equations and $s - r$ algebraic equations. Therefore the reduction to the orbit space leads to an algebro–differential equation, a feature which has not yet been exploited.

The reduction to the orbit space gives some extra tools, which we want to describe by the way of an example, compare [38]. If we look at the local bifurcation for the natural action of the group D_3 on $\mathbf{R}^2$ (which is the same theory as the local bifurcation theory for the 5-dimensional irreducible representation of $\mathbf{O}(3)$, see [22]) then one finds two algebraic independent generators of the algebra on invariant functions, i.e. $r = s = 2$, the orbit space is a subset of $\mathbf{R}^2$. Since one of the invariants can always be chosen as $\pi_1(v) = \|v\|^2$, the range of Π is in the right half plane. It is easy to check, that the invariants can be chosen of degree 2 and 3. Up to a scaling of the invariants one has the following lemma

Lemma 3.8 *The range of* Π *is equal to* $\Delta(\pi_1, \pi_2) \geq 0$ *with* $\Delta(\pi_1, \pi_2) = \pi_1^3 - 27\pi_2^2$.

Proof: See [38], or compute the invariants and check.   □

Now this function Δ satisfies a differential equation, which can be easily seen:

$$\dot{\Delta} = 6f_1(\pi_1, \dots, \pi_s, \lambda)\Delta. \tag{14}$$

It is a general fact, that the algebraic relations describing the boundary of the range of Π give rise to differential equations. In our D_3 example this equation can be used to derive *global* information in a bifurcation problem with topological codimension 3, see [38].

Example 3.9 *Let us look at a* D_3-*equivariant problem on* $\mathbf{R}^2$ *(or at the five dimensional absolutely irreducible representation of* $\mathbf{O}(3)$*) and let us write the equation in form of equation (11). For both cases* D_3 *or* $\mathbf{O}(3)$ *this equation has the same form, this is why these two theories are the same. We choose* $f_1(\pi_1, \pi_2, \lambda) = \lambda + B_1\pi_2$ *and* $f_2(\pi_1, \pi_2, \lambda) = A_2\pi_1$. *Using the notion of contact equivalence this problem has* C^∞-*codimension 5 and topological codimension 3. For the computation of these codimensions one can follow the line of [6] or one uses a direct computation to compute the relevant modules using the Gröbner package in some computer algebra system. We do not attempt to describe the behaviour for an unfolding, we just describe an interesting region in parameter space. From the computation of the codimension we find an unfolding of the form*

$$
\begin{aligned}
f_1 &= \lambda + a_1\pi_1 + \varepsilon_1\pi_2 \\
f_2 &= c + a_2\pi_1 + \varepsilon_2\pi_2.
\end{aligned}
$$

where ε_1 is near B_1, a_2 is near A_2 and c, ε_2 and a_1 are close to 0. Choosing the parameters such that $a_1\varepsilon_2 - a_2\varepsilon_1 > 0$ and $12a_1^2 + a_2c < 0$ is satisfied, one finds

Theorem 3.10 *At* $\lambda = 0$ *the trivial solution* $v = 0$ *looses stability and a transcritical bifurcation takes place. In the orbit space, we find a secondary bifurcation to steady states and tertiary Hopf branch.*

The proof of this result follows classical lines and is omitted here. Concerning the global behaviour of the branch of periodic solutions we use the global Hopf bifurcation theorem by ALEXANDER & YORKE [1]. This theorem tells us that one of the following is true

1. the amplitude goes to infinity, or

2. the the branch is unbounded in parameter space, or

3. the period goes to infinity, or finally

4. the the closure of the connected component of periodic solutions emanating at our Hopf point contains another Hopf point.

In our example the third alternative is true, in fact we can show

Theorem 3.11 *Along the connected component of periodic solutions in the orbit space bifurcating at the tertiary Hopf point the minimal period goes to infinity, in fact the closure contains a heteroclinic cycle with two equilibria. The two equilibria have isotropy* $\mathbf{Z}_2$, *there is one connection in the space with isotropy type* $\mathbf{Z}_2$, *one connection in the space with trivial isotropy.*

Proof: We prove the part where we have to use the differential equation for Δ. This is the part where we show that the amplitude of the periodic solutions in the connected component containing the Hopf point in its closure does not approach infinity. Suppose it did. From our assumption concerning f_1 we conclude that there is a $\pi_1^0 > 0$ such that f_1 is of one sign in the domain $\pi_1 > \pi_1^0$. Due to the equation for Δ we conclude that Δ is a Lyapunov function in the domain where $\pi_1 > \pi_1^0$. Hence there exists a number $c_0 > 0$ such that the domains $\Delta > c > c_0$ are positively or negatively invariant (depending on the sign of f_1). From the fact that the curves $\Delta = 0$ and $\Delta = c > 0$ are asymptotically equal, we find that the amplitudes of periodic solutions have to be uniformly bounded. $\qquad\square$

Remark 3.12   1. *Here the result for the* $\mathbf{O}(3)$ *case looks slightly different, the isotropy* $\mathbf{Z}_2$ *is replaced by* $\mathbf{O}(2)$ *and the trivial isotropy by* D_2 *(which is the principal isotropy in this example)*

2. *Since* D_3 *is a finite group it is easy to see that equilibria in the orbit space correspond to equilibria in the state space and periodic orbits correspond to periodic orbits. In the case of the continuous group* $\mathbf{O}(3)$ *this also true but less trivial to see. This property is very specific to the case of the 5-dimensional irreducible representation of* $\mathbf{O}(3)$.

3. Another difference between the two cases occurs if we allow perturbations which destroy the equivariance property, such that the perturbed equation is only equivariant with respect to a subgroup. In the D_3-case the periodic solutions will lead to periodic solutions in the perturbed equation near the original periodics. In the continuous case a very complicated dynamical behaviour is expected near the manifold of equilibria. This has not yet been completely studied, however it is clear that this question leads to interesting topological and dynamical problems.

4   Forced symmetry breaking

As before we consider a domain or an embedded compact manifold without boundary $\Omega \subset \mathbf{R}^n$ with a partial differential equation on Ω of the form

$$\frac{\partial u}{\partial t} + Lu = f(u, \lambda), \tag{15}$$

where L is a sectorial operator and f is sufficiently smooth. In the case of a domain Ω we also require boundary values, let us say of the form

$$Bu = \phi,$$

where B is a boundary operator of the form

$$Bu = au + b\frac{\partial u}{\partial n}, \tag{16}$$

with functions a, b, ϕ on the boundary $\partial\Omega$ and n denotes the outer normal unit vector. As said before, if a, b, ϕ are constant (homogeneous boundary conditions) and if the coefficients of L do not depend explicitly on $\omega \in \Omega$ then we have a G_Ω equivariant equation. Forced symmetry breaking may occur through several mechanism which may differ in the physical mechanism, but which can lead to similar mathematical problems. We classify according to the mathematical effects.

Let us first mention some physical situations as perturbations of boundary conditions, perturbations by adding some terms depending explicitly on the state variable or introducing drift. The first example might be physically the most important one, when outside influences perturb the boundary conditions. We might think of non homogeneous temperature distribution in the spherical Bénard problem or to speak about more recent problems the phase locking of high frequency pulses in DFB-lasers to periodic outside signals, compare WÜNSCHE, BANDELOW, FEISTE ET AL. [4, 16, 42].

Associated to Ω we have $G = G_\Omega$ the symmetry group of Ω. We assume that the equation (15) and the boundary operator B, both are equivariant under the group action of G. Let H be a (closed) subgroup of G. The forced symmetry breaking perturbations which we want to study in our mathematical framework are the following

1. Add a function of the form $\varepsilon h(x, u)$ in the equation. More specifically we add terms of the form $h_1(x)g(u)$, where g is a (non-)linear function of u and h_1 is invariant under the action of H. We refer to these perturbations as *class I* perturbations.

2. Add a term of the form $h_1(x) < e(x), g(u)\nabla u >$, where h_1 is H invariant and e is an H–equivariant mapping $\Omega \to \mathbf{R}^n$. These perturbations will be called class II perturbations.

Note that these two classes of perturbations of the symmetry lead to H–equivariant equations. If we perturb in a similar fashion the boundary operator B or the prescribed function ϕ, we can reduce these perturbations to perturbations of the equation on Ω. This does not remain true if we perturb the type of boundary condition, like Dirichlet to mixed or Neumann to mixed, by adding small terms. Then some functional analytic problems arise which have not yet been solved in general, compare ASHWIN & MEI [3].

A specific situation arises when we look at the effects of forced symmetry breaking near a bifurcation point. We begin with the discussion of a steady state bifurcation point. In principle we have different ways to proceed. We could compute the effect of the perturbation on the center manifold and then discuss the finite dimensional problem. There are two main difficulties involved with this approach. First of all symmetry in general leads to multiple eigenvalues. Perturbing the symmetry may split (some of) the eigenvalues and we have several bifurcation points. If we discuss the behaviour near the full set of bifurcation points we run into extremely messy calculations. Near the bifurcation point we will see the branches as they come out of the perturbed points, further away, when the effects of the forced symmetry breaking become smaller (compared with the hyperbolic structure of manifolds of equilibria) we see a slightly distorted picture of the original bifurcation problem. There is very complex recombination of branches and lots of secondary bifurcations going on. Even in the simple example of a spherical problem with the $\ell = 2$–representation on the kernel a perturbation to axisymmetric symmetry leads to almost unsurmountable computational difficulties. This may reflect the following fact. If we consider the G-equivariant problem within the class of H–equivariant problem we could use a singularity theory approach in the sense of [22] to classify these problems. However if G is not a finite group and if the dimension of $\dim(G/H)$ (as a homogeneous manifold) exceeds 0, then any G-equivariant bifurcation problem has codimension infinity, compare GOLUBITSKY & SCHAEFFER [20]. Therefore we study a more specific question, than describing the perturbed flow in a complete neighborhood of the bifurcation point. In order to describe the principal ideas of our approach, let us consider a G–equivariant ODE

$$\dot{x} = f(x, \lambda) \tag{17}$$

and suppose a H–equivariant vector field $h(x)$ is given. Consider

$$\dot{x} = f(x, \lambda) + \varepsilon h(x). \tag{18}$$

Furthermore suppose x_0 is a steady state solution of (17). Then the orbit

$$\mathcal{O}(x_0) = Gx_0$$

is contained in the set of equilibria of (17). Let K denote the isotropy of x_0. Then $\mathcal{O}(x_0)$ is diffeomorphic to G/K. We impose the following hypotheses

H1) $\mathcal{O}(x_0)$ is isolated in the set of equilibria.

H2) $\mathcal{O}(x_0)$ is a normally hyperbolic manifold.

HIRSCH. PUGH & SUB [26] give a detailed theory of normally hyperbolic manifolds. Here, we just need, that normally hyperbolic invariant manifolds are persistent, i.e. if M is such a manifold, then for any vector field sufficiently close to (17) there exists a unique invariant, normally hyperbolic manifold $\tilde{M}$ near M which is diffeomorphic to M, i.e. there exists a diffeomorphism

$$\Psi : M \rightarrow \tilde{M}. \tag{19}$$

For a manifold of equilibria to be normally hyperbolic it is necessary and sufficient that at each point $x \in M$ the linearization of the vector field has precisely $\dim(M)$ eigenvalues on the imaginary axis and all the others off the imaginary axis. Applying this concept to our present situation. LAUTERBACH & ROBERTS [36] have shown, that for each sufficiently small H–equivariant perturbation h (18) of (17) and for each normally hyperbolic manifold of equilibria there exists a unique invariant manifold $\tilde{M}$ for (18) near $\mathcal{O}(x_0)$ which is H–equivariantly diffeomorphic to $\mathcal{O}(x_0)$. i.e. the diffeomorphism Ψ is H–equivariant. We follow the exposition in [36]) and start with the observation that there is an action of H on $\tilde{M}$, and since $\tilde{M}$ is H–equivariantly diffeomorphic to G/K. and this homogeneous space is a H–space. i.e. there is a natural action of H on G/K we find that the a-priori unknown manifold $\tilde{M}$ is diffeomorphic to G/K, with the natural action of H on G/K. This action is given by multiplication

$$h[g]_K = [hg]_K. \tag{20}$$

In general this manifold $\tilde{M}$ does not consist of equilibria, but it carries a nontrivial flow. Our aim is to describe some properties of this flow. Now it is possible to classify H–equivariant flows on G/K, a program which was initiated and carried through in [36] for some examples with $G = \mathbf{SO}(3)$ and H, K closed subgroups of G. In that paper possible flows for $H = \mathbf{T}$ and $K = \mathbf{SO}(2)$ (or vice versa) were classified and in the case of ODE's it was possible to construct flows with heteroclinic cycles. The main observation is a description of the precise location of the fixed point space for the action of K on G/H. The fixed point space of a subgroup H_1 of this action is given by the set of points where

$$h[g]_K = [g]_K \ \forall h \in H_1. \tag{21}$$

This is satisfied if and only if $g^{-1}hg \in K$ for all $h \in H_1$, or $g^{-1}H_1g \subset K$. Therefore we see

$$\mathrm{Fix} H_1 = \{g \in G \mid g^{-1}H_1g \subset K\}/K. \qquad (22)$$

We denote this set $\{g \in G \mid g^{-1}H_1g \subset K\}$ by $N(H_1, K)$. This set was introduced by IHRIG & GOLUBITSKY [28]. Some properties of $N(H_1, K)$ were derived in [28]. In the spherical case the computations of $N(H_1, K)$ for all pairs of subgroups was started in [33], and continued in [11, 36]. Now all fixed point spaces for actions of groups H on G/K for $G = \mathbf{O}(3)$ and closed subgroups H, K are available [35, 43]. From (22) the fixed point spaces for subgroups can be characterized in a purely algebraic fashion. These fixed point spaces are flow invariant which gives severe restrictions on the flow. Pictures of the geometry of some of these spaces can be found in [36, 35, 39]. The main idea in [35] is to give group theoretic conditions for heteroclinic cycles in problems with forced symmetry breaking. This is translated into a graph theoretical problem using the stratification of the double quotient $H \backslash G / K$ into orbit types for the action of H on G/K. In this context we find a notion which is similar to Krupa's notion of a robust heteroclinic cycle [29].

Definition 4.1 *A point $[g]_K$ which is isolated in its stratum is called a group theoretic equilibrium. A group theoretic connection of two equilibria is a one dimensional fixed point space, containing both equilibria ξ_1, ξ_2 and an arc with endpoints ξ_1, ξ_2 containing no other group theoretic equilibria. A collection of group theoretic equilibria $\xi_1, \ldots, \xi_m$ and of one dimensional fixed point spaces $V_1 \ldots, V_k$ is called a group theoretic cycle if we can find a directed closed path consisting of group theoretic equilibria and of arcs on group theoretic connections.*

An application of the theoretical results to problems with spherical symmetry yields

Theorem 4.2 *Given an ODE of the form (17) which is equivariant with respect to $\mathbf{O}(3)$. Suppose a normally hyperbolic orbit of equilibria with isotropy type K is given. A necessary condition for the occurrence of group theoretical cycles is that either $H = \mathbf{T}$ or $K = \mathbf{T}$.*

In [43, 35] all graphs associated to forced symmetry breaking in problems with spherical symmetry are computed. From this one gets a complete list of group theoretical cycles. There is duality between the dynamics associated to the pair (H, K) and the one corresponding to the pair (K, H). Here again lifting theorem (SCHWARZ [46]) is used.

In order to apply this to PDE's we do not use center manifold reductions, but we compute a approximation to the group theoretic cycle in the Banach space and determine the flow on this cycle. Let us first define the notation: suppose a G–equivariant equation (2) is given and defines a semidynamical system on $H^1(\Omega)$. Assume

1. $u = 0$ is a solution for all $\lambda \in \mathbf{R}$.

2. For λ_0 the linearization at the trivial solution has a nontivial kernel V.

3. V is an absolutely irreducible representation of G.

4. K is an isotropy subgroup for this action on V and has a one-dimensional fixed point space.

5. The hypotheses of the equivariant branching lemma are satisfied.

6. The bifurcating branch of steady states with isotropy K is normally hyperbolic. Observe that this is a generic property, compare FIELD [17].

Suppose that we perturb equation (2) by an H-equivariant term as described above. We would like to compute the group orbit of bifurcating solution and then the nearby invariant manifold for the perturbed system. However the second step is very difficult. As an approximation we compute a group orbit in the kernel V of a point with isotropy K. Observe, that all these points lie in a one dimensional subspace. Up to a scaling by a real parameter s we get a unique group orbit, of the form sGv_0, where s is the real parameter and v_0 is a unit vector with isotropy K. On this orbit we can compute the group theoretical cycle. This is a purely group theoretic data and does not depend on the equation or on its perturbation. For each s, sufficiently small, we find a unique orbit M_s of steady states of (2), just use the mapping $\sigma : V \to V^\perp$ describing the center manifold, here $V^\perp$ denotes a closed complement to V in the Banach space. For the class of problems we have studied it is possible to prove the existence of a closed invariant complement. This mapping transports the group theoretic cycle onto M_s as well. Finally we use the mapping (19) Ψ constructed via normal hyperbolicity to transport all the information to M_s. It can be shown (LAUTERBACH & ROBERTS, [37]), that for additive perturbations the flow on the group theoretic cycle can be computed, by computing the scalar product between the tangent vector to the one-dimensional pieces of the cycle and the perturbations, i.e. let $v(\tau)$ be a parametrization of an arc in the group theoretic cycle and let

$$t_\tau = \frac{d}{d\tau}v(\tau)$$

be the tangent vector to the arc at τ. We have

Theorem 4.3 *There exists some $\varepsilon_0 > 0$, such that for $|s| < \varepsilon_0, s \neq 0$ the direction of the flow on the group theoretic cycle on $\tilde{M}_s$ at $v(\tau)$ is given by the scalar product*

$$\langle t_\tau, h \rangle_{L_2(\Omega)}, \tag{23}$$

where εh denotes perturbation, of either form.

Using this result one can study what kind of perturbations h lead to heteroclinic cycles (on the group theoretic cycle). In MAIER–PAAPE & LAUTERBACH [39] this result is used to investigate forced symmetry breaking near the $\ell = 2$ bifurcation for a problem with spherical symmetry.

Theorem 4.4 *Consider a PDE of form (2) which is equivariant with respect to* $\mathbf{O}(3)$. *Suppose for all* $\lambda \in \mathbf{R}$ $u = 0$ *is a solution which changes stability at* λ_0. *Suppose moreover that the kernel* V *of the linearization at* $u = 0$, λ_0 *is the* $\ell = 2$ - *representation of* $\mathbf{O}(3)$. *Then there exists a branch of axisymmetric solutions. We consider perturbations with* $H = \mathbf{T} \oplus \mathbf{Z}_2^-$ -*equivariance. Then there exists an open set of perturbations (in the space of* H-*equivariant perturbations of class 1 and class 2 in the* $C(\overline{\Omega})$ *topology) which lead to heteroclinic cycles.*

Proof: The main difficulty is to study the type of perturbations leading to heteroclinic cycles. The classification is based on a detailed study of the invariant theory for the exceptional subgroups of $\mathbf{O}(3)$. The details can be found in [39]. □
A similar theory can be developed for the perturbations of Hopf bifurcations, this is work in progress. Different techniques to investigate perturbations of Hopf branches were developed by CHOSSAT and FIELD [9], for applications in physics, see DANGELMAYR & KNOBLOCH [13], and HIRSCHBERG & KNOBLOCH [27].

So far we have looked at forced symmetry breaking of continuous groups. In the case of finite groups these techniques cannot work. SANDSTEDE & SCHEEL [44] look at the problem of forced symmetry breaking for finite groups. By projecting on the orbit space they find a codimension 2 homoclinic bifurcation. This leads to various periodic orbits, heteroclinic cycles and even geometric Lorenz attractors.

5   Heteroclinic cycles and invariant planes

A typical scenario for the creation of heteroclinic cycles in equivariant systems is the following (in the simplest possible case). Assume that G contains subgroups H_0, H_1 and H_2 with

1. $H_0 \supset H_1$ and $H_0 \supset H_2$, and

2. (a) $\dim\mathrm{Fix}(H_0) = 1$,
   (b) $\dim\mathrm{Fix}(H_1) = 2$ and
   (c) $\dim\mathrm{Fix}(H_2) = 2$.

Moreover, we assume that there are two nontrivial hyperbolic fixed points in $\mathrm{Fix}(H_0)$, say v_1, v_2 such that the unstable manifold of v_1 intersects $\mathrm{Fix}(H_1)$ in a one dimensional manifold, and so does the stable manifold of v_1 with $\mathrm{Fix}(H_2)$. For stable and unstable manifolds of the point v_2 we require the opposite inclusions, i.e. we have

$$\dim\left(W^u(v_1) \cap \mathrm{Fix}(H_1)\right) = \dim\left(W^s(v_1) \cap \mathrm{Fix}(H_2)\right) =$$
$$\dim\left(W^u(v_2) \cap \mathrm{Fix}(H_2)\right) = \dim\left(W^s(v_2) \cap \mathrm{Fix}(H_1)\right) = 1,$$

see figure 1.

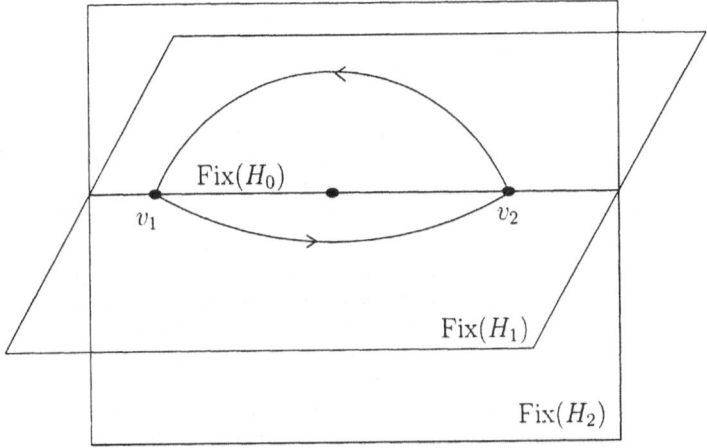

Figure 1: The geometry of the fixed point planes with a heteroclinic cycle.

This type of heteroclinic cycle is called *robust heteroclinic cycle* in KRUPA [29].

Of course. there might be several groups conjugate to H_1 or H_2 contained in H_0. In [11] the number $n(H, K)$ is introduced as the number of conjugate copies of H contained in K. CHOSSAT & GUYARD [10] make a distinction between the two cases

1. $n(H_j. H_0) = 1$ for $j = 1, 2$ and

2. $n(H_j. H_0) > 1$ for $j = 1$ or $j = 2$.

A nice and simple example for this scenario is due to GUCKENHEIMER & HOLMES [23]. They give a vector field on $\mathbf{R}^3$ which is equivariant with respect to the group $\mathbf{T} \oplus \mathbf{Z}_2^r$ of all rigid motions of a regular tetrahedron $\mathbf{T}$ together with $\mathbf{Z}_2^r$ which acts as a reflection at one of the coordinate planes. The subgroups of the form $\mathbf{Z}_2 \oplus \mathbf{Z}_2^r$ have a one dimensional fixed point subspace. They contain two subgroups of order 2. Consider the vector field

$$\begin{aligned}
\dot{x} &= \lambda x + x(ax^2 + by^2 + cz^2) \\
\dot{y} &= \lambda y + y(ay^2 + bz^2 + cx^2) \\
\dot{z} &= \lambda z + z(az^2 + bx^2 + cy^2).
\end{aligned}$$

This vector field has the right equivariance property. therefore we find the three coordinate planes as invariant subspaces. Choosing the parameter values $a < 0, , \lambda > 0$ and either $b < a < c$ or $c < a < b$ we obtain a pair of nontrivial equilibria on each coordinate line and a heteroclinic orbit connected them, compare [23, 44]. This gives a heteroclinic cycle involving 3 equilibria. This example is slightly more complicated than the scenario shown in figure 1.

An application of this technique to problems with spherical symmetry was given in CHOSSAT & GUYARD [10]. It can be shown that in irreducible representations of $O(3)$ there is no possibility of a local steady state bifurcation giving rise to a heteroclinic cycle through this scenario. In fact the bifurcation equations have some variational structure ([45] to prohibit heteroclinic cycles. The interest in heteroclinic cycles in spherical problems comes partly from geophysics. Such cycles could be a model for the change of orientation of the earths magnetic field. There are some indications that the relevant bifurcations come from mode interactions involving several irreducible representations of $O(3)$. A systematic investigation of the scenario described in [41] in mode interactions for problems with spherical symmetry was done by CHOSSAT & GUYARD [10, 24]. They study two types of mode interactions, the $\ell = 2, \ell = 6$ mode interaction and the interactions of type $\ell, \ell + 1$. Here ℓ stands for the $2ell + 1$–dimensional representation of $O(3)$. These studies follow some earlier work of CHOSSAT & ARMBRUSTER [2, 8], where heteroclinic cycles in the $(1, 2)$ mode interaction were found. Concerning the $(\ell. \ell + 1)$ mode interaction CHOSSAT & GUYARD [10] give a complete list of heteroclinic cycles which can be constructed with the invariant planes scenario. To describe the results concerning the spherical symmetric case we follow the notation in [11, 22, 28].

Theorem 5.1 *Consider the spherical Bénard problem and let $\ell > 1$. If the loss of stability of the purely heat conducting solution leads to a kernel with the $(\ell, \ell + 1)$ mode interaction, then there exists an open region in parameter space and an open neighborhood U of the bifurcation point such that for each parameter value in the open region there exists at least one heteroclinic cycle in U connecting two $O(2) \oplus Z_2^c$ symmetric points. If $\ell = 8$ then besides this heteroclinic cycle there is another one connecting two $O \oplus Z_2^c$ symmetric points.*

Observe that the functions invariant under the group $O(2)$ are axisymmetric, the ones invariant under O have the symmetry of a cube. Z_2^c stands for the group generated by $x \mapsto -x$ in R^3.

Proof: The proof consists of two parts. We begin with a group theoretic verification of the geometry of the fixed point subspaces. There is a necessary condition on the partial ordered set (po-set) of isotropy subgroups, namely the occurrence of a subgraph of the form indicated in figure 2. This gives the possibility of a heteroclinic connection in $\mathrm{Fix}(H_1)$ and in $\mathrm{Fix}(H_2)$ and in the fixed point subspaces of groups $H_k \subset H_0$ conjugate to H_1 or to H_2. The number of such conjugate subgroups plays a crucial rôle. In order to establish the existence of heteroclinic connections one has to look at the vector fields restricted to these subspaces. For the genericity statement one has to show that for open regions in parameter space the equations give rise to a steady state bifurcation of a pair of points in $\mathrm{Fix}(H_0)$ which have the correct stability assignments within $\mathrm{Fix}(H_{1,2})$ and moreover one has to show that the stable or unstable manifold cannot go off to infinity.   □

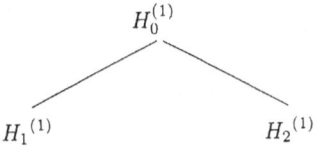

Figure 2: In order to find heteroclinic cycles a part of the po-set of isotropy subgroups has to have the indicated form The numbers indicate the dimension of the corresponding fixed point subspace.

Remark 5.2 *For a proof that the generic hypotheses are satisfied in a given system one has to study the specific equation. Here, the Clebsch-Gordan coefficients allow to gather sufficient information to prove the existence of the heteroclinic cycles as asserted.*

Remark 5.3 *This result cannot be directly applied to the geophysical problem, one reason is the earths rotation. Taking it into account the problem can be treated as a forced symmetry breaking to a $SO(2)$-equivariant problem. Some work in this direction has been done by CHOSSAT [7]. The study of the behaviour of the heteroclinic cycles under this symmetry breaking perturbations is under way and promises some interesting dynamical effects.*

Remark 5.4 *It is amusing to note that the group theoretic computations to verify the necessary condition in the mode interaction case are very similar to the group theoretic computations for the forced symmetry breaking analysis, compare GUYARD [24], LAUTERBACH & ROBERTS [36], LAUTERBACH, MAIER & REISSNER [35] and REISSNER [43].*

References

[1] J. C. ALEXANDER & J. YORKE. Global bifurcation of periodic orbits. *Amer. J. Math.* 100, 263–292, 1978.

[2] D. ARMBRUSTER & P. CHOSSAT. Heteroclinic cycles in a spherically invariant system. *Physica 50D*, 155–176, 1991.

[3] P. ASHWIN & Z. MEI. A Hopf bifurcation with Robbin boundary conditions. *J. Dyn. Diff. Equat.*, 6(3), 487–505, 1994.

[4] U. BANDELOW, H. J. WÜNSCHE & H. WENZEL. Theory of selfpulsation in two-section DFB lasers. *IEEE Photonics Technol Lett.*, 5, 1176–1179, 1993.

[5] G. E. BREDON. *Introduction to Compact Transformation Groups.* Academic Press, 1972.

[6] E. BUZANO, G. GEYMONAT & T. POSTON. Post–buckling behavior of a nonlinear-lyhyperelastic thin rod with cross-section invariant under the dihedral group D_n. *Arch. Rat. Mech. Anal.*, 89(4), 307–388, 1985.

[7] P. CHOSSAT. *Le Problème de Bénard dans une Couche Sphérique.* PhD thesis, Nice, 1981.

[8] P. CHOSSAT & D. ARMBRUSTER. Structurally stable heteroclinic cycles in a system with $O(3)$-symmetry. In M. Roberts & I. Stewart, editors, *Singularity Theory and Its Applications, Warwick 1989, Part II*, 38–62. Springer Verlag. 1991. Lecture Notes in Mathematics 1463.

[9] P. CHOSSAT & M. FIELD. Geometric analysis of the effect of symmetry breaking on an $O(2)$-invariant homoclinic cycle. *Preprint*, 1992.

[10] P. CHOSSAT & F. GUYARD. Heteroclinic cycles in bifurcation problems with $O(3)$ symmetry and the spherical Bénard problem. *Preprint*, 1995.

[11] P. CHOSSAT, R. LAUTERBACH & I. MELBOURNE. Steady-state bifurcation with $O(3)$-symmetry. *Arch. Rat. Mech. Anal.*, 113(4), 313–376, 1991.

[12] G. CICOGNA. Symmetry breakdown from bifurcation. *Lettere al Nuovo Cimento*, 31, 600–602, 1981.

[13] G. DANGELMAYR & E. KNOBLOCH. Hopf bifurcation with broken circular symmetry. *Nonlinearity*, 4, 399–428, 1991.

[14] B. DIONNE, M. GOLUBITSKY & I. STEWART. Coupled cells with internal symmetry part i: Wreath products. *Preprint*, (to appear).

[15] B. DIONNE, M. GOLUBITSKY & I. STEWART. Coupled cells with internal symmetry part ii: Direct products. *Preprint*, (to appear).

[16] U. FEISTE, D. J. AS & A. ERHARD. 18 ghz all-optical frequency locking and clock recovery using a self-pulsating two section DFB laser. *IEEE Photonics Technol. Lett.*, 6, 106–108, 1994.

[17] M. FIELD. *Symmetry Breaking for Compact Lie Groups.* Preprint. 1994.

[18] Gap. Program.

[19] K. GATERMANN & R. LAUTERBACH. Automatic classification of normal forms. *Preprint*, 1995.

[20] M. GOLUBITSKY & D. G. SCHAEFFER. A discussion of symmetry and symmetry breaking. *Proc. Symp. Pure Math.*, 40, 499–515, 1982.

[21] M. GOLUBITSKY & I. STEWART. Hopf bifurcation in presence of symmetry. *Arch. Rat. Mech. Anal.*, 87(2), 107–165, 1985.

[22] M. GOLUBITSKY. I. STEWART & D. G. SCHAEFFER. *Singularities and Groups in Bifurcation Theory*, Vol. II. Springer Verlag, 1988.

[23] J. GUCKENHEIMER & P. HOLMES. Structurally stable heteroclinic cycles. *Math. Proc. Cambridge Phil. Soc.*, 103, 189–192, 1988.

[24] F. GUYARD. *Interactions de mode dans les problèmes de bifurcation avec symétry sphérique.* PhD thesis. Université de Nice Sophia – Antipolis. 1994.

[25] E. HEWITT & K. A. ROSS. *Abstract Harmonic Analysis I*, volume 115 of *Grundl. d. math. Wiss.* Springer Verlag, 1963.

[26] M. W. HIRSCH, C. C. PUGH & M. SHUB. *Invariant Manifolds*, volume 583 of *Lecture Notes in Mathematics*. Springer Verlag. 1977.

[27] P. HIRSCHBERG & E. KNOBLOCH. Šil'nikov-Hopf bifurcation. *Physica D*, 62, 202–216, 1993.

[28] E. IHRIG & M. GOLUBITSKY. Pattern selection with O(3)-symmetry. *Physica 13D*, 1–33, 1984.

[29] M. KRUPA. Robust heteroclinic cycles. *Preprint*. 1994.

[30] M. KRUPA & I. MELBOURNE. Asymptotic stability of heteroclinic cycles in systems with symmetry. *Preprint*, 1–50, 1991.

[31] M. KRUPA & I. MELBOURNE. Nonasymptotically stable attractors in O(2) mode interactions. *Preprint*, 1993.

[32] S. LANG. *Algebra*. Addison–Wesley, 1970.

[33] R. LAUTERBACH. *Problems with Spherical Symmetry - Studies on O(3)-Equivariant Equations*. Habilitationsschrift, Univ. Augsburg. 1988.

[34] R. LAUTERBACH. Äquivariante dynamische Systeme. Vorlesung FU Berlin, SS 1995.

[35] R. LAUTERBACH, S. MAIER & E. REISSNER. A systematic study of heteroclinic cycles in dynamical system with broken symmetries. *Proc. Roy. Soc. Edinb.*, to appear.

[36] R. LAUTERBACH & M. ROBERTS. Heteroclinic cycles in dynamical systems with broken spherical symmetry. *J. Diff. Equat.*, 100, 428–448, 1992.

[37] R. LAUTERBACH & M. ROBERTS. (In preparation.)

[38] R. LAUTERBACH & J. SANDERS. Bifurcation analysis for spherically symmetric systems using invariant theory. *Preprint*, 1994.

[39] S. MAIER-PAAPE & R. LAUTERBACH. Reaction diffusion systems on the 2–sphere and forced symmetry breaking. *Preprint*, in preparation.

[40] I. MELBOURNE. An example of a non-asymptotically stable attractor. *Nonlinearity*, 4, 835–844, 1991.

[41] I. MELBOURNE, P. CHOSSAT & M. GOLUBITSKY. Heteroclinic cycles involving periodic solutions in mode interactions with O(2)-symmetry. *Proc. Roy. Soc. Edinb.*, 113(5), 315–345, 1989.

[42] M. MÖHRLE, U. FEISTE, J. HÖRER, R. MOLT & B. SATORINS. Gigahertz self-pulsation in 1.5 μm wavelength multisection DFB lasers. *IEEE Photonics Technol. Letter*. 4, 976–979, 1992.

[43] E. REISSNER. Dynamische Systeme und erzwungene Symmetriebrechung am Beispiel sphärischer Probleme. Master's thesis, Univ. Augsburg. 1993.

[44] B. SANDSTEDE & A. SCHEEL. Forced symmetry breaking of homoclinic cycles. *Nonlinearity*, 8, 333–365, 1994.

[45] D. H. SATTINGER. *Group Theoretic Methods in Bifurcation Theory*. volume 762 of *Lecture Notes in Mathematics*. Springer Verlag. 1978.

[46] G. SCHWARZ. Lifting smooth homotopies. *IHES*. 51, 37–135, 1980.

[47] J.-P. SERRE. *Représentations Linéaires des Groupes Finis*. Herrmann, 1978.

[48] A. VANDERBAUWHEDE. *Local Bifurcation and Symmetry*, volume 75 of *Research Notes in Mathematics*. Pitman, 1982.

Progress in Nonlinear Differential Equations
and Their Applications, Vol. 19
© 1996 Birkhäuser Verlag Basel/Switzerland

Invariant C^j functions and center manifold reduction*

Matthias Rumberger   Jürgen Scheurle[†]

1 Introduction

Recently the orbit space reduction method has been successfully applied to study complex dynamical systems with symmetry (see e.g. Dias and Chossat [3], Marsden [5], and Scheurle [9]). In particular, combined with some kind of center manifold reduction procedure, that method has been used to analyze bifurcation and stability properties of relative equilibrium solutions of mechanical systems. Those solutions become genuine equilibria on the orbit space.

A key feature of orbit space reduction is to identify whole orbits of the symmetry group of the system in phase space with single points of the so-called orbit space. Thus, modulo the motion on group orbits, the original dynamical system induces a dynamical system on the orbit space, the *reduced dynamical system*. Relative equilibrium solutions have the property that their dynamic orbit in phase space is also an orbit of a subgroup of the symmetry group. So, they become fixed points under orbit space reduction. In general, this simplifies their analysis to a large extent.

A common way to represent the orbit space and the reduced dynamical system is to use a Hilbert basis of the ring of polynomials on the original phase space, that are invariant under the symmetry group. Let the symmetry group Γ be a compact Lie group which acts orthogonally on $\mathbb{R}^m$. Then a well-known theorem says that the ring $\mathcal{P}(\Gamma)$ of Γ-invariant polynomials over $\mathbb{R}^m$ is finitely generated, i.e., there exist finitely many polynomials $\rho_j \in \mathcal{P}(\Gamma)$ $(j = 1, \ldots, k)$ such that any $p \in \mathcal{P}(\Gamma)$ can be represented in the form $p = P(\rho_1, \ldots, \rho_k)$, where P is a polynomial over $\mathbb{R}^k$. In other words, there exists a finite Hilbert basis for $\mathcal{P}(\Gamma)$ (see e.g. Weyl [12]). Moreover, $\mathcal{P}(\Gamma)$ separates orbits of Γ in $\mathbb{R}^m$. Hence, an embedding into $\mathbb{R}^k$ of the orbit space $\mathbb{R}^m/\Gamma$ of Γ is given by $\rho(\mathbb{R}^m)$ with $\rho = (\rho_1, \ldots, \rho_k)$, i.e., by functional relations of the polynomials ρ_j that generate $\mathcal{P}(\Gamma)$.

---

*This work has been supported by the DFG under the contract Sche233/3-1

[†] Institut für Angewandte Mathematik, Universität Hamburg. Bundesstrasse 55,D-20146 Hamburg. Germany

We call a dynamical system on $\mathbb{R}^m$ Γ-*symmetric*, if it is given by a Γ-*equivariant* vector field X on $\mathbb{R}^m$, i.e., if for all $x \in \mathbb{R}^m$ and $\gamma \in \Gamma$, $X(\gamma x) = \gamma X(x)$ holds. Obviously, such a system induces a flow on the orbit space. This flow is given by the evolution of the values of the polynomials ρ_j along X-orbits. If X is a polynomial vector field, then it follows from the theorem mentioned above, that the reduced dynamical system is generated by a polynomial vector field $\tilde{X}$ on $\mathbb{R}^k$ (cf. section 3).

In Schwarz [10], the above mentioned theorem has been generalized to (germs of) C^∞ functions. In fact, if $C^\infty(\mathbb{R}^m)$ denotes the Frechét space of arbitrarily often continuously differentiable functions on $\mathbb{R}^m$ equipped with the usual topology (analogously, if $\mathbb{R}^m$ is replaced by some subset of $\mathbb{R}^m$) and if a superscript Γ denotes the corresponding subspace of Γ-invariant functions, then for any $f \in C^\infty(\mathbb{R}^m)^\Gamma$ there exists an $F \in C^\infty(\mathbb{R}^k)$ such that $f = F(\rho_1, \ldots, \rho_k)$, where as before the ρ_j are elements of a given Hilbert basis of $\mathcal{P}(\Gamma)$. Furthermore, F can be chosen such that the map

$$I : f \mapsto F = I(f); \ C^\infty(\mathbb{R}^m)^\Gamma \to C^\infty(\mathbb{R}^k)$$

is linear and continuous (cf. Mather [6, Theorem 1]). Moreover, $F_{|_{\rho(\mathbb{R}^m)}}$ is uniquely determined by f, and the map I from $C^\infty(\mathbb{R}^m)^\Gamma$ onto its range in $C^\infty(\mathbb{R}^k)$ is a homeomorphism. Accordingly, given any equivariant C^∞ vector field X on $\mathbb{R}^m$, the corresponding reduced dynamical system is generated by a C^∞ vector field $\tilde{X}$ on $\mathbb{R}^k$.

Let us now assume that locally in $\mathbb{R}^m$, say near $x = 0$, the vector field X represents a possibly infinite dimensional dynamical system on a center manifold. As is well-known, in general such a vector field is not C^∞ even if the original vector field is analytic. On the other hand, it is of class C^j, provided that the original vector field has that property. Thus, the question arises whether in the C^j case, there is an analogue to Schwarz's theorem and the orbit space reduction procedure as outlined above.

In this note we want to make the point that the answer to that question is 'yes'. In fact, the following theorem holds true, where the spaces C^j of j-times continuously differentiable functions are defined analogously to the C^∞ case.

Theorem 1.1 *Let* Γ *be a compact Lie group acting orthogonally on* $\mathbb{R}^m$, *let* $\{\rho_1, \ldots, \rho_k\}$ *be a Hilbert basis of* $\mathcal{P}(\Gamma)$, *and let* U *be some* Γ-*invariant neighborhood of* $x = 0$ *in* $\mathbb{R}^m$. *Then there is an integer* q *and a neighborhood* $V \subset U$ *of* $x = 0$ *such that for all* $n \in \mathbb{N}$, *there exists a continuous, linear map*

$$I : f \mapsto F = I(f); \ C^{qn}(U)^\Gamma \to C^n(\mathbb{R}^k)$$

with

$$f = F(\rho_1, \ldots, \rho_k) \quad on \ V$$

for any $f \in C^{qn}(U)^\Gamma$.

See Whitney [13] for the special case of this theorem for even functions, where $q = 2$.

It is to be noted that there are differences compared to the C^∞ case. First, elementary examples show that there is a loss of smoothness between f and F in general. According to Theorem 1.1, an upper bound for this loss of smoothness is given by $(n-1)q$. It turns out that the integer q can be estimated from above by 4^a times the maximal degree of the polynomials ρ_j, where a denotes the maximal length of direct paths through the lattice of isotropy subgroups of Γ. We point out, though, that this is a rough bound only. In particular, this bound does not depend on the functional relations of the polynomials ρ_j.

Second, the map I from $C^{qn}(U)^\Gamma$ onto its range in $C^n(\mathbb{R}^k)$ is not a homeomorphism in general, not even in the case $U = \mathbb{R}^m$, where V can be set equal to $\mathbb{R}^m$. In turn, the proof in Schwarz [10] does not fully carry over to the present case. Especially, several nice abstract arguments from that paper do not work here. See, in particular, the proof of the Lemma in §2 of [10].

We shall outline a proof of Theorem 1.1 in section 2. In particular, this theorem justifies an application of the method of orbit space reduction on center manifolds for sufficiently smooth symmetric dynamical systems. To illustrate this briefly, we consider the Swift-Hohenberg equation as a simple example in section 3. This equation is supposed to model convective instabilities of the basic motion of certain fluids.

2 Outline of a proof of Theorem 1.1

In this section we mainly want to give the reader a rough idea of how the loss of a finite number of derivatives for the function F in Theorem 1.1 comes about. Basically, we follow the ideas of Schwarz [10]. For technical details in the present case we refer to Rumberger [8]. Throughout this section, n is chosen to be a fixed integer.

The proof proceeds by mathematical induction with respect to the lattice of isotropy subgroups of the symmetry group Γ acting on $\mathbb{R}^m$. Note that those are again compact Lie groups and there are only finitely many of them. If $\Gamma = \{id\}$ is the trivial group, then the theorem is clearly true with $k = m$, $q = 1$, and for example, $\rho_j(x) = x_j$ $(j = 1, \ldots, m)$. We now assume, as inductive hypothesis, that the theorem has been proved for appropriate representations (see below) of any proper isotropy subgroup of Γ. Of course, if the theorem is true for one Hilbert basis of $\mathcal{P}(\Gamma)$, then it is also true for any other one. We can assume that the polynomials ρ_j are homogeneous and of degrees $d_j > 0$. Also, first we consider the case that the fixed point subspace of Γ in $\mathbb{R}^m$ is equal to $\{0\}$. Moreover, since both the Taylor polynomial and its remainder are Γ-invariant for a Γ-invariant function f, and Theorem 1.1 is true for polynomials of a fixed maximal degree by the classical result, it suffices to prove the theorem for the remainder. Thus, we assume that f and all its derivatives up through order qn vanish at $x = 0$. Let us denote the corresponding subspace of $C^{qn}(\mathbb{R}^m)^\Gamma$ by $C^{qn}(\mathbb{R}^m)_0^\Gamma$.

The basic idea of the induction argument is the decompostion of I into $I_3 \circ I_2 \circ I_1 \circ I_0$, to treat the point of maximal isotropy in $\mathbb{R}^m$, i.e. the origin, and the points of less isotropy separately. Starting with I_0, the maps I_j will be constructed one after the other in four steps. First one performs a blow-up of the singularity of f at $x = 0$ to separate the "radial" part and the "angular" part. The second step is the main step of the induction argument. There one proves the claim of the theorem for the "angular" part. Finally, in the last two steps one deals with the "radial" part to capture the origin.

To perform a blow-up of the singularity of f at $x = 0$, let us now introduce the map

$$\alpha : (r, \theta) \mapsto r\theta \; ; \quad \mathbb{R} \times S^{m-1} \to \mathbb{R}^m ,$$

where S^{m-1} denotes the unit sphere in $\mathbb{R}^m$. Also, we introduce the subspaces E_j of functions $f \in C^j(\mathbb{R} \times S^{m-1})^\Gamma$ such that $f(-r, -\theta) = f(r, \theta)$ holds for all (r, θ) and such that f and all its derivatives up through order j vanish for $r = 0$. Here and subsequently, the action of Γ on S^{m-1} is given by the restriction of the Γ-action on $\mathbb{R}^m$, and Γ acts trivially on $\mathbb{R}$. Using Whitney's extension theorem one easily finds a continuous linear map $I_0 : C^{qn}(U)_0^\Gamma \to E_{qn}$ such that

$$I_0(f)|_{(-\delta, \delta) \times S^{m-1}} = f \circ \alpha|_{(-\delta, \delta) \times S^{m-1}}$$

holds for all $f \in C^{qn}(U)_0^\Gamma$. Here $\delta > 0$ is chosen such that the closed ball $V := \overline{B_\delta(0)}$ in $\mathbb{R}^m$ is contained in U.

In the second step, we prove the assertion of Theorem 1.1 for functions $f \in C^{\tilde{q}n}(S^{m-1})^\Gamma$. Choose $p \in S^{m-1}$ arbitrarily. According to our assumptions above, the isotropy subgroup Γ_p of p, i.e. the subgroup of all $\gamma \in \Gamma$ that fix p, is different from Γ. Therefore we can apply the inductive hypothesis to that symmetry group. Let N_p denote the manifold given by the intersection of S^{m-1} with that subspace of the affine space $p + \mathbb{R}^m$ that is orthogonal to the Γ-orbit at p. Introducing a coordinate chart, we can identify this manifold with a neighborhood of 0 in $\mathbb{R}^{\tilde{m}}$ for some $\tilde{m} < m$. Obviously, N_p is Γ_p-invariant, and Γ_p acts orthogonally on it. Choose any minimal Hilbert basis $\{\sigma_1, \ldots, \sigma_\ell\}$ of the corresponding ring $\mathcal{P}(\Gamma_p)$. (In view of the indicated upper bound for q, we mention that this Hilbert basis can be chosen such that the maximal degree of the polynomials σ_j is less than or equal to the maximal degree of the polynomials ρ_j.) Then, by the inductive hypothesis, locally in N_p near p, and provided that $\tilde{q}$ is sufficiently large, any function $f \in C^{\tilde{q}n}(S^{m-1})^\Gamma$ can be represented as $f = F(\sigma_1, \ldots, \sigma_\ell)$ where $F = I_p(f)$ is given by a continuous linear map $I_p : C^{\tilde{q}n}(N_p)^{\Gamma_p} \to C^n(\mathbb{R}^\ell)$. Moreover, since different Γ_p-orbits in N_p are contained in different Γ-orbits in $\mathbb{R}^m$ (cf. Bredon [1. p. 85]). the polynomials ρ_j restricted to N_p separate Γ_p-orbits there. Consequently, the Γ_p-invariants σ_j are functions of the ρ_j on N_p. In fact, using the property that the Hilbert basis $\{\sigma_1, \ldots, \sigma_\ell\}$ is minimal, one can prove that those functions are smooth. Hence, there exists a C^n function $\tilde{F}$ such that $f = \tilde{F} \circ \rho$ holds locally on N_p near p. In fact, by the Γ-invariance of f and the ρ_j, this relation is even valid on S^{m-1} locally near p. Moreover, $\tilde{F}$ can be extended to a C^n function on an open neighborhood, say W_p, of $\rho(p)$ in $\mathbb{R}^k$, e.g. by extending the relation that gives σ as a function of ρ on N_p, to all of $\mathbb{R}^k$. Finally, one uses a partition of unity on S^{m-1} that is subordinated to the covering $\{W_p\}_{p \in S^{m-1}}$ of $\rho(S^{m-1})$, to establish the existence of a continuous linear map $I_{S^{m-1}} : C^{\tilde{q}n}(S^{m-1})^\Gamma \to C^n(\mathbb{R}^k)$ such that

$$I_{S^{m-1}}(f) \circ \rho_{|S^{m-1}} = f$$

holds for all $f \in C^{\tilde{q}n}(S^{m-1})^\Gamma$. Since Γ acts trivially on the $\mathbb{R}$-component of functions in E_j, a similar Lemma as in Schwarz [10,§2] shows that the latter result implies the existence of a continuous linear map $I_1 : E_{2\tilde{q}n} \to F_n$ such that

$$I_1(f)(id, \rho) = f$$

holds for all $f \in E_{2\tilde{q}n}$. Here we assume that, in particular,

$$\tilde{q} \geq d := \max_{1 \leq j \leq k} \{d_j\}$$

holds, and we use the notation F_n for the function space

$$F_n = \{ \quad f : \mathbb{R} \times \mathbb{R}^k \to \mathbb{R} \mid f \circ A = f, \text{ all derivatives of } f \text{ that are at most}$$
$$\text{of order } dn \text{ with respect to the } \mathbb{R}\text{-argument and of order } n \text{ with}$$
$$\text{respect to the } \mathbb{R}^k\text{-argument, exist, are continuous, and vanish}$$
$$\text{together with } f \text{ along } \{0\} \times \mathbb{R}^k \}.$$

The map A is defined by

$$A : (t, y_1, \ldots, y_k) \mapsto (-t, (-1)^{d_1} y_1, \ldots, (-1)^{d_k} y_k) : \mathbb{R} \times \mathbb{R}^k \to \mathbb{R} \times \mathbb{R}^k,$$

where d_j is the degree of the polynomial ρ_j.

Next, in the third step of the proof, we introduce a map

$$\beta : (t, y_1, \ldots, y_k) \mapsto (t^2, t^{d_1} y_1, \ldots, t^{d_k} y_k) : \mathbb{R} \times \mathbb{R}^k \to \mathbb{R} \times \mathbb{R}^k.$$

Obviously, we have the relation

$$\beta \circ (id, \rho) = (r^2, \rho) \circ \alpha$$

where $r^2(x) = x_1^2 + \cdots + x_m^2$ is the square of the radius function in $\mathbb{R}^m$. Then, using again Whitney's extension theorem, one shows that there is a continuous linear map $I_2 : F_n \to C^n(\mathbb{R} \times \mathbb{R}^k)_0$ such that

$$I_2(f) \circ \beta = f \quad \text{on } \mathbb{R} \times \rho(S^{m-1})$$

holds for all $f \in F_n$. Here one needs to employ the property that sufficiently many derivatives of f vanish along $\{0\} \times \mathbb{R}^k$.

Finally, in the last step of the induction argument, we observe that there is a polynomial R such that

$$r^2 = R \circ \rho$$

holds for the square r^2 of the radius function, because r^2 is Γ-invariant. The map $I_3 : C^n(\mathbb{R} \times \mathbb{R}^k)_0 \to C^n(\mathbb{R}^k)_0$ given by

$$I_3(f) = f \circ (R, id)$$

is clearly linear and also continuous. Therefore, the map $I : C^{\tilde{q}n}(U)^\Gamma \to C^n(\mathbb{R}^k)$ that we are looking for is given by the composition

$$I = I_3 \circ I_2 \circ I_1 \circ I_0 \,,$$

with $\tilde{\tilde{q}} = 2\tilde{q}$. Thus the induction argument to prove Theorem 1.1 in the case that the fixed point subspace $Fix(\Gamma) \subset \mathbb{R}^m$ is trivial, is finished.

To complete the proof, we point out that Γ acts trivially on its fixed point subspace $Fix(\Gamma)$, and orthogonally on the orthogonal complement $Fix(\Gamma)^\perp$ in $\mathbb{R}^m$. Thus, after the assertion of Theorem 1.1 has been proved as above, with $\mathbb{R}^m$ replaced by $Fix(\Gamma)^\perp$, we can again use the above mentioned Lemma to extend I to a map from $C^{qn}(U)^\Gamma$ into $C^n(Fix(\Gamma) \times \mathbb{R}^k)$ with $q = 2\tilde{\tilde{q}}$, such that

$$I(f) \circ (id, \rho) = f$$

holds in a neighborhood of $x = 0$ in $\mathbb{R}^m$.

Remark: Accordingly, there is a version of Poènaru's normal form theorem for equivariant vector fields [7] in the C^j-case. In fact, we can use Theorem 1.1 to establish such a result. Namely, if X is a Γ-equivariant vector field on $\mathbb{R}^m$ of class C^{qn} with q sufficiently large, then we define a function $f \in C^{qn}(\mathbb{R}^m \times (\mathbb{R}^m)^*)^\Gamma$ by $f(x, w) = w(X(x))$. Here Γ acts on the second variable w by $(\gamma w)(x) = w(\gamma^{-1}x)$ for all $\gamma \in \Gamma$, $x \in \mathbb{R}^m$. Let $\{\rho_1, \ldots, \rho_k\}$ be a Hilbert basis of the corresponding ring $\mathcal{P}(\Gamma)$. By Theorem 1.1, there exists a function $Q \in C^n(\mathbb{R}^k)$ such that $f = Q \circ \rho$. Now, differentiating f with respect to w at $w = 0$ yields

$$X(x) = D_w f(x, 0) = \sum_{j=1}^{k} \partial_{y_j} Q(\rho(x, 0)) D_w \rho_j(x, 0).$$

Obviously, for all j, $D_w \rho_j(x, 0)$ can be identified with a Γ-equivariant polynomial vector field on $\mathbb{R}^m$, and the functions $\partial_{y_j} Q(\rho(x, 0))$ are of class $C^{n-1}(\mathbb{R}^m)$ and Γ-invariant.

3   Example

In this section we briefly illustrate the usefulness of Theorem 1.1 in the context of orbit space reduction on center manifolds. To that end we consider the Swift-Hohenberg equation. As pointed out in the introduction, the Swift-Hohenberg equation models certain hydrodynamical instability phenomena. It reads

$$\frac{\partial u}{\partial t} = -(1 + \frac{\partial^2}{\partial x^2})^2 u + \mu u - u^3$$

where $x \in \mathbb{R}$ is a one-dimensional spatial variable. $t \in \mathbb{R}$ denotes time, and $\mu \in \mathbb{R}$ is supposed to be a bifurcation parameter. Furthermore, we impose the periodic boundary condition

$$u(x + 2\pi, t) = u(x, t)$$

for the unknown function $u = u(x, t) \in \mathbb{R}$. It is well-known, that this equation generates a semiflow in an appropriate Banach space H of 2π-periodic functions (cf. Collet and Eckmann [2], Temam [11]). As a consequence of the periodic boundary condition, $\Gamma = O(2)$ is a spatial symmetry group of the system. That means, given any solution, the image of this solution under the action of any element of Γ is again a solution. Here Γ acts by translations $T_a : x \mapsto x + a$ with $a \in [0, 2\pi)$ as well as by the flip $S : x \mapsto -x$ on the x-axis. The action of Γ on u is given by $(T_a u)(x, t) = u(x + a, t)$ and $(Su)(x, t) = u(-x, t)$. Note that an additional symmetry of the present equation is given by $Ru = -u$.

Obviously, the Swift-Hohenberg equation has the trivial solution $u \equiv 0$ for all values of μ. Therefore, the first step in a stability and bifurcation analysis of that model consists in determining the stability properties of the trivial solution depending on μ. Since the partial differential equation is parabolic, we can employ the principle of linearized stability (cf. e.g. Henry [4]). So we have to study the location of the spectrum of the linear operator

$$L(\mu) = -(1 + \frac{\partial}{\partial x^2})^2 + \mu$$

in the space H. But $L(\mu)$ has only eigenvalues, since its resolvent is compact. Those are easily computed to be $\lambda(\mu) = \mu - (1 - k^2)^2$ $(k = 0, 1, 2, \ldots)$, the corresponding normalized eigenfunctions being $\zeta_k^\pm = e^{\pm ikx}$. Hence, the eigenvalue $\lambda_0(\mu)$ is simple, while for $k > 1$, the eigenvalues $\lambda_k(\mu)$ are double. For $\mu < 0$, all eigenvalues are negative, i.e. the solution $u = 0$ is asymptotically stable. Furthermore, if μ increases through 0, then the double eigenvalue $\lambda_1(\mu)$ crosses zero from left to right, while all other eigenvalues strictly remain negative. Consequently, $u = 0$ becomes unstable for $\mu > 0$, and locally near $\mu = 0, u = 0$, the system has a two-dimensional center manifold which can be invoked to perform a local stability and bifurcation analysis. In fact, center manifolds are invariant under the semiflow in H, and contain all the relevant local dynamics including stability properties in the present case.

In complex coordinates $\zeta, \bar{\zeta} \in \mathbb{C}$, one has the following representation of such a center manifold M:

$$u = \zeta \zeta_1^+ + \bar{\zeta} \zeta_1^- + h(\zeta, \bar{\zeta}, \mu)$$

Here, for any $j \in \mathbb{N}$, h considered as a function with values in H is of class C^j in a neighborhood of the point $B : \zeta = \bar{\zeta} = 0, \mu = 0$, where the diameter of that neighborhood might shrink to zero, as j tends to infinity. Also. h and its first derivatives with respect to ζ and $\bar{\zeta}$ are zero at B. Furthermore, there is an induced action of the symmetry group on M, given by

$$T_a\zeta = e^{ia}\zeta , \quad T_a\bar{\zeta} = e^{-ia}\bar{\zeta}$$
$$S\zeta = \bar{\zeta} , \quad S\bar{\zeta} = \zeta .$$

The reduced dynamics on M is governed by a vector field $X(\zeta.\bar{\zeta},\mu)$ which is as smooth as h, and equivariant with respect to the induced Γ-action. Also, $X(0,0,\mu) = 0$ for all μ, and the derivative of X with respect to ζ and $\bar{\zeta}$ is zero at B.

Now, in order to achieve a further reduction of the dynamics on M, we use the orbit space reduction method with respect to the maximal continuous subgroup S^1 of Γ, that is represented by the operators T_a. Obviously, the corresponding ring $\mathcal{P}(S^1)$ is generated by the polynomial

$$\rho(\zeta,\bar{\zeta}) = \zeta\bar{\zeta}.$$

Thus, using ρ as a coordinate, the corresponding orbit space M/S^1 can be identified with the non-negative real axis $\mathbb{R}_0^+$. The flow induced on M/S^1 by the flow on M, is determined by the evolution of ρ along orbits on M, i.e. by the equation

$$\dot{\rho} = \nabla\rho \cdot X$$

But $f = \nabla\rho\cdot X$ is a C^j function of $\zeta,\bar{\zeta}$ and μ, that is invariant under the S^1-action on M. This follows by a straight-forward calculation using the invariance of ρ and the commutativity of X with respect to the operators T_a. Thus. if $j = qn$ for some $n \in \mathbb{N}$ with an appropriate $q \in \mathbb{N}$, then Theorem 1.1 applies to establish the existence of a C^n function F with $f = F(\rho,\mu)$. Here we let the group S^1 act trivially on the parameter μ, so that, in addition to Theorem 1.1, we can use the same Lemma that we used twice in section two, to take care of the parameter μ. Since $\rho = 0$ corresponds to $\zeta = \bar{\zeta} = 0$, we have $F(0,\mu) = 0$ for all μ. Also, since the derivative of X with respect to ζ and $\bar{\zeta}$ vanishes at B, we have $\partial_\rho F(0,0) = 0$. In fact, a more careful analysis shows that

$$F(\rho,\mu) = \mu\rho + a\rho^2 + O(\rho^3), \quad \text{as } \rho \to 0,$$

holds uniformly near $\mu = 0$, where $a = a(\mu) < 0$. We note that the extra symmetry generated by S acts trivially on the orbit space, since S maps any S^1-orbit on M into itself.

Finally, we analyze the reduced system on the orbit space which now reads ($\rho > 0$):

$$\dot{\rho} = F(\rho,\mu) .$$

According to the properties of F mentioned above, there is a supercritical steady state bifurcation of the trivial solution $\rho = 0$ at $\mu = 0$.

Asymptotically, as $\mu \downarrow 0$, the nontrivial steady state solutions are given by

$$\rho = -\frac{\mu}{a(0)} + O(\mu^2), \quad \mu > 0.$$

By the standard exchange of stability principle, those nontrivial solutions are asymptotically stable near $\mu = 0$. The corresponding solutions in the original phase space are relative equilibria (cf. the characterization of relative equilibria in the introduction) that are orbitally asymptotically stable. In fact, those are contained in a S^1-orbit given by the corresponding value of ρ and could be either steady state or time-periodic solutions. In the former case, the whole S^1-orbit is filled by S^1-conjugated steady state solutions. This non-generic case actually occurs in the present problem. The reason is the additional flip symmetry in the phase space generated by S.

References

[1] G.E. BREDON, Introduction to Compact Transformation Groups. Academic Press, New York 1972.

[2] P. COLLET AND J.-P. ECKMANN, The time dependent amplitude equation for the Swift-Hohenberg problem, Comm. Math. Phys. 132 (1990), 139–153.

[3] F. DIAS AND P. CHOSSAT, The 1:2 resonance wit $O(2)$ symmetry and its applications in Hydrodynamics, J. Nonlinear Sci. 5(2) (1995), 105–129.

[4] D. HENRY, Geometric Theory of Semilinear Parabolic Equations, Lect. Notes in Math. Vol 840, Springer-Verlag 1981.

[5] J.E. MARSDEN, Lectures on Mechanics, London Math. Soc. Lect. Note Ser. 174, Cambridge Univ. Press 1992.

[6] N. MATHER, Differentiable invariants, Topology 16 (1977). 145–155.

[7] V. POÈNARU, Singularités C^∞ en Présence de Symétrie. Springer Lect. Notes in Math. Vol 510, Springer-Verlag, Berlin 1976.

[8] M. RUMBERGER, C^r-Funktionen mit kompakter Liescher Symmetriegruppe dargestellt in Invarianten, Diploma thesis, Univ. Hamburg 1994.

[9] J. SCHEURLE, Some aspects of successive bifurcations in the Couette-Taylor problem, to appear in Fields Inst. Comm. Vol 4, AMS.

[10] G.W. SCHWARZ, Smooth functions invariant under the action of a compact Lie group, Topology 14 (1975), 63–68.

[11] R. TEMAM, Infinite-Dimensional Dynamical Systems in Mechanics and Physics, Springer-Verlag, New York 1988.

[12] H. WEYL, The Classical Groups (2nd edn.), Princeton Univ. Press 1946.

[13] H. WHITNEY, Differentiable even functions, Duke Math. 10 (1943), 159–160.

Progress in Nonlinear Differential Equations
and Their Applications, Vol. 19
© 1996 Birkhäuser Verlag Basel/Switzerland

Hopf bifurcation at k-fold resonances in conservative systems

Jürgen Knobloch*   André Vanderbauwhede†

Abstract

We show that the bifurcation picture for families of periodic orbits at a k-fold resonance in systems with a first integral is the same as in the reversible case studied in [4]. The approach is based on the general reduction method introduced in [2] and a careful reduction to normal form of the linear part of the equation.

1 Introduction

Consider a parameter-dependent autonomous system

$$\dot{x} = f(x, \lambda) \tag{1.1}$$

where $f : \mathbb{R}^n \times \mathbb{R}^m \to \mathbb{R}^n$ is smooth and such that $f(0, \lambda) = 0$ for all $\lambda \in \mathbb{R}^m$. We will assume that (1.1) has a first integral, i.e. there exists a smooth function $K : \mathbb{R}^n \times \mathbb{R}^m \to \mathbb{R}$ such that

$$D_x K(x, \lambda) \cdot f(x, \lambda) = 0, \quad \forall (x, \lambda) \in \mathbb{R}^n \times \mathbb{R}^m. \tag{1.2}$$

We will be concerned with the existence of periodic orbits near the equilibrium $x = 0$ in such *conservative systems*. One of the main results in this context is the Liapunov Center Theorem which we can formulate as follows. Suppose that for some value of the parameter, say $\lambda = \lambda_0$, the following holds:

(i) the linearization $A_0 := D_x f(0, \lambda_0) \in \mathcal{L}(\mathbb{R}^n)$ has a pair of simple purely imaginary eigenvalues $\pm i\omega_0$ ($\omega_0 > 0$), with (real) eigenspace $U := \ker(A_0^2 + \omega_0^2 I)$;

(ii) A_0 has no other eigenvalues of the form $\pm i\ell\omega_0$, with $\ell \in \mathbb{N}$, $\ell \neq 1$;

(iii) the quadratic form $Q_0(u) := D_x^2 K(0, \lambda_0) \cdot (u, u)$ on U is non-degenerate.

---

*Department of Mathematics, Technische Universität Ilmenau, PSF 327, D-98684 Ilmenau, Germany

†Department of Pure Mathematics and Computer Algebra, University of Gent, Krijgslaan 281, B-9000 Gent, Belgium

Then the equation (1.1) has, for each λ sufficiently close to λ_0. a smooth one-parameter family of periodic orbits $\gamma_\rho^{(\lambda)}$ ($0 \leq \rho < \rho_0$), with $\gamma_{\rho=0}^{(\lambda)} = \{0\}$, and with minimal period $T_\rho^{(\lambda)}$ converging to $2\pi/\omega_0$ as $(\rho, \lambda) \to (0, \lambda_0)$.

This theorem associates a one-parameter family of periodic orbits to each pair of purely imaginary eigenvalues of A_0 for which the non-resonance and non-degeneracy conditions are satisfied; so if A_0 has several such pairs of eigenvalues, then several local families of periodic orbits will emanate from the equilibrium. In particular one might think about a situation where (for some $k \geq 2$) there are k pairs of simple purely imaginary eigenvalues close to each other; by changing the parameters these eigenvalues may move even closer and finally coalesce. Generically such scenario requires at least $k - 1$ parameters, and at the critical value of these parameters (that is, when all k pairs coalesce) the linearization will have a non-semisimple pair of purely imaginary eigenvalues, with geometric multiplicity one but algebraic multiplicity k; we call this a k-fold resonance. Moreover, exploring the full neighborhood of such resonance, one finds that for certain parameter values some or all of the critical eigenvalues will be off the imaginary axis (see Section 3 and [4] for details). So, as one moves around in parameter space the number of one-parameter families of periodic orbits as given by the Liapunov Center Theorem will change, i.e. bifurcations will occur. The aim of this paper is to describe these bifurcations; more in particular, we want to determine the bifurcation set in parameter space and the changes which the set of small periodic orbits of (1.1) undergoes as the parameter crosses this bifurcation set.

In an earlier paper [4] we have studied the same problem for reversible systems; it will appear that the bifurcation results for the conservative case considered here are completely analogous to those found in [4] for the reversible case. So this is one further example of something which has been noticed in many other studies: periodic orbits of Hamiltonian or more generally conservative systems behave mostly in a way which is similar to the behaviour of *symmetric* periodic orbits of reversible systems. This does not mean that the analysis leading to the results is the same for both types of systems: in most cases one can use parallel arguments, but at crucial points one has to distinguish between reversible systems at one side, and conservative systems at the other side. Therefore we will in this paper outline the reduction procedure for the study of the k-fold resonance in conservative systems, and work out in detail those parts of the reduction which are specific for such conservative systems (this includes in particular the normal form reduction of the linear part of the system). For details on a number of arguments and calculations which are common for the conservative and the reversible case we will refer to [4]; also the discussion of the final bifurcation picture has been given in [4]. Where possible we will also use the same notation as in [4].

We start by making our hypotheses precise. We notice that by a parameter change we can assume that the critical parameter value is at $\lambda = 0$, while by a time rescale we can make the critical eigenvalues equal to $\pm i$; this allows us to formulate our main hypothesis as follows:

(**H**) The smooth mapping $f : \mathbb{R}^n \times \mathbb{R}^m \to \mathbb{R}^n$ appearing in (1.1) is such that:

    (i) $f(0, \lambda) = 0$ for all $\lambda \in \mathbb{R}^m$;

    (ii) $D_x K(x, \lambda) \cdot f(x, \lambda) = 0$ for some smooth $K : \mathbb{R}^n \times \mathbb{R}^m \to \mathbb{R}$;

    (iii) the operator $A_0 := D_x f(0, 0) \in \mathcal{L}(R^n)$ has the eigenvalues $\pm i$, and no other eigenvalues of the form $\pm \ell i$, with $\ell \in \mathbb{N}, \ell \neq 1$;

    (iv) the subspace $\ker (A_0^2 + I)$ is irreducible under A_0;

    (v) the quadratic form $Q_0(u) := D_x^2 K(0, 0) \cdot (u, u)$ is non-degenerate on the space $U := \ker (S_0^2 + I)$; here S_0 is the semisimple part of A_0.

The hypothesis (**H**)(iv) means that the eigenvalues $\pm i$ of A_0 have geometric multiplicity equal to 1; their algebraic multiplicity is given by the smallest integer k such that $\ker (A_0^2 + I)^{k+1} = \ker (A_0^2 + I)^k$. We will denote by $A_0 = S_0 + N_0$ the (unique) semisimple-nilpotent decomposition of A_0; this means that S_0 is semisimple (i.e. complex diagonalizable), N_0 is nilpotent, and $S_0 N_0 = N_0 S_0$. The subspace U appearing in (**H**)(v) is then the *generalized* eigenspace corresponding to the eigenvalues $\pm i$ of A_0; it is also given by $U = \ker (A_0^2 + I)^k$. It is clear that U is invariant under each of the operators A_0, S_0 and N_0; we denote the restrictions of these operators to U by respectively A, S and $N \in \mathcal{L}(U)$. It follows from $U = \ker (S^2 + I_U)$ that S generates an S^1-action on U, given by

$$\varphi \in S^1 \cong \mathbb{R}/2\pi\mathbb{Z} \longmapsto \exp(\varphi S) \in \mathcal{L}(U). \tag{1.3}$$

Also, k is the smallest integer such that $N^k = 0$; although our hypotheses do not exclude that $k = 1$ we are mainly interested in the case $k \geq 2$.

Under the hypothesis (**H**) we want to study the following problem:

    (**P**) Describe, for all sufficiently small $\lambda \in \mathbb{R}^m$, all small periodic solutions of (1.1) with period T near 2π.

As in [4] we will use the general reduction results of [2] to immediately reduce this problem to a similar one on the *reduced phase space* U. Indeed, it follows from the part of [2] that deals with conservative systems that the solutions of (**P**) are in one-to-one relation with the small T-periodic solutions of a *reduced system*

$$\dot{u} = g(u, \lambda), \tag{1.4}$$

where the reduced vector field $g : U \times \mathbb{R}^m \to U$ can be chosen to be of class C^q for any $q \geq 2$ (see [2] and [4] for more details), can be determined up to any order by bringing the original vector field f in normal form with respect to S_0 (again, see [2]), and has the following properties:

    (a) $g(0, \lambda) = 0$ for all $\lambda \in \mathbb{R}^m$ and $D_u g(0, 0) = A = S + N$;

    (b) g is S^1-equivariant:

$$g(\exp(\varphi S)u, \lambda) = \exp(\varphi S)g(u, \lambda). \quad \forall \varphi \in S^1; \tag{1.5}$$

    (c) g has an S^1-invariant first integral; more precisely, there exists a smooth mapping $M : U \times \mathbb{R}^m \to \mathbb{R}$, with $D_u^2 M(0, 0) \cdot (u, u) = D_x^2 K(0, 0) \cdot (u, u)$ and

$$M(\exp(\varphi S)u, \lambda) = M(u, \lambda), \quad \forall \varphi \in S^1, \tag{1.6}$$

such that

$$D_u M(u, \lambda) \cdot g(u, \lambda) = 0, \quad \forall (u, \lambda) \in U \times \mathbb{R}^m. \tag{1.7}$$

Concerning property (b) we should stress the fact that the S^1-equivariance of g is exact, and not just formal as with the normal form; this exact S^1-equivariance of the reduced vector field is due to the fact that we concentrate on periodic solutions, forgetting the other dynamics (see [2] for more details). It follows from the S^1-equivariance that for all (λ, T) near $(0, 2\pi)$ all sufficiently small T-periodic solutions of $(1.4)_\lambda$ necessarily have the form

$$\tilde{u}(t) = \exp((1 + \sigma) S t) u \tag{1.8}$$

for some small $u \in U$ and with $\sigma \in \mathbb{R}$ given from $T = 2\pi / (1 + \sigma)$: these periodic solutions can therefore be obtained by solving the *determining equation*

$$(1 + \sigma) S u = g(u, \lambda) \tag{1.9}$$

for (u, λ, σ) near $(0, 0, 0)$ in $U \times \mathbb{R}^m \times \mathbb{R}$. The solutions of (1.9) come in S^1-orbits; each such solution orbit corresponds to a periodic orbit of our original equation (1.1).

In the next section we will show how by parameter-dependent linear transformations we can bring the linear part $D_u g(0, \lambda)$ of the reduced vector field g in an explicit normal form which will allow us to determine the bifurcation set for (1.9).

2 The linear normal form

For each λ we define the linear operator $A_\lambda := D_u g(0, \lambda) \in \mathcal{L}(U)$ and the quadratic form $Q_\lambda(u) := D_u^2 M(0, \lambda) \cdot (u, u)$ ($u \in U$); referring to the S^1-action on U given by (1.3) one shows that these have the following properties (for small λ):

(1) $A_{\lambda=0} = A = S + N$, where $S^2 + I_U = 0$, S generates the S^1-action on U, $SN = NS$, $N^k = 0$ but $N^{k-1} \neq 0$, and $\ker(A^2 + I) = \ker N$ is irreducible under S;

(2) $A_\lambda S = S A_\lambda$ (i.e. A_λ is S^1-equivariant);

(3) $Q_{\lambda=0} = Q_0$ is non-degenerate;

(4) $Q_\lambda(\exp(S\varphi)u) = Q_\lambda(u)$ for all $\varphi \in S^1$;

(5) $DQ_\lambda(u) \cdot A_\lambda u = 0$ for all $u \in U$.

One obtains (5) by differentiating the identity (1.7) at $u = 0$: a first differentiation combined with the fact that A_λ is non-singular for small λ (this follows from (1)) shows that $D_u M(0, \lambda) = 0$; a further differentiation then gives (5).

Our aim in this section is to bring the family of linear operators A_λ ($\lambda \in \mathbb{R}^m$) in a simple and explicit form. As a preliminary step we first prove the existence of an appropriate scalar product on the space U.

Lemma 1 *Under the foregoing hypotheses there exists a scalar product $\langle \cdot, \cdot \rangle$ on U such that the following properties hold (we denote by $L^T \in \mathcal{L}(U)$ the transpose of $L \in \mathcal{L}(U)$ with respect to this scalar product):*

(i) the S^1-action on U is orthogonal, i.e. $S^T = -S = S^{-1}$;

(ii) there exists an orthogonal and S^1-invariant decomposition $U = U_1 \oplus \cdots \oplus U_k$ of the space U such that each subspace U_j ($1 \leq j \leq k$) is S^1-irreducible, $U_1 = \ker N$, and N is an isomorphism from U_j onto U_{j-1} for $2 \leq j \leq k$;

(iii) we can write $Q_0(u)$ in the form $Q_0(u) = \langle u, B_0 u \rangle$, with $B_0 \in \mathcal{L}(U)$ such that $B_0^T = B_0$ and $B_0^2 = I_U$.

Moreover, these properties also imply the following:

(iv) $N^T S = S N^T$, $U_k = \ker N^T$ and

$$N^T(U_j) = U_{j+1}, \qquad N^T|_{U_j} = \left(N|_{U_{j-1}} \right)^{-1}. \qquad 1 \leq j \leq k-1; \quad (2.1)$$

(v) any S^1-equivariant linear operator $L_j \in \mathcal{L}(U_j)$ ($1 \leq j \leq k$) necessarily has the form

$$L_j = \alpha I_{U_j} + \beta S|_{U_j}$$

for some $\alpha, \beta \in \mathbb{R}$;

(vi) $B_0 S = S B_0$ and $N^T B_0 + B_0 N = 0$.

Proof For the first part of the proof we refer to [4, Section 3] where the details of the following construction are given. Let U_k be any S^1-invariant complement of $\ker((A^2 + I_U)^{k-1})$ in U, and set $U_j := N^{k-j}(U_k)$ for $1 \leq j < k$. Then $U = U_1 \oplus \cdots \oplus U_k$, each of the subspaces U_j ($1 \leq j \leq k$) is invariant and irreducible under the S^1-action, $U_1 = \ker N$, and N maps U_j isomorphically onto U_{j-1} for $2 \leq j \leq k$. Choose a scalar product $\langle \cdot, \cdot \rangle_1$ on U_1 for which the S^1-action on U_1 is orthogonal, and define a scalar product on U by

$$\left\langle \sum_{j=1}^k u_j, \sum_{j=1}^k u_j' \right\rangle := \sum_{j=1}^k \left\langle N^{j-1} u_j, N^{j-1} u_j' \right\rangle_1. \qquad \forall u_j, u_j' \in U_j, \ 1 \leq j \leq k. \quad (2.2)$$

Then the conditions (i) and (ii) are satisfied, and as a consequence also (iv) and (v) hold (see [4, Lemma 6] for the proof of (v)). Moreover, there exists a unique linear operator $B_0 \in \mathcal{L}(U)$ such that $Q_0(u) = \langle u, B_0 u \rangle$ and $B_0^T = B_0$. Since Q_0 is non-degenerate B_0 will be invertible, and the condition $DQ_0(u) \cdot Au = 0$ translates into

$$A^T B_0 + B_0 A = 0. \qquad (2.3)$$

This implies $A = -B_0^{-1} A^T B_0 = B_0^{-1} S B_0 - B_0^{-1} N^T B_0$, and the uniqueness of the semisimple-nilpotent decomposition of A then proves that $B_0^{-1} S B_0 = S$ and $-B_0^{-1} N^T B_0 = N$, i.e. we have (vi).

It remains to show that the foregoing construction can be done in such a way that also $B_0^2 = I_U$ holds. To prove this we will (in an indirect way) use the fact that in our construction we can choose for U_k any S^1-invariant complement of $\ker\left((A^2 + I_U)^{k-1}\right)$: although formulated differently, the proof which follows in fact shows that U_k can be chosen such that $B_0^2 = I_U$. To start, fix a particular such complement U_k and carry out the construction just described: this leads to a particular decomposition $U = U_1 \oplus \cdots \oplus U_k$ and to a particular scalar product $\langle \cdot, \cdot \rangle$ on U, which we keep fixed from now on. In this framework also the symmetric operator $B_0 \in \mathcal{L}(U)$ is uniquely determined and satisfies (vi); we will use (vi) to get more details on the possible form of B_0.

Using the S^1-equivariance of B_0 and the properties (ii) and (v) it is easily seen that there must exist numbers $\alpha_j, \beta_j \in \mathbb{R}$ $(1 \leq j \leq k)$ such that

$$B_0(u_k) = \sum_{j=1}^{k} \left(\alpha_j S^{k-j} N^{k-j} u_k + \beta_j S^{k-j+1} N^{k-j} u_k \right), \quad \forall u_k \in U_k. \qquad (2.4)$$

(We have included the operators S^{k-j} in order to simplify some of the expressions which follow). From (2.4) we see that for all $u_k, \tilde{u}_k \in U_k$ and all $j \in \{1, \ldots, k\}$ we have

$$\left\langle N^{k-j}\tilde{u}_k, B_0 u_k \right\rangle = \alpha_j \left\langle \tilde{u}_k, S^{k-j} u_k \right\rangle + \beta_j \left\langle \tilde{u}_k, S^{k-j+1} u_k \right\rangle.$$

From the other side it follows from (vi) and the symmetry of B_0 that

$$\begin{aligned}
\left\langle N^{k-j}\tilde{u}_k, B_0 u_k \right\rangle &= \left\langle B_0 N^{k-j}\tilde{u}_k, u_k \right\rangle = (-1)^{k-j} \left\langle (N^T)^{k-j} B_0 \tilde{u}_k, u_k \right\rangle \\
&= (-1)^{k-j} \left\langle B_0 \tilde{u}_k, N^{k-j} u_k \right\rangle \\
&= (-1)^{k-j} \alpha_j \left\langle S^{k-j}\tilde{u}_k, u_k \right\rangle + (-1)^{k-j} \beta_j \left\langle S^{k-j+1}\tilde{u}_k, u_k \right\rangle \\
&= \alpha_j \left\langle \tilde{u}_k, S^{k-j} u_k \right\rangle - \beta_j \left\langle \tilde{u}_k, S^{k-j+1} u_k \right\rangle.
\end{aligned}$$

Comparing the two expressions we conclude that $\beta_j = 0$ $(1 \leq j \leq k)$, such that (2.4) simplifies to

$$B_0(u_k) = \sum_{j=1}^{k} \alpha_j S^{k-j} N^{k-j} u_k, \quad \forall u_k \in U_k. \qquad (2.5)$$

A further calculation (using (iv) and (vi)) then gives

$$\begin{aligned}
B_0(S^i N^i u_k) &= (-SN^T)^i B_0(u_k) = (-SN^T)^i \sum_{j=1}^{k} \alpha_j S^{k-j} N^{k-j} u_k \\
&= (-S)^i \sum_{j=1}^{k} \alpha_j S^{k-j} (N^T)^i N^{k-j} u_k = \sum_{j=1}^{k-i} \alpha_j S^{k-i-j} N^{k-i-j} u_k \qquad (2.6)
\end{aligned}$$

valid for all $u_k \in U_k$ and for $0 \leq i \leq k-1$. In particular we find for $i = k-1$ that $B_0(S^{k-1} N^{k-1} u_k) = \alpha_1 u_k$ for all $u_k \in U_k$. Since $S^{k-1} N^{k-1}(U_k) = U_1$ and B_0 is non-singular we conclude that $\alpha_1 \neq 0$; in fact, multiplying the first integral M of (1.4) by α_1^{-1} we can without loss of generality assume that $\alpha_1 = 1$.

Now *suppose* that

$$\alpha_2 = \cdots = \alpha_k = 0. \tag{2.7}$$

Then (2.6) gives

$$B_0(S^i N^i u_k) = S^{k-i-1} N^{k-i-1} u_k, \qquad \forall u_k \in U_k, \quad 0 \le i \le k-1, \tag{2.8}$$

which implies $B_0^2(S^i N^i u_k) = S^i N^i u_k$. Since the subspaces $S^i N^i(U_k) = U_{k-i}$ ($0 \le i \le k-1$) generate U we conclude that (2.7) implies that $B_0^2 = I_U$; also the opposite is true, as one can easily show. Therefore, to conclude the proof of the lemma we have to find a way to satisfy (2.7); we will do this by applying an appropriate linear transformation to the equation (1.4).

Let $T \in \mathcal{L}(U)$ be of the form $T = \sum_{i=0}^{k-1} \beta_i S^i N^i$, with $\beta_0 = 1$ and $\beta_i \in \mathbb{R}$ ($1 \le i \le k-1$); then T is an S^1-equivariant isomorphism of U, with an inverse of the same form. Now replace the phase variable u in (1.4) by Tu; then the vector field $g(u, \lambda)$ is replaced by $\hat{g}(u, \lambda) := T^{-1} g(Tu, \lambda)$, the first integral $M(u, \lambda)$ by $\hat{M}(u, \lambda) := M(Tu, \lambda)$, and the quadratic form $Q_\lambda(u)$ by $\hat{Q}_\lambda(u) := Q_\lambda(Tu)$. It follows that for $\lambda = 0$ the linearization $D_u \hat{g}(0,0) = T^{-1} AT = A = S + N$ remains unchanged, while B_0 is replaced by $\hat{B}_0 := T^T B_0 T$; as before we can write $\hat{B}_0(u_k) = \sum_{j=1}^k \hat{\alpha}_j S^{k-j} N^{k-j} u_k$ for all $u_k \in U_k$ and for certain coefficients $\hat{\alpha}_j$ ($1 \le j \le k$). We will now show that it is possible to choose the constants β_i ($1 \le i \le k-1$) in the definition of T such that (2.7) is satisfied for $\hat{B}_0$, i.e. such that $\hat{\alpha}_2 = \cdots = \hat{\alpha}_k = 0$.

We have $(SN)^T B_0 = B_0(SN)$, and hence $\hat{B}_0 = B_0 T^2$. The operator T^2 has the form $T^2 = \sum_{i=0}^{k-1} \gamma_i S^i N^i$, with $\gamma_0 = \beta_0^2 = 1$ and $\gamma_i = 2\beta_0 \beta_i + \zeta_i(\beta_0, \ldots, \beta_{i-1})$ for some quadratic polynomials ζ_i ($1 \le i \le k-1$); it follows that the mapping $(\beta_1, \ldots, \beta_{k-1}) \longmapsto (\gamma_1, \ldots, \gamma_{k-1})$ is a bijection on $\mathbb{R}^{k-1}$. Using (2.6) and rearranging terms one finds then that

$$
\begin{aligned}
\hat{B}_0 u_k &= B_0 \left(\sum_{i=0}^{k-1} \gamma_i S^i N^i \right) u_k = \sum_{i=0}^{k-1} \sum_{j=1}^{k-i} \gamma_i \alpha_j S^{k-j-i} N^{k-j-i} u_k \\
&= \sum_{j=1}^k \left(\sum_{i=0}^{j-1} \gamma_i \alpha_{j-i} \right) S^{k-j} N^{k-j} u_k,
\end{aligned}
$$

i.e. the new coefficients $\hat{\alpha}_j$ are given by $\hat{\alpha}_j = \sum_{i=0}^{j-1} \gamma_i \alpha_{j-i}$ ($1 \le j \le k$). Using $\gamma_0 = \alpha_1 = 1$ it is then an easy matter to show that the set of equations

$$\hat{\alpha}_j = \sum_{i=0}^{j-1} \gamma_i \alpha_{j-i} = 0, \quad (2 \le j \le k) \tag{2.9}$$

can be solved for $\gamma_1, \ldots, \gamma_{k-1}$ in function of $\alpha_2, \ldots, \alpha_k$: hence also the coefficients $\beta_1, \ldots, \beta_{k-1}$ appearing in the definition of T can be chosen such that (2.9) holds.

By our earlier results. this implies $\hat{B}_0^2 = I_U$, which completes the proof of the lemma. Observe that the linear transformation which we used in fact amounts to replacing the originally chosen subspace U_k by $\hat{U}_k := T^{-1}(U_k)$. which is again (because of the form of T) an S^1-invariant complement of $\ker((A^2 + I_U)^{k-1})$ in U. ∎

From now on we fix a scalar product on U such that all the properties of Lemma 1 are satisfied; each of the properties given by the Lemma will be used at various places in the reduction which follows. The first step in this reduction consists in making the quadratic form $Q_\lambda(u)$ independent of the parameter λ.

Lemma 2 *There exists a smooth mapping* $\Psi : \mathbb{R}^m \to \mathcal{L}(U), \lambda \mapsto \Psi_\lambda$, *with* $\Psi_0 = I_U$, *and such that we have for all sufficiently small* λ *that*

$$\Psi_\lambda S = S\Psi_\lambda \tag{2.10}$$

and

$$Q_\lambda(\Psi_\lambda u) = Q_0(u), \quad \forall u \in U. \tag{2.11}$$

Proof We write $Q_\lambda(u)$ in the form $Q_\lambda(u) = \langle u, B_\lambda u \rangle$, with $B_\lambda \in \mathcal{L}(U)$ symmetric ($B_\lambda^T = B_\lambda$), commuting with S and depending smoothly on λ; for $\lambda = 0$ this operator is precisely the operator B_0 considered in Lemma 1. Then $Q_\lambda(\Psi u) = \langle u, \Psi^T B_\lambda \Psi u \rangle$ for $\Psi \in \mathcal{L}(U)$, and we have to show that the equation

$$\Psi^T B_\lambda \Psi = B_0 \tag{2.12}$$

has for each sufficiently small λ a solution $\Psi = \Psi_\lambda \in \mathcal{L}(U)$ which depends smoothly on λ, reduces to I_U for $\lambda = 0$, and commutes with S. Such solution can be obtained from an application of the implicit function theorem, as follows. Let $X := \{\Psi \in \mathcal{L}(U) \mid \Psi S = S\Psi\}$, $Y := \{\Phi \in \mathcal{L}(U) \mid \Phi S = S\Phi$ and $\Phi^T = \Phi\}$. and define $F : X \times \mathbb{R}^m \to Y$ by $F(\Psi, \lambda) := \Psi^T B_\lambda \Psi - B_0$. Then $F(I_U, 0) = 0$ and $D_\Psi F(I_U, 0) \cdot \Psi = \Psi^T B_0 + B_0 \Psi$; the result will follow if we can show that $D_\Psi F(I_U, 0) \in \mathcal{L}(X, Y)$ is surjective, i.e. that the equation

$$\Psi^T B_0 + B_0 \Psi = \Phi \tag{2.13}$$

has for each $\Phi \in Y$ a solution $\Psi \in X$. Since B_0 is non-singular the equation (2.13) has, by a classical result. for each symmetric $\Phi \in \mathcal{L}(U)$ a solution $\Psi \in \mathcal{L}(U)$ — in fact, (2.13) has a unique solution Ψ which is lower-triangular with respect to an orthonormal basis of U which consists of eigenvectors of B_0. If Φ belongs also to Y (i.e. commutes with S), and if $\Psi \in \mathcal{L}(U)$ is *any* solution of (2.13). then also

$$\widehat{\Psi} := \frac{1}{2\pi} \int_0^{2\pi} \exp(-S\varphi) \Psi \exp(S\varphi) \, d\varphi \in X$$

is a solution (here one uses the orthogonality of the S^1-action). This shows that $D_\Psi F(I_U, 0)$ is indeed surjective from X onto Y, which completes the proof. ∎

Using this lemma and replacing u by $\Psi_\lambda u$ in (1.4) we find a new equation which is still S^1-equivariant, and which has the S^1-invariant first integral $M_1(u, \lambda) := M(\Psi_\lambda u, \lambda)$: the quadratic part of this first integral is given by $\frac{1}{2}Q_\lambda(\Psi_\lambda u) = \frac{1}{2}Q_0(u)$, i.e. it is independent of the parameter λ. Denoting the new vector field again by $g(u, \lambda)$ and the new first integral by $M(u, \lambda)$ this means that after this first linear transformation we can replace the property (5) above by

(5′) $DQ_0(u) \cdot A_\lambda u = 0$ for all $u \in U$.

Next we define for each $\Phi \in \mathcal{L}(U)$ linear operators $\operatorname{Ad}(\Phi)$ and $\operatorname{ad}(\Phi)$ on $\mathcal{L}(U)$ by

$$\operatorname{Ad}(\Phi) \cdot \Psi := e^{-\Phi} \Psi e^{\Phi} \quad \text{and} \quad \operatorname{ad}(\Phi) \cdot \Psi := \Psi\Phi - \Phi\Psi, \quad \forall \Psi \in \mathcal{L}(U). \quad (2.14)$$

We also introduce the vector space

$$Z := \{\Phi \in \mathcal{L}(U) \mid \Phi S = S\Phi \text{ and } DQ_0(u) \cdot \Phi u = 0, \forall u \in U\}. \quad (2.15)$$

Using $Q_0(u) = \langle u, B_0 u\rangle$ the second condition in (2.15) can be rewritten as

$$\Phi^T B_0 + B_0 \Phi = 0. \quad (2.16)$$

and is also equivalent to

$$Q_0\left(e^{\Phi t} u\right) = Q_0(u), \quad \forall t \in \mathbb{R}, \forall u \in U. \quad (2.17)$$

It follows from the properties (2) and (5′) above that the operators A_λ ($\lambda \in \mathbb{R}^m$) belong to the space Z. The following lemma summarizes the main properties of Z.

Lemma 3 *The space Z defined by (2.15) has the following properties:*

(i) $\Phi \in Z$ implies $\Phi^T \in Z$;

(ii) if $\Phi \in Z$ then $\operatorname{Ad}(\Phi)$ and $\operatorname{ad}(\Phi)$ map Z into itself, i.e.

$$\operatorname{Ad}(\Phi) \cdot \Psi \in Z \quad \text{and} \quad \operatorname{ad}(\Phi) \cdot \Psi \in Z, \quad \forall \Phi, \Psi \in Z; \quad (2.18)$$

(iii) if $\Phi = S_\Phi + N_\Phi$ is the semisimple-nilpotent decomposition of $\Phi \in Z$, then both S_Φ and N_Φ belong to Z.

Proof Property (i) follows from $S^T = -S$, $B_0^2 = I_U$ and (2.16). To prove (ii) observe first that both $\operatorname{Ad}(\Phi) \cdot \Psi$ and $\operatorname{ad}(\Phi) \cdot \Psi$ commute with S if Φ and Ψ do. Also, if Φ belongs to Z then we have by (2.17) that $Q_0(e^\Phi u) = Q_0(u)$ for all $u \in U$, and hence

$$DQ_0(e^\Phi u) \cdot e^\Phi \tilde{u} = DQ_0(u) \cdot \tilde{u}, \quad \forall u, \tilde{u} \in U.$$

This implies that for all $\Psi \in Z$ and all $u \in U$ we have

$$DQ_0(u) \cdot (\operatorname{Ad}(\Phi) \cdot \Psi) \cdot u = DQ_0(u) \cdot e^{-\Phi} \Psi e^\Phi u = DQ_0(e^\Phi u) \cdot \Psi e^\Phi u = 0,$$

i.e. $\text{Ad}(\Phi) \cdot \Psi$ belongs indeed to Z. The same conclusion then also holds for $\text{ad}(\Phi) \cdot \Psi$, as follows from the relation

$$\text{ad}(\Phi) \cdot \Psi = \left.\frac{d}{ds}\right|_{s=0} \text{Ad}(\Phi s) \cdot \Psi, \tag{2.19}$$

which is valid for all $\Phi, \Psi \in \mathcal{L}(U)$. Finally, if $\Phi = S_\Phi + N_\Phi$ is the semisimple-nilpotent decomposition of $\Phi \in Z$, then it follows from (2.16) that

$$\Phi = -B_0^{-1}\Phi^T B_0 = -B_0^{-1}S_\Phi^T B_0 - B_0^{-1}N_\Phi^T B_0,$$

which gives us a second semisimple-nilpotent decomposition of Φ. The uniqueness of this decomposition then implies that $S_\Phi^T B_0 + B_0 S_\Phi = 0$ and $N_\Phi^T B_0 + B_0 N_\Phi = 0$. Since obviously S_Φ and N_Φ commute with S, this proves (iii).   ∎

Since $A = S + N$ belongs to Z it follows from this lemma that also S, N and N^T belong to Z; this allows us to formulate the main result of this section in the following form.

Theorem 4 *There exist a neighborhood Ω of A in Z and a smooth mapping Φ^* : $\Omega \to Z$ with $\Phi^*(A) = 0$ and such that for all $\Psi \in \Omega$ we have*

$$\text{Ad}(\Phi^*(\Psi)) \cdot \Psi - A \in Z \cap \ker(\text{ad}(N^T)). \tag{2.20}$$

Proof Define $F : Z \times Z \to Z$ by $F(\Phi, \Psi) := \text{Ad}(\Phi) \cdot \Psi - A$. Then $F(0, A) = 0$ and $D_\Phi F(0, A) \cdot \tilde{\Phi} = A\tilde{\Phi} - \tilde{\Phi}A = N\tilde{\Phi} - \tilde{\Phi}N$ for all $\tilde{\Phi} \in Z$, i.e.

$$D_\Phi F(0, A) = -\left.\text{ad}(N)\right|_Z \in \mathcal{L}(Z). \tag{2.21}$$

We will show further on that

$$Z = \text{Im}\left(\left.\text{ad}(N)\right|_Z\right) \oplus \ker\left(\left.\text{ad}(N^T)\right|_Z\right). \tag{2.22}$$

Denote by π the projection in Z onto $Z_1 := \text{Im}\left(\left.\text{ad}(N)\right|_Z\right)$ and parallel to $\ker\left(\left.\text{ad}(N^T)\right|_Z\right)$, and define $F_1 : Z \times Z \to Z_1$ by $F_1(\Phi, \Psi) := \pi F(\Phi, \Psi)$. Then $F_1(0, A) = 0$, and (2.21) implies that $D_\Phi F_1(0, A) = \pi \circ D_\Phi F(0, A) \in \mathcal{L}(Z; Z_1)$ is surjective. So we can apply the implicit function theorem to conclude that there exist a neighborhood Ω of A in Z and a smooth mapping $\Phi^* : \Omega \to Z$ such that $\Phi^*(A) = 0$ and $F_1(\Phi^*(\Psi), \Psi) = 0$ for all $\Psi \in \Omega$; going back through the definitions this means precisely that (2.20) holds.

It remains to prove (2.22). To do so we introduce a scalar product $\ll \cdot, \cdot \gg$ on Z by

$$\ll \Psi_1, \Psi_2 \gg := \text{trace}(\Psi_1^T \Psi_2), \quad \forall \Psi_1, \Psi_2 \in Z. \tag{2.23}$$

A direct calculation then shows that for all $\Psi_1, \Psi_2 \in Z$ we have

$$\begin{aligned}
\ll \text{ad}(N) \cdot \Psi_1, \Psi_2 \gg &= \ll \Psi_1 N - N\Psi_1, \Psi_2 \gg = \text{trace}((\Psi_1 N - N\Psi_1)^T \Psi_2) \\
&= \text{trace}(N^T \Psi_1^T \Psi_2 - \Psi_1^T N^T \Psi_2) \\
&= \text{trace}(\Psi_1^T(\Psi_2 N^T - N^T \Psi_2)) \\
&= \ll \Psi_1, \text{ad}(N^T) \cdot \Psi_2 \gg.
\end{aligned}$$

So the adjoint $\left(\operatorname{ad}(N)\big|_Z\right)^* \in \mathcal{L}(Z)$ of $\operatorname{ad}(N)\big|_Z \in \mathcal{L}(Z)$ with respect to $\ll \cdot, \cdot \gg$ is given by $\left(\operatorname{ad}(N)\big|_Z\right)^* = \operatorname{ad}(N^T)\big|_Z$, and (2.22) follows from the classical result which says that $Z = \operatorname{Im}(L) \oplus \ker(L^*)$ for any $L \in \mathcal{L}(Z)$. ∎

Again we use this theorem to perform a linear transformation on the equation (1.4); more precisely, we replace u by $\Phi^*(A_\lambda)u$. Keeping the old notation for the new equation one can then easily verify that all properties which we had before remain unchanged, but that moreover the linearization A_λ has the special form $A_\lambda = A + \Psi_\lambda$, with $\Psi_\lambda \in Z \cap \ker(\operatorname{ad}(N^T))$ (see (2.20)). Therefore our next task consists in determining the explicit form of the elements of $Z \cap \ker(\operatorname{ad}(N^T))$; this is done in the next lemma.

Lemma 5 A linear operator $\Psi \in \mathcal{L}(U)$ belongs to $Z \cap \ker(\operatorname{ad}(N^T))$ if and only if it has the form

$$\Psi = \sum_{j=1}^{k} \delta_j S^j (N^T)^{j-1} \tag{2.24}$$

for some $\delta_j \in \mathbb{R}$ $(1 \leq j \leq k)$.

Proof It is clear that if $\Psi \in \mathcal{L}(U)$ has the form (2.24) then it commutes with S and N^T, and using Lemma 1(vi) one can also directly verify that $\Psi^T B_0 + B_0 \Psi = 0$; hence, if Ψ has the form (2.24) then it belongs to $Z \cap \ker(\operatorname{ad}(N^T))$. To prove the converse let us assume that Ψ belongs to $Z \cap \ker(\operatorname{ad}(N^T))$; in particular, Ψ commutes with S and N^T. By [4, Theorem 7] this implies that Ψ must have the form

$$\Psi = \sum_{j=1}^{k} (\alpha_j I_U + \beta_j S)(N^T)^{j-1} = \sum_{j=1}^{k} (\gamma_j S^{j-1} + \delta_j S^j)(N^T)^{j-1} \tag{2.25}$$

for some constants $\alpha_j, \beta_j, \gamma_j, \delta_j \in \mathbb{R}$ $(1 \leq j \leq k)$. Using again Lemma 1(vi)) the condition $\Psi^T B_0 + B_0 \Psi = 0$ then gives

$$\sum_{j=1}^{k} \gamma_j B_0 (S N^T)^{j-1} = 0. \tag{2.26}$$

Applying $(SN)^{k-1}$ to (2.26) shows that $\gamma_1 B_0(SN^T)^{k-1} = 0$, from which we get $\gamma_1 = 0$. since $B_0(SN^T)^{k-1} \neq 0$. Next we apply $(SN)^{k-2}$ to (2.26) and obtain in the same way that $\gamma_2 = 0$; the argument can be repeated to show that all γ_j $(1 \leq j \leq k)$ must be zero. Hence each element $\Psi \in Z \cap \ker(\operatorname{ad}(N^T))$ must have the form (2.24). ∎

Using Lemma 5 in combination with the remark after Theorem 4 we see that we can assume that the linearization A_λ has the form

$$A_\lambda = S + N + \sum_{j=1}^{k} \delta_j(\lambda) S^j (N^T)^{j-1} = (1 + \delta_1(\lambda))S + N + \sum_{j=2}^{k} \delta_j(\lambda) S^j (N^T)^{j-1}, \tag{2.27}$$

for some sufficiently smooth functions $\delta_j : \mathbb{R}^m \to \mathbb{R}$ $(1 \le j \le k)$ satisfying $\delta_j(0) = 0$. There are two further simplifications which we can carry out. First, we can make $\delta_1(\lambda) \equiv 0$ by a parameter-dependent rescaling of both the time and the phase variables (see [4] for the details); that is, we can assume that

$$A_\lambda = S + N + \sum_{j=2}^{k} \delta_j(\lambda) S^j (N^T)^{j-1}. \tag{2.28}$$

Observe that after such rescaling we will no longer have that the quadratic form $Q_\lambda(u)$ is independent of λ; however, at this point of the reduction this is no longer important, since we have already achieved our goal, namely to bring A_λ in a simple explicit form. The second simplification requires an additional hypothesis, motivated by (2.28): we assume the following *transversality condition*:

(**T**) The mapping $\Delta : \mathbb{R}^m \longrightarrow \mathbb{R}^{k-1}$, $\lambda \longmapsto (\delta_2(\lambda), \ldots, \delta_k(\lambda))$ is transversal to the origin at $\lambda = 0$, i.e. $D\Delta(0) \in \mathcal{L}(\mathbb{R}^m, \mathbb{R}^{k-1})$ is surjective.

This implies that $m \ge k - 1$, and without loss of generality we can assume that

$$\frac{\partial(\delta_2, \ldots, \delta_k)}{\partial(\lambda_1, \ldots, \lambda_{k-1})}(0) \ne 0.$$

By a change of parameters we can then put the functions δ_j in the explicit form $\delta_j(\lambda) = \lambda_{j-1}$ $(2 \le j \le k)$, which gives us our final form for A_λ:

$$A_\lambda = S + N + \sum_{j=1}^{k-1} \lambda_j S^{j+1} (N^T)^j. \tag{2.29}$$

This is precisely the same form as the one obtained in [4] for the reversible case.

3   The bifurcation analysis

With the explicit form (2.29) of the linearization A_λ at hand we can now turn our attention to solving the determining equation (1.9); we write this equation in the form

$$\Theta(u, \lambda, \sigma) := -(1 + \sigma) S u + g(u, \lambda) = 0. \tag{3.1}$$

We have to solve this equation for $(u, \lambda, \sigma) \in U \times \mathbb{R}^m \times \mathbb{R}$ near $(0.0.0)$, and from the foregoing section we know that the vector field $g(u, \lambda)$ appearing in (3.1) has the following properties:

(I) $g : U \times \mathbb{R}^m \to U$ is sufficiently smooth and S^1-equivariant. $g(0, \lambda) = 0$ for all $\lambda \in \mathbb{R}^m$, and $A_\lambda := D_u g(0, \lambda)$ has the explicit form (2.29);

(II) $D_u M(u, \lambda) \cdot g(u, \lambda) = 0$ for some smooth S^1-invariant $M : U \times \mathbb{R}^m \to \mathbb{R}$ with $D_u M(0, \lambda) = 0$ for all λ, and such that $Q_0(u) := D_u^2 M(0,0) \cdot (u, u)$ has the form $Q_0(u) = \langle u, B_0 u \rangle$, with $B_0 \in \mathcal{L}(U)$ as in Lemma 1.

It follows from the S^1-invariance of M that $D_u M(u,\lambda) \cdot Su = 0$ for all (u,λ), such that in fact we have

$$D_u M(u,\lambda) \cdot \Theta(u,\lambda,\sigma) = 0, \quad \forall (u,\lambda,\sigma) \in U \times \mathbb{R}^m \times \mathbb{R}, \qquad (3.2)$$

or equivalently,

$$\langle \Gamma(u,\lambda), \Theta(u,\lambda,\sigma) \rangle = 0, \quad \forall (u,\lambda,\sigma) \in U \times \mathbb{R}^m \times \mathbb{R}, \qquad (3.3)$$

where $\Gamma(u,\lambda) := \nabla_u M(u,\lambda)$ is the gradient (in the u-variable) of M with respect to the scalar product $\langle \cdot, \cdot \rangle$. The mapping $\Gamma : U \times \mathbb{R}^m \to U$ is smooth and S^1-equivariant, $\Gamma(0,\lambda) = 0$ for all λ, and $D_u\Gamma(0,0) = B_0$. The relation (3.3) will play an important role when we solve (3.1); in fact, (3.3) will take over the role of the reversibility which we had in [4] but which we do not assume here. For a more general discussion of bifurcation problems satisfying a *constraint* such as (3.3) see [1].

In order to solve (3.1) we proceed as in [4]. Setting $\hat{U} := \ker((N^T)^{k-1})$ it follows from Lemma 1 that $U = U_1 \oplus \hat{U}$; also, the linear mapping $u \mapsto (S^{-k}N^{k-1}u, N^T u)$ is an S^1-equivariant isomorphism from U onto $U_1 \times \hat{U}$, with inverse $(u_1, \hat{u}) \mapsto N\hat{u} + S^k(N^T)^{k-1}u_1$ (for $u_1 \in U_1$ and $\hat{u} \in \hat{U}$). Writing $u \in U$ as $u = u_1 + \hat{u}$, with $u_1 \in U_1$ and $\hat{u} \in \hat{U}$, it follows that (3.1) is equivalent to the system of equations

$$\begin{cases} \Theta_1(u_1, \hat{u}, \lambda, \sigma) := S^{-k}N^{k-1}\Theta(u_1 + \hat{u}, \lambda, \sigma) = 0, \\ \hat{\Theta}(u_1, \hat{u}, \lambda, \sigma) := N^T\Theta(u_1 + \hat{u}, \lambda, \sigma) = 0. \end{cases} \qquad (3.4)$$

Since $\Theta(u_1 + \hat{u}, \lambda, \sigma) = N\hat{\Theta}(u_1, \hat{u}, \lambda, \sigma) + S^k(N^T)^{k-1}\Theta_1(u_1, \hat{u}, \lambda, \sigma)$ we can rewrite (3.3) as

$$\langle \Gamma_1(u_1, \hat{u}, \lambda), \Theta_1(u_1, \hat{u}, \lambda, \sigma) \rangle + \left\langle \hat{\Gamma}(u_1, \hat{u}, \lambda), \hat{\Theta}(u_1, \hat{u}, \lambda, \sigma) \right\rangle = 0, \qquad (3.5)$$

where $\Gamma_1 : U_1 \times \hat{U} \times \mathbb{R}^m \to U_1$ and $\hat{\Gamma} : U_1 \times \hat{U} \times \mathbb{R}^m \to \hat{U}$ are given by

$$\Gamma_1(u_1, \hat{u}, \lambda) := S^{-k}N^{k-1}\Gamma(u_1 + \hat{u}, \lambda) \quad \text{and} \quad \hat{\Gamma}(u_1, \hat{u}, \lambda) := N^T\Gamma(u_1 + \hat{u}, \lambda). \quad (3.6)$$

As in [4] the second equation in (3.4) can be solved by the implicit function theorem for $\hat{u} = \hat{u}^*(u_1, \sigma)$; the mapping $\hat{u}^* : U_1 \times \mathbb{R}^m \times \mathbb{R} \to \hat{U}$ is uniquely defined, S^1-equivariant and as smooth as the mapping g for (u_1, λ, σ) sufficiently small; also $\hat{u}^*(0, \lambda, \sigma) = 0$ for all (λ, σ), and $D_1\hat{u}^*(0,0,0) = 0$. Bringing this solution in the first equation of (3.4) gives us the *bifurcation equation*

$$\Theta_0(u_1, \lambda, \sigma) := \Theta_1(u_1, \hat{u}^*(u_1, \lambda, \sigma), \lambda, \sigma) = 0. \qquad (3.7)$$

The bifurcation mapping $\Theta_0 : U_1 \times \mathbb{R}^m \times \mathbb{R} \to U_1$ is as smooth as g, S^1-equivariant, and satisfies $\Theta_0(0, \lambda, \sigma) = 0$ for all (λ, σ), and $D_1\Theta_0(0,0,0) = 0$. As in [4] it is possible to do more detailed calculations based on (2.29): these show that

$$\hat{u}^*(u_1, \lambda, \sigma) = \sum_{i=1}^{k-1} h_i(\sigma, \lambda) S^i (N^T)^i u_1 + O(\|u_1\|^2) \qquad (3.8)$$

and

$$\Theta_0(u_1, \lambda, \sigma) = -h_k(\sigma, \lambda)u_1 + O(\|u_1\|^2), \tag{3.9}$$

where the functions $h_i(\sigma, \lambda)$ are polynomials given by the following iteration scheme:

$$\begin{cases} h_0(\sigma, \lambda) & := \quad 1, \\ h_1(\sigma, \lambda) & := \quad \sigma, \\ h_i(\sigma, \lambda) & := \quad \sigma h_{i-1}(\sigma, \lambda) - \sum_{j=1}^{i-1} \lambda_j h_{i-1-j}(\sigma, \lambda) \quad (2 \le i \le k). \end{cases} \tag{3.10}$$

These polynomials only depend on the first $k - 1$ components $(\lambda_1, \ldots, \lambda_{k-1})$ of $\lambda \in \mathbb{R}^m$; we call these the *essential parameters*. Also, as shown in [4, Remark 5], we have $h_k(\sigma, \lambda) = 0$ if and only if $\pm(1 + \sigma)i$ are eigenvalues of the linear operator A_λ given by (2.29). Finally, the bifurcation mapping $\Theta_0(u_1, \lambda, \sigma)$ has one further important property which is related to the fact that our original system had a first integral. Namely, when we set $\hat{u} = \hat{u}^*(u_1, \lambda, \sigma)$ in (3.5) we get the identity

$$\langle \Gamma_0(u_1, \lambda, \sigma), \Theta_0(u_1, \lambda, \sigma) \rangle = 0, \quad \forall (u_1, \lambda, \sigma) \in U_1 \times \mathbb{R}^m \times \mathbb{R}. \tag{3.11}$$

where the S^1-equivariant mapping $\Gamma_0 : U_1 \times \mathbb{R}^m \times \mathbb{R} \to U_1$ is given by $\Gamma_0(u_1, \lambda, \sigma)$ $:= \Gamma_1(u_1, \hat{u}^*(u_1, \lambda, \sigma), \lambda)$. It follows that $\Gamma_0(0, \lambda, \sigma) = 0$ for all (λ, σ). and that

$$D_1\Gamma_0(0, 0, 0) = S^{-k}N^{k-1}B_0|_{U_1} = (-1)^k S|_{U_1} . \tag{3.12}$$

(We have used (2.8) to obtain the last equality in (3.12)).

It remains to solve the bifurcation equation (3.7): since $\dim U_1 = 2$ this forms strictly speaking a two-dimensional problem. However, because of the S^1-equivariance of this equation it's solutions (and hence also those of (1.9)) come in S^1-orbits, which, according to the reduction method of [2], correspond to periodic orbits of the original equation (1.1). Therefore it is sufficient to find one point on each solution orbit; we do this as follows. Choose a non-zero vector $u_1^0 \in U_1$, normalized such that $\langle u_1^0, u_1^0 \rangle = 1$. Then $\{u_1^0, Su_1^0\}$ forms an orthonormal basis of U_1, and hence we can find for each $u_1 \in U_1$ some $\rho \in \mathbb{R}$ and some $\varphi \in S^1$ such that $u_1 = (\rho \cos \varphi)u_1^0 + (\rho \sin \varphi)Su_1^0 = \exp(\varphi S)(\rho u_1^0)$; we conclude that each S^1-orbit in U_1 contains a point of the form ρu_1^0, and hence we can also put $u_1 = \rho u_1^0$ in (3.7). In fact, even this still contains some redundancy, since together with ρu_1^0 also $-\rho u_1^0$ will be a solution of (3.7), both corresponding to the same periodic orbit of (1.1); so we can restrict to $\rho \ge 0$, but for convenience of formulation we allow for the moment any $\rho \in \mathbb{R}$. Since $\Theta_0(0, \lambda, \sigma) = \Gamma_0(0, \lambda, \sigma) = 0$ we can then find scalar functions $\theta_i(\rho, \lambda, \sigma)$ and $\gamma_i(\rho, \lambda, \sigma)$ $(i = 0, 1)$ which are even in ρ and such that

$$\Theta_0(\rho u_1^0, \lambda, \sigma) = \rho\theta_0(\rho, \lambda, \sigma)u_1^0 + \rho\theta_1(\rho, \lambda, \sigma)Su_1^0 \tag{3.13}$$

and

$$\Gamma_0(\rho u_1^0, \lambda, \sigma) = \rho\gamma_0(\rho, \lambda, \sigma)u_1^0 + \rho\gamma_1(\rho, \lambda, \sigma)Su_1^0. \tag{3.14}$$

For non-zero solutions of the form $u_1 = \rho u_1^0$ the equation (3.7) reduces to

$$\theta_0(\rho, \lambda, \sigma) = \theta_1(\rho, \lambda, \sigma) = 0, \tag{3.15}$$

while (3.11) gives us the relation

$$\gamma_0(\rho, \lambda, \sigma)\theta_0(\rho, \lambda, \sigma) + \gamma_1(\rho, \lambda, \sigma)\theta_1(\rho, \lambda, \sigma) \equiv 0 \tag{3.16}$$

between the two components of (3.15). From (3.12) and the definitions we find

$$\gamma_0(0,0,0)u_1^0 + \gamma_1(0,0,0)Su_1^0 = D_1\Gamma_0(0,0,0) \cdot u_1^0 = (-1)^k Su_1^0,$$

such that $\gamma_0(0,0,0) = 0$ and $\gamma_1(0,0,0) = (-1)^k \neq 0$. It follows then from (3.16) that for sufficiently small (ρ, λ, θ) we have $\theta_1(\rho, \lambda, \sigma) = 0$ if $\theta_0(\rho, \lambda, \sigma) = 0$, i.e. instead of solving the system (3.15) it is sufficient to solve the scalar bifurcation equation

$$\theta_0(\rho, \lambda, \sigma) = 0. \tag{3.17}$$

From (3.9). (3.13) and the fact that $\theta_0(\rho, \lambda, \sigma)$ is even in ρ we conclude that there exists another scalar function $\zeta_0(\rho, \lambda, \sigma)$ which is even in ρ and such that

$$\theta_0(\rho, \lambda, \sigma) = -h_k(\sigma, \lambda) + \zeta_0(\rho, \lambda, \sigma)\rho^2. \tag{3.18}$$

Now we introduce a final hypothesis; namely we assume the non-degeneracy condition
(N-D) $\zeta_0(0,0,0) \neq 0$.
Since $\zeta_0(0,0,0)$ can (in principle) be calculated from the Taylor expansion up to order 3 of the vector field $f(x, \lambda_0)$ this is a generically satisfied condition on the coefficients appearing in this Taylor expansion. Under this condition we can set $\epsilon := \text{sgn}(\zeta_0(0,0,0))$ and perform a simple transformation (see [4]) which puts the equation (3.17) in the explicit polynomial form

$$\epsilon\rho^2 - h_k(\sigma, \lambda) = 0. \tag{3.19}$$

This equation describes the complete bifurcation behavior for the problem we were interested in. Also, this equation is exactly the same as the one obtained in [4] for the reversible case.

In [4] we have shown that the bifurcation set is diffeomorphic to the standard cuspoid of order k (see [7]); as the parameter crosses the different strata of this cuspoid in the appropriate direction one or more pairs of purely imaginary eigenvalues of the linearization at the equilibrium coalesce and split off the imaginary axis. The periodic orbits themselves appear in so-called *local loops*, *global branches* or *detached branches*; when the parameter crosses the bifurcation set one can see transitions between those different types of branches via either *elliptic* or *hyperbolic* bifurcations, similar to those which appear at a Krein instability in Hamiltonian systems (see e.g. [6]). For more details we refer to [4].

References

[1] M. Golubitsky, J. Marsden, I. Stewart and M. Dellnitz. The constrained Liapunov-Schmidt procedure and periodic orbits. Preprint 1994.

[2] J. Knobloch and A. Vanderbauwhede. A general reduction method for periodic solutions in conservative and reversible systems. *TU Ilmenau Preprint No M 11/94*. To appear in Journal of Dynamics and Differential Equations.

[3] J. Knobloch and A. Vanderbauwhede. Hopf bifurcation at k-fold resonances in equivariant reversible systems. In P. Chossat, editor, *Dynamics, Bifurcation and Symmetry. New Trends and New Tools*, NATO ASI Series C, Vol. 437, pages 167–179, Dordrecht 1994. Kluwer Acad. Publ.

[4] J. Knobloch and A. Vanderbauwhede. Hopf bifurcation at k-fold resonances in reversible systems. *TU Ilmenau Preprint No M 16/95*.

[5] A. Vanderbauwhede. Hopf bifurcation for equivariant conservative and time-reversible systems. *Proceedings of the Royal Society of Edinburgh. 116A*, pages 103–128, 1990.

[6] J.-C. van der Meer. *The Hamiltonian Hopf Bifurcation*. Lect. Notes in Math. 1160, Springer-Verlag, Berlin, 1986.

[7] A.E.R. Woodcock and T. Poston. *A geometrical study of the elementary catastrophes*. LNM 373. Springer-Verlag, Berlin, 1989.

Progress in Nonlinear Differential Equations
and Their Applications, Vol. 19
© 1996 Birkhäuser Verlag Basel/Switzerland

Families of Quasi-Periodic Motions in Dynamical Systems Depending on Parameters

H.W. Broer[*]   G.B. Huitema[†]   M.B. Sevryuk[‡]

1   Introduction and set-up

One of the central topics in the qualitative theory of differential equations is the study of invariant submanifolds. A number of general theorems establishing the existence and/or persistence and describing the properties of those submanifolds play a fundamental rôle in the analysis of nonlinear dynamical systems [23, 64, 13]. A particular example of an invariant submanifold is an invariant torus with parallel dynamics.

Definition 1.1 *Let X be a smooth vector field on a smooth real finite-dimensional manifold M. An invariant n-torus T of vector field X is a submanifold of M invariant under the flow of X and diffeomorphic to the standard n-torus $\mathbb{T}^n = (\mathbb{R}/2\pi\mathbb{Z})^n$. An invariant n-torus T is said to carry parallel dynamics if there exist coordinates $\varphi \in \mathbb{T}^n$ on T in which the restriction of X to T gets the form $\omega \, \partial/\partial\varphi$ for some vector $\omega \in \mathbb{R}^n$. This vector is called the frequency vector of invariant torus T, and the components of ω are called the frequencies (or internal frequencies) of T.*

The frequency vector is determined uniquely up to changes of the form $\omega \mapsto A\omega$, where A is an $n \times n$ matrix with integer entries and determinant ± 1.

Invariant tori with parallel dynamics are of great importance in the theory of dynamical systems which stems, in the long run, from the fact that *any finite-dimensional connected compact abelian Lie group is a torus* [34, 2]. However, the dynamical properties of the flow on those tori are very different for frequency vectors ω with different number-theoretical properties.

Definition 1.2 *An invariant n-torus T with parallel dynamics is said to be resonant if its frequencies $\omega_1, \ldots, \omega_n$ are linearly dependent over the field $\mathbb{Q}$ of rational numbers. Otherwise, such a torus T is said to be nonresonant or quasi-periodic.*

[*]Department of Mathematics, University of Groningen. P.O.Box 800, 9700 AV Groningen, The Netherlands
[†]KPN Research, P.O. Box 15.000, 9700 CD Groningen. The Netherlands
[‡]Institute of Energy Problems of Chemical Physics. Lenin prospect 38, Bldg. 2, 117829 Moscow, Russia

Each orbit on a quasi-periodic invariant torus T is everywhere dense in T, and the whole motion on T is ergodic. On the other hand, a resonant invariant torus T is smoothly foliated into invariant tori of smaller dimensions. According to the Kupka–Smale theorem and its generalizations [37], a *generic* vector field possesses no resonant invariant tori, and consequently, all invariant tori *with parallel dynamics* of a generic dynamical system are quasi-periodic. It turns out that all the invariant tori with parallel dynamics in the phase space of a generic vector field are usually not only nonresonant but, moreover, their frequencies satisfy certain Diophantine conditions.

Before formulating these conditions, we introduce some notations to be used in the sequel. By the angle brackets, we will denote the standard inner product of two vectors, so that

$$\langle a, b \rangle = \sum_{j=1}^{N} a_j b_j \quad \text{for} \quad a \in \mathbb{R}^N, b \in \mathbb{R}^N.$$

The symbols $|a|$ and $\|a\|$ will denote the l_1-norm and l_2-norm (Euclidean norm), respectively, of vector a:

$$|a| = \sum_{j=1}^{N} |a_j| \quad \text{and} \quad \|a\|^2 = \sum_{j=1}^{N} |a_j|^2 \quad \text{for} \quad a \in \mathbb{C}^N.$$

We will write $O_\ell(u)$ instead of $O(|u|^\ell)$. Finally, $\mathbb{N}$ is the set of positive integers, and $\mathbb{Z}_+ = \mathbb{N} \cup \{0\}$ is the set of non-negative integers.

Definition 1.3 *A quasi-periodic invariant n-torus is said to be Diophantine if its frequency vector ω satisfies the infinite system of inequalities*

$$|\langle \omega, k \rangle| \geq \gamma |k|^{-\tau} \quad \forall k \in \mathbb{Z}^n \setminus \{0\}$$

for some constants $\tau > 0$ and $\gamma > 0$.

Diophantine invariant tori of dynamical systems are the subject of the Kolmogorov–Arnol'd–Moser (KAM) theory.

The variational equations along invariant tori with parallel dynamics can sometimes be reduced to a constant coefficient form.

Definition 1.4 *An invariant n-torus T with parallel dynamics of a vector field X on a K-dimensional manifold M is said to be Floquet if there exist such coordinates $(x \in \mathbb{T}^n, w \in \mathbb{R}^{K-n})$ on M near T in which the torus T itself takes the form $\{w = 0\}$ while the vector field X takes the Floquet form*

$$[\omega + O(w)]\frac{\partial}{\partial x} + [\Omega w + O_2(w)]\frac{\partial}{\partial w} \tag{1}$$

where $\omega \in \mathbb{R}^n$ is the frequency vector of T and $\Omega \in \mathrm{gl}(K - n, \mathbb{R})$ is a constant matrix called the Floquet matrix of T.

The Floquet matrix is determined up to similarity. i.e., up to changes of the form $\Omega \mapsto A^{-1}\Omega A$, where $A \in GL(K - n, \mathbb{R})$.

Example 1.5 Any equilibrium of a vector field is a Floquet Diophantine invariant 0-torus. Its frequency vector is $0 \in \mathbb{R}^0 = \{0\}$ whereas its Floquet matrix is just the linearization matrix of the vector field. Any closed trajectory of a vector field is a Diophantine invariant 1-torus. Its frequency is $\omega = 2\pi/S$ where S is the period. Vector fields around closed trajectories are not necessarily reducible to the form (1), the sufficient condition for this reducibility being the absence of negative real eigenvalues of the monodromy matrix [3].

The properties of invariant tori in dynamical systems are very sensitive to what structures (symmetries) the system in question is assumed to preserve or be compatible with. For instance, Hamiltonian and reversible vector fields often exhibit Cantor-like families of Floquet Diophantine invariant tori [10, 49]. On the other hand, invariant n-tori of a vector field that possesses no symmetry and satisfies no conservation law (such vector fields are sometimes said to be dissipative [12]) are generically isolated in the phase space and do not carry parallel dynamics for $n > 1$ [3], but dissipative vector fields *depending on external parameters* often admit Floquet Diophantine invariant tori (also isolated in the phase space).

So. consider a real finite-dimensional connected manifold M equipped with some structure $\mathfrak{S}$ (e.g., a tensor field on M, the action of a Lie group, etc.). We will assume the manifold M itself and the structure $\mathfrak{S}$ to be analytic. Our aim is to study l-parameter families of Floquet Diophantine invariant n-tori in s-parameter families of vector fields on M compatible (in the sense to be made precise) with structure $\mathfrak{S}$. The tori we look for are also supposed to be compatible with structure $\mathfrak{S}$ in an appropriate sense.

Thus. we will encounter mainly two different kinds of objects: families of vector fields and families of quasi-periodic invariant tori of those vector fields. An s-parameter family X^μ of vector fields on M is just a vector field on M which depends on a parameter $\mu \in P \subset \mathbb{R}^s$ (P being an open domain). An l-parameter family of quasi-periodic invariant n-tori of X^μ is a much more complicated object whose precise definition will be given below in Section 3. For the time being, one can think of a certain set of n-tori in $M \times P$ such that the $(n + l)$-dimensional Lebesgue measure $meas_{n+l}$ of the union of all the tori is finite (neither 0 nor $+\infty$), and each torus lies in one of the "fibers" $M \times \{\mu_0\}$, is invariant under the flow of field X^{μ_0} and carries quasi-periodic dynamics.

To fix thoughts, let C^r, $r \in \mathbb{N} \cup \{\infty, \omega\}$, be some smoothness class (ω is for real analyticity). Denote by $\mathfrak{X}_s^r$ the space of all s-parameter families of C^r-smooth vector fields on M compatible with the structure $\mathfrak{S}$. the dependence of these fields on the s-dimensional parameter being also of class C^r. Equip the space $\mathfrak{X}_s^r$ with an appropriate C^r weak or strong topology [22] (in fact. the difference between these two standard topologies is entirely irrelevant for our purposes). The main question this paper deals with is as follows:

For what values of $n \in \mathbb{Z}_+$ and $l \in \mathbb{Z}_+$, does a *typical* family of vector fields in $\mathfrak{X}_s^r$ possess l-parameter families of Floquet Diophantine invariant n-tori compatible with the structure $\mathfrak{S}$? What are the properties of these families of tori?

The word "*typical*" above means that families of vector fields possessing l-parameter families of Floquet Diophantine invariant n-tori constitute an *open* set in $\mathfrak{X}_s^r$ (to be more precise, a set with non-empty interior). In other words, we are looking for some persistent phenomena in families of vector fields.

The easiest way to study quasi-periodic invariant tori of typical families of vector fields we know is to perturb *integrable* vector fields, i.e.. vector fields equivariant with respect to the free action of the standard n-torus $\mathbb{T}^n$. Integrable systems are quite exceptional, but their perturbations, no matter how small the perturbation size is, are already typical. If we succeed in proving that *any* sufficiently small perturbation of a family X_0^μ of integrable vector fields admits an l-parameter family of Floquet Diophantine invariant n-tori, then we will be able to conclude that such families of tori occur typically, namely, in a neighborhood of X_0^μ in $\mathfrak{X}_s^r$. We can impose some further restrictions on the unperturbed family X_0^μ of integrable vector fields (e.g., some nondegeneracy and nonresonance conditions). In fact, we can choose the unperturbed family as special as we please, and this would not affect our final conclusion. The only important restriction is that *perturbations* of that initial family should not be subject to any conditions (of course, within the space $\mathfrak{X}_s^r$).

From now on, we will confine ourselves with the real analytic category $r = \omega$, i.e., we will consider real analytic families of real analytic vector fields. The structures $\mathfrak{S}$ we will be interested in are a volume element, a symplectic structure, and an involution. Accordingly, we will distinguish four "contexts" of our theory (cf. [25, 12, 11]):

1) the dissipative context, where no structure on the phase space is present;

2) the volume preserving context, where the structure on M is a volume element, i.e., a differential form of the maximal degree which vanishes nowhere;

3) the Hamiltonian context, where the structure on M is a symplectic structure, i.e., a closed nondegenerate differential 2-form;

4) the reversible context, where the structure on M is an involution $G : M \to M$ (a mapping whose square is the identity transformation).

In the volume preserving context, the vector fields in question are globally divergence-free. Recall that if σ is a volume element on an N-dimensional manifold M then the divergence $\mathrm{div}X$ of a vector field X on M is a real-valued function on M defined as

$$d(i_X\sigma) = (\mathrm{div}X)\sigma.$$

Here $i_X\sigma$ is the $(N-1)$-form whose value at the vectors $X_1,\dots.X_{N-1}$ is equal to the value of σ at the vectors $X, X_1,\dots,X_{N-1}$. Thus, divergence-free vector fields X are those for which the form $i_X\sigma$ is closed. A divergence-free vector field X is said to be *globally divergence-free* if the form $i_X \sigma$ is not only closed but even exact (cf. [1]).

In the Hamiltonian context, the vector fields in question are Hamiltonian, and their invariant n-tori with parallel dynamics are assumed to be either isotropic (for $n \leq N$) or coisotropic (for $n \geq N$), where N is the number of degrees of freedom (so that the phase space M is of dimension $2N$). Recall that a submanifold L of a symplectic $(2N)$-dimensional manifold M is said to be *isotropic* [*coisotropic*] if the tangent space $T_u L$ to L at each point $u \in L$ lies in its skew-orthogonal complement $(T_u L)^{\perp}$ [respectively, contains $(T_u L)^{\perp}$]. see [1, 4]. For any isotropic submanifold L, one has $\dim L \leq N$, while for any coisotropic submanifold L, one has $\dim L \geq N$. Any submanifold of dimension 0 or 1 is isotropic, while any submanifold of codimension 0 or 1 is coisotropic. For N-dimensional submanifolds the concepts of isotropicity and coisotropicity coincide. and an N-dimensional submanifold $L \subset M$ for which $(T_u L)^{\perp} = T_u L$ at each point $u \in L$ is called Lagrangian.

In the reversible context, the vector fields X in question are reversible with respect to a fixed involution G of the phase space M. This means that G transforms the field X into the opposite field $-X$. In other words. a vector field X [and the corresponding differential equation $du/dt = X(u)$] is said to be reversible with respect to G if $G(u(-t))$ is a solution of that equation whenever $u(t)$ is a solution. This definition is applicable to any diffeomorphism $G : M \to M$ but we will consider only the case $G^2 = \mathrm{id}$. Besides, n-tori with parallel dynamics in the reversible context are assumed to be invariant not only under the flow of the field itself, but also under the reversing involution. It turns out that on any of such tori with a nonresonant frequency vector, one can choose a coordinate system $\varphi \in \mathbb{T}^n$ which normalize both the dynamics and the involution.

Lemma 1.6 [49] *If the differential equation $d\phi/dt = \omega$ on the torus $\mathbb{T}^n$ with nonresonant frequency vector ω is reversible with respect to a diffeomorphism $G : \mathbb{T}^n \to \mathbb{T}^n$ then this diffeomorphism is an involution and has the form $G(\phi) = c - \phi$ for some constant $c \in \mathbb{T}^n$.*

Proof. Choose an arbitrary point $\phi_0 \in \mathbb{T}^n$. Then $G(\phi_0 + \omega t) = G(\phi_0) - \omega t$ for each real t. In other words, $G(\phi) = G(\phi_0) + \phi_0 - \phi$ for all $\phi \in \mathbb{T}^n$ of the form $\phi_0 + \omega t$. But ω is nonresonant, and points of this form are everywhere dense in $\mathbb{T}^n$. Therefore $G(\phi) = c - \phi$ for all the points $\phi \in \mathbb{T}^n$ with $c = G(\phi_0) + \phi_0$. $\square$

Note that the coordinate change $\varphi = \phi - c/2$ retains the differential equation (which takes the form $d\varphi/dt = \omega$) but puts the involution into the standard form $G : \varphi \mapsto -\varphi$.

The nonresonance condition on ω in Lemma 1.6 is very essential. For instance, the system $d\phi_1/dt = 1$, $d\phi_2/dt = 0$ on $\mathbb{T}^2$ is reversed by any diffeomorphism of the form $G : (\phi_1, \phi_2) \mapsto (a(\phi_2) - \phi_1, b(\phi_2))$.

Recall that the action of any compact Lie group on an arbitrary manifold M around a fixed point $u \in M$ is linear in an appropriate local coordinate system on M centered at u (the Bochner theorem [34]). Since an involution generates the group $\mathbb{Z}_2$ of two elements, it is always conjugate around a fixed point to its linear part. Consequently, the fixed points of any analytic involution $G : M \rightarrow M$ constitute an analytic submanifold Fix G of M. However, this submanifold can be disconnected and its connected components can be of different dimensions, even in the case where the manifold M itself is connected [43].

Definition 1.7 *An involution $G : M \rightarrow M$ of a $(q_+ + q_-)$-dimensional manifold M is said to be of type (q_-, q_+) if all the connected components of its submanifold Fix G of fixed points are of dimension q_+.*

Lemma 1.6 implies that the restriction of an involution G to a quasi-periodic invariant n-torus T of a G-reversible vector field is always of type $(n, 0)$, and the set $(\text{Fix } G) \cap T$ consists of 2^n isolated points. Hence, if involution G itself is of type (q_-, q_+) then $q_- \geq n$.

The reader is referred to works [36, 16, 5, 49, 51, 44, 43, 26, 55] for a survey of the main properties of reversible dynamical systems, examples, and physical applications (the papers [51, 44, 43, 26, 55] contain also an extended bibliography).

Now let us consider in detail the integrable set-up in all the four contexts above. In other words, we will deal with s-parameter families X^μ of vector fields on M under the assumption that these vector fields as well as structure $\mathfrak{S}$ are equivariant with respect to the free action of the standard n-torus $\mathbb{T}^n$. We will also suppose that structure $\mathfrak{S}$ is "linear" in a certain sense although this often leads to no additional restrictions due to normalization theorems like the Darboux or the Bochner theorem. The parameter μ varies in an open domain $P \subset \mathbb{R}^s$, and we will always assume that $0 \in P$ and that the field X^0 possesses an invariant n-torus with parallel dynamics. This torus is one of the orbits of the action of $\mathbb{T}^n$ on M.

Dissipative (n, p, s) context: invariant n-tori of s-parameter families of vector fields on an $(n + p)$-dimensional manifold. In this context, we have a family of vector fields

$$X = \{X^\mu\}_\mu = \varpi(w, \mu) \frac{\partial}{\partial x} + W(w, \mu) \frac{\partial}{\partial w} \tag{2}$$

where $x \in \mathbb{T}^n$, $w \in \mathbb{R}^p$ and $W(w_0, 0) = 0$ (the equivariance just means that the components of these vector fields do not depend on x). Generically the Jacobi matrix $D_w W(w_0, 0)$ is nondegenerate, and the equation $W(w, \mu) = 0$ can be solved with respect to w as $w = a(\mu)$, $a(0) = w_0$. Having introduced new coordinate $z = w - a(\mu)$, we arrive at the family of vector fields

$$[\omega(\mu) + O(z)] \frac{\partial}{\partial x} + [\Omega(\mu) z + O_2(z)] \frac{\partial}{\partial z} \tag{3}$$

where z varies in a neighborhood of $0 \in \mathbb{R}^p$ and matrix $\Omega(\mu)$ is nondegenerate for all μ. For each value of μ, the torus $\{z = 0\}$ is invariant under the flow of X^μ and carries parallel dynamics with frequency vector $\omega(\mu)$. So, we obtain an s-parameter analytic family of invariant n-tori with parallel dynamics (with no restrictions on the non-negative integers n, p, s).

Volume preserving (n, p, s) context: invariant n-tori of s-parameter families of globally divergence-free vector fields on an $(n + p)$-dimensional manifold. In this context, we have the same family (2) of vector fields which, however, now are supposed to be globally divergence-free with respect to volume element $dx \wedge dw$. First of all, observe that this requirement excludes the case $p = 0$ since the constant vector field $\omega \partial/\partial \varphi$ on the torus $\mathbb{T}^n$ with volume element $d\varphi$ is always divergence-free but never *globally* divergence-free (except for the trivial case $\omega = 0$). For $p = 1$, one can easily verify that vector fields (2) are divergence-free if and only if W does not depend on w, and they are globally divergence-free if and only if $W \equiv 0$. For $p > 1$, the notion of being divergence-free and that of being globally divergence-free for vector fields on $M = \mathbb{T}^n \times \mathbb{R}^p$ are equivalent (because the cohomology $H^{n+p-1}(M, \mathbb{R}) = 0$ in this case), and fields (2) are divergence-free (and globally divergence-free) with respect to $dx \wedge dw$ if and only if vector fields $W(w, \mu)\partial/\partial w$ are divergence-free (and globally divergence-free) with respect to dw (i.e., $\operatorname{Tr} D_w W(w, \mu) \equiv 0$). The cases $p = 1$ and $p > 1$ therefore turn out to be drastically different in the volume preserving context [25, 12].

If $p = 1$, we arrive at the family of vector fields

$$\omega(y, \mu)\frac{\partial}{\partial x} \tag{4}$$

where $y \in \mathbb{R}$ (to achieve the consistency with the notations in the Hamiltonian and reversible contexts discussed below we prefer here to write y instead of w and ω instead of ϖ). For each value of μ, each torus $\{y = const\}$ is invariant under the flow of X^μ and carries parallel dynamics with frequency vector $\omega(y, \mu)$. So, we obtain an $(s + 1)$-parameter analytic family of invariant n-tori with parallel dynamics.

If $p > 1$, we can proceed in exactly the same way as in the dissipative context. We will obtain the family (3) of vector fields and an s-parameter analytic family of invariant n-tori $\{z = 0\}$ with parallel dynamics. The only difference is that now $\Omega(\mu) \in sl(p, \mathbb{R})$ for each μ (i.e., $\operatorname{Tr} \Omega(\mu) \equiv 0$).

Hamiltonian isotropic (n, p, s) context: isotropic invariant n-tori of s-parameter families of Hamiltonian vector fields with $n + p$ degrees of freedom ($n \geq 0$, $p \geq 0$, $n + p \geq 1$). The codimension of the tori in the phase space is equal to $n + 2p$. We suppose that field X^0 possesses an isotropic invariant n-torus T and near this torus, there exists a coordinate system ($\varphi \in \mathbb{T}^n$, $w \in \mathbb{R}^{n-2p}$) in which T takes the form $\{w = 0\}$, the Hamilton function does not depend on φ and the symplectic structure has constant coefficients. It is not hard to verify that after an appropriate coordinate change $\varphi = x + Ay + Bz$, $w = Cy + Dz$, where $x \in \mathbb{T}^n$, $y \in \mathbb{R}^n$, $z \in \mathbb{R}^{2p}$, and A, B, C, D are constant matrices of suitable sizes. this symplectic structure will get the form

$$\sum_{\iota=1}^{n} dy_\iota \wedge dx_\iota + \sum_{j=1}^{p} dz_j \wedge dz_{j+p}. \tag{5}$$

In fact, the assumption of constant coefficients is not really necessary for such normalization of the symplectic structure because the latter can always be reduced to form (5) around an isotropic torus (the generalized Darboux theorem [1]). The Hamiltonian $H^\mu(y, z)$ determines the family of vector fields

$$X = \{X^\mu\}_\mu = \frac{\partial H}{\partial y}\frac{\partial}{\partial x} + \left(J\frac{\partial H}{\partial z}\right)\frac{\partial}{\partial z}$$

with

$$J = \begin{pmatrix} 0 & -I \\ I & 0 \end{pmatrix}, \quad I = diag(1,\ldots,1) \in SL(p, \mathbb{R}). \tag{6}$$

Since the torus $\{y = 0, z = 0\}$ is invariant under the field X^0. the origin $z = 0$ is a critical point of the function $\mathbb{R}^{2p} \to \mathbb{R}$, $z \mapsto H^0(0, z)$. Generically this critical point is nondegenerate, i.e., $\det \partial^2 H^0(0,0)/\partial z^2 \neq 0$, in which case the equation $D_z H^\mu(y, z) = 0$ determines an analytic surface $z = Z(y, \mu)$, $Z(0.0) = 0$. We obtain an $(n + s)$-parameter analytic family of isotropic invariant n-tori with parallel dynamics for X^μ. Namely, for each value of μ, each torus $\{y = const, z = Z(y, \mu)\}$ is isotropic, invariant under the flow of X^μ and carries parallel dynamics with frequency vector $\partial H^\mu(y, Z(y, \mu))/\partial y$.

For simplicity, in the sequel we will always assume that $Z(y. \mu) \equiv 0$ (recall that we are allowed to choose the integrable vector fields arbitrarily special). In this case the Hamiltonian has the form

$$H = H^\mu(y, z) = F(y, \mu) + \tfrac{1}{2}\langle z, K(y, \mu)z \rangle + O_3(z) \tag{7}$$

(the $2p \times 2p$ matrix $K(y, \mu)$ is symmetric for all values of y and μ). The corresponding family of vector fields has the form

$$X = \{X^\mu\}_\mu = [\omega(y, \mu) + O_2(z)]\frac{\partial}{\partial x} + [\Omega(y, \mu)z + O_2(z)]\frac{\partial}{\partial z} \tag{8}$$

where $\omega = \partial F/\partial y$ and $\Omega = JK$. For each value of μ, each torus $\{y = const, z = 0\}$ is isotropic, invariant under the flow of X^μ and carries parallel dynamics with frequency vector $\omega(y, \mu)$.

Hamiltonian coisotropic (n, p, s) context: coisotropic invariant n-tori of s-parameter families of Hamiltonian vector fields with $n - p$ degrees of freedom $(n \geq 3, 0 < p < n/2)$. The codimension of the tori in the phase space is equal to $n - 2p$. The case $p = n/2$ is excluded because invariant tori in the Hamiltonian context lie in the energy levels of the Hamilton function and have therefore positive codimension. For n odd and $p = (n-1)/2$, each invariant torus is a connected component of an energy level. We consider the space $M = \mathbb{T}^n \times \mathbb{R}^{n-2p}$ with coordinates (x, y) and a symplectic structure ω^2 with constant coefficients, each torus $\{y = const\}$ being coisotropic with respect to ω^2. Identify the tangent spaces $T_{x_0} \mathbb{T}^n = U \cong \mathbb{R}^n$ to $\mathbb{T}^n$ at all the points $x_0 \in \mathbb{T}^n$ and the tangent spaces $T_{y_0} \mathbb{R}^{n-2p} = V \cong \mathbb{R}^{n-2p}$ to $\mathbb{R}^{n-2p}$ at all the points $y_0 \in \mathbb{R}^{n-2p}$. The structure ω^2 can be treated as a nondegenerate skew-symmetric bilinear form on $U \oplus V$ such that the $(n-2p)$-dimensional skew-orthogonal complement $U^\perp$ to plane U lies in U. We will denote this bilinear form by the same symbols ω^2. The bilinear form $\omega^2(\mathbf{v}, \mathbf{u})$, where $\mathbf{u} \in U^\perp$ and $\mathbf{v} \in V$, is nondegenerate (otherwise the whole form ω^2 would be degenerate).

Define the linear mapping $\mathbf{f} : V^* \to U^\perp$ (here V^* is the space of linear functions on V) as

$$\omega^2(\mathbf{v}, \mathbf{f}(\xi)) = \xi(\mathbf{v}) \quad \forall \mathbf{v} \in V$$

for $\xi \in V^*$. Now let

$$H = H^\mu(y) \tag{9}$$

be a family of integrable Hamiltonians on M and $X = \{X^\mu\}_\mu$, the corresponding family of Hamiltonian vector fields.

Lemma 1.8 *For each value of μ and at each point of M*

$$X = \mathbf{f}\,(dH|_V) \in U^\perp.$$

Proof. Let $\mathbf{f}\,(dH|_V) = \mathbf{a}$. Then $\forall \mathbf{u} \in U \ \forall \mathbf{v} \in V$

$$dH(\mathbf{u} + \mathbf{v}) = dH(\mathbf{v}) = \omega^2(\mathbf{v}, \mathbf{a}) = \omega^2(\mathbf{u} + \mathbf{v}, \mathbf{a})$$

(the first equality in this chain follows from $\partial H / \partial x \equiv 0$ and the last one, from $\mathbf{a} \in U^\perp$). Consequently, $\mathbf{a} = X$. $\qquad\square$

Thus, all the tori $\{y = const\}$ are invariant under the flow of X^μ for each μ and carry parallel dynamics with frequency vectors $\omega(y, \mu) = \mathbf{f}\,(dH^\mu(y)|_V)$ (after the identification of U and $\mathbb{R}^n$). So, we obtain an $(n - 2p + s)$-parameter analytic family of coisotropic invariant n-tori with parallel dynamics (in fact, the space $M \times P$ is foliated into those tori). On the other hand, all the frequency vectors of the tori lie in the fixed $(n - 2p)$-dimensional subspace $U^\perp$ of $\mathbb{R}^n$ (any vector in this subspace can be realized as the frequency vector for a suitable Hamiltonian). The subspace $U^\perp$ is determined by the symplectic structure ω^2 only and does not depend on the Hamilton function H. The behavior of Hamiltonian systems on M is very sensitive to the arithmetical properties of the arrangement of the plane $U^\perp$ with respect to the lattice $\mathbb{Z}^n$. For instance, it is possible that all the vectors in $U^\perp$ are resonant (e.g., if $U^\perp$ is one of the coordinate planes). The alternative possibility for $1 \leq p < (n-1)/2$ is that both resonant vectors and nonresonant ones constitute everywhere dense subsets of $U^\perp$. For $p = (n-1)/2$ [of course, in this case n is odd] the alternative possibility is that all the nonzero vectors in $U^\perp$ are *nonresonant* [$U^\perp$ is a straight line for $p = (n-1)/2$]. We see that the cases $1 \leq p < (n-1)/2$ and $p = (n-1)/2$ within the Hamiltonian coisotropic (n, p, s) context are rather different.

As the arithmetical properties of the space $U^\perp$ are different for different forms ω^2, there is no universal normal form for the symplectic structure in the Hamiltonian coisotropic context [like (5) in the Hamiltonian isotropic context].

Coisotropic invariant tori of Hamiltonian systems are encountered, e.g., in the quasiclassical theory of motion of a conduction electron [39]. It turns out that the motion of a conduction electron in an electric and magnetic field can proceed along four-dimensional coisotropic invariant tori in the six-dimensional phase space $T^* \mathbb{R}^3$. Here $n = 4$, $p = 1$. For homogeneous fields, an open domain in the phase space is smoothly foliated into such tori.

Reversible context: invariant n-tori of s-parameter families of reversible vector fields. We consider the space $M = \mathbb{T}^n \times \mathbb{R}^{u+v}$ with coordinates (x, w^+, w^-) where $x \in \mathbb{T}^n$, $w^+ \in L^+ = \mathbb{R}^u$, $w^- \in L^- = \mathbb{R}^v$, and the involution

$$G : (x, w^+, w^-) \mapsto (-x, w^+, -w^-) \tag{10}$$

of type $(n + v, u)$. An s-parameter family of integrable G-reversible vector fields on M has the form

$$X = \{X^\mu\}_\mu = \varpi(w^+, w^-, \mu)\frac{\partial}{\partial x} + W^+(w^+, w^-, \mu)\frac{\partial}{\partial w^+} + W^-(w^+, w^-, \mu)\frac{\partial}{\partial w^-} \tag{11}$$

where functions ϖ and W^- are even in w^- whereas function W^+ is odd in w^- [and therefore $W^+(w^+, 0, \mu) \equiv 0$]. We suppose that the torus $\{w^+ = 0, w^- = 0\}$ is invariant under the flow of field X^0, i.e., $W^-(0, 0, 0) = 0$. An n-torus $\{w^+ = w_0^+, w^- = w_0^-\}$ is invariant under involution (10) if and only if $w_0^- = 0$, and for $w_0^- = 0$ it is invariant under the flow of X^{μ_0} if and only if $W^-(w_0^+, 0, \mu_0) = 0$. Consider the mapping

$$\mathbb{R}^{u+s} \to \mathbb{R}^v, \quad (w^+, \mu) \mapsto W^-(w^+, 0, \mu). \tag{12}$$

The preimage of zero under this mapping generically is empty for $u + s < v$. This means that for $u + s < v$ a generic s-parameter family of integrable G-reversible vector fields on M admits no invariant n-tori: even if some particular family possesses such tori, the latter can be destroyed by an arbitrarily small perturbation of the family (within the integrable realm!). Now consider the case $u + s \geq v$ which, in turn, splits into two quite different subcases: $u \geq v$ and $s \geq v - u > 0$. These will be referred to as the reversible context 1 and the reversible context 2, respectively. In the first subcase, we will write $v = p$, $u = m + p$. In the second subcase, we will write $u = p$, $v = m + p$ ($1 \leq m \leq s$). We now consider these subcases separately.

Reversible (n, m, p, s) context 1: invariant n-tori of s-parameter families of vector fields reversible with respect to involutions of type $(n + p, m + p)$, $m \geq 0$,

$p \geq 0, s \geq 0$. Here $w^+ \in L^+ = \mathbb{R}^{m+p}$, $w^- \in L^- = \mathbb{R}^p$. For each fixed value of μ, one can generically introduce a new coordinate system (y, z^+) in $L^+ = \mathbb{R}^{m+p}$ ($y \in \mathbb{R}^m$, $z^+ \in \mathbb{R}^p$) via the formula $w^+ = Q(y, z^+, \mu)$ in such a way that $Q(0, 0, 0) = 0$ and $W^-(Q(y, z^+, \mu), 0, \mu) \equiv z^+$. We will also write z^- instead of w^-. Then involution (10), the family of differential equations on M determined by vector fields (11), and mapping (12) take the form

$$G : (x, y, z^+, z^-) \mapsto (-x, y, z^+, -z^-), \tag{13}$$

$$\begin{aligned} \dot{x} &= \omega(y, \mu) + O(z^+, z^-), & \dot{z}^+ &= O(z^-) = a(y, \mu)z^- + O_2(z^+, z^-), \\ \dot{y} &= O(z^-), & \dot{z}^- &= z^+ + O_2(z^-), \end{aligned} \tag{14}$$

and

$$\mathbb{R}^{m+p+s} \to \mathbb{R}^p, \quad (y, z^+, \mu) \mapsto z^+, \tag{15}$$

respectively. For each value of μ, each torus $\{y = const, z^+ = z^- = 0\}$ is invariant under both involution G and the flow of X^μ and carries parallel dynamics with frequency vector $\omega(y, \mu)$. We thus obtain an $(m + s)$-parameter analytic family of invariant n-tori with parallel dynamics for X^μ.

In the sequel, we will prefer to consider involutions G and families of G-reversible differential equations of a slightly more general form

$$G : (x, y, z) \mapsto (-x, y, Rz) \tag{16}$$

and

$$\dot{x} = \omega(y, \mu) + O(z), \quad \dot{y} = O(z), \quad \dot{z} = \Omega(y, \mu)z + O_2(z), \tag{17}$$

where $z \in \mathbb{R}^{2p}$, R is an arbitrary fixed involutive $(2p) \times (2p)$ matrix whose 1- and (-1)-eigenspaces are p-dimensional, and $\Omega(y, \mu)R + R\Omega(y, \mu) \equiv 0$. Expressions (13) and (14) correspond to

$$R = \begin{pmatrix} I & 0 \\ 0 & -I \end{pmatrix}, \quad \Omega(y, \mu) = \begin{pmatrix} 0 & a(y, \mu) \\ I & 0 \end{pmatrix}$$

[I being defined in (6)].

Reversible (n, m, p, s) context 2: invariant n-tori of s-parameter families of vector fields reversible with respect to involutions of type $(n + m + p, p)$, $m \geq 1$, $p \geq 0$, $s \geq m$. Here $w^+ \in L^+ = \mathbb{R}^p$, $w^- \in L^- = \mathbb{R}^{m+p}$. Generically one can split the coordinates $\mu_1, \ldots, \mu_s$ in the parameter space $\mathbb{R}^s$ into two groups $\mu = (\mu^1, \mu^2)$, $\mu^1 \in \mathbb{R}^m$, $\mu^2 \in \mathbb{R}^{s-m}$ in such a way that for each fixed value of μ^2, the mapping

$$\mathbb{R}^{p+m} \to \mathbb{R}^{m+p}, \quad (w^+, \mu^1) \mapsto W^-(w^+, 0, \mu^1, \mu^2) \tag{18}$$

is a local diffeomorphism near the point $(w^+ = 0, \mu^1 = 0)$, the preimage of zero being $(w^+ = \xi(\mu^2), \mu^1 = \zeta(\mu^2))$, $\xi(0) = 0$, $\zeta(0) = 0$. Shifting the variables w^+

and μ^1, if necessary. one can achieve $\xi \equiv 0$, $\zeta \equiv 0$. For each μ^2. introduce the new coordinate system (z^-, y) in $L^- = \mathbb{R}^{m+p}$ ($z^- \in \mathbb{R}^p$, $y \in \mathbb{R}^m$) via the formula

$$w^- = Q(z^-, y. \mu^2) = \frac{\partial W^-(0,0,0,\mu^2)}{\partial w^+} z^- + \frac{\partial W^-(0.0.0.\mu^2)}{\partial \mu^1} y.$$

Note that the coordinates z^- and y depend on the initial coordinate w^- *linearly*, so that multiplying w^- by -1 is equivalent to multiplying z^- and y simultaneously by -1. In the new coordinate system, mapping (18) takes the form

$$\begin{aligned}
(w^+, \mu^1) &\mapsto (z^-(w^+, \mu^1, \mu^2), y(w^+, \mu^1, \mu^2)) \\
&= (w^+ + A(w^+, \mu^1, \mu^2), \mu^1 + B(w^+. \mu^1, \mu^2)).
\end{aligned}$$

where $A = O_2(w^+, \mu^1)$. $B = O_2(w^+, \mu^1)$. Let also $z^+ = w^+ + A(w^+, \mu^1, \mu^2)$ be the new coordinate in $L^+ = \mathbb{R}^p$. Then involution (10), the family of differential equations on M determined by vector fields (11), and mapping (12) take the form

$$G : (x, z^+, z^-, y) \mapsto (-x, z^+, -z^-, -y). \tag{19}$$

$$\begin{aligned}
\dot{x} &= \omega(\mu^2) + O(z^+, z^-, y, \mu^1), \\
\dot{z}^+ &= O(z^-, y) = a(\mu)z^- + b(\mu)y + O_2(z^+, z^-. y). \\
\dot{z}^- &= z^+ + O_2(z^-, y), \\
\dot{y} &= \mu^1 + O_2(z^+, z^-, y, \mu^1),
\end{aligned} \tag{20}$$

and

$$\mathbb{R}^{p+s} \to \mathbb{R}^{m+p}, \quad (z^+, \mu) \mapsto (z^+, \mu^1 + O_2(z^+. \mu^1)). \tag{21}$$

respectively. For $\mu^1 = 0$ and each value of μ^2, torus $\{z^+ = z^- = 0, y = 0\}$ is invariant under both involution G and the flow of $X^\mu = X^{(0,\mu^2)}$ and carries parallel dynamics with frequency vector $\omega(\mu^2)$. We thus obtain an $(s - m)$-parameter analytic family of invariant n-tori with parallel dynamics for X^μ.

2 Summary of the results

To summarize, in all the contexts we have found l-parameter *analytic* families of invariant n-tori with parallel dynamics for some $l \in \mathbb{Z}_+$ provided that the vector fields X^μ are *integrable*. The question is what will happen to these families if one perturbs the vector fields. Attempting to answer this question. suppose first that the perturbation is *still integrable*, so that it just slightly shifts the initial family of tori. In essence, to perturb X^μ within the integrable realm means to change $\omega(\mu)$ and $\Omega(\mu)$ in (3). $\omega(y, \mu)$ in (4), $F(y, \mu)$ and $K(y, \mu)$ in (7). $H^\mu(y)$ in (9), $\omega(y, \mu)$ and $\Omega(y, \mu)$ in (17), or $\omega(\mu^2)$, $a(\mu)$ and $b(\mu)$ in (20). At first glance, this would not lead to anything interesting. However, when examining the behavior of the frequency vectors of the tori under the perturbation, three following different situations can be met:

(a) all the tori in the unperturbed family are nonresonant, and this property is preserved by perturbations;

(b) for a generic unperturbed family of vector fields. some tori are resonant, some are not. and this property is preserved by perturbations;

(c) by an arbitrarily small perturbation, one can make all the tori resonant.

Situation (a) always takes place for $n = 0$ and $n = 1$, since equilibria and closed trajectories treated as invariant tori with parallel dynamics of dimensions 0 and 1, respectively, are always nonresonant. This situation also occurs in the Hamiltonian coisotropic (n, p, s) context whenever $p = (n-1)/2$ and the straight line $U^\perp$ does not lie in any resonant hyperplane in $\mathbb{R}^n$.

Situation (c) corresponds to $n > 1$ and $l = 0$, i.e.. to the cases where $n > 1$ and the family X^μ of integrable vector fields possesses a single n-torus [this takes place in the dissipative (n, p, s) context for $s = 0$. in the volume preserving (n, p, s) context for $p > 1$, $s = 0$, in the reversible (n, m, p, s) context 1 for $m = s = 0$, and in the reversible (n, m, p, s) context 2 for $s = m$]. By an arbitrarily small perturbation, it is possible to make this single torus resonant. However, situation (c) is also realized in the Hamiltonian coisotropic (n, p, s) context (with $l = n - 2p + s \geq 1$) when the subspace $U^\perp$ lies in one of the resonant hyperplanes. In the latter case, not only one can make all the tori resonant by arbitrarily small perturbations, but also all the tori of the unperturbed family of vector fields are always resonant.

The "most familiar" situation is (b). It occurs in all the cases except those indicated above as pertaining to situations (a) or (c).

Now we can formulate the following heuristic principle describing the fate of analytic families of invariant tori with parallel dynamics of the initial family of vector fields under *nonintegrable* perturbations:

A small *generic* perturbation: preserves the initial family of tori and leaves it analytic in situation (a); preserves the family of tori but makes it Cantor-like in situation (b); destroys the initial family of tori completely in situation (c). In the first two situations, the unperturbed family of vector fields is assumed to satisfy some nondegeneracy and nonresonance conditions. Also, in the first two situations, all the invariant tori with parallel dynamics of the perturbed family of vector fields are nonresonant (i.e., quasi-periodic), and even Diophantine.

A generic dynamical system admits no resonant tori (the Kupka–Smale theorem [37]). so it is not surprising that generically perturbed families of vector fields no longer possess invariant tori with parallel dynamics in situation (c) or l-parameter *analytic* families of invariant tori with parallel dynamics in situation (b). Note. however, that in situation (c) with $l = 0$, i.e.. when the unperturbed family of vector fields has a single n-torus with parallel dynamics, a perturbation does not destroy this torus as an invariant submanifold: a perturbed family still has an invariant torus close to the unperturbed one. but the dynamics on the perturbed torus is generically no longer parallel. What is highly unexpected is that a perturbed family of vector fields does possess many quasi-periodic invariant tori in situations (a) and (b). However, the principle above is indeed true "in the first approximation". To make the latter statement more precise, let us find out what this principle tells in each context for nontrivial torus dimensions $n > 1$:

(α) Dissipative (n, p, s) context: a *generic* family of vector fields has no invariant n-tori with parallel dynamics for $s = 0$ (i.e., when there are no parameters) while a *typical* family of vector fields possesses s-parameter Cantor-like families of Diophantine invariant n-tori for $s \geq 1$.

(β) Volume preserving (n, p, s) context with $p > 1$: a *generic* family of globally divergence-free vector fields has no invariant n-tori with parallel dynamics for $s = 0$ (i.e., when there are no parameters) while a *typical* family of globally divergence-free vector fields possesses s-parameter Cantor-like families of Diophantine invariant n-tori for $s \geq 1$.

(γ) Volume preserving (n, p, s) context with $p = 1$: a *typical* family of globally divergence-free vector fields possesses $(s + 1)$-parameter Cantor-like families of Diophantine invariant n-tori.

(δ) Hamiltonian isotropic (n, p, s) context: a *typical* family of Hamiltonian vector fields possesses $(n + s)$-parameter Cantor-like families of Diophantine isotropic invariant n-tori.

(ϵ) Hamiltonian coisotropic (n, p, s) context: depending on the global properties of the symplectic structure on the phase space, either 1) a *generic* family of Hamiltonian vector fields has no coisotropic invariant n-tori with parallel dynamics, or 2) a *typical* family of Hamiltonian vector fields possesses $(n - 2p + s)$-parameter Cantor-like [for $1 \leq p < (n-1)/2$] or analytic [for $p = (n-1)/2$] families of Diophantine coisotropic invariant n-tori.

(ζ) Reversible (n, m, p, s) context 1: a *generic* family of reversible vector fields has no invariant n-tori with parallel dynamics for $m = s = 0$ while a *typical* family of reversible vector fields possesses $(m + s)$-parameter Cantor-like families of Diophantine invariant n-tori for $m + s \geq 1$.

(η) Reversible (n, m, p, s) context 2: a *generic* family of reversible vector fields has no invariant n-tori with parallel dynamics for $s = m$ while a *typical* family of reversible vector fields possesses $(s - m)$-parameter Cantor-like families of Diophantine invariant n-tori for $s > m$.

In all the cases above, the word "*generic*" means that families of vector fields with the property indicated constitute an open everywhere dense set in the space $\mathfrak{X}_s^\omega$ of all the families of vector fields, whereas the word "*typical*" means that families of vector fields with the property indicated constitute a set in the space $\mathfrak{X}_s^\omega$ with non-empty interior.

Are these seven statements (α)–(η) actually true? We start with the least studied cases.

To the best of the authors' knowledge, the reversible context 2 has not been considered yet in the literature, and nothing can be said about quasi-periodic invariant tori in this context [although statement (η) above seems very likely].

Statement (ϵ) above concerning the Hamiltonian coisotropic context is true (however, that the space $U^\perp$ does not lie in any resonant hyperplane in $\mathbb{R}^n$ does not guarantee the presence of many quasi-periodic coisotropic invariant n-tori in small perturbations of generic integrable Hamiltonians: one should also impose some Diophantine conditions on $U^\perp$). More precisely, the following two theorems hold (in the formulations of these theorems, we will suppose for simplicity that $s = 0$, i.e., there are no external parameters). Let Y be some finite open domain in $\mathbb{R}^{n-2p}$.

Theorem 2.1 [38, 39, 65] *Let $n \geq 4$, $1 \leq p \leq (n-2)/2$, the symplectic structure ω^2 with constant coefficients on $M = \{(x,y)\} = \mathbb{T}^n \times Y \subset \mathbb{T}^n \times \mathbb{R}^{n-2p}$ satisfy some Diophantine conditions (the set of structures which do not meet those conditions is of measure zero), and the unperturbed integrable Hamiltonian $H(y)$ satisfy some nondegeneracy conditions. Then any Hamiltonian vector field on M with Hamilton function $H(y) + \Delta(x,y)$ sufficiently close to $H(y)$ possesses Floquet Diophantine coisotropic invariant analytic n-tori, the $2(n-p)$-dimensional Lebesgue measure of the union of these tori tending to $\mathrm{meas}_{2(n-p)} M = (2\pi)^n \mathrm{meas}_{n-2p} Y$ as the perturbation magnitude tends to 0. The Floquet $(n-2p) \times (n-2p)$ matrix of each of the tori is zero.*

Theorem 2.2 [21, 20] *Let $n \geq 3$, $p = (n-1)/2$, the symplectic structure ω^2 with constant coefficients on $M = \{(x,y)\} = \mathbb{T}^n \times Y \subset \mathbb{T}^n \times \mathbb{R}$ satisfy some Diophantine conditions (the set of structures which do not meet those conditions is of measure zero), and the unperturbed integrable Hamiltonian $H(y)$ satisfy the nondegeneracy condition $dH/dy \neq 0$. Then each energy level of any Hamilton function $H(y) + \Delta(x,y)$ sufficiently close to $H(y)$ is a Floquet Diophantine coisotropic invariant n-torus of the corresponding Hamiltonian vector field. The frequency vectors of these tori are proportional to one and the same vector ω^0 which depends on neither the torus nor the Hamiltonian but is determined by the symplectic structure only. The Floquet 1×1 matrix of each of the tori is zero.*

In the context of Theorem 2.2, we have the following amazing picture [21, 20]. The energy levels of the unperturbed Hamiltonian $H(y)$ are n-tori with Diophantine motion, all the frequency vectors being proportional to some fixed vector ω^0. Now we perturb this Hamiltonian arbitrarily. Obviously, the energy levels of a perturbed Hamiltonian will be still n-tori close to the unperturbed ones, but it turns out that the motion on those perturbed tori *will be still Diophantine* with frequency vectors proportional to the same vector ω^0!. In fact, all the Hamilton functions close to $H(y)$ are integrable. This phenomenon is a direct consequence of the fact that the Hamiltonian nature of a vector field imposes very severe restrictions on the motion on invariant tori of small codimensions.

For exact formulations of Diophantine conditions for ω^2 and nondegeneracy conditions for $H(y)$ and the proofs, we refer the reader to the original papers [38, 39, 21, 20] (see also [65] for the case $p = (n - 2)/2$).

In the rest of this article, we will no longer consider the Hamiltonian coisotropic context and reversible context 2. So, our topics will be the dissipative context, both volume preserving contexts, Hamiltonian isotropic context, and reversible context 1. All the statements (α), (β), (γ), (δ), (ζ) above concerning these five contexts are true (and all the tori can be required to be Floquet as well). However, in these contexts, much more is known about the properties of l-parameter Cantor-like families of Floquet Diophantine invariant n-tori than in the Hamiltonian coisotropic context described by Theorem 2.1. Namely, not only the union of the tori is of positive $(n + l)$-dimensional Lebesgue measure, but also the tori depend on the labeling l-dimensional parameter (ranging in some Cantor-like subset of $\mathbb{R}^l$) *in the Whitney-smooth way*. The expression "a Whitney-smooth function on a closed set $\Xi \subset \mathbb{R}^l$" means a function on Ξ that can be extended to a smooth function defined on the whole space $\mathbb{R}^l$ [63, 40]. Now we would like to give a precise definition of a "Whitney-smooth l-parameter family of Floquet Diophantine invariant n-tori" (cf. [55, 58]).

3   Whitney-smooth families of invariant tori: definition

Definition 3.1 *Let* $X = \{X^\mu\}_\mu$ *be an analytic s-parameter family of analytic vector fields on a K-dimensional manifold M, parameter μ varying in an open domain $P \subset \mathbb{R}^s$. A Whitney-smooth l-parameter family of Floquet Diophantine invariant analytic n-tori of X is the image*

$$\mathcal{F}(\mathbb{T}^n \times \{0\} \times \Xi)$$

of the set $\mathbb{T}^n \times \{0\} \times \Xi$ ($\Xi \subset \mathbb{R}^l$) under a map

$$\mathcal{F} : \mathbb{T}^n \times \mathcal{O} \times \mathbb{R}^l \to M \times P, \quad 0 \in \mathcal{O} \subset \mathbb{R}^{K-n}$$

possessing the following properties (below x, w, and ξ are the coordinates in $\mathbb{T}^n$, $\mathbb{R}^{K-n}$, and $\mathbb{R}^l$, respectively).

a) $\mathcal{O}$ is a neighborhood of the origin in $\mathbb{R}^{K-n}$ while Ξ is a closed subset of $\mathbb{R}^l$ of positive Lebesgue measure.

b) The map $\mathcal{F}$ is analytic in $x \in \mathbb{T}^n$ and $w \in \mathcal{O}$ and is of class C^∞ in $\xi \in \mathbb{R}^l$. The restriction of $\mathcal{F}$ to $\mathbb{T}^n \times \mathcal{O} \times \Xi$ is injective and the inverse map

$$\mathcal{F}^{-1} : \mathrm{Image}(\mathbb{T}^n \times \mathcal{O} \times \Xi) \to \mathbb{T}^n \times \mathbb{R}^{K-n} \times \mathbb{R}^l$$

can be extended to a C^∞-map defined in $M \times P$.

c) For any $\xi \in \Xi$, the set $\mathcal{F}(\mathbb{T}^n \times \mathcal{O} \times \{\xi\})$ lies in one of the "fibers" $M \times \{\Lambda(\xi)\}$ (the function $\Lambda : \xi \mapsto \Lambda(\xi) \in P$ is defined for $\xi \in \Xi$ only). Thus, for $\xi \in \Xi$ the *restriction* $\mathcal{F}^\xi$,

$$\mathcal{F}^\xi : \mathbb{T}^n \times \mathcal{O} \to M, \quad \mathcal{F}(x, w, \xi) = (\mathcal{F}^\xi(x, w), \Lambda(\xi)),$$

is well defined.

d) For any $\xi \in \Xi$, $\mathcal{F}^\xi$ *is a diffeomorphism of* $\mathbb{T}^n \times \mathcal{O}$ *onto its image, and the vector field* $(\mathcal{F}^\xi)_*^{-1} X^{\Lambda(\xi)}$ *has the form*

$$[\omega(\xi) + O(w)]\frac{\partial}{\partial x} + [\Omega(\xi)w + O_2(w)]\frac{\partial}{\partial w} \tag{22}$$

where $\omega(\xi)$ *is some constant vector in* $\mathbb{R}^n$ *and* $\Omega(\xi)$ *is some constant matrix in* $gl(K - n, \mathbb{R})$.

e) Moreover, for all $\xi \in \Xi$, *the vectors* $\omega(\xi)$ *are uniformly Diophantine nonresonant, i.e., there exist constants* $\tau > 0$ *and* $\gamma > 0$ *independent of* ξ *and such that*

$$|\langle \omega(\xi), k \rangle| \geq \gamma |k|^{-\tau} \quad \forall k \in \mathbb{Z}^n \setminus \{0\}.$$

Condition d) implies that each set

$$T_\xi = \mathcal{F}^\xi(\mathbb{T}^n \times \{0\}) \subset M$$

($\xi \in \Xi$) is an invariant n-torus of the field $X^{\Lambda(\xi)}$ carrying parallel dynamics, and, moreover, this torus is Floquet. Condition e) implies that the parallel dynamics on each torus T_ξ is in fact Diophantine. Since $meas_l \Xi > 0$ [condition a)] and the inverse of the restriction of $\mathcal{F}$ to $\mathbb{T}^n \times \{0\} \times \Xi$ is Whitney-smooth [condition b)], the $(n+l)$-dimensional Lebesgue measure of the union of all the tori T_ξ is positive. However, a Whitney-smooth l-parameter family of Floquet Diophantine invariant n-tori is not just a collection of Floquet Diophantine invariant n-tori such that the $(n+l)$-dimensional measure of the union of the tori is positive. We also require that the tori depend on the labeling l-dimensional parameter ξ in a Whitney-smooth way and that the normalizing coordinate system around each torus [in which the corresponding vector field takes form (22)] can be also chosen to depend on ξ in a Whitney-smooth manner.

If one deals with Hamiltonian or reversible vector fields, the following additional conditions are imposed on Whitney-smooth families of invariant tori.

In the Hamiltonian isotropic context, we also require each torus T_ξ *to be isotropic.*

In the Hamiltonian coisotropic context, we also require each torus T_ξ *to be coisotropic.*

In the reversible context (where all the vector fields X^μ *are reversible with respect to involution* $G : M \to M$ *of the phase space), we also require that for each* $\xi \in \Xi$, *the involution* $(\mathcal{F}^\xi)^{-1} G \mathcal{F}^\xi$ *has the form*

$$(x, w) \mapsto (-x, Rw) \tag{23}$$

with ξ-independent involutive matrix R. In particular, each torus T_ξ is invariant not only under the flow of field $X^{\Lambda(\xi)}$, but also under the reversing involution G.

Whereas the components of vector $\omega(\xi)$ in (22) are called *internal frequencies* of torus T_ξ, the positive imaginary parts of the eigenvalues of matrix $\Omega(\xi)$ are called *normal frequencies* of this torus [25, 12, 11]. They constitute the *normal frequency vector* that will be denoted in the sequel by $\omega^N(\xi)$. Thus, if, e.g., the eigenvalues of $\Omega(\xi)$ are $\delta_1, \ldots, \delta_{N_1}, \alpha_1 \pm i\beta_1, \ldots, \alpha_{N_2} \pm i\beta_{N_2}$ where $N_1 + 2N_2 = K - n$ and $\delta \in \mathbb{R}^{N_1}$, $\alpha \in \mathbb{R}^{N_2}$, $\beta \in \mathbb{R}^{N_2}$, $\beta_j > 0$, all the numbers $\beta_1, \ldots, \beta_{N_2}$ being distinct, then $\omega^N(\xi) = \beta$.

The differentiability of Cantor-like families of invariant tori in dynamical systems was first established by Lazutkin [27, 28, 29, 30] for mappings of the plane possessing the so called intersection property (such mappings are slight generalizations of exact symplectic diffeomorphisms). Lazutkin's results were generalized to higher dimensions by Svanidze [61]. Analogous theorems for Hamiltonian vector fields were first proven by Pöschel [40] and Chierchia and Gallavotti [15], and those for reversible vector fields, by Pöschel [40]. For a very recent and detailed exposition of these results concerning the case where external parameters are absent, the reader is referred to Lazutkin's book [31]. Whitney-smooth families of Floquet Diophantine invariant tori in dynamical systems depending on external parameters were obtained in a general set-up by Broer, Huitema, and Takens [25, 12, 11]. Dissipative, globally divergence-free, and Hamiltonian vector fields were considered in [25, 12] while reversible vector fields were examined in [25] {the reversible $(n, m, 0, s)$ context 1} and [11] {the reversible (n, m, p, s) context 1 for arbitrary $p \geq 0$}.

4   Quasi-periodic stability

The main idea of the multiparameter KAM theory as developed in [25, 12, 11] is to consider integrable vector fields depending on a parameter of dimension *large enough* to guarantee the existence of a Floquet invariant torus with parallel dynamics possessing any collection (ω, Ω) of internal and normal data. From the practical viewpoint, this requirement means *the submersivity* of the mapping

the parameter labeling the tori $\mapsto (\omega$, the spectrum of $\Omega)$

[cf. mappings (25), (33), (34), (35), (37), (48), (59) below]. In a sufficiently small nonintegrable perturbation of this multiparameter family of integrable vector fields, one then looks for Floquet invariant tori with parallel dynamics whose internal ω and normal ω^N frequencies satisfy certain Diophantine conditions, and it turns out to be possible to find tori with *all* the collections (ω, Ω) of internal and normal data meeting those conditions (the so called quasi-periodic stability [12]).

In this section, we give precise formulations of quasi-periodic stability theorems for all the five contexts under consideration. We will call these theorems "main theorems", because all the results concerning vector fields depending on a smaller number of parameters can be relatively easily deduced from these theorems by some parameter reduction techniques.

In the sequel, parameter μ labeling the vector fields is always assumed to vary in an open domain $P \subset \mathbb{R}^s$, $s \geq 0$. All the quantities δ_j, ε_j, α_j, β_j are supposed to be real. The symbols $\mathcal{O}(0)$, $\widehat{\mathcal{O}(0)}$, and $\underline{\mathcal{O}(0)}$ denote neighborhoods of the origin in the Euclidean spaces $\mathbb{R}^d$ of dimensions d indicated below. The letter Y denotes an open domain in $\mathbb{R}^d$. Also, $\mathbb{RP}^d$ is the d-dimensional real projective space ($d \in \mathbb{Z}_+$) and $\Pi : \mathbb{R}^d \setminus \{0\} \to \mathbb{RP}^{d-1}$ for $d \in \mathbb{N}$ denotes the natural projection.

In all the theorems below (except for Theorem 4.3 pertaining to the volume preserving context with $p = 1$), the unperturbed vector fields X^μ are not, strictly speaking, integrable [equivariant with respect to the free action of $\mathbb{T}^n$]. The family X of vector fields is just assumed to possess an *analytic* family of Floquet invariant analytic n-tori with parallel dynamics.

4.1 Dissipative context

Consider an analytic family of analytic vector fields on $\mathbb{T}^n \times \mathbb{R}^p$ ($n \geq 1$, $p \geq 0$):

$$X = \{X^\mu\}_\mu = [\omega(\mu) + f(x, z, \mu)]\frac{\partial}{\partial x} + [\Omega(\mu)z + h(x, z, \mu)]\frac{\partial}{\partial z} \qquad (24)$$

[cf. (3)], where $x \in \mathbb{T}^n$, $z \in \mathcal{O}(0) \subset \mathbb{R}^p$, $\mu \in P \subset \mathbb{R}^s$, $\omega : P \to \mathbb{R}^n$, $\Omega : P \to gl(p, \mathbb{R})$, $f = O(z)$, $h = O_2(z)$. Let for $\mu \in \Gamma \subset P$ [Γ being diffeomorphic to a closed s-dimensional ball]

1) all the eigenvalues

$$\delta_1, \ldots, \delta_{N_1}, \quad \alpha_1 \pm i\beta_1, \ldots, \alpha_{N_2} \pm i\beta_{N_2}$$

($\beta_j > 0$) of matrix $\Omega(\mu)$ are simple and other than zero ($N_1 + 2N_2 = p$);

2) the mapping

$$\mathbb{R}^s \ni \mu \mapsto (\omega, \delta, \alpha, \beta) \in \mathbb{R}^{n+p} \qquad (25)$$

is submersive (so that $s \geq n + p$).

Fix $\tau > n - 1$. Set $\omega^N = \beta \in \mathbb{R}^r$, where $r = N_2$. By Γ_γ, where $\gamma > 0$, denote the set

$$\Gamma_\gamma = \left\{ \mu \in \Gamma : \forall k \in \mathbb{Z}^n \setminus \{0\} \; \forall \ell \in \mathbb{Z}^r, |\ell| \leq 2, \; |\langle \omega, k \rangle + \langle \omega^N, \ell \rangle| \geq \gamma|k|^{-\tau} \right\}. \qquad (26)$$

The set (26) is Cantor-like for $n \geq 2$. Since mapping (25) is submersive, this set consists of $(s-n-r+1)$-dimensional analytic surfaces Γ_γ^ι (maybe, with boundary), each surface being a part of the preimage of some point $\iota \in \mathbb{RP}^{n+r-1}$ under the mapping

$$\mathbb{R}^s \ni \mu \mapsto \Pi(\omega, \omega^N) \in \mathbb{RP}^{n+r-1}. \qquad (27)$$

Note that $\operatorname{meas}_s \Gamma_\gamma / \operatorname{meas}_s \Gamma \to 1$ as $\gamma \to 0$.

Theorem 4.1 (Main theorem in the dissipative context) [25, 12] *Then for any* $\gamma > 0$ *and any neighborhood* $\mathfrak{O}$ *of zero in the space of all* C^∞-*mappings*

$$\mathbb{T}^n \times P \to \mathbb{R}^n \times \mathbb{R}^p \times gl(p, \mathbb{R}) \times \mathbb{R}^s, \quad (x, \mu) \mapsto (\chi, \zeta_1, \zeta_2, \lambda), \qquad (28)$$

analytic in x *and such that* λ *does not depend on* x, *there exists a neighborhood* $\mathcal{X}$ *of the family* X *in the space of all analytic families of analytic vector fields*

$$\widetilde{X} = \{\widetilde{X}^\mu\}_\mu = [\omega(\mu) + f(x, z, \mu) + \widetilde{f}(x, z, \mu)]\frac{\partial}{\partial x} + [\Omega(\mu)z + h(x, z, \mu) + \widetilde{h}(x, z, \mu)]\frac{\partial}{\partial z}$$
$$\tag{29}$$

such that for any $\widetilde{X} \in \mathcal{X}$ *there is a mapping in* $\mathfrak{O}$

$$(x, \mu) \mapsto (\chi(x, \mu), \zeta_1(x, \mu), \zeta_2(x, \mu), \lambda(\mu)) \qquad (30)$$

with the following property: for each $\mu_0 \in \Gamma_\gamma$ *the vector field* $(\Phi_{\mu_0})_*^{-1} \widetilde{X}^{\mu_0 + \lambda(\mu_0)}$, *where*

$$\Phi_{\mu_0} : (\bar{x}, \bar{z}) \mapsto (\bar{x} + \chi(\bar{x}, \mu_0), \bar{z} + \zeta_1(\bar{x}, \mu_0) + \zeta_2(\bar{x}, \mu_0)\bar{z}). \qquad (31)$$

has the form

$$[\omega(\mu_0) + O(\bar{z})]\frac{\partial}{\partial \bar{x}} + [\Omega(\mu_0)\bar{z} + O_2(\bar{z})]\frac{\partial}{\partial \bar{z}}. \qquad (32)$$

Moreover, the mapping Φ_{μ_0} *depends on* μ_0 *analytically when* μ_0 *varies on each of the surfaces* Γ_γ^ι.

In a "less topological" language, this theorem runs as follows: for any $\gamma > 0$, $N^\star \in \mathbb{N}$ and $\epsilon^\star > 0$, there exists $\delta^\star > 0$ such that the following holds. For any analytic family of analytic vector fields (29), where $|\widetilde{f}(x, z, \mu)|$ and $|\widetilde{g}(x, z, \mu)|$ are less than $\delta^\star$ in a fixed (independent of γ, $N^\star$, $\epsilon^\star$) complex neighborhood of $\mathbb{T}^n \times \{0\} \times \Gamma \subset \mathbb{T}^n \times \mathbb{R}^p \times \mathbb{R}^s$, there exists a C^∞ mapping

$$\mathbb{T}^n \times P \to \mathbb{R}^n \times \mathbb{R}^p \times gl(p, \mathbb{R}) \times \mathbb{R}^s, \quad (x, \mu) \mapsto (\chi(x, \mu), \zeta_1(x, \mu), \zeta_2(x, \mu), \lambda(\mu))$$

analytic in x and possessing the following properties: **(1)** all the partial derivatives of the functions χ, ζ_1, ζ_2, λ of orders $0, 1, \ldots, N^\star$ are less than $\epsilon^\star$ in $\mathbb{T}^n \times P$ and **(2)** for each $\mu_0 \in \Gamma_\gamma$ the vector field $(\Phi_{\mu_0})_*^{-1} \widetilde{X}^{\mu_0 + \lambda(\mu_0)}$ [Φ_{μ_0} being defined by (31)] has the form (32). Moreover, the mapping Φ_{μ_0} depends on μ_0 analytically when μ_0 varies on each of the surfaces Γ_γ^ι.

4.2   Volume preserving context ($p \geq 2$)

Consider an analytic family of analytic vector fields (24) on $\mathbb{T}^n \times \mathbb{R}^p$ ($n \geq 1, p \geq 2$) globally divergence-free with respect to the volume element $dx \wedge dz$. where $x \in \mathbb{T}^n$, $z \in \mathcal{O}(0) \subset \mathbb{R}^p$, $\mu \in P \subset \mathbb{R}^s$, $\omega : P \to \mathbb{R}^n$, $\Omega : P \to sl(p, \mathbb{R})$, $f = O(z)$, $h = O_2(z)$. Here one should distinguish the cases $p = 2$ and $p > 2$. The reason is that the presence of purely imaginary eigenvalues of a 2×2 real matrix with trace zero is a typical possibility whereas for $p > 2$, all the eigenvalues of a generic $p \times p$ real matrix with trace zero have nonzero real parts. Assume that for $\mu \in \Gamma \subset P$ [Γ being diffeomorphic to a closed s-dimensional ball] the following holds.

For $p = 2$ (hyperbolic case): 1) the eigenvalues

$$\delta_1, \delta_2, \quad \delta_1 + \delta_2 = 0$$

of matrix $\Omega(\mu)$ are simple;

2) the mapping

$$\mathbb{R}^s \ni \mu \mapsto (\omega, \delta) \in \mathbb{R}^n \times L \cong \mathbb{R}^{n+1} \tag{33}$$

is submersive, where

$$L = \left\{ \delta \in \mathbb{R}^2 : \delta_1 + \delta_2 = 0 \right\} \cong \mathbb{R}$$

(so that $s \geq n + 1$).

Set $r = 0$, $\omega^N = 0 \in \mathbb{R}^0 = \{0\}$.

For $p = 2$ (elliptic case): 1) the eigenvalues

$$\pm i\varepsilon$$

of matrix $\Omega(\mu)$ are simple (and $\varepsilon > 0$);

2) the mapping

$$\mathbb{R}^s \ni \mu \mapsto (\omega, \varepsilon) \in \mathbb{R}^{n+1} \tag{34}$$

is submersive (so that $s \geq n + 1$).

Set $r = 1$, $\omega^N = \varepsilon \in \mathbb{R}$.

For $p > 2$: 1) all the eigenvalues

$$\delta_1, \ldots, \delta_{N_1}, \quad \alpha_1 \pm i\beta_1, \ldots, \alpha_{N_2} \pm i\beta_{N_2}, \quad \delta_1 + \ldots + \delta_{N_1} + 2(\alpha_1 + \ldots + \alpha_{N_2}) = 0$$

($\beta_j > 0$) of matrix $\Omega(\mu)$ with trace zero are simple and other than zero ($N_1 + 2N_2 = p$);

2) the mapping

$$\mathbb{R}^s \ni \mu \mapsto (\omega, \delta, \alpha, \beta) \in \mathbb{R}^n \times L \times \mathbb{R}^{N_2} \cong \mathbb{R}^{n+p-1} \tag{35}$$

is submersive, where

$$L = \left\{ (\delta, \alpha) \in \mathbb{R}^{N_1 + N_2} : \delta_1 + \ldots + \delta_{N_1} + 2(\alpha_1 + \ldots + \alpha_{N_2}) = 0 \right\} \cong \mathbb{R}^{N_1 + N_2 - 1}$$

(so that $s \geq n + p - 1$).

Set $r = N_2$, $\omega^N = \beta \in \mathbb{R}^r$.

Fix $\tau > n - 1$. By Γ_γ, where $\gamma > 0$, denote the set (26). As in the dissipative context, this set consists of $(s - n - r + 1)$-dimensional analytic surfaces Γ_γ^ι (maybe, with boundary), each surface being a part of the preimage of some point $\iota \in \mathbb{RP}^{n+r-1}$ under mapping (27). Also, $\text{meas }_s \Gamma_\gamma / \text{meas }_s \Gamma \to 1$ as $\gamma \to 0$.

Theorem 4.2 (Main theorem in the volume preserving context with $p \geq 2$) [25, 12] *Then for any $\gamma > 0$ and any neighborhood $\mathfrak{O}$ of zero in the space of all C^∞-mappings (28) analytic in x and such that λ does not depend on x. there exists a neighborhood $\mathcal{X}$ of the family X in the space of all analytic families of analytic globally divergence-free vector fields (29) such that for any $\widetilde{X} \in \mathcal{X}$ there is a mapping (30) in $\mathfrak{O}$ with the following property: for each $\mu_0 \in \Gamma_\gamma$ the vector field $(\Phi_{\mu_0})_*^{-1} \widetilde{X}^{\mu_0 + \lambda(\mu_0)}$ [Φ_{μ_0} being defined by (31)] has the form (32). Moreover, the mapping Φ_{μ_0} depends on μ_0 analytically when μ_0 varies on each of the surfaces Γ_γ^ι. Finally, all the mappings Φ_μ are volume preserving.*

4.3 Volume preserving context ($p = 1$)

Consider an analytic family of analytic vector fields on $\mathbb{T}^n \times \mathbb{R}$ ($n \geq 1$) globally divergence-free with respect to the volume element $dx \wedge dy$:

$$X = \{X^\mu\}_\mu = \omega(y, \mu)\frac{\partial}{\partial x} \tag{36}$$

[cf. (4)], where $x \in \mathbb{T}^n$, $y \in Y \subset \mathbb{R}$, $\mu \in P \subset \mathbb{R}^s$, $\omega : Y \times P \to \mathbb{R}^n$. Let for $(y, \mu) \in \Gamma \subset Y \times P$ [Γ being diffeomorphic to a closed $(s + 1)$-dimensional ball] the mapping

$$\mathbb{R}^{s+1} \ni (y, \mu) \mapsto \omega \in \mathbb{R}^n \tag{37}$$

is submersive (so that $s \geq n - 1$).

Fix $\tau > n - 1$. By Γ_γ, where $\gamma > 0$, denote the set

$$\Gamma_\gamma = \left\{ (y, \mu) \in \Gamma \ : \ \forall k \in \mathbb{Z}^n \setminus \{0\} \ |\langle \omega, k \rangle| \geq \gamma |k|^{-\tau} \right\}. \tag{38}$$

The set (38) is Cantor-like for $n \geq 2$. Since mapping (37) is submersive, this set consists of $(s - n + 2)$-dimensional analytic surfaces Γ_γ^ι (maybe, with boundary), each surface being a part of the preimage of some point $\iota \in \mathbb{RP}^{n-1}$ under the mapping

$$\mathbb{R}^{s+1} \ni (y, \mu) \mapsto \Pi\omega \in \mathbb{RP}^{n-1}. \tag{39}$$

Note also that $\mathrm{meas}_{s+1}\Gamma_\gamma / \mathrm{meas}_{s+1}\Gamma \to 1$ as $\gamma \to 0$.

Theorem 4.3 (Main theorem in the volume preserving context with $p = 1$) [25, 12] *Then for any $\gamma > 0$ and any neighborhood $\mathfrak{O}$ of zero in the space of all C^∞-mappings*

$$\mathbb{T}^n \times Y \times P \to \mathbb{R}^n \times \mathbb{R} \times \mathbb{R} \times \mathbb{R}^s, \quad (x, y, \mu) \mapsto (\chi, \eta_1, \eta_2, \lambda), \tag{40}$$

analytic in x and such that λ does not depend on x, there exists a neighborhood $\mathcal{X}$ of the family X in the space of all analytic families of analytic globally divergence-free vector fields

$$\widetilde{X} = \{\widetilde{X}^\mu\}_\mu = [\omega(y, \mu) + \widetilde{f}(x, y, \mu)]\frac{\partial}{\partial x} + \widetilde{g}(x, y, \mu)\frac{\partial}{\partial y} \tag{41}$$

such that for any $\widetilde{X} \in \mathcal{X}$ there is a mapping in $\mathfrak{D}$

$$(x, y, \mu) \mapsto (\chi(x, y, \mu), \eta_1(x, y, \mu), \eta_2(x, y, \mu), \lambda(y, \mu)) \qquad (42)$$

with the following property: for each $(y_0, \mu_0) \in \Gamma_\gamma$ the vector field
$(\Phi_{y_0\mu_0})_*^{-1} \widetilde{X}^{\mu_0 + \lambda(y_0, \mu_0)}$, *where*

$$\Phi_{y_0\mu_0} : (\bar{x}, \bar{y}) \mapsto (\bar{x} + \chi(\bar{x}, y_0, \mu_0), \bar{y} + \eta_1(\bar{x}, y_0, \mu_0) + \eta_2(\bar{x}, y_0, \mu_0)(\bar{y} - y_0)), \quad (43)$$

has the form

$$[\omega(y_0, \mu_0) + O(\bar{y} - y_0)]\frac{\partial}{\partial \bar{x}} + O_2(\bar{y} - y_0)\frac{\partial}{\partial \bar{y}}. \qquad (44)$$

Moreover, the mapping $\Phi_{y_0\mu_0}$ depends on (y_0, μ_0) analytically when the point (y_0, μ_0) varies on each of the surfaces Γ_γ^ι.

4.4 Hamiltonian isotropic context

Consider an analytic family of vector fields on $\mathbb{T}^n \times \mathbb{R}^{n+2p}$ ($n \geq 1$, $p \geq 0$) which are Hamiltonian with respect to the symplectic structure (5) with the analytic Hamilton function

$$H^\mu(x, y, z) = F(y, \mu) + \tfrac{1}{2}\langle z, K(y, \mu)z \rangle + \Delta(x, y, z, \mu) \qquad (45)$$

[cf. (7)], where $x \in \mathbb{T}^n$, $y \in Y \subset \mathbb{R}^n$, $z \in \mathcal{O}(0) \subset \mathbb{R}^{2p}$, $\mu \in P \subset \mathbb{R}^s$, $F : Y \times P \to \mathbb{R}$, $K : Y \times P \to gl(2p, \mathbb{R})$, $K(y, \mu)$ is symmetric for all values of y and μ, $\Delta = O_3(z)$. This family has the form

$$X = \{X^\mu\}_\mu = [\omega(y, \mu) + O_2(z)]\frac{\partial}{\partial x} + O_3(z)\frac{\partial}{\partial y} + [\Omega(y, \mu)z + O_2(z)]\frac{\partial}{\partial z} \qquad (46)$$

[cf. (8)], where $\omega = \partial F/\partial y$ and $\Omega = JK$, the $2p \times 2p$ matrix J being defined by (6). Let for $(y, \mu) \in \Gamma \subset Y \times P$ [Γ being diffeomorphic to a closed $(n + s)$-dimensional ball]

1) all the eigenvalues

$$\pm \delta_1, \ldots, \pm \delta_{N_1}, \quad \pm i\varepsilon_1, \ldots, \pm i\varepsilon_{N_2}, \quad \pm \alpha_1 \pm i\beta_1, \ldots, \pm \alpha_{N_3} \pm i\beta_{N_3} \qquad (47)$$

($\delta_j > 0$, $\varepsilon_j > 0$, $\alpha_j > 0$, $\beta_j > 0$) of Hamiltonian matrix $\Omega(y, \mu)$ are simple ($N_1 + N_2 + 2N_3 = p$);

2) the mapping

$$\mathbb{R}^{n+s} \ni (y, \mu) \mapsto (\omega, \delta, \varepsilon, \alpha, \beta) \in \mathbb{R}^{n+p} \qquad (48)$$

is submersive (so that $s \geq p$).

Fix $\tau > n - 1$. Set $\omega^N = (\varepsilon, \beta) \in \mathbb{R}^r$, where $r = N_2 + N_3$. By Γ_γ, where $\gamma > 0$, denote the set

$$\Gamma_\gamma = \left\{ (y, \mu) \in \Gamma : \forall k \in \mathbb{Z}^n \setminus \{0\} \; \forall \ell \in \mathbb{Z}^r, |\ell| \leq 2. \; |\langle \omega, k \rangle + \langle \omega^N, \ell \rangle| \geq \gamma |k|^{-\tau} \right\}. \qquad (49)$$

The set (49) is Cantor-like for $n \geq 2$. Since mapping (48) is submersive, this set consists of $(s - r + 1)$-dimensional analytic surfaces Γ_γ^ι (maybe, with boundary), each surface being a part of the preimage of some point $\iota \in \mathbb{R}P^{n+r-1}$ under the mapping

$$\mathbb{R}^{n+s} \ni (y, \mu) \mapsto \Pi(\omega, \omega^N) \in \mathbb{R}P^{n+r-1}. \tag{50}$$

Note also that $\operatorname{meas}_{n+s} \Gamma_\gamma / \operatorname{meas}_{n+s} \Gamma \to 1$ as $\gamma \to 0$.

Theorem 4.4 (Main theorem in the Hamiltonian isotropic context) [25, 12] *Then for any $\gamma > 0$ and any neighborhood $\mathfrak{O}$ of zero in the space of all C^∞-mappings*

$$\mathbb{T}^n \times Y \times \widehat{\mathcal{O}(0)} \times \mathcal{O}(0) \times P \to \mathbb{R}^n \times \mathbb{R}^n \times \mathbb{R}^{2p} \times \mathbb{R}^s, \quad (x, y, \hat{y}, z, \mu) \mapsto (\chi, \eta, \zeta, \lambda), \tag{51}$$

affine in $\hat{y}$ and z, analytic in x and such that χ does not depend on $\hat{y}$ and z while λ does not depend on x, $\hat{y}$ and z (here and henceforth in the Hamiltonian isotropic context, $\widehat{\mathcal{O}(0)} \subset \mathbb{R}^n$), there exists a neighborhood $\mathcal{X}$ of the Hamilton function H^μ in the space of all analytic and analytically μ-dependent Hamiltonians

$$\tilde{H}^\mu(x, y, z) = F(y, \mu) + \tfrac{1}{2} \langle z, K(y, \mu) z \rangle + \Delta(x, y, z, \mu) + \tilde{\Delta}(x, y, z, \mu) \tag{52}$$

(Hamilton function $\tilde{H}^\mu$ determines the family $\{\tilde{X}^\mu\}_\mu$ of vector fields) such that for any $\tilde{H}^\mu \in \mathcal{X}$ there is a mapping in $\mathfrak{O}$

$$(x, y, \hat{y}, z, \mu) \mapsto (\chi(x, y, \mu), \eta(x, y, \hat{y}, z, \mu), \zeta(x, y, \hat{y}, z, \mu), \lambda(y, \mu)) \tag{53}$$

with the following property: for each $(y_0, \mu_0) \in \Gamma_\gamma$ the vector field $(\Phi_{y_0 \mu_0})_*^{-1} \tilde{X}^{\mu_0 + \lambda(y_0, \mu_0)}$, where

$$\begin{aligned}
&\Phi_{y_0 \mu_0} : (\bar{x}, \bar{y}, \bar{z}) \mapsto \\
&(\bar{x} + \chi(\bar{x}, y_0, \mu_0), \bar{y} + \eta(\bar{x}, y_0, \bar{y} - y_0, \bar{z}, \mu_0), \bar{z} + \zeta(\bar{x}, y_0, \bar{y} - y_0, \bar{z}, \mu_0)),
\end{aligned} \tag{54}$$

has the form

$$\begin{aligned}
&[\omega(y_0, \mu_0) + O(|\bar{y} - y_0| + |\bar{z}|)]\frac{\partial}{\partial \bar{x}} + O_2(|\bar{y} - y_0| + |\bar{z}|)\frac{\partial}{\partial \bar{y}} + \\
&[\Omega(y_0, \mu_0)\bar{z} + O_2(|\bar{y} - y_0| + |\bar{z}|)]\frac{\partial}{\partial \bar{z}}.
\end{aligned} \tag{55}$$

Moreover, the mapping $\Phi_{y_0 \mu_0}$ depends on (y_0, μ_0) analytically when the point (y_0, μ_0) varies on each of the surfaces Γ_γ^ι. Finally, the invariant n-torus

$$\{\Phi_{y_0 \mu_0}(\bar{x}, y_0, 0) : \bar{x} \in \mathbb{T}^n\} \tag{56}$$

of the field $\tilde{X}^{\mu_0 + \lambda(y_0, \mu_0)}$ is isotropic.

Remark. The literature devoted to the Hamiltonian isotropic KAM theory is now immense. As far as the "classical" isotropic $(n, 0, 0)$ context is concerned, see, e.g., [10] for a review and a large bibliography. Some important references on the *lower-dimensional* invariant tori in Hamiltonian vector fields [the Hamiltonian isotropic (n, p, s) context with $p \geq 1$ in our terminology] are [32, 33, 35, 36, 8, 18, 66, 25, 17, 41, 14, 12, 62, 9].

4.5 Reversible context 1

Consider an analytic family of analytic G-reversible vector fields on $\mathbb{T}^n \times \mathbb{R}^{m+2p}$ ($n \geq 1$, $m \geq 0$, $p \geq 0$):

$$
\begin{aligned}
X = \{X^\mu\}_\mu &= [\omega(y, \mu) + f(x, y, z, \mu)]\frac{\partial}{\partial x} + g(x, y, z, \mu)\frac{\partial}{\partial y} + \\
&\quad [\Omega(y, \mu)z + h(x, y, z, \mu)]\frac{\partial}{\partial z}
\end{aligned} \tag{57}
$$

[cf. (17)],

$$
G : (x, y, z) \mapsto (-x, y, Rz) \tag{58}
$$

[cf. (16)], where $x \in \mathbb{T}^n$, $y \in Y \subset \mathbb{R}^m$, $z \in \mathcal{O}(0) \subset \mathbb{R}^{2p}$, $\mu \in P \subset \mathbb{R}^s$, R is an involutive $(2p) \times (2p)$ real matrix whose 1- and (-1)-eigenspaces are p-dimensional, $\omega : Y \times P \to \mathbb{R}^n$, $\Omega : Y \times P \to \{L \in sl(2p, \mathbb{R}) : LR + RL = 0\}$, $f = O(z)$, $g = O_2(z)$, $h = O_2(z)$. The reversibility with respect to G imposes the following conditions on the terms f, g, and h:

$$
f(-x, y, Rz, \mu) \equiv f(x, y, z, \mu), \qquad g(-x, y, Rz, \mu) \equiv -g(x, y, z, \mu),
$$
$$
h(-x, y, Rz, \mu) \equiv -Rh(x, y, z, \mu).
$$

Let for $(y, \mu) \in \Gamma \subset Y \times P$ [Γ being diffeomorphic to a closed $(m + s)$-dimensional ball]

1) all the eigenvalues (47) of infinitesimally R-reversible matrix $\Omega(y, \mu)$ are simple;

2) the mapping

$$
\mathbb{R}^{m+s} \ni (y, \mu) \mapsto (\omega, \delta, \varepsilon, \alpha, \beta) \in \mathbb{R}^{n+p} \tag{59}
$$

is submersive (so that $s \geq n - m + p$).

Fix $\tau > n - 1$. Set $\omega^N = (\varepsilon, \beta) \in \mathbb{R}^r$, where $r = N_2 + N_3$. By Γ_γ, where $\gamma > 0$, denote the set (49). Since mapping (59) is submersive, this set consists of $(s + m - n - r + 1)$-dimensional analytic surfaces Γ_γ^ι (maybe, with boundary), each surface being a part of the preimage of some point $\iota \in \mathbb{R}\mathrm{P}^{n+r-1}$ under the mapping

$$
\mathbb{R}^{m+s} \ni (y, \mu) \mapsto \Pi(\omega, \omega^N) \in \mathbb{R}\mathrm{P}^{n+r-1}. \tag{60}
$$

Also, $\mathrm{meas}\,_{m+s}\Gamma_\gamma / \mathrm{meas}\,_{m+s}\Gamma \to 1$ as $\gamma \to 0$.

Theorem 4.5 (Main theorem in the reversible context 1) [11] *Then for any $\gamma > 0$ and any neighborhood $\mathfrak{O}$ of zero in the space of all C^∞-mappings*

$$\mathbb{T}^n \times Y \times \widehat{\mathcal{O}(0)} \times \mathcal{O}(0) \times P \to \mathbb{R}^n \times \mathbb{R}^m \times \mathbb{R}^{2p} \times \mathbb{R}^s, \quad (x, y, \hat{y}, z, \mu) \mapsto (\chi, \eta, \zeta, \lambda), \quad (61)$$

affine in $\hat{y}$ and z, analytic in x and such that χ does not depend on $\hat{y}$ and z while λ does not depend on x, $\hat{y}$ and z (here and henceforth in the reversible context 1, $\widehat{\mathcal{O}(0)} \subset \mathbb{R}^m$), there exists a neighborhood $\mathcal{X}$ of the family X in the space of all analytic families of analytic G-reversible vector fields

$$\begin{aligned}
\widetilde{X} = \{\widetilde{X}^\mu\}_\mu \quad = \quad & [\omega(y, \mu) + f(x, y, z, \mu) + \tilde{f}(x, y, z, \mu)]\frac{\partial}{\partial x} + \\
& [g(x, y, z, \mu) + \tilde{g}(x, y, z, \mu)]\frac{\partial}{\partial y} + \\
& [\Omega(y, \mu)z + h(x, y, z, \mu) + \tilde{h}(x, y, z, \mu)]\frac{\partial}{\partial z}
\end{aligned} \quad (62)$$

such that for any $\widetilde{X} \in \mathcal{X}$ there is a mapping (53) in $\mathfrak{O}$ with the following properties:

$$\chi(-x, y, \mu) \equiv -\chi(x, y, \mu), \qquad \eta(-x, y, \hat{y}, Rz, \mu) \equiv \eta(x, y, \hat{y}, z, \mu),$$
$$\zeta(-x, y, \hat{y}, Rz, \mu) \equiv R\zeta(x, y, \hat{y}, z, \mu) \quad (63)$$

and for each $(y_0, \mu_0) \in \Gamma_\gamma$ the vector field $(\Phi_{y_0\mu_0})_*^{-1} \widetilde{X}^{\mu_0 + \lambda(y_0, \mu_0)}$ [$\Phi_{y_0\mu_0}$ being defined by (54)] has the form (55). Moreover, the mapping $\Phi_{y_0\mu_0}$ depends on (y_0, μ_0) analytically when the point (y_0, μ_0) varies on each of the surfaces Γ_γ^i.

Remark 1. If $g = O(z)$. then the statement of Theorem 4.5 is still true with the summand $O(|\bar{y}-y_0|+|\bar{z}|)\partial/\partial\bar{y}$ instead of $O_2(|\bar{y}-y_0|+|\bar{z}|)\partial/\partial\bar{y}$ in (55). Nevertheless, the invariant n-torus (56) of the field $\widetilde{X}^{\mu_0 + \lambda(y_0, \mu_0)}$ is still Floquet and its Floquet matrix is similar to $\mathbf{0}_m \oplus \Omega(y_0, \mu_0)$ where $\mathbf{0}_m$ is the $m \times m$ zero matrix.

Remark 2. For an extensive bibliography on the reversible KAM theory, see [51, 43, 55]. Here we confine ourselves by some references on the *lower-dimensional* invariant tori in reversible vector fields [the reversible (n, m, p, s) context 1 with $p \geq 1$], namely, [48, 50, 51] (for $m = n$), [53, 54] (for $m \geq n$), and [9, 11, 55] (for arbitrary n and m).

All the five Theorems 4.1–4.5 can be proven by the KAM technique of an infinite sequence of coordinate transformations within a unified Lie-algebraic approach. The proofs are given in [25, 12, 11].

5   The parameter reduction

The mappings Φ_{μ_0} or $\Phi_{y_0\mu_0}$ defined by (31), (43), (54) in Theorems 4.1–4.5 preserve all the internal and normal data of the unperturbed tori satisfying the appropriate

Diophantine conditions. However, the latter are formulated in terms of the internal ω and normal ω^N frequencies only [to be more precise, these conditions consist in that the point μ_0 or (y_0, μ_0) labeling the unperturbed tori belongs to the set Γ_γ defined by (26), (38), (49)]. Moreover, if some pair (ω^0, ω^{N0}) satisfies these conditions, so does any pair $(c\omega^0, c\omega^{N0})$ with $c \geq 1$. Consequently, it turns out to be possible to reduce the required number of parameters in Theorems 4.1–4.5, thereby weakening the results, namely, obtaining the preservation of the internal and normal frequencies only, and up to proportionality. In these "relaxed" counterparts of Theorems 4.1, 4.2, 4.3, 4.4, 4.5, the number s of external parameters is no less than $n+r-1, n+r-1, n-2, r-1, n-m+r-1$, respectively [25, 12, 11]. We will not present here the precise formulations of the "relaxed" theorems, because the lower bounds for s they provide are still too large compared to the heuristic predictions of Section 2. The reason is that in those theorems, although already "relaxed" with respect to the information on the perturbed tori (we no longer control the real parts of the eigenvalues of the Floquet matrices of the tori), there is still a correspondence between the perturbed tori and unperturbed ones: to each perturbed torus, there corresponds an unperturbed torus with the same collection of the internal and normal frequencies (up to proportionality). If we get rid of any intention to connect the perturbed tori and unperturbed ones and wish to prove just that a perturbed system possesses a Whitney-smooth family of Floquet Diophantine invariant n-tori (and that the relative measure of the union of these tori tends to unit as the perturbation size tends to zero), then we would be able to reduce the number of parameters to the heuristic values [$s = 1$ for the dissipative context and volume preserving context with $p \geq 2$, $s = 0$ for the volume preserving context with $p = 1$ and Hamiltonian isotropic context, and, finally, $s = \max(1 - m, 0)$ for the reversible context 1]. This parameter reduction requires Diophantine approximations on submanifolds of the Euclidean space (*Diophantine approximations of dependent quantities* in Sprindžuk's terminology [60]).

Let $W \subset \mathbb{R}^d$ be an open domain ($d \in \mathbb{N}$) and $\Gamma \subset W$ a subset in W diffeomorphic to a closed d-dimensional ball. Let $\mathfrak{F} : W \to \mathbb{R}^n$ be a mapping of class C^Q, $Q \in \mathbb{N}$. We will write

$$D^q \mathfrak{F}(w) = \frac{\partial^{|q|} \mathfrak{F}(w)}{\partial w^q} \in \mathbb{R}^n \text{ for } q \in \mathbb{Z}_+^d, \ 0 \leq |q| \leq Q,$$

where $\partial w^q = \partial w_1^{q_1} \cdots \partial w_d^{q_d}$. Choose an arbitrary vector $e \in \mathbb{R}^n \setminus \{0\}$. It is obvious that if for some $w^0 \in W$ and $j \in \mathbb{Z}_+$, $0 \leq j \leq Q$, the equality

$$\sum_{|q|=j} \langle D^q \mathfrak{F}(w^0), e \rangle u^q = 0$$

holds for all $u \in \mathbb{R}^d$ (here $u^q = u_1^{q_1} \cdots u_d^{q_d}$), then all the $(d + j - 1)!/j!(d - 1)!$ vectors $D^q \mathfrak{F}(w^0)$, $|q| = j$, are orthogonal to e. Consequently, if for some $w \in W$ the collection of $(d + Q)!/d!Q!$ vectors

$$D^q \mathfrak{F}(w), \quad q \in \mathbb{Z}_+^d, \ 0 \leq |q| \leq Q \tag{64}$$

span $\mathbb{R}^n$, then the quantity

$$\rho_{\mathfrak{F}}^Q(w) = \min_{\|e\|=1} \max_{j=0}^Q \max_{\|u\|=1} \left| \sum_{|q|=j} \langle D^q \mathfrak{F}(w), e \rangle u^q \right|$$

is positive ($e \in \mathbb{R}^n$, $u \in \mathbb{R}^d$).

Let also $\mathfrak{G} : W \to \mathbb{R}$ be a function of class C^Q. For any $w \in W$ we will write

$$\Xi_{\mathfrak{G}}^Q(w) = \max_{j=0}^Q \max_{\|u\|=1} \left| \sum_{|q|=j} D^q \mathfrak{G}(w) u^q \right|$$

(here again $u \in \mathbb{R}^d$).

Lemma 5.1 (Diophantine approximations on submanifolds) [55] *Suppose that for each $w \in \Gamma$ the collection of $(d+Q)!/d!Q!$ vectors (64) span $\mathbb{R}^n$. Assume also that*

$$\langle k, \mathfrak{F}(w) \rangle \neq \mathfrak{G}(w) \tag{65}$$

for all $w \in \Gamma$ and $k \in \mathbb{Z}^n$, $0 < \|k\| \leq \Xi_{\mathfrak{G}}^Q(w)/\rho_{\mathfrak{F}}^Q(w)$.

Then there is such $\delta^* > 0$ that for any C^Q-mappings $\widetilde{\mathfrak{F}} : W \to \mathbb{R}^n$ and $\widetilde{\mathfrak{G}} : W \to \mathbb{R}$ subject to inequalities

$$\sup_{w \in W} \max_{|q| \leq Q} |D^q \widetilde{\mathfrak{F}}(w) - D^q \mathfrak{F}(w)| < \delta^*, \quad \sup_{w \in W} \max_{|q| \leq Q} |D^q \widetilde{\mathfrak{G}}(w) - D^q \mathfrak{G}(w)| < \delta^*$$

($q \in \mathbb{Z}_+^d$) the following holds. For any $\tau > nQ - 1$ and $\epsilon^* > 0$ there exists such $\gamma = \gamma(\tau, \epsilon^*) > 0$ that the Lebesgue measure of the set of those points $w \in \Gamma$ for which

$$|\langle k, \widetilde{\mathfrak{F}}(w) \rangle - \widetilde{\mathfrak{G}}(w)| \geq \gamma |k|^{-\tau} \quad \text{for all} \quad k \in \mathbb{Z}^n \setminus \{0\}$$

is greater than $(1 - \epsilon^*) \mathrm{meas}_d \Gamma$.

Note that $\Xi_{\mathfrak{G}}^Q(w)/\rho_{\mathfrak{F}}^Q(w)$ is bounded from above on Γ, and condition (65) involves therefore only finitely many resonances to be avoided.

The "homogeneous nonperturbative" analogue of Lemma 5.1 (where the function $\mathfrak{G}$ is absent and one has to estimate $\langle k, \mathfrak{F}(w) \rangle$ only) was obtained by Bakhtin [6, 7]. In turn, Bakhtin's theorem generalizes an earlier result by Pyartli [42] pertaining to the case $d = 1$.

Lemma 5.2 [45, 46] *For an analytic mapping $\mathfrak{F} : W \to \mathbb{R}^n$, the following two statements are equivalent:*

(1) there is a number $Q \in \mathbb{N}$ such that the collection of $(d+Q)!/d!Q!$ vectors (64) span $\mathbb{R}^n$ for each point $w \in \Gamma$;

(2) the image of the mapping $\mathfrak{F} : \Gamma \to \mathbb{R}^n$ does not lie in any linear hyperplane in $\mathbb{R}^n$ passing through the origin.

Proof. If the image $\mathfrak{F}(\Gamma)$ of set Γ lies in some hyperplane in $\mathbb{R}^n$ passing through the origin, then for any $w \in \Gamma$ and $q \in \mathbb{Z}_+^d$ the vector $D^q\mathfrak{F}(w)$ lies in this hyperplane. Thus, (1) $\Longrightarrow$ (2) (here we have not used the analyticity of $\mathfrak{F}$). On the other hand, suppose that $\mathfrak{F}(\Gamma)$ does not lie in any hyperplane in $\mathbb{R}^n$ passing through the origin. If for some point $w^0 \in \Gamma$ all the vectors $D^q\mathfrak{F}(w^0)$, $q \in \mathbb{Z}_+^d$, belong to some hyperplane in $\mathbb{R}^n$, i.e., all of them are orthogonal to some vector $e \in \mathbb{R}^n \setminus \{0\}$, then $\langle \mathfrak{F}(w), e \rangle \equiv 0$ for $w \in \Gamma$ due to the analyticity of $\mathfrak{F}$. Thus, for each point $w^0 \in \Gamma$, there exists a number $Q(w^0) \in \mathbb{N}$ such that the vectors $D^q\mathfrak{F}(w^0)$, $q \in \mathbb{Z}_+^d$, $|q| \leq Q(w^0)$, span $\mathbb{R}^n$. Now observe that the vectors $D^q\mathfrak{F}(w)$, $|q| \leq Q(w^0)$, will span $\mathbb{R}^n$ for each $w \in W$ sufficiently close to w^0. As Γ is compact, there is $Q \in \mathbb{N}$ such that the vectors $D^q\mathfrak{F}(w)$, $|q| \leq Q$, span $\mathbb{R}^n$ for each $w \in \Gamma$. Thus, (2) $\Longrightarrow$ (1). $\qquad\square$

For C^∞-mappings $\mathfrak{F} : W \to \mathbb{R}^n$, condition (1) above is much stronger than condition (2). In fact, there are C^∞-mappings $\mathfrak{F} : W \to \mathbb{R}^n$ such that for all the points w in some subset $\Xi \subset \Gamma$ of positive measure all the derivatives $D^q\mathfrak{F}(w)$, $q \in \mathbb{Z}_+^d$, vanish, but each point $w^0 \in W$ possesses a neighborhood $\mathcal{O}(w^0) \subset W$ whose image $\mathfrak{F}(\mathcal{O}(w^0))$ does not lie in any hyperplane in $\mathbb{R}^n$ passing through the origin.

Example 5.3 Let $d = 1$ and Ξ be a closed perfect nowhere dense subset of segment Γ of positive measure (the word "*perfect*" means that any neighborhood of each point of Ξ contains infinitely many points of Ξ). Let $\mathcal{F} : W \to \mathbb{R}$ be a C^∞-function such that $\{w \in W : \mathcal{F}(w) = 0\} = \Xi$. Then $\mathcal{F}^{(l)}(w) = 0$ for each point $w \in \Xi$ and integer $l \in \mathbb{Z}_+$. We can further require that the set of points $w \in W$ for which $\mathcal{F}'(w) = 0$ be nowhere dense. Now set $\mathfrak{F} = (\mathcal{F}, \mathcal{F}^2, \ldots, \mathcal{F}^n)$. This C^∞-mapping possesses the desired properties.

5.1 Dissipative context

Let $s \geq 1$. Introduce the following notations:

$$\rho^Q(\mu) = \min_{\|e\|=1} \max_{j=0}^{Q} \max_{\|u\|=1} \left| \sum_{|q|=j} \langle D^q\omega(\mu), e \rangle u^q \right| \tag{66}$$

($q \in \mathbb{Z}_+^s$, $e \in \mathbb{R}^n$, $u \in \mathbb{R}^s$) where $Q \in \mathbb{N}$,

$$\Xi_\ell^Q(\mu) = \max_{j=0}^{Q} \max_{\|u\|=1} \left| \sum_{|q|=j} \langle D^q\omega^N(\mu), \ell \rangle u^q \right| \tag{67}$$

($q \in \mathbb{Z}_+^s$, $u \in \mathbb{R}^s$) where $Q \in \mathbb{N}$ and $\ell \in \mathbb{Z}^r$.

Let all the conditions of Theorem 4.1 be met except for that the mapping $\mu \mapsto (\omega, \delta, \alpha, \beta)$ is assumed to possess the following properties instead of submersivity:

a) there exists $Q \in \mathbb{N}$ such that for any $\mu \in \Gamma$ the collection of $(s + Q)!/s!Q!$ vectors

$$D^q \omega(\mu) \in \mathbb{R}^n, \quad q \in \mathbb{Z}_+^s, \ 0 \leq |q| \leq Q \tag{68}$$

span $\mathbb{R}^n$ [in particular, $(s + Q)!/s!Q! \geq n$, whence $s \geq n - 1$ for $Q = 1$ but any value of $s \geq 1$ is allowed for $Q = n - 1$], or, equivalently, the image of the map $\omega : \Gamma \to \mathbb{R}^n$ does not lie in any linear hyperplane passing through the origin (see Lemma 5.2); this condition ensures that $\rho^Q(\mu) > 0$ for any $\mu \in \Gamma$.

b) for each $\mu \in \Gamma$, $\ell \in \mathbb{Z}^r, 1 \leq |\ell| \leq 2$, and $k \in \mathbb{Z}^n, 0 < \|k\| \leq \Xi_\ell^Q(\mu)/\rho^Q(\mu)$, the following inequality holds:

$$\langle \omega(\mu), k \rangle \neq \langle \omega^N(\mu), \ell \rangle. \tag{69}$$

Theorem 5.4 (Miniparameter theorem in the dissipative context) *Then for any $\tau > nQ - 1$ fixed, any $\gamma > 0$ and any neighborhood $\mathfrak{O}$ of zero in the space of all C^∞-mappings*

$$\mathbb{T}^n \times P \to \mathbb{R}^n \times \mathbb{R}^p \times gl(p, \mathbb{R}), \quad (x, \mu) \mapsto (\chi, \zeta_1, \zeta_2). \tag{70}$$

analytic in x, there exists a neighborhood $\mathcal{X}$ of the family X in the space of all analytic families of analytic vector fields (29) such that for any $\widetilde{X} \in \mathcal{X}$ there are a set $\mathcal{G} \subset \Gamma$ and a mapping in $\mathfrak{O}$

$$(x, \mu) \mapsto (\chi(x, \mu), \zeta_1(x, \mu), \zeta_2(x, \mu)) \tag{71}$$

with the following properties: $\operatorname{meas}_s \mathcal{G} \geq (1 - \gamma)\operatorname{meas}_s \Gamma$ and for each $\mu_0 \in \mathcal{G}$ the vector field $(\Phi_{\mu_0})_*^{-1} \widetilde{X}^{\mu_0}$ [Φ_{μ_0} being defined by (31)] has the form

$$[\omega' + O(\bar{z})]\frac{\partial}{\partial \bar{x}} + [\Omega' \bar{z} + O_2(\bar{z})]\frac{\partial}{\partial \bar{z}}, \tag{72}$$

where the eigenvalues of matrix Ω' are

$$\delta_1', \ldots, \delta_{N_1}', \quad \alpha_1' \pm i\beta_1', \ldots, \alpha_{N_2}' \pm i\beta_{N_2}'$$

and

$$\forall k \in \mathbb{Z}^n \setminus \{0\} \ \forall \ell \in \mathbb{Z}^r, |\ell| \leq 2, \ |\langle \omega', k \rangle + \langle \omega'^N, \ell \rangle| \geq \gamma |k|^{-\tau} \tag{73}$$

with $\omega'^N = \beta'$.

Here there is no correspondence between the perturbed tori and unperturbed ones, and we do not need to introduce the parameter shift $\lambda(\mu_0)$ which was present in Theorem 4.1.

5.2 Volume preserving context $(p \geq 2)$

Let $s \geq 1$ and all the conditions of Theorem 4.2 be met except for that the mapping $\mu \mapsto (\omega, \delta, \alpha, \beta)$ [for $p = 2$ (hyperbolic case) and $p > 2$] or $\mu \mapsto (\omega, \varepsilon)$ [for $p = 2$ (elliptic case)] is assumed to possess the following properties instead of submersivity:

a) there exists $Q \in \mathbb{N}$ such that for any $\mu \in \Gamma$ the collection of $(s+Q)!/s!Q!$ vectors (68) span $\mathbb{R}^n$ [in particular, $(s+Q)!/s!Q! \geq n$, whence $s \geq n-1$ for $Q = 1$ but any value of $s \geq 1$ is allowed for $Q = n-1$], or, equivalently, the image of the map $\omega : \Gamma \to \mathbb{R}^n$ does not lie in any linear hyperplane passing through the origin (see Lemma 5.2); this condition ensures that $\rho^Q(\mu) > 0$ for any $\mu \in \Gamma$, where the quantities $\rho^Q(\mu)$ and $\Xi_\ell^Q(\mu)$ are defined by (66) and (67), respectively,

b) for each $\mu \in \Gamma$, $\ell \in \mathbb{Z}^r, 1 \leq |\ell| \leq 2$, and $k \in \mathbb{Z}^n. 0 < \|k\| \leq \Xi_\ell^Q(\mu)/\rho^Q(\mu)$, inequality (69) holds.

Theorem 5.5 (Miniparameter theorem in the volume preserving context with $p \geq 2$) *Then for any $\tau > nQ - 1$ fixed, any $\gamma > 0$ and any neighborhood $\mathfrak{O}$ of zero in the space of all C^∞-mappings (70) analytic in x, there exists a neighborhood $\mathcal{X}$ of the family X in the space of all analytic families of analytic globally divergence-free vector fields (29) such that for any $\tilde{X} \in \mathcal{X}$ there are a set $\mathcal{G} \subset \Gamma$ and a mapping (71) in $\mathfrak{O}$ with the following properties: $\mathrm{meas}_s \mathcal{G} \geq (1-\gamma)\mathrm{meas}_s \Gamma$ and for each $\mu_0 \in \mathcal{G}$ the vector field $(\Phi_{\mu_0})_*^{-1} \tilde{X}^{\mu_0}$ [Φ_{μ_0} being defined by (31)] has the form (72), where the eigenvalues of matrix Ω' are*

$$\delta'_1, \ldots, \delta'_{N_1}, \quad \alpha'_1 \pm i\beta'_1, \ldots, \alpha'_{N_2} \pm i\beta'_{N_2}, \quad \delta'_1 + \ldots + \delta'_{N_1} + 2(\alpha'_1 + \ldots + \alpha'_{N_2}) = 0$$

[for $p = 2$ (hyperbolic case) and $p > 2$; set $\omega'^N = \beta'$] or

$$\pm i\varepsilon'$$

[for $p = 2$ (elliptic case); set $\omega'^N = \varepsilon'$] and inequalities (73) hold. Moreover, all the mappings Φ_μ are volume preserving.

5.3 Volume preserving context $(p = 1)$

Let all the conditions of Theorem 4.3 be met except for that the mapping $(y, \mu) \mapsto \omega$ is assumed to possess the following property instead of submersivity:

there exists $Q \in \mathbb{N}$ such that for any $(y, \mu) \in \Gamma$ the collection of $(s + Q + 1)!/(s+1)!Q!$ vectors

$$D^q \omega(y, \mu) \in \mathbb{R}^n, \quad q \in \mathbb{Z}_+^{s+1}, \ 0 \leq |q| \leq Q \tag{74}$$

span $\mathbb{R}^n$ [in particular, $(s + Q + 1)!/(s+1)!Q! \geq n$, whence $s \geq n-2$ for $Q = 1$ but any value of $s \geq 0$ is allowed for $Q = n-1$], or, equivalently, the image of the map $\omega : \Gamma \to \mathbb{R}^n$ does not lie in any linear hyperplane passing through the origin (see Lemma 5.2).

Theorem 5.6 (Miniparameter theorem in the volume preserving context with $p = 1$) *Then for any $\tau > nQ - 1$ fixed, any $\gamma > 0$ and any neighborhood $\mathfrak{O}$ of zero in the space of all C^∞-mappings*

$$\mathbb{T}^n \times Y \times P \to \mathbb{R}^n \times \mathbb{R} \times \mathbb{R}, \quad (x, y, \mu) \mapsto (\chi, \eta_1, \eta_2), \quad (75)$$

analytic in x, there exists a neighborhood $\mathcal{X}$ of the family X in the space of all analytic families of analytic globally divergence-free vector fields (41) such that for any $\widetilde{X} \in \mathcal{X}$ there are a set $\mathcal{G} \subset \Gamma$ and a mapping in $\mathfrak{O}$

$$(x, y, \mu) \mapsto (\chi(x, y, \mu), \eta_1(x, y, \mu), \eta_2(x, y, \mu)) \quad (76)$$

with the following properties: $\mathrm{meas}_{s+1}\mathcal{G} \geq (1-\gamma)\mathrm{meas}_{s+1}\Gamma$ and for each $(y_0, \mu_0) \in \mathcal{G}$ the vector field $(\Phi_{y_0\mu_0})_*^{-1} \widetilde{X}^{\mu_0}$ [$\Phi_{y_0\mu_0}$ being defined by (43)] has the form

$$[\omega' + O(\bar{y} - y_0)]\frac{\partial}{\partial \bar{x}} + O_2(\bar{y} - y_0)\frac{\partial}{\partial \bar{y}}, \quad (77)$$

and

$$\forall k \in \mathbb{Z}^n \setminus \{0\} \quad |\langle \omega', k \rangle| \geq \gamma |k|^{-\tau}. \quad (78)$$

5.4   Hamiltonian isotropic context

Introduce the following notations:

$$\rho^Q(y, \mu) = \min_{\|e\|=1} \max_{j=0}^{Q} \max_{\|u\|=1} \left| \sum_{|q|=j} \langle D^q\omega(y, \mu), e \rangle u^q \right| \quad (79)$$

$(q \in \mathbb{Z}_+^{n+s}, e \in \mathbb{R}^n, u \in \mathbb{R}^{n+s})$ where $Q \in \mathbb{N}$,

$$\Xi_\ell^Q(y, \mu) = \max_{j=0}^{Q} \max_{\|u\|=1} \left| \sum_{|q|=j} \langle D^q\omega^N(y, \mu), \ell \rangle u^q \right| \quad (80)$$

$(q \in \mathbb{Z}_+^{n+s}, u \in \mathbb{R}^{n+s})$ where $Q \in \mathbb{N}$ and $\ell \in \mathbb{Z}^r$.

Let all the conditions of Theorem 4.4 be met except for that the mapping $(y, \mu) \mapsto (\omega, \delta, \varepsilon, \alpha, \beta)$ is assumed to possess the following properties instead of submersivity:

a) there exists $Q \in \mathbb{N}$ such that for any $(y, \mu) \in \Gamma$ the collection of $(n + s + Q)!/(n + s)!Q!$ vectors

$$D^q\omega(y, \mu) \in \mathbb{R}^n, \quad q \in \mathbb{Z}_+^{n+s}, \ 0 \leq |q| \leq Q \quad (81)$$

span $\mathbb{R}^n$, or, equivalently, the image of the map $\omega : \Gamma \to \mathbb{R}^n$ does not lie in any linear hyperplane passing through the origin (see Lemma 5.2): this condition ensures that $\rho^Q(y, \mu) > 0$ for any $(y, \mu) \in \Gamma$,

b) for each $(y, \mu) \in \Gamma$, $\ell \in \mathbb{Z}^r$, $1 \le |\ell| \le 2$, and $k \in \mathbb{Z}^n$, $0 < \|k\| \le \Xi_\ell^Q(y, \mu)/\rho^Q(y, \mu)$, the following inequality holds:

$$\langle \omega(y, \mu), k \rangle \ne \langle \omega^N(y, \mu), \ell \rangle. \tag{82}$$

Remark. The hypotheses of this kind were first introduced by Rüssmann [45, 46, 47]. However, these papers contained no proofs.

Theorem 5.7 (Miniparameter theorem in the Hamiltonian isotropic context)
Then for any $\tau > nQ - 1$ fixed, any $\gamma > 0$ and any neighborhood $\mathfrak{O}$ of zero in the space of all C^∞-mappings

$$\mathbb{T}^n \times Y \times \widehat{\mathcal{O}(0)} \times \mathcal{O}(0) \times P \to \mathbb{R}^n \times \mathbb{R}^n \times \mathbb{R}^{2p}, \quad (x, y, \hat{y}, z, \mu) \mapsto (\chi, \eta, \zeta), \tag{83}$$

affine in $\hat{y}$ and z, analytic in x and such that χ does not depend on $\hat{y}$ and z, there exists a neighborhood $\mathcal{X}$ of the Hamilton function H^μ in the space of all analytic and analytically μ-dependent Hamiltonians (52) (Hamilton function $\tilde{H}^\mu$ determines the family $\{\tilde{X}^\mu\}_\mu$ of vector fields) such that for any $\tilde{H}^\mu \in \mathcal{X}$ there are a set $\mathcal{G} \subset \Gamma$ and a mapping in $\mathfrak{O}$

$$(x, y, \hat{y}, z, \mu) \mapsto (\chi(x, y, \mu), \eta(x, y, \hat{y}, z, \mu), \zeta(x, y, \hat{y}, z, \mu)) \tag{84}$$

with the following properties: $\text{meas}_{n+s} \mathcal{G} \ge (1-\gamma) \text{meas}_{n+s} \Gamma$ and for each $(y_0, \mu_0) \in \mathcal{G}$ the vector field $(\Phi_{y_0 \mu_0})_*^{-1} \tilde{X}^{\mu_0}$ [$\Phi_{y_0 \mu_0}$ being defined by (54)] has the form

$$[\omega' + O(|\bar{y} - y_0| + |\bar{z}|)]\frac{\partial}{\partial \bar{x}} + O_2(|\bar{y} - y_0| + |\bar{z}|)\frac{\partial}{\partial \bar{y}} + [\Omega' \bar{z} + O_2(|\bar{y} - y_0| + |\bar{z}|)]\frac{\partial}{\partial \bar{z}}, \tag{85}$$

where the eigenvalues of matrix Ω' are

$$\pm \delta_1', \ldots, \pm \delta_{N_1}', \quad \pm i\varepsilon_1', \ldots, \pm i\varepsilon_{N_2}', \quad \pm \alpha_1' \pm i\beta_1', \ldots, \pm \alpha_{N_3}' \pm i\beta_{N_3}' \tag{86}$$

and inequalities (73) hold with $\omega'^N = (\varepsilon', \beta')$. Moreover, the invariant n-torus (56) of the field $\tilde{X}^{\mu_0}$ is isotropic.

5.5 Reversible context 1

Let $m + s \ge 1$ and all the conditions of Theorem 4.5 be met except for that the mapping $(y, \mu) \mapsto (\omega, \delta, \varepsilon, \alpha, \beta)$ is assumed to possess the following properties instead of submersivity:

a) there exists $Q \in \mathbb{N}$ such that for any $(y, \mu) \in \Gamma$ the collection of $(m + s + Q)!/(m + s)!Q!$ vectors

$$D^q \omega(y, \mu) \in \mathbb{R}^n, \quad q \in \mathbb{Z}_+^{m+s}, \ 0 \le |q| \le Q \tag{87}$$

span $\mathbb{R}^n$ [in particular, $(m + s + Q)!/(m + s)!Q! \geq n$, whence $s \geq n - m - 1$ for $Q = 1$ but any value of $s \geq 1 - m$ is allowed for $Q \geq n - 1$], or, equivalently, the image of the map $\omega : \Gamma \to \mathbb{R}^n$ does not lie in any linear hyperplane passing through the origin (see Lemma 5.2); this condition ensures that $\rho^Q(y, \mu) > 0$ for any $(y, \mu) \in \Gamma$, where the quantities $\rho^Q(y, \mu)$ and $\Xi_\ell^Q(y, \mu)$ are defined by (79) and (80), respectively (with $u \in \mathbb{R}^{m+s}$ instead of $u \in \mathbb{R}^{n+s}$ and $q \in \mathbb{Z}_+^{m+s}$ instead of $q \in \mathbb{Z}_+^{n+s}$),

b) for each $(y, \mu) \in \Gamma$, $\ell \in \mathbb{Z}^r, 1 \leq |\ell| \leq 2$, and $k \in \mathbb{Z}^n, 0 < \|k\| \leq \Xi_\ell^Q(y, \mu)/\rho^Q(y, \mu)$, inequality (82) holds.

Theorem 5.8 (Miniparameter theorem in the reversible context 1) *Then for any* $\tau > nQ - 1$ *fixed, any* $\gamma > 0$ *and any neighborhood* $\mathfrak{O}$ *of zero in the space of all* C^∞*-mappings*

$$\mathbb{T}^n \times Y \times \widehat{\mathcal{O}(0)} \times \mathcal{O}(0) \times P \to \mathbb{R}^n \times \mathbb{R}^m \times \mathbb{R}^{2p}, \quad (x, y, \hat{y}, z, \mu) \mapsto (\chi, \eta, \zeta), \quad (88)$$

affine in $\hat{y}$ *and* z, *analytic in* x *and such that* χ *does not depend on* $\hat{y}$ *and* z, *there exists a neighborhood* $\mathcal{X}$ *of the family* X *in the space of all analytic families of analytic G-reversible vector fields (62) such that for any* $\widetilde{X} \in \mathcal{X}$ *there are a set* $\mathcal{G} \subset \Gamma$ *and a mapping (84) in* $\mathfrak{O}$ *satisfying identities (63) and possessing the following properties:* $\mathrm{meas}_{m+s}\mathcal{G} \geq (1 - \gamma)\mathrm{meas}_{m+s}\Gamma$ *and for each* $(y_0, \mu_0) \in \mathcal{G}$ *the vector field* $(\Phi_{y_0\mu_0})_*^{-1} \widetilde{X}^{\mu_0}$ *[$\Phi_{y_0\mu_0}$ being defined by (54)] has the form (85), the eigenvalues of matrix* Ω' *being of the form (86) and inequalities (73) holding with* $\omega'^N = (\varepsilon', \beta')$.

Remark. If $g = O(z)$, then the statement of Theorem 5.8 is still true with the summand $O(|\bar{y}-y_0|+|\bar{z}|)\partial/\partial\bar{y}$ instead of $O_2(|\bar{y}-y_0|+|\bar{z}|)\partial/\partial\bar{y}$ in (85). Nevertheless, the invariant n-torus (56) of the field $\widetilde{X}^{\mu_0}$ is still Floquet and its Floquet matrix is similar to $\mathbf{0}_m \oplus \Omega'$.

The "miniparameter" Theorems 5.4–5.8 can be obtained from the main Theorems 4.1–4.5 using Lemma 5.1. Below we demonstrate this reduction technique in the reversible context 1, the proofs for the other contexts are entirely similar.

Proof of Theorem 5.8. Let $\nu \in \mathcal{O}(0) \subset \mathbb{R}^t$ be an additional parameter and let analytic mappings

$$\omega^{\mathrm{new}} : Y \times P \times \underline{\mathcal{O}(0)} \to \mathbb{R}^n, \quad \Omega^{\mathrm{new}} : Y \times P \times \underline{\mathcal{O}(0)} \to \{L \in sl(2p, \mathbb{R}) : LR + RL = 0\}$$

possess the following properties:
  i) $\omega^{\mathrm{new}}(y, \mu, 0) \equiv \omega(y, \mu)$, $\Omega^{\mathrm{new}}(y, \mu, 0) \equiv \Omega(y, \mu)$;
  ii) the mapping

$$\mathbb{R}^{m+s+t} \ni (y, \mu, \nu) \mapsto (\omega^{\mathrm{new}}, \delta^{\mathrm{new}}, \varepsilon^{\mathrm{new}}, \alpha^{\mathrm{new}}, \beta^{\mathrm{new}}) \in \mathbb{R}^{n+p}$$

is submersive for $(y, \mu) \in \Gamma$, $\nu = 0$ (so that $t \geq n - m + p - s$).

The possibility of extending $\Omega(y,\mu)$ in this way follows from the theory of versal unfoldings for infinitesimally reversible matrices [52, 59, 24].

One can apply Theorem 4.5 to the family of G-reversible vector fields $\widetilde{X}_{\text{new}} = \left\{ \widetilde{X}_{\text{new}}^{\mu,\nu} \right\}_{\mu,\nu}$ having been obtained from the family $\widetilde{X} = \{\widetilde{X}^\mu\}_\mu$ (62) by replacing $\omega(y,\mu)$ and $\Omega(y,\mu)$ with $\omega^{\text{new}}(y,\mu,\nu)$ and $\Omega^{\text{new}}(y,\mu,\nu)$, respectively. One has $\widetilde{X}_{\text{new}}^{\mu,0} = \widetilde{X}^\mu$. By $\Gamma_{\gamma'}^{\text{new}}$, where $\gamma' > 0$, denote the set

$$\Gamma_{\gamma'}^{\text{new}} = \Big\{ (y,\mu,\nu) \in \Gamma \times \mathcal{O}(0) \;\; : \;\; \forall k \in \mathbb{Z}^n \setminus \{0\} \; \forall \ell \in \mathbb{Z}^r, |\ell| \leq 2,$$
$$|\langle \omega^{\text{new}}, k \rangle + \langle (\omega^N)^{\text{new}}, \ell \rangle| \geq \gamma' |k|^{-\tau} \Big\}$$

where $(\omega^N)^{\text{new}} = (\varepsilon^{\text{new}}, \beta^{\text{new}})$. Now for any $\gamma' > 0$ and any $(y_0, \mu_1, \nu_1) \in \Gamma_{\gamma'}^{\text{new}}$, Theorem 4.5 provides a mapping

$$\Phi_{y_0\mu_1\nu_1}^{\text{new}} : (\bar{x}, \bar{y}, \bar{z}) \;\; \mapsto \;\; (\bar{x} + \chi^{\text{new}}(\bar{x}, y_0, \mu_1, \nu_1),$$
$$\bar{y} + \eta^{\text{new}}(\bar{x}, y_0, \bar{y} - y_0, \bar{z}, \mu_1, \nu_1),$$
$$\bar{z} + \zeta^{\text{new}}(\bar{x}, y_0, \bar{y} - y_0, \bar{z}, \mu_1, \nu_1))$$

such that the vector field

$$\left(\Phi_{y_0\mu_1\nu_1}^{\text{new}}\right)_*^{-1} \widetilde{X}_{\text{new}}^{\mu_1 + \lambda^{\text{new}}(y_0,\mu_1,\nu_1), \nu_1 + \theta(y_0,\mu_1,\nu_1)}$$

has the form

$$[\omega^{\text{new}}(y_0, \mu_1, \nu_1) + O(|\bar{y} - y_0| + |\bar{z}|)]\frac{\partial}{\partial \bar{x}} + O_2(|\bar{y} - y_0| + |\bar{z}|)\frac{\partial}{\partial \bar{y}} +$$
$$[\Omega^{\text{new}}(y_0, \mu_1, \nu_1)\bar{z} + O_2(|\bar{y} - y_0| + |\bar{z}|)]\frac{\partial}{\partial \bar{z}}.$$

One can solve the equation $\nu + \theta(y,\mu,\nu) = 0$ with respect to ν and obtain $\nu = \xi(y,\mu)$ where the function ξ is C^∞-small. Set

$$\lambda(y,\mu) = \lambda^{\text{new}}(y, \mu, \xi(y,\mu)).$$

The equation $\mu_1 + \lambda(y_0, \mu_1) = \mu_0$ for each fixed y_0 can be solved with respect to μ_1 as $\mu_1 = \kappa(y_0, \mu_0)$ where the function $\kappa(y,\mu) - \mu$ is also C^∞-small. Define the set $\mathcal{G}$ as

$$\mathcal{G} = \{(y_0, \mu_0) \in \Gamma : (y_0, \kappa(y_0, \mu_0)) \in \Gamma$$
$$\text{and} \quad (y_0, \kappa(y_0, \mu_0), \xi(y_0, \kappa(y_0, \mu_0))) \in \Gamma_{\gamma'}^{\text{new}} \}.$$

According to Lemma 5.1, $\text{meas}_{m+s}\mathcal{G}/\text{meas}_{m+s}\Gamma \to 1$ as $\gamma' \to 0$, and one can choose $\gamma' \leq \gamma$ such that $\text{meas}_{m+s}\mathcal{G} \geq (1-\gamma)\text{meas}_{m+s}\Gamma$. It remains to set

$$\chi(x, y, \mu + \lambda(y,\mu)) = \chi^{\text{new}}(x, y, \mu, \xi(y,\mu)),$$
$$\eta(x, y, \hat{y}, z, \mu + \lambda(y,\mu)) = \eta^{\text{new}}(x, y, \hat{y}, z, \mu, \xi(y,\mu)),$$
$$\zeta(x, y, \hat{y}, z, \mu + \lambda(y,\mu)) = \zeta^{\text{new}}(x, y, \hat{y}, z, \mu, \xi(y,\mu)). \tag{89}$$

or, to be more "rigorous",

$$\chi(x,y,\mu) = \chi^{\text{new}}(x,y,\kappa(y,\mu),\xi(y,\kappa(y,\mu))),$$
$$\eta(x,y,\hat{y},z,\mu) = \eta^{\text{new}}(x,y,\hat{y},z,\kappa(y,\mu),\xi(y,\kappa(y,\mu))),$$
$$\zeta(x,y,\hat{y},z,\mu) = \zeta^{\text{new}}(x,y,\hat{y},z,\kappa(y,\mu),\xi(y,\kappa(y,\mu))).$$

Indeed, let $(y_0,\mu_0) \in \mathcal{G}$, $\mu_1 = \kappa(y_0,\mu_0)$, and $\nu_1 = \xi(y_0,\mu_1)$. Then $(y_0,\mu_1,\xi(y_0,\mu_1)) \in \Gamma^{\text{new}}_{\gamma'}$. Taking into account that $\nu_1 + \theta(y_0,\mu_1,\nu_1) = 0$ and

$$\mu_0 = \mu_1 + \lambda(y_0,\mu_1) = \mu_1 + \lambda^{\text{new}}(y_0,\mu_1,\nu_1),$$

we have for the mapping $\Phi_{y_0\mu_0}$ defined by (54) and (89):

$$\left(\Phi_{y_0\mu_0}\right)^{-1}_* \widetilde{X}^{\mu_0} = \left(\Phi^{\text{new}}_{y_0\mu_1\nu_1}\right)^{-1}_* \widetilde{X}^{\mu_1+\lambda^{\text{new}}(y_0,\mu_1,\nu_1),0}_{\text{new}} =$$

$$\left[\omega^{\text{new}}(y_0,\mu_1,\nu_1) + O(|\bar{y} - y_0| + |\bar{z}|)\right]\frac{\partial}{\partial\bar{x}} + O_2(|\bar{y} - y_0| + |\bar{z}|)\frac{\partial}{\partial\bar{y}} +$$

$$\left[\Omega^{\text{new}}(y_0,\mu_1,\nu_1)\bar{z} + O_2(|\bar{y} - y_0| + |\bar{z}|)\right]\frac{\partial}{\partial\bar{z}}.$$

□

This parameter reduction technique was invented by Herman [19] (see also [65]) and Sevryuk [57, 56, 55]. Talk [19] and paper [57] consider the Hamiltonian isotropic $(n,0,0)$ context, paper [56] examines the Hamiltonian isotropic $(n,0,0)$ context and reversible $(n,n,0,0)$ context 1, and paper [55] is devoted to more general reversible $(n,m,p,0)$ context 1 with arbitrary n, m, p. A similar technique is used in [65] for the Hamiltonian coisotropic $(n,p,0)$ context for $n \geq 4$, $p = (n-2)/2$ and for the volume preserving $(n,1,0)$ context.

6 Whitney-smooth families of invariant tori: results

Theorems 5.4–5.8 (together with the "relaxed" analogues of Theorems 4.1–4.5) present a complete description of Whitney-smooth families of Floquet Diophantine invariant tori in typical families of vector fields in various contexts. Recall that all the vector fields and their dependence on external parameters are assumed to be analytic. Let $n \geq 2$.

Theorem 6.1 (dissipative context) *For any $p \geq 0$ and $s \geq 1$, a typical s-parameter family of vector fields on an $(n+p)$-dimensional manifold possesses a Whitney-smooth s-parameter family of Floquet Diophantine invariant n-tori (at most one torus per each parameter value). If the number of the pairs of complex conjugate eigenvalues of the Floquet matrices of these tori is r and $s \geq n+r-1$, then the Whitney-smooth s-parameter family of n-tori consists generically of $(s-n-r+1)$-parameter analytic subfamilies.*

Theorem 6.2 (volume preserving context with $p \geq 2$) *For any $p \geq 2$ and $s \geq 1$, a typical s-parameter family of globally divergence-free vector fields on an $(n + p)$-dimensional manifold possesses a Whitney-smooth s-parameter family of Floquet Diophantine invariant n-tori (at most one torus per each parameter value). If the number of the pairs of complex conjugate eigenvalues of the Floquet matrices of these tori is r and $s \geq n + r - 1$, then the Whitney-smooth s-parameter family of n-tori consists generically of $(s - n - r + 1)$-parameter analytic subfamilies.*

Theorem 6.3 (volume preserving context with $p = 1$) *For any $s \geq 0$, a typical s-parameter family of globally divergence-free vector fields on an $(n + 1)$-dimensional manifold possesses a Whitney-smooth $(s + 1)$-parameter family of Floquet Diophantine invariant n-tori. The Floquet 1×1 matrix of each of these tori is zero. If $s \geq n - 2$, then the Whitney-smooth $(s + 1)$-parameter family of n-tori consists generically of $(s - n + 2)$-parameter analytic subfamilies.*

Theorem 6.4 (Hamiltonian isotropic context) *For any $p \geq 0$ and $s \geq 0$, a typical s-parameter family of Hamiltonian vector fields with $n + p$ degrees of freedom possesses a Whitney-smooth $(n + s)$-parameter family of Floquet Diophantine isotropic invariant n-tori. The Floquet $(n + 2p) \times (n + 2p)$ matrix of each of these tori has eigenvalue 0 of multiplicity n, while the remaining $2p$ eigenvalues occur in pairs $(\lambda, -\lambda)$. If the number of distinct values of the positive imaginary parts of the eigenvalues of the Floquet matrices of these tori is r and $s \geq r - 1$, then the Whitney-smooth $(n + s)$-parameter family of n-tori consists generically of $(s - r + 1)$-parameter analytic subfamilies.*

Theorem 6.5 (reversible context 1) *For any $m \geq 0$, $p \geq 0$ and $s \geq \max(1 - m, 0)$, a typical s-parameter family of vector fields reversible with respect to an involution of type $(n + p, m + p)$ on an $(n + m + 2p)$-dimensional manifold possesses a Whitney-smooth $(m + s)$-parameter family of Floquet Diophantine invariant n-tori. The Floquet $(m + 2p) \times (m + 2p)$ matrix of each of these tori has eigenvalue 0 of multiplicity m (if $m > 0$), while the remaining $2p$ eigenvalues occur in pairs $(\lambda, -\lambda)$. If the number of distinct values of the positive imaginary parts of the eigenvalues of the Floquet matrices of these tori is r and $s \geq n - m + r - 1$, then the Whitney-smooth $(m + s)$-parameter family of n-tori consists generically of $(s + m - n - r + 1)$-parameter analytic subfamilies.*

Of course, all the invariant tori in Theorems 6.1–6.5 are analytic.

If $n = 1$ or $n = 0$ (when no small divisors are present), then no restrictions on s are needed, and Whitney-smooth families of invariant n-tori are in fact analytic. So, if $n \leq 1$, then for any $s \geq 0$ typical s-parameter families of vector fields possess analytic l-parameter families of invariant n-tori, where l is equal to s, s, $s + 1$, $n + s$. $m + s$ in the contexts of Theorems 6.1, 6.2, 6.3. 6.4, 6.5, respectively.

7 Acknowledgments

We are grateful to V.I. Arnol'd, M.R. Herman, A.I. Neĭshtadt, and J. Pöschel for long time discussions on the KAM theory and related mathematical topics. The third author (M.B.S.) thanks also all his colleagues of the University of Groningen for their hospitality.

References

[1] R.H. Abraham and J.E. Marsden. *Foundations of Mechanics (2nd edition)*. Benjamin/Cummings, 1978.

[2] J.F. Adams. *Lectures on Lie Groups*. Benjamin, New York, 1969.

[3] V.I. Arnol'd. *Geometrical Methods in the Theory of Ordinary Differential Equations*. Springer-Verlag, 1983.

[4] V.I. Arnol'd and A.B. Givental'. Symplectic geometry. In V.I. Arnol'd and S.P. Novikov, editors, *Encyclopædia of Mathematical Sciences, Vol. 4, Dynamical Systems IV*, pages 1–136. Springer-Verlag, 1990.

[5] V.I. Arnol'd and M.B. Sevryuk. Oscillations and bifurcations in reversible systems. In R.Z. Sagdeev, editor, *Nonlinear Phenomena in Plasma Physics and Hydrodynamics*, pages 31–64. Mir, Moscow, 1986.

[6] V.I. Bakhtin. Averaging in multifrequency systems. *Funct. Anal. Appl.*, 20(2): 83–88, 1986.

[7] V.I. Bakhtin. Diophantine approximations on images of mappings. *Dokl. Akad. Nauk Beloruss. SSR*, 35(5): 398–400, 1991 [in Russian].

[8] Yu.N. Bibikov. A sharpening of a theorem of Moser. *Sov. Math. Dokl.*, 14(6): 1769–1773, 1973.

[9] Yu.N. Bibikov. *Multifrequency Nonlinear Oscillations and their Bifurcations*. Leningrad Univ. Press, 1991 [in Russian].

[10] J.-B. Bost. Tores invariants des systèmes dynamiques hamiltoniens (d'après Kolmogorov, Arnol'd, Moser, Rüssmann, Zehnder, Herman, Pöschel, ...). In *Séminaire Bourbaki, Vol. 639, 1984–1985*, pages 113–157. Astérisque, 133–134, 1986.

[11] H.W. Broer and G.B. Huitema. Unfoldings of quasi-periodic tori in reversible systems. *J. Dyn. Differ. Eq.*, 7(1): 191–212, 1995.

[12] H.W. Broer, G.B. Huitema, and F. Takens. Unfoldings of quasi-periodic tori. *Mem. Amer. Math. Soc.*, 83(421): 1–81, 1990.

[13] I.U. Bronstein and A.Ya. Kopanskiĭ. *Smooth Invariant Manifolds and Normal Forms*. World Scientific, 1994.

[14] A.D. Bruno. *Local Methods in Nonlinear Differential Equations*. Springer-Verlag, 1989.

[15] L. Chierchia and G. Gallavotti. Smooth prime integrals for quasi-integrable Hamiltonian systems. *Nuovo Cimento B*, 67(2): 277–295, 1982.

[16] R.L. Devaney. Reversible diffeomorphisms and flows. *Trans. Amer. Math. Soc.*, 218: 89–113, 1976.

[17] L.H. Eliasson. Perturbations of stable invariant tori for Hamiltonian systems. *Ann. Sc. Norm. Super. Pisa, Cl. Sci., IV Ser.*, 15(1): 115–147, 1988.

[18] S.M. Graff. On the conservation of hyperbolic invariant tori for Hamiltonian systems. *J. Differ. Eq.*, 15(1): 1–69, 1974.

[19] M.R. Herman. *Talk held on the International Conference on Dynamical Systems* (Lyons), 1990.

[20] M.R. Herman. Différentiabilité optimale et contre-exemples à la fermeture en topologie C^∞ des orbites récurrentes de flots hamiltoniens. *C. R. Acad. Sci. Paris, Série I*, 313(1): 49–51, 1991.

[21] M.R. Herman. Exemples de flots hamiltoniens dont aucune perturbation en topologie C^∞ n'a d'orbites périodiques sur un ouvert de surfaces d'énergies. *C. R. Acad. Sci. Paris, Série I*, 312(13): 989–994, 1991.

[22] M.W. Hirsch. *Differential Topology*. Springer-Verlag, 1976.

[23] M.W. Hirsch, C.C. Pugh, and M. Shub. *Invariant Manifolds*. Lect. Notes Math., Vol. 583. Springer-Verlag, 1977.

[24] I. Hoveijn. Versal deformations and normal forms for reversible and Hamiltonian linear systems. *J. Differ. Eq. (to appear)*.

[25] G.B. Huitema. *Unfoldings of quasi-periodic tori*. PhD thesis, University of Groningen, 1988.

[26] J.S.W. Lamb. *Reversing symmetries in dynamical systems*. PhD thesis, University of Amsterdam, 1994.

[27] V.F. Lazutkin. The existence of a continuum of closed invariant curves for a convex billiard. *Uspekhi Mat. Nauk*, 27(3): 201–202, 1972 [in Russian].

[28] V.F. Lazutkin. The existence of caustics for a billiard problem in a convex domain. *Math. USSR Izv.*, 7(1): 185–214, 1973.

[29] V.F. Lazutkin. Concerning Moser's theorem on invariant curves. In *Problems in the Dynamical Theory of Seismic Waves Propagation, Vol. 14*, pages 109–120. Nauka, Leningrad, 1974 [in Russian].

[30] V.F. Lazutkin. *Convex Billiard and Eigenfunctions of the Laplace Operator*. Leningrad Univ. Press, 1981 [in Russian].

[31] V.F. Lazutkin. *KAM Theory and Semiclassical Approximations to Eigenfunctions*. Springer-Verlag, 1993.

[32] V.K. Mel'nikov. On some cases of conservation of conditionally periodic motions under a small change of the Hamilton function. *Sov. Math. Dokl.*, 6(6): 1592–1596, 1965.

[33] V.K. Mel'nikov. A family of conditionally periodic solutions of a Hamiltonian system. *Sov. Math. Dokl.*, 9(4): 882–886, 1968.

[34] D. Montgomery and L. Zippin. *Topological Transformation Groups*. Interscience, New York, 1955.

[35] J. Moser. Convergent series expansions for quasi-periodic motions. *Math. Ann.*, 169(1): 136–176, 1967.

[36] J. Moser. *Stable and Random Motions in Dynamical Systems, with Special Emphasis on Celestial Mechanics.* Ann. Math. Studies, Vol. 77. Princeton Univ. Press, 1973.

[37] J. Palis and W.C. de Melo. *Geometric Theory of Dynamical Systems.* Springer-Verlag, 1982.

[38] I.O. Parasyuk. Conservation of multidimensional invariant tori in Hamiltonian systems. *Ukrain. Math. J.*, 36(4): 380–385, 1984.

[39] I.O. Parasyuk. Coisotropic invariant tori of Hamiltonian systems in the quasiclassical theory of motion of a conduction electron. *Ukrain. Math. J.*, 42(3): 308–312, 1990.

[40] J. Pöschel. Integrability of Hamiltonian systems on Cantor sets. *Comm. Pure Appl. Math.*, 35(5): 653–696, 1982.

[41] J. Pöschel. On elliptic lower dimensional tori in Hamiltonian systems. *Math. Z.*, 202(4): 559–608, 1989.

[42] A.S. Pyartli. Diophantine approximations on submanifolds of Euclidean space. *Funct. Anal. Appl.*, 3(4): 303–306, 1969.

[43] G.R.W. Quispel and M.B. Sevryuk. KAM theorems for the product of two involutions of different types. *Chaos*, 3(4): 757–769, 1993.

[44] J.A.G. Roberts and G.R.W. Quispel. Chaos and time-reversal symmetry. Order and chaos in reversible dynamical systems. *Phys. Rep.*, 216(2–3): 63–177, 1992.

[45] H. Rüssmann. Non-degeneracy in the perturbation theory of integrable dynamical systems. In M.M. Dodson and J.A.G. Vickers, editors, *Number Theory and Dynamical Systems*, London Math. Soc. Lect. Note Ser., Vol. 134, pages 5–18. Cambridge Univ. Press, 1989.

[46] H. Rüssmann. Nondegeneracy in the perturbation theory of integrable dynamical systems. In S. Albeverio, Ph. Blanchard, and D. Testard, editors, *Stochastics, Algebra and Analysis in Classical and Quantum Dynamics*, Math. and its Appl., Vol. 59, pages 211–223. Kluwer Academic, Dordrecht, 1990.

[47] H. Rüssmann. On twist-Hamiltonians. *Talk held on the Colloque international: Mécanique céleste et systèmes hamiltoniens* (Marseille), 1990.

[48] J. Scheurle. Bifurcation of quasi-periodic solutions from equilibrium points of reversible dynamical systems. *Arch. Rat. Mech. Anal.*, 97(2): 103–139, 1987.

[49] M.B. Sevryuk. *Reversible Systems.* Lect. Notes Math., Vol. 1211. Springer-Verlag, 1986.

[50] M.B. Sevryuk. Invariant m-tori of reversible systems with the phase space of dimension greater than $2m$. *J. Soviet Math.*, 51(3): 2374–2386, 1990. [Russian original: *Trudy Sem. I.G. Petrovskogo*, 14: 109–124, 1989].

[51] M.B. Sevryuk. Lower-dimensional tori in reversible systems. *Chaos*, 1(2): 160–167, 1991.

[52] M.B. Sevryuk. Linear reversible systems and their versal deformations. *J. Soviet Math.*, 60(5): 1663–1680, 1992. [Russian original: *Trudy Sem. I.G. Petrovskogo*, 15: 33–54, 1991].

[53] M.B. Sevryuk. Invariant tori of reversible systems of intermediate dimensions. *Russ. Acad. Sci. Dokl. Math.*, 47(1): 129–133, 1993.

[54] M.B. Sevryuk. New results in the reversible KAM theory. In S.B. Kuksin, V.F. Lazutkin, and J. Pöschel, editors, *Seminar on Dynamical Systems*, pages 184–199. Birkhäuser, Basel, 1994.

[55] M.B. Sevryuk. The iteration-approximation decoupling in the reversible KAM theory. *Chaos*, 5(3): 552–565, 1995.

[56] M.B. Sevryuk. KAM-stable Hamiltonians. *J. Dyn. Control Syst. (to appear)*, 1995.

[57] M.B. Sevryuk. Invariant tori of Hamiltonian systems nondegenerate in the sense of Rüssmann. *Dokl. Akad. Nauk (to appear)*, 1995 [in Russian, to be translated into English in *Russ. Acad. Sci. Dokl. Math.*].

[58] M.B. Sevryuk. Some problems of the KAM theory: quasi-periodic motions in typical systems. *Uspekhi Mat. Nauk*, 50(2): 111–124, 1995 [in Russian, to be translated into English in *Russian Math. Surveys*].

[59] Ch.-W. Shih. Normal forms and versal deformations of linear involutive dynamical systems. *Chinese J. Math.*, 21(4): 333–347, 1993.

[60] V.G. Sprindžuk. *Metric Theory of Diophantine Approximations*. John Wiley, New York, 1979.

[61] N.V. Svanidze. Small perturbations of an integrable dynamical system with an integral invariant. *Proc. Steklov Inst. Math.*, 2: 127–151, 1981.

[62] D.V. Treshchëv. A mechanism for the destruction of resonant tori in Hamiltonian systems. *Math. USSR Sbornik*, 68(1): 181–203, 1991.

[63] H. Whitney. Analytic extensions of differentiable functions defined in closed sets. *Trans. Amer. Math. Soc.*, 36(1): 63–89, 1934.

[64] S. Wiggins. *Normally Hyperbolic Invariant Manifolds in Dynamical Systems*. Springer-Verlag, 1994.

[65] J.-C. Yoccoz. Travaux de Herman sur les tores invariants. In *Séminaire Bourbaki, Vol. 754, 1991-1992*, pages 311–344. Astérisque, 206, 1992.

[66] E. Zehnder. Generalized implicit function theorems with applications to some small divisor problems, I and II. *Comm. Pure Appl. Math.*, 28(1): 91–140, 1975 and 29(1): 49–111, 1976.

Progress in Nonlinear Differential Equations
and Their Applications, Vol. 19
© 1996 Birkhäuser Verlag Basel/Switzerland

Towards a Global Theory of Singularly Perturbed Dynamical Systems

John Guckenheimer*

Dynamical systems with multiple time scales arise naturally in many domains. Models of neural systems provide the principal motivation for this paper. Most of the previous mathematical analysis of qualitative properties of multiple time scale systems has dealt with local phenomena that occur in low dimensions. The neural system models raise questions that lie beyond the scope of existing work. Our aim here is to outline a theory that extends the local theory to a description of qualitative features of the global dynamics for systems with two time scales. We fill in portions of this outline within the context of systems that have two slow variables and two fast variables. Even within this low dimensional setting, most basic questions about global properties remain unanswered.

The setting within which we work is a system of differential equations in R^{m+n} of the form

$$\begin{array}{ccc} \dot{x} & = & f(x,y) \\ \dot{y} & = & \epsilon g(x,y) \end{array} \quad \text{or} \quad \begin{array}{ccc} \epsilon x' & = & f(x,y) \\ y' & = & g(x,y) \end{array}$$

Such systems are called *singularly perturbed* differential equations or *fast-slow* vector fields. For brevity, we say that these equations define an (m,n) dimensional SP system. Here $\epsilon \geq 0$ is a small parameter, the x variables are called *fast variables*, the y variables are called *slow variables* and the two systems have time variables t and T with $dT/dt = \epsilon$. The system for $\epsilon = 0$ will be called the *singular* system or SP_0 field of the corresponding SP system. We have written the equations of an SP system in two different forms to emphasize that it will be studied from two points of view. When we examine the fast time scale t, the system appears as a perturbation of a family of vector fields parametrized by y. The perturbation parameter induces a slow variation of the parameters, producing a *slowly varying system*. Bifurcations in the family of vector fields induce transitions of trajectories from the neighborhood of one family of attractors to another on the fast time scale. The asymptotic properties of the simplest types of such transitions have been studied extensively. The second system has as its limit a *differential algebraic equation* in which the differential equations for the evolution of x become algebraic equations. This limit makes good sense as a singularly perturbed system for describing trajectories that track regular portions of the 0 level set of g. In these regions, one

*Department of Mathematics, Cornell University, White Hall, Ithaca, NY 14853-7901, USA

can solve the equation $g = 0$ for x using the implicit function theorem and obtain a system of equations that describes the evolution of y. At places on the 0 level set of g where g_y is singular, the differential algebraic system may not have solutions at all. To make sense of the solutions, it is necessary to follow the evolution of the fast vector field from these points. The fast evolution of y can be viewed as "jumps" from one slow manifold to another. Piecing together trajectories from the continuous evolution along the 0 level set of g with these jumps produces a *hybrid system* of the sort studied by Back, Guckenheimer and Myers [2].

Generic properties of singularly perturbed systems have been studied previously, but much of this work has been devoted to local properties. the study of particular classes of solutions and phenomena that involve only one or two transitions between attractors of the fast subsystems. Arnold et al. [1], Grasman [7] and Mishchenko [12] give surveys of much of this theory. We note that the analysis of the forced van der Pol equation by Cartwright and Littlewood [3] can be viewed as a seminal example of a long time analysis of a system with three time scales. Takens [15, 16, 17, 18] was perhaps the first to consider the development of a systematic theory, mainly in the context of systems for which the fast subsystems had gradient structures. Rinzel [14] has used systems with two time scales as models for bursting oscillations of electrical activity in biological systems. His work does not attempt to build a comprehensive theory for singularly perturbed systems. Here, we take initial steps to construct a systematic global theory of singularly perturbed systems, including those that display oscillations on fast time scales. We sketch topological aspects of the theory and touch briefly on the asymptotic analysis of various phenomena. To make our task more manageable. we focus our attention to phenomena that occur in SP fields with $m = 2$ and $n = 2$. This allows us to draw upon the relatively complete theory of codimension two bifurcations of two dimensional vector fields and to avoid considerations of chaotic dynamics occurring within fast subsystems.

1 Preliminaries

Our goal is to characterize the qualitative properties that occur in "generic" SP systems. Our interpretation of genericity will be partly dependent upon context. Let X_ϵ be a system in R^{m+n} of the form

$$\dot{x} = f(x, y) \qquad \dot{y} = \epsilon g(x, y) \tag{1}$$

with two time scales. We regard X_ϵ as a smooth (C^∞) map $X_\epsilon : R^{m+n} \to R^m \times R^n$. The range of X is written as a product to reflect the roles of the two factors in the product. Since the singular limit X_0 is a family of vector fields in R^m parametrized by R^n, the most stringent definition of equivalence of two systems will be that there is a smooth coordinate change of R^{m+n} that commutes with the projection $R^{m+n} \to R^n$ and maps one vector field to another. These coordinate changes have the form $(h(x, y), k(y))$. The local classification of normal forms given by Arnold

et al. [1] is based upon this definition of equivalence. The limit $\epsilon = 0$ is indeed
singular. The solutions to the equation defining an SP_0 field do not approximate
the solutions of an SP system for $\epsilon > 0$ since the slow variables of an SP_0 field
remain constant. Therefore, we make the following definition.

Definition: Let X_ϵ be the SP vector field (1). A curve $\gamma_0 \subset R^m \times R^n$ is a trajectory
of the SP_0 field X_0 if it is the limit of a one parameter family of trajectories γ_ϵ for
$\epsilon > 0$. The limit is assumed to be the limit of the curves as subsets of $R^m \times R^n$
with respect to the Hausdorff metric on subsets.

The dynamics of a vector field can be decomposed into a phase portrait
consisting of chain recurrent invariant sets and sets of trajectories that connect
pairs of chain recurrent sets. Trajectories of the SP_0 field can be decomposed into
segments during which the trajectories remain in an invariant set of the fast subsystem and segments in which the trajectories makes transitions between invariant
sets. We call these segments (slow) S segments and (fast) F segments respectively.
The time scales associated with S segments and F segments of the SP_0 field trajectories have an infinite separation. Trajectories that lie on equilibrium manifolds
of the fast subsystem can be regarded as evolving on the slow time scale. If M is
an equilibrium manifold on which $D_y g$ is nonsingular, then the slow vector field
that describes this evolution is called the *reduced* field. The reduced field at $x \in M$
is the unique vector v tangent to M at x with the property that v projects onto
$f(x)$ by the projection of $R^m \times R^n$ onto its first factor R^m. We distinguish FS
transitions where an S segment precedes a F segment from SF transitions in which
an S segment follows a F segment. The local theory is primarily an asymptotic
analysis of the generic SF transitions that one expects to find from equilibrium
manifolds of the SP_0 field to F segments.

The *long time scale* will refer to times that are long in the slow time scale
t. On the long time scale, we expect that there will be trajectories that pass
through many transitions. A global theory of SP systems will segment trajectories
so that normal forms can be fit to each segment uniformly as $\epsilon \to 0$. Through the
analysis of the normal forms, this leads to a matched asymptotic expansion for
the trajectories. For the long time scale, we want to perform an *epoch analysis* in
which the slow time evolution of trajectories is collapsed to a sequence of discrete
maps that send points at the onset of an FS transition to the onset of the next FS
transition. We call these maps induced maps. They need not be invertible like the
cross-section maps of a flow, and they may be multivalued.

2 Transitions based on bifurcations of two parameter families of vector fields

Bifurcation theory of dynamical systems provides a guide to the types of FS transitions that we expect to find in SP_0 fields. Classification of FS transitions in (m, n)
dimensional SP systems can therefore be based upon the classification of bifurcations of n parameter families of m dimensional vector fields. In this section, we

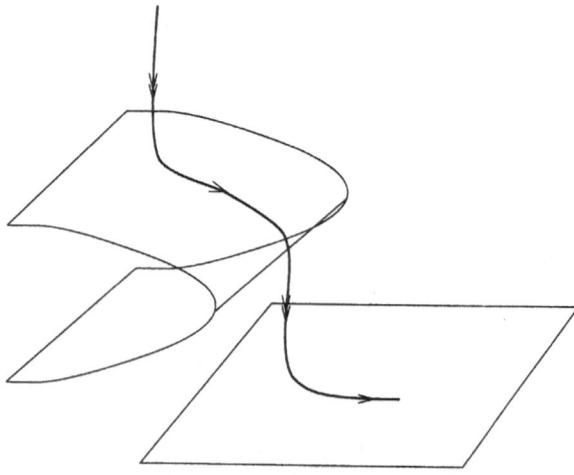

Figure 1: A trajectory of a (1,2) dimensional SP system approaches a manifold of stable equilibria, leaves this manifold in an SF transition at a fold of the manifold and then approaches a second equilibrium manifold in an FS transition. The trajectory is plotted in a heavier line style than the equilibrium manifolds.

review the classification of codimension one and two bifurcations of two parameter families of two dimensional vector fields and discuss their relationship with FS transitions of $(2,2)$ dimensional SP systems.

Bifurcations are classified by codimension [8], and the codimension reflects the dimension of the set of trajectories undergoing an FS transition. Codimension one bifurcations of attractors give rise to FS transitions that are found in open subsets of trajectories of an SP_0 field, while higher codimension bifurcations give rise to FS transitions that are found only on nowhere dense sets of trajectories of SP_0 fields. In the setting of $(2,2)$ dimensional SP systems, this means that codimension two bifurcations occur only at isolated trajectories of an SP_0 field. Moreover, the invariant sets of the fast subsystem are all equilibrium points, periodic orbits or polycycles composed of closed curves containing equilibrium points.

The codimension one bifurcations of two dimensional vector fields are

1. Saddle-nodes equilibria or folds: These bifurcations give rise to trajectories that make jump transitions from an equilibrium manifold of an SP_0 field. The asymptotic properties of SP trajectories near folds have been described in substantial detail [1]. The vector field

$$\dot{x} = \epsilon$$
$$\dot{y} = -y^2 - x$$

is a normal form for a$(1,1)$ dimensional SP system with a fold. The matched asymptotic expansions for trajectories of this system are based upon a partition of the plane into three regions: an ϵ-dependent neighborhood U_ϵ of the origin, a neighborhood of the regular portion of the curve defined by $y^2 + x = 0$, and regions of the plane where $y >> \epsilon$. Scaling the system with the coordinate transformation $y = \epsilon^{1/3}Y$, $x = \epsilon^{2/3}X$ and $t = \epsilon^{-1/3}T$ produces the system

$$\begin{aligned} X' &= 1 \\ Y' &= -Y^2 - X \end{aligned}$$

which is independent of ϵ. Therefore, trajectories of this system can be used as a scaled model for the behavior of the vector field in U_ϵ. Explicit solutions of the equation $Y' = -Y^2 - T$ can be expressed in terms of Airy functions [7].

2. Hopf bifurcations: Hopf bifurcations are characterized by an equilibrium manifold of an SP_0 field along which complex eigenvalues of the fast field at the equilibrium cross the imaginary axis transversally. In an SP system, trajectories have difficulty tracking an emerging branch of periodic orbits at Hopf bifurcations. Consider the case of a supercritical Hopf bifurcations in which a family of stable limit cycles emerges from stable equilibria. The trajectory of the SP system comes so close to the equilibria prior to the the Hopf bifurcation that it remains close to the branch of post-critical unstable equilibria for a substantial period of time. The asymptotics of this "delayed loss of stability" have been analyzed by Nejshtadt [13]. The smoothness or analyticity of the vector field significantly affects the asymptotics of delayed loss of stability in generic SP systems. With analytic systems, Nejshtadt [13] calculates of the length of delay following bifurcation in terms of a complex-time extension of the system.

3. Saddle-connections: Homoclinic and heteroclinic bifurcations are global and have not been studied extensively within the context of SP systems. Homoclinic bifurcations appear in generic families as one mechanism for the termination of a family of stable limit cycles. Rinzel [14] observed the corresponding phenomenon in SP systems at the termination of spiking in models for bursting oscillations of neural oscillators. The asymptotics of this transition have not been analyzed beyond the dynamics of the SP_0 field itself close to its homoclinic bifurcation. Here are a few preliminary comments about the analysis of the periods of oscillation in a trajectory passing through a homoclinic bifurcation. In a generic reduced system with a homoclinic orbit, there is a manifold E of equilibria for the fast system that are hyperbolic saddles. These saddles have stable and unstable manifolds that intersect transversally at a set of slow variables that form a codimension one submanifold H of E. Fenichel [5] proves that the hyperbolic structure of the manifold E persists

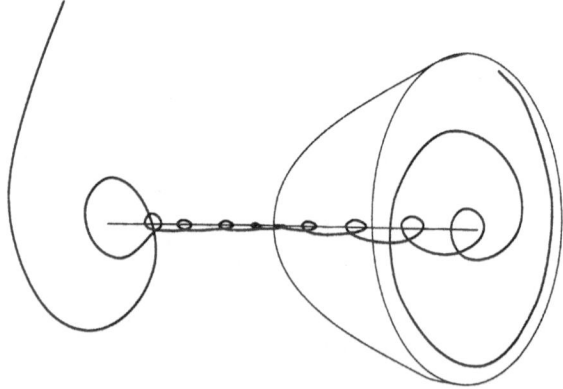

Figure 2: A trajectory of a (2,1) dimensional SP system passes through a Hopf bifurcation and then exhibits a delay in tracking the manifold of periodic orbits emerging from the Hopf bifurcation.

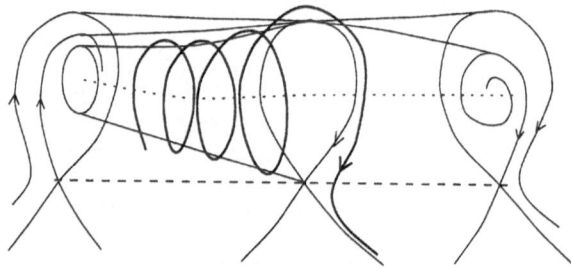

Figure 3: A trajectory of a (2,1) dimensional SP system passes a homoclinic orbit. A curves of sources is depicted by a dotted line and a curve of saddles is depicted by a a dashed line. The surface of periodic orbits and the stable and unstable manifolds of the saddles are shown. The homoclinic orbit is contained in the closure of all three of these surfaces.

for $\epsilon > 0$. The manifold E is the limit of a family of normally hyperbolic manifolds E_ϵ as $\epsilon > 0$. Its stable and unstable manifolds continue to intersect transversally. In a neighborhood of E, Fenichel defines coordinates for which the families of strong stable and unstable manifolds are invariant.

Periodic orbits of a reduced vector field close to a homoclinic orbit can be decomposed into a pair of segments, one lying in a neighborhood U of the manifold E and the second lying in the complement of U. The portion of trajectories lying outside U are regular orbits that vary smoothly with x and have bounded duration as $\epsilon \to 0$. The dominant portion of the increase in period of the reduced system comes from the passage of trajectories through U. In the reduced system, the length of time that a trajectory spends in U is determined by the smallest magnitude of its coordinates along the unstable manifold of the saddle point. Assume that there are coordinates for the reduced system in which the fast equations are defined by $\dot{x}_i = \lambda_i(y)x_i$. Assume that $\lambda_i(y) > 0$ for $i = 1, \ldots, u$ and that $\lambda_i(x) < 0$ for $i = u+1, \ldots, n$ (For generic systems, we can expect that coordinate systems of this form can at best be determined with finite degrees of smoothness [10].) In each unstable direction, the x_i increase exponentially in magnitude, implying that the time a trajectory spends in U is comparable to $\max(-\ln(x_i)/\lambda_i, \, i = 1, \ldots, u$. If the slow variable evolves at a constant rate, then each cycle of the oscillation of the trajectory will carry it closer to the stable manifold of E. The cycles of the oscillation get longer, resulting in a larger change in the slow variable from one cycle to the next. Thus the growth of the period of the oscillations is faster than they would be if there were a constant change in parameter values between cycles. We conjecture that the period between cycles should be asymptotic to a function of the form $a_1 \ln(1/\epsilon) + a_2 \ln(\ln(1/\epsilon)) + a_3$.

Heteroclinic bifurcations of fast systems can be expected to lead to discontinuities in the fast transition maps that occur along fold singularities, though we are unaware of any examples of this phenomenon in the literature.

4. Saddle-nodes of periodic orbits: Manifolds of periodic orbits may have folds in generic one parameter families of vector fields. These folds are singularities for the projection map of the manifold onto the parameter space of the family. The return map for the family of periodic orbits has a discrete time saddle-node bifurcation. The asymptotic analysis of the return map is analogous to that for saddle-nodes of equilibrium points. The behavior of the "phase" variable that is ignored by the return map has not been studied. Note that the isochrons (strong stable manifolds) of a limit cycle for a planar system make an angle with the cycles that is proportional to the characteristic exponent of the cycle. Semistable limit cycles typically do not have isochrons.

The consequences of codimension two bifurcations in the fast subsystems of a $(2, 2)$ dimensional SP field have hardly been studied. The only case that gives rise to a local bifurcation is the cusp [1]. The list of codimension two bifurcations is manageable for a case by case asymptotic analysis, so we present it here as a challenge to readers.

1. Cusps: Projection of the equilibrium manifold M of a generic SP_0 field onto a space of slow variables will have singularities that are folds (codimension 1) and cusps (codimension 2). A normal form for a mapping with a cusp is

$f(x_1, x_2) = (x_1^3 + x_2 x_1, x_2)$. An SP_0 vector field with a cusp is

$$
\begin{aligned}
\dot{x} &= -x^3 + y_1 x + y_2 \\
\dot{y_1} &= \epsilon c_1 \\
\dot{y_2} &= \epsilon c_2
\end{aligned}
$$

As with the fold bifurcation, there is a region near the origin in which the time scales for the dynamics of x differ from those when the slow variables y_i are far from the origin. A scaling argument similar to the one used for folds indicates the size of this region. We solve the equations for $\dot{y_i}$ and substitute these solutions into the equation for $\dot{x}$. We then look for a region of the phase space in which all of the terms have comparable magnitude. Such a region exists for $x \sim O(\epsilon^{1/5})$, $y_1 \sim O(\epsilon^{2/5})$, $y_2 \sim O(\epsilon^{3/5})$, and $t \sim O(\epsilon^{-2/5})$. Trajectories that track the equilibrium manifold in this region will relax towards the stable equilibrium surface on the slower time scale $t \sim O(\epsilon^{-2/5})$ than the unit time scale for the typical evolution of x.

2. Takens-Bogdanov bifurcations: Consider the equilibrium manifold M of a generic SP_0 field with more than $n > 1$ fast variables. The linearization of the fast subsystem along M defines a smooth map of M into the space of $m \times m$ matrices. There are no constraints on this map from the equilibrium conditions, so there will be a codimension two submanifold of M at which the linearization of the fast subsystem has zero as an eigenvalue of algebraic multiplicity two. The generic points of this submanifold are called Takens-Bogdanov points. They lie on the codimension one fold surface of M at places where the number of stable and unstable eigenvalues of the linearization of the fast subsystem changes along the fold surface. The unfolding of the Takens-Bogdanov bifurcation as a codimension two bifurcation of planar vector fields is well known. There are small amplitude periodic orbits and homoclinic bifurcations associated with the Takens-Bogdanov bifurcation.

3. Degenerate Hopf bifurcations: The normal form for Hopf bifurcations contains a "nonlinear" term that determines whether the equilibrium point is weakly stable or unstable in the marginal directions. Degenerate Hopf bifurcation occurs when this quantity vanishes. A point of degenerate Hopf bifurcation in a generic two parameter family of vector fields signals the change from subcritical to supercritical Hopf bifurcation for families that cross the bifurcation curve transversally.

4. Resonant homoclinic bifurcations: In generic one parameter families of vector fields, the stability of the periodic orbits near a homoclinic orbit are determined by the properties of the saddle point that is the limit of the homoclinic orbit. In two dimensional vector fields, the trace of the linearization of the saddle point determines whether the periodic orbits are attracting or repelling. Resonant homoclinic bifurcation occurs when the trace vanishes.

This occurs in generic multiparameter families of vector fields. Associated with the resonant homoclinic bifurcations in generic two parameter families are paths of saddle-node bifurcations of limit cycles that end at the codimension two point, meeting the curve of homoclinic bifurcations with a tangency that is flat.

5. Saddle-node loop: Consider the phase portrait of a two dimensional vector field near a point with a saddle-node equilibrium at which the linearization of the field has a negative eigenvalue along with its zero eigenvalue. The stable set of the equilibrium is diffeomorphic to a half plane, and the unstable set of the equilibrium is a single trajectory. If the unstable set is contained in the interior of the stable set, then the saddle-node bifurcation has been called a homoclinic saddle-node, "saddle-node in cycle" or SNIC. This is a codimension one bifurcation that persists with perturbation of the saddle-node. If the unstable set is contained in the boundary of the stable set, this is a codimension two bifurcation that has also been called a homoclinic saddle-node or saddle-node loop. Passing through the codimension two point in the parameter space of a generic two parameter family is a curve of saddle-node bifurcation, and there is a curve of homoclinic bifurcations that terminates at the saddle-node point. Algorithms for computing these codimension two bifurcations have been discussed by Schechter.

6. Triple cycles: Just as the equilibrium manifolds of SP_0 fields have projections with cusps as codimension two singularities, families of limit cycles for SP_0 fields have projections with codimension two singularities corresponding to cusps. The cusps appear by restricting the projection to the family of fixed points in a cross-section to the SP field.

7. Double saddle connections: There are several cases of transversal intersections of curves of independent codimension one bifurcations that occur in two parameter families of vector fields. The case of two intersecting curves of homoclinic bifurcations for the same saddle point leads to a more complicated situation with additional subsidiary bifurcation curves appearing. In two dimensional vector fields, this bifurcation has subsidiary homoclinic bifurcations with both "convex" and "non-convex" homoclinic orbits. In the parameter space, there are two curves of homoclinic bifurcation passing through the codimension two point for homoclinic orbits that enclose a convex angle at the saddle point. There are also two branches of homoclinic bifurcation for non-convex homoclinic orbits that terminate at the codimension two point.

3   The Long Time Scale: Induced Maps

We turn to the analysis of the long time scale dynamics of an SP system. In particular, we would like to describe the long time invariant sets of an SP system. This task is complex because the $\epsilon = 0$ limit of an SP system is indeed singular. For example, in a $(2, 2)$ dimensional system undergoing supercritical Hopf bifurcation, the transition from the equilibrium manifold of the SP_0 field to its manifold of limit cycles will have trajectories that connect every unstable point along the equilibrium manifold with every point on the limit cycle in the same fast subsystem. (This observation is a corollary of the theory of Nejshtadt [13] that analyzes the delay of SP trajectories in tracking the family of limit cycles following the Hopf bifurcation.) The pathology represented by this example is not pursued further here. Instead, we restrict our attention to the simpler setting for (2.2) dimensional systems with only equilibrium attractors.

Manifolds of stable equilibria for $(2, 2)$ dimensional systems are two dimensional and have one dimensional boundaries at which trajectories undergo FS transitions. If γ is a curve that is in the boundary of a manifold of stable equilibria, then we follow the evolution of γ through the FS transition until γ meets the next manifold M of stable equilibria and begins flowing once more. Singular phenomena may be encountered along the F segments emanating from γ or in the relationship of γ to the slow flow along M. The induced map of the system will be a one dimensional map from γ to the boundary of M.

The Poincaré-Bendixson Theorem implies that the limit set of a bounded trajectory of a planar vector field is a periodic orbit or contains an equilibrium point. This implies that the trajectories forming the boundary of the basin of attraction of a sink for a structurally stable vector field are either in the stable manifold of saddles, tend to infinity or emerge from sources that are equilibrium points or unstable limit cycles. This observation applies to systems with saddle-nodes if the saddle-node points and their strong stable manifolds are included in the list of trajectories that form boundaries of basins of attraction. If we follow the unstable separatrices of saddle-node points in a generic (2.2) dimensional SP system, discontinuities for the induced map will be created at separatrices that lie in the stable manifold of a saddle. Transversal crossing of two saddle-node bifurcation curves also leads to discontinuity of induced maps. When equilibrium points of a system disappear in a saddle-node of equilibria, there is a discontinuous change in a basin of attraction. This change creates the possibility for new global bifurcations involving saddle connections of the fast subsystems to appear. The phenomenon may be related to a codimension two bifurcation that can be detected by bifurcation analysis. For example, if the unstable separatrix of a saddle-node p connects to a saddle point q, there may be another saddle separatrix approaching p that connects to q when it "breaks through" the saddle-node region.

Although we don't discuss systems with limit cycle attractors in detail here, we remark that new saddle connections may be created by the destruction of the cycles without the involvement of a codimension two bifurcation. If γ is a periodic orbit undergoing a saddle-node of limit cycle bifurcation. and if there are saddle separatrices forward and backwards asymptotic to γ. then there can be infinite number of heteroclinic bifurcations that accumulate at the saddle-node bifurcation point.

Most of the attention that has been directed at analysis of singularly perturbed systems examines the geometry of how trajectories leave equilibrium manifolds of an SP_0 field. There has been little analysis of how these sets of trajectories approach a new equilibrium manifold. We regard a FS transition as having as limit a map $h : \partial M \to N$, with ∂M the boundary of a manifold of attractors M, and N another manifold of attractors for the SP_0 field. The geometry of $h(\partial M)$ relative to the slow flow on N is important in determining the singularities of the induced map from ∂M to ∂N. Since the slow flows are independent of the fast subsystems, there are no implicit constraints on the geometry of ∂M relative to the flow on N. The slow time scale dynamics of trajectories that make an infinite number of transitions can be analyzed in terms of the induced maps.

We describe three phenomena that occur in generic $(2, 2)$ dimensional SP systems following an FS transition between manifolds of equilibria. In this setting, ∂M is a curve, and the points of this curve arrive on the two dimensional manifold N in an arbitrary position relative to the phase portrait of the slow flow on N. There may be points where

1. $h(\partial M)$ is tangent to the slow flow on N. This leads to turning points of induced maps.

2. $h(\partial M)$ may intersect a limit cycle of the reduced flow on N transversally. This leads to a countable sequence of discontinuities of the induced map if the limit cycle is unstable and points near it do not remain on N forever under its slow flow.

3. $h(\partial M)$ may intersect a saddle separatrix for the slow flow. This produces discontinuities and singularities in induced maps. The singularities are described by power laws determined by the eigenvalues of the saddle.

Within the context of "hybrid systems", each of these phenomena has been discussed by Guckenheimer and Johnson [9].

Combining all of the induced maps for equilibrium manifolds of a generic $(2, 2)$ dimensional SP system yields a one dimensional mapping φ. The domain of φ is a one dimensional manifold that may well have several components. Furthermore φ can have discontinuities, singularities and critical points associated to the phenomena described above. The extensive theory of one dimensional iterations [4] can be used to characterize the types of attractors that may occur for the induced map. The examples of attractors for two dimensional hybrid systems described by Guckenheimer and Johnson [9] are readily translated into examples of SP systems whose long time dynamics display varied types of attractors.

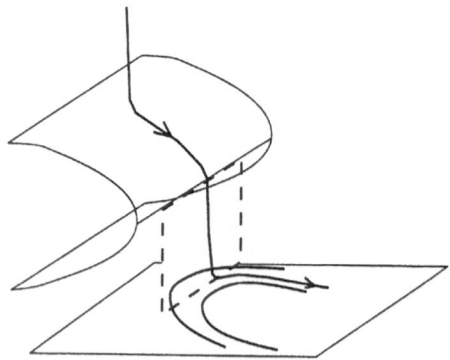

Figure 4: The trajectories of a fold curve on one manifold of equilibria flow to another manifold of equilibria. The image of the fold curve after the FS transition is tangent to the reduced vector field on the lower manifold of equilibria at an isolated point.

4 Concluding Remarks

Our outline of a global theory of singularly perturbed systems is incomplete and sketchy. Much more needs to be done to describe the important geometric structures that occur in generic SP systems (and families of SP systems) and to elucidate their properties with regard to transversality. Jones and Kopell's analysis of periodic orbits in generalizations of the Hodgkin-Huxley model [11] for the action potential is a good example of what needs to be done. The importance of the applications that give rise to SP systems justifies the effort required to flesh out this outline, even though the theory will necessarily be complicated and lack the coherent elegance of singularity theory [6]. The use of singularly perturbed systems to model neural systems is compelling. Neural systems operate with a hierarchy of time scales and SP systems provide a natural way to represent this hierarchy.

Dynamical systems analysis of applications usually requires simulation that relies upon numerical integration. For SP systems, numerical integration presents the problem that "standard" numerical integration techniques require time steps that are small with respect to the fast time scale in the system. Numerical integration techniques for stiff systems and differential algebraic equations address the problem of computing slow-time trajectories of a reduced system, but the problems associated with integration along non-equilibrium attractors of an SP system and with SF transitions have not been studied numerically. We contend that effective solutions to these problems can be constructed through the further development of the global theory outlined in this paper, together with algorithms that exploit the geometric structures found in generic SP systems.

Acknowledgments. This research was partially supported by grants by the Department of Energy, the Office of Naval Research and the National Science foundation.

References

[1] V. I. Arnold, V. S. Afrajmovich, Yu. S. Il'yashenko, L. P. Shil'nikov, Dynamical Systems V, Springer-Verlag, 1994.

[2] A. Back, J. Guckenheimer and M. Myers, A Dynamical Simulation Facility for Hybrid Systems, Springer Lecture notes in Comp. Sci., 736, 255–267, 1993.

[3] M. L. Cartwright and J. E. Littlewood, On nonlinear differential equations of second order II, Ann. of Math. 48, 472–494, 1947.

[4] W. de Melo and S. van Strien, One-dimensional dynamics, Springer-Verlag 1993.

[5] N. Fenichel, Persistence and smoothness of invariant manifolds for flows, Indiana Univ. Math. J. 21, 193–226, 1971

[6] M. Golubitsky and V. Guillemin, Stable mappings and their singularities, Springer-Verlag, 1973.

[7] J. Grasman, Asymptotic methods for relaxation oscillations and applications, Springer-Verlag, 1987.

[8] J. Guckenheimer and P. Holmes, Nonlinear Oscillations, Dynamical Systems, and Bifurcation of Vector Fields, Springer-Verlag, 1983.

[9] J. Guckenheimer and S. Johnson, Planar hybrid systems, to appear in Hybrid Systems II, Springer Lecture Notes in Computer Science, 1995.

[10] Yu. S. Il'yashenko and S. Yu. Yakovenko, Finitely-smooth normal forms of local families of diffeomorphisms and vector fields, Russian Math Surveys, 46, 1–43, 1991.

[11] C. Jones and N. Kopell, Tracking invariant manifolds with differential forms, Journal of Differential Equations, 108, 64–88, 1994.

[12] E. F. Mishchenko and N. Kh. Rozov, Differential equations with a small parameter multiplying the highest derivative and relaxation oscillations, Plenum Press, 1980.

[13] A. I. Nejshtadt Asymptotic investigation of the loss of stability as apair of eigenvalues slowly cross the imaginary axis, Usp. mat. Nauk 40, 190–191, 1985.

[14] J. Rinzel, A formal classification of bursting mechansims in excitable systems, Proc. Intern. Congr. of Mathematicians (A.M. Gleason, ed.), Amer. Math. Soc., 1578–1594, 1987.

[15] F. Takens, Constrained differential equations, in: Dynamical systems-Warwick 1974, LNM **468**, Springer-Verlag, 1975.

[16] F. Takens, Constrained equations; a study of implicit differential equations and their discontinuous solutions, in: Structural stability, the theory of catastrophes, and applications in the sciences, LNM **525**, Springer-Verlag, 1976.

[17] F. Takens, Implicit differential equations; some open problems, in: Singularités D'applications différentiables, LNM **535**, Springer-Verlag, 1976.

[18] F. Takens, Transition from periodic to strange attractors in constrained equations, in: Dynamical systems and bifurcation theory, Longman, 1987.

Progress in Nonlinear Differential Equations
and Their Applications, Vol. 19
© 1996 Birkhäuser Verlag Basel/Switzerland

Equivariant Perturbations of the Euler Top

Heinz Hanßmann

Abstract: The motion of a rigid body in a small non-constant force field is
studied. A normal form approach yields a formal 2-torus symmetry, which
in turn allows to reduce to a one-degree-of-freedom system. The behaviour
of this system is used to identify quasi-periodic motions of the rigid body
with two or three independent frequencies.

The external force field is supposed to be invariant under two spacial reflec-
tions. This $\mathbb{Z}_2 \times \mathbb{Z}_2$-symmetry influences the distribution of invariant tori. To
simplify the necessary calculations, two of the principal moments of inertia
of the rigid body are set equal. Furthermore, the study is concentrated on
the case of an affine (constant+linear) force field.

1 Introduction

Of all integrable cases of the rigid body the free one, named after Euler, deserves
our special attention. Although the number of degrees of freedom is three, there are
four independent integrals : the energy and the three components of the angular
momentum — the Euler top is a *superintegrable* Hamiltonian system. Replacing
one of the components of the angular momentum by its length, two of the integrals commute with all four. In particular the general motion is restricted to a
2-torus. Possible generalizations of this situation are discussed in [Nehorošev;72],
in [Mishchenko,Fomenko;78] and in [Karasev,Maslov;93].

The other classical integrable cases of the rigid body, named after Lagrange and
Kovalevskaya, lead to a *ramified 3-torus bundle* : the phase space is fibrated into
three-parameter families of invariant 3-tori, while two-parameter families of invariant 2-tori and one-parameter families of periodic orbits give rise to singular fibres.
The superintegrable Euler top leads to 4-parameter families of invariant 2-tori and
3-parameter families of periodic orbits, a very degenerate situation.

On the other hand, the Euler top is the limit of any rigid body under forces,
integrable or not, when the potential governing the external force field goes to
zero. This allows us to treat the general problem of a rigid body subject to a
(weak) conservative force field as a perturbation of the Euler top. In this paper we
address the case that the perturbing force field admits a $\mathbb{Z}_2 \times \mathbb{Z}_2$-symmetry, being
invariant under two reflections with respect to two orthogonal planes in space. For
a similar study of non-symmetric forces the reader is referred to [Hanßmann;95].

Perturbations of the non-degenerate integrable cases of the rigid body allow an immediate application of KAM-theory, confer [Broer, Huitema, Takens;90] and references therein. While the resonant tori break up, most of the 3-tori survive the perturbation and are only slightly deformed. They are parametrised over a Cantor set, and their relative measure tends to 1 as the perturbing forces tend to zero. Similarly, the two-parameter families of normally hyperbolic and normally elliptic invariant 2-tori persist, parametrised by Cantor sets of (Hausdorff-)dimension 2 . Periodic orbits persist by means of the implicit mapping theorem. in particular they survive as entire one-parameter families.

We study small perturbations of the superintegrable Euler top, so we have to answer the question : *"What happens to the 4-parameter family of invariant 2-tori under this perturbation ?"* Already the other classical integrable cases, the Lagrange top and the Kovalevskaya case, show that most of the *Eulerian 2-tori* are destroyed, giving rise to a slow motion of the angular momentum. But those 2-tori that do persist may help to understand the global flow of the perturbed system.

The above discussion generalizes to integrable systems with n degrees of freedom. When non-degenerate integrable systems are perturbed, the families of maximal and lower dimensional tori persist on Cantor sets. This is the contents of KAM-theory — for k-tori with $2 \leq k \leq n-2$ one has to ressort to the techniques of [Broer, Huitema, Sevryuk;95] or [Eliasson;88] and [Pöschel;89].

The abstract pendant of the Euler top is a *minimal superintegrable system*, in which the general motion takes place on $(n-1)$-tori. Again these lower dimensional tori form $(n+1)$-parameter families, and the question remains how this very degenerate situation behaves under perturbation. In [Arnol'd;63] KAM-theory is adapted to show that a measure-theoretically large part of the phase space becomes filled by invariant n-tori. Unlike the non-degenerate case, where the geometry of the Cantor-fibration into n-tori is imposed by the unperturbed integrable system, the distribution of invariant n-tori in the perturbation of a superintegrable system is determined by the perturbation itself.

The aim of this paper is to exemplify in the rigid body context how the perturbation determines the geometry of the Cantor-fibration into 3-tori. In particular we are interested in the singular fibres, *i.e.* in invariant 2-tori.

In order to identify the surviving Eulerian 2-tori, we express the perturbed Hamilton function in action-angle variables of the Euler top. Because of its superintegrability the kinetic energy only depends on two of the actions. By means of a normal form theorem we make the perturbed Hamilton function independent of the two angles conjugate to these actions — up to higher order terms. Then we truncate the Hamilton function. The two actions become integrals, and we can treat them as (internal) parameters. In this fashion we may reduce to a one-degree-of-freedom problem. For the integrable cases of Lagrange and Kirchhoff the results one obtains with this perturbative approach are compared in [Hanßmann;93] with the known behaviour of the top.

The following three-step program will guide us.

First the one-degree-of-freedom problem has to be understood. Here we see whether the normal form is of sufficiently high order. We require the one-degree-of-freedom problem to be *structurally stable* as a family, parametrised by the two actions. At this point the $\mathbb{Z}_2 \times \mathbb{Z}_2$-symmetry becomes important : structural stability is meant within the class of $\mathbb{Z}_2 \times \mathbb{Z}_2$-symmetric Hamiltonian systems. We use equivariant singularity theory to study the equilibria and their normal behaviour.

In the second step we reconstruct the dynamics of the normal form on the full phase space from the family of one-degree-of-freedom systems. In this way the equilibria give rise to invariant 2-tori. The *energy-momentum mapping* makes the phase space a ramified torus bundle, the regular fibres being invariant 3-tori. The invariant 2-tori are singular fibres of this bundle, in the normally hyperbolic case the fibre includes the stable and unstable manifolds. These fibres are attached to critical values of the energy-momentum mapping. In this way the relevant information is collected in the set of critical values.

The third and final step has to show what the dynamics defined by the normal form has to do with the original rigid body motion. This is actually a perturbation problem in itself. If the potential energy is a perturbation of the kinetic energy of order ε, we want the normal form to be ε^2-close to the original system. From KAM-theory, see [Arnol'd;63] and [Pöschel;82], we know that most of the invariant 3-tori survive the perturbation. These 3-tori survive as a Cantor family of (Hausdorff-) dimension 3. Using a centre manifold one can similarly show that normally hyperbolic invariant 2-tori survive the perturbation as a 2-dimensional Cantor family. See also [Graff;74] and [Zehnder;75,76]. For normally elliptic invariant 2-tori we use the results of [Moser;67] and [Broer, Huitema, Takens;90]. It turns out that also the normally elliptic invariant 2-tori survive the perturbation as a 2-dimensional Cantor family. However, the parametrising Cantor sets for normally hyperbolic and for normally elliptic 2-tori have different structures.

This paper is organized as follows. In the remaining part of this introduction we give a description of the possible motions of the free rigid body and formulate the normal form theorem. We assume the body to be *dynamically symmetric*, by this we mean that two of its principal moments of inertia are equal. In this case the action-angle co-ordinates are the Andoyer variables. confer [Andoyer;23] or [Deprit;67]. The force field we consider is affine (constant+linear), this renders the necessary computations feasible and does not require moments of the mass distribution other than the well-known centre of mass and the moments of inertia. Section 2 contains the treatment of the one-degree-of-freedom problem. We identify the equilibria of a two-parameter family of Hamiltonian vector fields on the sphere S^2, as well as their bifurcations. This is used in Section 3 where we calculate the set of critical values of the energy-momentum mapping.

In the final Section 4 we see how the flow of the normal form is perturbed when the higher order terms are added. We consider both the cases that the rigid body is symmetric and that the centre of mass of the rigid body lies off the figure axis.

1.1   The Euler top

We perturb the free rigid body, fixed at one point. Therefore, we now recapitulate the main facts about it. The reader may find a comprehensive introduction in [Abraham,Marsden;78], [Arnol'd;74], [Cushman,Bates;96] or [Goldstein;50].

We choose a set of axes $\vec{e}_x, \vec{e}_y, \vec{e}_z$ fixed in space and a body set of axes $\vec{e}_1, \vec{e}_2, \vec{e}_3$ along the principal axes of inertia. The configuration space is the group $SO(3)$ of orientation preserving orthogonal three-by-three matrices. The elements of $SO(3)$ specify how to transform $\vec{e}_x, \vec{e}_y, \vec{e}_z$ into $\vec{e}_1, \vec{e}_2, \vec{e}_3$.

The phase space is the cotangent bundle $T^*SO(3)$, the space of positions and (angular) momenta. This space comes equipped with the canonical one-form ϑ , allowing to define the equations of motion by means of a Hamilton function, the total energy of the system.

The kinetic energy of the rigid body is defined by means of a left-invariant Riemannian metric $\langle .. \,|\, .. \rangle$, derived from the mass distribution. The relation to the bi-invariant Killing metric $(.. \,|\, ..)$ is given by the inertia tensor $I = \begin{pmatrix} I_1 & 0 & 0 \\ 0 & I_2 & 0 \\ 0 & 0 & I_3 \end{pmatrix}$,

where $I_1 = I_2$ as we assume the body to be dynamically symmetric. An element $\alpha \in T^*SO(3)$ yields the components ℓ_1, ℓ_2, ℓ_3 of the angular momentum with respect to the body set of axes $\vec{e}_1, \vec{e}_2, \vec{e}_3$. Then the kinetic energy is expressed as $T(\alpha) = \frac{1}{2}\langle \alpha \,|\, \alpha \rangle = \frac{\ell_1^2 + \ell_2^2}{2I_1} + \frac{\ell_3^2}{2I_3}$.

Using the Killing metric we construct an S^1-action on $T^*SO(3)$, the *central action*. To this end define the function

$$|\mu| : \quad T^*SO(3) \quad \longrightarrow \quad \mathbb{R}$$

by $|\mu|(\alpha) := \|\alpha\| := \sqrt{(\alpha \,|\, \alpha)}$, measuring the length of the angular momentum vector. On the complement $T^*SO(3)^{\backslash SO(3)}$ of the zero-section $SO(3) \subseteq T^*SO(3)$ we obtain a Hamiltonian vector field $X_{|\mu|}$, the trajectories of which are periodic with common period 2π . Thus the flow ψ_t of $X_{|\mu|}$ yields the central action

$$\begin{array}{rccc} \Gamma : & S^1 \times T^*SO(3)^{\backslash SO(3)} & \longrightarrow & T^*SO(3)^{\backslash SO(3)} \\ & (\xi, \alpha) & \mapsto & \psi_\xi(\alpha) \end{array} .$$

The orbits of the central action are related to a pure precession of the rigid body about the angular momentum. Therefore we call ξ the precession angle.

The kinetic energy does not depend on the precession angle. Hence, we can use the central symmetry to reduce the phase space. Fixing the length $|\mu| \neq 0$ of the angular momentum vector, the orbit space $\{\alpha \in T^*SO(3) \mid \|\alpha\| = |\mu|\}_{/S^1}$ can be identified with

$$S^2_{|\mu|} \times S^2_{|\mu|} = \{ (\mu, \ell) \in \mathbb{R}^3 \times \mathbb{R}^3 \mid \|\mu\| = |\mu| = \|\ell\| \} .$$

see [Hanßmann;95]. Here μ_1, μ_2, μ_3 are the components of the angular momentum with respect to the set of axes $\vec{e}_x, \vec{e}_y, \vec{e}_z$ fixed in space (and ℓ_1, ℓ_2, ℓ_3 those with respect to $\vec{e}_1, \vec{e}_2, \vec{e}_3$). The bundle $\{\alpha \in T^*SO(3) \mid \|\alpha\| = |\mu|\} \longrightarrow S^2_{|\mu|} \times S^2_{|\mu|}$ is related to the Hopf fibration, confer [Abraham,Marsden;78]. In particular, this bundle does not have a global (continuous) section.

The reduced Hamilton function is $H(\mu, \ell) = \frac{\ell_1^2 + \ell_2^2}{2I_1} + \frac{\ell_3^2}{2I_3}$ and the symplectic structure reduces to $\dfrac{\sigma_L}{|\mu|} - \dfrac{\sigma_R}{|\mu|}$, where σ_L (σ_R) denotes the area element on the left (right) hand sphere. Hence, the equations of motion are

$$\dot{\mu} = \frac{\partial T}{\partial \mu} \times \mu = 0$$

$$\dot{\ell} = -\frac{\partial T}{\partial \ell} \times \ell = \ell \times I^{-1}(\ell) \quad .$$

In particular the equations decouple. On the right hand sphere we have Euler's equations ; since $I_1 = I_2$ these are

$$\dot{\ell}_1 = -\frac{I_3 - I_1}{I_1 I_3} \ell_2 \ell_3 , \quad \dot{\ell}_2 = \frac{I_3 - I_1}{I_1 I_3} \ell_1 \ell_3 , \quad \dot{\ell}_3 = 0 \quad .$$

The vector field on the left hand sphere vanishes identically, the (direction of the) angular momentum is conserved.

From this we reconstruct the flow on $T^*SO(3)$, confer *Figure 1.1*. While the length $|\mu|$ of the angular momentum is conserved, the precession angle varies according to $\dot{\xi} = \frac{\partial T}{\partial |\mu|}$. To compute this we express T in Andoyer variables $(\xi, \rho, q, |\mu|, \Im, p)$, the action-angle co-ordinates of the dynamically symmetric free rigid body. Here $\Im := \ell_3$ is the height on the right hand sphere, the component of the angular momentum along the *figure axis* $\vec{e}_3$, and ρ is the conjugate angle, measuring the rotation of the body about the figure axis. Similarly $p := \mu_3$ and q denote cylindrical co-ordinates on the left hand sphere. The kinetic energy now reads $T = \frac{|\mu|^2}{2I_1} - \frac{I_3 - I_1}{I_1 I_3} \cdot \frac{\Im^2}{2}$, and we conclude

$$\dot{\xi} = \frac{|\mu|}{I_1} \quad \text{and} \quad \dot{\rho} = -\frac{I_3 - I_1}{I_1 I_3} \Im \quad .$$

In the sequel it is preferable to (re)substitute (μ_1, μ_2, μ_3) for (q, p) , keeping in mind the relation $\mu_1^2 + \mu_2^2 + \mu_3^2 = |\mu|^2$ between the co-ordinates $(\xi, \rho, |\mu|, \Im, \mu_1, \mu_2, \mu_3)$ on $T^*SO(3)$. The Eulerian 2-tori $\{ (\xi, \rho, |\mu|, \Im, \mu_1, \mu_2, \mu_3) \mid \xi, \rho \in S^1 , \mu_1^2 + \mu_2^2 + \mu_3^2 = |\mu|^2 \}$ are parametrised by $|\mu|$, $\Im \in\,]-|\mu|, |\mu|[$ and the points on the left hand sphere.

We perturb the free rigid body by a small conservative affine (constant+linear) force field. The set of axes $\vec{e}_x, \vec{e}_y, \vec{e}_z$ fixed in space may be chosen to consist of eigenvectors of the linear part of the force field. These axes give rise to co-ordinates

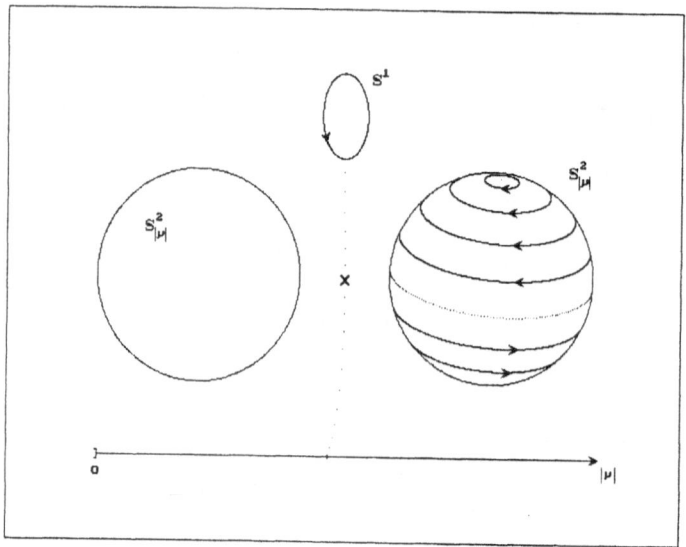

Figure 1 1: Schematic representation of the bundle structure of the phase space $T^*SO(3)\backslash SO(3)$. The Euler top gives rise to the flow depicted on the right hand sphere, while the left hand sphere consists of equilibria. The unreduced flow on $T^*SO(3)$ is equivariant with respect to the central action on the S^1-fibres ; and $|\mu|$ remains fixed.

x, y, z in space. The linear part of the force field has a $\mathbb{Z}_2 \times \mathbb{Z}_2 \times \mathbb{Z}_2$-symmetry, being equivariant with respect to the three reflections $x \mapsto -x$, $y \mapsto -y$ and $z \mapsto -z$. In this paper we let the constant part of the force field be parallel to the eigenvector $\vec{e}_y$ corresponding to the middle eigenvalue. This leads to a $\mathbb{Z}_2 \times \mathbb{Z}_2$-symmetry of the perturbing force field

$$\vec{G} \;=\; -\,\beta\,\vec{e}_y \;-\; 2ax\,\vec{e}_x \;-\; 2by\,\vec{e}_y \;-\; 2cz\,\vec{e}_z \quad . \tag{1}$$

From the potential $\beta y + ax^2 + by^2 + cz^2$ of the force field one can compute the potential energy $V_{\beta,a,b,c}$ of the rigid body, confer [Hanßmann;95]. The force field (1) is positional, *i.e.* the potential energy only depends on the configuration $g \in SO(3)$ of the system. As a result $V_{\beta,a,b,c}$ is homogeneous of degree zero in the momenta. Since an additive term $d \cdot (x^2 + y^2 + z^2)$ in the potential of the force field has no effect on the motion, we may assume $0 < a < b < c$.

1.2   The normal form

Let $\omega(|\mu|, \Im)$ denote the frequency vector of the parallel flow on the Eulerian 2-tori, *i.e.*

$$\omega(|\mu|, \Im) = \begin{pmatrix} \dfrac{\partial T}{\partial |\mu|} \\[2mm] \dfrac{\partial T}{\partial \Im} \end{pmatrix} = \begin{pmatrix} \dfrac{1}{I_1} \cdot |\mu| \\[2mm] -\dfrac{I_3 - I_1}{I_1 I_3} \cdot \Im \end{pmatrix} \quad .$$

For resonant 2-tori already a small perturbation usually leads to frequency locking. This results in some isolated periodic orbits, but the 2-torus breaks up. Therefore we single out a $(|\mu|_0, \Im_0)$ with (even) a *diophantine* frequency vector $\omega(|\mu|_0, \Im_0)$ and concentrate on the situation around the associated quasi-periodic tori.

We simplify and unify the notation by setting $y_1 := |\mu| - |\mu|_0$, $y_2 := \Im - \Im_0$, $x_1 := \xi$, $x_2 := \rho$, $z := (\mu_1, \mu_2, \mu_3)$ and $\varepsilon := (\beta, a, b, c)$. Recall that $x = (x_1, x_2)$ takes its values on a 2-torus T, $y = (y_1, y_2)$ takes its values in an open part $\mathsf{Y} = \{(y_1, y_2) \mid |\Im_0 + y_2| < |\mu|_0 + y_1\}$ of $\mathbb{R}^2$ and z takes its values in the Poisson space $\mathsf{P} = \{(z_1, z_2, z_3) \in \mathbb{R}^3 \mid z_1^2 + z_2^2 \neq 0\}$. Instead of the third axis we also could have taken away any other line through the origin out of $\mathbb{R}^3$, but we cannot work globally on $\mathbb{R}^3 \backslash \{0\}$ as the precession angle $\xi = x_1$ is only locally defined.

In these co-ordinates we may write our Hamilton function as

$$H(x, y, z, \varepsilon) = t + (\omega \mid y) + \tau(y) + V_\varepsilon(x, y, z) \quad .$$

Here $t := T(|\mu|_0, \Im_0)$ is the constant part of the kinetic energy and $\omega = \omega(|\mu|_0, \Im_0)$ is the frequency vector. Since we are in the dynamically symmetric case, $\tau(y) = \frac{y_1^2}{2I_1} - \frac{I_3 - I_1}{I_1 I_3} \cdot \frac{y_2^2}{2}$ is exactly the quadratic part of the kinetic energy.

We expand our Hamilton function as a Taylor series in (y, ε) . The first term of the perturbation V_ε is of order one in ε . Note that we do not expand our Hamilton function in z as well, our construction is global with respect to P. We introduce the following *gradation* of the ring of all (=formal) power series in y and ε with coefficients in $C^\infty(\mathsf{T} \times \mathsf{P})$ that vanish for $\varepsilon = 0$,

$$\mathcal{A}_k := \{ f \in C^\infty(\mathsf{T} \times \mathsf{P})[y, \varepsilon] \mid D^l f(0,0) = 0 \ \forall_{l \neq k}, \ f(y, 0) \equiv 0 \}$$

together with the *filtration* $\mathcal{F}_l := \prod_{k \geq l} \mathcal{A}_k$. This terminology replaces expressions like " $+ O(\varepsilon \cdot (y + \varepsilon)^k)$ uniformly as $y, \varepsilon \to 0$ " by " $\mod \mathcal{F}_{k+1}$ ", confer [Broer;81]. With $\pi_{\mathcal{A}_k}$ we denote the projection of $\prod_{l \geq 1} \mathcal{A}_l$ onto $\mathcal{A}_k$, and π_{Y} denotes the projection of $\mathsf{T} \times \mathsf{Y} \times \mathsf{P}$ onto the second component.

Theorem 1.1   *Let T be an n-torus, Y an open neighbourhood of the origin in $\mathbb{R}^n$ and P a Poisson space. Generate a Poisson bracket on $\mathsf{T} \times \mathsf{Y}$ by $\{x_i, x_j\} =$*

$0 = \{y_i, y_j\}$ and $\{x_i, y_j\} = \delta_{ij}$. Supply $(\mathsf{T} \times \mathsf{Y}) \times \mathsf{P}$ with a Poisson structure that extends the brackets on $\mathsf{T} \times \mathsf{Y}$ and P and satisfies $\{y_i, z_j\} = 0$. Let $H(x, y, z, \varepsilon) = T(y) + V_\varepsilon(x, y, z)$ be a perturbed Hamilton function on $\mathsf{T} \times \mathsf{Y} \times \mathsf{P}$, depending on the perturbation parameter $\varepsilon \in \mathbb{R}^m$ and invariant under a $\mathbb{Z}_2 \times \mathbb{Z}_2$-action. Suppose that the frequency vector $\omega = \frac{\partial T}{\partial y}(0)$ is diophantine and fix an integer $k \in \mathbb{N}$. Then there is a $\mathbb{Z}_2 \times \mathbb{Z}_2$-equivariant Poisson co-ordinate transformation ψ such that $(H \circ \psi)(x, y, z, \varepsilon) = T(y) + W_\varepsilon(y, z) \pmod{\mathcal{F}_{k+1}}$, with $\mathbb{Z}_2 \times \mathbb{Z}_2$-invariant $T + W_\varepsilon$.

Proof See [Hanßmann;93,95], the $\mathbb{Z}_2 \times \mathbb{Z}_2$-symmetry is automatically preserved ;
$$\text{q.e.d.}$$

The terms up to order k in $T(y) + W_\varepsilon(y, z)$ are called the normal form (of order k). In the new co-ordinates, H becomes independent of x up to the prescribed order k . The normal form of order one $T + \bar{V}_\varepsilon$ simply arises averaging V_ε along $x \in \mathsf{T}$. For the rigid body in the $\mathbb{Z}_2 \times \mathbb{Z}_2$-symmetric affine force field (1) we get as normal form of order one

$$\bar{H}_{|\mu|, \Im}(\mu_1, \mu_2, \mu_3) = \frac{|\mu|^2}{2I_1} - \frac{I_3 - I_1}{I_1 I_3}\frac{\Im^2}{2} + \frac{a + b + c}{2}\left(I_1 + (I_3 - I_1)\frac{\Im^2}{|\mu|^2}\right)$$

$$+ s_3 \frac{\Im}{|\mu|} \beta \frac{\mu_2}{|\mu|} - \frac{I_3 - I_1}{2}\frac{3\Im^2 - |\mu|^2}{|\mu|^2}\left(a\frac{\mu_1^2}{|\mu|^2} + b\frac{\mu_2^2}{|\mu|^2} + c\frac{\mu_3^2}{|\mu|^2}\right) \quad .$$

Here s_3 is the component of the centre of mass along the figure axis $\vec{e}_3$. Note that the normal form $\bar{H}$ is globally defined on $T^*SO(3) \backslash SO(3)$, the result does not depend on the line through the origin we took out of $\mathbb{R}^3$ to define P . In [Fassò;91,95] the local expressions ψ are shown to represent a global co-ordinate transformation. The simplicity of the normal form — a quadratic polynomial in μ_1, μ_2, μ_3 — is due to the dynamical symmetry of the rigid body. The normal form in the general case with three different moments of inertia would involve $(\mathbb{Z}_2 \times \mathbb{Z}_2$-symmetric) elliptic functions.

Restricting the domain Y of y to an ε-neighbourhood of the origin in $\mathbb{R}^2$, we see that the difference between the transformed Hamilton function $H \circ \psi_1$ and the normal form of order one $\bar{H}$ is of order ε^2 . In case the centre of mass lies on the figure axis we can obtain this estimate on a much larger domain. Indeed, in this case the Hamilton function does not depend on ρ and Theorem 1.1 applies for $x = x_1 = \xi$, $y = y_1 = |\mu| - |\mu|_0$ and $z = (\mu_1, \mu_2, \mu_3, \ell_1, \ell_2, \ell_3)$. We can even simplify the calculations replacing the necessary integration by means of a Fourier series by an indefinite integral. In particular we have no problems with diophantine conditions, and this allows us to improve Theorem 1.1.

The domain of the Andoyer variables on $T^*SO(3)$ can be embedded in $\mathsf{T} \times \mathsf{Y} \times \mathsf{P}$ with $\mathsf{T} := S^1$, $\mathsf{Y} =]0, \infty[$ and $\mathsf{P} = \{z \in \mathbb{R}^6 \mid z_1^2 + z_2^2 \neq 0, z_4^2 + z_5^2 \neq 0\}$. Unlike Theorem 1.1 we do not have to restrict to a neighbourhood of 1-tori with frequency $\omega = \frac{|\mu|_0}{I_1}$, so we work with $y = |\mu|$ instead of $y = |\mu| - |\mu|_0$. With $\|..\|_A$ we denote the supremum norm on A .

Corollary 1.2 Let $H = T + V_\varepsilon$ be the Hamilton function of a symmetric rigid body in a $\mathbb{Z}_2 \times \mathbb{Z}_2$-symmetric conservative affine force field. Fix $\eta > 0$ and let the 1-torus T and the Poisson space P be as given above. Then there is a $\mathbb{Z}_2 \times \mathbb{Z}_2$-equivariant Poisson co-ordinate transformation ψ_1 on $\mathsf{T} \times [\eta, \infty[\times \mathsf{P}$ and a constant c_η such that

$$\| H \circ \psi_1 - \bar{H} \|_{\mathsf{T} \times [\eta, \infty[\times \mathsf{P}} < c_\eta \, \varepsilon^2 \quad .$$

Here $\bar{H}$ denotes the normal form of order one.

Proof See [Hanßmann;95], Corollary 6.1 ;   q.e.d.

If averaging along two (or more) angles is required, the necessary diophantine conditions allow estimates $|H \circ \psi_1 - \bar{H}| \sim \varepsilon^2$ only on y-domains of order ε . But in the particular situation of a dynamically symmetric rigid body in an affine force field only finitely many harmonics of the Fourier series (which give rise to the small denominators) are involved. Therefore we can estimate $|H \circ \psi_1 - \bar{H}|$ on a large domain even in the case that the centre of mass is in general position. We work again with $\mathsf{T} = \{(\xi, \rho) \mid \xi \in S^1, \rho \in S^1\}$ and $\mathsf{P} = \{(\mu_1, \mu_2, \mu_3) \in \mathbb{R}^3 \mid \mu_1^2 + \mu_2^2 \neq 0\}$, but let y vary in the whole subset $\{(|\mu|, \Im) \mid \Im \in [-|\mu|, |\mu|]\}$ of $\mathbb{R}^2$.

Corollary 1.3 Let $H = T + V_\varepsilon$ be the Hamilton function of a rigid body, fixed at one point, with principal moments of inertia $I_1 = I_2 \neq I_3$ and centre of mass in general position, and moving in a conservative affine $\mathbb{Z}_2 \times \mathbb{Z}_2$-symmetric force field. Denote by Y a set of $(|\mu|, \Im) \in \mathbb{R}^2$ with $\Im \in [-|\mu|, |\mu|]$ for which the frequency vector

$$\omega(|\mu|, \Im) = \begin{pmatrix} \dfrac{1}{I_1} \cdot |\mu| \\[2ex] -\dfrac{I_3 - I_1}{I_1 I_3} \cdot \Im \end{pmatrix}$$

is bounded away from resonances in $\{0, \pm 1, \pm 2\}^2$ and let the 2-torus T and the Poisson space P be as given above. Then there is a $\mathbb{Z}_2 \times \mathbb{Z}_2$-equivariant Poisson co-ordinate transformation ψ_1 on $\mathsf{T} \times \mathsf{Y} \times \mathsf{P}$ such that

$$\| H \circ \psi_1 - \bar{H} \|_{\mathsf{T} \times \mathsf{Y} \times \mathsf{P}} \leq c_\mathsf{Y} \, \varepsilon^2$$

for some constant c_Y depending on the distance to the above resonances.

Proof See [Hanßmann;95], Corollary 7.2 ;   q.e.d.

In the sequel we also need the normal form of order two. It is not the precise expression of the normal form itself that we use, but the fact that the difference between the transformed Hamilton function $H \circ \psi_1 \circ \psi_2$ and the normal form of order two $\overline{H \circ \psi_1}$ is of order ε^3 . The co-ordinate transformation ψ_1 introduces some higher resonances, but there are still only finitely many harmonics involved. This allows again to obtain the estimate on a large domain (i.e. on a domain that does not shrink to a submanifold $\{|\mu| = |\mu|_0, \Im = \Im_0\}$ as $\varepsilon \to 0$).

Corollary 1.4   *Under the assumptions of Corollary 1.3, let* Y *denote a set of* $(|\mu|, \Im) \in \mathbb{R}^2$ *with* $\Im \in [-|\mu|, |\mu|]$ *for which the frequency vector* $\omega(|\mu|, \Im)$ *is bounded away from resonances in* $\{0, \pm 1, \ldots, \pm 4\}^2$. *Then there is a* $\mathbb{Z}_2 \times \mathbb{Z}_2$-*equivariant Poisson co-ordinate transformation* ψ_2 *on* $\mathsf{T} \times \mathsf{Y} \times \mathsf{P}$ *such that*

$$\| H \circ \psi_1 \circ \psi_2 - \overline{H \circ \psi_1} \|_{\mathsf{T} \times \mathsf{Y} \times \mathsf{P}} \leq c_{\mathsf{Y}} \varepsilon^3$$

for some constant c_{Y} *depending on the distance to the above resonances.*

Proof   See [Hanßmann;95], Theorem 7.4 ;   q.e.d.

Let us close this section with a remark concerning the general situation that the perturbation is a full Fourier series. With an 'ultra-violet cutoff' we may reduce to a Fourier polynomial and obtain estimates similar to those above for some normal form, the tail of the Fourier series allowing a separate estimate. In fact, this procedure is used in [Arnol'd;63] to show the existence of a large set of maximal invariant tori in the perturbation of superintegrable Hamiltonian systems.

Our main aim is to understand how these maximal tori are distributed in phase space, how they are separated by stable and unstable manifolds of normally hyperbolic 2-tori and how they shrink to normally elliptic 2-tori.

2   The one-degree-of-freedom system

The averaged Hamilton function $\bar{H}$ on $T^*SO(3) \backslash SO(3)$ is invariant under the central action Γ and under rotations about the figure axis. These two S^1-actions commute and define a 2-torus action. The corresponding momentum mapping is

$$(|\mu|, \Im) : \quad T^*SO(3) \quad \longrightarrow \quad \mathbb{R}^2 \quad .$$

The level sets of the momentum mapping are invariant manifolds of the Hamiltonian vector field $X_{\bar{H}}$. These level sets are also invariant under the 2-torus action. In the schematic representation of *Figure 1.1* the level sets correspond to fixed radius $|\mu|$ of the two spheres and fixed height $\Im$ on the right hand sphere $S^2_{|\mu|}$. The 2-torus action rotates along the S^1-fibres and along circles of height $\Im$ on the right hand sphere $S^2_{|\mu|}$. Dividing out this action leads to the reduced phase space $S^2_{|\mu|} \times \{\Im\}$. The reduced Hamilton function $\bar{H}_{|\mu|, \Im}$ defines a one-degree-of-freedom system on this reduced phase space. Recall that $S^2_{|\mu|} \times \{\Im\}$ comes equipped with the symplectic structure $\dfrac{\sigma_L}{|\mu|}$, where σ_L denotes the area element on the sphere.

The first step of our program is to study the reduced Hamilton function $\bar{H}_{|\mu|, \Im}$. We consider the external parameters β, a, b, c as fixed constants. The internal or distinguished parameters $|\mu|$ and $\Im$ on the other hand are allowed to vary. We want to show that this family of one-degree-of-freedom systems is structurally stable in the subclass of (families of) $\mathbb{Z}_2\times\mathbb{Z}_2$-equivariant Hamiltonian systems on $S^2_{|\mu|} \times \{\Im\}$.

Let us start with a rough description of the expected phase portraits. For an open and dense part of the parameter plane the corresponding system $X_{\bar{H}_{|\mu|,\Im}}$ itself is structurally stable. The flow on $S^2_{|\mu|} \times \{\Im\}$ mostly consists of periodic orbits and can be characterized by the distribution of its equilibria. We have several cases of structurally stable systems with, respectively, four centres and two saddles, or three centres and one saddle, or two centres. In the cases where saddles are present their stable and unstable manifolds consist of homoclinic orbits. In the Hamiltonian context at hand homoclinic connections are generic.

To understand the whole family $\bar{H}_{|\mu|,\Im}$ we also study the bifurcations between these structurally stable cases. Although we have two parameters, only bifurcations of codimension one occur. To pass from one of the above cases to another a centre has to split into a saddle and two centres in a Hamiltonian pitchfork bifurcation. We also encounter a *connection bifurcation* where the two saddles have the same energy.

To be more explicit, the orbits of the Hamiltonian vector field $X_{\bar{H}_{|\mu|,\Im}}$ are the intersections of the level sets $\{\bar{H}_{|\mu|,\Im}(\mu_1,\mu_2,\mu_3) = h\} \subseteq \mathbb{R}^3$ with $S^2_{|\mu|}$ (from now on we drop the factor $\{\Im\}$ in $S^2_{|\mu|} \times \{\Im\}$). Since $\bar{H}_{|\mu|,\Im}$ is a quadratic polynomial in μ_1, μ_2, μ_3 these level sets are quadrics. Using this the phase portraits are easily obtained.

If $3\Im^2 = |\mu|^2$ the level sets are planes perpendicular to $\begin{pmatrix} 0 \\ 1 \\ 0 \end{pmatrix}$. We get 2 centres, surrounded by periodic orbits, see *Figure 2.1a*. In this case the quadratic part of the potential energy has no influence.

For $3\Im^2 \neq |\mu|^2$ the level sets of $\bar{H}_{|\mu|,\Im}$ are triaxial ellipsoids with centre

$$\frac{s_3 \, \Im \, |\mu|}{(I_3 - I_1)(3\Im^2 - |\mu|^2)} \begin{pmatrix} 0 \\ \beta/b \\ 0 \end{pmatrix} \quad ,$$

which tends to infinity as $(|\mu|, \Im) \longrightarrow \{3\Im^2 = |\mu|^2\}$. Let us consider values $(|\mu|, \Im)$ near $\{3\Im^2 = |\mu|^2\}$. When it comes to an intersection of the ellipsoid $\{\bar{H}_{|\mu|,\Im} = h\}$ with the sphere $S^2_{|\mu|}$, the curvature of the ellipsoid is so small that locally, *i.e.* on the sphere, it resembles a plane. Qualitatively the flow does not change, confer *Figure 2.1b* ; we get two centres, surrounded by periodic orbits.

When the centre of the ellipsoid approaches the sphere $S^2_{|\mu|}$, a Hamiltonian pitchfork bifurcation takes place, see *Figures 2.1a-c*. The centre at $\begin{pmatrix} 0 \\ 1 \\ 0 \end{pmatrix}$ turns into a saddle, the stable and unstable manifolds of which form a $\mathbb{Z}_2\times\mathbb{Z}_2$-invariant 'figure eight', encircling two newly born centres.

As the centre of the ellipsoid further approaches the origin, a second Hamiltonian pitchfork bifurcation takes place, leading to a second saddle and a fourth centre.

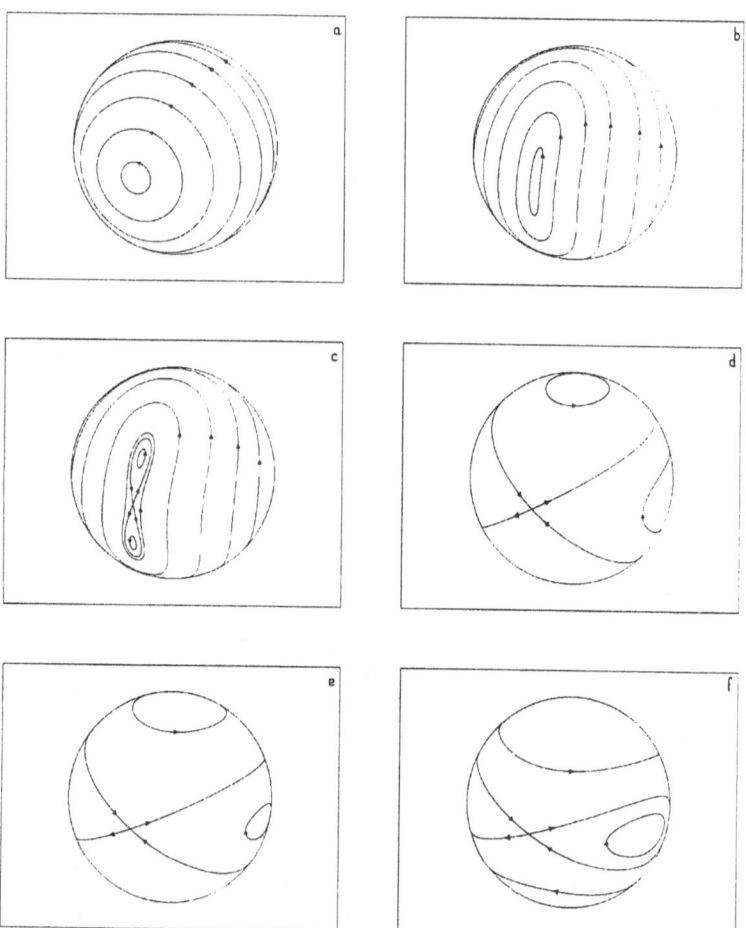

Figure 2.1: Phase portraits of the one-degree-of-freedom system.

confer *Figure 2.1d*. If $\Im = 0$ the ellipsoid is centered at the origin. The family of periodic orbits between the stable and unstable manifolds of the two saddles vanishes and the saddles become connected by heteroclinic orbits, see *Figure 2.1e*. In this case the linear part of the potential energy has no influence. When $\Im$ has passed through 0 , the scenario reverses, starting with *Figure 2.1f*.

The vector field $X_{\bar{H}_{|\mu|,\Im}}$ on $S^2_{|\mu|}$ depends only on the last two terms of $\bar{H}_{|\mu|,\Im}$. Furthermore, these terms are invariant under the transformation $\mu \mapsto \frac{\mu}{|\mu|}$ (which turns $\Im$ into $\frac{\Im}{|\mu|}$). Multiplying these last two terms of $\bar{H}_{|\mu|,\Im}$ by $\frac{1}{I_3-I_1} \cdot \frac{|\mu|^2}{|\mu|^2-3\Im^2}$ we are led to

$$\mathcal{H}_{\mathcal{I}}(x,y,z) \;=\; \mathcal{I}y + \frac{1}{2}(ax^2 + by^2 + cz^2) \tag{2}$$

where $x = \frac{\mu_1}{|\mu|}$, $y = \frac{\mu_2}{|\mu|}$, $z = \frac{\mu_3}{|\mu|}$ and $\mathcal{I} = \frac{s_3\beta}{I_3 - I_1} \cdot \frac{\Im|\mu|}{|\mu|^2 - 3\Im^2}$. This shows that $\bar{H}_{|\mu|,\Im}$ is induced by a one-parameter family, the parameters $|\mu|$ and $\Im$ enter only as quotient $\frac{\Im}{|\mu|}$. We see that for (generic) positional forces only bifurcations of codimension one can occur.

Proposition 2.1 Let $X_{\mathcal{H}_{\mathcal{I}}}$ be the family of Hamiltonian vector fields on S^2 given by (2). Then $X_{\mathcal{H}_{\mathcal{I}}}$ has phase portraits as shown in Figure 2.1. As $\mathcal{I}$ varies, the flow undergoes four Hamiltonian pitchfork bifurcations and one connection bifurcation.

Proof To find the equilibria of $X_{\mathcal{H}_{\mathcal{I}}}$ we have to solve the equation

$$\begin{pmatrix} \frac{\partial \mathcal{H}_{\mathcal{I}}}{\partial x} \\ \frac{\partial \mathcal{H}_{\mathcal{I}}}{\partial y} \\ \frac{\partial \mathcal{H}_{\mathcal{I}}}{\partial z} \end{pmatrix} \;=\; \lambda \cdot \begin{pmatrix} x \\ y \\ z \end{pmatrix}$$

under the condition $x^2 + y^2 + z^2 = 1$. For all values of $\mathcal{I}$ we have the solutions $x = 0 = z$, $y = \pm 1$ (with $\lambda = b \pm \mathcal{I}$). Putting $\lambda = c$ yields the solutions $x = 0, y = \frac{\mathcal{I}}{c-b}, z = \pm\sqrt{1-y^2} = \frac{\pm\sqrt{(c-b)^2-\mathcal{I}^2}}{c-b}$, while $\lambda = a$ leads to $z = 0, y = \frac{\mathcal{I}}{a-b}$, $x = \pm\sqrt{1-y^2} = \frac{\pm\sqrt{(b-a)^2-\mathcal{I}^2}}{b-a}$. In particular the 'permanent' equilibria bifurcate for $\mathcal{I} = a - b, c - b$ and for $\mathcal{I} = b - c, b - a$, respectively.

In the neighbourhood of the equilibria $(0, \epsilon, 0), \epsilon = \pm 1$ we choose local co-ordinates q and p on S^2 according to

$$\begin{pmatrix} x \\ y \\ z \end{pmatrix} \;\overset{!}{=}\; \sqrt{1-q^2-p^2} \begin{pmatrix} 0 \\ \epsilon \\ 0 \end{pmatrix} + q \begin{pmatrix} 0 \\ 0 \\ \epsilon \end{pmatrix} + p \begin{pmatrix} 1 \\ 0 \\ 0 \end{pmatrix} \quad .$$

Writing $\mathcal{I} = \epsilon(a - b) + \nu$ the 4-jet of $\mathcal{H}_{\mathcal{I}}$ in q and p reads

$$\begin{aligned}
\mathcal{H}^{\text{4-jet}}_{\nu,\epsilon}(q,p) \;=\;\; & a - \frac{b}{2} + \epsilon\nu + \frac{c-a-\epsilon\nu}{2}q^2 - \frac{\epsilon\nu}{2}p^2 \\
& + \frac{b-a-\epsilon\nu}{4}p^4 + \frac{b-a-\epsilon\nu}{2}q^2p^2 + \frac{b-a-\epsilon\nu}{4}q^4
\end{aligned}$$

and around $\mathcal{I} \approx \epsilon(c-a)$ we have

$$
\begin{aligned}
\mathcal{H}_{\nu,\epsilon}^{4\text{-jet}}(q,p) = \quad & c - \frac{b}{2} + \epsilon\nu - \frac{c-a+\epsilon\nu}{2}p^2 - \frac{\epsilon\nu}{2}q^2 \\
& - \frac{c-b+\epsilon\nu}{4}q^4 - \frac{c-b+\epsilon\nu}{2}q^2p^2 - \frac{c-b+\epsilon\nu}{4}p^4 \quad .
\end{aligned}
$$

This shows that the bifurcations at $\nu = 0$ are Hamiltonian pitchfork bifurcations, confer Theorem 4.1 below. In particular the equilibria at $(0, \epsilon, 0)$ change from centres to saddles as $|\mathcal{I}|$ decreases. The reader may easily check that the other equilibria remain centres (until they vanish in one of the other Hamiltonian pitchfork bifurcations).

At $\mathcal{I} = 0$ both equilibria $(0, -1, 0)$ and $(0, 1, 0)$ are saddles. Their energies coincide, giving rise to heteroclinic connections. As

$$
\frac{\mathrm{d}}{\mathrm{d}\mathcal{I}}\Big(\mathcal{H}_{\mathcal{I}}(0,1,0) - \mathcal{H}_{\mathcal{I}}(0,-1,0)\Big)\Big|_{\mathcal{I}=0} = 2
$$

the occurrence of this heteroclinic connection is robust ;   q.e.d.

3   Reconstruction of the averaged flow

In the previous section we gave a detailed description of the one-degree-of-freedom systems $X_{\bar{H}_{|\mu|,\Im}}$ on $S^2_{|\mu|} \times \{\Im\}$. The second step of our program consists in reconstructing the dynamics of $X_{\bar{H}}$ on $T^*SO(3)\backslash SO(3)$ from this. We have to solve two problems. On the one hand, we have to reconstruct the dynamics on the 4-dimensional invariant manifolds of constant $|\mu|$ and $\Im$. We obtain a ramified torus bundle, the regular fibres of which are invariant 3-tori. On the other hand, we want to understand how the global flow on $T^*SO(3)\backslash SO(3)$ is organized. We have to answer questions like : what is the topology of the bundle of invariant 3-tori ? how are the singularities of this bundle distributed ?

To perform the reconstruction we just have to attach a 2-torus to every point on the reduced phase space $S^2_{|\mu|} \times \{\Im\}$. The periodic orbits correspond to invariant 3-tori, while the equilibria give rise to invariant 2-tori. The normal behaviour of the latter is induced from the linearisation of the corresponding equilibrium. We get normally elliptic as well as normally hyperbolic and occasionally normally *parabolic* invariant 2-tori.

The motion of the rigid body corresponding to such an invariant 2-torus is the regular precession of the Euler top. The figure axis $\vec{e}_3$ precesses about the angular momentum (which is fixed in space), while the body rotates about $\vec{e}_3$. For an invariant 3-torus the slow motion of the angular momentum according to $X_{\bar{H}_{|\mu|,\Im}}$ has to be superposed. The same holds true for the motion reconstructed from the stable and unstable manifolds of normally hyperbolic invariant 2-tori.

When $\Im = \pm|\mu|$ the angular momentum is parallel to the figure axis and the 2-torus action degenerates. Indeed, the orbits of the action become circles S^1, coinciding with the orbits of the central action. Hence, the periodic orbits of the reduced Hamiltonian vector field $X_{\bar{H}_{|\mu|,\pm|\mu|}}$ give rise to invariant 2-tori.

However, these 2-tori are "not Eulerian". In the schematic representation of *Figure 1.1* we are in the limiting centres of the *right* hand sphere, while following a slow periodic orbit on the left hand sphere (with superposed fast precession). In particular these invariant 2-tori are normally elliptic.

From equilibria on $S^2_{|\mu|} \times \{\pm|\mu|\}$ we now reconstruct periodic orbits. The body rotates about the fixed angular momentum (which is parallel to the figure axis). Centres become elliptic periodic orbits, and saddles give rise to periodic orbits whose Floquet exponents are one real pair and one complex conjugate pair on the unit circle. We call such a periodic orbit *hypoelliptic*.

We also have to understand how the invariant level sets fit together to form the phase space $T^*SO(3)^{\backslash SO(3)}$. We can collect the relevant information in one picture, the set of critical values of the energy-momentum mapping

$$\mathcal{EM} := (\bar{H}, |\mu|, \Im) : \quad T^*SO(3)^{\backslash SO(3)} \quad \longrightarrow \quad \mathbb{R}^3 \quad .$$

The regular values of $\mathcal{EM}$ just correspond to invariant 3-tori. The critical values correspond to 2-tori (in the normally hyperbolic and parabolic cases together with their stable and unstable manifolds) and to elliptic and hypoelliptic periodic orbits, the latter again together with their stable and unstable manifolds. The critical values also mark the boundary of the image of the energy-momentum mapping. Since only the ratio $\frac{\Im}{|\mu|}$ is important we restrict to a level set, fixing $|\mu|$.

The critical values are symmetric with respect to the reflection $\Im \mapsto -\Im$. For external parameters β, a, b, c in general position the set of critical values has, for fixed value of $|\mu|$, one of the four forms shown in *Figures 3.1a-d*. As the energy $\bar{H}$ depends quadratically on $|\mu|$ (+ small perturbation, homogeneous of degree zero in the momenta), the whole set of critical values is a paraboloid, with diffeomorphic $|\mu|$-slices. In particular also a level set $\bar{H} = h$ is diffeomorphic to a $|\mu|$-slice.

Proposition 3.1 Let $\bar{H} : T^*SO(3)^{\backslash SO(3)} \longrightarrow \mathbb{R}$ be the normal form of order one of the Hamilton function describing a dynamically symmetric rigid body subject to a $\mathbb{Z}_2 \times \mathbb{Z}_2$-symmetric affine force field. Then the set $\Sigma \subseteq \mathbb{R}^3$ of critical values of the energy-momentum mapping $\mathcal{EM}$ is the union of six surfaces $\Sigma = \Sigma_+ \cup \Sigma_- \cup \Sigma_a \cup \Sigma_c \cup \Sigma_l \cup \Sigma_r$ that allow the following parametrisations :

$$\Sigma_+ = \{ (k(|\mu|,\Im) + s_3\beta\tfrac{\Im}{|\mu|} - l(|\mu|,\Im)b, |\mu|, \Im) \in \mathbb{R}^3 \mid |\mu| > 0, \Im \in [-|\mu|, |\mu|] \}$$

$$\Sigma_- = \{ (k(|\mu|,\Im) - s_3\beta\tfrac{\Im}{|\mu|} - l(|\mu|,\Im)b, |\mu|, \Im) \in \mathbb{R}^3 \mid |\mu| > 0, \Im \in [-|\mu|, |\mu|] \}$$

$$\Sigma_a = \{ (k(|\mu|,\Im) - l(|\mu|,\Im)a, |\mu|, \Im) \in \mathbb{R}^3 \mid |\mu| > 0, \tfrac{s_3\Im\Im}{2|\mu|l(|\mu|,\Im)} \in [a-b, b-a] \}$$

$$\Sigma_c = \{ (k(|\mu|,\Im) - l(|\mu|,\Im)c, |\mu|, \Im) \in \mathbb{R}^3 \mid |\mu| > 0, \tfrac{s_3\Im\Im}{2|\mu|l(|\mu|,\Im)} \in [b-c, c-b] \}$$

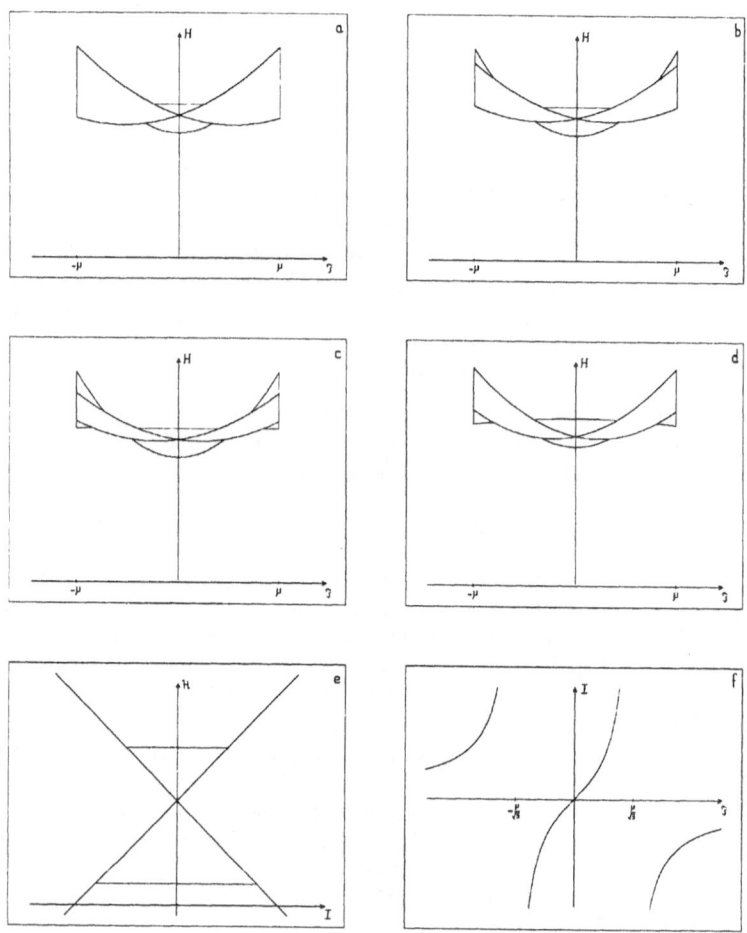

Figure 3.1:

a-d) Possible μ-slices of the set of critical values of the energy-momentum mapping.

e) Set of critical values of $(\mathcal{H}, \mathcal{I})$.

f) Relation between $\mathfrak{I}$ and $\mathcal{I}$.

$$\Sigma_l = \{ (h, |\mu|, -|\mu|) \in \mathbb{R}^3 \mid |\mu| > 0, \, h - k(|\mu|, -|\mu|) + (I_3 - I_1)b \in [-s_3\beta, s_3\beta] \}$$
$$\Sigma_r = \{ (h, |\mu|, |\mu|) \in \mathbb{R}^3 \mid |\mu| > 0, \, h - k(|\mu|, |\mu|) + (I_3 - I_1)b \in [-s_3\beta, s_3\beta] \}$$

with $k(|\mu|, \Im) := \frac{|\mu|^2}{2I_1} - \frac{I_3 - I_1}{I_1 I_3}\frac{\Im^2}{2} + \frac{a+b+c}{2}(I_1 + (I_3 - I_1)\frac{\Im^2}{|\mu|^2})$ and $l(|\mu|, \Im) = \frac{I_3 - I_1}{2}\frac{3\Im^2 - |\mu|^2}{|\mu|^2}$. The four possible cases are distinguished according to whether $|\frac{s_3\beta}{I_3 - I_1}| < \min(b - a, c - b)$, $b - a < |\frac{s_3\beta}{I_3 - I_1}| < c - b$, $c - b < |\frac{s_3\beta}{I_3 - I_1}| < b - a$ or $|\frac{s_3\beta}{I_3 - I_1}| > \max(b - a, c - b)$.

Proof The family (2) of Hamilton functions $\mathcal{H}_\mathcal{I}$ of Proposition 2.1 defines a two-degree-of-freedom system on the product $S^2 \times TS^1$ of the sphere with a cylinder. The energy-momentum mapping

$$(\mathcal{H}, \mathcal{I}): \quad S^2 \times TS^1 \quad \longrightarrow \quad \mathbb{R}^2$$

of this system has the set of critical values $\{(\frac{b}{2} \pm \mathcal{I}, \mathcal{I}) \mid \mathcal{I} \in \mathbb{R}\} \cup \{(\frac{a}{2}, \mathcal{I}) \mid \mathcal{I} \in \,]a - b, b - a[\,\} \cup \{(\frac{c}{2}, \mathcal{I}) \mid \mathcal{I} \in \,]b - c, c - b[\,\}$, see *Figure 3.1e*. To obtain Σ we have to invert the relation $\mathcal{I} = \frac{s_3\beta}{I_3 - I_1} \cdot \frac{\Im |\mu|}{|\mu|^2 - 3\Im^2}$ between $\mathcal{I}$ and $\Im$, see *Figure 3.1f*, and translate this back to

$$\Im = \frac{|\mu|}{6\mathcal{I}}\left(\frac{-s_3\beta}{I_3 - I_1} + \delta\sqrt{\frac{s_3^2\beta^2}{(I_3 - I_2)^2} + 12\mathcal{I}^2}\right) \quad, \delta = \pm 1 \quad .$$

Recall that multiplying $\mathcal{H}$ by $(I_3 - I_1)\frac{|\mu|^2 - 3\Im^2}{|\mu|^2} \overset{!}{=} \frac{s_3\Im\Im}{\mathcal{I}|\mu|}$ yields the two last terms of $\bar{H}$. As the first three terms of $\bar{H}$ yield $k(|\mu|, \Im)$, the four curves $(\frac{a}{2}, \mathcal{I}), (\frac{c}{2}, \mathcal{I}), (\frac{b}{2} + \epsilon\mathcal{I}, \mathcal{I}), \epsilon = \pm 1$ give rise to Σ_a, Σ_c and Σ_ϵ . To these we have to add the set $\Sigma_l \cup \Sigma_r$ where the mappings $|\mu|$ and $\Im$ are tangent ;   q.e.d.

4   Implications for the original system

In the previous section we could give a quite detailed description of the dynamics of $\bar{H}$ on $T^*SO(3)^{\backslash SO(3)}$. This is due to the 2-torus symmetry. The behaviour is governed by the reduced system with one degree of freedom. But unlike these one-degree-of-freedom systems the vector field $X_{\bar{H}}$ is not structurally stable (in the class of all Hamiltonian vector fields on $T^*SO(3)^{\backslash SO(3)}$). Under any small perturbation it can (and usually does) lose its integrability.

We have to keep track of several perturbation problems here. First, our "original Hamilton function" H is a perturbation (of order ε) of the Euler top. Using Theorem 1.1 we consider H as a perturbation (of order ε^2) of its normal form $\bar{H}$. Furthermore we know that perturbations of the one-degree-of-freedom systems defined by $\bar{H}$ do not lead to qualitative changes. This is what we want to use to gain information about the perturbation from $\bar{H}$ to H .

The (structural) stability of the one-degree-of-freedom systems concerns the Eulerian 2-tori. Indeed, the periodic orbits on $S^2_{|\mu|} \times \{\Im\}$ do not tell us whether the family $X_{\bar{H}_{|\mu|,\Im}}$ is structurally stable ; this question is decided by the behaviour of the equilibria. These equilibria give rise to invariant 2-tori on $T^*SO(3)\backslash SO(3)$, the Eulerian 2-tori that survive the perturbation from T to $\bar{H} = T + \bar{V}$.

The case that the 2-torus symmetry of the normal form is not completely broken by the step from $\bar{H}$ to H is considerably easier. Therefore we first investigate the case of a symmetric rigid body, where the 2-torus symmetry of the normal form is only partially broken, with an S^1-symmetry remaining. The persistence results we obtain will serve as a guideline when we consider the 'general case' of an 'unsymmetric dynamically symmetric rigid body'.

4.1   Motion of a symmetric rigid body

We investigate the motion of a rigid body in a $\mathbf{Z}_2 \times \mathbf{Z}_2$-symmetric affine force field with potential $\beta y + ax^2 + by^2 + cz^2$. The rigid body is dynamically symmetric, *i.e.* the principal moments of inertia satisfy $I_1 = I_2 \neq I_3$. In this subsection we also assume that the centre of mass lies on the figure axis.

The affine force field is not able to distinguish such a rigid body from a really symmetric one : the Hamilton function $H = T + V_\varepsilon$ does not depend on ρ. The co-ordinate free formulation of this is that H is invariant under the right S^1-action

$$R^{T^*} :\quad S^1 \times T^*SO(3) \quad \longrightarrow \quad T^*SO(3)$$
$$(\rho, \alpha) \quad \longmapsto \quad T^*R_{\exp_\rho}(\alpha) \quad .$$

Here $\exp_\rho$ stands for $\exp\begin{pmatrix} 0 & -\rho & 0 \\ \rho & 0 & 0 \\ 0 & 0 & 0 \end{pmatrix} = \begin{pmatrix} \cos\rho & -\sin\rho & 0 \\ \sin\rho & \cos\rho & 0 \\ 0 & 0 & 1 \end{pmatrix}$ and $R_{\exp_\rho} : SO(3) \longrightarrow SO(3)$ is defined by right multiplication $g \mapsto g \circ \exp_\rho$. The orbits of R^{T^*} are related to a rotation of the rigid body about the figure axis $\vec{e}_3$.

The regular reduction of this free and proper action is classically known as the "elimination of the cyclic Euler angle" and goes back to Poisson. The reduced phase space turns out to be $\mathbb{R}^3 \times S^2$, expressing that modulo rotations about $\vec{e}_3$ the state of the rigid body is given by the components μ_1, μ_2, μ_3 of the angular momentum and the components $\zeta_1, \zeta_2, \zeta_3$ of the figure axis with respect to the axes $\vec{e}_x, \vec{e}_y, \vec{e}_z$ fixed in space (note $\zeta_1^2 + \zeta_2^2 + \zeta_3^2 = 1$). The reduced Poisson bracket on $\mathbb{R}^3 \times S^2$ leads to the equations of motion

$$\dot{\mu} = \frac{\partial H}{\partial \mu} \times \mu + \frac{\partial H}{\partial \zeta} \times \zeta \qquad \dot{\zeta} = \frac{\partial H}{\partial \mu} \times \zeta \quad .$$

The momentum mapping to the right S^1-action $\Im : T^*SO(3) \longrightarrow \mathbb{R}$ turns on the reduced phase space into the Casimir element

$$\Im :\quad \mathbb{R}^3 \times S^2 \quad \longrightarrow \quad \mathbb{R}$$
$$(\mu, \zeta) \quad \longmapsto \quad (\mu \,|\, \zeta)$$

and the central action becomes

$$\Gamma: \quad \begin{array}{ccc} S^1 \times (\mathbb{R}^3\backslash\{0\} \times S^2) & \longrightarrow & \mathbb{R}^3\backslash\{0\} \times S^2 \\ (\xi,(\mu,\zeta)) & \longmapsto & (\mu, \exp(-\xi\frac{\mu}{\|\mu\|})(\zeta)) \end{array}$$

with momentum mapping $\quad \begin{array}{ccc} |\mu|: & \mathbb{R}^3 \times S^2 & \longrightarrow & \mathbb{R} \\ & (\mu,\zeta) & \longmapsto & \|\mu\| \end{array}$

The Eulerian 2-tori of the free motion (*i.e.* at $V_0 \equiv 0$) get reduced to 1-tori. These periodic orbits coincide with the orbits of the central action on $\mathbb{R}^3\backslash\{0\} \times S^2$. The equilibria $\zeta = \pm\frac{\mu}{\|\mu\|}$ correspond to steady rotations of the rigid body about the figure axis. The other steady rotations all get reduced to the periodic orbit at the equator $\zeta \perp \mu$.

Using Corollary 1.2 we consider H as a perturbation of the normal form $\bar{H}$ on $\mathbb{R}^3\backslash\{0\} \times S^2$. As we have seen in Section 3, the flow of the normal form can be understood from the distribution of invariant 2-tori. Since these are periodic orbits on $\mathbb{R}^3\backslash\{0\} \times S^2$ we expect them to survive the step from $\bar{H}$ to H by means of the implicit mapping theorem.

The distribution of periodic orbits of $\bar{H}$ on $\mathbb{R}^3\backslash\{0\} \times S^2$ (*i.e.* of the 2-tori on $T^*SO(3)$) is in turn organized by the equilibria, the connection bifurcation and the periodic Hamiltonian pitchfork bifurcations. These correspond to the singularities of the set of critical values of the energy-momentum mapping.

The equilibria are either elliptic (those corresponding to the edges of the set of critical values) or hypoelliptic. In both cases they persist by means of the implicit mapping theorem.

While $\Im$ passes through 0, the two hyperbolic periodic orbits interchange their energies. In particular they are (at $\Im = 0$) connected by heteroclinic orbits. In this formulation these are robust properties of the flow of $\bar{H}$.

However, the heteroclinic orbits form separatrices, and these may be expected to split, giving rise to transversal heteroclinic orbits. As a result heteroclinic orbits would exist for a whole interval of energy values. To show that the separatrices indeed split one has to study the Mel'nikov function, but this would be outside the scope of the present paper.

The robustness of the periodic Hamiltonian pitchfork bifurcation is due to the $\mathbb{Z}_2\times\mathbb{Z}_2$-symmetry in our problem. In fact a $\mathbb{Z}_2$-symmetry would be sufficient as the following theorem shows — but the occurring periodic Hamiltonian pitchfork bifurcations are governed by two different $\mathbb{Z}_2$-symmetries.

Theorem 4.1 *Let* $\mathsf{T} := S^1$, Y *be an open interval including 0 and* $\mathsf{S} \subseteq \mathbb{R}^2$ *an open neighbourhood of the origin. Supply* $\mathsf{T} \times \mathsf{Y} \times \mathsf{S}$ *with the symplectic structure* $dx \wedge dy + dq \wedge dp$, *where* $x \in \mathsf{T}$, $y \in \mathsf{Y}$ *and* $(q,p) \in \mathsf{S}$. *Consider a Hamilton function* K *on* $\mathsf{T} \times \mathsf{Y} \times \mathsf{S}$ *that does not depend on* x, *is invariant under the reflection* $q \mapsto -q$ *and has a Taylor series starting with*

$$\omega y + a p^2 + \frac{b}{4}q^4 - c y q^2 \quad .$$

where all four coefficients ω, a, b, c *are positive. Let* $H : \mathsf{T} \times \mathsf{Y} \times \mathsf{S} \longrightarrow \mathbb{R}$ *be* C^4-*close to* K *and invariant under* $q \mapsto -q$. *Then* X_H *undergoes near* $\mathsf{T} \times \{0\} \times \{(0,0)\}$ *a periodic Hamiltonian pitchfork bifurcation.*

Proof Because $a > 0$ we can use a preliminary translation $p \mapsto p + u(x)$ to get rid of the linear term of $\bar{H}$ in p . The linear term of $\bar{H}$ in q is already zero due to the symmetry $q \mapsto -q$, so $\{q = 0, p = 0\}$ is an $X_{\bar{H}}$-invariant submanifold, a one-parameter family of periodic orbits. On account of the frequency condition $\frac{\partial K}{\partial y} > 0$ we can replace the parameter y by the value h of the energy.

Since $a \cdot c > 0$ the periodic orbits are elliptic for $h \ll 0$ and hyperbolic for $h \gg 0$. In between, say at $h = h_0$, there must be a parabolic periodic orbit : here the iso-energetic Poincaré mapping has a Floquet multiplier 1 with algebraic multiplicity two and geometric multiplicity one. With the preliminary translation we already arranged the eigenvector to lie along the q-axis.

As $b \cdot c > 0$ the one-parameter family of periodic orbits tangent to this eigenvector lies in the half space $h \geq h_0$. In other words, on each energy shell $H = h > h_0$ we have two more periodic orbits next to the hyperbolic periodic orbit at the origin. Under the reflection $q \mapsto -q$ these periodic orbits get mapped onto each other, and from $a \cdot b > 0$ we conclude that both are elliptic ;   **q.e.d.**

Note that changing the signs of the coefficients a, b, c may lead to the situation where two hyperbolic periodic orbits split off at the bifurcating periodic orbit. In [Guckenheimer,Mahalov;92] a $\mathbb{Z}_2 \times \mathbb{Z}_2$-symmetric example is given where both bifurcations take place subsequently.

Our aim is to understand the motion of a symmetric rigid body subject to a $\mathbb{Z}_2 \times \mathbb{Z}_2$-symmetric affine force field. We used the S^1-symmetry of the Hamilton function H to reduce the phase space to $\mathbb{R}^3 \backslash \{0\} \times S^2$, and from Section 3 we know the dynamics of the integrable approximation $\bar{H}$ of H . We now extract from the robustness properties of the individual features of the flow to $\bar{H}$ a (partial) description of the flow defined by X_H .

Theorem 4.2 *Let* $H = T + V_\varepsilon$ *be the reduced Hamilton function of a symmetric rigid body, fixed at one point and moving in a generic small conservative* $\mathbb{Z}_2 \times \mathbb{Z}_2$-*symmetric affine force field. The principal moments of inertia* $I_1 = I_2$ *and* I_3 *satisfy* $I_1 \neq I_3$, *and the centre of mass does not coincide with the fixed point. Then the flow on* $\mathbb{R}^3 \backslash \{0\} \times S^2$ *has the following properties.*

In the phase space there are nested 3-dimensional Cantor families of invariant 2-tori. The motion on these 2-tori is quasi-periodic. In compact subsets of $\mathbb{R}^3 \backslash \{0\} \times S^2$ *the measure of their complement goes to zero as* $\varepsilon \to 0$.

Different Cantor families of invariant 2-tori are separated by the stable and unstable manifolds of 2-parameter families of hyperbolic periodic orbits. The latter arise in periodic Hamiltonian pitchfork bifurcations. Two subfamilies of the hyperbolic periodic orbits are connected by heteroclinic orbits. Depending on the affine force field, hyperbolic periodic orbits may also shrink down to one-parameter families of hypoelliptic equilibria.

The Cantor families of invariant 2-tori shrink down to 2-parameter families of elliptic periodic orbits. There are two types of elliptic periodic orbits. The "Eulerian" elliptic periodic orbits have a short period. These families originate from centres and vanish in the periodic Hamiltonian pitchfork bifurcations. The families of "slow" elliptic periodic orbits occur for $\Im$ close to $\pm|\mu|$. All elliptic periodic orbits and the centres are stable in the sense of Lyapunov.

Proof We write the coefficients of the affine force field as $(\beta, a, b, c) = \varepsilon \cdot (\tilde{\beta}, \tilde{a}, \tilde{b}, \tilde{c})$ with $\tilde{\beta}^2 + \tilde{a}^2 + \tilde{b}^2 + \tilde{c}^2 = 1$. From Corollary 1.2 we get

$$\|H \circ \psi_1 - \bar{H}\|_{A(h)} \leq C(h) \cdot \varepsilon^2 \quad ,$$

where $\|..\|_{A(h)}$ denotes the supremum norm on the union $A(h)$ of the energy shells $\{H = h'\}$ with $h' \geq h$ and $C(h)$ is a constant that only depends on h (and not on ε).

To apply the implicit mapping theorem on equilibria and periodic orbits (including the periodic Hamiltonian pitchfork bifurcation) we need estimates $|H \circ \psi_1 - \bar{H}| < \Gamma \cdot \varepsilon$ on some neighbourhood, with a constant Γ independent of ε . Such estimates hold true for $\varepsilon < \frac{\Gamma}{C(h)}$. The persistence of most invariant 2-tori finally follows from [Arnol'd;63]. Indeed, on the energy shell the frequency $\dot{\xi} \approx \frac{|\mu|}{I_1}$ stays bounded, while the frequency of the "slow" motion converges to zero as the invariant 2-tori approach a separatrix, showing that the (analytic) frequency ratio is not a constant function.

The momentum mapping $\Im$ is an integral of motion not only for $\bar{H}$, but also for H . The 2-dimensional invariant tori divide the 3-dimensional invariant submanifolds of constant $\Im$ and H and we obtain Lyapunov-stability ;   **q.e.d.**

From Theorem 4.2 we can reconstruct the motion of the rigid body in $T^*SO(3)$, attaching an S^1 to every point on the reduced phase space $\mathbb{R}^3 \times S^2$. Equilibria thereby turn into periodic orbits, while periodic orbits on $\mathbb{R}^3 \times S^2$ become invariant 2-tori on $T^*SO(3)$, under preservation of the normal behaviour. With the exception of normally elliptic 2-tori that are perturbed from the families at $\Im = \pm|\mu|$, these invariant 2-tori are the surviving Eulerian 2-tori. From invariant 2-tori on $\mathbb{R}^3 \times S^2$ with quasi-periodic motion we get invariant 3-tori that may be resonant, but do not foliate into periodic orbits.

In the next subsection we want to get a similar description of the dynamics when the centre of mass lies off the figure axis. In that case we have to dereduce from $\mathbb{R}^3 \times S^2$ to $T^*SO(3)$ *before* starting the perturbation analysis. Showing that invariant tori survive the step from $\bar{H}$ to H directly on $T^*SO(3)$ yields less information, for instance we will obtain Cantor families rather than smooth 2-parameter families of surviving Eulerian 2-tori.

4.2 Motion of a dynamically symmetric rigid body

We investigate the motion of a rigid body in a $\mathbb{Z}_2 \times \mathbb{Z}_2$-symmetric affine force field with potential $\beta y + ax^2 + by^2 + cz^2$. The rigid body is dynamically symmetric, but in this subsection the centre of mass does not lie on the figure axis. Unlike the previous subsection, the Hamilton function $H = T + V_\varepsilon$ is not invariant under the right S^1-action R^{T^*} .

Using Corollary 1.3 we consider H as a perturbation of the normal form $\bar{H}$, the flow of which is organized by "isolated" periodic orbits, the quasi-periodic Hamiltonian pitchfork bifurcations and the connection bifurcation. While the periodic orbits survive the step from $\bar{H}$ to H due to the implicit mapping theorem, the other organizing centres are not robust in the present situation. Let us discuss this in some more detail.

Although we cannot expect a result as sharp as Theorem 4.1 to hold, with all invariant 2-tori involved in the quasi-periodic Hamiltonian pitchfork bifurcation persisting, it seems reasonable that the normally hyperbolic and the normally elliptic 2-tori survive on 2-dimensional Cantor sets, defined by diophantine conditions on the internal and in the latter case also on the normal frequencies. The normally parabolic 2-tori, where the bifurcation takes place, should persist as a 1-dimensional Cantor family, defined by diophantine conditions of the ratio of the internal frequencies. None of this is proven yet, but in this fashion the quasi-periodic Hamiltonian pitchfork bifurcation may be conjectured to persist in a generic $\mathbb{Z}_2$-equivariant setting.

However, in the situation of our dynamically symmetric rigid body the normally parabolic invariant 2-tori are given by an equation of the form $\frac{\Im}{|\mu|} = const.$ and the value h of the energy. But the frequency ratio of the Eulerian tori is the quotient of $-\frac{I_3 - I_1}{I_1 I_3} \Im$ by $\frac{1}{I_1} |\mu|$, $i.e.$ the same equation $\frac{\Im}{|\mu|} = const.$ also fixes the frequency ratio.

Our situation is therefore degenerate, the internal parameters $|\mu|$ and $\Im$ do not unfold the bifurcation and the frequency ratio independently. Even if the conjecture mentioned above is true we cannot apply it here. We do not know anything about the persistence of this organizing centre.

The third organizing centre of the flow defined by $X_{\bar{H}}$ is the connection bifurcation of the stable and unstable manifolds of normally hyperbolic invariant 2-tori. In general one may expect these separatrices to split, while the 2-tori survive on Cantor sets ; their complement being of small relative measure this would yield many transversal heteroclinic orbits between normally hyperbolic invariant 2-tori. But for the normal form defined by our dynamically symmetric rigid body all normally hyperbolic 2-tori connected by heteroclinic orbits have the same frequency ratio $[\omega_1 : 0] = [1 : 0]$. Hence, we cannot expect these 2-tori to survive a small perturbation, the heteroclinic scenario "falls into a resonance hole".

We understand the dynamics of the integrable approximation $\bar{H}$ of H , but we have seen that we lack information about the robustness of its organizing centres.

However, many of the individual features of $\bar{H}$ persist, and we can give a partial description of the flow defined by X_H .

Theorem 4.3   Let $H = T + V_\varepsilon$ be the Hamilton function of a dynamically symmetric rigid body, fixed at one point and moving in a generic small conservative $\mathbb{Z}_2 \times \mathbb{Z}_2$-symmetric affine force field. The centre of mass has a nonzero component s_3 along the figure axis, and the principal moments of inertia $I_1 = I_2$ and I_3 satisfy $kI_1 \neq lI_3$ for $(k,l) \in \{(2,l)\,|\,l \in \mathbb{N}\} \cup \{(3,l)\,|\,l = 2,\ldots.7\}$. Then the flow on $T^*SO(3)^{\backslash SO(3)}$ has the following properties.

A measure-theoretically large part of the phase space is filled by 3-dimensional Cantor families of invariant 3-tori. In compact subsets of $T^*SO(3)^{\backslash SO(3)}$ the measure of their complement goes to zero as $\varepsilon \to 0$. The motion on these 3-tori is quasi-periodic, with diophantine frequencies. The rigid body motion along these invariant 3-tori is a superposition of a "fast" rotational-precessional motion with a "slow" nutation.

Close to the separatrices of the integrable approximation $X_{\bar{H}}$ there are stable and unstable manifolds of Cantor families of normally hyperbolic invariant 2-tori. The normally hyperbolic 2-tori are Eulerian 2-tori, giving rise to a "fast" motion of the figure axis. Depending on the affine force field, these invariant 2-tori may shrink down to one-parameter families of hypoelliptic periodic orbits.

The $X_{\bar{H}}$-invariant families of 2-tori at $\Im = \pm|\mu|$ give rise to a 2-dimensional Cantor family of X_H-invariant 2-tori. In compact subsets of $\{\Im = \pm|\mu| \neq 0\}$ the 4-dimensional measure of the complement of persisting 2-tori goes to zero as $\varepsilon \to 0$. The rigid body motion along these "non-Eulerian" invariant 2-tori consists of a "fast" rotation superposed by a "slow" movement of the figure axis.

Similarly, the persisting normally elliptic Eulerian 2-tori form 2-dimensional Cantor families. In compact subsets of the union of $X_{\bar{H}}$-invariant normally elliptic Eulerian 2-tori the 4-dimensional measure of the complement of persisting 2-tori goes to zero as $\varepsilon \to 0$. In phase space normally elliptic invariant 2-tori shrink down to one-parameter families of elliptic periodic orbits.

Proof   From Corollary 1.3 we get the inequality

$$\|H \circ \psi - \bar{H}\|_{A(h)} \leq c(h)\,\varepsilon^2 \quad .$$

where $A(h)$ denotes a subset of $T^*SO(3)$ on which the frequency vector $\omega(|\mu|,\Im)$ is bounded away from the "low order resonances". The elliptic and hypoelliptic periodic orbits persist by means of the implicit mapping theorem if $|H \circ \psi_1 - \bar{H}| < \Gamma \varepsilon$ on some neighbourhood, which holds true for $\varepsilon < \frac{\Gamma}{c(h)}$.

For the persistence of "Eulerian" invariant 2-tori we use Corollary 6.2 of .[Broer, Huitema, Takens;90]. In the normally elliptic case one has to show that the ratio $[\omega_1 : \omega_2 : \Omega]$ of the internal frequencies ω_1 and ω_2 and the normal frequency Ω has maximal rank as function of $|\mu|$ and $\Im$, while in the normally hyperbolic case it is sufficient that $\omega = (\omega_1, \omega_2)$ has maximal rank as function of $|\mu|$ and $\Im$. Furthermore a smallness condition $\|H \circ \psi_1 - \bar{H}\|_{A(h)} < \kappa \varepsilon \gamma^2 \delta$ has to be satisfied, where

$\gamma > 0$ is the constant appearing in the diophantine condition and κ is a lower bound of $\frac{1}{\varepsilon}|\Omega|$. The details are completely analogous to [Hanßmann;95], Theorem 7.3. The persistence of most invariant 3-tori follows again from [Arnol'd;63] as $\det D^2 T(|\mu|, \Im) = -\frac{I_3 - I_1}{I_1^2 I_3} \neq 0$ and the third frequency, related to the "slow" motion, has a nonzero derivative with respect to the corresponding action.

For the $X_{\bar{H}}$-invariant 2-tori at $\Im = \pm|\mu|$ we need the estimate

$$\|H \circ \psi_1 \circ \psi_2 - \overline{H \circ \psi_1}\|_{A(h)} \leq \hat{c}(h)\,\varepsilon^3$$

of Corollary 1.4. We do not have to compute the normal form of order two $\overline{H \circ \psi_1}$ as the corresponding flow has the same qualitative description as the flow of $X_{\bar{H}}$, due to the stability of the family $\bar{H}_{|\mu|,\Im}$.

Again one has to show that the diophantine conditions that ensure the persistence of the normally elliptic 2-tori are fulfilled on a 2-dimensional Cantor set and that the (in this case somewhat stronger) smallness condition $|H \circ \psi - \bar{H}| < \kappa \varepsilon^2 v^2 \delta$ is met, confer [Hanßmann;95], Theorem 7.4 for the details ;   q.e.d.

The normally elliptic Eulerian 2-tori in the symmetric case are stable in the sense of Lyapunov, see Theorem 4.2. In the dynamically symmetric case at hand the invariant 3-tori can no longer help to separate different parts of the phase space. However, the time it takes to leave neighbourhoods of persisting normally elliptic Eulerian 2-tori is very long, it increases exponentially with the smallness of the affine force field. The necessary Nehorošev estimates for perturbations of the free rigid body have been obtained in [Fassò;91] and [Benettin,Fassò;95].

Note that we do not claim that the Cantor set that parametrises the normally hyperbolic invariant 2-tori is nonempty. Indeed, if the linear part of the force field is very small compared to the constant part, these tori might all fall into the "resonance hole" around $\Im = 0$. In this situation we have the choice between two different sizes ε of the perturbing force field. For KAM-theory to work (and the conclusions of Theorem 4.3 to hold true) the size ε must already be rather small. Further restriction of ε shrinks the width of the excluded strip around $\Im = 0$, until for a distinctly smaller size ε of the perturbation the quasi-periodic Hamiltonian pitchfork bifurcations of the integrable approximation lie sufficiently far outside this strip. As expected, the normally hyperbolic 2-tori are then parametrised by a 2-dimensional Cantor set.

Conclusions

In Sections 2–4 we were able to identify periodic and quasi-periodic motions of a dynamically symmetric rigid body in a $\mathbf{Z}_2 \times \mathbf{Z}_2$-symmetric affine force field. Along the same lines one can treat perturbations by other ($\mathbf{Z}_2 \times \mathbf{Z}_2$-symmetric) force fields. We can make the reduced dynamics on $S^2_{|\mu|} \times \{\Im\}$ arbitrarily complex, just choosing the appropriate force field. However, if the nonlinear part of the force field is sufficiently small with respect to the affine part, Theorem 4.2 and 4.3 remain true. In this sense we described a robust situation.

In the dissipative context the quasi-periodic period-doubling bifurcation is shown to be robust on Cantor sets in [Braaksma, Broer, Huitema;90]. For $\mathbb{Z}_2$-symmetric Hamiltonian systems a similar approach should yield the persistence of the quasi-periodic Hamiltonian pitchfork bifurcation. In applications one often encounters symmetries, so it would be worthwile to proceed towards an equivariant KAM-theory. On the other hand, every symmetry raises the question : what happens if this symmetry is broken, how does the symmetric system unfold within the space of all systems? For the rigid body system discussed here the answer can be guessed from [Hanßmann;95].

Acknowledgements. I want to thank Henk Broer, Francesco Fassò, Gerton Lunter and Florian Wagener for valuable comments and remarks. The figures were drawn using the software package DYNPAO of Rense Posthumus, for a description see [Posthumus,Scholtmeijer;90]. Finally I would like to acknowledge the support of the Department of Mathematics at the Rijksuniversiteit Groningen where part of this paper was written.

Bibliography

R. Abraham, J.E. Marsden [78] *Foundations of Mechanics.* 2nd ed.; Benjamin (1978)

V.I. Arnol'd [63] Small denominators and problems of stability of motion in classical and celestial mechanics ; *Russ. Math. Surv.* **18**(6), p. 85–191 (1963)

V.I. Arnol'd [74] *Mathematical Methods of Classical Mechanics* ; Springer (1978)

G. Benettin, F. Fasso [95] Fast Rotations of the Rigid Body : A study by Hamiltonian Perturbation Theory. Part I ; preprint, University of Padova

B.J.L. Braaksma, H.W. Broer, G.B. Huitema [90] Toward a quasi-periodic bifurcation theory ; *Mem. AMS* **83** #421, p. 83–167 (1990)

H.W. Broer [81] Formal Normal Form Theorems for Vector Fields and some Consequences for Bifurcations in the Volume Preserving Case ; p. 54–74 in *Dynamical systems and turbulence, Warwick 1980* (eds. D. Rand, L-S. Young) LNM **898**, Springer (1981)

H.W. Broer, G.B. Huitema, M.B. Sevryuk [95] Families of quasi-periodic motions in dynamical systems depending on parameters — *this volume*

H.W. Broer, G.B. Huitema, F. Takens [90] Unfoldings of quasi-periodic tori ; *Mem. AMS* **83** #421, p. 1–82 (1990)

R. Cushman, L. Bates [96] *Classical Integrable Systems* — to appear

L.H. Eliasson [88] Perturbations of stable invariant tori for Hamiltonian systems ; *Ann. Sc. Norm. Sup. Pisa, Cl. Sci., Ser.IV* **15**(1), p. 115–147 (1988)

F. Fassò [91] *Fast rotations of the rigid body and Hamiltonian perturbation theory* ; Ph.D. thesis, SISSA, Trieste (1991)

F. Fassò [95] Hamiltonian perturbation theory on a manifold ; *Celest. Mech. Dyn. Astr.* **62**, p. 43–69 (1995)

H. Goldstein [50] *Classical Mechanics* ; Addison-Wesley (1950)

S.M. Graff [74] On the Conservation of Hyperbolic Invariant Tori for Hamiltonian Systems ; *J. Diff. Eq.* **15**, p. 1–69 (1974)

J. Guckenheimer, Mahalov [92] Instability Induced by Symmetry Reduction ; *Phys. Rev. Lett.* **68**(15), p. 2257–2260 (1992)

H. Hanßmann [93] Normal Forms for Perturbations of the Euler Top ; preprint W9309, Rijksuniversiteit Groningen — to appear in *Proceedings of the Workshop on Normal Forms and Homoclinic Chaos, Waterloo 1992* (eds. W.F. Langford, W. Nagata) Fields Institute Communications

H. Hanßmann [95] *Quasi-periodic Motions of a Rigid Body — A case study on perturbations of superintegrable systems* ; Ph.D. thesis, Rijksuniversiteit Groningen (1995)

M.V. Karasev, V.P. Maslov [93] *Nonlinear Poisson Brackets — Geometry and Quantization* ; Translations of the AMS **119** (1993)

A.S. Mishchenko, A.T. Fomenko [78] Generalized Liouville method of integration of Hamiltonian systems ; *Funct. Anal. Appl.* **12**(2), p. 113–121 (1978)

J. Moser [67] Convergent series expansion for quasi-periodic motions : *Math. Ann.* **169**, p. 136–176 (1967)

N.N. Nehorošev [72] Action-Angle Variables and their generalizations ; *Trans. Moscow Math. Soc.* **26**, p. 180–198 (1972)

J. Pöschel [82] Integrability of Hamiltonian Systems on Cantor Sets ; *Comm. Pure Appl. Math.* **35**, p. 653–696 (1982)

J. Pöschel [89] On Elliptic Lower Dimensional Tori in Hamiltonian Systems ; *Math. Z.* **202**, p. 559–608 (1989)

R.A. Posthumus, J. Scholtmeijer [90] *Chaos and Fractals. Practicumhandleiding bij* DYNPAO ; Rijksuniversiteit Groningen (1990)

E. Zehnder [75] Generalized Implicit Function Theorems with Applications to Some Small Divisor Problems I ; *Comm. Pure Appl. Math.* **28**, p. 91–140 (1975)

E. Zehnder [76] Generalized Implicit Function Theorems with Applications to Some Small Divisor Problems II ; *Comm. Pure Appl. Math.* **29**, p. 49–111 (1976)

Progress in Nonlinear Differential Equations
and Their Applications, Vol. 19
© 1996 Birkhäuser Verlag Basel/Switzerland

On stability loss delay for a periodic trajectory

A. I. Neishtadt, C. Simó and D. V. Treschev

1 Introduction

Stability loss delay is an interesting, important and so far not completely clear phenomenon. Its essence is as follows. Consider a system of differential equations depending on a slowly varying parameter. Suppose that the system has an equilibrium position or a periodic trajectory for any fixed value of the parameter. Suppose also that the parameter passes through a bifurcational value: the equilibrium (periodic trajectory) loses stability but remains nondegenerate. In the case of an equilibrium a pair of conjugate eigenvalues leaves the left half-plane not passing through zero. For a periodic trajectory either a pair of conjugate multipliers leaves the unit circle not passing through the point 1, or one real multiplier goes away from the unit circle through the point -1. If the system is analytic, a delay of stability loss takes place: phase points attracted to the equilibrium (periodic trajectory) long before the moment of the bifurcation remain close to the unstable equilibrium (periodic trajectory) until the change of the parameter is of order one. The velocity of the parameter changing can be arbitrary small. In non-analytic systems (even in the C^∞ case) in general there is no such a delay of stability loss.

In the present paper stability loss delay for periodic trajectories is considered. We obtain a lower estimate for the time of the delay. Under some assumptions this estimate gives also the asymptotics of the escape time. We illustrate analytic results by results of numerical experiments.

The phenomenon of stability loss delay for equilibria was discovered in [1] for one model system of ODE. In [2, 3, 4] it was shown that this phenomenon is unavoidable in analytic systems of differential equations and in analytic maps when a parameter passes slowly through a bifurcational value as was described above. For the case of equilibrium an asymptotic formula for the moment of departure from an unstable equilibrium is known under rather general conditions [4, 5]. Some aspects of stability loss delay for periodic trajectories are considered in [6]. There are examples of stability loss delay in some problems of laser physics, biophysics and chemical kinetics (references can be found in [7]).

2 Basic equations

We consider systems of the following form:

$$\dot{x} = f(x, \tau, \varepsilon), \quad \dot{\tau} = \varepsilon. \tag{2.1}$$

Here ε is a small nonnegative parameter, and x belongs to an open subset of $\mathbb{R}^{n+1}$. It is useful to consider the system (2.1) for $\varepsilon = 0$ separately. It is as follows:

$$\dot{x} = f(x, \tau, 0), \quad \tau = \text{const.} \tag{2.2}$$

Suppose that for any fixed $\tau \in [\tau_1, \tau_2]$ the system (2.2) has a periodic solution continuously depending on τ. Let L_τ be the trajectory of this solution and let $T(\tau)$, $\omega(\tau) = 2\pi/T$, $\rho_1(\tau), \ldots, \rho_n(\tau)$ be period, frequency and multipliers of L_τ respectively. Later we consider L_τ as sets in the space of the variables x, τ as well as in the space of the variables x.

We assume that for $\tau < \tau_* \in [\tau_1, \tau_2]$ the periodic solution L_τ is linearly asymptotically stable (all the multipliers are inside the unit circle) and it loses stability at $\tau = \tau_*$. Either a pair of complex conjugate multipliers ρ_1 and $\rho_2 = \bar{\rho}_1$ leaves the unit circle and $\rho_{1,2}(\tau_*) \neq 1$, or one real multiplier ρ_1 leaves the unit circle through the point -1. The other multipliers remain inside the unit circle.

If the right-hand sides of the equations (2.1) can be continued analytically in x, τ and smoothly in ε to a complex neighborhood N of the periodic trajectory L_{τ_*} then the stability loss delay occurs provided the neighborhood N can be chosen independent of ε [3].

3 Basic assumptions

Later on we assume that the periodic solutions L_τ exist for $\tau \in U$. where $U \subset \mathbb{C}$ is an open neighborhood of the point τ_*. Moreover, we need the following assumptions to hold:

C1. *In the domain U all the multipliers $\rho_l(\tau)$ are analytic in τ and distinct from one. The frequency ω is analytic and does not vanish in U.*
C2. *In the domain U the multipliers $\rho_l(\tau)$ are pairwise distinct.*

a. Escape of a pair of conjugate multipliers from the unit circle

Suppose that a pair of complex conjugate multipliers ρ_1 and $\rho_2 = \bar{\rho}_1$ leaves the unit circle at $\tau = \tau_*$ and $\rho_{1,2}(\tau_*) \neq 1$. In stationary bifurcation theory such a loss of stability of a periodic solution is called a Poincaré-Andronov-Hopf bifurcation.

Without loss of generality we may assume that

$$\operatorname{Im} \rho_1(\tau_*) = -\operatorname{Im} \rho_2(\tau_*) < 0.$$

We put

$$\lambda_l^0(\tau) = \frac{\log \rho_l(\tau)}{T(\tau)}, \quad \operatorname{Im} \lambda_l^0(\tau_*) T(\tau_*) \in (-\pi, \pi),$$
$$\lambda_l^k(\tau) = \lambda_l^0(\tau) - ik\omega(\tau), \quad k \in \mathbb{Z}. \tag{3.1}$$

Obviously, for any $k \in \mathbb{Z}$ the equality $\rho_l = \exp(T\lambda_l^k)$ holds. We have the following relations:

$$\lambda_1^k(\tau) = \bar{\lambda}_2^{-k}(\bar{\tau}), \quad \tau \in U, \quad k \in \mathbb{Z}, \tag{3.2}$$
$$\operatorname{Im} \lambda_1^0(\tau_*) T(\tau_*) \in (-\pi, 0), \quad \operatorname{Re} \lambda_1^k(\tau_*) = \operatorname{Re} \lambda_2^k(\tau_*) = 0. \tag{3.3}$$

Below an important role is played by the functions

$$\psi_l^k(\tau) = \int_{\tau_*}^{\tau} \lambda_l^k(s) \, ds, \quad \phi_l^k(\tau) = \operatorname{Re} \psi_l^k(\tau), \quad \phi^{\pm}(\tau) = \mp \operatorname{Re} \int_{\tau_*}^{\tau} i\omega(s) \, ds, \quad \tau \in U. \tag{3.4}$$

Obviously, $\phi^{\pm}(\tau) = 0$ for real τ. Later on we need the following assumption

C3. *The equation $\phi^{\pm}(\tau) = 0$, $\tau \in U$ has only real solutions.*

Let $\gamma_l^k(\sigma)$ be level lines of the functions $\phi_l^k(\tau)$: $\gamma_l^k(\sigma) = \{\tau \in \mathbb{C} : \phi_l^k(\tau) = \sigma\}$. From (3.3) it follows that the functions ϕ_1^k and ϕ_2^k, restricted to the real axis, have a critical point $\tau = \tau_*$. Consequently, the curves $\gamma_l^k(0)$. $l = 1, 2$ are tangent to the real axis at the point τ_*. Since the functions $\operatorname{Re} \lambda_l^k(\tau)$. $l = 1, 2$, $\tau \in \mathbb{R}$ change sign from $-$ to $+$ when τ passes through the point τ_*, the point τ_* is a local minimum of the functions ϕ_1^k and ϕ_2^k for real τ.

The relations (3.1)–(3.2) imply that for fixed $\sigma > 0$ the curves $\gamma_l^k(\sigma)$, $l = 1, 2$ intersect at the real points $\tau_{\pm}(\sigma)$, where

$$\tau_-(\sigma) < \tau_* < \tau_+(\sigma), \quad \int_{\tau_*}^{\tau_{\pm}(\sigma)} \operatorname{Re} \lambda_l^k(s) \, ds = \sigma, \quad s \in \mathbb{R}. \tag{3.5}$$

Note that due to (3.2) the curves $\gamma_1^k(\sigma)$ and $\gamma_2^{-k}(\sigma)$ are symmetric in the real axis.

Now consider the family $\gamma_1^k(\sigma)$, $\sigma > 0$ in more detail. Let $\Gamma^k(\sigma)$, $\sigma \geq 0$ be continuous in σ families of segments of the curves $\gamma^k(\sigma)$ satisfying the following conditions:

1. endpoints of $\Gamma^k(\sigma)$ are $\tau_{\pm}(\sigma)$,

2. the segments $\Gamma^k(0)$ degenerate to the point τ_*,

3. $\Gamma^k(\sigma) \subset U$.

Proposition 3.1. *The families $\Gamma^k(\sigma)$ are defined on an interval $[0, \sigma_0]$, $\sigma_0 > 0$.*

The geometry of the curves $\Gamma^k(\sigma)$ is described by the following

Proposition 3.2. *Suppose that the equation* $|\rho_1(\tau)| = 1$ *has on the line segment* $[\tau_-(\sigma_0), \tau_+(\sigma_0)] \subset \mathbb{R}$ *the unique solution* $\tau = \tau_*$. *Let the families* $\Gamma^0(\sigma)$ *and* $\Gamma^{-1}(\sigma)$ *be defined for* $0 \le \sigma \le \sigma_0$. *Then the following assertions hold.*
(a) *The curves* $\Gamma^0(\sigma)$ *(respect-break ively,* $\Gamma^{-1}(\sigma)$*),* $0 \le \sigma \le \sigma_0$ *lie in the upper (respectively, in the lower) half-plane.*
(b) *The families* $\Gamma^k(\sigma)$ *are also defined for* $0 \le \sigma \le \sigma_0$.
(c) *All the curves* $\Gamma^k(\sigma)$, $0 \le \sigma \le \sigma_0$ *lie in the closed domain* $\hat{D} \subset U$ *bounded by* $\Gamma^{-1}(\sigma_0)$ *and* $\Gamma^0(\sigma_0)$.

Let us fix σ_0 satisfying the conditions of Proposition 3.2. We denote by D^{-1} (respectively, D^0) the domain bounded on the complex plane $\mathbb{C}_\tau$ by the curve $\Gamma^{-1}(\sigma_0)$ (respectively, $\Gamma^0(\sigma_0)$) and the line segment $[\tau_-(\sigma_0), \tau_+(\sigma_0)]$. Let $D_s^{-1}, D_s^0 \subset \mathbb{C}_\tau$ be the domains symmetric in the real axis to D^{-1} and D^0 respectively. Our basic assumptions are as follows:

A1. *Either* $D_s^{-1} \subset D^0$ *or* $D_s^0 \subset D^{-1}$.

We put $D = D^0 \cup D_s^0$ in the first case and $D = D^{-1} \cup D_s^{-1}$ in the second one. Without loss of generality we may assume that the first case takes place.

A2. *For any* $3 \le l \le n$ *and* $\tau \in D \cap \mathbb{R}$ *the inequalities* $|\rho_l(\tau)| < 1$ *hold.*
A3. *Each of the functions* $\phi_l^k(\tau), \phi^\pm(\tau)$ *(except* $\phi_1^0(\tau)$ *and* $\phi_2^0(\tau)$*) restricted to the boundary* ∂D *of the domain* D *has a unique maximum and maxima of the functions* $\phi^\pm|_{\partial D}$ *are nondegenerate.*

Remark Since $\phi^\pm(\tau)$ vanish for real τ, the curve ∂D is smooth at the points of maximum of the functions $\phi^\pm|_{\partial D}$.

Proposition 3.3. *If the multipliers* $\rho_1(\tau), \rho_2(\tau)$, $\tau \in \mathbb{R}$ *cross the unit circle at* $\tau = \tau_*$ *with nonzero velocity then conditions A1–A3 are valid for small* $\sigma_0 > 0$.

b. Escape of a real multiplier from the unit circle through the point -1

In this subsection we assume that stability loss of the periodic solution L_τ occurs as a result of escape of the real multiplier ρ_1 from the unit circle through -1. This case turns out to be very similar to that considered in the previous subsection and all constructions are almost the same. Here we again assume that the conditions $C1$–$C2$ hold.

First, we introduce the characteristic exponents λ_l^k by using relations (3.1), where the condition $\operatorname{Im} \lambda_1^0(\tau_*) T(\tau_*) \in (-\pi, \pi)$ is replaced by the following one: $\lambda_1^0(\tau_*) T(\tau_*) = -i\pi$.

Second, we introduce the functions $\psi_l^k(\tau), \phi_l^k(\tau), \phi^\pm(\tau)$ and the curves $\gamma_l^k(\sigma)$ in the same way as in subsection **a**. We assume that the functions $\phi^\pm(\tau)$ satisfy condition $C3$. The point τ_* is a local minimum of the functions ϕ_1^k for real τ and the curves $\gamma_1^k(0)$ are tangent to the real axis at the point τ_*. For fixed $\sigma > 0$ the curves $\gamma_1^k(\sigma)$ intersect at the real points $\tau_\pm(\sigma)$ satisfying (3.5) for $l = 1$.

Now we can consider families of the segments $\Gamma^k(\sigma)$, $\sigma \geq 0$ defined by conditions (1)–(3). Propositions 3.1–3.2 obviously, remain valid. Let σ_0 satisfy conditions of Proposition 3.2 and let $D \subset \mathbb{C}_\tau$ be the domain bounded by the curves $\Gamma^0(\sigma_0)$ and $\Gamma^{-1}(\sigma_0)$. We assume that the following conditions are valid:

B1. For any $2 \leq l \leq n$ and $\tau \in D$ the inequalities $|\rho_l(\tau)| < 1$ hold.

B2. Each of the functions $\phi_l^k(\tau), \phi^\pm(\tau)$ (except $\phi_1^0(\tau)$ and $\phi_1^{-1}(\tau)$) restricted to the boundary ∂D of the domain D has a unique maximum and maxima of the functions $\phi^\pm|_{\partial D}$ are nondegenerate.

We have the following analog of Proposition 3.3:

Proposition 3.4. If the multiplier $\rho_1(\tau)$, $\tau \in \mathbb{R}$ crosses the unit circle at $\tau = \tau_*$ with nonzero velocity then conditions B1–B2 hold for small $\sigma_0 > 0$.

4 Estimates for the time of delay of stability loss

In this section we present theorems estimating the time of delay of stability loss. In two cases we deal with the following: when two complex conjugate multipliers leave the unit circle and when one real multiplier does the same. We call these two cases **A** and **B** respectively.

We define the functions

$$\Pi\,[\tau_-(\sigma_0), \tau_*] \to [\tau_*, \tau_+(\sigma_0)], \quad \hat{\Pi}\,[\tau_-(\sigma_0), \tau_*] \to [\tau_*, \tau_+(\sigma_0)]$$

by the following relations

$$\int_{\tau_*}^{\tau} \operatorname{Re} \lambda_1^0(s)\,ds = \int_{\tau_*}^{\Pi(\tau)} \operatorname{Re} \lambda_1^0(s)\,ds, \quad \int_{\tau_*}^{\tau} \mu(s)\,ds = \int_{\tau_*}^{\hat{\Pi}(\tau)} \operatorname{Re} \lambda_1^0(s)\,ds,$$
$$\mu(\tau) = \max_{1 \leq j \leq n} \operatorname{Re} \lambda_j^0(\tau).$$

The functions $\Pi(\tau)$ and $\hat{\Pi}(\tau)$ evidently coincide for τ close to τ_*. Consider a solution $x(t)$ of the system (2.1) with initial conditions at $t = t_0 = \tau_0/\varepsilon$, $\tau_0 \in [\tau_-(\sigma_0), \tau_*)$.

Theorem 4.1. Suppose that the point $x(t_0)$ is inside a c_1^{-1}-neighborhood of the periodic trajectory L_{τ_0}. Then for $\tau_0 + c_2\varepsilon|\log \varepsilon| \leq \varepsilon t \leq \hat{\Pi}(\tau_0) - c_3\varepsilon|\log \varepsilon|$ the solution $x(t)$ is inside a $c_4\varepsilon$-neighborhood of $L_{\varepsilon t}$.

Here and below c_j are positive constants. Appearance of a constant c_j in some assertion means that there exists a constant c_j satisfying this assertion.

According to Theorem 4.1 for any solution attracted to the family L_τ at $\tau \approx \tau_0$, the loss of stability delays at least up to the moment $\tau \approx \hat{\Pi}(\tau_0)$.

To calculate an asymptotics of the escape moment we need additional assumptions and notations. Let the following condition hold:

A4. $\operatorname{Re} \lambda_j^0(\tau) < \operatorname{Re} \lambda_1^0(\tau)$ for all $j = 3, \ldots, n$ and $\tau \in [\tau_-(\sigma_0), \tau_*]$ for the case **A** and

B3. $\operatorname{Re} \lambda_j^0(\tau) < \operatorname{Re} \lambda_1^0(\tau)$ for all $j = 2, \ldots, n$ and $\tau \in [\tau_-(\sigma_0), \tau_*]$ for the case **B**.

Obviously, $A4$ implies the identity $\hat{\Pi}(\tau) \equiv \Pi(\tau)$ in the case **A** and $B3$ implies the same identity in the case **B**.

If the point $x(t_0)$ is close to the trajectory L_{τ_0} then there exists a unique point $x_{00} \in L_{\tau_0}$ which is the closest to $x(t_0)$. Consider a hyperplane $\Theta_0 \subset \mathbb{R}^{n+1}$ containing the point x_{00} and transversal to L_{τ_0} at x_{00}. In a sufficiently small neighborhood of the point x_{00} on the hyperplane Θ_0 the first return map of the system (2.2) with $\tau = \tau_0$ can be defined. The multipliers $\rho_1(\tau_0), \ldots, \rho_n(\tau_0)$ are eigenvalues of the linearized at x_{00} first return map. Let Θ' and Θ'' be eigenspaces of the linearized first return map, which correspond to the multipliers ρ_1, ρ_2 and $\rho_3, \ldots, \rho_n$ respectively in case **A**. In case **B** they correspond to ρ_1 and $\rho_2, \ldots, \rho_n$. We denote by β' and β'' projections of the vector $x(t_0) - x_{00}$ to Θ' and Θ''. If $n = 2$ we put $\beta'' = 0$.

Theorem 4.2. *Suppose that the point $x(t_0)$ is inside a c_5^{-1}-neighborhood of the periodic trajectory L_{τ_0}. Let $|\beta'| > c_6 \varepsilon$ and $|\beta''| < c_7^{-1} \varepsilon$. Then*

(1) for $\tau_0 + c_2 \varepsilon |\log \varepsilon| \leq \varepsilon t \leq \Pi(\tau_0) - c_3 \varepsilon |\log \varepsilon|$ the solution $x(t)$ is inside a $c_4 \varepsilon$-neighborhood of $L_{\varepsilon t}$,

(2) the point $x(t_d)$ is outside a c_8^{-1}-neighborhood of the trajectory $L_{\varepsilon t_d}$ for some t_d, where $|\varepsilon t_d - \Pi(\tau_0)| < c_9 \varepsilon |\log \varepsilon|$.

Assertion (1) of Theorem 4.2 is just a repetition of Theorem 4.1. According to assertion (2), a solution attracted to the family L_τ at $\tau \approx \tau_0$ escapes from L_τ at $\tau \approx \Pi(\tau_0)$.

Introducing assumptions additional to $A1$–$A3$ or to $B1$–$B2$ one can get lower estimates for the delay time, which are better than the ones in Theorem 4.1 (see e.g. assumptions $A4, B3$ and Theorem 4.2). Below we formulate conditions $A5$ and $B4$, which as a rule are less restrictive compared with $A4$ and $B3$. Under condition $A5, B4$ the delay time turns out to be greater than in Theorem 4.1 but less than in Theorem 4.2 (see Theorem 4.3). We denote

$$\nu_j^k(\tau) = \lambda_j^k(\tau) - \lambda_1^0(\tau), \quad \kappa_j^k(\tau) = \operatorname{Re} \int_{\tau_*}^{\tau} \nu_j^k(s) \, ds. \qquad k \in \mathbb{Z}.$$

Here $j = 3, \ldots, n$ in case **A** and $j = 2, \ldots, n$ in case **B**. Let $\delta_j^k(\sigma) \subset \mathbb{C}_\tau$ be level lines of the functions $\kappa_j^k(\tau)$. Assume that the following condition holds.

$A5, B4$. *There exists a domain $\tilde{D} \subset D$ such that*
(a) the set $\tilde{D} \cap \mathbb{R} \subset \mathbb{C}$ coincides with the line segment $[\tau_-(\sigma_0), \tau_+(\sigma_0)]$,
(b) any curve $\delta_j^k(\sigma)$ has at most two points in common with the boundary of the domain $\tilde{D}$.

We define the function $\tilde{\Pi} [\tau_-(\sigma_0), \tau_*] \to [\tau_*, \tau_+(\sigma_0)]$ by the equality

$$\int_{\tau_*}^{\tau} \tilde{\mu}(s)\, ds = \int_{\tau_*}^{\tilde{\Pi}(\tau)} \operatorname{Re} \lambda_1^0(s)\, ds,$$
$$\tilde{\mu}(\tau) = \begin{cases} \max\{\operatorname{Re} \lambda_1^0(\tau), 2\operatorname{Re} \lambda_3^0(\tau), \ldots, 2\operatorname{Re} \lambda_n^0(\tau)\} & \text{in case } \mathbf{A}, \\ \max\{\operatorname{Re} \lambda_1^0(\tau), 2\operatorname{Re} \lambda_2^0(\tau), \ldots, 2\operatorname{Re} \lambda_n^0(\tau)\} & \text{in case } \mathbf{B}. \end{cases}$$

Obviously, $\hat{\Pi}(\tau) \leq \tilde{\Pi}(\tau) \leq \Pi(\tau)$. Under conditions $A5, B4$ we can improve the estimate of Theorem 4.1 in the following way.

Theorem 4.3. *Suppose that the point $x(t_0)$ is inside a c_1^{-1}-neighborhood of the trajectory L_{τ_0}. Then for $\tau_0 + c_2\varepsilon|\log\varepsilon| \leq \varepsilon t \leq \tilde{\Pi}(\tau_0) - c_3\varepsilon|\log\varepsilon|$ the solution $x(t)$ is inside a $c_4\varepsilon$-neighborhood of $L_{\varepsilon t}$.*

5 Discussion of results and conditions. Examples

When $\varepsilon \to 0$ the attraction of a solution to the family L_τ and the escape take place instantly in the slow time. The corresponding limit values of the slow time are called the moments of fall and escape. Dependence of the escape moment on the moment of fall, if it exists for a majority of initial conditions with the given moment of fall, determines an in-out function. In this definition one should take the initial conditions from a small $O(1)$-neighborhood of the periodic trajectory L_{τ_0}, where τ_0 is the moment of fall.

According to Theorem 4.2, under certain assumptions the in-out function coincides with the function Π introduced in Section 4. Theorems 4.1 and 4.3 give lower estimates for the moment of escape. Apparently, the estimate given in Theorem 4.1 is not sharp. (This estimate would be sharp if we did not assume that condition C2 holds, see Example 5.2, but without this condition the estimate has not been proved.) The estimate given by Theorem 4.3 is sharp (see Example 5.3).

Theorems 4.1 and 4.2 would be rather simple assertions if the surface $M = \bigcup_{\tau_1 \leq \tau \leq \tau_2} L_\tau$, formed by the periodic solutions L_τ, were invariant with respect to the phase flow of the system (2.1). This case can be called the case of a "trivial" delay of stability loss. In this case the solution falls by the moment $\tau = \tau_*$ to an exponentially narrow neighborhood of the surface M. Hence, the solution needs a long time ($\sim 1/\varepsilon$) to go aside from the surface M. Theorems 4.1 and 4.2 give an estimate of this time and asymptotics of the moment of escape.

As a rule the surface M is not invariant. In proofs of Theorems 4.1-4.2 a surface M_ε is constructed which is invariant with respect to the phase flow of the system (2.1) and $O(\varepsilon)$-close to the surface M for $\tau \in I$, where $I \subset \mathbb{R}$ is an interval containing τ_*. This interval is determined by the behaviour of the system for complex values of τ. When the surface M_ε is constructed the original problem is reduced to the problem of analysis of a "trivial" stability loss delay.

Remarks **1.** In general the surface M is not analytic in ε. The formal series in ε of the perturbation theory for M_ε diverges at $\tau = \tau_*$.

2. In the paper [3] a formal series of the perturbation theory for M_ε is cut at terms of order $\varepsilon^{\mathrm{const}/\varepsilon}$. By this method the existence of a surface invariant with a precision $\sim \exp(-c_1^{-1}/\varepsilon)$ has been proved but the constant c_1 has not been controlled. The existence of an invariant surface M_ε can be easily derived from this. But for the interval $I = [a, b]$ only the estimate $a < \tau_* - c_2^{-1}, b > \tau_* + c_3^{-1}$ with constants c_2, c_3 we can not control is obtained.

To prove Theorem 4.3 an additional analysis is needed after constructing the surface M_ε and reducing the problem to the case of "trivial" delay. We introduce for the system (2.1) new coordinates such that the matrix of the system linearized near M_ε has a nill corner: the equations corresponding to the multipliers $\rho_{1,2}$ separate (recall that the multipliers $\rho_{1,2}$ are responsible for the stability loss). To construct these coordinates it is necessary again to deal with complex values of τ. The idea is similar to that of constructing the surface M_ε. Estimates in the new variables lead to Theorem 4.3.

Below we present examples illustrating the phenomenon of stability loss delay and behavior of in-out function.

Example 5.1. *(Multiplier One)*. Consider the following linear non-homogeneous system written in a complex form:

$$\dot{z} = (\tau - \alpha i)z + \varepsilon h(\vartheta), \quad \dot{\vartheta} = 1, \quad \dot{\tau} = \varepsilon, \quad z = x_1 + ix_2.$$
$$h = \sum_{k=-\infty}^{\infty} h^k e^{ik\vartheta}, \quad h^k = e^{-|k|}, \quad 0 < \alpha < 1/2. \tag{5.1}$$

The variable ϑ is defined on the circle $\mathbb{R}/(2\pi\mathbb{Z})$. For $\varepsilon = 0$ the system has the periodic solution $z = 0$, $\dot{\vartheta} = 1$, $\tau = \mathrm{const}$. Its multipliers are $\rho_{1,2} = e^{2\pi(\tau \mp \alpha i)}$. The periodic solution is stable for $\tau < 0$ and unstable for $\tau > 0$. At $\tau = \tau_* = 0$ there are two conjugate multipliers on the unit circle. The eigenvalues introduced in Section 3 are $\lambda_{1,2}^k = \tau \mp i\alpha - ik$. The functions $\psi_{1,2}^k$ and $\phi_{1,2}^k$ are as follows:

$$\psi_{1,2}^k = (\tau \mp i\alpha - ik)^2/2 - (\mp i\alpha - ik)^2/2, \quad \phi_{1,2}^k = \operatorname{Re}\psi_{1,2}^k.$$

The eigenvalue λ_1^0 vanishes (and the multiplier ρ_1 becomes equal to one) for $\tau = i\alpha$. The level lines $\phi_1^0 = \mathrm{const}$ passing through the point $i\alpha$ are two straight lines intersecting the real axis at the points $-\alpha$ and α. The other level lines of ϕ_1^0 are hyperbolas (see Fig. 1).

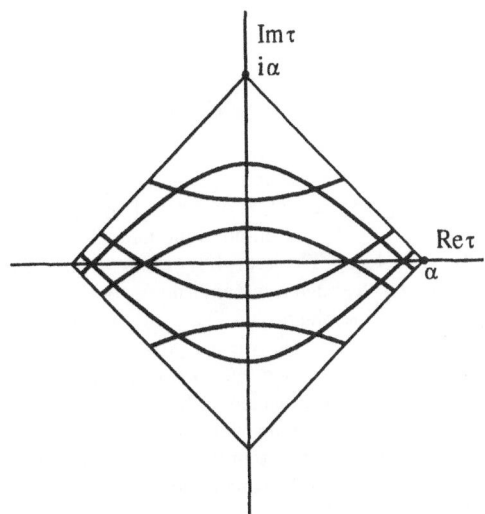

Figure 1

Level lines of the function ϕ_1^k can be obtained from level lines of ϕ_1^0 by the shift on k along the imaginary axis. Level lines of ϕ_2^k are symmetric to those of ϕ_1^{-k} in the real axis. Let D_* be the square bounded by the level (straight) lines of $\phi_{1,2}^0 = \text{const}$ passing through the points $\pm i$. We can choose as a domain D (see Section 3) any sub-domain of D_* bounded by the hyperbolas $\phi_1^0 = \phi_1^0(\sigma)$ and $\phi_2^0 = \phi_2^0(\sigma)$ which intersect the real axis at the points $\pm\sigma$, $0 < \sigma < \alpha$.

To investigate stability loss delay in the system (5.1) we construct an integral surface M_ε of the form

$$z = \hat{z}(\tau, \vartheta, \varepsilon) = \sum_{k=-\infty}^{\infty} z^k(\tau, \varepsilon)e^{ik\vartheta}. \tag{5.2}$$

The coefficients z^k satisfy the differential equations

$$\dot{z}^k = (\tau - \alpha i - ik)z^k + \varepsilon h^k. \tag{5.3}$$

Hence, we reduce the problem of stability loss of a periodic solution to the problem of stability loss of an equilibrium.

The behavior of solutions of equation (5.3) is well-known (see e.g. [4]). In particular let us choose initial conditions at $\tau = \tau_0 = -\alpha - 1/2$ in the following way: $z^k(\tau_0, \varepsilon) = h^k$. Then for $k \neq 0$ we have $|z^k| < c_1\varepsilon h^k$ for $\tau_0 + c_2\varepsilon|\log\varepsilon| \leq \tau \leq -\tau_0 - c_2\varepsilon|\log\varepsilon|$. The function z^0 satisfies the estimates

$$|z^0| \leq c_1\varepsilon \quad \text{for} \quad \tau_0 + c_2\varepsilon|\log\varepsilon| \leq \tau \leq \alpha - c_2\varepsilon|\log\varepsilon|,$$
$$|z^0| > 1 \quad \text{for} \quad \tau \geq \alpha + c_2\varepsilon|\log\varepsilon|.$$

Hence for the surface (5.2) we have:

$$|\hat{z}| = O(\varepsilon) \quad \text{for} \quad \tau_0 + c_2\varepsilon|\log\varepsilon| \le \tau \le \alpha - c_2\varepsilon|\log\varepsilon|.$$
$$|\hat{z}| > 1/2 \quad \text{for} \quad \tau = \alpha + c_2\varepsilon|\log\varepsilon|.$$

The magnitude $\xi = z - \hat{z}$ satisfies the homogeneous equation $\dot{\xi} = (\tau - \alpha i)\xi$.

Now it is easy to construct the in-out function for the system (5.1). Phase points attracted to the family L_τ at $\tau \approx \tau_- \in (-\alpha, 0)$ escape at $\tau \approx \tau_+ = -\tau_-$. Therefore, for $\tau \in (-\alpha, 0)$ the in-out function coincides with the function Π introduced in Section 4 (cf. Theorem 4.2). Phase points attracted to L_τ at $\tau \approx \tau_- < -\alpha$ escape at $\tau \approx \alpha$. Such a metamorphosis of the in-out function at the point $\tau = -\alpha$ happens because the Stokes line passes through this point. Here we mean by the Stokes line the curve $\phi_1^0 = \text{const}$ going through the point αi, where the multiplier ρ_1 becomes equal to one (cf. assumption C1).

Remark In the example we have considered above a periodic solution L_τ and its multipliers that are regular at the point where a multiplier becomes equal to one. This case is degenerate. In a typical system the periodic solution and its multipliers branch at the point where a multiplier one appears.

Example 5.2. (*Double multiplier*). Consider the system

$$\begin{aligned}
\dot{z} &= (\tau - \alpha i)z + \varepsilon w + \varepsilon h(\vartheta), & z &= x_1 + ix_2, \\
\dot{w} &= (-\mu - \beta i)w, & w &= x_3 + ix_4. \\
\dot{\vartheta} &= 1, & \dot{\tau} &= \varepsilon.
\end{aligned} \tag{5.4}$$

Here h and α are the same as in (5.1), $0 < \beta < 1$ and $0 < \mu < \alpha$. Multipliers in the system (5.4) are $\rho_{1,2} = e^{2\pi(\tau \mp \alpha i)}$ and $\rho_{3,4} = e^{2\pi(-\mu \mp \beta i)}$. The system (5.4) has a double multiplier at $\tau = -\mu \pm (\alpha - \beta)i + ik$ and $\tau = -\mu \pm (\alpha + \beta)i + ik$, $k \in \mathbb{Z}$. In the example in question only the pair with $k = 0$ is essential. If the condition

$$\mu + |\alpha - \beta| < \alpha \tag{5.5}$$

holds, the double multiplier is inside the square D_*, see Fig 1. In this example the in-out function F can be calculated by using an explicit solution of (5.4).

Proposition 5.1. (a). *Suppose that the condition* (5.5) *is valid. Then*

$$F(\tau) = \begin{cases} -\tau, & -\mu - |\alpha - \beta| \le \tau < 0, \\ \sqrt{(\alpha - \beta)^2 - \mu^2 - 2\tau\mu}, & -\dfrac{\alpha^2 + \mu^2 - (\alpha - \beta)^2}{2\mu} \le \tau < -\mu - |\alpha - \beta|, \\ \alpha, & \tau < -\dfrac{\alpha^2 + \mu^2 - (\alpha - \beta)^2}{2\mu}. \end{cases}$$

(b). *If the condition* (5.5) *does not hold then*

$$F(\tau) = \begin{cases} -\tau, & -\alpha \le \tau < 0, \\ \alpha & \tau < -\alpha. \end{cases}$$

Let condition (5.5) be satisfied. Theorem 4.1 gives us an estimate from below for the in-out function on the interval $(-\sqrt{\mu^2 + 2\alpha|\alpha - \beta| - (\alpha - \beta)^2}, 0)$ only. (The left endpoint of this interval is connected with the point τ where two multipliers coincide by a level line of the function ϕ_1^0.) Condition A2 does not allow us to enlarge the interval. The functions Π and $\hat{\Pi}$, introduced in the Section 4, are as follows:

$$\Pi(\tau) = -\tau, \qquad \hat{\Pi}(\tau) = \begin{cases} -\tau, & -\mu \leq \tau < 0, \\ \sqrt{-2\mu\tau - \mu^2}, & \tau < -\mu. \end{cases}$$

We see, that $\hat{\Pi} < F < \Pi$ for $-\alpha < \tau < -\mu - |\alpha - \beta|$ and $\alpha \neq \beta$. If $\alpha = \beta$ (the double multiplier is on the real axis), $F(\tau) \equiv \hat{\Pi}$. This example shows that first, one can not replace $\hat{\Pi}$ by Π in the Theorem 4.1 and second, the assertion of Theorem 4.1 may remain valid without assumption A2 about absence of double multipliers. We believe that also in the general case one can skip assumption A2 in Theorem 4.1.

Remark In Example 5.2 the multipliers remain regular at the point τ, where a double multiplier appears. In a general case the multipliers branch at this point.

Example 5.3. (*Resonance.*) Consider the system

$$\dot{z} = (\tau - \alpha i)z + \varepsilon w^2 + \varepsilon h(\vartheta), \quad \dot{w} = (-\mu - \beta i)w, \qquad (5.6)$$
$$\dot{\vartheta} = 1, \quad \dot{\tau} = \varepsilon.$$

Here $h, \alpha, \mu, \beta, z, w$ and multipliers $\rho_{1,2,3,4}$ are the same as in Example 5.2. There is the resonance $\rho_1 = \rho_2^2$ in the system (5.6) at $\tau = -2\mu \pm i(\alpha - 2\beta) + ik$, $\tau = -2\mu \pm i(\alpha + 2\beta) + ik$, $k \in \mathbb{Z}$. In Theorem 4.3 the in-out function is estimated on the interval $(-\alpha, 0)$ from below by the function $\tilde{\Pi}$. The exact in-out function for the system (5.6) is given by Proposition 5.1, where μ and β should be replaced by 2μ and 2β. In particular, if $\alpha = 2\beta$ (resonance on the real axis) the in-out function for $-\alpha < \tau < 0$ coincides with the function $\tilde{\Pi}$. This example shows that the estimate of the Theorem 4.3 cannot be improved.

6   Proof of Theorems 4.1–4.3

In this section we prove Theorems 4.1–4.3. We consider only case **A**. The proof in case **B** is the same. Below we assume that conditions $C1$–$C3$ and $A1$–$A3$ hold.

a. Basic Lemma

Since for any $\tau \in D$ all the multipliers $\rho_l(\tau)$ are pairwise distinct and distinct from one (see assumptions C1–C2), in the vicinity of the trajectories L_τ, $\tau \in D$ there exists an analytic change of the variables $(x_1, \ldots, x_{n+1}) \to (\hat{\xi}_1, \ldots, \hat{\xi}_n, \hat{\varphi})$ reducing the system (2.2) to the form

$$d\hat{\xi}/dt = \Lambda(\tau)\hat{\xi} + \Phi(\hat{\xi}, \hat{\varphi}, \tau), \quad d\hat{\varphi}/dt = \omega(\tau) + \Phi'(\hat{\xi}, \hat{\varphi}, \tau).$$
$$\Lambda = \mathrm{diag}(\lambda_1^0, \ldots, \lambda_n^0), \quad \Phi = (\Phi_1, \ldots, \Phi_n)^T = O(|\hat{\xi}|^2), \quad \Phi' = O(|\hat{\xi}|).$$

All the functions $\Phi', \Phi_1, \ldots, \Phi_n$ are 2π-periodic in $\hat{\varphi}$. The variables $(\hat{\xi}_1, \ldots, \hat{\xi}_n, \hat{\varphi})$ are usually called the Floquet coordinates.

The system (2.1) in the variables $\hat{\xi}, \hat{\varphi}, \tau$ has the form

$$
\begin{aligned}
d\hat{\xi}/dt &= [\Lambda(\tau) + \varepsilon\hat{B}(\hat{\varphi}, \tau, \varepsilon)]\hat{\xi} + \varepsilon\hat{h}(\hat{\varphi}, \tau, \varepsilon) + \hat{F}(\hat{\xi}, \hat{\varphi}, \tau, \varepsilon), \\
d\hat{\varphi}/dt &= \omega(\tau) + \varepsilon\hat{a}(\hat{\varphi}, \tau, \varepsilon) + \hat{F}'(\hat{\xi}, \hat{\varphi}, \tau, \varepsilon), \\
d\tau/dt &= \varepsilon.
\end{aligned} \tag{6.1}
$$

Here the functions $\hat{B} = (\hat{B}_{lk})$, $\hat{h} = (\hat{h}_1, \ldots, \hat{h}_n)^T$, $\hat{F} = (\hat{F}_1, \ldots, \hat{F}_n)^T$, $\hat{a}$, and $\hat{F}'$ are 2π-periodic in $\hat{\varphi}$ and the following conditions hold:

$$
\hat{F} = O(|\hat{\xi}|^2), \quad \hat{F}' = O(|\hat{\xi}|).
$$

By an analytic change of variables $\xi = \hat{\xi} + O(\varepsilon)$, $\varphi = \hat{\varphi} + O(\varepsilon)$, which is 2π-periodic in φ we can reduce the system (6.1) to the following form:

$$
\begin{aligned}
\dot{\xi} &= [\tilde{\Lambda}(\tau, \varepsilon) + \varepsilon^2 B_0(\varphi, \tau, \varepsilon)]\xi + \varepsilon^3 h_0(\varphi, \tau, \varepsilon) + F_0(\xi, \varphi, \tau, \varepsilon), \\
\dot{\varphi} &= \tilde{\omega}(\tau, \varepsilon) + \varepsilon^3 a_0(\varphi, \tau, \varepsilon) + F_0'(\xi, \varphi, \tau, \varepsilon), \\
\dot{\tau} &= \varepsilon,
\end{aligned} \tag{6.2}
$$

where for any $\tau \in D$

$$
\begin{aligned}
\tilde{\Lambda}(\tau, \varepsilon) &= \operatorname{diag}(\tilde{\lambda}_l(\tau, \varepsilon)), \quad \tilde{\lambda}_l(\tau, \varepsilon) = \lambda_l^0(\tau) + O(\varepsilon). \\
\tilde{\omega}(\tau, \varepsilon) &= \omega(\tau) + O(\varepsilon), \quad F_0 = O(|\xi|^2), \quad F_0' = O(|\xi|).
\end{aligned} \tag{6.3}
$$

We call a function bounded if its absolute value is bounded by a constant not depending on ε.

Basic Lemma. *There exists a change of the variables*

$$
\xi = \eta + \vartheta(\psi, \tau), \quad \varphi = \psi + \varkappa(\psi, \tau), \quad \vartheta = (\vartheta_1, \ldots, \vartheta_n)^T \tag{6.4}
$$

such that the functions

$$
\vartheta(\varphi, \tau)/\varepsilon^2, \quad \varkappa(\varphi, \tau)/\varepsilon \quad \tau \in D, \quad |\operatorname{Im}\psi| \le \beta_0/2
$$

are bounded and the system (6.2) in the variables η, ψ, τ has the form

$$
\begin{aligned}
\dot{\eta} &= [\tilde{\Lambda}(\tau, \varepsilon) + \varepsilon^2 B(\psi, \tau, \varepsilon)]\eta + F(\eta, \psi, \tau, \varepsilon). \\
\dot{\psi} &= \tilde{\omega}(\tau, \varepsilon) + F'(\eta, \psi, \tau, \varepsilon), \\
\dot{\tau} &= \varepsilon,
\end{aligned} \tag{6.5}
$$

Here the functions $B, \tilde{\omega}, F, F'$ are analytic and bounded for

$$
|\eta| \le \alpha_0/2, \quad \tau \in D, \quad |\operatorname{Im}\psi| \le \beta_0/2,
$$

$F = O(|\eta|^2)$, $F' = O(|\eta|)$ *and the constants α_0, β_0 are positive.*

We prove this Basic Lemma in Section 7.

b. Separation Lemma

To prove Theorem 4.3 we need one more lemma. Let $\mathfrak{M}$ and $\mathfrak{N}$ be the following linear subspaces of $L(n, \mathbb{R})$:

$$\mathfrak{M} = \{M = (M_{jk}) \in L(n, \mathbb{R})\, M_{jk} = M_{kj} = 0, \quad \text{for } j = 1, 2,\ k = 3, \dots, n\}$$
$$\mathfrak{N} = \{N = (N_{jk}) \in L(n, \mathbb{R})\, N_{jk} = N_{kj} = 0, \quad \text{for } j, k = 1, 2,\ \text{or } j, k = 3, \dots, n\}$$

Obviously, $L(n, \mathbb{R}) = \mathfrak{N} \oplus \mathfrak{M}$ and for any $N_1, N_2 \in \mathfrak{N}$, $M \in \mathfrak{M}$ we have:

$$N_1 N_2 \in \mathfrak{M}, \quad N_1 M \in \mathfrak{N}, \quad M N_1 \in \mathfrak{N}.$$

Separation Lemma. *Suppose that condition A5 holds. Then there exists a linear change of the variables*

$$\eta = (I + \varepsilon N(\psi, \tau, \varepsilon))\zeta, \qquad I = diag(1 \dots, 1) \tag{6.6}$$

such that the matrix function $N(\psi, \tau, \varepsilon) \in \mathfrak{N}$ is bounded in the complex domain $|Im\,\psi| \le \beta_0/4$, $\tau \in D$ and the system (6.5) in the variables ζ, ψ, τ takes the form

$$\begin{aligned}
\dot{\zeta} &= [\tilde{\Lambda}(\tau, \varepsilon) + \varepsilon^2 \tilde{B}(\psi, \tau, \varepsilon)]\zeta + \tilde{F}(\zeta, \upsilon, \tau, \varepsilon), \\
\dot{\psi} &= \tilde{\omega}(\tau, \varepsilon) + \tilde{F}'(\zeta, \psi, \tau, \varepsilon), \\
\dot{\tau} &= \varepsilon,
\end{aligned} \tag{6.7}$$

where $\tilde{B}(\psi, \tau, \varepsilon) \in \mathfrak{M}$, the functions $\tilde{B}, \tilde{F}, \tilde{F}'$ are analytic and bounded for

$$|\eta| \le \alpha_0/4, \quad \tau \in D, \quad |Im\,\psi| \le \beta_0/4,$$

$\tilde{F} = O(|\eta|^2)$ *and* $\tilde{F}' = O(|\eta|)$.

The Separation Lemma shows that condition $A5$ implies the possibility of a linear separation of the variables η_1, η_2 from $\eta_3, \dots, \eta_n$ by a linear change. In case **B** the corresponding statement asserts that the variable η_1 can be separated in linear terms from $\eta_2, \dots, \eta_n$ provided condition $B4$ holds.

c. Proof of Theorem 4.1

Consider the solution $\eta(t), \psi(t)$ of the system (6.5) with initial conditions at $t = t_0 = \tau_0/\varepsilon$.

Lemma 6.1. *Suppose that $|\eta(t_0)| < d_1^{-1}$. Then for $\tau_0 + d_2\varepsilon|\log \varepsilon| \le \varepsilon t \le \hat{\Pi}(\tau_0) - d_3\varepsilon|\log \varepsilon|$ we have $|\eta(t)| < d_4\varepsilon$.*

Theorem 4.1 is an obvious corollary of Lemma 6.1. One should consider the solution $x(t)$ in the variables η, ψ and use Lemma 6.1.

Proof of Lemma 6.1. Introduce the Lyapunov function

$$W(\eta) = \sum_{j=1}^{n} \eta_j \bar{\eta}_j.$$

Denote $w(t) = \sqrt{W(\eta(t))} \geq 0$. Then (6.2), (6.3) and the definition of $\mu(\tau)$ (Section 4.a) imply:

$$\dot{w} \leq [\mu(\tau) + O(\varepsilon) + O(w)]w. \tag{6.8}$$

If $w < a_1^{-1}$ then the expression in square brackets in (6.8) for $\tau = \tau_0$ is smaller than $\mu(\tau_0)/2$. Let $w(t_0) < a_1^{-1}/2$. Then (6.8) implies the estimate $w(t_1) < \varepsilon$ for $t_1 = t_0 + d_2|\log\varepsilon|$. On the time interval $[t_0, t_1]$ the function $w(t)$ decays exponentially: $w(t) < w(t_0)\exp(\mu(\tau_0)t/4)$. Denote by $\hat{t}_2$ the supremum of the time moments t_2, $\varepsilon t_2 < \hat{\Pi}(\tau_0) + 1$ such that for $t_1 \leq t \leq t_2$ the inequality $w(t) < 2\varepsilon$ is satisfied. Obviously, $\int_{t_0}^{\hat{t}_2} w(t)dt = O(1)$. Therefore, for $t_1 \leq t \leq \hat{t}_2$ equation (6.8) implies the estimate

$$w(t) \leq a_2 w(t_1) \exp\left(\frac{1}{\varepsilon}\int_{\varepsilon t_1}^{\varepsilon t} \mu(s)ds\right). \tag{6.9}$$

By using (6.9) we get that $\varepsilon\hat{t}_2 > \hat{\Pi}(\tau_0) - d_3\varepsilon|\log\varepsilon|$. We see also that $w(t) < \varepsilon$, $|\eta(t)| < d_4\varepsilon$ for $\tau_0 + d_2\varepsilon|\log\varepsilon| \leq \varepsilon t \leq \hat{\Pi}(\tau_0) - d_3\varepsilon|\log\varepsilon|$. Lemma 6.1 is proved.

d. Proof of Theorem 4.2

Denote for the system (6.5) by $\eta^{(1)}$ and $\eta^{(2)}$ the vectors, formed by the first two and the last $(n-2)$ components of the vector η respectively. Consider the solution $\eta(t), \psi(t)$ of the system (6.5) with initial conditions at $t = t_0 = \tau_0/\varepsilon$. Suppose, that assumption A4 is satisfied.

Lemma 6.2. *Suppose that $|\eta(t_0)| < d_5^{-1}$. Let $|\eta^{(2)}(t_0)| < d_6\varepsilon$ and $|\eta^{(1)}(t_0)| > d_7 d_6\varepsilon$. Then*

(1) for $\tau_0 + d_2\varepsilon|\log\varepsilon| \leq \varepsilon t \leq \Pi(\tau_0) - d_3\varepsilon|\log\varepsilon|$ we have $|\eta(t)| < d_4\varepsilon$.

(2) $|\eta(t_d)| > d_8^{-1}$ for some t_d, where $|\varepsilon t_d - \Pi(\tau_0)| < d_9\varepsilon|\log\varepsilon|$.

Theorem 4.2 is an obvious corollary of Lemma 6.2. One should consider the solution $x(t)$ in the variables η, ψ and use Lemma 6.1.

Proof of Lemma 6.2 Item (1) of Lemma 6.2 is just a repetition of Lemma 6.1. To prove item (2) introduce functions

$$W_1(\eta) = \sum_{j=1}^{2}\eta_j\bar{\eta}_j, \quad W_2(\eta) = \sum_{j=3}^{n}\eta_j\bar{\eta}_j.$$

Denote $w_l(t) = \sqrt{W_l(\eta(t))} \geq 0$, $l = 1, 2$. The constant d_6 in Lemma 6.2 can be defined arbitrarily; for example, $d_6 = 1$. Define d_7 in such a way, that $w_1(t_0) > 2w_2(t_0)$.

For any time moment t such that $w_1(t) \neq 0, w_2(t) \neq 0$ and $w_1(t) \geq w_2(t)$ the system (6.7) implies

$$\dot{w}_1 = [\mathrm{Re}\lambda_1^0(\tau) + O(\varepsilon) + O(w_1)]w_1. \tag{6.10}$$

$$\dot{w}_2 < [-\gamma(\tau) + O(\varepsilon) + O(w_1)]w_2 + O(w_1^2). \tag{6.11}$$

Here $-\gamma(\tau) = \max \mathrm{Re}\lambda_j^0$, where $3 \leq j \leq n$. Assumption A4 implies the estimate $-\gamma(\tau) < \mathrm{Re}\lambda_1^0$.

Due to (6.10), (6.11) for $w_1 < a_1^{-1}$ the following implication is valid: if $w_1 = w_2 \neq 0$ then $\dot{w}_1 > \dot{w}_2$. Also for $w_1 < a_1^{-1}$ the expressions in square brackets in (6.10) and (6.11) for $\tau = \tau_0$ are smaller, than $\mathrm{Re}\lambda_1^0(\tau)/2$ and $-\gamma(\tau)/2$ respectively, and the expression in square brackets in (6.10) for $\tau = \Pi(\tau_0)$ is bigger than $\mathrm{Re}\lambda_1^0(\tau)/2$. Let $w_1(t_0) < a_1^{-1}/2$. Denote by t_d the supremum of the time moments t_1, $\varepsilon t_1 < \Pi(\tau_0) + 1$ such that for $t_0 \leq t \leq t_1$ the inequality $w_1(t) < a_1^{-1}$ is satisfied. Then for $t_0 \leq t \leq t_d$ the inequality $w_1 \geq w_2$ is satisfied and, therefore, the estimates (6.10)–(6.11) are valid. For $t_0 \leq t \leq t_d$ equation (6.10) implies the estimate

$$a_2^{-1}w_1(t_0)\exp\left(\frac{1}{\varepsilon}\int_{\tau_0}^{\varepsilon t}\mathrm{Re}\lambda_1^0(s)ds\right) < w_1(t) < a_2 w_1(t_0)\exp\left(\frac{1}{\varepsilon}\int_{\tau_0}^{\varepsilon t}\mathrm{Re}\lambda_1^0(s)ds\right). \tag{6.12}$$

Inequalities (6.12) show that $|\varepsilon t_d - \Pi(\tau_0)| < d_9\varepsilon|\log\varepsilon|$. Lemma 6.2 is proved.

e. Proof of Theorem 4.3

Consider the solution $\eta(t), \psi(t)$ of the system (6.5) with initial conditions at $t = t_0 = \tau_0/\varepsilon$.

Lemma 6.3. *Suppose that $|\eta(t_0)| < d_1^{-1}$. Then for*

$$\tau_0 + d_2\varepsilon|\log\varepsilon| \leq \varepsilon t \leq \tilde{\Pi}(\tau_0) - d_3\varepsilon|\log\varepsilon|$$

we have $|\eta(t)| < d_4\varepsilon$.

Lemma 6.3 implies Theorem 4.3 when we consider the solution $x(t)$ in the variables η, v.

Proof of Lemma 6.3 Consider the functions

$$W_1(\eta) = \sum_{j=1}^{2}\eta_j\bar{\eta}_j, \quad W_2(\eta) = \sum_{j=3}^{n}\eta_j\bar{\eta}_j, \quad W(\eta) = W_1(\eta) + W_2(\eta)$$

and denote $w_l(t) = \sqrt{W_l(\eta(t))} \geq 0$, $l = 1, 2$, $w(t) = \sqrt{W(\eta(t))} \geq 0$.

For any time moment t such that $w_1(t) \neq 0, w_2(t) \neq 0$ the system (6.5) implies

$$\dot{w}_1 = [\mathrm{Re}\lambda_1^0(\tau) + O(\varepsilon) + O(w)]w_1 + O(w_2^2),$$

$$\dot{w}_2 < [-\gamma(\tau) + O(\varepsilon) + O(w)]w_2 + \varepsilon^2 O(w_1). \tag{6.13}$$

Here $-\gamma(\tau) = \max \operatorname{Re} \lambda_j^0$, where $3 \leq j \leq n$. Note that the first equation (6.13) does not contain terms of first order in w_2.

If $|\eta(\tau_0)| < d_1^{-1}$ then (6.13) implies the estimate $w(t_1) < \varepsilon^3$ for $t = t_1 = t_0 + d_2|\log \varepsilon|$. Let the functions $\tilde{w}_1(t)$ and $\tilde{w}_2(t)$ satisfy the relations

$$
\begin{aligned}
w_1(t) &= \tilde{w}_1(t) \exp\left(\frac{1}{\varepsilon} \int_{\varepsilon t_1}^\tau \tilde{\mu}(s)\, ds\right), \\
w_2(t) &= \tilde{w}_2(t) \exp\left(\frac{1}{\varepsilon} \int_{\varepsilon t_1}^\tau \frac{\tilde{\mu}(s)}{2}\, ds\right), \\
\tilde{\mu}(\tau) &= \max\{\operatorname{Re} \lambda_1^0(\tau), -2\gamma(\tau)\}.
\end{aligned}
\tag{6.14}
$$

Equations (6.13) can be transformed as follows:

$$
\begin{aligned}
d\tilde{w}_1/dt &= \operatorname{Re} \lambda_1^0(\tau) - \tilde{\mu}(\tau) + O(\varepsilon) + O(w)]\tilde{w}_1 + O(\tilde{w}_2^2). \\
d\tilde{w}_2/dt &< [-\gamma(\tau) - \tilde{\mu}(\tau)/2 + O(\varepsilon) + O(w)]\tilde{w}_2 + \varepsilon^2 O(\sqrt{w_1 \tilde{w}_1}).
\end{aligned}
\tag{6.15}
$$

Denote by $\hat{t}_2$ the supremum of the time moments t_2, $\varepsilon t_2 < \tilde{\Pi}(\tau_0) + 1$ such that for $t_1 \leq t \leq t_2$ the following inequalities hold:

$$
w(t) < \varepsilon^2, \quad \tilde{w}_1(t) < \varepsilon^2, \quad \tilde{w}_2(t) < \varepsilon^2.
$$

Now the relations (6.14)–(6.15) imply the inequality $\varepsilon \hat{t}_2 \geq \tilde{\Pi}(\tau_0) - d_3 \varepsilon |\log \varepsilon|$. For $t_0 + d_2|\log \varepsilon| \leq t \leq \hat{t}_2$ we have $|\eta(t)| < d_4 \varepsilon$. Lemma 6.3 is proved.

7   Proof of the Basic Lemma

We obtain the change (6.4) going to the limit in the following sequence of changes:

$$
\begin{aligned}
\xi^{(m)} &= \xi^{(m+1)} + \vartheta^{(m)}(\varphi^{(m+1)}, \tau), \quad \varphi^{(m)} = \varphi^{(m+1)} + \varkappa^{(m)}(\varphi^{(m+1)}, \tau), \\
\xi^{(0)} &= \xi, \quad \varphi^{(0)} = \varphi, \quad \xi^{(\infty)} = \eta, \quad \varphi^{(\infty)} = \psi.
\end{aligned}
\tag{7.1}
$$

In the variables $\xi^{(m)}, \varphi^{(m)}, \tau$ the system (6.2) has the form

$$
\begin{aligned}
\dot{\xi}^{(m)} &= [\tilde{\Lambda}(\tau) + \varepsilon^2 B_m(\varphi^{(m)}, \tau)]\xi^{(m)} + \varepsilon^3 h_m(\varphi^{(m)}, \tau) + F_m(\xi^{(m)}, \varphi^{(m)}, \tau), \\
\dot{\varphi}^{(m)} &= \tilde{\omega}(\tau) + \varepsilon^3 a_m(\varphi^{(m)}, \tau) + F_m'(\xi^{(m)}, \varphi^{(m)}, \tau), \quad \dot{\tau} = \varepsilon, \\
F_m &= O(|\xi^{(m)}|^2), \quad F_m' = O(|\xi^{(m)}|).
\end{aligned}
\tag{7.2}
$$

Here and henceforth for brevity we do not write ε among the arguments of the functions $\tilde{\Lambda}, B_m, h_m$ etc. We assume that the functions $\vartheta^{(m)}$ and $\varkappa^{(m)}$ satisfy the equations

$$
\begin{aligned}
\varepsilon \vartheta_\tau^{(m)}(\varphi, \tau) + \tilde{\omega}(\tau)\vartheta_\varphi^{(m)}(\varphi, \tau) &- \tilde{\Lambda}(\tau)\vartheta^{(m)}(\varphi, \tau) \\
&= \varepsilon^2 B_m(\varphi, \tau)\vartheta^{(m)}(\varphi, \tau) + \varepsilon^3 h_m(\varphi, \tau), \\
\varepsilon \varkappa_\tau^{(m)}(\varphi, \tau) + \tilde{\omega}(\tau)\varkappa_\varphi^{(m)}(\varphi, \tau) &= \varepsilon^3 a_m(\varphi, \tau) + (F_m')_\xi(0, \varphi, \tau)\vartheta^{(m)}(\varphi, \tau),
\end{aligned}
\tag{7.3}
$$

where the subscripts τ, φ, ξ denote partial derivatives and

$$(F'_m)_\xi \vartheta^{(m)} = \sum_{l=1}^n (F'_m)_{\xi_l} \vartheta_l^{(m)}.$$

We introduce the following notations:

(1) If $f(\varphi, \tau) = \sum_{k\in\mathbb{Z}} f^k(\tau) e^{ik\varphi}$ is an analytic scalar function and $\beta > 0$ then

$$||f||_\beta = \sup_{k\in\mathbb{Z}, \tau\in D} |f^k(\tau)| e^{\beta|k|}.$$

(2) For a vector $v = (v_1(\varphi, \tau), \ldots, v_n(\varphi, \tau))^T$ and a matrix $V = (V_{lj}(\varphi, \tau))$, $1 \le l, j \le n$ we put

$$||v||_\beta = \max_{1\le l\le n} ||v_l||_\beta, \quad ||V||_\beta = \max_{1\le l,j\le n} ||V_{l,j}||_\beta.$$

(3) For positive α and β let

$$E_{\alpha,\beta} = \{x \in \mathbb{C}^n, \tau \in D, \varphi \in \mathbb{C}/(2\pi\mathbb{Z})\, |x| \le \alpha, |\text{Im}\,\varphi| \le \beta\}.$$

Then for any function $F(x, \varphi, \tau)$ analytic in $E_{\alpha,\beta}$ we put

$$|F|_{\alpha,\beta} = \max_{E_{\alpha,\beta}} |F|.$$

Obviously, for some positive $\alpha_0, \beta_0, \chi_0, b_0$ and f_0 the following relations hold:

$$\begin{aligned}
||h_0||_{\beta_0} &\le \chi_0, \quad ||B_0||_{\beta_0} \le b_0, \quad ||a_0||_{\beta_0} \le \chi_0, \\
||F_0||_{\alpha_0,\beta_0} &\le f_0, \quad ||(F_0)_{\xi\xi}||_{\alpha_0,\beta_0} \le f_0, \\
||F'_0||_{\alpha_0,\beta_0} &\le f_0, \quad ||(F'_0)_\xi||_{\alpha_0,\beta_0} \le f_0.
\end{aligned} \tag{7.4}$$

For any $m \in \mathbb{N}$ we put

$$\begin{aligned}
\alpha_m &= \alpha_0(1 + 2^{-m})/2, \quad \beta_m = \beta_0(1 + 2^{-m})/2, \quad \beta'_m = (\beta_m + \beta_{m+1})/2, \\
\chi_{m+1} &= (\sqrt{\varepsilon}\chi_m)^{3/2}, \quad b_m = 2^m \chi_m, \quad f_m = f_0(2 - 2^{-m}).
\end{aligned}$$

Lemma 7.1. *Suppose that the following relations hold:*

$$\begin{aligned}
B_m &= \tilde{B}_0 + \tilde{B}_1 + \cdots + \tilde{B}_m, \quad ||\tilde{B}_j||_{\beta_j} \le b_j, \quad \tilde{B}_0 = B_0, \\
||h_m||_{\beta_m} &\le \chi_m, \quad ||a_m||_{\beta_m} \le \chi_m, \quad ||(F'_m)_\xi(0, \varphi, \tau)||_{\beta_m} \le f_m.
\end{aligned}$$

Then the equations (7.3) have a solution $\vartheta^{(m)}, \varkappa^{(m)}$ satisfying the estimates

$$||\vartheta^{(m)}||_{\beta'_m} \le P\varepsilon^2 \chi_m, \quad ||\varkappa^{(m)}||_{\beta'_m} \le Q2^m\varepsilon\chi_m. \qquad P, Q > 0. \tag{7.5}$$

We prove Lemma 7.1 in Section 8. Substituting into (7.2) the relations (7.1) and using (7.3) we get:

$$\dot{\xi}^{(m+1)} = (\tilde{\Lambda}(\tau) + \varepsilon^2 B_m(\varphi^{(m)}, \tau))\xi^{(m+1)} + U_m(\xi^{(m+1)}, \varphi^{(m+1)}, \tau),$$
$$\dot{\varphi}^{(m+1)} = \tilde{\omega}(\tau) + V_m(\xi^{(m+1)}, \varphi^{(m+1)}, \tau),$$

where

$$V_m = [1 + \varkappa_{\varphi}^{(m)}(\varphi^{(m+1)}, \tau)]^{-1} \big[\varepsilon^3 (a_m(\varphi^{(m)}, \tau) - a_m(\varphi^{(m+1)}, \tau))$$
$$+ F_m'(\xi^{(m)}, \varphi^{(m)}, \tau) - (F_m')_\xi(0, \varphi^{(m+1)}, \tau)\, \vartheta^{(m)}(\varphi^{(m+1)}, \tau) \big],$$
$$U_m = -V_m \vartheta_\varphi^{(m)}(\varphi^{(m+1)}, \tau) + \varepsilon^2 \big(B_m(\varphi^{(m)}, \tau) - B_m(\varphi^{(m+1)}, \tau) \big) \vartheta^{(m)}(\varphi^{(m+1)}, \tau)$$
$$+ \varepsilon^3 (h_m(\varphi^{(m)}, \tau) - h_m(\varphi^{(m+1)}, \tau)) + F_m(\xi^{(m)}, \varphi^{(m)}, \tau).$$

Therefore,

$$\varepsilon^3 h_{m+1}(\varphi^{(m+1)}, \tau) = U_m(0, \varphi^{(m+1)}, \tau),$$
$$\varepsilon^2 \tilde{B}_{m+1}(\varphi^{(m+1)}, \tau) = \varepsilon^2 B_m(\varphi^{(m)}, \tau) - \varepsilon^2 B_m(\varphi^{(m+1)}, \tau)$$
$$+ (U_m)_\xi(0, \varphi^{(m+1)}, \tau),$$
$$F_{m+1}(\xi^{(m+1)}, \varphi^{(m+1)}, \tau) = U_m(\xi^{(m+1)}, \varphi^{(m+1)}, \tau)$$
$$- U_m(0, \varphi^{(m+1)}, \tau) - (U_m)_\xi(\vartheta^{(m)}(0, \varphi^{(m+1)}, \tau)\, \xi^{(m+1)},$$
$$\varepsilon^3 a_{m+1}(\varphi^{(m+1)}, \tau) = V_m(0, \varphi^{(m+1)}, \tau),$$
$$F_{m+1}'(\xi^{(m+1)}, \varphi^{(m+1)}, \tau) = V_m(\xi^{(m+1)}, \varphi^{(m+1)}, \tau) - V_m(0, \varphi^{(m+1)}, \tau).$$

Now the Basic Lemma follows from

Lemma 7.2. *For all integer $m \geq 0$ the following estimates hold:*

$$\|h_m\|_{\beta_m} \leq \chi_m, \quad \|\tilde{B}_m\|_{\beta_m} \leq b_m, \quad \|a_m\|_{\beta_m} \leq \chi_m.$$
$$\|F_m\|_{\alpha_m, \beta_m} \leq f_m, \quad \|(F_m)_{\xi\xi}\|_{\alpha_m, \beta_m} \leq f_m.$$
$$\|F_m'\|_{\alpha_m, \beta_m} \leq f_m, \quad \|(F_m')_\xi\|_{\alpha_m, \beta_m} \leq f_m.$$

Lemma 7.2 is analogous to some technical assertions in KAM-theory [8, 9]. The proof of the lemma is quite standard. It is based on Lemma 7.1 and on quadratic convergence of our sequence of changes of variables. We use induction in m. For $m = 0$ assertions of the lemma hold (see (7.4)). Suppose that for $m \leq l$ Lemma 7.2 is valid. Then by using Lemma 7.1 and Cauchy estimates we obtain the following relations:

$$\|V_l(0, \varphi^{(l+1)}, \tau)\|_{\beta_l'} \leq c_1' 4^l \varepsilon^3 \chi_l^2, \quad \|U_l(0, \varphi^{(l+1)}, \tau)\|_{\beta_l'} \leq c_2' 4^l \varepsilon^4 \chi_l^2.$$

Now we obtain the following assertions:

(i) $\|\varepsilon^3 h_{l+1}(\varphi^{(l+1)}, \tau)\|_{\beta_{l+1}} \leq c_2' 4^l \varepsilon^4 \chi_l^2.$

(ii) $\|\varepsilon^2 \tilde{B}_{l+1}(\varphi^{(l+1)}, \tau)\|_{\beta_{l+1}} \leq 2^l \varepsilon^2 \chi_l.$

(iii) $\|\varepsilon^3 a_{l+1}(\varphi^{(l+1)}, \tau)\|_{\beta_{l+1}} \leq c_1' 4^l \varepsilon^3 \chi_l^2.$

(iv) The function $F'_{l+1}(\xi^{(l+1)}, \varphi^{(l+1)}, \tau)$ satisfies the relation

$$F'_{l+1} = F'_l(\xi^{(l+1)} + \vartheta^{(l)}(\varphi^{(l+1)}, \tau), \varphi^{(l)}, \tau) - F'_l(\vartheta^{(l)}(\varphi^{(l+1)}, \tau), \varphi^{(l)}, \tau).$$

Hence the derivatives of the functions F_l and F_{l+1} with respect to ξ^{m+1} have the same estimates. Since $F'_l = O(\xi^{(m)})$, the following estimate holds:

$$\|F'_{l+1}\|_{\alpha_{l+1}, \beta_{l+1}} \leq f_l(1 + c'_3 2^l \varepsilon^2 \chi_l.$$

The function $F_{l+1}(\xi^{(l+1)}, \varphi^{(l+1)}, \tau)$ and its derivatives can be estimated analogously with the help of the equality

$$\begin{aligned}
F_{l+1} &= -\vartheta^{(l)}_\varphi(\varphi^{(l+1)}, \tau)\big[F'_l(\xi^{(l+1)} + \vartheta^{(l)}(\varphi^{(l+1)}, \tau), \varphi^{(l)}, \tau) \\
&\quad - F'_l(\vartheta^{(l)}(\varphi^{(l+1)}, \tau), \varphi^{(l)}, \tau) - (F'_l)_\xi(\vartheta^{(l)}(\varphi^{(l+1)}, \tau), \varphi^{(l)}, \tau)\xi^{(l+1)}\big] \\
&\quad + F_l(\xi^{(l+1)} + \vartheta^{(l)}(\varphi^{(l+1)}, \tau), \varphi^{(l)}, \tau) \\
&\quad - F'_l(\vartheta^{(l)}(\varphi^{(l+1)}, \tau), \varphi^{(l)}, \tau) - (F'_l)_\xi(\vartheta^{(l)}(\varphi^{(l+1)}, \tau), \varphi^{(l)}, \tau)\xi^{(l+1)}
\end{aligned}$$

Lemma 7.2, obviously, follows from assertions (i)–(iv).

8 Proof of Lemma 7.1

In this section we write for brief $\vartheta, \varkappa$ instead of $\vartheta^{(m)}, \varkappa^{(m)}$. First, we expand the equations (7.3) in the Fourier series in φ. The Fourier coefficients $\vartheta^k(\tau), \varkappa^k(\tau)$ satisfy the system

$$\begin{aligned}
\varepsilon \vartheta^k_\tau + (ik\tilde{\omega}(\tau) - \tilde{\Lambda})\vartheta^k &= \sum_{j \in \mathbb{Z}} \varepsilon^2 B_m^{k-j} \vartheta^j + \varepsilon^3 h_m^k, \\
\varepsilon \varkappa^k_\tau + ik\tilde{\omega}\varkappa^k &= \varepsilon^3 a_m^k + \sum_{j \in \mathbb{Z}} (F'_m)_\xi^{k-j}(0, \tau)\vartheta^j,
\end{aligned} \tag{8.1}$$

where B_m^k, h_m^k, a_m^k, and $(F'_m)_\xi^k$ are Fourier coefficients of the functions B_m, h_m, a_m, and $(F'_m)_\xi$ respectively.

Let the function ϕ^+ (respectively ϕ^-), see (3.4), restricted to the domain D have a local maximum at the point τ_+ (respectively τ_-). Since $\phi^\pm$ are harmonic, $\tau_\pm$ lie on the boundary of D. According to A3 the maxima $\tau_\pm$ are unique (and consequently, are global on D). Since the function $\omega(\tau)$ is real and positive for real $\tau \in D$, the points $\tau_\pm$ are outside the real axis. Therefore, the boundary ∂D of the domain D is smooth at these points.

Proposition 8.1. *Suppose that the maxima $\tau_\pm$ of the functions $\phi^\pm$ restricted to ∂D are nondegenerate. Then there exist points τ_l^k and τ^k, where $k \in \mathbb{Z}$, $l = 1, \ldots, n$ such that for any $\tau \in D$*

$$\tilde{\varphi}_l^k(\tau) \leq 0, \quad \tilde{\varphi}^k(\tau) \leq 0.$$

where

$$\tilde{\varphi}_l^k(\tau) = \operatorname{Re} \int_{\tau_l^k}^{\tau} (\tilde{\lambda}_l^0(s) - ik\tilde{\omega}(s))\, ds, \quad \tilde{\varphi}^k(\tau) = -\operatorname{Re} \int_{\tau^k}^{\tau} ik\tilde{\omega}(s)\, ds.$$

Furthermore, there exists a system of curves $\mathfrak{C}_{kl}^\tau(\nu) \subset D$, $\mathfrak{C}_k^\tau(\nu) \subset D$, $\tau \in D$, $\nu \in [0,1]$ (here we regard ν as a parameter on a curve and τ, k, l as parameters labeling the curves) with the following properties.

(i) $\mathfrak{C}_{kl}^\tau(0) = \tau_l^k$, $\mathfrak{C}_k^\tau(0) = \tau^k$; $\mathfrak{C}_{kl}^\tau(1) = \tau$, $\mathfrak{C}_k^\tau(1) = \tau$.

(ii) *There exists a constant K such that for any $0 \le \nu_1 \le \nu_2 \le 1$*

$$\tilde{\varphi}_l^k(\mathfrak{C}_{kl}^\tau(\nu_2)) - \tilde{\varphi}_l^k(\mathfrak{C}_{kl}^\tau(\nu_1)) \le K\varepsilon, \quad \tilde{\varphi}^k(\mathfrak{C}_k^\tau(\nu_2)) - \tilde{\varphi}^k(\mathfrak{C}_k^\tau(\nu_1)) \le K\varepsilon.$$

(iii) *Lengths of the curves $\mathfrak{C}_{kl}^\tau$, $\mathfrak{C}_k^\tau$ are bounded from above by a constant L not depending on ε.*

It is easy to check that solutions of the following integral equation

$$(\vartheta, \varkappa) = P_m(\vartheta), \tag{8.2}$$

satisfy the system (8.1). In more detail,

$$\vartheta_l^k = [P_m^{(\vartheta)}(\vartheta)]_l^k, \quad \varkappa^k = [P_m^{(\varkappa)}(\vartheta)]^k,$$

where

$$[P_m^{(\vartheta)}(\vartheta)]_l^k = \frac{1}{\varepsilon} \int_{\tau_l^k}^{\tau} e^{\frac{1}{\varepsilon} \int_{\nu}^{\tau} (\tilde{\lambda}_l(s) - ik\tilde{\omega}(s))\, ds} \Big[\sum_{j \in \mathbb{Z}} \varepsilon^2 (B_m^{k-j}\vartheta^j)_l(\nu) + \varepsilon^3 (h_m^k)_l(\nu) \Big]\, d\nu,$$
$$\tag{8.3}$$

$$[P_m^{(\varkappa)}(\vartheta)]^k = \frac{1}{\varepsilon} \int_{\tau^k}^{\tau} e^{-\frac{1}{\varepsilon} \int_{\nu}^{\tau} ik\tilde{\omega}(s)\, ds} \Big[\varepsilon^3 a_m^k(\nu) + \sum_{j \in \mathbb{Z}} (F_m')_\xi^{k-j}(0,\nu)\vartheta^j(\nu) \Big]\, d\nu, \tag{8.4}$$

$$(B_m^{k-j}\vartheta^j)_l = \sum_{r=1}^{n} B_{m,l,r}^{k-j}\vartheta_r^j, \quad (F_m')_\xi^{k-j}(0,\nu)\vartheta^j = \sum_{r=1}^{n} (F_m')_{\xi_r}^{k-j}\vartheta_r^j,$$

$$P_m(\vartheta) = \sum_{k \in \mathbb{Z}} \Big(\big([P_m^{(\vartheta)}(\vartheta)]_1^k, \dots, [P_m^{(\vartheta)}(\vartheta)]_n^k\big)^T, \, [P_m^{(\varkappa)}(\vartheta)]^k \Big) e^{ik\varphi}.$$

Here we assume that the points τ_l^k and τ^k are determined by Proposition 8.1. The integrals in the right-hand sides of (8.3)–(8.4) can be taken along any paths (with the indicated endpoints) lying in the domain D. Since the integrands are holomorphic functions and D is simple-connected, the values of the integrals do not depend on the choice of paths.

To prove that the system (8.3)–(8.4) has a solution we show that the operator

$$P_m^{(\vartheta)} S_{\beta'_m} \to S_{\beta'_m}, \qquad S_{\beta_m} = \{\vartheta \mid \|\vartheta\|_{\beta'_m} < \infty\}$$

is contracting. We regard $S_{\beta'_m}$ as a Banach space with the norm $\|\vartheta\|_m = \|\vartheta\|_{\beta'_m}$.

We can assume that in equations (8.3) and (8.4) the integration takes place along the paths $\mathfrak{C}_{kl}^\mu$ and $\mathfrak{C}_k^\mu$ respectively. Then according to Proposition 8.1,

$$\mathrm{Re}\Big(\frac{1}{\varepsilon}\int_\nu^\tau (\tilde\lambda_l^0(s) - ik\tilde\omega(s))\,ds\Big) \le 2K, \quad -\mathrm{Re}\Big(\frac{1}{\varepsilon}\int_\nu^\tau ik\tilde\omega(s)\,ds\Big) \le 2K.$$

Hence the exponentials in (8.3)–(8.4) are uniformly bounded by the magnitude e^{2K}. Let the constant L be defined by Proposition 8.1 and $M = Le^{2K}$.

For any two functions $\vartheta, \vartheta_* \in S_{\beta'_m}$ we have:

$$\begin{aligned}
\big|[P_m^{(\vartheta)}(\vartheta)]_l^k - [P_m^{(\vartheta)}(\vartheta_*)]_l^k\big| &\le M\varepsilon \max_{\tau \in D}\Big|\sum_{j\in\mathbb{Z}}\big(B_m^{k-j}(\tau)(\vartheta^j(\tau) - \vartheta_*^j(\tau))\big)_l\Big| \\
&\le nM\varepsilon \sum_{r=0}^m \sum_{j\in\mathbb{Z}} \|\tilde B_r\|_{\beta_r}\|\vartheta - \vartheta_*\|_m e^{-|k-j|\beta_r - |j|\beta'_m} \\
&\le nM\varepsilon \sum_{r=0}^m b_r\Big(\frac{1}{1-e^{\beta'_m-\beta_r}} + \frac{2}{1-e^{-\beta'_m}}\Big)\|\vartheta-\vartheta_*\|_m e^{-|k|\beta'_m} \\
&\le \|\vartheta-\vartheta_*\|_m e^{-|k|\beta'_m}/2.
\end{aligned}$$

Therefore,

$$\|P_m^{(\vartheta)}(\vartheta) - P_m^{(\vartheta)}(\vartheta_*)\|_m \le \|\vartheta-\vartheta_*\|_m/2.$$

and the operator $P_m^{(\vartheta)}$ is contracting with the contraction coefficient not exceeding $1/2$. The existence of the solution $(\vartheta, \varkappa)$ is established.

Now we prove the estimate (7.5). The triangle inequality implies:

$$\|\vartheta\|_m = \|P_m^{(\vartheta)}(\vartheta)\|_m \le \|P_m^{(\vartheta)}(\vartheta) - P_m^{(\vartheta)}(0)\|_m + \|P_m^{(\vartheta)}(0)\|_m \tag{8.5}$$

Since $\|P_m^{(\vartheta)}(\vartheta) - P_m^{(\vartheta)}(0)\|_m \le \|\vartheta\|_m/2$, by using (8.5) we obtain the following estimate:

$$\|\vartheta\|_m \le 2\|P_m^{(\vartheta)}(0)\|_m.$$

The magnitude $\|P_m^{(\vartheta)}(0)\|_m$ can be estimated as follows:

$$\big|\big(P_m^{(\vartheta)}(0)\big)_l^k\big| \le M\varepsilon^2 \|h_m\|_{\beta_m} e^{-\beta_m|k|} \le M\varepsilon^2 \chi_m e^{-\beta'_m|k|}.$$

These inequalities imply the first estimate (7.5) with $P = 2M$. Now let us verify the second one:

$$\begin{aligned}
\big|\big(P_m^{(\varkappa)}(\vartheta)\big)^k\big| &\le \varepsilon^{-1}M\Big[\varepsilon^3\|a_m\|_{\beta_m} e^{-\beta_m|k|} + \max_{\tau\in D}\Big|\sum_{j\in\mathbb{Z}}(F_m')_\xi^{k-j}(0,\tau)\vartheta^j(\tau)\Big|\Big] \\
&\le \varepsilon^{-1}M\Big[\varepsilon^3\chi_m e^{-\beta_m|k|} + n\sum_{j\in\mathbb{Z}}f_m\|\vartheta\|_m e^{-|k-j|\beta_m - |j|\beta'_m}\Big] \\
&\le \varepsilon^{-1}M\Big[\varepsilon^3\chi_m + \frac{3nf_m}{1-e^{\beta'_m-\beta_m}}P\varepsilon^2\chi_m\Big]e^{-|k|\beta'_m}.
\end{aligned}$$

Hence, we can put $Q = 48Mnf_0P/\beta$. Lemma 7.1 is proved.

9   Proof of the Separation Lemma

The change (6.6) transforms the first equation (6.5) to

$$\dot{\zeta} = M\zeta + O(|\zeta|^2), \qquad M = (I + \varepsilon N)^{-1}(-\varepsilon N_\psi \tilde{\omega} - \varepsilon^2 N_\tau + (\tilde{\Lambda} + \varepsilon^2 B)(I + \varepsilon N)), \quad (9.1)$$

where we denote by $N_\psi \tilde{\omega}$ the sum $\sum_{k=1}^n N_{\psi_k} \tilde{\omega}_k$. The lemma asserts that we can take the function N with values in $\mathfrak{N}$ in such a way that $M \in \mathfrak{M}$. To prove this we study the equation

$$\varepsilon^2 N_\tau = -\varepsilon N_\psi \tilde{\omega} - (I + \varepsilon N)M + (\tilde{\Lambda} + \varepsilon^2 B)(I + \varepsilon N) \qquad (9.2)$$

which is a consequence of (9.1). We put

$$B = B_\mathfrak{N} + B_\mathfrak{M}, \qquad B_\mathfrak{N} \in \mathfrak{N}, \quad B_\mathfrak{M} \in \mathfrak{M},$$
$$M = \tilde{\Lambda} + \varepsilon^2 B_\mathfrak{M} + \varepsilon^3 B_\mathfrak{N} N \in \mathfrak{M}.$$

Equality (9.2) takes the form

$$\varepsilon N_\tau + N_v \tilde{\omega} - [\tilde{\Lambda}, N] = \varepsilon B_\mathfrak{N} + \varepsilon^2 [B_\mathfrak{M}, N] - \varepsilon^3 N B_\mathfrak{N} N. \qquad (9.3)$$

Here the brackets [.] denote the matrix commutator. Since all terms in (9.3) belong to $\mathfrak{N}$, we can regard this equation as a system for the coefficients N_{kl}, where $k = 1, 2, l = 3, \dots, n$ or $l = 1, 2, k = 3, \dots, n$:

$$\varepsilon(N_{kl})_\tau + (N_{kl})_\psi \tilde{\omega} - (\tilde{\lambda}_k - \tilde{\lambda}_l)N_{kl} = \varepsilon B_{kl} + \varepsilon^2 \big([B_\mathfrak{M}, N]\big)_{kl} - \varepsilon^3 (N B_\mathfrak{N} N)_{kl}. \quad (9.4)$$

This system is analogous to (7.2). It can be solved following the same line (see Section 7), i.e.

(1) We expand the system (9.4) into the Fourier series in ϕ.

(2) We search for a solution as the sum of a rapidly converging series (Newton method), where each term of the series satisfy certain linear system.

(3) We construct solutions of these linear systems by using the method of contracting mappings.

The details are omitted.

10   Proof of propositions

1. Proof of Proposition 3.1

Suppose for simplicity that $\tau_* = 0$ and expand the functions λ_1^k into Maclaurin series:

$$\lambda_1^k(\tau) = \lambda_1^0(0) + p\tau^l + O(\tau^{l+1}) - ik\omega(\tau), \quad l \geq 1. \qquad (10.1)$$

The magnitude $\lambda_1^0(0)$ is purely imaginary. If $\tau \in \mathbb{R}$ increases and passes through the origin, $\lambda_1^0(\tau)$ crosses the imaginary axis from the left to the right. Consequently, $\mathrm{Re}\, p > 0$ and l is odd. In a typical situation $l = 1$. The function $\omega(\tau)$ is real-analytic. We put $\tau = x + iy$. The functions ϕ_1^k have the form:

$$\phi_1^k(\tau) = \int_0^x \mathrm{Re}\,\lambda_1^0(s)\,ds + \big[-\mathrm{Im}\,\lambda_1^0(x) + k\omega(x)\big]y + O((|k|+1)y^2). \qquad (10.2)$$

The equations $\phi_1^k(\tau) = \sigma$, with small $\sigma > 0$ has the following solutions:

$$\text{If } y = 0 \text{ then } x = \tau_\pm(\sigma), \quad \tau_\pm(\sigma) = \pm\Big(\frac{(l+1)\sigma}{\mathrm{Re}\,p}\Big)^{1/(l+1)} + O(\sigma^{2/(l+1)}),$$

$$\text{If } x \in [\tau_-(\sigma), \tau_+(\sigma)] \text{ then } y = \frac{\sigma - \mathrm{Re}\,p\,x^{l+1}/(l+1) + O(\sigma x) + O(x^{l+2})}{k\omega(0) + i\lambda_1^0(0)}. \qquad (10.3)$$

Proposition 3.1 is proved.

2. Proof of Proposition 3.2

(a) Equations (10.3) show that for small $\sigma > 0$ the segments $\Gamma^0(\sigma)$ lie in the upper half-plane. Suppose that some segment $\Gamma^0(\sigma)$, $0 \leq \sigma \leq \sigma_0$ has a point with negative imaginary part. Then for some $\sigma = \sigma' \in (0, \sigma_0)$ the equation $\phi_1^0(x) = \sigma'$, $x \in U \cap \mathbb{R}$ has a solution distinct from $\tau_\pm(\sigma')$. But this is impossible since the function $\phi_1^0(x)$ is monotonic on the intervals $U \cap \mathbb{R}^+$, $U \cap \mathbb{R}^-$, where $\mathbb{R}^\pm = \{x \in \mathbb{R} : \pm x \geq 0\}$. Arguments concerning the curves $\Gamma^{-1}(\sigma)$ are analogous.

(b) There are two possible reasons because of which a family $\Gamma^k(\sigma)$ may be not defined for $0 \leq \sigma \leq \sigma_0$. Either the curves $\Gamma^k(\sigma)$ go out from the domain U, or there is a singularity for these curves inside U. The first possibility does not take place due to assertion (c). The second one is equivalent to the condition $\partial\phi_l^k/\partial x = \partial\phi_l^k/\partial y = 0$, for some $\tau = x + iy \in U$. This condition can be rewritten in the form $\lambda_l^k(\tau) = 0$ or $\rho_l(\tau) = 1$. The last equality holds nowhere in U due to assumption $C1$.

(c) First, note that the magnitude $-\mathrm{Im}\,\lambda_1^0(x)$ (respectively, $-\mathrm{Im}\,\lambda_1^0(x) - \omega(x)$) is positive (respectively, negative) for $x \in U \cap \mathbb{R}$. Indeed. for $x = 0$ (recall that we assume that $\tau_* = 0$) this assertion follows from definition of λ_1^0. If $-\mathrm{Im}\,\lambda_1^0(x)$ is negative for some $x \in U \cap \mathbb{R}$ then there exists $x' \in U \cap \mathbb{R}$ such that $-\mathrm{Im}\,\lambda_1^0(x') = 0$. Let us choose the closest to zero point x' satisfying this equation. Then at this point the relation $\rho_1 = \rho_2$ holds which contradicts to assumption $C2$.

The following inequalities hold for $x \in U \cap \mathbb{R}$:

$$\omega(x) > 0, \quad \mathrm{Re}\,\lambda_1^0(x) > 0 \text{ for } x > 0, \quad \mathrm{Re}\,\lambda_1^0(x) < 0 \text{ for } x < 0.$$

Using these remarks and the equality (10.2) we see that for small $\beta \geq 0$ the curves $\Gamma^k(\sigma_0)$ can be partially determined by the relations

$$
\begin{aligned}
x &= \tau_-(\sigma_0) + \beta, & y &= \frac{-\operatorname{Re}\lambda_1^0(\tau_-(\sigma_0))\beta}{-\operatorname{Im}\lambda_1^0(\tau_-(\sigma_0)) + k\omega(\tau_-(\sigma_0))} + O(\beta^2), \\
x &= \tau_+(\sigma_0) - \beta, & y &= \frac{\operatorname{Re}\lambda_1^0(\tau_+(\sigma_0))\beta}{-\operatorname{Im}\lambda_1^0(\tau_+(\sigma_0)) + k\omega(\tau_+(\sigma_0))} + O(\beta^2)
\end{aligned}
$$

which imply in particular, that for $k \geq 0$ ($k < 0$) the segments $\Gamma^k(\sigma_0)$ lie at least near the points $\tau_\pm(\sigma_0)$ in the upper (lower) half-plane. Moreover, these relations show that for any integer k distinct from 0 and -1 the curve $\Gamma^k(\sigma_0)$ lies between $\Gamma^0(\sigma_0)$ and $\Gamma^{-1}(\sigma_0)$ at least near the points $\tau_\pm(\sigma_0)$. Hence, to complete the proof of assertion (c) it is sufficient to verify that the curves $\Gamma^k(\sigma_0)$ and $\Gamma^l(\sigma_0)$, $l = 0, -1$ intersect only at the points $\tau_\pm(\sigma_0)$. The existence of other intersection point τ is equivalent to the equality $\phi_1^k(\tau) = \phi_1^l(\tau) = \sigma_0$ which cannot hold for real $\tau \neq \tau_\pm(\sigma_0)$. For $\tau \notin \mathbb{R}$ this equality contradicts to assumption $C3$. Proposition 3.2 is proved.

3. Proof of Proposition 3.3

Note that the two positive numbers

$$
i\lambda_1^0(0) \quad \text{and} \quad \omega(0) - i\lambda_1^0(0) \quad \text{are distinct} \tag{10.5}
$$

because otherwise $i\lambda_1^0(0) = -i\omega(0)/2$ and $\rho_1(0) = \rho_2(0) = -1$ (the last relation contradicts assumption $C2$). Now assumption $A1$ follows for small $\sigma_0 > 0$ from (10.3) and (10.5).

Assumption $A2$ holds for small $\sigma_0 > 0$ because the corresponding domain $D = D(\sigma_0)$ is a small neighborhood of the point $\tau = \tau_*$.

Now we verify assumption $A3$. According to the conditions of Proposition 3.3 we have $l = 2$ in the relations (10.1) and (10.3). Putting $\tau_* = 0$ and $\tau = x + iy$, we have the following expansions:

$$
\begin{aligned}
\phi^\pm(\tau) &= \omega(x)y + O(y^2), \\
\phi_j^k(\tau) &= \operatorname{Re}\lambda_j^0(0)x + (-\operatorname{Im}\lambda_j^0(x) + k\omega(x))y + O(x^2) + O((|k| + 1)y^2).
\end{aligned}
$$

First, take the function $\phi^+|_{\partial D}$. It has a maximum on the curve Γ^0. Since for small $\sigma_0 > 0$ the curve $\Gamma^0(\sigma_0)$ can be defined by the relations (10.3), the function $\phi^+|_{\Gamma^0}$ has the form

$$
\frac{\omega(0)(\sigma - \operatorname{Re}p\, x^2/2)}{i\lambda_1^0(0)} + O(\sigma x) + O(x^3), \qquad x \in [\tau_-(\sigma_0), \tau_+(\sigma_0)].
$$

This function has a maximum at a point $x = O(\sigma_0)$. This maximum is nondegenerate since $\omega(0)\operatorname{Re}p/(i\lambda_1^0(0)) > 0$. The function ϕ^- can be analyzed analogously.

For $x \in [\tau_-(\sigma_0), \tau_+(\sigma_0)]$ the functions $\phi_j^k|_{\Gamma_0}$, $j \geq 3$ can be presented in the following form:

$$\operatorname{Re} \lambda_j^0(0)x + (-\operatorname{Im} \lambda_j^0(0) + k\omega(x))\frac{\sigma - \operatorname{Re} px^2/2}{i\lambda_1^0(0)} + O(\sigma x) + O(x^2). \qquad (10.6)$$

Formulas determining the functions $\phi_j^k|_{\Gamma_{-1}}$, $j \geq 3$ are analogous.

Since $\tau_\pm(\sigma) = O(\sqrt{\sigma})$ and $\operatorname{Re} \lambda_j^k(0) < 0$, it is clear that for $|k| < (c\sqrt{\sigma_0})^{-1}$ (the constant c is sufficiently large) the functions $\phi_j^k|_{\partial D}$. $j \geq 3$ have maxima only at the point $\tau_-(\sigma_0)$. If $|k| \geq (c\sqrt{\sigma_0})^{-1}$, the functions $\phi_j^k(\tau)/k$ are small perturbations of the function $\phi^+(\tau)$:

$$\phi_j^k(\tau)/k = \phi^+(\tau) + \frac{1}{k}\int_0^\tau \lambda_j^0(s)\,ds.$$

Hence, in this case assumption $A3$ holds because it holds for the functions $\phi^\pm$.

For the functions $\phi_j^k|_{\partial D}$, $j = 1, 2$ and small σ_0 Assumption $A3$ can be verified analogously by using (10.5).

4. Proof of Proposition 8.1

Let τ^k be maxima of the functions

$$-\operatorname{Re}\int_0^\tau ik\tilde{\omega}(s)\,ds \qquad (10.7)$$

restricted to D. For small values of ε and $k \neq 0$ these maxima are unique according to A3 and due to the closeness of the functions $\tilde{\omega}$ and ω on D. (The point $\tau^0 \in D$ is arbitrary.) Note that $\tau^k \in \partial D$ for $k \neq 0$ and $\tau^k = \tau^l$ if k and l have the same sign. Moreover, for $k \neq 0$ restrictions of the functions (10.7) to the boundary ∂D have nondegenerate maxima at the points τ^k.

Analogously, let τ_l^k be global maxima of the functions

$$-\operatorname{Re}\int_0^\tau \left(\bar{\lambda}_l^0 - ik\tilde{\omega}(s)\right)ds \qquad (10.8)$$

restricted to D. Note that if $|k|$ is not very large, the functions (10.8) restricted to D can have other (local) maxima.

Inequalities $\tilde{\phi}_l^k(\tau) \leq 0$, $\tilde{\phi}^k(\tau) \leq 0$, $\tau \in D$ hold according to the definition of τ_l^k and τ^k. Now we construct the curves $\mathfrak{C}_{kl}^\tau$, $\mathfrak{C}_k^\tau$.

Let any of the curves $\mathfrak{C}_{kl}^\tau$ ($\mathfrak{C}_k^\tau$) consist of two parts. The second part links the boundary ∂D with the point τ along a level line of the function $\tilde{\phi}_l^k$ ($\tilde{\phi}^k$). Note that since the functions $\tilde{\phi}_l^k, \tilde{\phi}^k$ are harmonic, their level lines having common points with D, reach ∂D. The first part goes along the curve ∂D and links the point τ_l^k (τ^k) with the beginning of the second part. Obviously. these curve satisfies the properties (i)–(iii).

Acknowledgment. The research described in this publication was partially supported by Russian Foundation of Basic Research, International Science Foundation, American Mathematical Society (fSU Aid Fund), INTAS and Centre de Recerca Matematica, Institut D'Estudis Catalans.

References

[1] M.A. Shishkova, *Examination of one system of differential equations with a small parameter in highest derivatives*, Dokl. Akad. Nauk SSSR **209** (1973), no. 3, 576–579 (Russian); English transl. in Soviet Math. Dokl. **14** (1973), no. 2, 384–387.

[2] A.I. Neishtadt, *Asymptotic study of stability loss of equilibrium under slow transition of two eigenvalues through imaginary axis*, Uspekhi Mat. Nauk **40** (1985), no. 5, 300–301. (Russian).

[3] A.I. Neishtadt, *On stability loss delay for dynamical bifurcations I*. Differ. Uravn. **23** (1987), no. 12, 2060–2067 (Russian); English transl. in Differ. Equations **23** (1987), no. 12, 1385–1390.

[4] A.I. Neishtadt, *On stability loss delay for dynamical bifurcations II* Differ. Uravn. **24** (1988), no. 2, 226–233 (Russian); English transl. in Differ. Equations **24** (1988), no. 2, 171–176.

[5] A.I. Neishtadt, *On calculation of stability loss delay time for dynamical bifurcations*, XI International Congress of Mathematical Physics, in press (1994).

[6] J. Su, *Effects of periodic forcing on delayed bifurcations, I, II.* Preprint. Univ. of Texas at Arlington, (1993).

[7] E. Benoit (Ed.), *Dynamic Bifurcations, Lect. Notes in Math.*. vol. 1493, Springer, Berlin, 1991.

[8] V. I. Arnold, *Proof of a theorem of A. N. Kolmogorov on the invariance of quasi-periodic motions under small perturbations of the Hamiltonian*, Uspehi. Mat. Nauk **18** (1963), 13–40 (Russian); English transl. in Russ. Math. Surveys **18** (1963), 9–36.

[9] J. Moser, *Convergent series expansions for quasi-periodic motions.* Math. Ann. **169** (1967), 136–176.

Progress in Nonlinear Differential Equations
and Their Applications, Vol. 19
© 1996 Birkhäuser Verlag Basel/Switzerland

Parametric and autoparametric resonance

M. Ruijgrok   F. Verhulst*

1   Introduction

Parametric resonance may arise in a mechanical system. the Excited System, in which one of the forces is varying periodically. The classical example is a pendulum with a suspension point which moves harmonically in the vertical direction. We shall discuss a fairly general one degree of freedom. parametrically excited system in section 2. This system is a dissipative version of the study by Broer and Vegter (1992). In section 3 we consider an autoparametric two degrees of freedom system which is in some sense a generalisation of section 2. Such a system admits a richer bifurcation structure and chaotic dynamics.

Parametric excitation in a system involves a periodic force which, in its turn, is not affected by the vibrations of the system. This can be a reasonable approximation of the mechanical configuration, however, in general this is too crude. It is at this point where autoparametric excitation and resonance comes into focus.

Autoparametric systems are vibrating systems which consist of at least two constituing subsystems. One is an Oscillator which is vibrating according to its nature; this Oscillator can be externally forced, self-excited. parametrically excited or a combination of these. For instance in Figure 1.1 the Oscillator consists of a mass mounted on a spring with damping and external, periodic forcing; the Oscillator in this case has one degree of freedom. In general the Oscillator has N degrees of freedom and will be described by the coordinates $y_i, \dot{y}_i, i = 1, 2, \ldots, N$.

The second constituing subsystem is called the Excited System. Whether this subsytem is actually excited or not depends on certain frequency ratios and nonlinearities. as we shall see. The Excited System is coupled to the Oscillator in a nonlinear way but such that the Excited System can be at rest while the Oscillator is vibrating. We call this state the semi-trivial solution; in fact there exist an infinite number of them, for instance all the transient states to a periodic solution of the Oscillator, but in discussing semi-trivial solutions we shall ignore transient states.

*Mathematisch Instituut, Universiteit Utrecht. P.O. Box, 80010, 3508 TA Utrecht, The Netherlands

In Figure 1.1 the Excited System consists of a pendulum attached to the mass of the Oscillator. As the spring is mounted vertically we have as the semi-trivial solution that the mass is in forced periodic vibration while the pendulum is hanging vertically at rest.

In general the Excited System has n degrees of freedom and will be described by the coordinates $x_i, \dot{x}_i, i = 1, 2, \ldots, n$.

Autoparametric systems can be characterized as follows:

1. Autoparametric systems consist of at least an Oscillator (coordinates $y, \dot{y}$) and an Excited System (coordinates $x, \dot{x}$) which are coupled.

2. Autoparametric systems admit semi-trivial solutions where
   $\sum_{i=1}^{N}(y_i^2(t) + \dot{y}_i^2(t)) \neq 0, x_i(t) = \dot{x}_i(t) = 0, i = 1, \ldots, n$.

3. Semi-trivial solutions can be stable or unstable in certain frequency intervals.

4. In the instability intervals of the semi-trivial solution we have autoparametric resonance. The vibrations of the Oscillator act as parametric excitation of the Excited System; in this context this is called autoparametric excitation.

We note here already that the interval of autoparametric resonance, resulting in stable vibrations of the Excited System, can be larger than the instability interval of the semi-trivial solution. In such a case we have the possibility of a stable semi-trivial solution co-existent with a stable autoparametric resonant solution; they will have their own domains of attraction. Also, it is clear that in studying autoparametric systems the determination of stability and instability conditions of the semi-trivial solution is always the first step.

1.1   Examples

We have seen one example of an autoparametric system which has two degrees of freedom. It consists of a mass mounted on a spring (Oscillator) and of a pendulum (Excited System) attached to the mass; see Figure 1.1. The mass on the spring can move in the vertical direction only and it is periodically excited. The forces acting on the pendulum result from gravity but also from the spring.

A related simple model consists of a mass, supported by a spring and mounted on the end of a long elastic element, for instance a leaf spring: see Figure 1.1. This is also a simple model for a vertical, oblong body (a building or other construction) which is mounted on a coherent elastic foundation. Again the excited spring with mass is the Oscillator, the leaf spring represents the Excited System.

An important class of engineering problems consists of vibrations which are self-excited. In mechanically simple cases the self-excitation effect can be expressed in terms of positive and negative damping; a well known example is given by the van der Pol equation. More complicated cases are represented by rotors in which the self-excitation effect is provided either by the action of forces in certain bearings

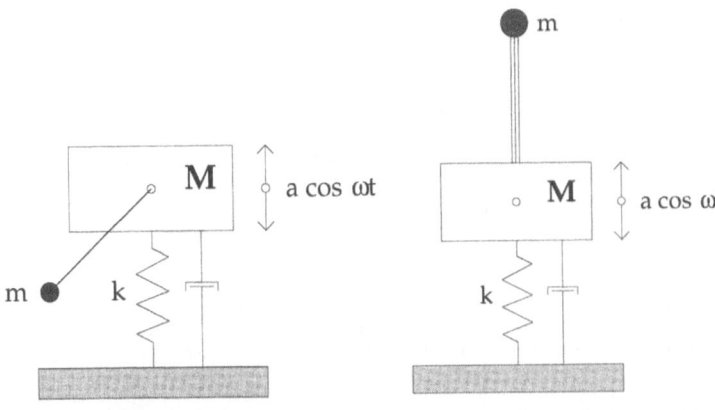

(a) Example of an autoparametric system consisting of a mass mounted on a spring (Oscillator) and of a pendulum attached to the mass (Excited System).

(b) Example of an autoparametric system where a mass m mounted on the end of a long elastic element (Excited System) is coupled to a forced, damped spring (the Oscillator).

Figure 1:

and forces arising in gap flows, or by the action of internal damping (Schmidt and Tondl, 1986).

Another important class consists of self-excited vibrations induced by cross-flow. A simple model is a spring-pendulum system as depicted in Figure 2 which is excited by cross-flow with constant velocity U. In this case the self-excitation is caused by hydrodynamic or aero-elastic forces which are very difficult to analyse. A simple model is discussed in section 3 and serves as a crude approximation. Note however that simple looking equations may lead to very complicated phenomena.

Another interesting type of autoparametric problems arises in dealing with nonlinear wave equations. In such problems one often applies a Galerkin-projection or finite Fourier-mode expansion which results in a finite chain of oscillators which is proposed as an approximating system of the original nonlinear wave problem. The question is then if higher order modes, which have been left out in the Galerkin-projection, can be excited by lower order modes. This would result in transfer of energy from lower order to higher order modes which destroys after some time the approximate character of the Galerkin-projection. In a number of cases this can be interpreted as autoparametric resonance.

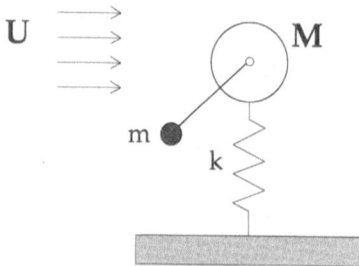

Figure 2: Example of an autoparametric system excited by flow consisting of a single-mass on a spring moving in one direction to which a pendulum is attached. The flow moves the mass M and so the spring (Oscillator) but not the pendulum (Excited System).

1.2   State of the art of autoparametric systems

Until recently, research in autoparametric systems in engineering was mainly concerned with the system consisting of a single-mass system with an attached pendulum, see for instance the books by Nayfeh and Mook (1979), Schmidt and Tondl (1986), and Cartmell (1990).

In the latter a comprehensive list of references is presented, and especially the papers of the following authors represent important contributions to this analysis: Barr, Bax, Cartmell, Ibrahim, Roberts. The last authors have directly investigated the autoparametric resonance of the previous system when it is tuned into internal resonance, i.e. for the case when the natural frequency of the harmonically excited mass-spring subsystem is approximately twice the natural frequency of the pendulum, and the excitation frequency is close to the natural frequency of the excited subsystem.

The stability investigation of the semi-trivial solution is presented in Tondl and Nabergoj (1990), Tondl (1991, 1992). In the approach of these papers one can determine all possible occurrences of autoparametric resonances.

In the literature up till 1992, the excitation of the Oscillator was considered to be external. In Tondl (1992) it was shown that not only external excitation but also parametric or self-excitation of the excited subsystem can be the source of parametric excitation for Excited Systems when certain conditions are met.

Mechanical models of ship behavior in longitudinal seas, developed to investigate the possibility of rolling motion instability, are analyzed in Tondl and Nabergoj (1990, 1992), Nabergoj and Tondl (1993). The analysis of the semi-trivial solution stability of more complicated models with a pendulum is given in Svoboda, Tondl and Verhulst (1992).

As remarked above, mainly external excitation for the Oscillator was considered in research up till 1992. The inclusion of other types of Oscillators in the autoparametric system as parametrically excited and self-excited Oscillators broadens the scope of autoparametric phenomena considerably.

In section 3 we shall present some of the results obtained by Ruijgrok (1995) and Verhulst and Tondl (1995).

2   Bifurcations of a parametrically forced nonlinear oscillator

Consider the following equation

$$\ddot{x} + k\dot{x} + (\alpha^2 + p(t))F(x) = 0 \tag{2.1}$$

where $k > 0$ is the damping coefficient, $F(x) = x + bx^2 + cx^3 + \cdots$, and time is scaled so that

$$p(t) = \sum_{l \in Z} a_{2l} e^{2ilt} \quad , \quad a_0 = 0 \quad , \quad a_{-2l} = \bar{a}_{2l} \tag{2.2}$$

is a π-periodic function with zero average. As is well known (see Yakubovitch and Starzhinskii, 1975), the trivial solution $x = 0$ is unstable when $k = 0$ and $\alpha^2 = n^2$, for all $n \in \mathbf{N}$. Fix a specific $n \in \mathbf{N}$ and assume that α^2 is close to n^2. We will study the bifurcations from the solution $x = 0$ in the case of primary resonance, which by definition occurs when the Fourier expansion of $p(t)$ contains nonzero terms $a_{2n}e^{2int}$ and $a_{-2n}e^{-2int}$. The parameters in this problem are the detuning $\sigma = \alpha^2 - n^2$, the damping coefficient k and the Fourier coefficients of $p(t)$. in particular a_{2n}. They are assumed to be small and of equal order of magnitude.

In Broer and Vegter (1992) the conservative case $k = 0$ was studied, here we consider the dissipative case $k > 0$. The analysis is based on Ruijgrok (1995).

2.1   Normal form equations

To find the time-periodic normal form of (2.1). we put $x = x_1$, $\dot{x} = x_2$ and write

$$\begin{aligned}
\dot{x}_1 &= x_2 \\
\dot{x}_2 &= -kx_2 - (n^2 + \sigma + p(t))F(x_1)
\end{aligned} \tag{2.3}$$

Equation (2.3) can be written in complex form. using $z = nx_1 - ix_2$ and expanding $F(x_1)$:

$$\dot{z} = inz - \tfrac{1}{2}k(z - \bar{z}) + \tfrac{1}{2n}i(\sigma + p(t))(z + \bar{z}) + \cdots \tag{2.4}$$

The equation for $\bar{z}$ has been omitted.

To equation (2.4) we apply the time-periodic normal form procedure as described in Hoveijn (1992). The righthand side of (2.4) is expanded in powers of

z, $\bar{z}$ and the parameters, which will be indicated by $\mu = (\sigma, k, a_2, a_4, \ldots)$. A long calculation yields the time-dependent normal form of (2.3), up to second order:

$$\dot{z} = inz + (-\tfrac{1}{2}k + \tfrac{1}{2n}i\sigma)z + \tfrac{1}{2n}ia_{2n}e^{2int}\bar{z} + L(z, \bar{z}, \mu, t) + K(z, \bar{z}, \mu, t) + igz|z|^2 + \mathcal{O}(|(z, \bar{z}, \mu)|^4) \qquad (2.5)$$

where

$$g = (\tfrac{3}{4}c - \tfrac{10}{3}b^2)$$

$$K(z, \bar{z}, \mu, t) = -\frac{ib}{6n^2}(a_{-n}e^{-int}z^2 + 2a_n e^{int}|z|^2 - 3a_{3n}e^{3int}\bar{z}^2) \qquad (2.6)$$

and $L(z, \bar{z}, \mu, t)$ contains terms which are linear in z and quadratic in the parameters. It can be assumed that after a suitable time translation, a_{2n} is real and positive. From this point on, it will be assumed that $g \neq 0$. This condition is satisfied by "almost all" choices of $F(x)$, and can therefore be called generic.

The normal form (2.5) can be made autonomous through the transformation $z = we^{int}$. After scaling time with a factor $\frac{1}{2n}$ and introducing $\kappa = nk$, the equation for w becomes:

$$\dot{w} = (-\kappa + i\sigma)w + ia_{2n}\bar{w} + K(w, \bar{w}, \mu) + igw|w|^2 + \hat{L}(w, \bar{w}, \mu) + \mathcal{O}(|(w, \bar{w}, \mu)|^4) \qquad (2.7)$$

where now:

$$K(w, \bar{w}, \mu) = -\frac{ib}{3n}(a_{-n}w^2 + 2a_n|w|^2 - 3a_{3n}\bar{w}^2) \qquad (2.8)$$

and

$$\hat{L}(w, \bar{w}, \mu) = 2nL(w, \bar{w}, \mu, 0) \qquad (2.9)$$

We now scale

$$\sigma = \varepsilon\hat{\sigma} \quad , \quad \kappa = \varepsilon\hat{\kappa} \quad , \quad a_{2j} = \varepsilon\hat{a}_{2j} \quad j = 1, 2, \ldots \qquad (2.10)$$

Following Broer and Vegter (1992) (where it is shown that all non-trivial fixed points of (2.7) are at $\mathcal{O}(\varepsilon^{\frac{1}{2}})$ of the origin) we also scale $w = \varepsilon^{\frac{1}{2}}\hat{w}$. Equation (2.7) becomes (dropping the hats and time scaling $\tau = \varepsilon t$)

$$\dot{w} = (-\kappa + i\sigma)w + ia_{2n}\bar{w} + \varepsilon^{\frac{1}{2}}K(w, \bar{w}, \mu) + igw|w|^2 + \mathcal{O}(\varepsilon) \qquad (2.11)$$

with $K(w, \bar{w}, \mu)$ as in (2.8). Note that the $\mathcal{O}(\varepsilon)$ estimate is valid, since it is easy to see that even terms in $(z, \bar{z})$ have coefficients of $\mathcal{O}(\varepsilon)$, so in particular terms of degree four will lead to $\mathcal{O}(\varepsilon)$ terms in the rescaled equation (2.11).

Equation (2.11) is invariant under $(w, \bar{w}) \rightarrow -(w, \bar{w})$ (up to $\mathcal{O}(\varepsilon)$ terms), if and only if $K(w, \bar{w}, \mu) = 0$. For sufficiently small ε, equation (2.11) can be treated as a perturbation of a symmetric system.

2.2   Dynamics and bifurcations of the symmetric system

In this section we assume that the autonomous normal form (2.7) is invariant under $(w, \bar{w}) \rightarrow -(w, \bar{w})$. This symmetry implies, amongst other things, that all fixed points come in pairs, and that bifurcations of the origin will be symmetric (such as pitchfork bifurcations). As was mentioned in the introduction, the normal form equation is symmetric when either $F(x)$ is odd in x or when n is odd. This is reflected by equation (2.11), which is invariant under $(x, y) \rightarrow -(x, y)$ only if the quadratic terms vanish. From (2.8) it is easy to see that this indeed is the case when $F(x)$ is odd, since then $b = 0$. Similarly, when n is odd, $p(t) = \sum_{l \in \mathbf{Z}} a_{2l} e^{2ilt}$ does not contain terms a_{-n}, a_n or a_{3n} and all the coefficients in (2.8) equal zero.

The symmetric equation, truncated at $\mathcal{O}(\varepsilon)$, is given by

$$\dot{w} = (-\kappa + i\sigma)w + ia_{2n}\bar{w} + igw|w|^2 \tag{2.12}$$

It is not difficult to show that, for sufficiently large R, the disc $|w| < R$ is invariant under the flow of (2.12), and that the only attractors in this area are fixed points. The dynamics of (2.12) can be summarized in Figure 3
Outside the hyperbola $\kappa^2 + \sigma^2 = a_{2n}^2$ (that is, outside area II) the trivial solution is stable. On the hyperbola a pitchfork bifurcation occurs, which is supercritical if $\sigma > 0$ and subcritical if $\sigma < 0$. On the half line $a_{2n} = \kappa$, $\sigma < 0$ there occurs a double saddle-node bifurcation, i.e. two simultaneous saddle-node's.

2.3   Bifurcations in the general case

As was remarked earlier, the general equation (2.11) can be seen as a non-symmetric $\mathcal{O}(\varepsilon^{\frac{1}{2}})$ perturbation of the symmetric case. For most values of the parameters σ and a_{2n}, the phase-portraits of the symmetric equation are structurally stable, so for sufficiently small ε, the perturbation will have no qualitative effect. There will still be zero, two or four (nontrivial) fixed points, respectively, and they will remain hyperbolic for values of (σ, a_{2n}) outside a neighbourhood of the boundaries in the bifurcation diagram (Figure 3). These fixed points will, however, no longer come in symmetric pairs.

For the half-line $a_{2n} = \kappa$, $\sigma < 0$ in Figure 3, we can make the following remark. In the symmetric case, two saddle-node bifurcations occur simultaneously. Since saddle-node bifurcations are generic, they will persist in the perturbed case. However, because of the symmetry breaking, they will, in general, no longer occur simultaneously. We therefore expect that the half line will break up into two curves of saddle-node bifurcations. These considerations hold outside a neighbourhood of the point $(\sigma, a_{2n}) = (0, \kappa)$. Near this point we will find more complicated behaviour.

It follows that we only have to consider values of (σ, a_{2n}) near the hyperbola $\kappa^2 + \sigma^2 = a_{2n}^2$. For these values of the parameters (that is, at points in parameter-space where the trivial solution loses stability) we will then apply center-manifold

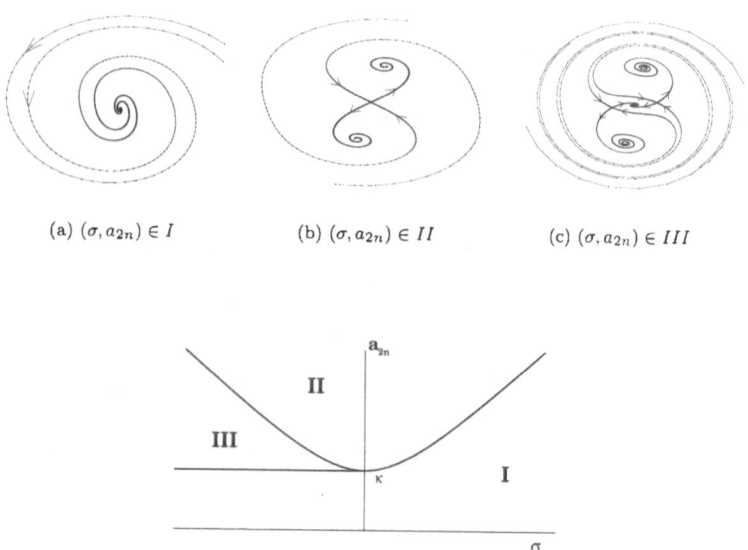

(a) $(\sigma, a_{2n}) \in I$ (b) $(\sigma, a_{2n}) \in II$ (c) $(\sigma, a_{2n}) \in III$

Figure 3: Bifurcation diagram in the (σ, a_{2n})-plane and phase-portraits of equation (2.12).

theory. Let $\lambda = -\kappa^2 - \sigma^2 + a_{2n}^2$ be the bifurcation-parameter. In Ruijgrok (1995) it is shown that the flow in the centre-manifold is given by

$$
\begin{aligned}
\dot{u} &= \mu_1 u + \varepsilon^{\frac{1}{2}}\mu_2 u^2 + \mu_3 u^3 + \varepsilon^{\frac{1}{2}}\mu_4 u^4 - u^5 \\
\dot{\lambda} &= 0
\end{aligned}
\tag{2.13}
$$

where μ_1 is proportional to λ and μ_3 is proportional to σ. The coefficients μ_2 and μ_4 are functions of λ, μ_3 and the Fourier-coefficients a_n and a_{3n}. The most degenerate member of the family of equations (2.13) is $\dot{u} = -u^5$, and it is not difficult to see that (2.13) defines a four-parameter unfolding of this degeneracy, such that $u = 0$ is always a solution (see Ruijgrok (1995) for details).

If σ is not small, the term u^3 in (2.13) dominates. We then have a symmetry breaking perturbation of the pitchfork-bifurcation (see Golubitsky and Schaeffer, 1983), leading to a saddle-node followed (or preceded) by a transcritical bifurcation.

We can now sketch the following partial bifurcation-diagram in (σ, a_{2n})-space, excluding a neighbourhood of the point $(\sigma, a_{2n}) = (0, \kappa)$ (see Figure 4).

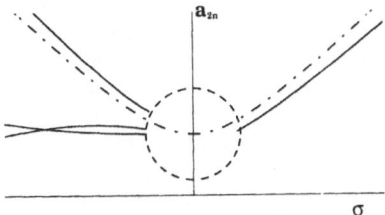

Figure 4: Partial bifurcation diagram in the (σ, a_{2n})-plane in the non-symmetric case.

It remains to analyse the case when σ is small. Rescale λ and σ through $\lambda = \varepsilon^{\frac{1}{2}}\hat{\lambda}$, $\sigma = \varepsilon^{\frac{1}{2}}\hat{\sigma}$. Truncating at $\mathcal{O}(\varepsilon^{\frac{1}{2}})$ and, as always, dropping the hats, yields the following equation for u

$$\dot{u} = \mu_1 u + \varepsilon^{\frac{1}{2}}\mu_2 u^2 + \mu_3 u^3 + \varepsilon^{\frac{1}{2}}\mu_4 u^4 - u^5$$

The translation $U = u + \frac{1}{5}\mu_4$ takes equation (2.14) to

$$\dot{U} = \nu_1 + \nu_2 U + \nu_3 U^2 + \nu_4 U^3 - U^5 \qquad (2.14)$$

where the ν_i, $i = 1, \ldots, 4$ are functions of the μ_i. The bifurcation-set of this equation (the "butterfly") is thoroughly examined in Poston and Stewart (1978) and Broecker and Lander (1975). It is not difficult to see that the map $(\nu_1, \nu_2, \nu_3, \nu_4) \rightarrow (\mu_1, \mu_2, \mu_3, \mu_4)$ is a diffeomorphism. The bifurcation-set in $(\mu_1, \mu_2, \mu_3, \mu_4)$ space is the image of the "butterfly" under this map.

A difficult problem is how to picture the bifurcation-set, since it lives in a four-dimensional space. Following an idea in Broecker and Lander (1975), we give a series of bifurcation pictures in the (μ_1, μ_3)-plane as the values of (μ_2, μ_4) are varied (see Figure 6). We choose (μ_1, μ_3), since we are interested in completing the bifurcation diagram Figure 4. The line $\mu_1 = 0$ corresponds to the hyperbola $\kappa^2 + \sigma^2 = a_{2n}^2$ and the line $\mu_3 = 0$ corresponds to the line $\sigma = 0$ (in Figure 6, we have actually reversed the direction of μ_3, so that now positive μ_3 corresponds to positive σ). It is therefore easy to transform a bifurcation-diagram in the (μ_1, μ_3)-plane to one in the (σ, a_{2n})-plane. As an example, consider Figure 5, which shows the bifurcation-diagram in the (μ_1, μ_3)-plane for $\mu_2 = 0$, $\mu_4 = 0$. This diagram, representing the symmetric case, is equivalent with Figure 3. On the line $\mu_1 = 0$, corresponding to the hyperbola in (σ, a_{2n})-space, there occurs a pitchfork bifurcation, which is supercritical when $\sigma < 0$ and subcritical when $\sigma > 0$. On the half-parabola $\mu_3^2 + 4\mu_1 = 0$, $-\mu_3 < 0$, corresponding to the half line $a_{2n} = \kappa$, $\sigma < 0$, there occur two saddle-nodes simultaneously. Figure 6 shows how this bifurcation diagram is perturbed when non-symmetric terms are added. In all cases the pitchfork-bifurcation is perturbed into a transcritical (the line $\mu_1 = 0$) and a saddle-node bifurcation. In all the bifurcation-diagrams in the (μ_1, μ_3)-plane we find cusp-points, i.e. points where the righthand side of (2.14) has a triple zero. For values of (μ_2, μ_4) on the cubic $\mu_4^2 - 16\mu_2 = 0$ (see Figure 6a and 6h), we have "swallowtail" points where the righthand side of (2.14) has a fourfold zero. Note that equation (2.14) is invariant under $(\mu_2, \mu_4) \rightarrow -(\mu_1, \mu_3)$, $u \rightarrow -u$. Therefore, going anti-clockwise in Figure 7 from $h)$ to $a)$ leads to the same sequence of bifurcation-diagrams. Finally, in Figure 7 we give the possible bifurcation diagrams in the (σ, a_{2n})-plane.

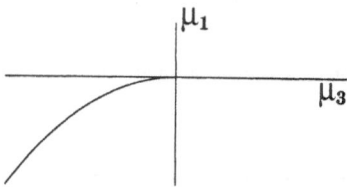

Figure 5: Bifurcation diagram in the (μ_1, μ_3)-plane in the symmetric case ($\mu_2 = \mu_4 = 0$).

2.4 Discussion

Consider again the original equation (2.3), written in complex form:

$$\dot{z} = F(z, \bar{z}, \mu, t) \quad z \in \mathbf{C} \tag{2.15}$$

and its normal form

$$\dot{\tilde{z}} = \tilde{F}(\tilde{z}, \bar{\tilde{z}}, \mu, t) \quad z \in \mathbf{C} \tag{2.16}$$

where $F(., t)$ and $\tilde{F}(., t)$ are π-periodic. The variables z and $\tilde{z}$ are related through $z = \tilde{z} + h(\tilde{z}, \bar{\tilde{z}}, \mu)$ with $h(\tilde{z}, \bar{\tilde{z}}, \mu)$ a C^∞ function whose Taylor expansion starts with quadratic terms.

One way to study the dynamics of (2.15) is through the Poincaré map $P_\mu :$ $\mathbf{C} \rightarrow \mathbf{C}$, defined by:

$$P_\mu(z) = X_\pi^\mu(z, 0) \tag{2.17}$$

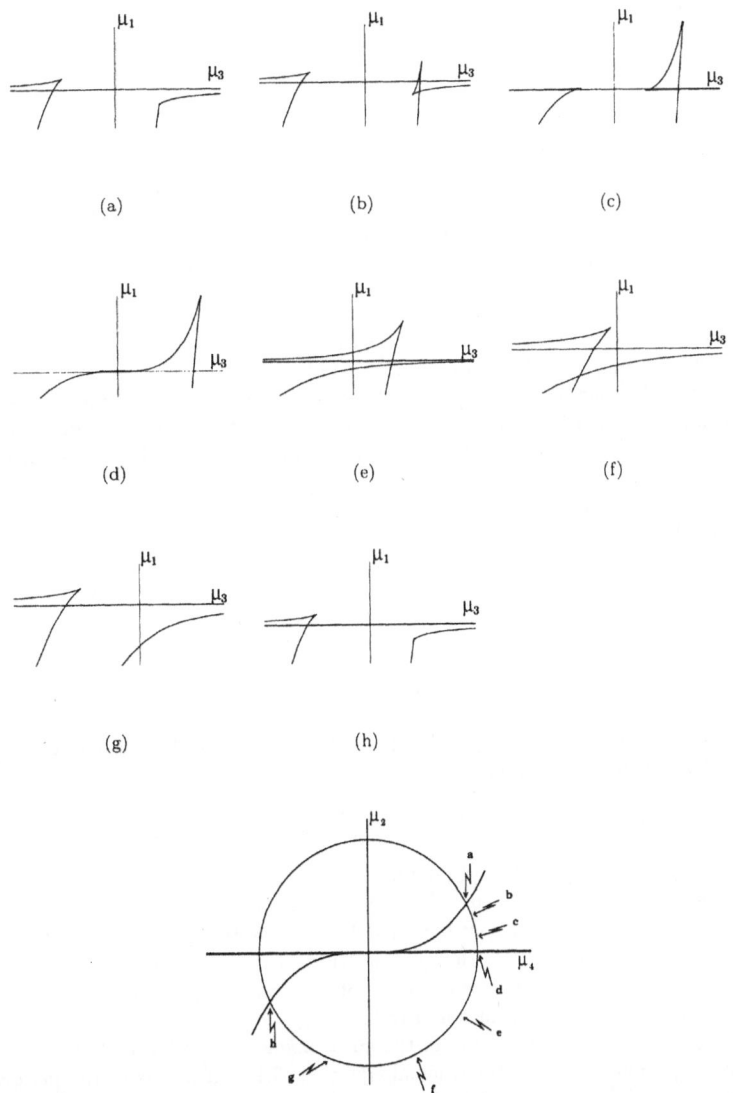

Figure 6: Bifurcation diagrams in the (μ_1, μ_3)-plane.

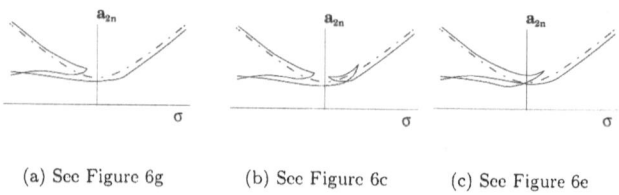

(a) See Figure 6g   (b) See Figure 6c   (c) See Figure 6e

Figure 7: Bifurcation diagrams in the (a_{2n}, σ)-plane.

where $X_t^\mu(v, t_0)$ is the solution of (2.15) with initial conditions $z(t_0) = v$, $\bar{z}(t_0) = \bar{v}$. The Poincaré map of (2.16) will be denoted by $\tilde{P}_\mu$.

A well known theorem on normal forms (see Hoveijn, 1992), implies that if $\tilde{z}_0$ is a hyperbolic fixed point of $\tilde{P}_\mu(\tilde{z})$, then there exists a nearby hyperbolic fixed point z_0 (of the same stability type) of $P_\mu(z)$. In this case, using the scalings of section 2.1, $z_0 = \tilde{z}_0 + \mathcal{O}(\varepsilon^{\frac{1}{2}})$.

Recall that the normal form (2.16) is made autonomous through the transformation

$$z = e^{int} w \tag{2.18}$$

leading to the equation

$$\dot{w} = \tilde{F}(w, \bar{w}, \mu, 0) \tag{2.19}$$

Now suppose that w_0 is a hyperbolic fixed point of (2.19). The image of w_0 under the Poincaré map $\tilde{P}_\mu$ is then

$$\tilde{P}_\mu(w_0) = e^{in\pi} w_0 \tag{2.20}$$

where we have used (2.17) and (2.18).

From (2.20) it follows that if n is even, w_0 is a hyperbolic fixed point of $\tilde{P}_\mu$, corresponding to a hyperbolic fixed point of P_μ, corresponding to a hyperbolic π-periodic solution of (2.15). However, when n is odd, we see that $\tilde{P}_\mu(w_0) = -w_0$, and that $(w_0, -w_0)$ is a hyperbolic orbit of period 2 of $\tilde{P}_\mu$, which corresponds to a hyperbolic 2π-periodic solution of (2.15).

The bifurcations occurring in the autonomous normal form (2.19) have the following implications for the original equation (2.15). If n is odd, the pitchfork bifurcation of the origin in (2.19) corresponds to a period-doubling (or *flip*) bifurcation of the solution $z = 0$ of the Poincaré map of (2.15). The normal form of the flip bifurcation is given by the 1-dimensional map $F(u, \lambda) = -u + \lambda u \pm u^3$, (see

Wiggins. 1990) where the sign of the cubic term determines whether this bifurcation is super- or subcritical. In general, a pair of fixed points $(w_0, -w_0)$ of (2.19) corresponds to one 2π-periodic solution of (2.15). Therefore, for parameter values in areas II and III in Figure 3, there are 1, respectively 2, 2π-periodic orbits. The symmetric saddle-node bifurcation corresponds with one saddle-node bifurcation of 2π-periodic orbits.

If n is even, but the function $F(x)$ in (2.15) is odd in x, the normal form (2.19) is the same as in the previous case. However, in this case the pitchfork bifurcation of the origin corresponds to a pitchfork bifurcation of π-periodic orbits in (2.15). The normal form of the pitchfork bifurcation is given by the 1 -dimensional map $F(u, \lambda) = u + \lambda u \pm u^3$. For parameter values in areas II and III in Figure 3, there are 2, respectively 4, π-periodic orbits. The symmetric saddle-node bifurcation corresponds to two saddle-node bifurcation of π-periodic orbits, occurring simultaneously.

If n is even, and $F(x)$ is not odd in x, there are only transcritical and saddle-node bifurcations occurring in (2.19), corresponding to transcritical and saddle-node bifurcations of π-periodic orbits. The normal forms are respectively $F(u, \lambda) = u + \lambda u \pm u^2$ and $F(u, \lambda) = u + \lambda \pm u^2$.

In Broer and Vegter (1992), the Hamiltonian case ($k=0$) of (2.1) is studied. Comparing with this study, we find similarities and differences.

The similarity is in the autonomous normal form (2.7), which for $k = 0$ is the same as calculated in Broer and Vegter (1992). Also. to the special (symmetric) cases of the present chapter, there correspond similar symmetric Hamiltonian cases, with similar codimension 1 bifurcations (transcritical, saddle-node, pitchfork or flip, depending on the specific symmetry). In the Hamiltonian case, however, there is an additional possibility of symmetry, namely when (2.1) is time-reversible (i.e. when $p(t) = p(-t)$ for all $t \in \mathbf{R}$). Reversibility does not occur in the present, dissipative, case.

The most important difference is that in the dissipative case, the bifurcation analysis can be reduced to studying a 1-dimensional system. Also, there are few difficulties in translating the bifurcation results for the autonomous normal form back to the original equation. The Hamiltonian case is rather more complicated. In Broer and Vegter (1992), the authors use singularity theory for families of planar Hamiltonian functions in their analysis. They reduce the normal form to a two parameter family of Hamiltonians (these parameters are roughly equivalent with σ and a_{2n}). In general, the analysis in the Hamiltonian case is more subtle than in the dissipative case.

3   Autoparametric resonance by self-excitation

3.1   The equations

Consider the following autoparametric version of (2.1)

$$\ddot{x} + \varepsilon k\dot{x} + (1 + \varepsilon\sigma + \varepsilon ay)F(x) = 0 \qquad (3.1)$$
$$\ddot{y} + 4y + \delta f(y,\dot{y}) + g(x,y) = 0 \qquad (3.2)$$

where $F(x)$ is a C^{∞} function whose Taylor expansion starts with linear terms. The coupling term in (3.2) has the form $g(x,y) = g_0(x) + \delta y g_1(x) + \delta^2 y^2 g_2(x) + \cdots$, and ε and δ are small parameters. It will be shown later that the most interesting case occurs when we take ε and δ to be of the same order of magnitude. It is further assumed that the subsystem

$$\ddot{y} + 4y + \delta f(y,\dot{y}) + g(0,y) = 0 \qquad (3.3)$$

defines a self-excited oscillator which has a stable π-periodic solution. Because (3.3) is autonomous, a closed orbit in the phase space actually corresponds to a one-parameter family of periodic solutions $(y,\dot{y}) = (y_p(t+\phi),\dot{y}_p(t+\phi))$. parametrized by the phase $\phi \in [0,\pi]$. From this point on we will identify a closed orbit with one element of this family. Analogously to previous cases, we expect that the semi-trivial solution $(x,\dot{x},y,\dot{y}) = (0,0,y_p(t),\dot{y}_p(t))$ loses stability for certain values of the parameters and that the x-mode then is excited. Also as in previous cases, it is expected that the resulting oscillations in the x-mode will have an amplitude of $\mathcal{O}(\varepsilon^{\frac{1}{2}})$.

In this section we present a summary of a study of system (3.1. 3.2). Details can be found in Ruijgrok (1995). We make the following assumptions:

1. y and time have been scaled so that subsystem (3.3) has a periodic solution of the form $y_p(t) = \cos(2t) + \mathcal{O}(\delta)$.

2. $f(y,\dot{y})$ has a Taylor expansion with non-trivial 3-jet, producing self-excitation. An example is the van der Pol case, where $f(y,\dot{y}) = \dot{y}(1-y^2)$.

3. ε and δ are small parameters such that $\varepsilon = \mathcal{O}(\delta)$ and $\delta = \mathcal{O}(\varepsilon)$. In the case that $\varepsilon = o(\delta)$, the equation for y in system (3.1, 3.2) decouples (to first order), and it can be described by equation (2.1) and an independent self-excited oscillator. If $\delta = o(\varepsilon)$, the selfexcitation term in the equation for y can be neglected with respect to the coupling term $g(x,y)$ and so system (3.1, 3.2) can in that case no longer be seen as an analogy of a parametrically excited system.

Then, after scaling x by $\sqrt{\varepsilon}$, writing $x = x_1$, $\dot{x} = x_2$ and introducing $z_1 = x_1 - ix_2$, $z_2 = 2y_1 - iy_2$, the normal form of system (3.1, 3.2), truncated at $\mathcal{O}(\varepsilon)$, becomes

$$\begin{aligned}
\dot{z}_1 &= iz_1 + \varepsilon((-k+i\sigma)z_1 + iA\bar{z}_1 z_2 + iBz_1|z_1|^2) \\
\dot{z}_2 &= 2iz_2 + \varepsilon(cz_2(1-|z_2|^2) + ic_0 z_1^2)
\end{aligned} \qquad (3.4)$$

with $z_1, z_2 \in \mathbf{C}$ (equations for $\bar{z}_1$ and $\bar{z}_2$ have been ommitted). The constants A, B, $k > 0$. σ. $c > 0$ and c_0 are real. It is this equation which will be studied in this section.

Note the following well known properties of truncated normal forms of the type (3.4): hyperbolic fixed points, closed orbits and invariant tori correspond to the same in the original system (3.1, 3.2). Also, the normal form (3.4) is invariant under the elements of the one-parameter group of linear transformations $\mathcal{G} \subset Gl(2, \mathbf{C})$, defined by:

$$\mathcal{G} = \{g | g = e^{L_0 s}, s \in [0, 2\pi], L_0 = \begin{pmatrix} i & 0 \\ 0 & 2i \end{pmatrix}\}$$

(see Iooss, 1988). This last property will be used in a following section to reduce the dimension of the phase space.

3.2 Stability and bifurcation of the semi-trivial solution

System (3.4) has a π-periodic solution given by $z_1 = 0$. $z_2 = e^{2it}$. To study the stability of this solution, we define $z_2 = e^{2it} + \hat{z}_2$. System (3.4), linearized near $z_1 = \hat{z}_2 = 0$ becomes: (dropping the hat)

$$\begin{aligned}
\dot{z}_1 &= iz_1 + \varepsilon((-k + i\sigma)z_1 + iAe^{2it}\bar{z}_1) \\
\dot{z}_2 &= 2iz_2 - c\varepsilon(z_2 + e^{4it}\bar{z}_2)
\end{aligned} \tag{3.5}$$

Let $z = (z_1, z_1, \bar{z}_1, \bar{z}_2)$. According to Floquet theory (see Verhulst, 1996), the solution of this π-periodic linear equation with initial condition $z = z_0$ can be written as $z = e^{Dt}P(t)z_0$ with D a constant 2×2 complex matrix and $P(t)$ a π-periodic 2×2 complex matrix. Let λ_i ($i = 1, \ldots, 4$) be the eigenvalues of $e^{\pi D}$. The λ_i are known as the characteristic multipliers. One of the multipliers will equal 1 (say $\lambda_4 = 1$), because we are linearising near a closed orbit of an autonomous equation. The closed orbit will be stable iff $|\lambda_i| < 1$ for $i = 1, 2, 3$. An exact solution of (3.5) can be found by transforming $z_1 = e^{it}w_1$, $z_2 = e^{i2t}w_2$, which yields the autonomous equation for w_1, w_2

$$\begin{aligned}
\dot{w}_1 &= \varepsilon((-k + i\sigma)w_1 + iA\bar{w}_1) & \dot{\bar{w}}_1 &= \varepsilon(-iAw_1 + (-k - i\sigma)\bar{w}_1) \\
\dot{w}_2 &= -c\varepsilon(w_2 + \bar{w}_2) & \dot{\bar{w}}_2 &= -c\varepsilon(w_2 + \bar{w}_2)
\end{aligned} \tag{3.6}$$

This equation has eigenvalues $\varepsilon(-k \pm \sqrt{\sigma^2 - A^2})$. $-2c\varepsilon$ and 0. The characteristic multipliers of D are then

$$\lambda_{1,2} = e^{(i + \varepsilon(-k \pm \sqrt{\sigma^2 - A^2}))\pi} \quad, \quad \lambda_3 = e^{(2i - 2\varepsilon c)\pi} \quad. \quad \lambda_4 = e^{i2\pi} \tag{3.7}$$

It then follows that the semi-trivial solution is stable iff:

$$A^2 < k^2 + \sigma^2 \tag{3.8}$$

From the expressions for the characteristic multipliers (3.7) it follows that at a point of bifurcation, i.e. when $A^2 = k^2 + \sigma^2$, one of the characteristic multipliers equals -1. This implies that the π-periodic semi-trivial solution undergoes a period-doubling bifurcation at this point (see Arnold, 1984).

To study this bifurcation, we first transform $z_1 = e^{it}w_1$, $z_2 = e^{2it}w_2$, yielding

$$
\begin{aligned}
\dot{w}_1 &= \varepsilon((-k + i\sigma)w_1 + iA\bar{w}_1 w_2 + iBw_1|w_1|^2) \\
\dot{w}_2 &= \varepsilon(cw_2(1 - |w_2|^2) + ic_0 w_1^2)
\end{aligned}
\tag{3.9}
$$

The fact that this transformation leads to an autonomous equation for w_1, w_2 is a consequence of the invariance of (3.4) under $z_1 \to e^{is}z_1$, $z_2 \to e^{2is}z_2$. It is easy to see that after a suitable scaling of w_1, w_2, A, B, k, σ and a time scaling, we can take $c = c_0 = 1$. In polar coordinates $w_1 = re^{i\phi_1}$, $w_2 = Re^{i\phi_2}$ and after a time scaling by a factor ε, (3.9) becomes

$$
\begin{aligned}
\dot{r} &= -kr + ArR\sin(2\phi_1 - \phi_2) \\
\dot{\phi}_1 &= \sigma + Br^2 + AR\cos(2\phi_1 - \phi_2) \\
\dot{R} &= R(1 - R^2) - r^2\sin(2\phi_1 - \phi_2) \\
\dot{\phi}_2 &= \frac{r^2}{R}\cos(2\phi_1 - \phi_2)
\end{aligned}
\tag{3.10}
$$

Because (3.9) is invariant under $w_1 \to e^{is}w_1$, $w_2 \to e^{2is}w_2$, it follows that (3.10) is invariant under $\phi_1 \to \phi_1 + s$, $\phi_2 \to \phi_2 + 2s$, which explains why only the combination-angle $2\phi_1 - \phi_2$ occurs in the righthandside of (3.10). Writing $\psi = 2\phi_1 - \phi_2$, the equations (3.10) can be reduced to the 3-dimensional system which will be central in the rest of this section:

$$
\begin{aligned}
\dot{r} &= -kr + ArR\sin\psi \\
\dot{R} &= R(1 - R^2) - r^2\sin\psi \\
\dot{\psi} &= \left(2AR - \frac{r^2}{R}\right)\cos\psi + 2\sigma + 2Br^2
\end{aligned}
\tag{3.11}
$$

Remark Let (r_0, R_0, ψ_0) be a fixed point of (3.11). System (3.10) then has a phase-locked solution of the form $r = r_0$, $R = R_0$, $\phi_1 = \alpha t + \phi_0$, $\phi_2 = 2\alpha t + 2\phi_0 - \psi_0$, where $\alpha = \sigma + AR_0\cos\psi_0 + br_0^2$. Therefore, (3.4) has a solution of the form

$$
z_1 = r_0 e^{i((1+\varepsilon\alpha)t + \phi_0)} \quad , \quad z_2 = R_0 e^{2i((1+\varepsilon\alpha)t + \phi_0 - \frac{1}{2}\psi_0)}
$$

This describes a periodic solution with period $\frac{2\pi}{1+\varepsilon\alpha}$, corresponding to a closed orbit in phase space. As was noted before, if this closed orbit is hyperbolic, then the phase space of the original system (3.1, 3.2) also contains a hyperbolic closed orbit. The constant ψ_0 is in a sense the phase difference between the z_1 and z_2 modes.

We will not go into the bifurcation analysis of the semi-trivial solution here (see Ruijgrok, 1995). As an illustration, we give a typical bifurcation diagram in the (A, σ)-plane (see Figure 8), obtained through a combination of center-manifold reduction and numerical analysis, using LOCBIF (Khibnik et al. 1992).

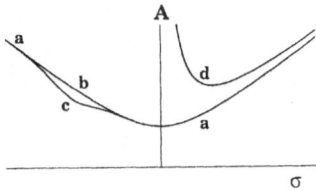

Figure 8: Typical bifurcation diagram in the (A, σ)-plane. On **a** (respectively on **b**) the origin undergoes a supercritical (respectively subcritical) pitchfork bifurcation. On **c** there is a saddle-node of non-trivial fixed points. On the curve inside the hyperbola, a secondary Hopf bifurcation occurs.

3.3 Chaotic dynamics

Even more interesting is that on the Hopf-curve in the (A, σ)-plane, there are isolated points where the eigenvalues of the linearised vectorfield are $\pm i\omega$ and 0. The bifurcations associated with this singularity have first been studied in Guckenheimer (1980) and Takens (1974).

In Ruijgrok (1995) it is shown that the normal form of (3.11) has a family of heteroclinic solutions, for certain values of the parameters near the above mentioned bifurcation-point. These solutions are not structurally stable. By adding a "flat" perturbation to (3.11), the family of heteroclinic solutions will in general break up and may yield an orbit homoclinic to a fixed point. If a technical condition on the eigenvalues of the fixed point is met, we have a Silnikov bifurcation. This is a global bifurcation for which it can be shown that near the bifurcation point chaotic dynamics, involving imbedded horseshoes, occur (see Silnikov, 1965 and Tresser, 1984). Phase-portraits of (3.11) illustrate some of the complex dynamics (see Figure 9).

Finally, a remark about the connection between the dynamics of (3.11), which after all is itself a truncated normal form equation, and the original system (3.1, 3.2). Solutions of the normal form equations are close approximations of solutions of the original equations, on a time scale of $\frac{1}{\varepsilon}$. In particular, if the normal form has chaotic solutions (for example near a Silnikov bifurcation point), then the original equations will show the same behaviour, at least on a time scale $\frac{1}{\varepsilon}$. Furthermore, hyperbolic fixed points and periodic solutions of (3.11) correspond to periodic solutions and invariant tori of the normal form of system (3.1, 3.2) and therefore persist for system (3.1, 3.2) itself.

When (3.11) undergoes a Silnikov bifurcation, it is known (see Wiggins, 1990) that a countable infinity of horseshoes is created. In Tresser (1984) it is shown

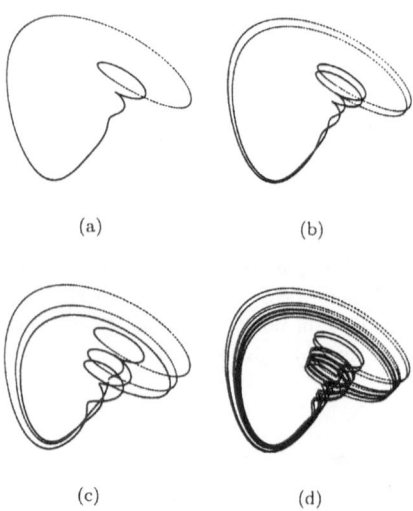

(a) (b)

(c) (d)

Figure 9: Some phase-portraits of (3.11), showing a sequence of period-doubling bifurcations.

that, under generic conditions, for a sufficiently small perturbation of (3.11) a finite number of these horseshoes persists. Because the original system (3.1, 3.2) is a small perturbation of its normal form (the size of the perturbation can be made arbitrarily small by normalizing to a sufficiently high degree), it is therefore conjectured that system (3.1, 3.2) possesses horseshoes. for most values of the parameters, when (3.11) undergoes a Silnikov bifurcation.

References

[1] Arnold V. I., *Geometrical Methods in the Theory of Ordinary Differential Equations*, Springer-Verlag, 1984.

[2] Broecker Th., Lander L., *Differentiable Germs and Catastrophes (LMS Lecture Notes vol 17)*, Cambridge University Press, 1975.

[3] Broer H.W., Vegter G., *Bifurcational aspects of parametric resonance, Dynamics reported: new series vol. 1: Expositions in dynamical systems*, eds. C.K.R.T. Jones, U. Kirchgraber and H.O. Walther, Springer, 1992.

[4] Cartmell, M., *Introduction to Linear, Parametric and Nonlinear Vibrations*, Chapman and Hall, London, 1990.

[5] Golubitsky M., Schaeffer D.G., *Singularities and Groups in Bifurcation Theory*, Springer-Verlag, New York, 1983.

[6] Guckenheimer J., Holmes P., *Nonlinear Oscillations, Dynamical Systems and Bifurcations of Vectorfields*, Springer-Verlag, New York, 1983.

[7] Guckenheimer J., On a codimension two bifurcation, *Dynamical Systems and Tubulence, Warwick 1980*, Springer LNM 898.

[8] Hoveijn I., *Aspects of Resonance in Dynamical Systems*, Thesis, Utrecht University, 1992.

[9] Iooss G., *Global characterization of the normal form for a vectorfield near a closed orbit.*, Journal of Differential Equations **76**, 47–76. (1988).

[10] Khibnik A.I., Kuznetsov Yu.A., Levitin V.V., Nikolaev E.V., LOCBIF, CAN Expertise Centre, Amsterdam, 1992.

[11] Nabergoj, R., Tondl, A., Simulation of parametric ship rolling: effects of hull bending and torsional elasticity, *Nonlinear Dynamics*, **5**, 1993.

[12] Nayfeh A.H., Mook D.T., *Nonlinear Oscillations*, Wiley Interscience, New York. (1979)

[13] Schmidt G., Tondl A., *Non-linear Vibrations*, Cambridge University Press, Cambridge, 1986.

[14] Poston T., Stewart I., *Catastrophe Theory*, Pitman, London, 1978.

[15] Ruijgrok M., *Studies in Parametric and Autoparametric Resonance*, Thesis, Utrecht University, 1995.

[16] Silnikov L.P., A case of the existence of a denumerable set of periodic motions, *Sov. Math. Dokl.*, **6**, 63–71, 1965.

[17] Svoboda, R.,Tondl, A., Verhulst, F., Autoparametric resonance by coupling of linear and nonlinear systems, *International Journal of Non-Linear Mechanics*, **123**, 1992.

[18] Takens F., Singularities of vectorfields, Publ. Math. I.H.E.S., **43**, 1974.

[19] Tondl, A., On the stability of a rotor system, *Acta Technica CSAV*, **36**, 331–338, 1991.

[20] Tondl, A., A contribution to the analysis of autoparametric systems, *Acta Technica CSAV*, **37**, 735–758, 1992.

[21] Tondl A., Elastically mounted body in cross flow with an attached pendulum, *Proceedings 14th Biennial ASME Conference Mechanical Vibration and Noise*, Albuquerque, 1993.

[22] Tondl, A., Nabergoj, R., Model simulation of parametrically excited ship rolling, *Nonlinear Dynamics*, **1**, 131–141, 1990.

[23] Tondl, A., Nabergoj, R., Simulation of parametric ship hull and twist oscillations, *Nonlinear Dynamics*, **3**, 41–56, 1992.

[24] Tresser C., About some theorems of L.P. Silnikov, *Ann. Inst. H. Poincare*, **40**, 440–461, 1984.

[25] Verhulst F., *Nonlinear Differential Equations and Dynamical Systems*, revised edition, Springer-Verlag, New York, (1996).

[26] Verhulst F. and Tondl A., Autoparametric resonance by self-excitation, to be published, 1995.

[27] Wiggins S., *Introduction to Applied Nonlinear Dynamical Systems and Chaos*, Springer-Verlag. New York, 1990.

[28] Yakubovich V.A., Starzhinskii V.M., *Linear differential equations with periodic coefficients*, Vols. I and II, Wiley, New York, 1975.

Progress in Nonlinear Differential Equations
and Their Applications, Vol. 19
© 1996 Birkhäuser Verlag Basel/Switzerland

Global attractors and bifurcations

Marcelo Viana*

Abstract

We present some recent developments in the study of attractors of smooth
dynamical systems, specially attractors whose basin has a global character.
A key point in our approach is to explore the relations between this study
and that of main bifurcation mechanisms.

1 Introduction

We consider both continuous time dynamical systems (flows) and discrete time
dynamical systems (smooth transformations, diffeomorphisms) on manifolds. In
the first setting we use $\varphi^t: M \longrightarrow M$, $t \in \mathbf{R}$, to denote the flow. In the second one
we let $\varphi: M \longrightarrow M$ be the transformation and denote its t-iterate $\varphi^t = \varphi \circ \cdots \circ \varphi$,
for each integer $t \geq 1$; if φ is invertible we also write $\varphi^{-t} = (\varphi^t)^{-1}$.

A main problem in Dynamics, which we want to address here, is to describe
the (typical) asymptotic behaviour of trajectories $\varphi^t(z)$. $z \in M$, as time t goes to
$+\infty$. Let an *attractor* be a (compact) subset A of the ambient manifold M such
that

- A is *invariant* under time evolution: $\varphi^t(A) = A$ for every $t > 0$;

- A is *dynamically indivisible*: it contains some dense orbit (alternatively, one
  may ask that A support an ergodic invariant measure);

- *the basin of A*, defined by $B(A) = \{z \in M: \varphi^t(z) \to A \text{ as } t \to +\infty\}$, *is a
  large set*: it contains a neighbourhood of A (weaker definitions are obtained
  by requiring $B(A)$ to have nonempty interior or even just positive Lebesgue
  measure).

Then this problem can be rephrased in terms of describing the properties of attractors, namely

- geometric and topological properties (fractional dimensions, topological invariants);

---

*Marcelo Viana, IMPA, Est. D. Castorina 110, Jardim Botânico, 22460-320 Rio de Janeiro,
Brazil, Tel: +55(21)294-9032   Fax: +55(21)512-4115   E-mail: viana@impa.br

- dynamical properties (symbolic dynamics, (non)hyperbolicity, Lyapunov exponents);

- ergodic properties (asymptotic measures, statistical parameters).

Of particular interest is to investigate the robustness (or *persistence*) of these features of the dynamics when the system is perturbed (either deterministically or randomly).

Besides the beautiful theory developed throughout the sixties and the seventies for the case of Axiom A systems, see e.g. [Sm], [Bo], a great deal of interest has been devoted in recent years to trying to provide a satisfactory answer to these questions for more general classes of attractors, lacking uniform hyperbolicity. Motivation comes both from the applications (models of natural phenomena are seldom uniformly hyperbolic) and from the intrinsic richness of such systems, which combine (structural) instability with some remarkable forms of persistence.

A fruitful approach, strongly advocated by J. Palis, has been to try and relate the study of (nonhyperbolic) attractors with that of the generic processes through which the dynamics varies as the initial system is modified (bifurcation processes). More precisely, one considers parametrized families of dynamical systems unfolding a given type of bifurcation (such as nontransverse homoclinic trajectories or nonstable cycles envolving periodic trajectories, for instance) and one tries to describe the presence and the properties of attractors in those families. Results such as [MV], [DRV], [Mo] or [MP], for instance, may be thought of from this perspective. A second, kind of converse, step has also been proposed by Palis: to show that generic dynamical systems with nonhyperbolic attractors (or other relevant unstable phenomena) can be approximated by others exhibiting one of a small number of bifurcation types. Results of this kind include e.g. [Ur]. [Ca].

Here we discuss a number of recent progresses in this general program. In Section 2 we analyse *the basin of Hénon-like attractors*, to prove that it *contains a neighbourhood of the attractor*, at least for a large set of parameters. This is well-known in the orientation-reversing case, but the, possibly even more relevant, orientation-preserving case seems to be new. We also announce a more quantitative result, of ergodic flavour, recently established by M. Benedicks, and myself: *almost every point in the basin of attraction is generic with respect to the Sinai-Ruelle-Bowen measure of the attractor*.

Section 3 corresponds to joint work with V. Baladi concerning the *ergodic properties of certain nonuniformly hyperbolic unimodal maps* of the interval. The main result asserts that those properties, including the fact that such maps are exponentially mixing (exponential decay of correlations), *are robust under random perturbations of the map* (stochastic stability).

Section 4 was written jointly with S. Luzzatto and contains a discussion of an extended geometric model for the behaviour of Lorenz equations. The goal of this model is to provide insight into the way the strange attractor is destroyed through the introduction of "folds", as the parameters are varied. The main statement is that *the attractor persists after the appearance of the folds. but only for a positive measure set of parameter values.*

In Section 5, we report on joint work with M. J. Pacifico and A. Rovella. We consider smooth flows in 3-dimensional manifolds exhibiting homoclinic connections associated to equilibrium points of saddle-focus type. Then we prove that *a new type of global attractor, with spiraling geometry. occurs (and is even a persistent phenomenon) in such families.*

2   The basin of Hénon-like attractors

In Section 2.1 we prove that *the basin of Hénon-like attractors contains a full neighbourhood of the attractor,* for a large set of parameter values, and we also state two related conjectures. Then, in Section 2.2, we discuss a substantial refinement of this result: *Lebesgue almost every orbit in the basin is (exponentially) asymptotic to some orbit in the attractor.* As a consequence. *almost every point in the basin is generic (in the sense of the ergodic theorem) with respect to the SRB-measure of the attractor.*

2.1   The topological basin

Let $(\Psi_a)_{1<a<2}$ denote the family of quadratic real maps $\Psi_a(x) = 1 - ax^2$ (this may replaced by much more general families of unimodal or multimodal maps of the interval, see e.g. [DRV, Section 5]). We also consider the corresponding family of endomorphisms ψ_a of the plane, given by $\psi_a(x,y) = (\Psi_a(x), 0)$. By a *Hénon-like map* we mean here any map φ on the plane which is close enough to some ψ_a in the C^r-sense

$$\|\varphi - \psi_a\|_{C^r} < b, \qquad b > 0 \quad \text{small}$$

(usually one assumes $r \geq 3$; just how small b should be depends on the context). In all that follows we suppose that φ is an orientation-preserving diffeomorphism but similar arguments apply in the orientation-reversing case.

It is straightforward to check that for every $a \in (1.2)$ the map Ψ_a has exactly two fixed points $Q_a < 0 < P_a$ and that these are both hyperbolic (repelling). Moreover. the unstable set of P_a is a compact interval contained in $(Q_a, -Q_a)$. Then a corresponding statement holds for ψ_a: it has exactly two fixed points $q_a = (Q_a, 0)$ and $p_a = (P_a, 0)$, which are hyperbolic saddles, and the unstable set of p_a lies inside $(Q_a, -Q_a) \times \{0\}$. Now let $p = p(\varphi)$. $q = q(\varphi)$ be the continuation of these fixed points for a nearby diffeomorphism φ. Then p and q are still hyperbolic saddles and the unstable manifold $W^u(p)$ is contained in a bounded region $(Q_a. -Q_a) \times (-b,b)$. It is well known (see e.g. [BC2]) that the compact set $A = A(\varphi) = $ closure $(W^u(p))$ has a basin $B(A)$ with nonempty interior. Moreover. [BC2]. [MV]. proved that very often (positive measure set of parameters) A contains a dense orbit with expanding behaviour (positive Lyapunov exponent).

Here we want to prove

Theorem 2.1 *There exist sequences $(I_j)_j$ of compact intervals converging to $a = 2$ and $(b_j)_j$ of positive numbers, such that, for any diffeomorphism φ satisfying $\|\varphi - \psi_a\|_{C^1} < b_j$ for some $a \in \cup I_j$, the basin $B(A)$ contains a neighbourhood of A.*

Proof: In order to exhibit the intervals I_j we go back to the quadratic family $\Psi_a(x) = 1 - ax^2$. An explicit calculation shows that if a is close to $a = 2$ then P_a is close to $x = 1/2$. Moreover, $\Psi_a^{-1}(P_a)$ consists of two points $P_{a,1} < 0 < P_{a,0} = P_a$ and $\Psi_a^{-2}(P_a)$ consists of four points $P_{a,2} < P_{a,1} < 0 < P_{a,0} < P'_{a,2}$. We denote $J_{a,0} = [P_{a,0}, P'_{a,2}]$ and let $J_{a,1} = [P_{a,1}, P'_{a,3}]$ be the connected component of $\Psi_a^{-1}(J_{a,0})$ situated to the left of zero (assuming once more that a is close enough to 2). More generally, for $j \geq 1$ we let $J_{a,j} = [P_{a,j}, P'_{a,j+2}]$ be the connected component of $\Psi_a^{-1}(J_{a,j-1})$ contained in $\{x < 0\}$. Observe that the $J_{a,j}$ converge to the repelling fixed point Q_a as $j \to +\infty$. Now we fix a slightly smaller compact interval $\tilde{J}_{a,j} \subset \mathrm{int}\,(J_{a,j})$ and define the parameter interval I_j by

$$a \in I_j \iff \Psi_a(1) \in \tilde{J}_{a,j}$$

(1 is the critical value of Ψ_a). Remarking that $\Psi_a^2(0) = Q_a$ when $a = 2$, one concludes without difficulty that the sets I_j defined in this way are indeed compact intervals accumulating on $a = 2$.

Now, for $a \in I_j$ we consider the endomorphism ψ_a. Note that the stable set $W^s(p_a) = \{z : \psi_a^n(z) \to p_a$ as $n \to +\infty\}$ consists of all the vertical lines of the form $\{(x, y) : \Psi_a^n(x) = P_a$ for some $n \geq 0\}$. In particular, it contains the vertical lines $P_{a,i} \times \mathbb{R}$, $P'_{a,i+2} \times \mathbb{R}$, for each $i \geq 0$.

Then we let φ be any orientation-preserving diffeomorphism sufficiently C^1-close to ψ_a. More precisely, we take φ to be defined in some large square $S = [-l, l]^2$, with $\|\varphi - \psi_a\|_{C^1(S)} < b_j$ for some small b_j. It is convenient to begin by extending φ to a (proper) diffeomorphism of the whole plane and from now on φ will denote such an extension (note that neither A nor the fact that $B(A)$ is a neighbourhood of it depend on the choice of the extension). Since local invariant manifolds of periodic points vary continuously with the the dynamics, see e.g. [Shu] or [PT, Appendix 1]. we get that, provided $b_j > 0$ is small enough.

1. $W^s(p)$ contains segments C^1-near each $P_{a,i} \times [-2l, 2l]$ and $P'_{a,i+2} \times [-l, l]$ with $0 \leq i \leq j$.

2. $W^u(p(\varphi))$ folds near $x = 1$; the first (resp. the second) image of this fold is near $x = \Psi_a(1)$ (resp. $x = \Psi_a^2(1)$) and hence it is contained in the interior of $J_{a,j} \times [-b, b]$ (resp. $J_{a,j-1} \times [-b, b]$).

In particular, the segments of $W^s(p)$ mentioned in 1 intersect $W^u(p)$ and $W^u(q)$; we denote by p_i, p'_i the points of intersection with the local unstable manifold of q, see Figure 1. Now, $W^s(p)$ is an immersed submanifold of the plane and so these segments must be connected in some way. Using the assumption that φ preserves orientation (hence both eigenvalues of $D\varphi(p)$ are negative) one checks

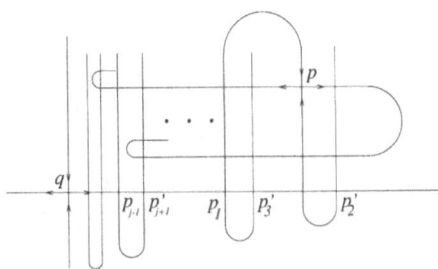

Figure 1: Invariant manifolds of p

easily that the segments passing through p, p'_2, p_1, must be connected as described in Figure 1. Moreover, iterating backwards we conclude that the segments passing through p_i and p'_{i+2}, $1 \leq i \leq j$, connect to each other as depicted. Observe here that each of these segments intersects the boundary of the horseshoe-shaped region $\varphi(S)$ in exactly four points, and so the corresponding preimage intersects ∂S also in four points.

At this point we fix some $\delta_j > 0$ and denote $K_j = [-2\delta_j, 2\delta_j] \times [-l, l]$. As long as δ_j and b_j are small enough, $\varphi(K_j)$, $\varphi^2(K_j)$, $\varphi^3(K_j)$ are small regions near $x = 1$, $x = \Psi_a(1)$, $x = \Psi_a^2(1)$, respectively. In particular, $\varphi^3(K_j)$ is contained in the interior of $J_{a,j-1} \times [-b, b]$ and, in fact, in the region Ω bounded by $W^u(p)$ and the piece of $W^s(p)$ connecting the nearly straight segments passing through p_{j-1} and p_{j+1}. It follows from well-known arguments (see e.g. Theorem 4 in [BC2]) that $\varphi^3(K_j) \subset \Omega$ is contained in the basin of the attractor $A = \text{closure}\,(W^u(p))$ and so the same holds for K_j.

We are left to consider those points in a neighbourhood of A whose forward orbit does not intersect K_j and we do this as follows. A result in [Ma] asserts that the set E_j of points in the interval whose forward trajectory is disjoint from $(-\delta_j, \delta_j)$ is a compact hyperbolic set for Ψ_a. Clearly, E_j contains the fixed point P_a and it is not difficult to deduce that $\Psi_a|E_j$ is transitive. Since hyperbolic sets are persistent under perturbations of the system, see e.g. [Shu] or [PT, Appendix 1], it follows that the set H_j of points whose full orbit remains outside $\tilde{K}_j = [-\delta_j, \delta_j] \times [-1, 1]$ is hyperbolic and transitive for φ. Moreover, the set of points whose forward orbit never enters K_j is contained in the stable set of H_j. In other words, all the points under consideration at this stage are attracted to H_j. On the other hand, H_j contains the fixed point p and so it is contained in $\text{closure}\,(W^u(p)) = A$. This completes our argument. $\square$

It is not difficult to check that the arguments in [MV] can be carried out within the parameter intervals I_j constructed above and so the present conclusions apply to the Hénon-like strange attractors found in there. Moreover, combining the present ideas with those in [Vi, Section 3] one obtains a similar conclusion for the quadratic-like attractors in higher-dimensional manifolds constructed there.

On the other hand, the previous result is somewhat unsatisfactory, in that one may expect the conclusion to hold for *all* values of a close to 2 (and small b), as happens in the orientation-reversing case. In this direction we state the following two conjectures.

1: There is a positive function $b(a)$, defined for a in a whole interval $(a_0, 2)$, such that the conclusion of the theorem holds if $\|\varphi - \psi_a\|_{C^1} < b(a)$ for some $a \in (a_0, 2)$.

2: Moreover, $b(a)$ may be taken so that if $\varphi(x, y) = (1 - ax^2 + by, \pm bx)$ (a Hénon map) with $b < b(a)$ then its nonwandering set $\Omega(\varphi)$ coincides with $Q(\varphi) \cup A(\varphi)$.

2.2　Exponential convergence and the ergodic basin

Another important problem is to characterize the ergodic basin of attraction of A. Let us formulate this more precisely. It was shown in [BY2] that for the parameter values such as in [BC2], [MV] the attractor A supports a measure of Sinai-Ruelle-Bowen. By such they mean an invariant probability measure μ supported on A, which is ergodic, has a positive Lyapunov exponent and, most important, induces absolutely continuous conditional measures along unstable manifolds (absolute continuity is with respect to the riemannian measure on the unstable manifold). Then standard arguments yield the following key property: μ *determines the asymptotic behaviour of time averages of all continuous functions* $g: B(A) \longrightarrow \mathbb{R}$

$$\frac{1}{t} \sum_{j=0}^{t-1} g(\varphi^j(x)) \to \int g \, d\mu \quad \text{as } t \to +\infty$$

for a positive (two-dimensional) Lebesgue measure set of points $x \in B(A)$. Of course, one would like to know whether this property holds for a *full measure* subset of the basin and this is the problem we are considering here.

Now, it is very easy to see that any (forward) asymptotic trajectories have the same (forward) limit time averages, for all continuous functions. Therefore, the previous problem is somewhat related to the question whether all (or almost all) the orbits in the basin are asymptotic to some orbit inside the attractor (as in the Axiom A case). Unfortunately, the elementary arguments we used in the proof of Theorem 2.1 seem of little help for solving this question. However, it turns out that the answer to both questions above is indeed positive, as shown recently by M. Benedicks and myself. More precisely, we prove that for the parameter values in [BC2] or in [MV], the attractor $A = A(\varphi)$ satisfies

Theorem 2.2 *Through Lebesgue almost every point in $B(A)$ there is a local stable manifold which intersects A. Moreover, we have*

$$\frac{1}{t} \sum_{j=0}^{t-1} g(\varphi^j(x)) \to \int g \, d\mu \quad as \ t \to +\infty,$$

for every continuous function g and for Lebesgue almost every $x \in B(A)$.

Recall that Lebesgue refers to the two-dimensional Lebesgue measure. Also, by a stable manifold we mean a curve which is exponentially contracted under all positive iterates of φ. The proof of Theorem 2.2 is to appear elsewhere.

3 Stochastic stability and exponential mixing

In this section we deal with nonuniformly hyperbolic unimodal maps of the interval $\varphi \colon I \longrightarrow I$, with $\varphi(I) \subset \text{int}(I)$. Our goal is to describe the main results and techniques in our paper with V. Baladi [BaV]. In an ongoing joint work with M. Benedicks we are extending part of these results (stochastic stability) to attractors of dissipative diffeomorphisms in higher-dimensional manifolds.

For simplicity we take φ to be quadratic, $\varphi(x) = a - x^2$, but our arguments hold for general unimodal maps with negative schwarzian derivative and nondegenerate critical point. We formulate the nonuniform hyperbolicity property in terms of the orbit of the critical point $c = 0$: let us assume that

1. $\left|(\varphi^k)'(\varphi(c))\right| \geq \lambda_c^k$ (positive Lyapunov exponent):

2. $\left|\varphi^k(c) - c\right| \geq e^{-\alpha k}$ (exponential recurrence bound)

for every $k \geq 1$ and for some constants $0 < \alpha \ll 1 < \lambda_c$. We also suppose that φ is topologically mixing (on the interval $\varphi^2(I)$).

This formulation is motivated by [BC1], [BC2], where it is proved that 1 and 2 above are satisfied by quadratic maps for a positive measure set of values of the parameter a. It follows from condition 1 and [Si] that φ can not have attracting periodic orbits. In contrast, as observed already by [BC1], [BC2], maps $\varphi_s(x) = \varphi(x) + s$ with small s may exhibit such periodic attractors. This means, in particular, that the dynamics of φ is very unstable under perturbations of the map. However, here we want to prove that *from a different, statistical, perspective the dynamics of such maps is, in fact, quite robust*. In order to state this in a precise way let us comment a bit more on conditions 1 and 2.

It is now well understood, [No], that condition 1 implies the conclusion of [Ja]: φ admits an invariant probability measure μ_0 which is absolutely continuous with respect to the Lebesgue measure m on I (even equivalent to m restricted to $\varphi^2(I)$). This measure μ_0 is unique, ergodic, and determines the asymptotics of typical orbits of φ:

$$\frac{1}{t}\sum_{j=0}^{t-1} g(\varphi^j(x)) \to \int g \, d\mu_0 \quad \text{as } t \to +\infty$$

for every continuous function g and m-almost all $x \in I$.

Now we want to consider the effect of adding random noise to the iteration of φ. More precisely, we want to compare the asymptotic behaviour of φ^t with that of $\varphi_{s_t} \circ \cdots \circ \varphi_{s_1}$, where $s_1, \ldots, s_t$ are chosen randomly and independently in some small interval $[-\varepsilon, \varepsilon]$ (we shall denote by θ_ε the corresponding probability distribution). Under general conditions, satisfied in our context, such a random scheme admits a stationary measure μ_ε, with

$$\frac{1}{t}\sum_{j=0}^{t-1} g(\varphi_{s_j} \circ \cdots \varphi_{s_1}(x)) \to \int g \, d\mu_\varepsilon \quad \text{as } t \to +\infty$$

for every continuous function g, m-almost all $x \in I$, and almost all choices of $(s_j)_{j\geq 1}$ (in the present context μ_ε is unique and absolutely continuous with respect to Lebesgue measure).

Then we say that φ is *(weakly) stochastically stable* if μ_ε is close to μ_0 (in the weak*-sense) when the noise level ε is close to zero.

Theorem 3.1 *[BaV]* φ *is strongly stochastically stable (hence stochastically stable), that is*

$$\frac{d\mu_\varepsilon}{dm} \to \frac{d\mu_0}{dm} \quad \text{in the } L^1\text{-sense,} \quad \text{as } \varepsilon \to 0.$$

That is, *small random noise has a neglectable effect on the asymptotic behaviour of the map.* Results such as this may be thought to provide some conceptual legitimacy to information concerning "chaotic" systems extracted from finite-precision numerical experiments (although round-off errors are not really random noise).

Before sketching the main points underlying Theorem 3.1, let us introduce another important, somewhat related, notion. We say that (φ, μ_0) is *exponentially mixing* (equivalently, has *exponential decay of correlations*), if there exists $\tau < 1$ such that, given test functions f and g,

$$\left| \int U_0^t(f) \cdot g \, d\mu_0 - \int f \, d\mu_0 \int g \, d\mu_0 \right| \leq C(f,g)\tau^t \quad \text{for all } t \geq 1$$

(U_0 denotes the spectral operator $U_0(f) = f \circ \varphi$). In other words, $f \circ \varphi^t$ and g, *viewed as random variables, become independent exponentially fast as $t \to +\infty$.* Formally speaking, f and g should be taken in some convenient Banach space, in our case this will be the space $BV(I)$ of functions of bounded variation on I. One can also define a notion of exponential mixing for the random scheme φ_s, $|s| \leq \varepsilon$, just by replacing above μ_0 by μ_ε, and U_0 by the perturbed spectral operator $U_\varepsilon(f)(x) = \int f(\varphi_s(x))\theta_\varepsilon(s)\,ds$.

Theorem 3.2 *[BaV] Both φ and its random perturbation schemes $(\varphi_s)_{|s|\leq\varepsilon}$, are exponentially mixing, with mixing rates τ, τ_ε, uniformly bounded away from 1.*

Not all the content of Theorems 3.1, 3.2 is new in [BaV]. Weak stochastic stability for quadratic maps was first proved by [KK], for uncountably many parameters, and by [BY1], for a positive measure set of parameters (but see also [Co], where strong stability was already considered). Exponential decay of correlations for (unperturbed) quadratic maps was proved independently by [KN] and [Yo]. See also [Ki] for many other references and general background.

3.1 Towers, co-cycles, and transfer operators

Now we outline the main ingredients in the proof of Theorems 3.1 and 3.2, referring the reader to [BaV] for details. The global strategy is inspired by [BaY], where similar results were obtained for certain uniformly hyperbolic systems.

Here we have to circumvent the lack of hyperbolicity and a first step in this direction is to construct a tower extension $\hat{\varphi}\colon \hat{I} \longrightarrow \hat{I}$ of $\varphi\colon I \longrightarrow I$. We fix positive constants $\beta \approx 2\alpha$ and $\delta \ll \alpha$ and then define

- $\hat{I} = \cup_{k\geq 0}(B_k \times \{k\})$, with $B_0 = I$ and B_k being the $e^{-\beta k}$-neighbourhood of $\varphi^k(c)$ for each $k \geq 1$.

- $\hat{\varphi}(x,k) = (\varphi(x), k+1)$ whenever $\varphi(x) \in B_{k+1}$ and either $k \geq 1$ or $k = 0$ with $|x| < \delta$; in all other cases $\hat{\varphi}(x,k) = (\varphi(x),0)$.

A main point in this construction is that return maps to the "ground floor" $E_0 = B_0 \times \{0\}$ are uniformly expanding:

(a) there is a constant $\lambda > 1$ such that $|(\hat{\varphi}^k)'(x)| \geq \lambda^{2k}$ whenever $(x,0) \in E_0$, $\hat{\varphi}^k(x,0) \in E_0$, and $\hat{\varphi}^i(x,0) \notin E_0$ for $0 < i < k$.

Note also that extensions $\hat{\varphi}_s\colon \hat{I} \longrightarrow \hat{I}$ of the perturbed maps $\varphi_s\colon \hat{I} \longrightarrow I$ can be defined in just the same way.

Next, we introduce a co-cycle $w_0\colon \hat{I} \longrightarrow [0,\infty)$, given by

- if $(y,k) = \hat{\varphi}^k(x,0)$ for some $x \in B_0$ then $w_0(y,k) = \lambda^k/|(\hat{\varphi}^k)'(x)|$ (in particular, $w_0 \equiv 1$ on E_0); otherwise $w_0(y,k) = 0$.

This definition ensures that the map $\hat{\varphi}$ is λ-expanding with respect to the metric $w_0 \, dx$ on the tower $\hat{I}$: this is automatic at points (x,k) with $\hat{\varphi}(x,k) = (\varphi(x),k+1)$ and (a) implies that it remains true when $\hat{\varphi}(x,k) = (\varphi(x),0)$. Then we also need a perturbed version w_ε of w_0, which we define by

$$w_\varepsilon(y,k) = \frac{1}{2} \int \frac{\lambda^k}{|(\varphi_{s_k} \circ \cdots \circ \varphi_{s_1})'(x_{s_1 \cdots s_k})|} \theta_\varepsilon(s_1) \, ds_1 \cdots \theta_\varepsilon(s_k) \, ds_k,$$

where the integral is taken over $(\hat{\varphi}_{s_k} \circ \cdots \circ \hat{\varphi}_{s_1})(x_{s_1 \cdots s_k}, 0) = (y,k)$ (the factor $1/2$ is introduced to compensate for the noninjectiveness of $\hat{\varphi}_{s_1}$ on E_0).

Now we define transfer operators $\mathcal{L}_0$ and $\mathcal{L}_\varepsilon$, associated to $\hat\varphi$ and its random perturbations $\hat\varphi_s$, $|s| \le \varepsilon$,

$$\mathcal{L}_0(\hat f)(y,k) = \sum_{\hat\varphi(x,l)=(y,k)} \frac{w_0(x,l)}{w_0(y,k)} \frac{\hat f(x,l)}{|\varphi'(x)|}$$

$$\mathcal{L}_\varepsilon(\hat f)(y,k) = \int \sum_{\hat\varphi_s(x,l)=(y,k)} \frac{w_\varepsilon(x,l)}{w_\varepsilon(y,k)} \frac{\hat f(x,l)}{|\varphi'_s(x)|} \, \theta_\varepsilon(s)\,ds$$

acting on the Banach space $BV(\hat I)$ of functions $\hat f\colon \hat I \longrightarrow \mathbb{R}$ such that

$$\left\| \hat f \right\|_{BV} = \operatorname{var} \hat f + \sup |\hat f| + \int |\hat f| w_0\,dx < \infty$$

(var $\hat f$ denotes the total variation of $\hat f$ on $\hat I$, that is, the sum of the variations on each $B_k \times \{k\}$). Note that (both for $\varepsilon = 0$ or for $\varepsilon > 0$) we have the duality relation

(b) $\int \hat U_\varepsilon(\hat f) \cdot g\, w_\varepsilon\,dx = \int \hat f \cdot \mathcal{L}_\varepsilon(\hat g)\, w_\varepsilon\,dx$

($\hat U_\varepsilon$ is defined in the same way as U_ε, with φ replaced by $\hat\varphi$). The main analytic step is to prove that

(c) $\mathcal{L}_0$ bounded and quasi-compact on $BV(\hat I)$;

(d) $\mathcal{L}_\varepsilon$ is "close" to $\mathcal{L}_0$ (in a convenient sense, which we borrow from [BaY]) if ε is small.

Here quasi-compactness means that the spectrum of $\mathcal{L}_0$ splits as $\sigma(\mathcal{L}_0) = \{1\} \cup S_0$ with S_0 contained in a disk of radius $\tau < 1$. Then the closeness in (d) implies that, for all small ε, the operator $\mathcal{L}_\varepsilon$ is also quasi-compact: $\sigma(\mathcal{L}_\varepsilon) = \{1\} \cup S_\varepsilon$ and S_ε contained in a disk of radius τ_ε bounded away from zero (uniformly in ε).

The proof of (c) and (d) relies on a delicate analysis of the action of the transfer operators on the L^1-norm, the supremum, and the variation (inequalities of Lasota-Yorke type), which falls outside the scope of this sketch. On the other hand, once these spectral properties have been derived the conclusions of our theorems follow through fairly standard arguments.

Indeed, if ρ_0 is an eigenfunction of $\mathcal{L}_0$ associated to the eigenvalue 1 then $\hat\mu_0 = \rho_0 w_0\,dx$ is an invariant measure for $\hat\varphi$ (use (b)). We normalize ρ_0 so that $\hat\mu_0$ be a probability and this defines ρ_0 uniquely. Then the absolutely continuous invariant probability measure of φ is given by $\mu_0 = p_*(\hat\mu_0)$, where $p\colon \hat I \longrightarrow I$ is the canonical projection. Similar statements hold for $\rho_\varepsilon, \mathcal{L}_\varepsilon, w_\varepsilon, \hat\mu_\varepsilon, \mu_\varepsilon$. Moreover, the fact that $\mathcal{L}_\varepsilon$ is close to $\mathcal{L}_0$ in the sense of [BaY] ensures that ρ_ε is close to ρ_0 in L^1-norm. From this one deduces that μ_ε is close to μ_0, as claimed in Theorem 3.1.

In order to prove Theorem 3.2 one uses the fact that (both for $\varepsilon = 0$ or $\varepsilon > 0$) the spectral projection π_ε associated to S_ε is given $\pi_\varepsilon(f) = f - \rho_\varepsilon \int f w_0\,dx$. As a consequence of this and the duality (b),

$$\int \hat{U}_\varepsilon^t(\hat{f}) \cdot \hat{g} \, d\hat{\mu}_\varepsilon - \int \hat{f} \, d\hat{\mu}_\varepsilon \int \hat{g} \, d\hat{\mu}_\varepsilon = \int \hat{f} \mathcal{L}_\varepsilon^t(\pi_\varepsilon(\hat{g}\rho_\varepsilon))w_\varepsilon \, dx$$

and, since $\mathcal{L}_\varepsilon$ acts as a τ_ε-contraction on $\pi_\varepsilon(BV(\hat{I}))$, this proves exponential mixing (with mixing rate bounded by τ_ε) at the tower level, for test functions in $BV(\hat{I})$. Finally, exponential mixing in $BV(I)$, as claimed in the theorem, is easily deduced by lifting bounded variation functions $f, g \colon I \longrightarrow \mathbb{R}$ to $BV(\hat{I})$ via the canonical projection $p \colon \hat{I} \longrightarrow I$.

4 Destruction of Lorenz attractors (joint with S. Luzzatto)

In this section we describe an extended geometric model for the dynamics of the system of differential equations

$$\begin{cases} \dot{x} = -\sigma x + \sigma y \\ \dot{y} = rx - y - xz \\ \dot{z} = -bz + xy \end{cases}$$

introduced by Lorenz [Lo]. Numerical analysis of this system for parameter values $\sigma \approx 10$, $b \approx 8/3$, and $r \approx 28$ led Lorenz to identify sensitive dependence to initial points as a main source of unpredictability in deterministic dynamical systems.

A rigorous description of the dynamics for these parameter values remains a challenging open problem to the present day, although some limited facts can be proved by classical methods. For instance, it is easy to see that, for all parameter values, there exists a singularity (equilibrium point) at the origin. Also, using the theory of Lyapunov functions one can find a (large) neighbourhood of this singularity which all trajectories enter and never leave. Since Lorenz equations are dissipative, this implies that there exists a compact invariant set Λ of zero Lebesgue measure containing the omega-limit sets of all trajectories. However it seems hard to prove any specific properties of this attractor (see [Sp] for a thorough discussion of numerical studies and classical approaches).

Results in this direction include [Ro], [Ry], where the existence of a "strange" attractor was proved for systems of (cubic) differential equations similar to that of Lorenz. Also, rigorous computer assisted proofs have been announced recently, see e.g. [HT]. concerning the existence of "chaotic" sets of trajectories in Lorenz equations. Such sets do not seem, in general, to be attracting and thus, even though their presence is very significant both from a mathematical and an experimental viewpoint, they only concern a set of trajectories of zero Lebesgue measure.

In fact, a large share of what we believe to know about chaotic behaviour in Lorenz equations comes from the study of geometric models. These were first introduced in [ABS], [GW], to try to describe the dynamics for the parameter values considered by Lorenz himself. Numerical studies of those equations indicate the presence of a nontrivial attractor containing the singularity at the origin and with strongly hyperbolic behaviour (exponentially contracting and exponentially expanding directions transverse to the flow). The papers mentioned above describe flows exhibiting attractors which do have these properties, and prove rigorous results on the corresponding topological, geometrical, and dynamical features. More recent results [BS], [Bu], [Pe1], [AP], [Pe2], [Sa] have built an extensive theory of such *generalized hyperbolic attractors*. Even more recently, [ACL], [Mo], [MP], provided a fairly detailed picture of the way these attractors can be formed already at the boundary of Morse-Smale flows.

Let us note that the methods used to study generalized hyperbolic systems are nontrivial generalizations of those developed for uniformly hyperbolic systems without singularities (e.g. geodesic flows on manifolds of negative curvature). Indeed the presence of a singularity constitutes an intrinsic obstruction to the existence of a uniform hyperbolic structure: since the hyperbolic decomposition $E^s \oplus E^u \oplus E^0$ of the tangent space at regular points must include a neutral direction E^0 tangent to the flow, which has no analog at singularities, such a decomposition can never be continuous on invariant sets containing regular trajectories accumulating at a singularity. This lack of a hyperbolic structure has more serious consequences than one might expect: the dynamics of flows is very often analysed in terms of Poincaré return maps to convenient cross-sections, however several features of smooth uniformly hyperbolic systems, like local product structure or continuous foliations by stable or unstable leaves, do not exist in general for such maps This is related with the fact that the presence of singularities in the vector field naturally translates in the form of discontinuities for the return maps (e.g. at the intersection of the cross-section with the local stable manifold of some singularity). As a consequence, global invariant (stable or unstable) sets are, in general, not connected and local invariant manifolds may have arbitrarily small size. An additional complication, which affects the ergodic properties of the attractor, is that the contraction and expansion rates are unbounded as the discontinuity is approached.

In the geometric models, and in most of the papers we mentioned above, these problems were partly overcome by assuming the existence of a smooth invariant stable foliation transverse to the flow. This is also the case in [Rv], which exhibited the first examples of attractors containing sungularities and with measure-theoretical (but not full) persistence. This hypothesis permits to reduce the analysis of the flow to that of a one-dimensional map and, in this way, to deduce several strong results (e.g. ergodicity of the attractor) from their one-dimensional analogs. However, this strategy breaks down in many other important situations, such as the one we want to consider here and which we describe in detail in the next section: the geometry of the problem (more precisely, the presence of criticalities) obstructs the existence of invariant foliations with any reasonable degree of regularity.

4.1 Critical and singular dynamics in Lorenz equations

The first part of Figure 2 is well-known: it describes the image of certain Poincaré return maps associated to the geometric models of [ABS], [GW]: the features of these return maps are coherent with the numerical data concerning Lorenz equations, for the original parameters of Lorenz. Subsequent numerical analysis of

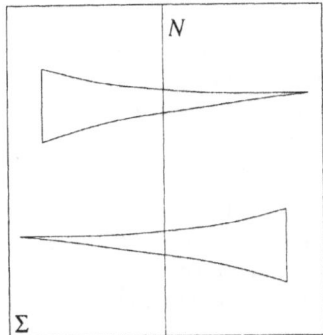 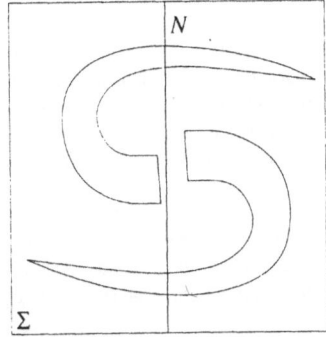

Figure 2: Formation of criticalities in the return map

these equations, [Sp], [GS], revealed that as the parameter r is increased to values of around 30 the flow begins to twist and fold in such a way that the image of the return map becomes as shown in the righthand half of Figure 2: it consists of two "hooks". See [GS] for an interpretation of this folding effect.

The new picture strongly suggests that the dynamics now contains *criticalities* (that is, nontransverse intersections between stable and unstable leaves) which, as we already mentioned, constitute a definite obstruction to the existence of regular invariant foliations (or of any uniform hyperbolic structure). The consequences of the loss of hyperbolicity due to the creation of such criticalities have been and continue to be the object of intense study, see [PT] and references therein for a presentation of the rich theory of homoclinic tangencies and a detailed study of the various dynamical phenomena occurring in their unfolding.

Flows with the characteristics above were first considered by [HP], [He], who introduced a family of smooth plane diffeomorphisms (the famous Hénon family) as a simplified model for the first return maps of the flow. These diffeomorphisms exhibit dynamical features arising from the presence of criticalities, without the additional complexity coming from the presence of a singularity. Notwithstanding this simplification, Hénon maps have been remarkably difficult to study rigorously. A major breakthrough occurred with the work of [BC2] in which new ideas and techniques were introduced to prove the existence and (measure-theoretical) persistence in the Hénon family of nontrivial attractors containing a dense orbit with a positive Lyapunov exponent. This result was generalized to strongly dissipative quadratic-like diffeomorphisms in [MV], [Vi] and the existence of SRB-measures for these attractors was proved in [BY2].

Our objective here is to recover the original project of Hénon-Pomeau and to develop a model for the dynamics exhibited by the Lorenz equation in the region of parameter values in which both criticalities and singularities are present. We define a class of one-parameter families of vector fields which exhibit, for a certain range of parameter values, generalized hyperbolic attractors as discussed above. As the parameter is varied a sequence of bifurcations takes place through which criticalities are formed. Beyond this sequence of bifurcations we encounter attractors in which features deriving from the presence of a singularity coexist with features related to the presence of criticalities. We study the way in which these two dynamical phenomena interact and show that a form of hyperbolicity remains, in a measure-theoretically persistent way.

Theorem 4.1 *There exists an open set of families $\{\mathcal{X}_a\}$ of smooth vector fields in $\mathbb{R}^3$ with the following properties. Let $(\varphi_a^t)_t$ denote the flow generated by the vector field $\mathcal{X}_a$. Then there exists a set $\mathcal{A}$ of positive Lebesgue measure in parameter space such that for each $a \in \mathcal{A}$ the flow $(\varphi_a^t)_t$ exhibits an attractor Λ_a and*

1. *Λ_a contains an equilibrium point with real eigenvalues and an infinite number of critical trajectories (consisting of criticalities).*

2. *Λ_a is transitive and (nonuniformly) hyperbolic in the following sense: there exists a point $z \in \Lambda_a$ and a vector $v \in T_z\mathbb{R}^3$ such that*

$$closure\left(\bigcup_{t \geq 0} \varphi_a^t(z)\right) = \Lambda_a \quad and \quad \limsup_{T \to \infty} \frac{1}{T} \log \|D\varphi_a^T(z)v\| > 0.$$

A final remark is in order, concerning an important difference between this theorem and the kind of results discussed above for generalized (or even uniformly) hyperbolic attractors. There, some form of hyperbolicity is assumed *a priori* and the effort then goes into proving that various dynamical properties follow from this hyperbolicity. In the presence of criticalities, however, one can expect a wide variety of dynamical behaviour, including periodic and quasi-periodic attractors, which occur intermittently alongside each other. For the theorem presented above we make some assumptions on the geometry of the flow and from this we draw the conclusion that there are indeed many parameter values for which a certain form of hyperbolicity exists. This is clearly a first fundamental step towards a more detailed analysis of the dynamical properties of the attractor.

4.2   Recurrence control yields positive Lyapunov exponent

The proof of Theorem 4.1 consists of two main parts. First we give a condition which implies the existence of an attractor with a positive Lyapunov exponent along certain "critical" orbits. Then we show that this condition is satisfied for a positive measure set of parameters and that for most of these parameters some critical orbit is dense. In this brief outline we shall concentrate on the first part, which already contains several interesting aspects from the point of view of the dynamics. A detailed proof is to appear in [LV1], [LV2].

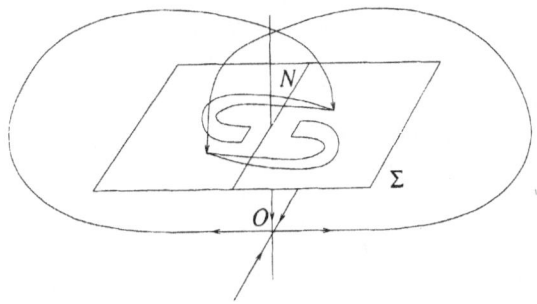

Figure 3: Defining the first-return map

Let $\{\mathcal{X}_a\}$ be a smooth family of vector fields in $\mathbf{R}^3$ with a singularity at the origin having real eigenvalues $\lambda_{ss} < \lambda_s < 0 < \lambda_u$ such that $|\lambda_s/\lambda_u| < 1/2$ and $|\lambda_{ss}/\lambda_u| > 1$. We define a first return map Φ_a to a horizontal cross-section $\Sigma \subset \{z = \epsilon\} \subset \mathbf{R}^3$ associating to each point $(x, y) \in \Sigma$ the point $\Phi_a(x, y) \in \Sigma$ given by the first intersection with Σ of the positive semi-trajectory $\{\varphi_a^t(x, y, \epsilon)\}_{t>0}$, see Figure 3. Notice that Φ_a is not defined on $N = \Sigma \cap W_{loc}^s(0)$, since points belonging to this set (which we assume to be given by $\{x = 0\} \subset \Sigma$) never leave a neighbourhood of the singularity and thus never return to intersect Σ. On the other hand, we always assume that $\Phi_a(\Sigma \setminus N) \subset \Sigma$, to guarantee the existence of some attracting set to which most orbits converge. Then we can write $\Phi_a = \Psi_a \circ P$ where Ψ_a is a diffeomorphism, corresponding to the holonomy of the flow far from the singularity, and P describes the holonomy of the flow near the origin. Ψ_a is responsable for the "folding" which we discussed above and which gives rise to the formation of criticalities. If one takes the flow to be locally linearizable then P has a simple explicit expression

$$P(x, y) = (|x|^{|\lambda_s/\lambda_u|} sgn(x), y|x|^{|\lambda_{ss}/\lambda_u|})$$

Note that the presence of the singularity yields some strong (local) hyperbolicity. Indeed, $|\lambda_s/\lambda_u| < 1$ gives $|\partial_x P_1| \approx |x|^{(|\lambda_s/\lambda_u|-1)} \to \infty$ as $x \to 0$, corresponding to a powerful horizontal expansion. Similarly, $|\partial_y P_2| \approx |x|^{|\lambda_{ss}/\lambda_u|-1} \to 0$ as $x \to 0$, corresponding to a strong contraction in the vertical direction.

Our strategy to obtain measure-theoretical persistence of positive Lyapunov exponents is very much inspired by [BC2], but we also have to deal with a difficulty which has no analog in the smooth case: controlling the recurrence of trajectories to the vicinity of the discontinuity N of Φ_a. Let us explain how this is done.

For the time being. the parameter a is fixed. Suppose that a certain set $\mathcal{C}$ of critical points is well defined for Φ_a (these lie roughly in the preimages of the folds and correspond to special points of nontransverse intersection between stable and unstable leaves). We fix some small neighbourhood $\Delta = \Delta_c \cup \Delta_0$ of the critical set $\mathcal{C}$ and of the line of discontinuity $\{x = 0\}$. Let $\{c_i\}_{i=0}^{\infty}$ be the orbit of some critical value $c_0 \in \Phi_a(\mathcal{C})$. According to a procedure which will be briefly recalled below, every iterate i is either *free* or *bound*. We say that ν is a *return* if $c_\nu \in \Delta$. A return is either free or bound according as to whether ν is a free iterate or a bound iterate. If ν is a return and $c_\nu \in \Delta_0$ we let $|||c_\nu|||$ denote the distance between c_ν and the line of discontinuity, if $c_\nu \in \Delta_c$ then we let $|||c_\nu|||$ denote the distance between c_ν and some critical point $\tilde{c} \in \mathcal{C}$ chosen so that the straight line joining c_ν and $\tilde{c}$ is essentially horizontal ("tangential position" [BC2]).

The key condition for controlling the recurrence to both the critical region and the singular region is the following. We fix some small $\alpha > 0$ and then we say that a critical point c satisfies condition $(*)$ if

$$\sum_{\nu \leq n} -\log |||c_\nu||| \leq \alpha n \quad \text{for all } n \geq 1,$$

where $c_\nu = \Phi_a^{\nu+1}(c)$ and the sum is taken over all *free* returns $\nu \leq n$. We also denote $w_j(c_0) = D\Phi_a^j(c_0) \cdot (1,0)$. Then we have, if $\alpha > 0$ is small enough,

Proposition 4.2 *If all the critical points $c \in \mathcal{C}$ satisfy condition $(*)$, then there exists a constant $\lambda > 0$ such that*

$$\|w_n(c_0)\| \geq e^{\lambda n} \quad \text{for all } n \geq 1 \text{ and all } c_0 \in \Phi_a(\mathcal{C}).$$

The proof of this fact relies on the decomposition of the orbit of each critical point into free and bound iterates mentioned above. Let a critical point c be fixed and ν be the first time that $c_\nu \in \Delta_c$. Then the first ν iterates are free, by definition. A crucial lemma says that during these iterates the vectors $w_j(c_0)$ are growing exponentially fast: there is $\lambda_0 > 0$ such that $\|w_j(c_0)\| \geq e^{\lambda_0 j}$ for all $j \leq \nu$. This expresses the fact that Φ_a has some hyperbolicity outside Δ_c: vectors which are roughly horizontal remain roughly horizontal and are expanded exponentially. The problems begin precisely when points fall into Δ_c, since there Φ_a is strongly contracting in all directions and, possibly even worse, rotates tangent vectors. This last fact means that $w_{\nu+1}(c_0)$ is likely to have large slope, making it difficult to control its growth during following iterates. Indeed, the hyperbolicity of Φ_a in the complement of Δ_c means that nearly horizontal vectors are expanded, but vectors which are almost vertical tend to be sharply contracted. The key to bypassing this effect is condition $(*)$, which implies $|||c_\nu||| \geq e^{-\alpha \nu}$, thence prevents the return from occurring too close to the critical set $\mathcal{C}$. This allows us to control the norm and the slope of $w_{\nu+1}(c_0)$ and to prove the following estimate:

Proposition 4.3 *There exist $\beta > 0$ and $p \approx \log |||c_\nu|||$ such that*
$$\|w_{\nu+p}(c_0)\| = \|D\Phi_a^p(c_\nu)w_\nu(c_0)\| \geq e^{\beta p}\|w_\nu(c_0)\|.$$
Moreover, the vector $w_{\nu+p}$ is roughly horizontal.

All iterates contained in the interval $[\nu+1, \nu+p)$ are called bound iterates. Starting at $\nu + p$ we then have a sequence of free iterates which continues until the next return to Δ_c. After this return another interval of bound iterates starts, followed by more free iterates, and so on. The above estimates are actually valid for each interval of free or bound iterates, respectively, and we get $\|w_n(c_0)\| \geq e^{\lambda_0 Q + \beta P}$ for $n \geq 1$, where P is the total number of bound iterates less than n and $Q = n - P$ is the number of free iterates. If the bound iterates correspond to returns $\nu_1, \ldots, \nu_s$ then by Proposition 4.3 and (∗) we have $P = \sum_{i=1}^{s} p_i \approx \sum_{i=1}^{s} - \log \|\|c_{\nu_i}\|\| \leq \alpha n$, which implies that $Q = n - P \geq (1 - \mathrm{const}\,\alpha)n$. From this we easily get

$$\|w_n(c_0)\| \geq e^{\lambda_0 Q + \beta P} \geq e^{\lambda_0 (1 - \mathrm{const}\,\alpha)n} \geq e^{\lambda n}$$

for some $0 < \lambda < (1 - \mathrm{const}\,\alpha)$, if α is small enough. This completes a heuristic outline of the proof of Proposition 4.2.

The second part of the proof of the theorem consists of an algorithm for excluding parameters for which condition (∗) fails. Though we shall not go into any detail concerning this algorithm, let us make a couple of brief remarks. Our estimates to guarantee that a positive measure set of parameters survives all the exclusions depend heavily on a global uniform bound on the distortion. This can be obtained only by controlling the recurrence of critical points near the discontinuity as well as in the critical region and that is one of the reasons why returns to Δ_0 must also be taken into account in (∗). A second remark is that, just as in the Hénon case, the set of critical points which we have implicitly assumed in the proposition is not given a priori but rather needs to be constructed by successive approximations. This construction proceeds alongside the algorithm for excluding parameters and even depends on it in the sense that it can be carried out at each iteration n only for those parameters which are not excluded up to that time. So, eventually, a set $\mathcal{C}$ of critical points is well defined only for those parameters for which all critical points satisfy (∗) at all times $n \geq 1$.

Finally, to obtain the statement in the theorem we need to show that the exponential growth of the vectors $w_j(c_0)$ for the map Φ_a implies exponential growth of the $w_j(c_0)$ also with respect to the (continuous) flow $(\varphi_a^t)_t$. The problem here is that return times to Σ are unbounded as points approach the discontinuity and so a given expansion may be distributed along longer and longer time intervals. In principle, this could give rise to a strictly subexponential growth for the flow but the control over the recurrence of critical points near the discontinuity provided by condition (∗) allows us to show that this is not the case. Some difficulty arises from the fact that we have to worry about all (both free and bound) returns to Δ_0, since all of them give rise to large return times, whereas (∗) only commits explicitly the free returns. However, using the fact that every bound return to Δ_0 occurs during bound periods associated to free returns to Δ_c, one can show that

$$\sum_{i:c_i \in \Delta_0} - \log \|\|c_i\|\| \leq \mathrm{const} \sum - \log \|\|c_i\|\| \leq \mathrm{const}\,\alpha n$$

where the second sum is taken over all free returns. Now this allows us to deduce that the total contribution to the return times corresponding to bound returns is dominated by that of the free returns. Indeed, let $t(z)$ denote the return time for the point z and let t_0 be the supremum of return times for points outside Δ_0. If $|z|$ denotes the distance of z to the discontinuity, then $t(z) \approx \log|z|$, by straightforward computation. Thus, for each $c_0 \in \Phi_a(\mathcal{C})$, and corresponding to each iterate $n \geq 1$ of the return map, we have a "continuous flow time"

$$T_n = \sum_{i=0}^{n-1} t(\Phi_a^i(c_0)) \leq \sum_{i:c_i \notin \Delta_0} t_0 + \sum_{i:c_i \in \Delta_0} t(c_i) \leq nt_0 + \text{const}\,\alpha n \leq \gamma n$$

for some constant $\gamma > 0$. Then $T_n^{-1} \log \|w_n\| \geq (\lambda n/\gamma n) = (\lambda/\gamma) > 0$, which implies the desired result.

5 Global spiral attractors

Our goal in this section is to prove that "chaotic" attractors with spiraling geometry occur, even in a measure-theoretically persistent way, in certain families of vector fields. This corresponds to very recent joint work with M. J. Pacifico and A. Rovella. The possible existence of spiral attractors seems to have been first mentioned by Ya. Sinai. Our results are motivated by the observations in [ACT] for which, in particular, they provide rigorous confirmation.

5.1 Saddle-focus connections

We consider smooth flows $(\varphi^t)_{t \in \mathbb{R}}$ in 3-dimensional space exhibiting a double saddle-focus homoclinic connection. By this we mean the following, see Figure 4. The flow has an equilibrium point O, at the origin say, which is of saddle-focus type: one expanding eigenvalue $\theta > 0$ and two complex contracting eigenvalues $\lambda \pm \omega i$, where $\lambda < 0$ and $\omega \neq 0$. Moreover, both unstable separatrices of O are contained in the stable manifold of O, that is, they are homoclinic trajectories.

For simplicity we assume that the flow is symmetric with respect to the origin, i.e. invariant under $(x, y, x) \longrightarrow (-x, -y, -z)$, but this is not strictly necessary for what follows. Furthermore, a convenient reformulation of our results holds when there is a single homoclinic connection, cf. comments below. These results also extend in straightforward way to general 3-dimensional manifolds. Generalization to higher dimensions was not yet carried out but seems a realistic task and is, certainly, an interesting one.

We want to describe the typical asymptotic behaviour of points in a neighbourhood of the homoclinic connections, not only for the initial flow $(\varphi^t)_t$ but also for "generic" nearby flows. More precisely, we consider smooth parametrized families of flows $(\varphi_\mu^t)_{t \in \mathbb{R}}$, $\mu \in (-\delta, \delta)$, generically unfolding the homoclinic connections: $(\varphi_0^t)_t = (\varphi^t)_t$ and as the parameter μ varies the unstable separatrices

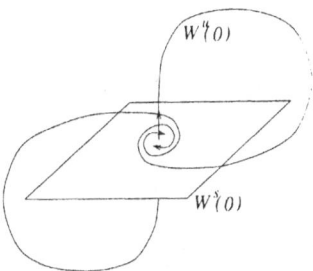

Figure 4: Double saddle-focus connections

move with nonzero speed with respect to the (local) stable manifold. Then we want to describe the attractors of such flows, close to the unstable separatrices, for a sizable portion of parameter values.

The answer to this problem depends crucially on the relative strength of the contracting and the expanding eigenvalues, that is on the value of $\alpha = -\lambda/\theta$. The *contracting case* $\alpha > 1$ is comparatively simple. The union A_0 of the two homoclinic connections is an attracting set for the unperturbed flow $(\varphi_0^t)_t$, with basin containing a neighbourhood of A_0. Moreover, this attracting set is, in some sense, persistent: varying μ leads to the formation of attracting periodic orbits close to (and with basins containing) the unstable separatrices. Therefore, periodic asymptotic behaviour is typical for small parameter values.

The dynamics is much richer in the *expanding case* $\alpha < 1$, as was already attested by the pioneer work of Shil'nikov [Shi]: he proved that infinitely many periodic orbits of saddle type (contained in suspended horseshoes) coexist in this situation. The main result in the present section states that, under convenient assumptions to be described below, *for a large (positive Lebesgue measure) set of values of μ the flow $(\varphi_\mu^t)_t$ admits a unique (global) attractor A_μ, close to A_0.* Moreover, this attractor is *chaotic* (sensitive dependence on initial conditions) and *singular* (contains the equilibrium point as well as regular trajectories) and has an intricate *spiraling geometry*.

Before going into discussing this result, let us point out that some complex dynamical phenomena are also present in the case $\alpha = 1$. Recently, Pumariño [Pu] used this context to give examples of coexistence of suspended Hénon-like attractors: he even finds parameter values for which infinitely many such attractors occur simultaneously.

From now on we restrict to the case $\alpha < 1$. In order to analyse the dynamics of the flow in the vicinity of the homoclinic connections we follow the standard procedure of considering the first-return map F to some convenient cross-section Σ. We take Σ intersecting one of the homoclinic trajectories and the local stable manifold of O; we also consider auxiliary cross-sections Σ^-, symmetric to Σ, and

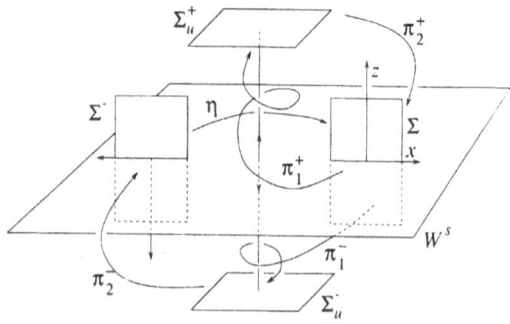

Figure 5: Constructing the first-return map

$\Sigma_u^{\pm}$, intersecting the two local unstable separatrices. Then $\pi_1^{\pm}$, $\pi_2^{\pm}$, and η denote corresponding Poincaré maps, as indicated in Figure 5, and we let

$$F(x,z) = \begin{cases} \pi_2^+ \circ \pi_1^+(x,z) & \text{if } z > 0 \\ \eta \circ \pi_2^- \circ \pi_1^-(x,z) & \text{if } z < 0 \end{cases}$$

(F is not defined on the line $\{z = 0\}$ of intersection between Σ and the local stable manifold of the singularity). We assume our flows to be linearizable near the equilibrium (a generic condition), so that it is easy to calculate η, $\pi_1^{\pm}$ explicitly and to see that the images of $\pi_1^{\pm}$ are spiraling regions in $\Sigma_u^{\pm}$ accumulating at the points $W_{loc}^u(O) \cap \Sigma_u^{\pm}$. On the other hand, $\pi_2^{\pm}$ are diffeomorphisms. Then, under open conditions on $\pi_2^{\pm}$, the image of Σ under F consists of two spiraling regions contained in the interior of Σ, see Figure 6.

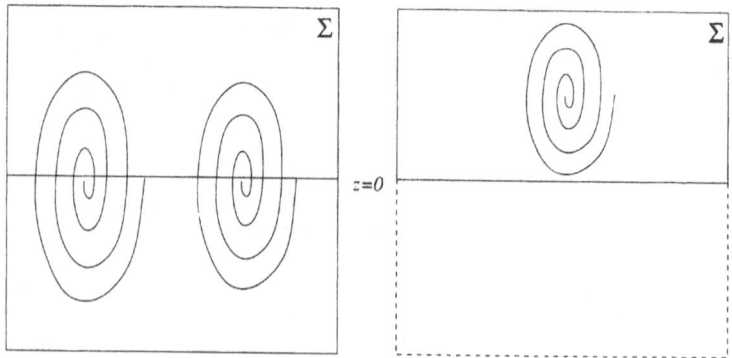

Figure 6: Trapping regions for double and for single connections

Furthermore, this implies that $F_\mu(\Sigma) \subset \Sigma$ for all small μ, where F_μ is the return map to Σ associated with the flow $(\varphi^t_\mu)_\mu$. Then all the asymptotic dynamics of points in this "trapping region" Σ is concentrated inside the maximal invariant set $\Lambda_\mu = \cap_{n\geq 0} F^n_\mu(\Sigma)$. In general, Λ_μ need not be dynamically indivisible, indeed it may contain several different types of dynamical behaviour. However, we have

Theorem 5.1 *For families of vector fields $\{\mathcal{X}_\mu\}$ unfolding double saddle-focus connections as above, there exists a positive Lebesgue measure set $S \subset (-\delta, \delta)$ such that Λ_μ is a (global) chaotic attractor of F_μ for all $\mu \in S$.*

This means, in particular, that for these parameter values Λ_μ contains dense orbits with positive Lyapunov exponent. Of course, the basin of Λ_μ contains the whole section Σ. Successive images $F^n_\mu(\Sigma)$ give increasingly better approximations to the remarkably complex geometry of Λ_μ, recall Figure 6. Note that the suspension $\hat{\Lambda}_\mu = $ closure $(\cup_{t\in\mathbb{R}} \varphi^t_\mu(\Lambda_\mu))$ contains the equilibrium point of the flow, in particular it can not be uniformly hyperbolic; $\hat{\Lambda}_\mu$ also contains criticalities and so it is not even a generalized uniformly hyperbolic attractor, recall Section 4.

Closing this section, we point out that our arguments to prove Theorem 5.1 apply also to the unfolding of flows with a unique saddle-focus connection, e.g. as in [Shi]. A main difference is that in this case one must consider large parameter values, i.e. far from the one for which the homoclinic connection occurs, in order to enforce invariance of Σ under the return map, cf. Figure 6.

5.2   Interval maps with infinitely many critical points

In this final section we give a brief discussion of the difficulties involved in the proof of Theorem 5.1 and of the methods we use to overcome them. A detailed presentation is to appear in [PRV1], [PRV2].

Explicit calculation of the return map along the lines sketched above leads to $F(x, z) = (1 + xg(z), xf(z))$ where

$$f(z) = \begin{cases} b_+|z|^\alpha \sin(\beta \log \frac{1}{|z|}) & \text{if } z > 0 \\ b_-|z|^\alpha \sin(\beta \log \frac{1}{|z|}) & \text{if } z < 0 \end{cases}$$

(we write $\alpha = -\lambda/\theta$, $\beta = \omega/\theta$, and the constants $b_\pm$ depend only on the flow) and g has a similar expression, with sin replaced by cos and $b_\pm$ replaced by constants $a_\pm$. Actually, these statements are accurate only for z close to zero (f must be modified away from the origin to create the trapping region), and then again only as a first-order approximation, but here we will allow ourselves this technical simplification.

A first difficulty arises from the fundamentally higher-dimensional nature of the system. It is not difficult to convince oneself, e.g. observing Figure 6, that F can not admit smooth invariant foliations, and so it can not be reduced in this way to a one-dimensional system; recall Section 4.1. To try to bypass this difficulty we assume the constants $|a_\pm|$ to be small. This is related with the small jacobian hypothesis in [BC2], but we observe that in our context having $a_\pm$ close to zero does not imply volume-dissipativeness (at least not if $\alpha < 1/2$). Then we have $F(1, z) \approx (1, f(z))$, which suggests that the dynamics of F may, to some extent, be mimed by that of the one-dimensional map f. This turns out to be only very roughly true, but it is indeed useful to study a version of our initial problem for such one-dimensional maps. In what follows we concentrate on discussing this simpler version, which is also interesting in itself, without further discussing the (considerable) work required to extend our conclusions back to the original setting to get Theorem 5.1.

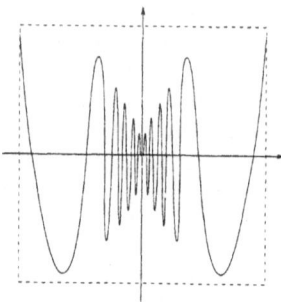

Figure 7: Maps with infinitely many critical points

More precisely, we want to consider unfoldings of f by parametrized families of transformations of the interval $I = [-1, 1]$ to itself, of the form

$$f_\mu(z) = \begin{cases} f(z) + c_+\mu & \text{if } z > 0 \\ f(z) - c_-\mu & \text{if } z < 0 \end{cases}$$

where $c_\pm$ are positive constants and μ is a parameter close to zero. The way f_μ depends on this parameter is meant to emulate the unfolding of the homoclinic connections by the flows $(\varphi_\mu^t)_t$ or, more precisely, the way the second coordinate of $F_\mu(x, z)$ depends on μ. It is easy to check that $f = f_0$ has two sequences of critical points accumulating at zero, see Figure 7; we denote them by $x_k^\pm$, with $x_k^- < 0 < x_k^+$. Of course, every f_μ has the same set of critical points, and we denote the corresponding critical values by $z_k^\pm(\mu) = f_\mu(x_k^\pm)$. Then we prove that the maps f_μ have global chaotic behaviour in a measure-theoretically persistent way, in the sense of the following theorem.

Theorem 5.2 *There exist $\sigma > 1$ and a positive Lebesgue measure set S of values of μ for which*
1. *$|(f_\mu^n)'(z_k^\pm)| \geq \sigma^n$ for all k and all $n \geq 1$;*
2. *almost every $z \in I$ has positive Lyapunov exponent.*

An important difference between this and similar results for quadratic maps of the interval, [Ja], [BC1], lies on the nature of the initial parameter $\mu = 0$. Indeed, persistent chaotic behaviour for smooth unimodal maps is usually found at parameter values close to one for which the critical point is nonrecurrent (e.g. preperiodic): (almost) nonrecurrence allows the critical orbit to build-up expansion during initial iterates ant then one proceeds by induction to prove that this initial expanding behaviour is preserved in the subsequent iterates, as long as the parameter is chosen conveniently. In contrast, here $\mu = 0$ corresponds to the origin being fixed under the map, and the origin is a particularly nasty point: not only it is a point of nonsmoothness/discontinuity of the dynamics, it is also accumulated by critical points of f, f_μ.

This means that a first main step in the proof of Theorem 5.2 must be devoted to proving that *all the critical orbits do exhibit initial expansion, at least for a large set of parameters*. We fix constants $\tau > 0$, $\varepsilon > 0$, $\gamma \in (\alpha, 1)$ and for each $\mu \in (-\varepsilon, \varepsilon)$ and $z \in (-\varepsilon, \varepsilon)$ we let $j(\mu, z)$ be the smallest iterate j for which $f_\mu^j(z) \notin (-\varepsilon, \varepsilon)$. Then the main ingredient is to show that the set G of parameter values $\mu \in (-\varepsilon, \varepsilon)$ for which every critical value $z = z_k^\pm(\mu)$ satisfies

1. $|f_\mu^{i+1}(z)| \geq |f_\mu^i(z)|$ for $0 \leq i < j(\mu, z)$;

2. $|f_\mu^i(z) - x_l^\pm \geq \tau|x_l^\pm|$ for $0 \leq i < j(\mu, z)$ and every critical point $x_l^\pm$

has almost full measure in $(-\varepsilon, \varepsilon)$ if ε and τ are small. Condition 1 implies that the orbit of $z = z_k^\pm$ moves away from the origin very fast and condition 2 means that while doing it it avoids the neighbourhood of the critical points. We prove that under these assumptions the orbit of z is expanding during the time interval $[0, j(\mu, z))$ it spends near zero.

In a second step we proceed from this set of parameters G, using arguments inspired in [BC1], [BC2]. The main difficulties at this point, with respect to the quadratic case, come from the nonsmoothness of f_μ and, most important, from the fact that it has infinitely many critical points. However, we are able to prove that, for parameters in a positive measure subset of G, all these critical orbits exhibit exponential growth of the derivative at all times, as stated in the theorem.

Acknowledgements. Discussions with V. Baladi, S. Luzzatto, C. Morales, M. J. Pacifico, J. Palis, and E. Pujals played an important role in the making of this paper. I am also most grateful to the hospitality of the University of Paris-Orsay and the University of Porto during part of the time this work was written.

References

[ABS]   V.S. Afraimovich, V.V. Bykov, and L.P. Shil'nikov, *On the appearence and structure of the Lorenz attractor*, Dokl. Acad, Sci. USSR 234-2 (1977), 336-339.

[ACL]   V.S. Afraimovich, S.-N. Chow, and W. Liu, *Lorenz type attractors from codimension one bifurcation*, J. Dynam. & Diff. Equ. 7-2 (1995), 375-407.

[AP]   V.S. Afraimovich and Ya. B. Pesin, *The Dimension of Lorenz type attractors*, Sov. Math.Phys. Rev., vol. 6, Gordon and Breach Harwood Academic, 1987.

[ACT]   A. Arneodo. P. Coullet, and C. Tresser, *Possible new strange attractors with spiral structure*. Comm. Math. Phys. 79 (1981), 573-579.

[BaY]   V. Baladi and L.-S. Young, *On the spectra of randomly perturbed expanding maps*, Comm. Math. Phys. 156 (1993), 355-385.

[BaV]   V. Baladi and M. Viana *Strong stochastic stability and rate of mixing for unimodal maps*. preprint IMPA 1995, to appear Annales E.N.S..

[BC1]   M. Benedicks and L. Carleson, *On iterations of* $1 - ax^2$ *on* (-1.1), Ann. Math. 122 (1985), 1-25.

[BC2]   M. Benedicks and L. Carleson, *The dynamics of the Hénon map*, Annals of Math. 133 (1991), 73-169.

[BY1]   M. Benedicks and L.-S. Young, *Absolutely continuous invariant measures and random perturbations for certain one-dimensional maps*. Ergod. Th. & Dynam. Sys. 12 (1992). 13-37

[BY2]   M. Benedicks and L.-S. Young, *SBR-measures for certain Hénon maps*, Invent. Math. 112-3 (1993), 541-576.

[Bo]   R. Bowen, *Equilibrium states and the ergodic theory of Anosov diffeomorphisms*, Lect. Notes in Math. 470 (1975), Springer Verlag.

[Bu]   L.A. Bunimovich, *Statistical properties of Lorenz attractors*. in *Nonlinear dynamics and turbulence*, 71-92, Longman publ., London, 1983.

[BS]   L.A. Bunimovich and Ya.G. Sinai, *Stochasticity of the attractor in the Lorenz model*, in *Nonlinear Waves*, Proc. Winter School, Moscow. 212-226, Nauka publ., Moscow. 1980.

[Ca]   E. Catsigeras. *Cascades of period-doubling of stable codimension one*, thesis IMPA 1994 and to appear.

[Co]   P. Collet, *Ergodic properties of some unimodal mappings of the interval*, Preprint 11, Institute Mittag Leffler 1984.

[DRV]   L.J. Díaz, J. Rocha, and M. Viana, *Strange attractors in saddle-node cycles: prevalence and globality*, preprint IMPA 1993 and to appear.

[GS]   P. Glendinning and C. Sparrow, *T-points: A codimension two heteroclinic bifurcation*, Jour. Stat. Phys. 43 (1986), 479-488.

[GW]   J. Guckenheimer and R.F. Williams, *Structural stability of Lorenz attractors*, Publ. Math. IHES 50 (1979), 307-320.

[HT]   S.P. Hastings and W.C. Troy, *A shooting approach to the Lorenz equations*, Bull. A.M.S. 27 (1992). 298-303.

[He]   M. Hénon, *A two dimensional mapping with a strange attractor*, Comm. Math. Phys. 50 (1976), 69–77.

[HP]   M. Hénon and Y. Pomeau, *Two strange attractors with a simple structure*, Lect. Notes Math. 565 (1976), 29–68.

[Ja]   M. Jakobson, *Absolutely continuous invariant measures for one-parameter families of one-dimensional maps*, Comm. Math. Phys. 81 (1981), 39–88.

[KK]   A. Katok and Y. Kifer *Random perturbations of transformations of an interval*, J. Analyse Math. 47 (1986), 193–237.

[KN]   G. Keller and T. Nowicki *Spectral theory, zeta functions and the distribution of periodic points for Collet-Eckmann maps*, Comm. Math. Phys. 149 (1992), 31–69.

[Ki]   Y. Kifer, *Random Perturbations of Dynamical Systems*, Birkhäuser. 1988.

[Lo]   E.N. Lorenz, *Deterministic nonperiodic flow*, J. Atmosph. Sci. 20 (1963), 130–141.

[LV1]   S. Luzzatto and M. Viana, *Positive Lyapunov exponents for Lorenz-like maps with criticalities*, preprint 1995.

[LV2]   S. Luzzatto and M. Viana, *Lorenz-like attractors without invariant foliations*, in preparation.

[Ma]   R. Ma né, *Hyperbolicity, sinks and measure in one-dimensional dynamics*, Comm. Math. Phys. 100 (1985), 495–524.

[MV]   L. Mora and M. Viana, *Abundance of strange attractors*, Acta Mathematica 171 (1993), 1–71.

[Mo]   C.A. Morales, *Lorenz attractors through saddle-node bifurcations*, thesis IMPA 1995, to appear Annales I.H.P., Analyse non linéaire.

[MP]   C.A. Morales and E. Pujals, *Singular attractors on the boundary of Morse-Smale systems*, preprint 1995.

[No]   T. Nowicki, *A positive Lyapunov exponent for the critical value of an S-unimodal mapping implies uniform hyperbolicity*, Ergod. Th. & Dynam. Sys. 8 (1988), 425–435.

[PRV1] M.J. Pacifico, A. Rovella, and M. Viana, *Global attractors in saddle-focus bifurcations: a one-dimensional model*, preprint 1995.

[PRV2] M.J. Pacifico, A. Rovella, and M. Viana, *Persistence of global spiraling attractors*, in preparation.

[PT]   J. Palis and F. Takens, *Hyperbolicity and sensitive-chaotic dynamics at homoclinic bifurcations*, Cambridge University Press, 1993.

[Pe1]   Ya.B. Pesin, *Ergodic properties and dimensionlike characteristics of strange attractors that are close to hyperbolic*, in Procs. ICM 1986.

[Pe2]   Ya.B. Pesin, *Dynamical systems with generalized hyperbolic attractors; hyperbolic, ergodic and topological properties*, Erg. Th. & Dyn. Syst. 12 (1992), 123–151.

[Pu]   A. Pumariño, *Coexistence with positive probability of strange attractors in ho-moclinic connections of saddle-focus type*, thesis Department of Mathematics, University of Oviedo, Spain, 1994.

[Ro]   C. Robinson, *Homoclinic bifurcation to a transitive attractor of Lorenz type*, Nonlinearity 2 (1989), 495-518.

[Rv]   A. Rovella, *The dynamics of perturbations of the contracting Lorenz attractor*, Bull. Braz. Math. Soc. 24 (1993), 233-259.

[Ry]   M. Rychlik, *Lorenz attractors through Shil'nikov-type bifurcation. Part 1*, Erg. Th. & Dyn. Syst. 10 (1989), 793-821.

[Sa]   E.A. Sataev, *Invariant measures for hyperbolic maps with singularities*, Russ. Math. Surveys **471** (1992), 191-251.

[Shu]   M. Shub, *Global stability of dynamical systems*, Springer Verlag. 1987.

[Shi]   L.P. Shil'nikov, *A case of the existence of a denumerable set of periodic motions*, Sov. Math. Dokl. 6 (1965), 163-166.

[Si]   D. Singer, *Stable orbits and bifurcations of maps of the interval*. SIAM J. Appl. Math. 35 (1978), 260-267.

[Sm]   S. Smale, *Differentiable dynamical systems*, Bull. Am. Math. Soc. 73 (1967), 747-817.

[Sp]   C. Sparrow, *The Lorenz equations: bifurcations, chaos and strange attractors*, Applied Mathematical Sciences, vol. 41, Springer-Verlag, 1982.

[Ur]   R. Ures, *On the approximation of Hénon-like attractors by homoclinic tangencies*, thesis IMPA 1993, to appear Ergod. Th. & Dynam. Sys..

[Vi]   M. Viana, *Strange attractors in higher dimensions*, Bull. Braz. Math. Soc. 24 (1993), 13-62.

[Yo]   L.-S. Young, *Decay of correlations for certain quadratic maps*. Comm. Math. Phys. 146 (1992), 123-138.

Progress in Nonlinear Differential Equations
and Their Applications, Vol. 19
© 1996 Birkhäuser Verlag Basel/Switzerland

Modulated waves in a perturbed Korteweg-de Vries equation

Stephan A. van Gils[*]   Edy Soewono[†]

Abstract

We show the existence of a global branch of modulated waves in a two-mode
approximation of a perturbed Korteweg-de Vries equation.

1 Introduction

In this paper we analyse the appearance of modulated traveling waves in perturbed
Korteweg-de Vries (KdV) equations. We consider the one-dimensional perturbed
KdV-equation:

$$u_t = \partial_x \delta H(u) + \epsilon(u_{xx} + \mu u). \tag{1.1}$$

with periodic boundary conditions and mean value. The Hamiltonian is

$$H(u) = \int_{-\pi}^{\pi} \frac{1}{2}(u_x^2 - u^3) \, dx. \tag{1.2}$$

We refer to equation (1.1) with $\epsilon = 0$ as the unperturbed equation. The unperturbed equation has, besides the Hamiltonian, another first integral, the so called
momentum:

$$I(u) = \int_{-\pi}^{\pi} u^2 \, dx. \tag{1.3}$$

In [DvG93] we have analyzed (1.1) with respect to traveling wave solutions. It is a
necessary condition for the perturbed equation to have a traveling wave solution
that

$$\int_{-\pi}^{\pi} u_x^2 \, dx - \mu \int_{-\pi}^{\pi} u^2 \, dx = 0. \tag{1.4}$$

It is proved in [DvG93] that if we take for u a stable traveling wave solution of
the unperturbed system, i.e. for a given value of γ, $u = u_\gamma$ is a minimizer of the

---

[*]Faculty of Applied Mathematics, University of Twente. P.O. Box 217, 7500 AE Enschede,
The Netherlands
[†]FMIPA. Institut Teknologi Bandung, Jalan Ganesha 10. Bandung 40132. Indonesia

constrained variational problem $\min\{H(u) \mid I(u) = \gamma\}$, then (1.4) determines μ uniquely as a function of γ, taking values in the interval $(1, \infty)$.

A relative equilibrium can Hopf-bifurcate. This has already been observed in [Der92]. This leads to the bifurcation of quasi periodic solutions. We analyse this in detail for the two-mode approximation. This should be considered as a first step towards the understanding of this bifurcation in the infinite dimensional system. The analysis is also interesting in its own right. Similar bifurcations occur in $1 : 1$ and $1 : 2$ resonance problems and have not been analyzed to our knowledge.

We compute for the two-mode approximation the Hopf bifurcation variety in the reduced phase space. For this reduced system we prove a *global* Hopf bifurcation result. This result implies that for ϵ small there exists, in the original phase space, a branch of quasi period solutions. We also show that no secondary bifurcations from this branch occur.

2 Preliminaries

2.1 N-mode approximation

In the N-mode approximation we let

$$u^{(N)} = \sum_{k=1}^{N} z_k e^{ikx} + \sum_{k=1}^{N} \zeta_k e^{-ikx}, \tag{2.1}$$

where $\zeta_k = \bar{z}_k$. For f and g in $L^2([-\pi, \pi], \mathbb{C})$ we let

$$(f, g) = \frac{1}{2\pi} \int_{-\pi}^{\pi} f(x)\overline{g(x)} \, dx \tag{2.2}$$

With u as in (2.1) we let

$$\begin{aligned} \widetilde{H}(z) &= \widetilde{H}(z_1, \ldots, z_N, \zeta_1, \ldots, \zeta_N) \\ &= H(u^{(N)}). \end{aligned} \tag{2.3}$$

It then follows that

$$\frac{\partial \widetilde{H}}{\partial z_k} = DH(u^{(N)})\frac{\partial u^{(N)}}{\partial z_k} = DH(u^{(N)})e^{ikx} = \left(\delta H(u^{(N)}), \, e^{-ikx}\right). \tag{2.4}$$

The dynamical equations for z_k are given by

$$\dot{z}_k = ik\frac{\partial \widetilde{H}}{\partial \zeta_k} + \epsilon(-k^2 z_k + \mu z_k). \tag{2.5}$$

For a traveling wave solution of (2.5), the discrete version of condition (1.4) reads

$$\sum_{k=-N}^{N} -k^2 z_k \zeta_k + \mu z_k \zeta_k = 0. \tag{2.6}$$

2.2 2-mode approximation

In the two-mode approximation we have

$$\tilde{H}(z) = z_1\zeta_1 - 3z_1^2\zeta_2 + 4z_2\zeta_2 - 3z_2\zeta_1^2. \tag{2.7}$$

Hence the system of differential equations is given by

$$\begin{aligned}
\dot{z}_1 &= iz_1 - 6iz_2\zeta_1 - \epsilon z_1 + \epsilon z_1\mu \\
\dot{z}_2 &= -6iz_1^2 + 8iz_2 - 4\epsilon z_2 + \epsilon z_2\mu.
\end{aligned} \tag{2.8}$$

The translation symmetry of system (1.1) induces equivariance of the system (2.5) with respect to the 4-dimensional representation of S^1: $\theta \mapsto S(\theta)$ where

$$S(\theta)(z_1, z_2, \zeta_1, \zeta_2) = (e^{i\theta_1}z_1, e^{2i\theta}z_2, e^{-i\theta_1}\zeta_1, e^{-2i\theta}\zeta_2). \tag{2.9}$$

There are various ways to exploit the symmetry. One way would be to introduce polar coordinates, $z_k = \rho_k e^{ik\theta_k}$ and to consider the reduced system in the variables ρ_1, ρ_2, $2\theta_1 - \theta_2$. We will not proceed in this direction, but we will use the so called Hilbert invariants.

2.3 Hilbert invariants

We introduce the 4 invariants for the symmetry (2.9)

$$\begin{aligned}
\pi_1 &= z_1\zeta_1 \\
\pi_2 &= z_2\zeta_2 \\
\pi_3 &= \mathrm{Re}(z_1^2\zeta_2) \\
\pi_4 &= \mathrm{Im}(z_1^2\zeta_2).
\end{aligned} \tag{2.10}$$

The invariants satisfy the relation

$$\pi_1^2\pi_2 = \pi_3^2 + \pi_4^2. \tag{2.11}$$

It is a straightforward calculation to compute the system of differential equations for the invariants:

$$\begin{aligned}
\dot{\pi}_1 &= -12\pi_4 + 2\epsilon(\mu - 1)\pi_1 \\
\dot{\pi}_2 &= 12\pi_4 + 2\epsilon(\mu - 4)\pi_2 \\
\dot{\pi}_3 &= 6\pi_4 + 3\epsilon(\mu - 2)\pi_3 \\
\dot{\pi}_4 &= -6\pi_3 - 12\pi_1\pi_2 + 6\pi_1^2 + 3\epsilon(\mu - 2)\pi_4,
\end{aligned} \tag{2.12}$$

which leaves the variety (2.11) invariant. It follows from (2.7) that the Hamiltonian is given in the invariants as

$$H = 4K - 3\pi_1 - 6\pi_3. \tag{2.13}$$

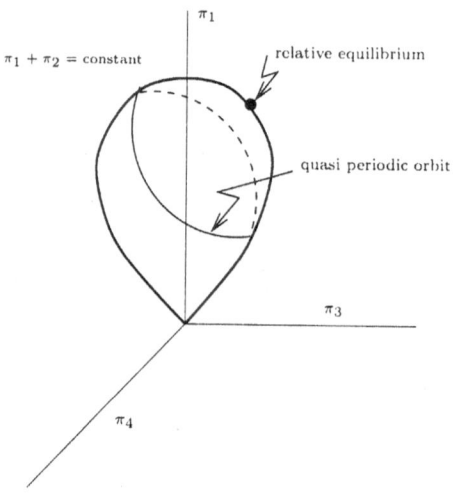

Figure 1: The orbits in the reduced phase space.

where,

$$K = \pi_1 + \pi_2. \tag{2.14}$$

Note that K, which is just the 2-mode approximation of the momentum, is a conserved quantity for the unperturbed system. The unperturbed equation is integrable. Using the fact that the Hamiltonian is linear in the invariants we can easily depict the orbits. We fix a value of K. The orbits are obtained by intersecting the Hamiltonian (2.13) with the variety (2.11). Except for the trivial solution, there are four types of orbits. Fixed points, which are traveling waves for the unperturbed un-reduced equation, periodic orbits, which correspond to quasi-periodic orbits in the un-reduced system, a homoclinic orbit. corresponding to a pinched torus in the un-reduced system, and finally the pure mode-2 solution. This is a solution for which $\pi_1 = 0$ and π_2 is constant. In the figure below we depict the two fixed points that are not pure mode-2 solutions. There are two curves in the $H - K$ plane, which have a quadratic tangency at $K = -1/12$. These are depicted in Figure 2.a. We have to keep in mind that there are two restrictions. Both K and π_1 must be positive. The values of π_1 along the branches is depicted in Figure 2.b. Positivity of π_1 reduces the curves to the ones depicted in the Figure 2.c.

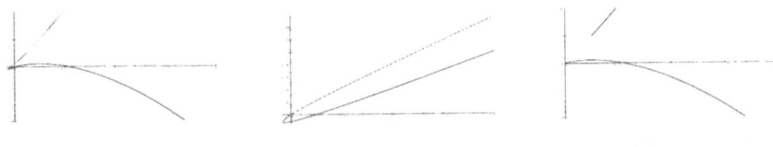

Figure 2.a   Figure 2.b   Figure 2.c

Figure 2: In each figure, K is on the horizontal axes. In figures (a) and (c), H is on the vertical axis, in figure (b) the value of π_1 is on the vertical axis. Leaving out the part of the curve in figure (a) with corresponding π_1-value less or equal to 0 reduces the graph in (a) to the graph in (c). See also formula 2.15 below.

Explicitly the relative equilibria are given as a function of k as

$$
\begin{aligned}
\pi_1 &= \frac{-1 + 12\,k - \sqrt{1 + 12\,k}}{18}, & \pi_1 &= \frac{-1 + 12\,k + \sqrt{1 + 12\,k}}{18} \\
\pi_2 &= \frac{6\,k + 1 + \sqrt{1 + 12\,k}}{18}, & \pi_2 &= \frac{6\,k + 1 - \sqrt{1 + 12\,k}}{18} \\
\pi_3 &= \frac{-6\,k\sqrt{1 + 12\,k} + 1 + \sqrt{1 + 12\,k}}{54}, & \pi_3 &= \frac{6\,k\sqrt{1 + 12\,k} + 1 - \sqrt{1 + 12\,k}}{54} \\
\pi_4 &= 0, & \pi_4 &= 0 \\
h &= \frac{1 + 36\,k + (12\,k + 1)^{\frac{3}{2}}}{18}, & h &= \frac{1 + 36\,k - (12\,k + 1)^{\frac{3}{2}}}{18} \\
k &\geq \frac{1}{4}, & k &\geq 0.
\end{aligned}
$$

$$(2.15)$$

Definition 2.1 *We define the* centerline *as the branch of relative equilibria for which*

$$
h = \frac{1 + 36\,k + (12\,k + 1)^{\frac{3}{2}}}{18}, \quad k \geq \frac{1}{4}.
$$

3 Traveling waves

3.1 Existence of traveling waves

It is easy to compute the fixed points of the system (2.12). We find that

$$
\begin{aligned}
\pi_1 &= -\frac{(\mu-1)(\mu-4)\left(\epsilon^2\mu^2 - 4\,\epsilon^2\mu + 4 + 4\,\epsilon^2\right)}{36\,(\mu-2)^2} \\
\pi_2 &= \frac{(\mu-1)^2\left(\epsilon^2\mu^2 - 4\,\epsilon^2\mu + 4 + 4\,\epsilon^2\right)}{36\,(\mu-2)^2} \\
\pi_3 &= \frac{(\mu-1)^2(\mu-4)\left(\epsilon^2\mu^2 - 4\,\epsilon^2\mu + 4 + 4\,\epsilon^2\right)}{108\,(\mu-2)^3} \\
\pi_4 &= -\frac{\epsilon\,(\mu-1)^2(\mu-4)\left(\epsilon^2\mu^2 - 4\,\epsilon^2\mu + 4 + 4\,\epsilon^2\right)}{216\,(\mu-2)^2}.
\end{aligned}
$$

$$(3.1)$$

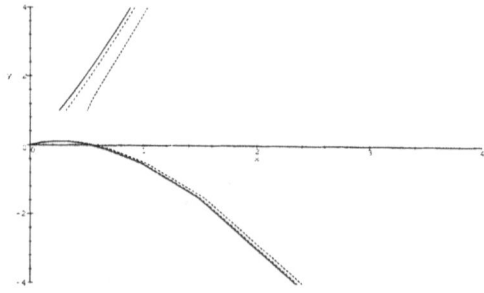

Figure 3: Relative equilibria for different values of ϵ. The dashed orbits are drawn with $\epsilon = 1$ and $\epsilon = 0.5$ (in the middle). The solid curve is with $\epsilon = 0.1$. K is on the horizontal axis and H is on the vertical axis.

As π_1 and π_2 must be positive we conclude that we find solutions for $\mu \in (1,2) \cup (2,4)$ and ϵ arbitrary. In addition we find the pure mode-2 solution: $\pi_1 = \pi_3 = \pi_4 = 0$, $\pi_2 \geq 0$, ϵ and μ arbitrary. If we let $\epsilon \downarrow 0$ in (3.1) and take μ in the set $(1,2) \cup (2,4)$, then the solution in (3.1) approaches the branch of stable relative equilibria of the unperturbed equation. This is depicted in the Figure 3. For $\mu \in (1,2)$ the lower branch in Figure 2.c is approached, whereas for $\mu \in (2,4)$ the upper branch is approached.

3.2   Stability of traveling waves

Lemma 3.1 *The Hopf bifurcation variety in the $\epsilon - \mu$ plane for the equilibria (3.1) is given by given by*

$$\epsilon^2 = \frac{(\mu - 1 - \sqrt{3})(\mu - 1 + \sqrt{3})}{(5\,\mu^2 - 22\,\mu + 26)\,(\mu - 2)^2},$$ (3.2)

$\mu \in (1 + \sqrt{3}, 4)$. *There are no saddle node bifurcations.*

Proof. If we linearize about the traveling wave solution (3.1) we find that the linearization is given by

$$\begin{bmatrix} 2(\mu - 1)\epsilon & 0 & 0 & -12 \\ 0 & 2(\mu - 4)\epsilon & 0 & 12 \\ 0 & 0 & 3(\mu - 2)\epsilon & 6 \\ -\frac{(\mu - 1)^2}{3(\mu - 2)^2}f & -\frac{(\mu - 1)(\mu - 4)}{3(\mu - 2)^2}f & \frac{(\mu - 1)(\mu - 4)}{3(\mu - 2)^2}f & 3(\mu - 2)\epsilon \end{bmatrix},$$ (3.3)

where $f = \epsilon^2\mu^2 - 4\epsilon^2\mu + 4 + 4\epsilon^2$. Hence the characteristic polynomial is

$$
\begin{aligned}
p(\lambda) = & -(-\lambda - 12\epsilon + 6\epsilon\mu)\left(6\epsilon^3\mu^5 - 66\epsilon^3\mu^4 + \lambda\epsilon^2\mu^4 + 276\epsilon^3\mu^3 - 12\lambda\epsilon^2\mu^3 - 4\lambda^2\epsilon\mu^3 \right. \\
& + 24\epsilon\mu^3 - 552\epsilon^3\mu^2 + 52\lambda\epsilon^2\mu^2 + 26\lambda^2\epsilon\mu^2 - 168\epsilon\mu^2 + \lambda^3\mu^2 - 12\lambda\mu^2 + 528\epsilon^3\mu \\
& - 96\lambda\epsilon^2\mu + 336\epsilon\mu - 56\lambda^2\epsilon\mu - 4\lambda^3\mu + 48\lambda\mu - 192\epsilon^3 + 64\lambda\epsilon^2 - 192\epsilon + 40\lambda^2\epsilon \\
& \left. + 4\lambda^3\right)/(\mu - 2)^2 .
\end{aligned}
$$

$$(3.4)$$

From this we conclude that the condition for an eigenvalue at zero is

$$
0 = -36\,\epsilon^2\,(\mu - 1)\,(\mu - 4)\,\left(\epsilon^2\mu^2 - 4\epsilon^2\mu + 4 + 4\epsilon^2\right) \tag{3.5}
$$

This shows that when $\epsilon \neq 0$, and $\mu \in (1,2) \cup (2,4)$ this condition is never satisfied.

Next we look at the possibility for purely imaginary eigenvalues. Recall that the equation

$$
A\lambda^3 + B\lambda^2 + C\lambda + D = 0
$$

has purely imaginary roots $\pm i\omega$ iff

$$
AD - BC = 0, \qquad \frac{C}{A} > 0.
$$

Applying this to (3.4) we conclude that the first condition is equivalent to either $\epsilon = 0$ or, more interestingly,

$$
0 = -5\,\epsilon^2\mu^4 + 42\,\epsilon^2\mu^3 - 134\,\epsilon^2\mu^2 + 12\,\mu^2 - 24\,\mu + 192\,\epsilon^2\mu - 104\,\epsilon^2 - 24. \quad (3.6)
$$

The second condition reads in this case

$$
\frac{(\mu - 2)^2}{(\mu - 4)\,(\epsilon^2\mu^3 - 8\,\epsilon^2\mu^2 - 12\,\mu + 20\,\epsilon^2\mu - 16\,\epsilon^2)} > 0. \tag{3.7}
$$

The denominator can never be zero for μ in the interval $(1,2) \cup (2,4)$, and hence (3.7) is satisfied for these values of μ. It remains to analyse (3.6). It follows from this equation that (3.2) holds. The interval in μ for which $\epsilon \geq 0$ is thus $(1 + \sqrt{3}, 4)$.
$\square$

Note that at $\epsilon = 0$ and $\mu = 1 + \sqrt{3}$ we have

$$
\pi_1 = \frac{1 + \sqrt{3}}{6} \quad \pi_2 = \frac{2 + \sqrt{3}}{6} \quad \pi_3 = -\frac{1}{6} - \frac{\sqrt{3}}{9} \quad \pi_4 = 0 \tag{3.8}
$$

We will refer to this point as the *Hopf point*. This is the traveling wave of the unperturbed equations that can bifurcate to modulated waves in the perturbed equation.

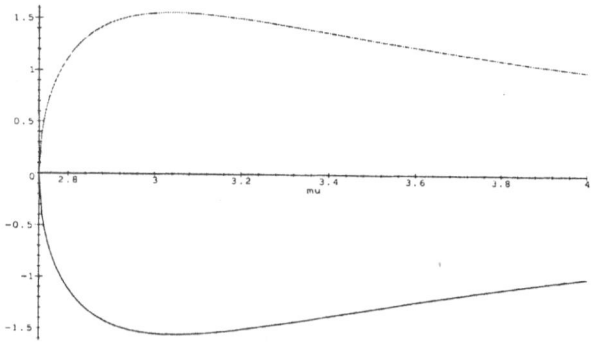

Figure 4: The Hopf bifurcation curve. On the vertical axis is ϵ and μ is on the horizontal axis.

4 Hopf bifurcation to modulated waves

4.1 The averaged equations transformed to standard form

In this section will determine the conditions for the existence of periodic solutions of equation (2.12). We compute averaged equations and by transformations put these equations in a certain standard form.

The unperturbed equation, (2.12) with $\epsilon = 0$, has two conserved quantities:

$$I_1 = \pi_1 + 2\pi_3$$
$$K = \pi_1 + \pi_2. \tag{4.1}$$

The values of I_1 and K will be denoted by h_1 and k respectively. The Hamiltonian can be expressed as

$$H = -3I_1 + 4K. \tag{4.2}$$

Let $t \mapsto p(t)$ be a periodic orbit of the unperturbed equation. So p depends on h_1 and k. Let $\Sigma = \Sigma_{h_1,k}$ be a neighborhood of the hyperplane transverse to $\dot{p}(0)$ through $p(0)$. A necessary and sufficient condition for the orbit of (2.12) through $p(0)$ to be periodic is that along the orbit

$$\int_0^T dI_1 = \int_0^T dK = 0, \tag{4.3}$$

where $I_1 = I_1(\epsilon, \mu, h_1, k)(t)$, $K = K(\epsilon, \mu, h_1, k)(t)$, and T is time of first return to Σ. This leads us to define

$$F(\epsilon, \mu, h_1, k) = \begin{pmatrix} \int_0^T dI_1 \\ \int_0^T dK, \end{pmatrix} \tag{4.4}$$

where we have suppressed in the right hand side the dependence on the variables. Using (2.12) we find that

$$
F = \left(
\begin{array}{c}
\epsilon \oint_{\gamma(h_1,k)} \frac{(\mu-4)\pi_1 + 3h_1(2-\mu)}{12\pi_4}\, d\pi_1 + o(\epsilon) \\
\epsilon \oint_{\gamma(h_1,k)} \frac{-3\pi_1 + (\mu-4)k}{6\pi_4}\, d\pi_1 + o(\epsilon)
\end{array}
\right)
\tag{4.5}
$$

where

$$
\gamma(h_1,k) = \{(\pi_1,\pi_2,\pi_3,\pi_4) \mid \pi_1^2\pi_2 = \pi_3^2 + \pi_4^2,\ I_1 = h_1,\ K = k\},
\tag{4.6}
$$

or, equivalently

$$
\gamma(h_1,k) = \{(\pi_1,\pi_4) \mid \pi_1^3 + (1/4 - h_2)\,\pi_1^2 - \frac{h_1\pi_1}{2} + \frac{h_1^2}{4} + \pi_4^4 = 0\}.
\tag{4.7}
$$

Our first goal is to obtain first precise information about the lowest order part of F. We thus consider the set of equations

$$
\begin{cases}
\oint_{\gamma(h_1,k)} (\mu - 4)\dfrac{\pi_1}{\pi_4} + 3h_1(2-\mu)\dfrac{1}{\pi_4}\, d\pi_1 = 0 \\[2mm]
\oint_{\gamma(h_1,k)} -3\dfrac{\pi_1}{\pi_4} + (\mu-4)k\dfrac{1}{\pi_4}\, d\pi_1 = 0.
\end{cases}
\tag{4.8}
$$

The following lemma is an immediate consequence of (4.4) and (4.8).

Lemma 4.1 (i). $F(0,\mu,h_1,k) = 0$

(ii). $D_\epsilon F(0,\mu,h_1,k) = 0$ *if and only if*

$$
\begin{aligned}
& -\frac{\mu^2 h_2}{72} + \frac{\mu h_2}{9} - \frac{2\,h_2}{9} - \frac{\mu h_1}{8} + \frac{h_1}{4} = 0 \\
& \oint_{\gamma(h_1,k)} -3\frac{\pi_1}{\pi_4} + (\mu-4)k\frac{1}{\pi_4}\, d\pi_1 = 0.
\end{aligned}
\tag{4.9}
$$

We apply a number of transformations that transforms (4.7) to the compact connected level sets of the standard form

$$
\frac{\pi_4^2}{2} + \frac{\pi_1^3}{3} - \frac{\pi_1^2}{2} + h = 0.
\tag{4.10}
$$

The price we will pay for this simplification is that h will depend in a complicated way on h_1 and k. In order to be able to transform to the standard form (4.10) we must impose a condition that guarantees that

$$
V(\pi_1) = \pi_1^3 + (1/4 - k)\,\pi_1^2 - \frac{h_1\pi_1}{2} + \frac{h_1^2}{4}
\tag{4.11}
$$

has two positive critical points.

Lemma 4.2 $V' = 0$ *has two nonnegative roots iff* $(1/4 - k)^2 + 3/2h_1 \geq 0$, $h_1 \leq 0$ *and* $k \geq 1/4$.

Proof. The solutions to the equation

$$\frac{3\pi_1{}^2}{2} + (1/4 - k)\pi_1 - \frac{h_1}{4} = 0 \tag{4.12}$$

are $\pi_1 = -\frac{1}{12} + \frac{k}{3} \pm \frac{\sqrt{1 - 8k + 16k^2 + 24h_1}}{12}$. From this explicit formula the proof is obvious. $\quad\square$

Introduce the variable x by

$$k = \frac{x^2}{12} - \frac{1}{12}. \tag{4.13}$$

Definition 4.3 $\mathcal{U}$ *is the region in the* $x - h_1$ *plane determined by the conditions*

$$-\frac{1}{216}(x - 2)^2(x + 2)^2 \leq h_1 \leq 0, \text{ and } x \geq 2.$$

Lemma 4.4 *Solving the equation* $D_\epsilon F(0, \mu, h_1, k) = 0$ *is equivalent to solving in* U *the equation*

$$Q \circ f = q, \tag{4.14}$$

where

$$Q(h) = \frac{\oint_{\Gamma(h)} \frac{\pi_4}{\pi_4} d\pi_1}{\oint_{\Gamma(h)} \frac{1}{\pi_4} d\pi_1}$$

$$f(x, h_1) = \frac{1}{12} + \frac{5832 h_1{}^2 - 324 h_1 x^2 + 1296 h_1 - x^6 + 12 x^4 - 48 x^2 + 64}{12 (16 - 8 x^2 + x^4 + 216 h_1)^{3/2}} \tag{4.15}$$

$$q(x, h_1) = \frac{1}{2} - \frac{x^2 - 4 - 54 h_1 - 6\sqrt{3}\sqrt{-h_1 (2 x^2 - 2 - 27 h_1)}}{2\sqrt{16 - 8 x^2 + x^4 + 216 h_1}}.$$

Here

$$\Gamma(h) = \{ (\pi_1, \pi_4) \mid \frac{\pi_4^2}{2} + \frac{\pi_1^3}{3} - \frac{\pi_1^2}{2} + h = 0, \ \pi_1 \geq 0 \}. \tag{4.16}$$

Proof. It follows from (2.12), (4.1) and (4.15) that π_1 and π_4 satisfy the system of equations

$$\begin{cases} \dot{\pi}_1 = -12\pi_4 \\ \dot{\pi}_4 = -3 h_1 + 3\pi_1 - 12\pi_1 \left(\frac{x^2}{12} - \frac{1}{12} - \pi_1 \right) + 6\pi_1{}^2. \end{cases} \tag{4.17}$$

We let

$$q = \frac{x^2}{36} - \frac{1}{9} - \frac{\sqrt{16 - 8 x^2 + x^4 + 216 h_1}}{36}$$

$$r = \left(1/3 - \frac{x^2}{12} \right)^2 + \frac{3 h_1}{2}. \tag{4.18}$$

Applying the transformation $\tilde{t} = 12t$, $\pi_1 = \tilde{\pi}_1 + q$ yields. dropping the tildes, we obtain

$$\begin{cases} \dot{\pi}_1 = -\pi_4 \\ \dot{\pi}_4 = \dfrac{\pi_1 \left(18\,\pi_1 - \sqrt{16 - 8\,x^2 + x^4 + 216\,h_1}\right)}{12}, \end{cases} \quad (4.19)$$

and the integral condition in (4.9) transforms to

$$\oint_{\tilde{\gamma}(x,k)} -\frac{\left(36\,\pi_1 - 3\,x^2 - \sqrt{16 - 8\,x^2 + x^4 + 216\,h_1} + \mu\,x^2 - \mu\right) d\pi_1}{72\,\pi_4} = 0. \quad (4.20)$$

Here $\tilde{\gamma}(x,k)$ is obtained from $\gamma(h_1, k)$ by the transformation. Next we rescale $\pi_k = \frac{2}{3}\tilde{\pi}_k\sqrt{r}$. $k = 1, 4$, which yields, dropping the tildes

$$\begin{cases} \dot{\pi}_1 = -\pi_4 \\ \dot{\pi}_4 = \dfrac{\pi_1\sqrt{16 - 8\,x^2 + x^4 + 216\,h_1}\,(\pi_1 - 1)}{12}, \end{cases} \quad (4.21)$$

while (4.20) transforms to

$$\oint_{\hat{\gamma}(x,k)} -\frac{\left((2\,\pi_1 - 1)\sqrt{16 - 8\,x^2 + x^4 + 216\,h_1} - 3\,x^2 + \mu\,x^2 - \mu\right) d\pi_1}{72\,\pi_4} = 0. \quad (4.22)$$

Here $\hat{\gamma}(x,k)$ is obtained from $\tilde{\gamma}(x,k)$ by the transformation. Finally we rescale once more (and drop the tildes) by $\pi_4 = \tilde{\pi}_4 r^{\frac{1}{4}}$, $\tilde{t} = r^{\frac{1}{4}}$. which gives

$$\begin{cases} \dot{\pi}_1 = -\pi_4 \\ \dot{\pi}_4 = \pi_1\,(\pi_1 - 1). \end{cases} \quad (4.23)$$

while (4.22) transforms to

$$\oint_{\Gamma(x,k)} -\frac{1}{72}\frac{\left((2\,\pi_1 - 1)\sqrt{16 - 8\,x^2 + x^4 + 216\,h_1} - 3\,x^2 + \mu\,x^2 - \mu\right) d\pi_1}{\pi_4\sqrt[4]{\frac{1}{9} - \frac{x^2}{18} + \frac{x^4}{144} + \frac{3\,h_1}{2}}} = 0. \quad (4.24)$$

Here $\Gamma(x,k)$ is obtained from $\hat{\gamma}(x,k)$ by the transformation. A straightforward, Maple checked, computation gives that

$$\Gamma(x,k) = \{(\pi_1, \pi_4) \mid \frac{\pi_4^2}{2} + \frac{\pi_1^3}{3} - \frac{\pi_1^2}{2} + f(x.k) = 0\}, \quad (4.25)$$

with

$$f(x,k) = \frac{1}{12} + \frac{5832\,h_1{}^2 - 324\,h_1 x^2 + 1296\,h_1 - x^6 + 12\,x^4 - 48\,x^2 + 64}{12\left(16 - 8\,x^2 + x^4 + 216\,h_1\right)^{3/2}}. \quad (4.26)$$

Next we eliminate μ using the first equation in (4.9). There are two solutions, but as we restrict to μ in the interval $(1, 4)$ it is not difficult to check that

$$\mu = -\frac{-4\,x^2 + 4 + 54\,h_1 + 6\sqrt{-3\,h_1\,(2\,x^2 - 2 - 27\,h_1)}}{(x - 1)\,(x + 1)} \quad (4.27)$$

From (4.24) and (4.27) the formula for q follows.

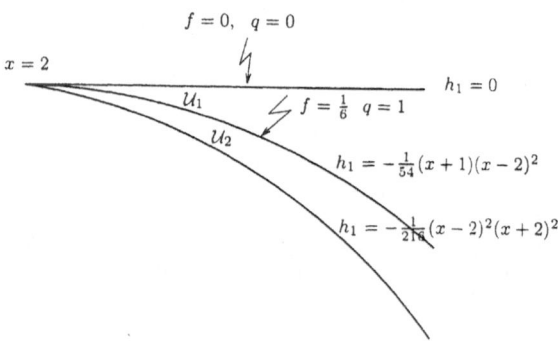

Figure 5: Illustration for Definition 4.5 and Lemma 4.6.

Definition 4.5 *We let* $\mathcal{U}_1$ *be the region in the* $x - h_1$ *plane determined by the conditions*

$$\mathcal{U}_1 = \{(x, h_1) \mid -\frac{1}{216}(x-2)^2(x+2)^2 \le h_1 \le -\frac{1}{54}(x+1)(x-2)^2, \text{ and } x \ge 2\}$$

$$\mathcal{U}_2 = \{(x, h_1) \mid -\frac{1}{54}(x+1)(x-2)^2 \le h_1 \le 0, \text{ and } x \ge 2\}$$

To make the formulas for f and q a bit simpler we introduce the new variable z by

$$h_1 = \frac{(-1+z^2)(-2+x)^2(2+x)^2}{216}. \tag{4.28}$$

Note that on $\mathcal{U}$, $z \in [0, 1]$ and that the centerline is given in the new coordinates by

$$centerline = \{(x, z) \mid z = \frac{x}{2+x}\} = \{(x, h_1) \mid h_1 = -\frac{1}{54}(x+1)(x-2)^2\}. \tag{4.29}$$

The mappings f and q. denoted by the same symbol, are given by

$$f(x, z) = \frac{(z-1)^2(2z+xz+x)(-2z+xz+x)}{96z^3}$$

$$q(x, z) = \frac{-x^2 + z^2x^2 + 4z}{8z}$$

$$+ \frac{\sqrt{(z-1)(1+z)(-8x^2 - x^4 + 16z^2 - 8z^2x^2 + z^2x^4)} - 4z^2}{8z} \tag{4.30}$$

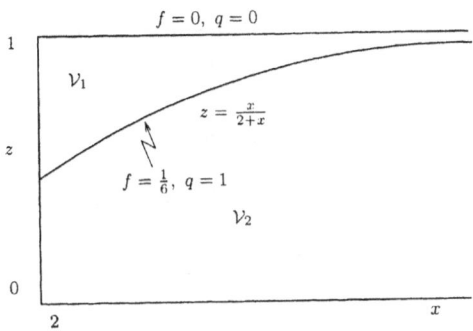

Figure 6: Blowing up Figure 5 in the h_1 direction . $\mathcal{V}_i$. $i = 1, 2$ is obtained from $\mathcal{U}_i$ by the transformation (4.28).

It is tempting to use q as one of the coordinates. From Figure 6 it would be reasonable to use q and x as coordinates. It seems difficult to perform the required transformation explicitly. On the other hand it is relatively simple to use q and z as variables. This transformation is singular as $q(x, 1) = 1$. It is straightforward to check that

$$q(x, z) = q. \quad x \geq 0$$
$$\Rightarrow x = \frac{2 z \sqrt{(z - 1)(1 + z)(2qz - z - 1)(1 + 2qz - z - 2q + 2q^2)}}{(z - 1)(1 + z)(2qz - z - 1)} \tag{4.31}$$

So, if we use q and z as variables then the half-line line $z = 1$, $x \geq 2$ is transformed to the point $q = z = 1$. This is depicted in Figure 7.

Lemma 4.6 *(i). At the centerline we have $f = \frac{1}{6}$ and $q = 1$. On the line $h_1 = 0$ we have $f = 0$ and $q = 0$.*

(ii). In $\mathcal{V}_1$ we have $0 \leq f \leq \frac{1}{6}$ and $0 \leq q \leq 1$.

(iii). In $\mathcal{V}_2$ we have $f \geq \frac{1}{6}$ and $q \geq 1$.

Proof. Direct computation shows that $\frac{\partial f}{\partial z}$ and $\frac{\partial q}{\partial z}$ are both negative in the interior of $\mathcal{V}$. □

Corollary 4.7 *If we let*

$$\mathcal{F}(x, z) = Q \circ f(x. z) - q(x. z). \tag{4.32}$$

then $\mathcal{F}$ vanishes identically at the centerline and at the half-line $z = 1$, $x \geq 2$.

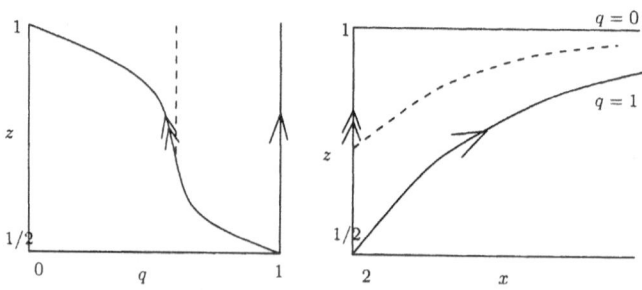

Figure 7: The transformation from x to q.

4.2 Global Hopf bifurcation

In this section we will prove that there exists a global branch of nontrivial solutions to equation (4.14) in the region $\mathcal{V}_1$. We need a number of lemmas which we state and prove below. But first we formulate the main theorem.

Theorem 4.8 *There exist a smooth mapping $k \mapsto \epsilon_0(k)$ from $(\frac{1}{3}\sqrt{3} + \frac{1}{2}, \infty)$ into $\mathbb{R}_+$ and a smooth mapping $(\epsilon, k) \mapsto \mu^*(\epsilon, k)$ defined for $0 < |\epsilon| \le \epsilon_0(k)$ and $k > \frac{1}{3}\sqrt{3} + \frac{1}{2}$ satisfying $\mu^*(0, \frac{\sqrt{3}}{3} + \frac{1}{2}) = 1 + \sqrt{3}$ such that for $0 < |\epsilon| \le \epsilon_0(k)$ and $k > \frac{1}{3}\sqrt{3} + \frac{1}{2}$ (2.12) has a periodic solution if and only if $\mu = \mu^*(\epsilon, k)$.*

Lemma 4.9 *There is a unique branch point on the centerline (4.29).*

Proof. At the centerline $\mathcal{F} \equiv 0$ and $f \equiv \frac{1}{6}$. Therefore $Q' = 1$ at the centerline, see Lemma 6.3, and the condition for bifurcation $D_x \mathcal{F}(x, \frac{x}{2+x}) = 0$ leads to the equation

$$\frac{x^2 - 4x + 1}{3x^2(x+2)} = 0. \tag{4.33}$$

This equation has a unique solution at $x = 2 + \sqrt{3}$ in the domain $x \ge 1$. $\qquad\square$

A local analysis shows that the branch bifurcates to the right. It will be shown later on that the branch extends to infinity without turning points. First we will show that this is the only branch.

Lemma 4.10 *There is no branch point on the line-segment $\{(x, z) \mid z = 1, \ x \ge 2\}$.*

Proof. Let I be a compact interval contained in $[2, \infty)$. $\mathcal{F}$ vanishes identically on the set $I \times 1$ in $\mathcal{V}_1$. We have the asymptotic expansions

$$f(x, z) = \frac{(x-1)(x+1)}{24}(z-1)^2 + \mathcal{O}\left((z-1)^3\right)$$

$$q(x, z) = \frac{\sqrt{-2x^2+2}\sqrt{z-1}}{2} + \left(\frac{x^2}{4} - \frac{1}{2}\right)(z-1) + \mathcal{O}((z-1)^{3/2})$$

$$Q'(h) = \frac{6}{h(\ln h)^2} + \mathcal{O}(\frac{1}{h(\ln h)^3}),$$

uniformly for $x \in I$. From these expansions we derive that

$$D_x\mathcal{F}(x, z) = \frac{12}{(z-1)(\ln \frac{(x-1)(x+1)(z-1)^2}{24})^2}(1+o(1)), \qquad (4.34)$$

uniformly for $x \in I$. Hence, there is a neighborhood of the set $I \times 1$ in $\mathcal{V}_1$ such that there are no other zeros of $\mathcal{F}$ other then the ones on $I \times 1$. $\qquad \square$

Lemma 4.11 $\mathcal{F}$ has no zero on the set $\{(2, z) \,|\, z \in (\frac{1}{2}, 1)\}$.

Proof. Using the inequality $Q(h) \geq (6h)^{\frac{1}{6}}$, see Lemma 6.4, it follows from (4.30) that it is sufficient to show that

$$\frac{2^{2/3}\sqrt[6]{2\ z^2 + z}\sqrt[3]{-z+1}}{2\ z^{2/3}} - \frac{-1 + z + \sqrt{3 - 3\ z^2}}{2\ z} \geq 0. \qquad (4.35)$$

Writing $z = \frac{w+1}{2}$, it is sufficient to show that

$$2^{5/6}\sqrt[6]{w+2}\sqrt{w+1} + (1-w)^{2/3} - \sqrt{3}\sqrt[6]{1-w}\sqrt{w+3} \geq 0 \qquad (4.36)$$

on the interval $(0, 1)$. Using the estimates, on the same interval, $(w+2)^{\frac{1}{6}} \geq 2^{\frac{1}{6}}$, $\sqrt{w+1} \geq 1 + \frac{w}{2} - \frac{w^2}{8}$, $(1-w)^{\frac{2}{3}} \geq 1 - w$, $(1-w)^{\frac{1}{6}} \leq 1 - \frac{1}{6}w - \frac{5}{72}w^2$, and finally $\sqrt{w+3} \leq \sqrt{3}(1 + \frac{1}{6}w)$, one shows that

$$2^{5/6}\sqrt[6]{w+2}\sqrt{w+1} + (1-w)^{2/3} - \sqrt{3}\sqrt[6]{1-w}\sqrt{w+3} \geq \frac{w^2}{24} + \frac{5\ w^3}{144} \qquad (4.37)$$

which is indeed positive on the interval $(0, 1)$. $\qquad \square$

Lemma 4.12 The only solution of the set of equations

$$\mathcal{F} = 0 \qquad (4.38)$$

$$D_q\mathcal{F} = 0 \qquad (4.39)$$

on the domain $\{(q, z) \,|\, 0 \leq q \leq 1, \ \frac{1}{2} \leq z \leq 1\}$ is the point $P = (1, \frac{5}{13} + \frac{2}{13}\sqrt{3})$.

Using the Riccati equation (6.12), we may replace (4.39) by

$$\mathcal{G} = 0, \tag{4.40}$$

where

$$\begin{aligned}
\mathcal{G} = {} & 13\,z^3 + 3\,z^2 - 9\,z + 1 + \left(18\,z^3 + 54\,z^2 - 14\,z - 2\right)(q-1) \\
& + \left(54\,z^2 - 24\,z^3 + 12\,z - 2\right)(q-1)^2 + \left(8\,z - 16\,z^3\right)(q-1)^3
\end{aligned} \tag{4.41}$$

$\mathcal{G}$ is a third order polynomial in q with coefficients depending smoothly on z. It is not difficult to show that

(i). When $z = \frac{1}{2}$ there are three real roots, one negative, one in the interval $(0,1)$ and one larger than 1.

(ii). $\mathcal{G}^{-1}(0) \cap \{(0,z) \mid z \in [\frac{1}{2}, 1]\} = \{(0,1)\}$.

(iii). $\mathcal{G}^{-1}(0) \cap \{(\frac{1}{2}, z) \mid z \in [\frac{1}{2}, 1]\} = \{\emptyset\}$.

(iv). $\mathcal{G}^{-1}(0) \cap \{(1, z) \mid z \in [\frac{1}{2}, 1]\} = \{P\}$.

(v). At $z = 1$ there are three real roots, one at $q = 0$, one at $q = 4 - \sqrt{10}$ and one larger than 1.

(vi). At $z = \frac{1}{2}\sqrt{2}$ there are two roots, which are real. One root approaches $-\infty$ as $z \uparrow \frac{1}{2}\sqrt{2}$ and one root approaches ∞ as $z \downarrow \frac{1}{2}\sqrt{2}$.

The conclusion is that there exists a smooth function $q_1(z)$, defined for $z \in [\frac{1}{2}, 1]$ such that

$$l_1 = \{\, (q_1(z), z) \mid z \in [\frac{1}{2}, 1] \,\} \tag{4.42}$$

is the zero set of $\mathcal{G}$ in the domain $\frac{1}{2} \le q \le 1$, $\frac{1}{2} \le z \le 1$. See Figure 8.

We need to demonstrate that $\mathcal{G}^{-1}(0) \cap \mathcal{F}^{-1}(0) = P$. From lemma 6.4 we conclude that it is sufficient to show that $\overline{\mathcal{F}} < 0$ on the set $l_1 \setminus P$, where

$$\overline{\mathcal{F}}(q, z) = 1 + (f(q, z) - \frac{1}{6}) - \frac{25}{12}(f(q, z) - \frac{1}{6})^2 - q. \tag{4.43}$$

Similar analysis as for $\mathcal{G}$ shows that there exists a smooth function $q_2(z)$, defined for $z \in [\frac{5}{13} + \frac{2}{13}\sqrt{3}, 1]$, and taking values in the interval $[\frac{1}{2}, 1]$, such that

$$l_2 = \{\, (q_2(z), z) \mid z \in [\frac{1}{2}, 1] \,\} \tag{4.44}$$

is the zero set of $\overline{\mathcal{F}}$ in the domain $0 \le q \le 1$, $\frac{1}{2} \le z \le 1$. See Figure 9.

Let $T_P l_1$ be the tangent line to l_1 at the point P. To complete the proof we will demonstrate that

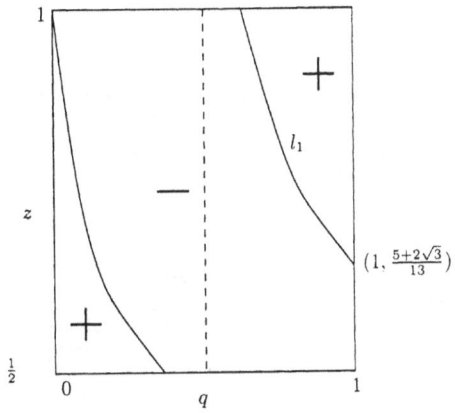

Figure 8: The zero set $\mathcal{G}$.

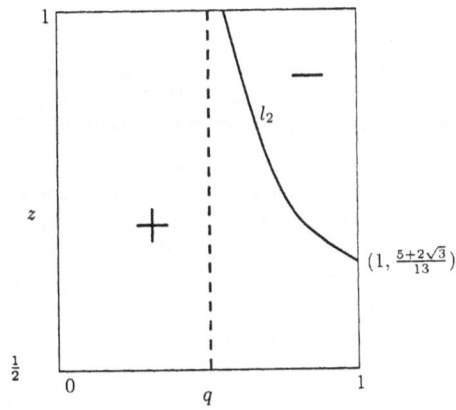

Figure 9: The zero set $\overline{\mathcal{F}}$.

(i). $q_2(z) \leq T_p l_1(z)$,

(ii). $T_p l_1(z) \leq q_1(z)$.

A straightforward computation shows that the tangent line is given by

$$z = -\frac{155}{169}q - \frac{160\sqrt{3}}{507}q + \frac{220}{169} + \frac{238\sqrt{3}}{507}. \tag{4.45}$$

To prove (i), we evaluate $\mathcal{G}$ for these values of z. The result is

$$\mathcal{G}]_{l_1}(q) = \frac{5}{57820345011}(309536\sqrt{3} + 553815)$$
$$(532400\,q^4 - 1940840\,q^3 - 58080\sqrt{3}q^3 + 3360456\,q^2 - 390192\sqrt{3}q^2 -$$
$$2712019\,q + 1233060\sqrt{3}q + 1028449 - 1142094\sqrt{3})$$
$$(q-1)^2$$

$$\tag{4.46}$$

It is straightforward to prove that the right hand side is negative on the interval $[4 - \sqrt{10}, 1)$ and vanishes at $q = 1$. To prove (ii), we evaluate $\overline{\mathcal{F}}$ for these values of z. The result is

$$\overline{\mathcal{F}}]_{l_1}(q) = \frac{5}{107488021375449}(27073417 + 15502976\sqrt{3})$$
$$(45753125\,q^6 + 11686845500\,q^5 - 6655000\,q^5\sqrt{3} - 64540389650\,q^4$$
$$- 1635532800\sqrt{3}q^4 + 155617401500\,q^3 - 822008000\sqrt{3}q^3 +$$
$$13948281280\sqrt{3}q^2 - 200612740091\,q^2 - 24107473896\sqrt{3}q +$$
$$142914707936\,q - 46013184784 + 13182109568\sqrt{3})$$
$$(q-1)$$

$$\tag{4.47}$$

To show that the right hand side is negative on the interval $[4 - \sqrt{10}, 1)$ and vanishes at $q = 1$ it is sufficient to show that the 6-th order polynomial in q is positive on the same interval. It is easy to see that this is indeed the case. This completes the proof.   □

Proof of Theorem 4.8. It is a consequence of Lemma 4.9 and Lemma 4.12 that there exist smooth mappings $k \mapsto \hat{\mu}(k)$ and $k \mapsto \hat{h}_1(k)$, defined for $k \geq \frac{1}{2} + \frac{1}{3}\sqrt{3}$, satisfying $\hat{\mu}(\frac{1}{2} + \frac{1}{3}\sqrt{3}) = 1 + \sqrt{3}$, $\hat{h}_1(\frac{1}{2} + \frac{1}{3}\sqrt{3}) = \frac{5}{2} + \frac{3}{2}\sqrt{3}$, such that $D_\epsilon F(0, \hat{\mu}(k), \hat{h}_1(k), k) = 0$. From Lemma 4.12 it also follows that along this branch $\det(D_\mu D_\epsilon F, D_{h_1} D_\epsilon F) \neq 0$. The existence part follows from the implicit function theorem. The uniqueness follows from Lemmas 4.9, 4.10, 4.12.

5 Numerical results

We have used the software package AUTO [Doe81] to follow the periodic orbits for large values of the momentum. We did start at the equilibrium point $\pi_1 = 23.222$, $\pi_2 = 13.444$. $\pi_3 = -85.148$ and $\pi_4 = 0.425$ with parameter values $\epsilon = 0.1$ and $\mu = 2.1$. Varying μ, a Hopf bifurcation has been detected at $\mu = 2.73$. At this point $\pi_1 = 0.45$, $\pi_2 = 0.62$, $\pi_3 = -0.35$ and $\pi_4 = 0.01$. So we are indeed very close to the Hopf point (3.8). Next we have followed the periodic solution inceasing μ until the L_2-norm was 500. At this point $\mu = 3.30$ The period of the orbit decreases monotonically, the momentum increases monotonically. It is not clear whether there is a limiting value for μ. The periodic solutions are far away in phase space from the branch of unstable relative equilibria and also they stay far away from the level-set $H = 0$, as is indicated by the fact that the period decreases. To get a precise insight in the limiting behavior it would be necessary to analyse the solutions of (4.16) for large values of x. We haven't pursued this analysis.

6 Appendix

Given the family of curves

$$\frac{y^2}{2} + \frac{x^3}{3} - \frac{x^2}{2} + h = 0 \tag{6.1}$$

we let

$$I_0 = I_0(h) = \oint_{\gamma(h)} y\,dx, \quad I_1 = I_1(h) = \oint_{\gamma(h)} xy\,dx, \tag{6.2}$$

where $\gamma(h)$ is the compact connected set that satisfies (6.1).

Lemma 6.1 I_0 and I_1 satisfy the differential equation

$$h\left(h - \frac{1}{6}\right)(I)'_0 I'_1 = \begin{pmatrix} \frac{5h}{6} - \frac{1}{6} & \frac{7}{36} \\ -\frac{h}{6} & \frac{7h}{6} \end{pmatrix}(I)_0 I_1 \tag{6.3}$$

Proof. Differentiation of (6.1) with respect to h, fixing x. gives

$$h\frac{dy}{dh} + 1 = 0. \tag{6.4}$$

Differentiation of (6.1) with respect to x, fixing h. gives

$$y\frac{dy}{dx} + x^2 - x = 0. \tag{6.5}$$

Multiplication of (6.5) by $\frac{x^m}{y}$ and integration over $\gamma(h)$ yields

$$-m\oint_{\gamma(h)} x^{m-1}y\,dx + \oint_{\gamma(h)} \frac{x^{m+2}}{y}\,Dx - \oint_{\gamma(h)} \frac{x^{m+1}}{y}\,dx = 0.$$

We compute the differential equation satisfied by I_0.

$$
\begin{aligned}
h\, I_0' &= -h \oint_{\gamma(h)} \frac{1}{y}\, dx \\
&= \oint_{\gamma(h)} \left(\frac{y^2}{2} + \frac{x^3}{3} - \frac{x^2}{2} \right) \frac{1}{y}\, dx \\
&= \frac{1}{2} \oint_{\gamma(h)} y\, dx + \frac{1}{3} \oint_{\gamma(h)} \frac{x^3}{y}\, dx - \frac{1}{2} \oint_{\gamma(h)} \frac{x^2}{y}\, dx \\
&\overset{RE_1}{=} \frac{5}{6} I_0 + \frac{1}{6} I_1'.
\end{aligned}
\tag{6.6}
$$

Similarly we compute

$$
\begin{aligned}
h\, I_1' &= -h \oint_{\gamma(h)} \frac{x}{y}\, dx \\
&= \frac{1}{2} \oint_{\gamma(h)} xy\, dx + \frac{1}{3} \oint_{\gamma(h)} \frac{x^4}{y}\, dx - \frac{1}{2} \oint_{\gamma(h)} \frac{x^3}{y}\, dx \\
&\overset{RE_2}{=} \frac{7}{6} I_1 - \frac{1}{6} \oint_{\gamma(h)} \frac{x^3}{y}\, dx \\
&\overset{RE_1}{=} \frac{7}{6} I_1 - \frac{1}{6} I_0 + \frac{1}{6} I_1'
\end{aligned}
\tag{6.7}
$$

The result follows now combining (6.6) and (6.7). $\square$

I_0 and I_1 are analytic at $h = \frac{1}{6}$ but are singular at $h = 0$. The singularity is of logarithmic type, see [BK81]. In the following Lemma we give the expansion of I_0 and I_1 at both endpoints. In the following lemma we give the asymptotic expansions at the two singular points. The proof is straightforward and omitted.

Lemma 6.2 Let $k = h - \frac{1}{6}$.

$$
\begin{aligned}
I_0(k) &= -2\pi k + \frac{5\pi k^2}{6} - \frac{385\pi k^3}{216} + \frac{85085\pi k^4}{15552} + \mathcal{O}(k^5) & k \uparrow 0 \\
I_1(k) &= -2\pi k - \frac{\pi k^2}{6} + \frac{35\pi k^3}{216} - \frac{5005\pi k^4}{15552} + \mathcal{O}(k^5) & k \uparrow 0 \\
I_0(h) &= -\frac{6}{5} + \frac{459h}{70} + \frac{5h^2}{14} - h\ln h - \frac{5h^3 \ln h}{12} + \mathcal{O}(h^4 \ln h) & h \\
I_1(h) &= -\frac{36}{35} + 6h + \frac{37h^2}{70} - \frac{h^2 \ln(h)}{2} + \mathcal{O}(h^3 \ln h) & h \downarrow 0
\end{aligned}
\tag{6.8}
$$

We introduce the quotient

$$
Q = Q(h) = \frac{I_1'(h)}{I_0'(h)}.
\tag{6.9}
$$

It is a corollary of Lemma 6.2 that we have the following asymptotic expansion for Q at the singular points:

Lemma 6.3 *Let* $k = h - \frac{1}{6}$.

$$Q(k) = 1 + k - \frac{25}{12}k^2 + \frac{775}{108}k^3 + O\left(k^4\right) \qquad k \uparrow 0$$
$$Q(h) = -\frac{6}{\ln h} + \mathcal{O}\left(\frac{1}{(\ln h)^2}\right) \qquad h \downarrow 0 \tag{6.10}$$

By differentiation of (6.3) we obtain the differential equation satisfied by I_0' and I_1'.

$$h(h - \frac{1}{6})\,(\,I\,)_0''\, I_1'' = \begin{pmatrix} -\frac{h}{6} & \frac{1}{36} \\ -\frac{h}{6} & \frac{h}{6} \end{pmatrix}(\,I\,)_0'\, I_1' \tag{6.11}$$

Hence the Riccati equation satisfied by Q reads

$$h(h - \frac{1}{6})Q' = -\frac{1}{36}Q^2 + \frac{h}{3}Q - \frac{h}{6}. \tag{6.12}$$

In the next lemma we state properties of the mapping Q that we have used. The proof is rather standard, and we only give the details of one inequality. For the proof of similar statements see for instance [CLW94]

Lemma 6.4 *On the interval* $[0, \frac{1}{6}]$, Q *has the following properties:*

(i). Q *is strictly increasing.*

(ii). $(6h)^{\frac{1}{6}} \leq Q(h) \leq \frac{335}{432} + \frac{61\,h}{36} - \frac{25\,h^2}{12}$.

Proof. We prove the first inequality in (ii). It follows from the asymptotic expansion of Q in Lemma 6.3 that there exists $\bar{h}$ such that $(6h)^{\frac{1}{6}} < Q(h)$ for $h \in (\bar{h}, \frac{1}{6}]$. Let $\bar{h}$ be the smallest element of the interval $(0, \frac{1}{6}$ with this property. Using the Riccati equation it follows that at this point

$$Q'(\bar{h}) = -\frac{1}{6}\left(\bar{h}^{7/6}\sqrt[3]{6} - 6\,\bar{h}^2\sqrt[6]{6} + 6\,\bar{h}^{\frac{11}{6}} - \sqrt[6]{6}\bar{h}\right)\bar{h}^{-\frac{11}{6}}\left(6\,\bar{h} - 1\right)^{-1}. \tag{6.13}$$

Writing $\bar{h} = x/6$ the right hand side takes the form

$$-\frac{\sqrt[6]{x} - x + x^{5/6} - 1}{x^{5/6}\,(x - 1)}$$

and this is negative for $x \in (0, 1)$, contradicting the choice of $\bar{h}$.   □

Acknowledgement. Part of the research is sponsored by (1) Commision of the European Communities, Directorate General XII-B. Joint Project CI1*-CT93-0018 between the Department of Mathematics, Institut Teknologi Bandung, Indonesia, and the Faculty of Applied Mathematics, University of Twente, The Netherlands, and (2) Hibah Tim URGE Project No. 007/HTPP/URGE/1995, DIKTI.

References

[BK81]   E. Brieskorn and H. Knörrer. *Ebene algebraische Kurven.* Birkhäuser, Basel, 1981.

[CLW94] S.-N. Chow, C. Li, and D. Wang. *Normal forms and bifurcations of planar vector fields.* Cambridge University Press, Cambridge, 1994.

[Der92] G. Derks. *Coherent structures in the dynamics of perturbed Hamiltonian systems.* PhD thesis, University of Twente, 1992.

[Doe81] E.J. Doedel. Auto: A program for the automatic bifurcation analysis of autonomous systems. *Cong. Num.*, 30:265–284, 1981.

[DvG93] G.L.A. Derks and S.A. van Gils. On the uniqueness of traveling waves in perturbed Korteweg-de Vries equations. *Japan Journal of Industrial and Applied Analysis*, 10(3):413–431, 1993.

Progress in Nonlinear Differential Equations
and Their Applications, Vol. 19
© 1996 Birkhäuser Verlag Basel/Switzerland

Hamiltonian Perturbation Theory for Concentrated Structures in Inhomogeneous Media

E.R. Fledderus*   E. van Groesen[†]

Abstract

We consider spatially inhomogeneous Hamiltonian systems for which the rate of change of the inhomogeneity is small. Connected to these systems is a 1-parameter family of homogenized versions, for which spatial variations vanish. Special solutions of these homogenized systems are relative equilibrium solutions: a 2-parameter manifold of solutions which are translations of an extremizer of the energy constrained to levelsets of momentum. A solution of the inhomogeneous system which describes the distortion of such a relative equilibrium solution is approximated using relative equilibrium states with the 3 parameters evolving in time in a way to be specified. The dynamics of the parameters is obtained using (i) a geometrically motivated projection argument, (ii) a dynamical consistent evolution of global quantities (energy and momentum), and (iii) a Fredholm-type of argument from a mathematical investigation of the error. The results are shown to be equivalent. The Fredholm-argument implies that the approximation is valid on spatial-temporal scales on which deformations are of order one, thereby justifying the physically more attractive method of consistent evolution. All results are illustrated to the motion of a Bloch wall in an inhomogeneous ferro-magnetic material.

Keywords: inhomogeneous hamiltonian systems, perturbation theory, quasi-homogeneous approximation, sine-Gordon.

AMS subject classification: 35B20, 35Q53, 47A55, 49D05, 78A25.

1 Introduction

In this paper we study spatially inhomogeneous dynamical systems. In particular, we derive an approximative description how structures that are characteristic for

---

*This research has been supported by the Netherlands Organization for Scientific Research, NWO, by contract 620–61–249.

[†]Part of the research is sponsored by the Commission of the European Communities, Directorate General XII-B, Joint Research Project CI1*-CT93-0018 between the Department of Mathematics, Institut Teknologi Bandung, Indonesia, and the Faculty of Applied Mathematics, University of Twente, The Netherlands.

homogeneous media deform as a consequence of the inhomogeneity. Under the assumption that the rate of change of the inhomogeneity is small. $\mathcal{O}(\varepsilon)$, it is shown that an $\mathcal{O}(\varepsilon)$ correct approximation can be found that describes changes of the structures of order one. This approximation will be a quasi-homogeneous evolution: a succession of structures corresponding to different homogeneous media. Restricting to concentrated structures, for which the spatial structure is localized in a sense that its "position" can be defined in a meaningful way[1], the succession will be determined by specifying at each moment one member out of a family of possible structures that belong to a homogeneous medium that resembles the local properties of the inhomogeneous medium at the position of the structure at the specific instant of time.

Before specifying in more detail the class of systems that can be treated in a unified way, we mention a few particular cases for which the theory applies.

The example that will be treated in detail in this paper is the motion of a Bloch wall in an inhomogeneous ferro-magnetic crystal. Denoting by $\theta = \theta(x,t)$ the angle of spin vectors along the x-axis, the governing equation is the inhomogeneous sine-Gordon equation (de Leeuw et al. (1980)):

$$\theta_{tt} = \theta_{xx} - K(x)\sin 2\theta. \tag{1.1}$$

(The spin vector can only take positions in a plane perpendicular to the x-axis so only one angle is needed to specify the state of the system.) Here. $K = K(x)$ is the positive magnetic anisotropy function. Along with this equation, we consider the family of uniform crystals, the homogeneous systems, for which the magnetic anisotropy is constant. Hence, for $\kappa \in \mathbb{R}_+$

$$\theta_{tt} = \theta_{xx} - \kappa \sin 2\theta. \tag{1.2}$$

In such a uniform medium, "kink-solutions" exist that are monotone transitions from "spin-up" ($\theta = 0$) to "spin-down" ($\theta = \pi$). The transition region, the Bloch wall, is essentially concentrated to a small interval; the width is related to the value of the horizontal momentum (which is a conserved quantity). and determines the (constant) speed at which the wall translates. In an anisotropic medium, when $K(x)$ is not constant, such solutions will deform since the momentum is not conserved any more. When the material properties change slowly. it is appealing to approximate the deforming profile at each instant by a uniform kink solution that "feels" the material property at the actual position of the transition region. Adapting the material parameter κ of the homogeneous system in time to take into account the changing value of the material properties at the transition region, a so-called *quasi-homogeneous approximation* will be obtained. The changing value of the horizontal momentum has to be adapted to the changing position; energy conservation will produce the governing equations for position and momentum.

---

[1]In a next paper we also study spatially extended structures and derive modulation equations in a unified same approach.

Just like the original equation, this parameter dynamics is a Hamiltonian system, which leads to a particle-like description (with K in the role of a potential energy function) for the moving Bloch wall.

Intuitively it is clear that for such an approximation to be valid, the assumption of slow variations in material properties is necessary. However, the resulting approximation may well describe large deformations. We will give an example below that illustrates this as follows. While in a uniform crystal, the boundary moves at constant speed with a fixed width, due to an inhomogeneity described by a convex-like function K, the boundary will experience a periodic motion, adapting at each time its velocity to the local value of the anisotropy. The change of K during this motion is of order one. This example illustrates at the same time that *any approximation that is based on taking some averaged value of K cannot produce the correct result on the large space and time scale of the periodic motion.* Stated differently, the use of a *family* of homogeneous systems is required to study the deformation in the inhomogeneous system.

This example will be used in this paper as a model to illustrate the mathematical methods. The methods are more generally applicable, however. Two other problems to which the same theory has been applied recently are surface waves above a varying bottom, and swirling flows in expanding pipes.

For surface waves, looking at a localized, solitary. wave that deforms due to bottom variations, the deformed wave is approximated at each instant by a solitary wave above a flat bottom at a depth equal to the depth below the top of the deformed wave. The quasi-homogeneous description of the deformation uses the amplitude (or horizontal momentum) of the homogeneous wave as a function of position, or actually, amplitude and position (depth) as a function of time. Energy conservation is the basic condition that relates the parameters and governs the dynamics: see Pudjaprasetya & Van Groesen (1995). The same method can be extended to describe the splitting, due to bottom variation, of a single wave into two or more waves.

For 3D Eulerian fluid dynamics in radially symmetric tubes with constant radius, various families of swirling flows can be found as extremizers of the energy at given helicity and axial flux. The deformation of such a swirling flow in a slowly expanding tube can also be studied along the lines of this paper. In this case, more parameters are needed in the quasi-homogeneous description, and the theory has to be extended to include Casimir functionals. This has been done in Van Groesen, Van der Fliert & Fledderus (1995) and Fledderus & Van Groesen (1995). One result is that a simple model is obtained that describes recirculation areas in waste burners: this is closely related to the phenomenon of vortex breakdown.

In each example above, the (inhomogeneous) dynamical system has a Poisson structure. The inhomogeneity destroys (translation) symmetry that is related to a conserved quantity (the momentum). This symmetry is present in the homogeneous systems and is used to find the basic structures of the approximation. In fact, the uniform structures are the relative equilibria of the underlying homogeneous

Poisson systems. One parameter measures the value of the homogeneity; fixing it, another one, related to the value of the conserved quantity, determines the various spatial structures in the homogeneous medium. Energy conservation will be decisive to find the actual parameter dynamics that specifies the inhomogeneous approximation. Since we are after a description that is correct for changes of order one, the actual choice of the parameter dynamics is crucial. In the presentation to follow this choice will be motivated in various ways. The justification will be based on an abstract analysis of the error. We now describe the essential ingredients for this analysis.

For any given approximation, the error is determined in lowest order by the linearized equation. In our approach, this equation is ("almost") degenerate because of the fact that for homogeneous media a symmetry is present. As a consequence, for the desired boundedness of the error on the time and length scales we are interested in, certain solvability conditions have to be satisfied as a consequence of the Fredholm alternative. Satisfying these solvability conditions then determines the parameter dynamics of the approximation.

Resuming it can be said that with the manifolds of relative equilibria chosen as the sets in which the approximation is sought, the dynamics of the actual evolution of the approximation in this set is found from the Fredholm alternative that determines the way how to restrict the original equation to this set.

It will be shown that the projection of the equation determined by the Fredholm solvability conditions to the manifolds of relative equilibria of the relevant homogeneous media has a clear geometrical interpretation at each point during the evolution. Furthermore, it turns out that, under an additional condition, the correct dynamics can equivalently be found by requiring that the physically relevant observables evolve in a natural, consistent way. This last result is often the easiest way to produce the governing quasi-homogeneous dynamics.

A final remark concerns the possibility to obtain the dynamics in an explicit way. In fact, the solvability conditions can be written down provided that the elements from the kernel of the adjoint linearized operator are available. In general, the elements form this kernel are difficult to find explicitly. However, for the systems considered here, owing to the underlying Poisson structure, the kernel of the adjoint operator can be expressed in terms of the (known) kernel of the linearized operator (Van Groesen (1995)) and can therefore be written down in an explicit way.

The arguments based on the Fredholm-alternative are not new of course. The classical text by Nayfey (1993), for instance, contains many examples for ODE's; applied to avaraging. see e.g. Hale (1969), and Sanders and Verhulst (1985). Some applications in averaging in pde's can be found in Buitelaar (1993) and Krol (1990) and the references mentioned there. Solvability conditions are also found in many examples with center manifold reduction, see e.g. Carr (1981) and multiple-scaling problems, see e.g. Aceves et al. (1986), Calogero & Eckhaus (1988). Connected to the sine-Gordon example, we mention the work by Kivshar and Malomed (1989),

Karpman and Solov'ev (1981) and a series of papers by Olsen and Samuelsen (1983), Salerno et al. (1985) and Sakai et al. (1987). They use the complete integrability of sine-Gordon in order to construct the dynamics of the scattering data with (Inverse) Scattering Theory. Sakai et al. (1987) study the inhomogeneous sine-Gordon equation and produce numerical results. Their treatment of the inhomogeneity different: the inhomogeneous equation is viewed as an externally perturbed system by means of a scaling of the spatial variable x. Using the Inverse Scattering Theory, they end up with the dynamics of the soliton parameters.

The organization of the paper is as follows. In section 2 we introduce the notation and derive the family of manifolds of relative equilibria for the homogeneous systems. In section 3, the quasi-homogeneous approximation is constructed for the inhomogeneous equation and it is shown that three methods to obtain the parameter dynamics are equivalent. In both sections the example of the Bloch wall illustrates the general theory. Finally, in section 4 we conclude with some remarks.

2 Homogeneous systems

We start to introduce the notation and the main notions from Hamiltonian dynamics that will be used in the sequel. Then relative equilibria are characterized in a variational way, and the geometry of the manifold of relative equilibria is described. The last subsection specializes the general notions to the sine-Gordon equation.

2.1 Notation

In this paragraph we introduce the notation and repeat the main definitions for Poisson systems.

With the state evolving in a state space denoted by $u \in \mathcal{U}$, the evolution equation for a Poisson system has a specific form that can be described as follows. Let Γ be a so-called "structure map" (possibly depending on the state), which means that

$$\{F, G\}(u) := \langle \delta F(u), \Gamma(u)\delta G(u)\rangle$$

defines a Poisson bracket (skew-symmetric and satisfying Jacobi's condition). Here and in the following δF denotes the variational derivative of the (density) functional F with respect to the spatial inner product or duality map $\langle \, , \, \rangle$; this derivative is an element of the cotangent space to $\mathcal{U}$: $\delta F(u) \in T_u^*\mathcal{U}.$, and $\Gamma(u)$ is a mapping from $T_u^*\mathcal{U}$ into the tangent space $T_u\mathcal{U}$. Then a Poisson system is described for a certain (autonomous) Hamiltonian $H(u)$ by

$$\partial_t u = \Gamma\delta H(u). \tag{2.1}$$

In the following we will use this notation for the inhomogeneous system for which the Hamiltonian is a density functional with density $h(u; K(x))$: $H(u) =$

$\int h(u; K(x)) \, dx$. The spatial variable x will enter only via some function K, $K = K(x) = \overline{K}(\varepsilon x)$ that measures some, slowly varying, scalar quantity of the underlying physical system (ε is small).

For simplicity we will actually restrict in the following to Hamiltonian systems, i.e., to Poisson systems for which the structure map is invertible (the method can be generalized to non-Hamiltonian Poisson systems, as is necessary in the problem for swirling flows).

The homogeneous version of (2.1) is obtained by taking a constant quantity for K, say $\kappa = $ constant. In doing this, the homogeneous Hamiltonian will be denoted by $\hat{H}(u; \kappa)$ and the homogeneous system then reads

$$\partial_t u = \hat{\Gamma} \delta \hat{H}(u; \kappa). \tag{2.2}$$

Here $\hat{\Gamma} = \hat{\Gamma}(u; \kappa)$ when we take into account that also the structure map may depend on the function K.

The vectorfields $X_H(u; K(x))$ and $\hat{X}_H(u; \kappa)$ will denote the Hamiltonian vectorfield of the inhomogeneous and of homogeneous Poisson system respectively:

$$X_H(u; K(x)) \equiv \Gamma(u; K(x)) \delta H(u), \quad \hat{X}_H(u; \kappa) \equiv \hat{\Gamma}(u; \kappa) \delta \hat{H}(u; \kappa). \tag{2.3}$$

In a homogeneous system the spatial translation symmetry implies the existence of a first integral, $\hat{I}$, describing the momentum in some spatial direction (Noether's theorem). In fact, also $\hat{I}$ may depend on κ. $\hat{I}$ is a first integral iff it Poisson commutes with $\hat{H}$, i.e., $\{\hat{H}, \hat{I}\} = 0$; it is a standard result (see e.g. Van Groesen and De Jager (1994)) that in this case the corresponding Hamiltonian flows commute.

Finally, when dealing with the arguments from the Fredholm alternative, the dynamic equation will be viewed as an operator equation. The following notation will then be useful. We will denote by $\mathcal{E}(u; K(x))$ and by $\hat{\mathcal{E}}(u; \kappa)$ the operators

$$\mathcal{E}(u; K(x)) := \partial_t u - X_H(u; K(x)) \tag{2.4}$$

and

$$\hat{\mathcal{E}}(u; \kappa) \equiv \partial_t u - \hat{X}_H(u; \kappa) \tag{2.5}$$

respectively. A solution of $\mathcal{E}(u; K(x)) = 0$ with initial value u_0 will be denoted by $u(t) = \Phi_t^H(u_0)$; a solution of $\hat{\mathcal{E}}(u; \kappa) = 0$ with initial value u_0 by $u(t) = \hat{\Phi}_t^H(u_0)$.

2.2 Extremal characterizations for relative equilibria

We now consider the homogeneous problems in which κ appears as a (fixed) parameter. However, for the application in the next section, we make the dependence on κ explicit.

The simplest solutions of a Poisson system are equilibria: stationary solutions $\bar{U}$ that satisfy

$$0 = \hat{X}_H(\bar{U}; \kappa). \tag{2.6}$$

A generalization of equilibria are relative equilibria. To motivate it, note that equilibria are extrema of $\hat{H}$. The value of the momentum $\hat{I}$ is 'slaved' in (2.6); that is, we cannot exert any influence on $\hat{I}(u;\kappa)$, its value being given by $\hat{I}(\hat{U};\kappa)$ that can be calculated once the equilibrium is found. We can change this situation since $\hat{I}$ is a first integral. So instead of looking for extrema of $\hat{H}$ in $\mathcal{U}$, we restrict the extremization problem to a subset of $\mathcal{U}$, namely a levelset of $\hat{I}$:

$$\text{Extr}\,\{\,\hat{H}(u;\kappa)\mid\hat{I}(u;\kappa)=\gamma\,\}. \tag{2.7}$$

Let us denote a typical solution of (2.7) by $\tilde{U}(\gamma;\kappa)$. Lagrange's multiplier rule implies that $\tilde{U}(\gamma;\kappa)$ satisfies for some multiplier λ the equation

$$\delta\hat{H}(U;\kappa)=\lambda\delta\hat{I}(U;\kappa). \tag{2.8}$$

Since both $\hat{H}$ and $\hat{I}$ are invariant for the $\hat{I}$-flow, the trajectory obtained by applying the $\hat{I}$-flow consists of critical points:

$$U(\gamma,\varphi_0;\kappa)\equiv\hat{\Phi}_{\varphi_0}^I\tilde{U}(\gamma;\kappa)\in\arg\text{Extr}\,\{\,\hat{H}(u;\kappa)\mid\hat{I}(u;\kappa)=\gamma\,\}; \tag{2.9}$$

all these elements satisfy the same multiplier equation.

For fixed κ and varying φ_0 and γ these functions $U(\gamma,\varphi_0;\kappa)$ form the MRE $\mathcal{M}_\kappa$, the manifold of relative equilibria at κ. When varying κ we get a family of MRE's (see also figure 1):

$$\mathcal{M}:=\underset{\kappa}{\cup}\mathcal{M}_\kappa,\qquad\mathcal{M}_\kappa=\{U(\gamma,\varphi_0,\kappa)|\gamma,\varphi_0\}$$

So far the relative equilibria are constructed from a variational problem, without

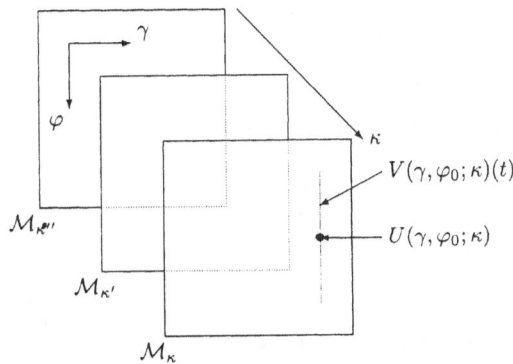

Figure 1: The MRE's $\mathcal{M}_\kappa$ embedded in $\mathcal{M}\subset\mathcal{U}$

direct reference to the dynamic equation. The reason for the MRE's to be important for the dynamical problem $\hat{\mathcal{E}}(u;\kappa)=0$ lies in the fact that when a trajectory in $\mathcal{M}_\kappa$ is transversed with a certain speed λ, that depends on γ and κ, the evolution $\hat{\Phi}^I_{\lambda t} U(\gamma, \varphi_0; \kappa) \equiv U(\gamma, \varphi_0 + \lambda t; \kappa)$ is a solution, a so-called *relative equilibrium solution*. These relative equilibrium solutions will be denoted by

$$V(\gamma, \varphi_0; \kappa)(t) = \hat{\Phi}^I_{\lambda(\gamma, \kappa) t} U(\gamma, \varphi_0; \kappa) = U(\gamma, \varphi_0 + \lambda(\gamma, \kappa) t; \kappa). \qquad (2.10)$$

The speed λ is in fact precisely the Lagrange multiplier that appears in the equation satisfied by the constrained extremizers $U(\gamma, \varphi_0; \kappa)$.

2.3   Geometry of the MRE and decomposition of $T_U \mathcal{U}$

We will specify the geometry of $\mathcal{M}_\kappa$ by describing its tangent and cotangent space at a certain point $U(\gamma, \varphi; \kappa)$. This defines two projection operators. $\mathbb{P}_1$ and $\mathbb{P}_2$, that will be used to split any perturbation on the MRE $\mathcal{M}_\kappa$ into a part tangential to $\mathcal{M}_\kappa$ and a remaining part transversal to it. The projection depends on the choice of the cotangent space: when chosen as below, it will be called the *'natural' projection (splitting)*. since it will be motivated by the argument from the Fredholm alternative.

The tangent space $T_U \mathcal{M}_\kappa$ is spanned by two vectors that are found by differentiating $U(\gamma, \varphi_0; \kappa)$ with respect to the two coordinates γ and φ_0:

$$T_U \mathcal{M}_\kappa = \text{span} \left\{ \hat{X}_I(U; \kappa), U_\gamma \right\}$$

with

$$\hat{X}_I(U; \kappa) \equiv \hat{\Gamma}(U; \kappa) \delta \hat{I}(U; \kappa) = \frac{\partial U}{\partial \varphi_0}, \quad U_\gamma = \frac{\partial U}{\partial \gamma}. \qquad (2.11)$$

Since $\hat{\Gamma}(U; \kappa)$ is assumed to be invertible, the 'natural' choice for a dual basis of $T^*_U \mathcal{M}_\kappa$ is

$$T^*_U \mathcal{M}_\kappa = \hat{\Gamma}^{-1}(U; \kappa) T_U \mathcal{M}_\kappa = \text{span} \left\{ \delta \hat{I}(U; \kappa), \hat{\Gamma}^{-1}(U; \kappa) U_\gamma \right\}.$$

This set is dual since

$$\begin{aligned}
\langle \delta \hat{I}(U; \kappa), \hat{X}_I(U; \kappa) \rangle = 0, &\qquad \langle \hat{\Gamma}^{-1}(U; \kappa) U_\gamma, \hat{X}_I(U; \kappa) \rangle = -1. \\
\langle \delta \hat{I}(U; \kappa). U_\gamma \rangle = 1, &\qquad \langle \hat{\Gamma}^{-1}(U; \kappa) U_\gamma, U_\gamma \rangle = 0.
\end{aligned} \qquad (2.12)$$

The extension to the case when more integrals (whether or not in involution) are present is immediate (see Van Groesen and De Jager (1994)).

Connected with the orthogonality relations in (2.12) we can define projection operators $\mathbb{P}_1$ and $\mathbb{P}_2$. They act on vector fields ξ defined on a subset $\mathcal{M}$ (the MRE) of $\mathcal{U}$:

$$\xi : \mathcal{M} \subset \mathcal{U} \to T\mathcal{U}, \quad \mathcal{U} \supset \mathcal{M} \ni U(\gamma, \varphi_0; \kappa) \mapsto \xi(U) \in T_U \mathcal{U}.$$

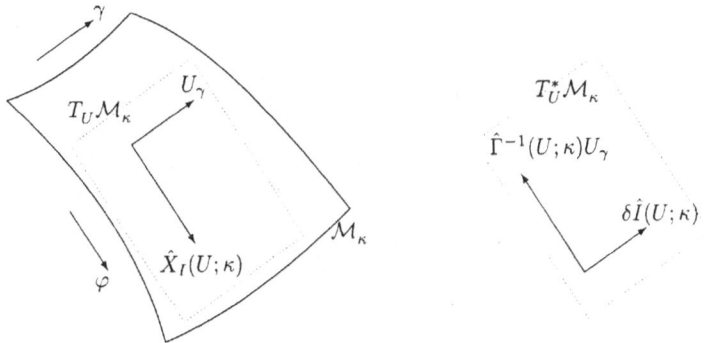

Figure 2: The tangent and cotangent space

These projection operators split the vector field in a part tangent to $\mathcal{M}_\kappa \subset \mathcal{M}$ and a part in the orthogonal complement:

$$T_U \mathcal{U} = T_U \mathcal{M}_\kappa \oplus T_U \mathcal{M}_\kappa^\perp.$$

$\mathbb{P}_1$ and $\mathbb{P}_2$ are given by

$$\mathbb{P}_1 \xi = \langle \delta \hat{I}(U;\kappa), \xi \rangle U_\gamma, \quad \mathbb{P}_2 \xi = \langle -\hat{\Gamma}^{-1}(U;\kappa)U_\gamma, \xi \rangle \hat{X}_I(U;\kappa).$$

Hence,

$$T_U \mathcal{M}_\kappa^\perp := \{\, u \in T_U \mathcal{U} \mid \mathbb{P}_1 u = \mathbb{P}_2 u = 0 \,\}.$$

2.4 Example: Motion of Bloch walls

With $\partial_t \theta = p$, the sine-Gordon equation (1.1) can be written as a Hamiltonian system with θ and p as a pair of canonically conjugate variables:

$$\partial_t \begin{pmatrix} \theta \\ p \end{pmatrix} = \begin{pmatrix} 0 & 1 \\ -1 & 0 \end{pmatrix} \begin{pmatrix} \delta_\theta H \\ \delta_p H \end{pmatrix}. \tag{2.13}$$

and with H the total energy,

$$H(\theta, p) = \int \left(\frac{1}{2} p^2 + \frac{1}{2} \theta_x^2 + K(x) \sin^2 \theta \right) dx.$$

Using the notation J_2 for the standard symplectic matrix in 2 dimensions, we rewrite (2.13) as

$$\partial_t \begin{pmatrix} \theta \\ p \end{pmatrix} = J_2 \delta_{\theta, p} H.$$

When K does not depend on x we are dealing with the homogeneous version of (2.13). The Hamiltonian for this case is denoted by

$$\hat{H}(\theta, p; \kappa) = \int \left(\frac{1}{2} p^2 + \frac{1}{2} \theta_x^2 + \kappa \sin^2 \theta \right) \, \mathrm{d}x.$$

We restrict ourselves to the homogeneous system in this paragraph. The translation symmetry that is then present is connected to the conserved momentum,

$$\hat{I}(\theta, p) = I(\theta, p) = - \int \theta_x p \, \mathrm{d}x, \qquad (2.14)$$

since its Hamiltonian flow is translation:

$$\partial_t \begin{pmatrix} \theta \\ p \end{pmatrix} = J_2 \delta_{\theta, p} I = -\partial_x \begin{pmatrix} \theta \\ p \end{pmatrix} \quad \Rightarrow \quad \Phi_t^I \begin{pmatrix} \theta \\ p \end{pmatrix} (x) = \begin{pmatrix} \theta \\ p \end{pmatrix} (x - t). \qquad (2.15)$$

The coherent/concentrated structures that are observed in this system are the so-called Bloch walls, a transition between domains with antiparallel orientation of the magnetisation. For the following we use as conditions at infinity (see also figure 3):

$$\begin{cases} \theta(-\infty, \cdot) &= 0 \ (\text{spin-up}) \\ \theta(+\infty, \cdot) &= \pi \ (\text{spin-down}). \end{cases} \qquad (2.16)$$

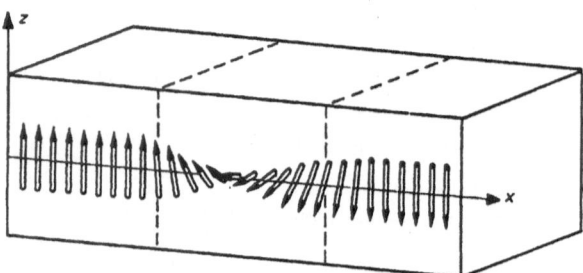

Figure 3: Structure of a Bloch wall in a uniaxial crystal. The magnetisation vector remains always normal to the x direction. Taken from de Leeuw et al. (1980).

These transition profiles, called kinks, are the relative equilibria in this case, i.e., they follow from the minimization problem

$$\underset{\theta, p}{\mathrm{Min}} \left\{ \hat{H}(\theta, p; \kappa) \mid I(\theta, p) = \gamma \right\}. \qquad (2.17)$$

If we denote a solution to (2.17) by $(\Theta(\gamma, \varphi_0; \kappa), P(\gamma, \varphi_0; \kappa))$, then Θ satisfies the Lagrange multiplier $(= \lambda)$ equation:

$$(\lambda^2 - 1)\Theta_{xx} + \kappa \sin 2\Theta = 0, \tag{2.18}$$

and $P = -\lambda \Theta_x$. Note that this equation is also found when looking for a travelling wave solution with speed λ.

The solution Θ of (2.18) can be written down explicitly when noting that besides (2.18) also

$$\frac{1}{2}(\lambda^2 - 1)\Theta_x^2 + \kappa \sin^2 \Theta = 0 \tag{2.19}$$

holds. This yields

$$\Theta(\gamma, \varphi_0; \kappa) = 2 \arctan(\exp(B(x - \varphi_0))). \tag{2.20}$$
$$P(\gamma, \varphi_0; \kappa) = -\lambda B \cosh^{-1} B(x - \varphi_0) \tag{2.21}$$

Re-expressing all quantities in γ and κ we find

$$\lambda(\gamma, \kappa) = \frac{\gamma}{\sqrt{8\kappa + \gamma^2}}, \quad B = \frac{1}{2}\sqrt{8\kappa + \gamma^2}. \tag{2.22}$$

The solutions $(\Theta(\gamma, \varphi_0; \kappa), P(\gamma, \varphi_0; \kappa))$ form together the family of MRE's,

$$\mathcal{M} = \bigcup_{\kappa > 0} \mathcal{M}_\kappa, \qquad \mathcal{M}_\kappa = \{(\Theta(\gamma, \varphi_0; \kappa), P(\gamma, \varphi_0; \kappa)) | \gamma, \varphi_0\}$$

If we let the I-flow act on an arbitrary element of the MRE with an amount $\lambda(\gamma, \kappa)t$ we have a solution to the dynamic problem, equation (1.1). These are the relative equilibrium solutions: kinks translated undeformed in shape at constant speed λ.

3 The quasi-homogeneous approximation

In this section we develop the perturbation theory for inhomogeneous systems.

As an approximation for a solution u of $\mathcal{E}(u; K(x)) = 0$ that describes the distortion of a relative equilibrium solution, we consider an evolution $\breve{u}$ in the family of MRE's that were constructed in section 2. Stated differently, $\breve{u}$ is a succession of relative equilibria with the parameters adjusted in a manner that has to be specified. When $\varepsilon = 0$ we recover the relative equilibrium solutions; the errors that are introduced vanish in the limit $\varepsilon \to 0$, and hence the theory produces results for slowly varying media.

The parameters that have to be adjusted in this proces of taking successive relative equilibria are γ, φ and κ. Thus,

$$t \mapsto U(\gamma(t), \varphi(t); \kappa(t))) = \breve{u}(t)$$

So the approximation $\check{u}(t)$ 'feels' at some specific time t a material 'constant' $\kappa(t)$, to approximate the inhomogeneous material at that time.

This section is arranged as follows. First we investigate the choice of κ, as a function of γ and φ. Then we determine the dynamics of $\varphi(t)$ and $\gamma(t)$ from three points of view. One is connected with the natural splitting of the perturbation P from the inhomogeneity, into an active part, $P_\parallel$, and a passive part. $P_\perp$. Letting the parameters move under the influence of the active part $P_\parallel$ is shown to be equivalent to a consistent evolution of the energy, and also equivalent to satisfying the necessary solvability conditions for the equation that determines the error between the approximation and an exact solution. The motion of the Bloch wall finishes the section. In order to simplify the formula we drop the dependence of I on κ: $\hat{I}(u) = I(u)$.

3.1 Pointwise homogenization

For the homogeneous medium, let $\tilde{U}(\gamma, \kappa)(x)$ be the relative equilibrium that is 'concentrated' around $x = 0$; then $U(\gamma, \varphi; \kappa)(x) \equiv \tilde{U}(\gamma; \kappa)(x - \varphi)$ is 'concentrated' around $x = \varphi$. In a moment we will specify in more detail what should be understood by 'concentrated': for now it suffices to interpret it as the position where the relative equilibrium is approximately confined to a small region. Then it is natural to take for κ the local value of the inhomogeneous medium. viz. $\kappa = K(\varphi)$; we will now investigate this in more detail.

The main property to determine the approximation will be energy conservation, also for the correct choice of κ. In contrast with the dissipation-like perturbations treated in Van Groesen (1995) that destroy the conservative character of the homogeneous system. the energy is conserved for solutions of the inhomogeneous system: $H(u)$ is conserved also when K varies with x. However, for any approximation $\check{u}$, the energy $H(\check{u})$ will in general not be constant. Consider for $U(\gamma, \varphi; \kappa)$ the difference between the inhomogeneous energy and the homogeneous energy at level κ, i.e.,

$$
\begin{aligned}
\Delta \quad &:= \quad H(U(\gamma, \varphi; \kappa)) - \hat{H}(U(\gamma, \varphi; \kappa); \kappa) \\
&= \quad \int \left[h(U(\gamma, \varphi; \kappa)(x); K(x)) - h(U(\gamma, \varphi; \kappa)(x); \kappa) \right] \mathrm{d}x.
\end{aligned}
$$

Introducing $y = x - \varphi$. and $\overline{\varphi} = \varepsilon\varphi$, so that $K(x) = \overline{K}(\overline{\varphi} + \varepsilon y)$, the difference will be viewed as a function of the variables γ, $\overline{\varphi}$, κ and ε:

$$
\Delta(\kappa; \gamma, \bar{\varphi}, \varepsilon) = \int \left[h(\tilde{U}(\gamma; \kappa)(y), \overline{K}(\overline{\varphi} + \varepsilon y)) - h(\tilde{U}(\gamma; \kappa)(y); \kappa) \right] \mathrm{d}y.
$$

Observe that this difference vanishes for $\varepsilon = 0$ for the choice $\kappa = K(\varphi) = \overline{K}(\overline{\varphi})$:

$$
\Delta(\overline{K}(\overline{\varphi}); \gamma, \overline{\varphi}, 0) = 0,
$$

and we look for the correct choice of κ for small ε.

One special case will be dealt with first. Sufficient for our purposes would be that

$$\Delta(\overline{K}(\overline{\varphi}); \gamma, \overline{\varphi}, \varepsilon) = \mathcal{O}(\varepsilon^2),$$

which is the case if

$$\frac{\partial \Delta}{\partial \varepsilon}\bigg|_{\varepsilon=0} \equiv \overline{K}'(\overline{\varphi}) \int \frac{\partial h}{\partial \kappa}(\tilde{U}(\gamma; \overline{K}(\overline{\varphi}))(y); \overline{K}(\overline{\varphi})) y \, dy$$

vanishes. This turns out to be the case in the example of the Bloch wall, and in several other examples, from the special symmetry property of the relative equilibrium, namely that $\frac{\partial h}{\partial \kappa}(\tilde{U}(\gamma; \overline{K}(\overline{\varphi}))(y); \overline{K}(\overline{\varphi}))$ is an even fuction.

In the more general case, $\frac{\partial \Delta}{\partial \varepsilon}\big|_{\varepsilon=0} \neq 0$. In fact, we have to require that this quantity is finite, which can only be expected when the relative equilibrium is sufficiently concentrated. *Hence, the finiteness of this expression is the explicit statement for the required 'concentratedness'.*

In this general case, we apply the implicit function theorem to the equation

$$\Delta(\kappa; \gamma, \overline{\varphi}, \varepsilon) = 0.$$

Since

$$\frac{\partial \Delta}{\partial \kappa}\bigg|_{\kappa=\overline{K}(\overline{\varphi}), \varepsilon=0} \neq 0,$$

this theorem leads to the result that in some neighbourhood M of 0 and some open set $N \subset \mathbb{R}^2$, there exists a unique, C^1, mapping $\hat{\kappa} = \hat{\kappa}(\gamma, \overline{\varphi}, \varepsilon)$ such that

- $\hat{\kappa}(\gamma, \overline{\varphi}, 0) = \hat{K}(\overline{\varphi}) = K(\varphi)$,

- $\Delta(\hat{\kappa}(\gamma, \overline{\varphi}, \varepsilon); \gamma, \overline{\varphi}, \varepsilon) = 0$, for all $(\gamma, \overline{\varphi}, \varepsilon) \in N \times M$.

Hence, for this choice of $\hat{\kappa}$, the energy difference vanishes. It holds that

$$\hat{\kappa} = \overline{K}(\overline{\varphi}) + \mathcal{O}(\varepsilon),$$

and so

$$\frac{\partial \hat{\kappa}}{\partial \gamma} = \mathcal{O}(\varepsilon), \qquad \frac{\partial \hat{\kappa}}{\partial \varphi} = \varepsilon \frac{\partial \hat{\kappa}}{\partial \overline{\varphi}} = \mathcal{O}(\varepsilon)$$

This choice for κ will be used in the sequel in the quasi-homogeneous approximation.

Notation. The quasi-homogeneous approximation will be based on the functions $U(\gamma, \varphi; \hat{\kappa}(\gamma, \overline{\varphi}, \varepsilon))$ and denoted by

$$\hat{u}(t) = U(\gamma(t), \varphi(t); \kappa(t))$$

where $\kappa(t) = \hat{\kappa}(\gamma(t), \overline{\varphi}(t), \varepsilon)$. Note that differentiating $\hat{u}(t)$ with respect to γ it holds that

$$\frac{\partial \hat{u}}{\partial \gamma} = \frac{\partial \check{u}}{\partial \gamma} + \frac{\partial \check{u}}{\partial \kappa}\frac{\partial \hat{\kappa}}{\partial \gamma} =: \check{u}_\gamma + \check{u}_\kappa \hat{\kappa}_\gamma = \check{u}_\gamma + \mathcal{O}(\varepsilon),$$

and similarly for φ:

$$\frac{\partial \hat{u}}{\partial \varphi} = \frac{\partial \hat{u}}{\partial \varphi} + \varepsilon \frac{\partial \hat{u}}{\partial \hat{\kappa}} \frac{\partial \hat{\kappa}}{\partial \overline{\varphi}} =: \hat{u}_\varphi + \varepsilon \hat{u}_\kappa \hat{\kappa}_{\overline{\varphi}} = \hat{u}_{\overline{\varphi}} + \mathcal{O}(\varepsilon).$$

As a final remark, observe the following. If one would like to identify the position ψ such that $\hat{\kappa} = K(\psi)$, upon introducing the slow variable $\overline{\Psi} = \varepsilon \Psi$, this quantity is found by inverting the relation

$$\hat{K}(\overline{\Psi}) = \hat{\kappa}(\gamma, \overline{\varphi}, \varepsilon).$$

Since in general this inverse is multiply defined, we take that Ψ that is closest to φ, the position of $\hat{u}$. Further, we have to stay away from an ε-neighbourhood of the extrema of K. This results in $\overline{\Psi} = \overline{\varphi} + \mathcal{O}(\varepsilon)$, and so $\psi = \varphi + \mathcal{O}(1)$. Consequently, when the value of $\hat{\kappa}$ changes order ε from $K(\varphi)$, the change in distance will be of order one. In the special case mentioned above, $\Delta(K(\varphi); \gamma, \overline{\varphi}, \varepsilon) = \mathcal{O}(\varepsilon^2)$, the situation is better in the sense that then $\psi - \varphi = \mathcal{O}(\varepsilon)$.

3.2   Parameter dynamics

3.2.1   A geometrical viewpoint

When approximating a solution u to a differential equation, $\mathcal{E}(u) = 0$, by $\hat{u}$, one very often studies the error $\eta = u - \hat{u}$ by considering the residual, i.e., the approximation $\hat{u}$ substituted in the equation. An exact solution satisfies $\mathcal{E}(u) = 0$, which is equivalent to requiring that the inner product of $\mathcal{E}(u)$ with any direction $\xi^* \in T_u^* \mathcal{U}$ equals zero:

$$\langle \mathcal{E}(u), \xi^* \rangle = 0, \quad \forall \xi^* \in T_u^* \mathcal{U} \iff \mathcal{E}(u) = 0. \tag{3.1}$$

For an approximation $\hat{u}$, the residual is by definition $\mathcal{E}(\hat{u})$ and will not vanish identically. When, as in our case, $\hat{u}$ still depends on parameters that have to be specified, these can be determined in a best possible way. The evolution of the parameters $t \mapsto (\gamma(t), \varphi(t))$ will be determined by requiring the residual to vanish in two directions. Like in Van Groesen (1995) we choose the residual, $\mathcal{E}(\hat{u}; K(x))$, to vanish in the directions $\delta I(\hat{u})$ and $\Gamma^{-1}(\hat{u}; K(x))\hat{u}_\gamma$, or, using our natural projections from the previous section, and extend these in a consistent(!) way by requiring that

$$\mathbb{P}_1 \mathcal{E}(\hat{u}; K(x)) = \mathbb{P}_2 \mathcal{E}(\hat{u}; K(x)) = 0.$$

The motivation is that when the perturbation is such that it leaves the MRE's invariant, the approximation is in fact an exact solution. Various examples support the idea that the MRE's are important for the perturbed dynamics: see, e.g., Van Groesen et al. (1990), Derks (1992), Derks and Van Groesen (1995), Van Groesen, Van de Fliert and Fledderus (1994) and Pudjaprasetya and Van Groesen (1995).

First we rewrite $\mathbb{P}_1 \mathcal{E}(\hat{u}; K(x)) = 0$ (we suppress the t-dependence).

$$
\begin{aligned}
0 &= \langle \delta I(\hat{u}), \mathcal{E}(\hat{u}; K(x)) \rangle \\
&= \langle \delta I(\hat{u}), \partial_t \hat{u} \rangle - \langle \delta I(\hat{u}), X_H(\hat{u}; K(x)) \rangle \\
&= \partial_t I(\hat{u}) - \langle \delta I(\hat{u}), X_H(\hat{u}; K(x)) \rangle \\
&= \partial_t \gamma - \langle \delta I(\hat{u}), X_H(\hat{u}; K(x)) \rangle,
\end{aligned}
$$

leading to

$$
\dot{\gamma} = \langle \delta I(\hat{u}), X_H(\hat{u}; K(x)) \rangle. \tag{3.2}
$$

Next, in order to find the dynamics for φ, we require $\mathbb{P}_2 \mathcal{E}(\hat{u}; K(x))$ to be zero. Let us abbreviate $\Gamma^{-1}(\hat{u}; K(x)) \hat{u}_\gamma$ with ζ_2 (the '2' corresponding with $\mathbb{P}_2$). Then

$$
\begin{aligned}
0 &= \langle \zeta_2, \mathcal{E}(\hat{u}; K(x)) \rangle \\
&= \langle \zeta_2, \partial_t \hat{u} \rangle - \langle \zeta_2, X_H(\hat{u}; K(x)) \rangle \\
&= -\dot{\varphi} \left(1 - \langle \zeta_2, \hat{X}_I(\hat{u}; \hat{\kappa}) - X_I(\hat{u}; K(x)) \rangle - \varepsilon \hat{\kappa}_{\overline{\varphi}} \langle \zeta_2, \hat{u}_\kappa \rangle \right) + \\
&\quad + \dot{\gamma} \hat{\kappa}_\gamma \langle \zeta_2, \hat{u}_\kappa \rangle + \langle \hat{u}_\gamma, \delta H(\hat{u}) \rangle \\
&= -\dot{\varphi} \left(1 - \langle \zeta_2, \hat{X}_I(\hat{u}; \hat{\kappa}) - X_I(\hat{u}; K(x)) \rangle - \varepsilon \hat{\kappa}_{\overline{\varphi}} \langle \zeta_2, \hat{u}_\kappa \rangle \right) + \\
&\quad + \dot{\gamma} \hat{\kappa}_\gamma \langle \zeta_2, \hat{u}_\kappa \rangle + \lambda(\gamma, \hat{\kappa}) + \langle \hat{u}_\gamma, \delta H(\hat{u}) - \delta \hat{H}(\hat{u}; \hat{\kappa}) \rangle
\end{aligned}
$$

leading to

$$
\dot{\varphi} = \frac{\lambda(\gamma, \hat{\kappa}) + \dot{\gamma} \hat{\kappa}_\gamma \langle \zeta_2, \hat{u}_\kappa \rangle + \langle \hat{u}_\gamma, \delta H(\hat{u}) - \delta \hat{H}(\hat{u}; \hat{\kappa}) \rangle}{1 - \langle \zeta_2, \hat{X}_I(\hat{u}; \hat{\kappa}) - X_I(\hat{u}; K(x)) \rangle - \varepsilon \hat{\kappa}_{\overline{\varphi}} \langle \zeta_2, \hat{u}_\kappa \rangle} \tag{3.3}
$$

From (3.2,3.3) one can observe that the dynamics of the parameters is driven by (i) the dependence of the relative equilibria (solutions) on the material quantity and (ii) the dependence of the Hamiltonian vectorfield on the material quantity. Moreover, when the structure map Γ does not depend on $K(x)$, the numerator in (3.3) reduces to $1 - \varepsilon \hat{\kappa}_{\overline{\varphi}} \langle \zeta_2, \hat{u}_\kappa \rangle$.

In the next two subsections we will show that the requirement that the residual $\mathcal{E}(\hat{u}; K(x))$ vanishes in the two directions $\delta I(\hat{u})$ and $\Gamma^{-1}(\hat{u}; K(x)) \hat{u}_\gamma$ is equivalent with two other requirements. Calling the projection argument geometrically inspired, the other two requirements are more physically and functional-analytic based, respectively.

3.2.2 Consistent evolution

For an exact solution the evolution of the momentum is given by

$$
\partial_t I(u) = \langle \delta I(u), X_H(u; K(x)) \rangle \equiv \{I, H\}(u).
$$

The immediate conclusion from (3.2) is that the same also holds for the approximation $\hat{u}$ as a consequence from the requirement that the residual vanishes in the direction $\delta I(\hat{u})$. That is, the approximation $\hat{u}$ is constructed in such a way that the momentum evolves *consistently*. Conversely, having started with the requirement of consistent evolution of I would have led directly to the equation (3.2). This is one of the reasons to use the value of a globally defined quantity, viz. the momentum functional I, as one of the parameters in the approximation.

It is more difficult to give a same argument for the evolution equation for φ. We will now show that it can be obtained in a same manner from energy conservation. Therefore, let us come back for a moment to the construction of $\hat{\kappa}$. There we required the inhomogeneous energy of the approximation, $H(\hat{u})$. to be equal to the homogeneous energy of $\check{u}$, $\check{H}(\check{u}; \kappa)$. From this requirement we constructed $\hat{\kappa} = \hat{\kappa}(\gamma, \overline{\varphi}, \varepsilon)$ and then $\hat{u}$. Now, since the energy is constant in time for an exact solution of the inhomogeneous system, it is tempting to demand the same for the approximation. That is, denote the energy of $\hat{u}$ by $\mathcal{H}(\gamma, \varphi)$:

$$\mathcal{H}(\gamma, \varphi) := \hat{H}(\hat{u}; \hat{\kappa}(\gamma, \overline{\varphi}, \varepsilon))) = H(\hat{u}),$$

energy conservation for the approximation is stated as

$$\frac{\mathrm{d}}{\mathrm{d}t}\mathcal{H}(\gamma, \varphi) = 0,$$

and means that the energy evolves consistently for the approximation.

In the rest of this subsection we will show the following result.

Proposition 3.1 *Under the technical assumption 3.2, the requirements that the momentum and energy evolve consistently are equivalent with the two requirements from the geometric projection argument:*

$$\mathbb{P}_1\mathcal{E}(\hat{u}; K(x)) = \mathbb{P}_2\mathcal{E}(\hat{u}; K(x)) = 0.$$

Proof Let us analyse the derivative of $\mathcal{H}(\gamma, \varphi)$ with respect to time:

$$\frac{\mathrm{d}}{\mathrm{d}t}\mathcal{H}(\gamma, \varphi) = \dot{\varphi}\mathcal{H}_\varphi + \dot{\gamma}\mathcal{H}_\gamma. \tag{3.4}$$

The first term at the righthandside can be rewritten:

$$\mathcal{H}_\varphi = \frac{\partial H(\hat{u})}{\partial \varphi} = \langle \delta H(\hat{u}), \check{u}_\varphi \rangle + \varepsilon \langle \delta H(\hat{u}), \check{u}_\kappa \rangle \hat{\kappa}_{\overline{\varphi}}. \tag{3.5}$$

The second term on the righthandside of (3.4) can be written as

$$\mathcal{H}_\gamma = \frac{\partial H(\hat{u})}{\partial \gamma} = \langle \delta H(\hat{u}), \check{u}_\gamma \rangle + \langle \delta H(\hat{u}), \check{u}_\kappa \rangle \hat{\kappa}_\gamma. \tag{3.6}$$

Let us first take care of the term $\langle \delta H(\hat{u}), \breve{u}_\kappa \rangle$ which appears in both (3.5) and (3.6). Decomposing $\delta H(\hat{u})$ yields

$$
\begin{aligned}
\langle \delta H(\hat{u}), \breve{u}_\kappa \rangle &= \langle \delta H(\hat{u}) - \delta \hat{H}(\hat{u}; \hat{\kappa}), \breve{u}_\kappa \rangle + \langle \delta \hat{H}(\hat{u}; \hat{\kappa}), \breve{u}_\kappa \rangle \\
&= \langle \delta H(\hat{u}) - \delta \hat{H}(\hat{u}; \hat{\kappa}), \breve{u}_\kappa \rangle + \lambda \langle \delta I(\hat{u}), \breve{u}_\kappa \rangle \\
&= \langle \delta H(\hat{u}) - \delta \hat{H}(\hat{u}; \hat{\kappa}), \breve{u}_\kappa \rangle, \quad (3.7)
\end{aligned}
$$

Assumption 3.2, (A1) implies that the second term on the righthandside of (3.5) and (3.6) is of the order $\mathcal{O}(\varepsilon^2)$. Note that in order to obtain this result the assumption is too strong: we only need $\delta H(\hat{u}) - \delta \hat{H}(\hat{u}; \hat{\kappa})$ to be $\mathcal{O}(\varepsilon)$ in the direction $\breve{u}_\kappa$.

Now we take a closer look at $\langle \delta H(\hat{u}), \breve{u}_\varphi \rangle$ and $\langle \delta H(\hat{u}), \breve{u}_\gamma \rangle$. The first term can easily be rewritten to

$$
\begin{aligned}
\langle \delta H(\hat{u}), \breve{u}_\varphi \rangle &= \langle \delta H(\hat{u}), \hat{X}_I(\hat{u}; \hat{\kappa}) - X_I(\hat{u}; K(x)) + X_I(\hat{u}; K(x)) \rangle \\
&= \langle \mathcal{E}(\hat{u}; K(x)) - \partial_t \hat{u}, \delta I(\hat{u}) \rangle + \langle \delta H(\hat{u}), \hat{X}_I(\hat{u}; \hat{\kappa}) - X_I(\hat{u}; K(x)) \rangle \\
&= \langle \mathcal{E}(\hat{u}; K(x)), \delta I(\hat{u}) \rangle - \langle \partial_t \hat{u}, \delta I(\hat{u}) \rangle + \\
&\quad + \langle \delta H(\hat{u}) - \delta \hat{H}(\hat{u}; \hat{\kappa}) + \delta \hat{H}(\hat{u}; \hat{\kappa}), (\hat{\Gamma}(\hat{u}; \hat{\kappa}) - \Gamma(\hat{u}; K(x))) \delta I(\hat{u}) \rangle \\
&= \langle \mathcal{E}(\hat{u}; K(x)), \delta I(\hat{u}) \rangle - \langle \partial_t \hat{u}, \delta I(\hat{u}) \rangle + \\
&\quad + \langle \delta H(\hat{u}) - \delta \hat{H}(\hat{u}; \hat{\kappa}), \hat{X}_I(\hat{u}; \hat{\kappa}) - X_I(\hat{u}; K(x)) \rangle \quad (3.8)
\end{aligned}
$$

whereas $\langle \delta H(\hat{u}), \breve{u}_\gamma \rangle$ can be reduced in the same way to

$$
\langle \delta H(\hat{u}), \breve{u}_\gamma \rangle = \langle \mathcal{E}(\hat{u}; K(x)), \Gamma^{-1}(\hat{u}; K(x)) \breve{u}_\gamma \rangle - \langle \partial_t \hat{u}, \Gamma^{-1}(\hat{u}; K(x)) \breve{u}_\gamma \rangle. \quad (3.9)
$$

Assumption 3.2, (A1) and (A2), assure that the last term in (3.8) is $\mathcal{O}(\varepsilon^2)$. Note again that this assumption is too strong: we only need an estimate for the specific inner product.

Finally, $\langle \partial_t \hat{u}, \delta I(\hat{u}) \rangle$ reduces immediately to $\dot{\gamma}$ whereas in the deduction of (3.3) we saw that

$$
\begin{aligned}
\langle \partial_t \hat{u}, \Gamma^{-1}(\hat{u}; K(x)) \breve{u}_\gamma \rangle &= -\dot{\varphi} + \dot{\varphi} \langle \zeta_2, X_I(\hat{u}; \hat{\kappa}) - X_I(\hat{u}; K(x)) \rangle + \\
&\quad + (\dot{\varphi} \varepsilon \hat{\kappa}_{\overline{\varphi}} + \dot{\gamma} \hat{\kappa}_\gamma) \langle \zeta_2, \breve{u}_\kappa \rangle \quad (3.10) \\
&\stackrel{(A1,A2)}{=} -\dot{\varphi} + \mathcal{O}(\varepsilon) \quad (3.11)
\end{aligned}
$$

and hence, (3.4) can be rewritten as

$$
\frac{d}{dt} \mathcal{H}(\gamma, \varphi) = \dot{\varphi} \langle \mathcal{E}(\hat{u}; K(x)), \delta I(\hat{u}) \rangle + \dot{\gamma} \langle \mathcal{E}(\hat{u}; K(x)), \Gamma^{-1}(\hat{u}; K(x)) \breve{u}_\gamma \rangle + \mathcal{O}(\varepsilon^2).
$$

(Note that $\dot{\gamma} = \mathcal{O}(\varepsilon)$ under assumption 3.2.) This finishes the proof. $\square$

Assumption 3.2 *The following technical conditions are sufficient in order for proposition 3.1 to hold true.*

(A1) $\|\delta H(\hat{u}) - \delta\hat{H}(\hat{u};\hat{\kappa})\| = \mathcal{O}(\varepsilon)$.

(A2) $\|X_I(\hat{u}; K(x)) - \hat{X}_I(\hat{u};\hat{\kappa})\| = \mathcal{O}(\varepsilon)$.

In the case that the structure map does not depend on $K(x)$, the second condition (A2) is met immediately.

These conditions specify in more detail what we mean by 'concentratedness of relative equilibria'. An illuminating example can be found in the next section where we treat the Bloch wall. In (3.23) we check condition (A1) and it is easily observed that, although the material 'constant' $K(x)$ can deviate by a large amount from its pointwise (in time) homogenized version $\kappa(t)$, the resulting energy from these tails is very small since the growth of $K(x)$ is compensated by a decay of—in this case—$\sin^2\hat{\theta}$.

3.2.3 Analytic solvability conditions

The consistent evolution of the momentum and the energy as considered above is an attractive formulation of the way how to derive the parameter dynamics. Moreover, it has a clear physical interpretation. Nevertheless, this formulation, nor the equivalent geometric projection described above, provide any information about the actual quality of the approxmation that results. To that end one has to consider the difference between the approximation and the solution that is to be approximated. The essence of the main result of this subsection will be that for the two (equivalent) methods above it holds that the error is small over long spatial and/or time-scales. This will follow by showing that the dynamics is also be obtained from Fredholm solvability conditions.

With u an exact solution of $\mathcal{E}(u; K(x)) = 0$, which describes the distortion of a relative equilibrium solution, introduce the error η as $\eta = u - \hat{u}$.

Notation. The Fréchet derivative of $\mathcal{E}(u; K(x))$ at u_0 is denoted by $\mathcal{E}'(u_0; K(x))$. Similarly, the Fréchet derivative of $\hat{\mathcal{E}}(u;\kappa)$ at u_0 is denoted by $\hat{\mathcal{E}}'(u_0;\kappa)$.

On substituting $u = \hat{u} + \eta$ in the inhomogeneous equation and using that u is an exact solution we derive

$$0 = \mathcal{E}(u; K(x)) = \mathcal{E}(\hat{u}; K(x)) + \mathcal{E}'(\hat{u}; K(x))\eta + \|\eta\|^2.$$

Assuming $\|\eta\| = \mathcal{O}(\varepsilon)$ (we have to check that afterwards), we have as a first order equation for the error

$$\mathcal{E}'(\hat{u}; K(x))\eta = -\mathcal{E}(\hat{u}; K(x)). \qquad (3.12)$$

When analysing (3.12) we note that when $\mathcal{E}'(\hat{u}; K(x))$ has a bounded inverse we can solve for η. In order to make sense, $\|\mathcal{E}(\hat{u}; K(x))\|$ should be $\mathcal{O}(\varepsilon)$ (this is equivalent with saying that $\hat{u}$ is an approximation), so when $\mathcal{E}'(\hat{u}; K(x))$ is regular we have the immediate result that $\|\eta\| = \mathcal{O}(\varepsilon)$. However, the presence of symmetry in the homogeneous case causes the linearized operator to be singular. This is clarified when we note that for the homogeneous problem it holds that

$$\hat{\mathcal{E}}(V(\gamma, \varphi_0; \kappa)(t); \kappa) \equiv 0, \quad \forall\gamma, \varphi_0,$$

and hence, by differentiating with respect to the two parameters γ and φ,

$$\hat{\mathcal{E}}'(V(\gamma, \varphi_0; \kappa)(t); \kappa) \frac{\partial V(\gamma, \varphi_0; \kappa)(t)}{\partial \varphi_0} = 0, \quad (3.13)$$

$$\hat{\mathcal{E}}'(V(\gamma, \varphi_0; \kappa)(t); \kappa) \left[t \frac{d\lambda}{d\gamma} \frac{\partial V(\gamma, \varphi_0; \kappa)(t)}{\partial \varphi_0} + \frac{\partial V(\gamma, \varphi_0; \kappa)(t)}{\partial \gamma} \right] = 0. \quad (3.14)$$

In most applications the linearized operator is a Fredholm operator, which has an adjoint with a non-trivial kernel too. (In most cases, the index is zero, meaning that the dimension of the kernel of the adjoint $\mathcal{E}'(u; K(x))^*$ equals the dimension of the kernel of the linearization $\mathcal{E}'(u; K(x))$). Here and in the following, the adjoint has to be understood with respect to the innerproduct over space and (arbitrary) time-interval. When the ajoint kernel is non-trivial, solvability conditions have to be satisfied in order that equation (3.12) has a solution. In fact, it then should hold that

$$\int_{t_1}^{t_2} \langle \zeta^*, \mathcal{E}(u; K(x)) \rangle \, dt = 0$$

for all ζ^* in the adjoint kernel: $\mathcal{E}'(u; K(x))^* \zeta^* = 0$. This necessary condition is immediately derived by taking the innerproduct of (3.12) with ζ^*.

In the following we shall exploit this reasoning in a perturbative way; see the classical monograph by Kato (1966) for more details.

In order to have a bounded solution η of (3.12), i.e., $\|\eta\| = \mathcal{O}(\varepsilon)$, it is necessary that the residual should be orthogonal to the kernel of the approximation $\mathcal{E}'(u; K(x))^*$.

This is the motivation for inspecting the kernel of the adjoint linearised operator, $\hat{\mathcal{E}}'(V; \kappa)$ and $\mathcal{E}'(\hat{u}; K(x))$. We have the following important theorem (Van Groesen (1995)).

Theorem 3.3 (Kernel Theorem)
The operators $\hat{\mathcal{E}}'(u; \kappa)\hat{\Gamma}(u; \kappa)$ and $\mathcal{E}'(u; K(x))\Gamma(u; K(x))$ are symmetric:

$$\hat{\mathcal{E}}'(u; \kappa)\hat{\Gamma}(u; \kappa) = \left[\hat{\mathcal{E}}'(u; \kappa)\hat{\Gamma}(u; \kappa) \right]^* \equiv -\hat{\Gamma}(u; \kappa)\hat{\mathcal{E}}'(u; \kappa)^*, \quad (3.15)$$

$$\mathcal{E}'(u; K(x))\Gamma(u; K(x)) = [\mathcal{E}'(u; K(x))\Gamma(u; K(x))]^* \equiv -\Gamma(u; K(x))\mathcal{E}'(u; K(x))^*. \quad (3.16)$$

for all u with $\hat{\mathcal{E}}(u; \kappa) = 0$ and $\mathcal{E}(u; K(x)) = 0$ respectively.

Proof See Van Groesen (1995). □

Hence, since $\hat{\Gamma}(V; \kappa)$ is invertible, it follows from the Kernel Theorem and (3.13, 3.14) that

$$T_V^* \mathcal{M}_\kappa \subset \ker[\hat{\mathcal{E}}'(V; \kappa)^*]. \quad (3.17)$$

The immediate candidates for elements from the kernel of $\mathcal{E}'(\hat{u}; K(x))^*$ are then $\delta I(\hat{u})$ and $t\dfrac{\mathrm{d}\lambda}{\mathrm{d}\gamma}\delta I(\hat{u}) + \Gamma^{-1}(\hat{u}; K(x))\check{u}_\gamma = \zeta_2$. This results in the necessary solvability conditions

$$\int_{t_1}^{t_2} \langle \mathbb{P}_1 \mathcal{E}(\hat{u}; K(x)), \delta I(\hat{u}) \rangle \, \mathrm{d}t = 0 \tag{3.18}$$

and

$$
\begin{aligned}
0 &= \int_{t_1}^{t_2} \left[\langle \mathbb{P}_2 \mathcal{E}(\hat{u}; K(x)), -\zeta_2 \rangle + \langle t\frac{\mathrm{d}\lambda}{\mathrm{d}\gamma}\delta I(\hat{u}), \mathcal{E}(\hat{u}; K(x)) \rangle \right] \mathrm{d}t \\
&= \int_{t_1}^{t_2} \langle \mathbb{P}_2 \mathcal{E}(\hat{u}; K(x)), -\zeta_2 \rangle \, \mathrm{d}t + \int_{t_1}^{t_2} t\frac{\mathrm{d}\lambda}{\mathrm{d}\gamma} \langle \mathbb{P}_1 \mathcal{E}(\hat{u}; K(x)), \delta I(\hat{u}) \rangle \, \mathrm{d}t.
\end{aligned}
\tag{3.19}
$$

This shows that requiring $\mathbb{P}_1 \mathcal{E}(\hat{u}; K(x)) = \mathbb{P}_2 \mathcal{E}(\hat{u}; K(x)) = 0$ at each instant (as was suggested from the projection argument), is equivalent to require the necessary conditions (3.18,3.19) to be fulfilled for every time interval $[t_1, t_2]$. Hence, the Fredholm alternative leads to the parameter dynamics, i.e., to the quasi-homogeneous approximation as constructed above.

3.3   Motion of the Bloch wall in inhomogeneous media

In a spatial inhomogeneous crystal, i.e., equation (1.1) with $K = K(x)$ slowly varying like

$$K(x) = \overline{K}(\varepsilon x),$$

with ε a small parameter, we look for a distortion, $\theta(x, t)$, of an initial kink solution as a quasi-homogeneous approximation, $\check{\theta}(x, t)$, by a specific succession of kinks,

$$\check{\theta}(t) = \Theta(\gamma(t), \varphi(t); \kappa(t)).$$

So the approximation $\check{\theta}(t)$ 'feels' at some specific time t an isotropy 'constant' $\kappa(t)$. After giving a computational example of finding $\hat{\kappa}$ as a function of γ, $\overline{\varphi}$ and ε, we will give the dynamics of the parameters obtained via the consistent evolution of the momentum and the energy.

Example 3.4 *In order to give a computional example of how to find $\hat{\kappa}$, we consider for $K(x)$ the function*

$$K(x) = K_0 + K_1(\varepsilon x)^2, \quad \overline{K}(x) = K_0 + K_1 x^2, \quad K_0, K_1 \in \mathbb{R}.$$

The quantity $\Delta(\kappa; \gamma, \overline{\varphi}, \varepsilon) = H(\check{\theta}, \check{p}) - \hat{H}(\check{\theta}, \check{p}; \kappa)$ reduces to

$$\Delta(\kappa; \gamma, \overline{\varphi}, \varepsilon) = \int (K_0 + K_1(\overline{\varphi} + \varepsilon y)^2 - \kappa)\sin^2 \check{\theta}(y, t) \, dy, \quad (y = x - \varphi),$$

which, after some maple-assisted manipulations reduces to

$$\Delta(\kappa:\gamma,\overline{\varphi},\varepsilon) = \frac{4}{3}\frac{\pi^2 K_1 \varepsilon^2}{(8\kappa+\gamma^2)^{3/2}} + 4\frac{K_1\overline{\varphi}^2 + K_0}{\sqrt{8\kappa+\gamma^2}} - \frac{4\kappa}{\sqrt{8\kappa+\gamma^2}}. \tag{3.20}$$

Taking $\varepsilon = 0$ we immediately have the result that $\kappa = K_0 + K_1\overline{\varphi}^2 = \overline{K}(\overline{\varphi})$. Moreover, differentiation of Δ with respect to κ and evaluating at $(\varepsilon,\kappa) = (0, K_0 + K_1\overline{\varphi}^2)$ yields

$$\left.\frac{\partial\Delta}{\partial\kappa}\right|_{(\varepsilon,\kappa)=(0,K_0+K_1\overline{\varphi}^2)} = -\frac{4}{\sqrt{8K_0 + 8K_1\overline{\varphi}^2 + \gamma^2}} \neq 0.$$

Hence, solving for κ, we find up to $\mathcal{O}(\varepsilon^4)$

$$\hat{\kappa} = K_0 + K_1\overline{\varphi}^2 + \frac{1}{3}\frac{\varepsilon^2\pi^2 K_1^2}{8K_0 + 8K_1\overline{\varphi}^2 + \gamma^2}. \tag{3.21}$$

Notice that the choice $\kappa = \overline{K}(\overline{\varphi}) = K_0 + K_1\overline{\varphi}^2$ results in a Δ that is $\mathcal{O}(\varepsilon^2)$ uniform for all φ.

Solving $\hat{K}(\overline{\Psi}) = \kappa(\gamma,\overline{\varphi},\varepsilon)$ for $\overline{\Psi}$ yields

$$\overline{\Psi}^2 = \overline{\varphi}^2 + \frac{1}{3}\frac{\varepsilon^2\pi^2 K_1}{8K_0 + 8K_1\overline{\varphi}^2 + \gamma^2} + \mathcal{O}(\varepsilon^4),$$

that is (outside an ε-neighbourhood of 0 where K assumes its minimum)

$$\overline{\Psi} = \overline{\varphi} + \frac{1}{6}\frac{\varepsilon^2\pi^2 K_1}{\overline{\varphi}(8K_0 + 8K_1\overline{\varphi}^2 + \gamma^2)}.$$

This result is even better than the general, expected result. It is due to symmetry in the underlying kink-solution.

Motivated by this example we take in the following $\kappa(t) = \overline{K}(\overline{\varphi}) = K(\varphi)$. Computing for this choice the homogeneous energy at level $K(\varphi)$. $\hat{H}(\hat{\theta},\hat{p};K(\varphi)) = \mathcal{H}(\gamma,\varphi)$, we find

$$\begin{aligned}
\mathcal{H}(\gamma,\varphi) &= \int\left\{\frac{1}{2}\hat{p}^2 + \frac{1}{2}\hat{\theta}_x^2 + K(\varphi(t))\sin^2\hat{\theta}\right\}\,dx \\
&= \int\left\{\frac{1}{2}(\lambda^2+1)\hat{\theta}_x^2 - \frac{1}{2}(\lambda^2-1)\hat{\theta}_x^2\right\}\,dx \\
&= \int\hat{\theta}_x^2\,dx = 2B \\
&= \sqrt{8K(\varphi(t)) + \gamma(t)^2} \tag{3.22}
\end{aligned}$$

At this point it is best to check the assumption 3.2. i.e., whether the norm of $\delta H(\hat{\theta},\hat{p}) - \delta\hat{H}(\hat{\theta},\hat{p};\kappa(t))$ is of the order ε (note that (A2) is trivially satisfied).

Calculating this quantity, using (2.18) with $\kappa(t) = K(\varphi(t))$, we are lead to check whether

$$\int_{\mathcal{R}} \left| 4e^{By} \frac{e^{2By} - 1}{(e^{2By} + 1)^2} (K(\varphi) - K(\varphi + y)) \right| \, dy \tag{3.23}$$

is of the required $\mathcal{O}(\varepsilon)$. This is the case, since making an expansion of the difference $K(\varphi) - K(\varphi + y)$ around $y = 0$, the norm of $\delta H(\hat{\theta}, \hat{p}) - \delta \dot{H}(\hat{\theta}, \hat{p}; K(\varphi))$ is easily shown to be

$$\| \delta H(\hat{\theta}, \hat{p}) - \delta \dot{H}(\hat{\theta}, \hat{p}; K(\varphi)) \| = \frac{2\pi \varepsilon}{B^2} + \mathcal{O}(\varepsilon^2).$$

Hence, the consistent dynamics is equivalent with the projected dynamics and with the 'Fredholm' dynamics.

If we now require the energy to evolve in a consistent way, i.e.,

$$\frac{d}{dt} \mathcal{H}(\gamma, \varphi) = \mathcal{H}_\gamma \dot{\gamma} + \mathcal{H}_\varphi \dot{\varphi} = 0,$$

we find that up to a scaling factor $\sigma(t)$, the dynamics of the variables constitute a Hamiltonian system themselves (see also the discussion in section 4):

$$\partial_t \begin{pmatrix} \varphi \\ \gamma \end{pmatrix} = \sigma(t) J_2 \begin{pmatrix} \mathcal{H}_\varphi \\ \mathcal{H}_\gamma \end{pmatrix}. \tag{3.24}$$

The factor $\sigma(t)$ follows from the requirement that also the momentum I evolves consistently:

$$\begin{aligned} \partial_t \gamma = \partial_t I(\hat{\theta}, \hat{p}) &= \left\langle \delta I(\hat{\theta}, \hat{p}), J_2 \delta H(\hat{\theta}, \hat{p}) \right\rangle \\ &= \int \left\{ \hat{p}\hat{p}_x - \hat{\theta}_x \hat{\theta}_{xx} + \hat{\theta}_x K(x) \sin 2\hat{\theta} \right\} \, dx \\ &\overset{(2.19)}{=} -\int K'(x) \sin^2 \hat{\theta} \, dx. \end{aligned} \tag{3.25}$$

Expanding $K'(x)$ around $x = \varphi$ yields

$$\begin{aligned} \partial_t \gamma &= -\int K'(\varphi) \sin^2 \hat{\theta} \, dx + \mathcal{O}(\frac{\varepsilon^3}{B^3}) \\ &= -\mathcal{H}_\varphi + \mathcal{O}(\varepsilon^3), \end{aligned} \tag{3.26}$$

and hence $\sigma(t) = 1$, correct up to $\mathcal{O}(\varepsilon^2)$. With this information (3.24) reduces, up to $\mathcal{O}(\varepsilon^2)$, to

$$\begin{cases} \dot{\gamma} = -\mathcal{H}_\varphi = -\dfrac{4K'(\varphi)}{2B}, \\ \dot{\varphi} = \mathcal{H}_\gamma = \dfrac{\gamma}{2B} = \lambda(\gamma(t), K(\varphi(t))) \end{cases} \tag{3.27}$$

Remarks

1. Considering (3.24) the motion of the wall is analogous to the dynamics of a particle from classical mechanics, with a potential function defined by the magnetic anisotropy function, K; see (3.22). This means that for convex-like potentials K, for instance $K(x) = K_0 + K_1(\varepsilon x)^2$, the motion will be a periodic one between extreme points $\varphi_\pm$, determined by $E_0 = \sqrt{8K(\varphi_\pm)}$; this describes a 'bound state' for the wall. See figure 4.

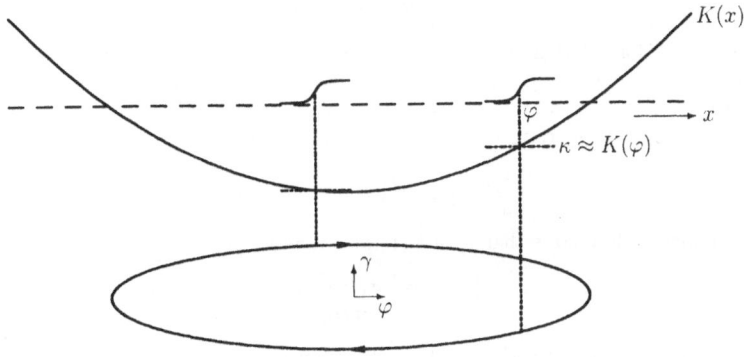

Figure 4: The particle analogy for the motion of a Bloch wall: trapped in a well of the magnetic anisotropy constant.

2. Numerical calculations (Blommers and Booij (1992) and Fledderus and Van Groesen (1995)) confirmed that the approximation is very accurate, even when the disturbances are large: during the evolution the value of K may change by order 1.

4 Interpretations and Conclusions

For a general inhomogeneous medium, with slowly varying inhomogeneity, we have constructed a quasi-homogeneous approximation, based on special solutions (relative equilibrium solutions) of a homogenized medium, where the value measuring the inhomogeneity changes with the propagation of the relative equilibrium solution. To be successful, the relative equilibria are supposed to be concentrated in space, a condition that was specified in the process of construction. In section 3 it was shown that there are three equivalent ways to obtain the parameter dynamics for the quasi-homogeneous approximation.

The derivation using the Fredholm alternative gives the desirable mathematical result that the approximation is valid on time (and space) scales in which the slowly varying material properties change a unit order of magnitude. so the result is valid on time and length scales of order $\mathcal{O}(1/\varepsilon)$.

Assuming the conditions (A1) and (A2) to hold, which turns out to be a specification of the concentratedness condition imposed on the relative equilibria, the dynamics could also, and in fact more simply, be derived from the requirement of energy conservation. That is, from the requirement that the energy of the inhomogeneous system, when restricted to the relative equilibria, is constant in time (in the required order):

$$\frac{d}{dt}\mathcal{H}(\gamma,\varphi) = \mathcal{O}(\varepsilon^2).$$

This last formulation has as another nice property that it allows to conclude that the parameter dynamics of the approximation has a Hamiltonian structure, just like the complete inhomogeneous system for which the approximation is constructed. Let us investigate this structure in some detail. From

$$\frac{d}{dt}\mathcal{H}(\gamma,\varphi) = \dot{\varphi}\mathcal{H}_\varphi + \dot{\gamma}\mathcal{H}_\gamma = \mathcal{O}(\varepsilon^2)$$

it immediately follows that with a suitable function σ

$$\begin{cases} \dot{\varphi} &= \sigma\mathcal{H}_\gamma + \mathcal{O}(\varepsilon^2), \\ \dot{\gamma} &= -\sigma\mathcal{H}_\varphi + \mathcal{O}(\varepsilon^2). \end{cases}$$

Hence, apart from a scaling in time determined by σ, the structure is that of a canonical Hamiltonian system with structure map $J_2 = \begin{pmatrix} 0 & 1 \\ -1 & 0 \end{pmatrix}$:

$$\begin{pmatrix} \dfrac{\partial\varphi}{\partial\tau} \\ \dfrac{\partial\gamma}{\partial\tau} \end{pmatrix} = \begin{pmatrix} 0 & 1 \\ -1 & 0 \end{pmatrix}\begin{pmatrix} \delta_\varphi H \\ \delta_\gamma H \end{pmatrix}.$$

We will now show that in general $\sigma = 1 + \mathcal{O}(\varepsilon)$, so that, in general, the scaling cannot be neglegted (in the required order).

To see that $\sigma = 1 + \mathcal{O}(\varepsilon)$, note that relations (3.5), (3.7) and (3.8) imply that

$$\mathcal{H}_\varphi = \langle \mathcal{E}(\hat{u}; K(x)), \delta I(\hat{u})\rangle - \dot{\gamma} + \mathcal{O}(\varepsilon^2) \tag{4.1}$$

when assumptions 3.2, (A1) and (A2), are satisfied. That is,

$$\dot{\gamma} = -\mathcal{H}_\varphi + \mathcal{O}(\varepsilon^2). \tag{4.2}$$

In the same way, relations (3.6), (3.7), (3.9) and (3.11) yield

$$\mathcal{H}_\gamma = \langle \mathcal{E}(\hat{u}; K(x)), \Gamma^{-1}(\hat{u}; K(x))\hat{u}_\gamma\rangle + \dot{\varphi} + \mathcal{O}(\varepsilon), \tag{4.3}$$

hence,

$$\dot{\varphi} = \mathcal{H}_\gamma + \mathcal{O}(\varepsilon), \tag{4.4}$$

which implies $\sigma = 1 + \mathcal{O}(\varepsilon)$.

Be aware that if one would aim to have the same order of accuracy in the equations for $\dot{\gamma}$ and $\dot{\varphi}$, and writing

$$\begin{cases} \dot{\varphi} &= \sigma\mathcal{H}_\gamma + \mathcal{O}(\varepsilon^2), \\ \dot{\gamma} &= -\mathcal{H}_\varphi + \mathcal{O}(\varepsilon^2), \end{cases}$$

where $\sigma = 1 + \mathcal{O}(\varepsilon)$ has to be retained to have the correct order $\mathcal{O}(\varepsilon)$ term in the righthandside of $\dot{\varphi}$, but can be neglected in the righthandside of $\dot{\gamma}$ since $\mathcal{H}_\varphi = \mathcal{O}(\varepsilon)$. the Hamiltonian character would be lost. This is a property that is quite common (and sometimes overlooked) when dealing with order-arguments in originally Hamiltonian systems.

In special cases, the scaling factor can be left out completely. This is the case if $\sigma = 1 + \mathcal{O}(\varepsilon^2)$, which turns out to be true for the Bloch walls for instance. In fact, some conditions can be formulated that guarantee $\sigma = 1 + \mathcal{O}(\varepsilon^2)$. First of all. when the structure map does not depend on $K(x)$, i.e., $\|X_I(\hat{u}; K(x)) - \hat{X}_I(\hat{u}; \hat{\kappa})\| = 0$, and $\langle \delta H(\hat{u}), \check{u}_\kappa \rangle$ in (3.7) is of the order $\mathcal{O}(\varepsilon^2)$, the $\mathcal{O}(\varepsilon^2)$-term in (4.1) turns into an $\mathcal{O}(\varepsilon^3)$-term. The $\mathcal{O}(\varepsilon)$ contribution in (3.11) is then reduced to $(\dot{\varphi}\varepsilon\hat{\kappa}_{\overline{\varphi}} + \dot{\gamma}\hat{\kappa}_\gamma) \langle \zeta_2, \check{u}_\kappa \rangle$, so imposing the condition $\langle \zeta_2. \check{u}_\kappa \rangle = \mathcal{O}(\varepsilon)$ yields $\dot{\varphi} = \mathcal{H}_\gamma + \mathcal{O}(\varepsilon^2)$ and thus improved result $\sigma = 1 + \mathcal{O}(\varepsilon^2)$. Recapitulating the sufficient conditions to obtain this result:

(A3) $\|X_I(\hat{u}: K(x)) - \hat{X}_I(\hat{u}; \hat{\kappa})\| = 0$,

(A4) $\langle \delta H(\hat{u}), \check{u}_\kappa \rangle = \mathcal{O}(\varepsilon^2)$,

(A5) $\langle \zeta_2. \check{u}_\kappa \rangle = \mathcal{O}(\varepsilon)$.

A final remark concerns an extension of the ideas above. As became clear from the example. for the theory above to apply the conditions on concentratedness are essential. However, in a future paper we will study problems for which the relative equilibria are extended (e.g. periodic) in space. Then looking for approximations that deviate slowly from the manifold, the argument of the Fredholm alternative, as well as a modified version of energy conservation. can still be used to find slowly varying approximations. This theory for extended structures is particularly suited to study equations that describe pattern formation; the resulting equation for the parameter dynamics may be Nonlinear Schrödinger type of equations or phase-diffusion type of equations.

References

[1] Aceves. A., Adachihara, H., Jones, C., Lerman, J.C., McLaughlin, D.W., Moloney, J.V.. and Newell, A.C. Chaos and coherent structures in partial differential equations. Physica 18(D), 85–112, 1986.

[2] Blommers. B., and Booij, W. Periodic solition solutions for a perturbed sine-Gordon equation (in Dutch). Technical report, University of Twente, 1992.

[3] Buitelaar, R.P. The method of averaging in Banach spaces: theory and application. PhD-thesis, Univ. of Utrecht, 1993.

[4] Calogero, F., and Eckhaus, W. Nonlinear evolution equations, rescalings, model PDE's and their integrability: II. Inverse Problems, 4, 11–33, 1988.

[5] Carr, J. *Applications of Centre Manifold Theory*. Springer-Verlag, New York, 1981.

[6] Derks, G. Coherent structures in the dynamics of perturbed Hamiltonian systems. PhD-thesis, University of Twente, 1992.

[7] Derks, G., and Groesen, E. van. Dissipation in Hamiltonian systems: Decaying cnoidal waves. To be pubished, SIAM J. Math. Anal., 1995.

[8] Fledderus, E.R., and Groesen, E. van. Critical swirling flows in expanding pipes, Part II. To be published, SIAM J. Math. An. Appl., 1995.

[9] Fledderus, E.R., and Groesen, E. van. Deformation of coherent structures in inhomogeneous media. Rep. Progr. Phys., forthcoming, 1995.

[10] Groesen, E. van, Beckum, F.P.H. van, and Valkering, T.P. Decay of travelling waves in dissipative Poisson systems. ZAMP, 41, 501-523, 1990.

[11] Groesen, E. van, Fliert, B.W. van de, and Fledderus, E. Quasi-homogeneous critical swirling flows in expanding pipes, I. SIAM J. Math. An. Appl., 192. 764-788, 1995.

[12] Groesen, E. van. A Hamiltonian Perturbation Theory for Coherent Structures illustrated to wave problems. Proceedings of the IUTAM/ISIMM Symposium, Hannover, Germany, 1995.

[13] Leeuw F.H. de, Doel, R. van den, and Enz, U. Dynamic properties of magnetic domain walls and magnetic bubbles. Rep. Prog. Phys. 43, 689-783, 1980.

[14] Hale, J.K. *Ordinary differential equations*. Wiley-Interscience, New York, 1969.

[15] Karpman, V.I., and Solov'ev, V.V. The influence of perturbations on the shape of a sine-Gordon soliton. Phys. Lett., 84A(2), 39–41, 1981.

[16] Kato, T. *Perturbation Theory for Linear Operators*. Springer-Verlag, New York, 1966.

[17] Kivshar, Y.S., and Malomed, B.A. Dynamics of solitons in nearly integrable systems. Rev. Mod. Phys. 61(4), 763–915, 1989.

[18] Krol, M.S. The Method of Averaging in Partial Differential Equations. PhD-thesis, Univ. of Utrecht, 1990.

[19] Nayfeh, A.H. *Introduction to Perturbation Techniques*. Wiley, New York, 1993.

[20] Olsen, O.H., and Samuelsen, M.R. Sine-Gordon 2π-kink dynamics in the presence of small perturbations. Phys. Rev. B, 28(1), 210–217, 1983.

[21] Pudjaprasetya, S.R.. and Groesen, E. van. Uni-directional waves over slowly varying bottom. Part II: Quasi-homogeneous approximation of distorting waves. To be published, Wave Motion, 1995.

[22] Sakai, S., Samuelsen. M.R., and Olsen, O.H. Perturbation analysis of a parametrically changed sine-Gordon equation. Phys. Rev. B, 36(1), 217-225, 1987.

[23] Salerno, M, Samuelsen. M.R., Lomdahl, P.S., and Olsen O.H. Non-dissipative perturbations in the sine-Gordon system. Phys. Lett. 108A, 241-244, 1985.

[24] Sanders, J.A., and Verhulst, F. *Averaging Methods in Nonlinear Dynamical Systems*. Springer-Verlag, New York, 1985.

Progress in Nonlinear Differential Equations
and Their Applications, Vol. 19
© 1996 Birkhäuser Verlag Basel/Switzerland

On instability of minimal foliations for a variational problem on T^2

Xianhua Huang*

Abstract

In this paper, we study the variational problem $\int F(t, x, x_t) dt$, where $F(t, x, p) \in C^r(S^1 \times S^1 \times R)$. Suppose $x = U(t, \alpha t + \beta)$ is its T-invariant minimal foliation and α does not satisfy Diophantine condition, then we can destroy this minimal foliation by a small C^r-perturbation of $F(t, x, p)$.

Key Words: Minimal foliation, instability, variational problems.

1 Introduction

The variational problem on a torus T^d has been studied extensively, see e.g. [2, 9, 10, 11] and references therein. An important topic is that of the foliations whose leaves are minimum surface with respect to a given metric. J. Moser showed the existence of such a minimum surface under certain hypotheses about the integrand [9] and proved the stability for some minimum surfaces, i.e. under small perturbation of the variational problem, there is a new foliation which is conjugate to the given one under a diffeomorphism close to identity.

In order to state the problems precisely, we describe the set-up and the leaves of the foliations generally.

Let $T^d = \mathbb{R}^d / Z^d$, set $d = n + 1$, $s = (s_1, \ldots, s_n)$, $\bar{s} = (s_1, \ldots, s_n, x)$, $\pi(s_1, \ldots, s_n, x, p_1, \ldots, p_n) = (s_1, \ldots, s_n)$. Suppose

i) $F \in C^r(\Omega)$, where $r \geq 2$ and Ω is an open domain in $T^{n+1} \times \mathbb{R}^n$ with the property that $\pi(\Omega) = T^d$;

ii) for all $\xi \in \mathbb{R}^n$, $(\bar{s}, p) \in \Omega$,

$$\sum_{\nu, \mu=1}^n F_{p_\nu p_\mu}(\bar{s}, p) \xi_\nu \xi_\mu \geq \lambda \|\xi\|^2 \tag{1}$$

where λ is a positive constant.

*Faculty of Technical Mathematics and Informatics, Delft University of Technology, 2628 CD Delft, The Netherlands, FAX: +31 15 2787209, EMAIL: X.Huang@twi.tudelft.nl

Consider the variational problem

$$\int F(s, x, x_s) ds, \tag{2}$$

where $ds = ds_1 \ldots ds_n$, $x_s = (\frac{\partial x}{\partial s_1}, \ldots, \frac{\partial x}{\partial s_n})$ and $F = F(\bar{s}, p)$ is a smooth function of period 1 in the first $d = n + 1$ variables, while $p = (p_1, \ldots, p_n)$ varies in an open subset of $\mathbb{R}^n$. The function x representing the leaves of the foliation are required to satisfy the Euler equation

$$\sum_\nu^n \frac{\partial}{\partial s_\nu} F_{p_\nu}(\bar{s}, x_s) = F_x(s, x, x_s), \tag{3}$$

which is a nonlinear elliptic differential equation. The solutions of (3) will be called *extremals*.

For example, consider a variational problem on T^2 by assuming that $F(t, x, \dot{x}) = \dot{x}^2$. For this trivial variational problem

$$\int \dot{x}^2 dt,$$

the Euler equation is given by $\ddot{x} = 0$ and the minimal solutions are straight lines given by $x = \alpha t + \beta$.

To define a minimal foliation on a torus T^{n+1}, we consider its lift on $\mathbb{R}^{n+1}$, taking account of the Z^{n+1}-action.

Definition 1 *For a given variational problem (2), we define a Z^{n+1}-invariant C^m-minimal foliation as a function $x \in C^m(\mathbb{R}^n \times \mathbb{R}, \mathbb{R})$ taking $(s, \lambda) \to x = x(s, \lambda)$ such that:*

  i) for any $\lambda \in \mathbb{R}$, $x(s, \lambda)$ is an extremal of (3);

  ii) for fixed $s \in \mathbb{R}^n$, $\lambda \to x(s, \lambda)$ is a C^m-homeomorphism of $\mathbb{R}$ onto $\mathbb{R}$ with $x(s, \lambda) \leq x(s, \lambda')$ for $\lambda \leq \lambda'$; and

  iii) the foliation given by the leaves $x = x(s, \lambda)$ for $(s, \lambda) \in \mathbb{R}^{n-1}$, is invariant under the Z^{n+1}-action.

The extremals representing the leaves of such a minimal foliation are special solutions of the Euler equation. For the leaves of a minimal foliation, minimal solutions of (2) without self-intersections qualify, as was pointed out in [10, 11]. From the theory developed there, it follows that, for a given Z^{n+1}-invariant minimal foliation, there exists a unique $\bar{\alpha} = (\alpha_1, \ldots, \alpha_n, -1)$ such that for every leaf

$$\sup_s |x(s, \lambda) - (\alpha, s)| < \infty,$$

where $\alpha = (\alpha_1, \ldots, \alpha_n)$. Moreover, there is a function $U(s, \theta) \in C^m(\mathbb{R}^n \times \mathbb{R}, \mathbb{R})$, where m depends on both the smoothness of F and the property of the minimal foliation itself, such that $U(s,\theta) - \theta \in C^m(T^d)$, $\frac{\partial}{\partial\theta}U(x,\theta) > 0$ if $m \geq 1$ such that the leaves of foliation take the form $x = U(s,(\alpha,s) + \beta)$. β= constant.

In particular, when $d = 2$(i.e. $n = 1$), we have that $F(t,x,p) \in C^r(S^1 \times S^1 \times \mathbb{R})$ ($r \geq 2$) and $0 < \delta \leq F_{pp}(t,x,p)$ accordingly. Without loss of generality, we may suppose that $0 < \delta \leq F_{pp}(t,x,p)\delta^{-1}$, since our problem can be restricted to a compact subset of $S^1 \times S^1 \times \mathbb{R}$. The following variational problem is considered

$$\int_{t_1}^{t_2} F(t,x,x_t)dt. \tag{4}$$

If $|F_{pp}| + |F_{tp}| = O(|p|), F_x = O(p^2)$, then, for any given number α, there exists a function $U(t,\theta)$ such that $U(t,\theta) - \theta$ has period 1 in t,θ, and $U(t,\theta)$ is strictly monotone increasing in θ, and such that for any β, $x(t) = U(t,\alpha t + \beta)$ is a solution of the Euler equation

$$\frac{\partial}{\partial t}F_p(t,x(t),\dot{x}(t)) = F_x(t,x(t),\dot{x}(t)). \tag{5}$$

Moreover, if $U(t,\theta)$ is continuous, we get a Z^2-invariant C^m-minimal foliation for some $m \geq 0$([11]).

In [9]. Moser proved that if $\alpha = (\alpha_1,\ldots,\alpha_n)$ is given such that

$$\sum_{\nu=1}^{n}(\alpha_\nu j_{n+1} + j_\nu)^2 \geq \gamma(1 + j_{n+1}^2)^{-\tau} \tag{6}$$

holds for fixed $\tau(\alpha),\gamma(\alpha)$ and arbitrary $\bar{j} = (j_1,\ldots,j_n,j_{n+1}) \in Z^{n+1} \setminus \{0\}$, then the above Z^{n+1}-invariant C^∞-minimal foliation $x = U(s,\alpha s + \beta)$ is stable under the perturbation of $F(s,x,p)$.

In this paper, we investigate the converse for $d = 2$(i.e. $n = 1$). Precisely, we show that, for $d = 2$, if α does not satisfy (6), i.e. for any positive numbers γ,τ there exist $p,q \in Z \setminus \{0\}$ such that $|q\alpha - p| < \gamma q^{-\tau}$, the Z^2-invariant C^m-minimal foliation $x = U(t,\alpha t + \beta)$ can be destroyed by a small perturbation.

In order to prove it, we first transform the Euler equation (5) into a Hamiltonian system

$$\begin{aligned} \dot{x} &= & H_y(t,x,y) \\ \dot{y} &= & - & H_x(t,x,y) \end{aligned} \tag{7}$$

by the Legendre transformation

$$y = F_p(t,x,p), \quad H(t,x,y) = yp - F(t,x,p) \tag{8}$$

then $0 < \delta < F_{pp} \leq \delta^{-1}$ implies $0 < \delta \leq H_{yy} \leq \delta^{-1}$. because $F_{pp}H_{yy} = 1$ and $H(t,x,y)$ has period 1 in x and t. Moreover, a Z^2-invariant C^m-minimal foliation given by $x = U(t,\alpha t + \beta)$ of (4) corresponds to an invariant circle of the "time-one" map of (7) with rotation number $\alpha([1, 3, 5, 11])$. By (8), a perturbation of $F(t,x,p)$ corresponds to one of the Hamiltonian function $H(t,x,y)$ in $C^r(\Omega)$, so in order to destroy a minimal foliation we only need to destroy the corresponding invariant circle.

Our main results can be stated as follows

Theorem A *Let $F(t,x,p) \in C^r(\Omega)$ $(r \geq 2)$ with $\pi(\Omega) = T^2$ and $0 < \delta \leq F_{pp}(t,x,p) \leq \delta^{-1}$. Suppose $x = U(t, \alpha t + \beta)$ is a Z^2-invariant C^m-minimal foliation for some $m \geq 0$. If α does not satisfy the Diophantine condition (i.e. (6)), then we can destroy this minimal foliation by a small perturbation of the variational problem in $C^r(\Omega)$.*

As we stated above, the pertubations of a variational problem are equivalent to those of the corresponding Hamiltonian system. In fact, we will prove the following equivalent form of Theorem A instead of proving Theorem A directly.

Theorem B *Let $H(t,x,y) \in C^r(T^2 \times \mathbb{R})$ with $r \geq 2$ and $0 < \delta \leq H_{yy}(t,x,y) \leq \delta^{-1}$. Suppose the "time-one" map of (5) has an invariant circle Γ_α with rotation number α. If α does not satisfy the Diophantine condition, then we can destroy Γ_α by a sufficiently small perturbation of $H(t,x,y)$.*

2   Composited twist maps and Peierl's barrier

Let $f : \mathbb{R}^2 \to \mathbb{R}^2$ be a C^r map $(r \geq 1)$ and $\Pi_i : \mathbb{R}^2 \to \mathbb{R}, i = 1.2$ is a projection to the ith component. f is called **a monotone twist map**, if it satisfies the following

   i) $f(x+1, y) = f(x, y)$;

   ii) $\delta^{-1} > \frac{\partial}{\partial y} \Pi_1 f(x,y) > \delta$;

   iii) f is homotopic to identity;

   iv) f is area-preserving.

f is a monotone map, then we can represent f by a generating function $h(x, x_1) \in C^{r+1}(\mathbb{R}^2)$ with $h(x+1, x_1+1) = h(x, x_1)$ and $\delta \leq -h_{xx_1} \leq \delta^{-1}$. We have $(x_1, y_1) = f(x, y)$ iff $y = -h_x(x, x_1), y_1 = h_{x_1}(x, x_1)$. We call h **the first generating function**. Also we can represent f by another generating function $w(x_1, y)$, where $w(x_1, y) \in C^{r+1}(\mathbb{R}^2)$ and $-\delta^{-1} \leq w_{yy}(x_1, y) \leq -\delta$ and $w(x_1 + 1.y) = w(w_1, y)$. We have $(x_1, y_1) = f(x, y)$ iff $x = x_1 + w_y(x_1, y), y_1 = y + w_{x_1}(x_1.y)$. We call w **the second generating function**.

Definition 2 *If $f : \mathbb{R}^2 \to \mathbb{R}^2$ is a map homotopic to identity and there are monotone twist maps $f_1, \ldots, f_k$ with the same twist direction such that $f = f_k \circ \ldots \circ f_1$, then we call f a composited twist map.*

Now we generalize the concept of energy in [1] to the above composited twist map $f = f_k \circ \ldots \circ f_1$.

Definition 3 *A* **configuration** *is a bi-infinite sequence* $\{x_n\}_{n \in Z}$. *Suppose* f_i *has the first generating function* $h_i(x, y)$ *for* $i = 1, \ldots, k$, *respectively, and set* $\phi(\{x_n\}) = \sum_{n=N'}^{N} h_n(x_{n-1}, x_n)$, *where* $h_n = h_i$ *when* $n = i (\mathrm{mod}\ k)$. *We call* ϕ **the energy function of** f **with respect to** $h_i(x, y), i = 1, \ldots, k$ *and* N', N. *(Abbrev. the energy function of* f.*)*

Definition 4 A minimum energy configuration*(abbrev. m.e. configuration) is a configuration* $\{x_n\}$ *such that a change of a finite set of atoms will necessarily increase the energy, i.e.*

$$\sum_{n=N'}^{N} h_n(x_{n-1} + \delta_{n-1}, x_n + \delta_n) - h_n(x_{n-1}, x_n) \geq 0$$

for $N', N, \{\delta_n\}$ *with* $N' < N$, *whereas* $\sum_{n=N'}^{N} |\delta_n| > 0$: $\delta_n = 0$ *for* $n < N'$ *or* $n > N$.

Clearly, if $\{x_n\}$ is an m.e. configuration, then there is a $\{y_n\}$ given by $y_{n-1} = -\frac{\partial h_n}{\partial x_{n-1}}(x_{n-1}, x_n), y_n = \frac{\partial h_n}{\partial x_n}(x_{n-1}, x_n)$ such that $f_n(x_{n-1}, y_{n-1}) = (x_n, y_n)$, where $f_n = f_i$ when $n = i (\mathrm{mod}\ k)$. For $k = 1$, the m.e. configuration has been studied extensively in e.g. [1, 6]. Here a more general case is considerated. Some results about m.e. configurations in [1] can be generalized to the above case.

We state some results similar to those in [1].

Proposition 1 *Let* $\{x_n\}$ *and* $\{x_n'\}$ *be two m.e. configurations with respect to* $h_1, \ldots, h_k$, *then the sequence* $\{x_{kn} - x_{kn}'\}(-\infty < n < \infty)$ *has at most one node. (i.e.* $\{x_{kn} - x'_{kn}\}$ *changes sign only once. If* $\{x_{kn}\}$ *and* $\{x'_{kn}\}$ *are asymptotic at infinity as* n *goes to* $+\infty$ *or* $-\infty$ *or both, the point at infinity must be considered as a uniqe node.)*

Proposition 2 *Suppose* $\{x_n\}$ *is an m.e. configuration, then there exists a* α *such that*

$$\lim_{|n-n'| \to +\infty} \frac{x_{kn} - x_{kn'}}{n - n'} = \alpha.$$

Proposition 3 *For any value* α, *there exists an m.e. configuration* $\{x_n\}$ *such that*

$$\lim_{|n-n'| \to +\infty} \frac{x_{kn} - x_{kn'}}{n - n'} = \alpha.$$

We call the above α the **rotation number** of m.e. configuration $\{x_n\}$.

Proposition 4 *Suppose* $\{x_n\}$ *and* $\{x_n'\}$ *are two m.e. configurations with the same rotation number* α.

i) *If* $\alpha \in \mathbb{R} \setminus Q$, *then either* $x_n > x_n'$ *or* $x_n = x_n'$ *or* $x_n < x_n'$:

ii) If $\alpha = p/q$ a rational number with $(p,q) = 1$ and $x_{n+kq} = x_n + p, x'_{n+kq} = x'_n + p$, then either $x_n > x'_n$ or $x_n = x'_n$ or $x_n < x'_n$:

iii) If $\alpha = p/q$, then either $x_{n+kq} > x_n + p$ for any n or $x_{n+kq} = x_n + p$ for any n or $x_{n+kq} < x_n + p$ for any n.

By the above proposition, we can define the rotation symbol of m.e. configurations as follows

Definition 5 *Suppose $\{x_i\}$ is an m.e. configuration, if it has rotation number $\alpha \in \mathbb{R} \setminus Q$, then we say that it has the rotation symbol α; if it has a rational rotation number $\alpha = p/q$, then we define the rotation symbol as follows: the rotation symbol is $p/q+$, if $x_{i+qk} > x_i + p$ for any i; p/q, if $x_{i+qk} = x_i + p$; $p/q-$, if $x_{i+qk} < x_i + p$.*

Proposition 5 *Suppose $\Gamma \subset S^1 \times \mathbb{R}$ is an invariant curve of the composited twist map $f = f_k \circ \ldots \circ f_1$. which is not homotopic to zero in $S^1 \times \mathbb{R}$. then Γ can be parametrized as $\Gamma = \{(x,y) : y = u(x)$ where $u(x+1) = u(x)\}$.*

The proofs of the above propositions refer readers to [1, 5]. and a little change is needed in some cases.

To define Peierl's barrier, we need to recall the relevant definitions from [1, 6].

Define a symbol space $S = (R \setminus Q) \bigcup (Q+) \bigcup Q \bigcup (Q-)$. where $Q+$ ($Q-$) denotes the set of all symbols $p/q+$ ($p/q-$). Let M_α be the set of m.e. configuration $\{x_n\}$ with rotation symbol α of the composited twist map $f = f_n \circ f_{n-1} \circ \ldots \circ f_1$. By Proposition 4, M_α is totally ordered. Define A_α to be the projection of M_α by $pr : \{x_n\} \to x_0$. Then A_α is closed, moreover, A_α is invariant under shift $x \to x + 1$. In addition, suppose that $\bar{x}$ belongs to a complementary interval of A_α, then there exist two m.e. configurations $\{u_{n-}\}$ and $\{u_{n+}\}$ with the same rotation symbol α and $u_{n-} < u_{n+}$ such that $\bar{x} \in (u_{0-}, u_{0+})$, $u_{kn-}. u_{kn+} \in A_\alpha$ and $(u_{kn-}, u_{kn+}) \bigcap A_\alpha = \phi \; \forall n \in Z$.

Definition 6 *For any given rotation symbol $\alpha \in (R \setminus Q) \bigcup (Q+) \bigcup Q \bigcup (Q-)$, we define Peierl's barrier P_α as follows:*

- *when $\bar{x} \in A_\alpha$, we define $P_\alpha(\bar{x}) = 0$;*

- *when $\bar{x} \notin A_\alpha$, thus there exist two m.e. configurations $\{u_{n-}\}$ and $\{u_{n+}\}$ with the same rotation symbol α such that $\bar{x} \in \{u_{0-}, u_{0+}\}$. $u_{kn-}. u_{kn+} \in A_\alpha$ and $(u_{kn-}, u_{kn+}) \bigcap A_\alpha = \phi \; \forall n \in Z$, then we define*

$$P_\alpha(\bar{x}) = \min\{\sum_{i \in I} h_n(x_i, x_{i+1}) - h_n(u_{i-}. u_{i+1-}) : x_i \in (u_{i-}, u_{i+}),$$
$$x_0 = \bar{x}, \text{ and } x_{i+kq} = x_i + p. \text{ if } \alpha = p/q\},$$

where $h_n = h_i$ for $i = 1, 2, \cdots, k$ when $n = i(\bmod k)$. and the index set I is Z if $\alpha \in (R \setminus Q) \bigcup (Q+) \bigcup (Q-)$; I is $\{0, 1, \ldots, kq - 1\}$ if $\alpha \in Q$.

Proposition 6 *For a given rotation symbol α, $P_\alpha(x) = 0$ for any x iff f has an invariant circle with rotation number $\alpha*$, where $\alpha*$ is the number underlying $\alpha([6])$.*

Proof Refer to [6]. The details are omitted.   □

3   Destruction of invariant circles for composited twist maps

Let $f_1, \ldots, f_k$ be monotone twist maps with the first generating functions $h_1, \ldots, h_k$, respectively, which satisfying $\delta \leq -\frac{\partial^2}{\partial x_1 \partial x} h_i(x, x_1) \leq \delta^{-1}$ for $i = 1, \ldots, k$. We define the conjunction of $h_1, \ldots, h_k$ by

$$h_1 * h_2(x, x_1) = \min_\xi \{h_1(x, \xi) + h(\xi, x_1)\}$$

then $(h_1 * h_2) * h_3 = h_1 * (h_2 * h_3)$, so we can represent it by $h_1 * h_2 * h_3$. We define $h = h_1 * \ldots * h_k$.

It can be proved that h_i for $i = 1, \ldots, k$, along with h, satisfy

H1: $h(x, x_1) = h(x + 1, x_1 + 1)$ for any $x, x_1 \in \mathbb{R}$ and $h \in C(\mathbb{R}^2, \mathbb{R})$;

H2: $\lim_{\xi \to \infty} h(x, x + \xi) = \infty$ uniformly in x;

H3: There is a positive function $\rho \in C(\mathbb{R}^2, \mathbb{R})$ such that

$$h(\xi, x_1) + h(x, \xi_1) - h(x, x_1) - h(\xi, \xi_1) \geq \int_x^\xi \int_{x_1}^{\xi_1} \rho,$$

if $x < \xi$ and $x_1 < \xi_1$;

H4: There is a $\theta > 0$ such that $x \to \frac{1}{2}\theta x^2 - h(x, x_1)$ is convex for any x_1 and $x_1 \to \frac{1}{2}\theta x_1^2 - h(x, x_1)$ is convex for any x.

Also, by the definition of conjunction and its properties. we can prove that

i) If $\{x_n\}_{n \in Z}$ is an m.e. configuration of $h_1, \ldots, h_k$. then $\{x_{kn}\}_{n \in Z}$ is an m.e. configuration of h;

ii) If $\{u_n\}_{n \in Z}$ is an m.e. configuration of h. then there is an m.e. configuration $\{x_n\}_{n \in Z}$ of $h_1, \ldots, h_k$ such that $u_n = \{x_{kn}\}$ $\forall n \in Z$;

iii) $P_\alpha(x) = \bar{P}_\alpha(x)$, where $\bar{P}_\alpha(x)$ is the Peierl's barrier with respect to h.

Because h satisfies (H1) through (H4), so [7. Theorem 2.2] holds. We can state

Theorem 3.1 *Suppose $h_1, \ldots, h_k$ satisfy (H1) through (H4), then there is a real positive number $c > 0$ such that, if p/q is a rational number and ω is a rotation symbol, $|P_{p/q+} - P_\omega(x)| \leq c\theta|\omega^* q - p|$ in the case $\omega \geq p/q+$; $|P_{p/q-} - P_\omega(x)| \leq c\theta|\omega^* q - p|$ in the case $\omega \leq p/q-$, here ω^* is the number underlying ω ([7]).*

Next. we state the main theorem in this section. which shows how to destroy the invariant circle for composited maps. In fact. it is a generalization of [7, Theorem 2.1].

Set $DC = \{\omega : \omega$ satisfies the Diophantine condition $\}$. i.e. $\omega \in DC$ iff there are $c, N > 0$ such that $|q\omega - p| > cq^{-N}$ for all $p, q \in Z \setminus \{0\}$. An irrational number ω, which does not satisfy the Diophantine condition, is called a *Liouville number*, i.e. for any $c, N > 0$ there are $p, q \in Z \setminus \{0\}$ such that $|q\omega - p| < cq^{-N}$.

Theorem 3.2 *Suppose* $\alpha \in \mathbb{R} \setminus DC$, *and* $f \in C^r (r \geq 1)$ *is a composited twist map with* $f = f_k \circ \ldots \circ f_1$, *where* $f_1, \ldots f_k \in C^r$ *are monotone twist maps, then there is a* $f_1^\epsilon \in C^r$, *which is arbitrarily close to* f_1 *with respect to* C^r *topology such that there is no homotopically non-trivial g-invariant circle with rotation number* α *for* $g = f_k \circ \ldots \circ f_2 \circ f_1^\epsilon$.

Proof Let $\alpha \in \mathbb{R} \subset DC$. We want to show that for a positive r. $h_i \in C^r$ and any $\epsilon > 0$, there is a $h_1^\epsilon \in C^r$ with $\|h_1^\epsilon - h_1\|_{C^r} < \epsilon$ and $h_i(i = 2\ldots, k)$ remains the same, such that the Peierl's barrier $P_\alpha(x)$ for $g = f_k \circ \ldots \circ f_1^\epsilon$ does not vanish identically.

Note that $h_1, \ldots, h_k$ satisfy (H1) through (H4), so does h.

For given α, we choose p/q close to α with $p/q < \alpha$, then we choose an m.e. configuration $\{u_i\}$ of $h_1, \ldots, h_k$ with rotation symbol p/q. and $u_{i+qk} = u_i + p$. Also, we choose a complementary interval J of length $\geq q^{-1}$ to the set $\{u_{ik} + j : i, j \in Z\}$.

Let $\delta(x) \in C^\times(\mathbb{R}, \mathbb{R})$ such that

i) $Supp(\delta) = \bigcup \{J + i : i \in Z\}$;

ii) $\delta(x) = u(x)$;

iii) $\delta(x)$ is less than $\frac{1}{2}\epsilon$ in C^r-norm;

iv) $\delta(x) \geq c_r \epsilon / q^r$ for $x \in J_1$, where J_1 is the middle third of J and c_r is a constant depending on r only.

Let $\tilde{h}_1(x, x') = h_1(x, x') + \delta(x)$, then the Peierl's barrier $\tilde{P}_{p/q}$ associated with $\tilde{h} = \tilde{h}_1 * h_2 * \ldots * h_k$ satisfies $\tilde{P}_{p/q}(x) = P_{p/q}(x) + \delta(x)$ for $x \in J$, where $P_{p/q}(x)$ is the Peierl's barrier associated with $h = h_1 * \ldots * h_k$. By performing this procedure, we can actually destroy an invariant circle with rotation number p/q for any p/q.

Let $\tilde{h} = \tilde{h}_1 * h_2 * \ldots * h_k$. Set $\tilde{H}(x, x') = \tilde{h}(x, \xi_1) * \ldots * \tilde{h}(\xi_{i-1}, \xi_i) * \ldots * \tilde{h}(\xi_{q-1}, x' - p) + const$. Since $\{u_{kn}\}$ is an m.e. configuration of rotation symbol p/q of h, it also is one for $\tilde{h}$, then $\{u_{kn}\}_{n \in Z}$ is an m.e. configuration of rotation symbol 0 for $\tilde{H}(x, x')$. Without loss of generality, we may choose the constant such that $\tilde{H}(u_{kn}, u_{kn}) = 0$. By the choice of $\delta(x)$, there is an m.e. configuration $\{v_i\}$ with rotation symbol 0+ for H such that $v_i \to J^\pm$(the ends of the interval J) as $i \to \pm\infty$.

In order to obtain a suitable perturbation of h_1, we perform as follows.

By [6], there is a $\theta' > 0$ such that

$$c_r \epsilon / q^r \leq \theta' |v_{i+1} - v_{i-1}|$$

for any i, therefore, either $|v_{i+1} - v_i|$ or $|v_i - v_{i-1}|$ is greater than $c_r\epsilon/(2\theta'q^r)$ for any i, hence $\max|v_{i+1} - v_i| \geq c_r\epsilon/(2\theta'q^r)$. Choose i such that $|v_{i+1} - v_i| \geq c_r\epsilon/(2\theta'q^r)$, and denote the interval $[v_i, v_{i+1}]$ by I.

Let $\beta \in C^\infty(\mathbb{R}, \mathbb{R})$ satisfy

i) $Supp(\beta) \subset \bigcup\{I + i : i \in Z\}$;

ii) $\beta(x + 1) = \beta(x)$;

iii) $\beta(x)$ is less than $\frac{1}{2}\epsilon$ in C^r-norm;

iv) $\max \beta \geq c_r\epsilon(c_r\epsilon/(2\theta'q^r))^r$.

We define $\bar{h}_1(x, x') = \tilde{h}_1(x, x') + \beta(x)$, and $h_i(x, x'), i = 2, \ldots, k$ remain the same. for $i = 2, \ldots, k$. Let $\bar{P}_{p/q}(x)$ be the Peierl's barrier associated with $g = \bar{h}_1 \circ h_2^* \circ \ldots \circ h_k^*$. Then, we have $\bar{P}_{p/q}(x) = \check{P}_{p/q}(x) + \beta(x)$. And if $x \in \{x : \beta(x) \geq c_r\epsilon(c_r\epsilon/(2\theta'q^r))^r\}$, then

$$\bar{P}_{p/q+}(x) \geq \beta(x) \geq c_r\epsilon(c_r\epsilon/(2\theta'q^r))^r = c_r^{r+1}\epsilon^{r+1}q^{-r^2}(2\theta')^{-r}.$$

Hence by the above procedure we have destroyed the invariant circle with rotation number p/q. Further, for $\alpha \in \mathbb{R} \setminus (DC \bigcup Q)$, we can choose p/q close to α enough so that $c_r\theta'|\alpha q - p| \leq c_r^{r+1}\epsilon^{r+1}q^{-r^2}(2\theta')^{-r}$. Suppose $\alpha > p/q$, we can perform the procedure as before, getting $f_1^\epsilon = \bar{f}_1$ such that $\|f_1^\epsilon - \bar{f}_1\|_{C^r} < \epsilon$ with

$$\bar{P}_{p/q+}(x) \geq c_r^{r+1}\epsilon^{r+1}q^{-r^2}(2\theta')^{-r}.$$

Hence, $\bar{P}_\alpha(x) \geq \bar{P}_{p/q+} - c\theta'|\alpha q - p| \geq c_r^{r+1}\epsilon^{r+1}q^{-r^2}(2\theta')^{-r} - c\theta'|\alpha q - p| > 0$.

For the case $p/q < \alpha$, we can give the proof similarly. So we can destroy the invariant circles with rotation number $\alpha \in \mathbb{R} \setminus (DC \bigcup Q)$. □

Theorem 3.3 *Suppose $\alpha \in \mathbb{R} \setminus DC$, and $f \in C^r(r \geq 1)$ is a composited twist map, then there is a $f^\epsilon \in C^r$, which is arbitrarily close to f with respect to C^r topology such that there is no homotopically non-trivial f^ϵ-invariant circle with rotation number α.*

Proof By the assumption, there are monotone twist maps $f_1, \ldots f_k$ such that $f = f_k \circ \ldots \circ f_1$. By Theorem 3.2, we can choose f_1^ϵ arbitrarily close to f_1 with respect to C^r-topology such that $f^\epsilon = f_k \circ \ldots \circ f_2 \circ f_1^\epsilon$ has no homotopically non-trivial f^ϵ-invariant circle with rotation number α. By the properties of composited functions, f^ϵ can be arbitrarily close to f with respect to C^r-topology. □

4   Proofs of the main theorems

In this section, we give the proof of Theorem B by applying Theorem 3.2.

Let $H(t,x,y) \in C^r(T^2 \times \mathbb{R})$, $H_{yy} > 0$. Consider the differential equation

$$\begin{aligned}
\dot{x} &= \quad H_y(t,x,y), \\
\dot{y} &= - \quad H_x(t,x,y).
\end{aligned} \tag{9}$$

Denote $x = \phi(t,t_0,x_0,y_0), y = \psi(t,t_0,x_0,y_0)$ is the solution of (9) taking the initial condition: $t = t_0, x = x_0, y = y_0$. Suppose Γ is an invariant circle of map $x = \phi(1,0,x_0,y_0), y = \psi(1,0,x_0,y_0)$ with rotation number α, then Γ can be represented by $\{(x,y(x))\}$, where $y(x)$ is continuous and of period 1.

By the hypothese about $H(t,x,y)$, for N large enough, $(x,y) \to (\phi(i/N,(i-1)/N,x,y), \psi(i/N,(i-1)/N,x,y))$, $i = 1,2,\ldots,N$ are monotone twist maps with variable boundaries. Suppose that they are well defined on the bounded annulus $A_i \equiv \{(x,y) : x \in \mathbb{R}, y \in [a_i,b_i]\}, i = 1,\ldots,N$, respectively. Without loss of generality, we can choose $A_i, i = 1,\ldots,N$ such that $f_i(A_i) \subset A_{i+1}$ and $A_i \subset A_{i+1}$ for $i = 1,\ldots,N$ and, of course, $\Gamma \subset A_1$, where $f_i(x,y) = (\phi(i/N,(i-1)/N,x,y), \psi(i/N,(i-1)/N,x,y))$. For convenience, we extend $f_i(x,y) : A_i \to A_{i+1}$ to $\tilde{f}_i(x,y) : \mathbb{R}^2 \to \mathbb{R}^2$ for $i = 1,\ldots,N$, respectively, such that $\tilde{f}_i(x,y)$ is a monotone twist map and $\tilde{f}_i|_{A_i} = f_i$ for $i = 1,\ldots,N$. We still denote $\tilde{f}_i$ by f_i. Note that we may set $a_1 = -n, b_1 = n$ for sufficiently large $n > 0$.

Let $h_i(x,x')$ be the generating function of f_i for $i = 1,\ldots,N$, respectively, then each h_i satisfies (H1) through (H4).

Since $f_N \circ \ldots \circ f_i \circ \ldots \circ f_1$ has an invariant circle with rotation number α, by using Theorem 3.2, we can choose f_1^ϵ with the generating function h_1^ϵ, which is close to f_1 enough in C^r-norm such that $g = f_N \circ \ldots \circ f_i \circ \ldots \circ f_2 \circ f_1^\epsilon$ has no invariant circle with rotation number α. And f_1^ϵ can be chosen such that $h_1^\epsilon(x,x') - h_1(x,x') \equiv v(x)$.

Let $\eta(t) : \mathbb{R} \to [0,1]$ be a C^∞-function satisfying $\eta(t) = 0$ for $t \in [0,1/(3N)]$ and $\eta(t) = 1$ for $t \in [2/(3N),1/N]$, and $\eta(t+1) = \eta(t)$. Here by the choice of $v(x)$, we may assume that $\eta(t)v(x)$ is small enough in C^r-norm. Let $h_1(t,x,x')$ be the first generating function of $(\phi,\psi)(t,0,x,y)$ for $t \in [0,1/N]$, then $h_1(t,x,x') + \eta(t)v(x)$ also is a generating function for some monotone twist maps. We denote the new first generating function by $h_1^\epsilon(t,x,x')$.

We transform $h_1(t,x,x')$ into the second generating function $x'y + w_1(t,x',y)$ for $t \in (0,1/N]$, where, for fixed t, $x = x' + \frac{\partial}{\partial y}w(t,x',y), y' = y + \frac{\partial}{\partial x'}w_1(t,x',y)$. Let $\tilde{H}_1(t,x',y') = -\frac{\partial}{\partial t}w_1(t,x',y)$ for $t \in [0,1/N]$. If $x = x(t,x_0,y_0), y = y(t,x_0,y_0)$ is the solution of equation $\dot{x} = \tilde{H}_{1y}(t,x,y), \dot{y} = -\tilde{H}_{1x}(t,x,y)$ satisfying the initial condition $t = 0, x = x_0, y = y_0$, then $(x(1/N,x_0,y_0), y(1/N,x_0,y_0)) = f_1^\epsilon(x_0,y_0)$.

Now we perturb $H(t,x,y)$ for $(x,y) \in A_1$ as follows

$$\tilde{H}(t,x,y) = \begin{cases} \tilde{H}_1(t,x,y) & \text{if } t \in [0,1/N]; \\ H(t,x,y) & \text{if } t \in (1/N,1]. \end{cases}$$

and extend $\tilde{H}(t, x, y)$ periodically in t. Further, we perturb the Hamiltonian function globally. Note that A_1 may be $S^1 \times [-n, n]$ for an n large enough and A_2 may be close to A_1, let us choose $\xi(y) \in C^\infty(\mathbb{R}, \mathbb{R})$ with $\xi(y) = 1$ for $|y| \leq \frac{1}{2}n$ and $\xi(y) = 0$ for $|y| \geq n$. we define

$$\bar{H}(t, x, y) = \tilde{H}(t, x, y)\xi(y) + H(t, x, y)(1 - \xi(y))$$

Then, for the new Hamiltonian $\bar{H}(t, x, y)$, there is no invariant circle with rotation number α. This means that for suitable A_1 and A_2, $n, N, \xi(y)$, $\bar{H}(t, x, y)$ is close to $H(t, x, y)$ in C^r-norm. Thus we have completed the proof of Theorem B. $\quad\square$

Acknowledgments. The author would like to express his gratitude to the referee for valuable suggestions which helped to improve the formulation.

References

[1] S. AUBRY, P.Y. LA DAERON, The discrete Frenkel-Kontorova model and its extension. Physica 8D (1983) 381–422.

[2] V. BANGERT, An uniqueness theorem for Z^n-periodic variational problems, *Comm. Math. Helv.* **62** (1987) 511–531.

[3] X. HUANG, Composition of twist maps and its application to $\ddot{x} + f(x, t) = 0$, Ann. of Diff. Eqs. **8** (1) (1992) 25–32.

[4] X. HUANG, A remark on fixed points and Morse decomposition of twist maps, *Mathematical Biquarterly, J. Nanjing University*, Vol.9. No. 2 (1992) 216–221.

[5] J. N. MATHER, More Denjoy minimal sets for area preserving diffeomorphism, *Comm. Math. Helv.* **60** (1985) 508–557.

[6] J. N. MATHER, Modules of continuity for Peierl's barrier, in *"Periodic solutions of Hamiltonian systems and related topics*, ed. P.H. Robinnowitz et al, *NATO ASI Series C209*, D. Reidel, Dordrecht (1987) 177–202.

[7] J.N. MATHER, Destruction of invariant circles, *Ergod. Th. Dyn. Sys.* **8*** (1988) 199–214.

[8] L. MORA, Birkhoff-Henon attractors for dissipative perturbations of area-preserving twist maps, *Ergod. Th. Dyn. Sys.* **14** (1994) 807–815.

[9] J. MOSER, A stability theorem for minimal foliations on torus, *Ergod. Th. Dyn. Sys.* **8*** (1988) 251–281.

[10] J. MOSER, Minimal solutions of variational problems, *Ann. Inst. H. Poincaré Anal. Nonlinéaire* **3**(1986) 229–272.

[11] J. MOSER, Recent developments in the theory of Hamiltonian systems, *SIAM Review* **28** (1986) 459–485.

[12] W. RUDIN, "Real and Complex Analysis", McGraw Hill, New York, 1966.

Progress in Nonlinear Differential Equations
and Their Applications, Vol. 19
© 1996 Birkhäuser Verlag Basel/Switzerland

Local and Global Existence of Multiple Waves Near Formal Approximations

Xiao-Biao Lin*

Keywords: singular perturbation, asymptotic expansion, reaction-diffusion system, internal layers, spatial shadowing lemma

Introduction

The formation of multiple wave fronts is important in applications of singularly perturbed parabolic systems. These solutions can be effectively constructed by formal asymptotic methods. When truncated to a certain order in ϵ, they become formal approximations of solutions to the given system. The precision of a formal approximation is judged by the smallness of the residual error in each regular and singular layer and the jump error between adjacent layers.

The purpose of this paper is to introduce *Spatial Shadowing Lemmas* that help to construct exact solutions near the formal approximations. The *Shadowing Lemma* was first developed for discrete mappings in $\mathbb{R}^n$, see [8]. It has been extended to continuous flows governed by ODEs [11], and semiflows governed by abstract parabolic equations [2]. We will call these *Temporal Shadowing Lemmas* since the dynamical systems considered there evolve in time.

The *Temporal Shadowing Lemma* has been used to construct exact solutions for singularly perturbed ODEs [11]. However, it cannot be applied directly to singularly perturbed parabolic equations. For formal approximations of multiple waves, the jumps between adjacent layers are functions of t and they occur along the x-direction. Since parabolic equations cannot be solved in the x-direction, therefore, they do not define a dynamical system in the spatial direction.

To solve this problem, we use an idea motivated by the works of Kirchässner and Renardy [10, 15]. We find stable and unstable subspaces of the trace space so that the parabolic system can be solved forward and backward in the spatial direction. Thus the jumps along the lateral common boundaries can be corrected using the technique of the usual *Shadowing Lemma* in abstract spaces.

---

*Department of Mathematics, North Carolina State University, Raleigh, North Carolina 27695–8205, USA, email xblin@xblsun.math.ncsu.edu, Research partially supported by NSF grant DMS9002803 and DMS9205535

This paper is divided into two parts. In the Part I. we show that for a general parabolic system, if a formal approximation is precise enough. then there is an exact solution near the formal approximation for at least a short time. The result obtained here applies to various systems including reaction-diffusion equations, Cahn-Hilliard equations [1], and viscous profile of conservation laws. In Part II, we show that with additional restrictions, the process in Part I can be repeated to obtain global solutions if the formal approximation is a global one. Examples include reaction-diffusion equations and phase field equations [7].

Part I. Local Existence of Multiple Waves

1. Consider a general singularly perturbed parabolic system,

$$\epsilon u_t + (-\epsilon^2)^m D_x^{2m} u = f(u, \epsilon u_x, \cdots, (\epsilon D_x)^{2m-1} u, x, \epsilon). \quad u \in \mathbb{R}^n. \; x \in \mathbb{R}, \quad (1)$$

where f is C^∞ with bounded derivatives in all the variables. Assume that the system has regular and internal layers located alternatively along the x axis. For simplicity, we solve the system for $x \in \mathbb{R}$, with no boundary conditions other than $u \in H^{2m,1}$. Assume that there are curves $\Gamma^i = \{(x,t) : x = x^i(t). \; t \in [0, \Delta t]\}$, $i \in \mathbb{Z}$, that divide the domain $x \in \mathbb{R}$, $t \in [0, \Delta t]$ into regular or singular layers Σ^i, each is between Γ^{i-1} and Γ^i.

Assume that a formal approximation is given, piecewise continuous, with $u = \tilde{u}^i(x, t, \epsilon)$ in Σ^i. Using the stretched variables $\xi = x/\epsilon. \; \tau = t/\epsilon$. let $w^i(\xi, \tau, \epsilon) = \tilde{w}^i(\epsilon\xi, \epsilon\tau, \epsilon)$ which depends slowly on τ. Assume that $w^i \in H^{2m,1}(\Sigma^i)$. Let $W^i = (w^i, D_\xi w^i, \cdots, D_\xi^{2m-1})^\tau$. The error terms in the followings are $-g^i$ and $-\delta^i$,

$$w_\tau^i + (-1)^m D_\xi^{2m} w^i - f(w^i, \cdots, \epsilon\xi, \epsilon) = -g^i, \quad \text{in } \Sigma^i. \quad (2)$$

$$W^i(\xi^i, \tau) - W^{i+1}(\xi^i, \tau) = -\delta^i(\tau), \quad \text{at } \Gamma^i. \quad (3)$$

Let an exact solution be $u^i + w^i$ in Σ^i. Let $U^i = (u^i, D_\xi u^i, \cdots, D_\xi^{2m-1} u^i)^\tau$. Let $\tau \in I = [0, \Delta\tau]$ where $\Delta\tau$ is independent of ϵ. The domain $[0. \Delta\tau]$ corresponds to a short time interval $[0, \epsilon\Delta\tau]$ in the t variable. Assume that ϵ is small so that a near identity change of coordinates in Σ^i can be used to straighten the boundaries Γ^{i-1} and Γ^i. In the following, we assume that $\xi^i = x^i/\epsilon$ is independent of τ, but may depend on ϵ, $\Omega^i = (\xi^{i-1}, \xi^i)$, and $\Sigma^i = \Omega^i \times I$. With the new coordinates introduced above, linearizing (1) around w^i at the fixed time $\tau = 0$. we have

$$u_\tau^i + (-1)^m D_\xi^{2m} u^i - \sum_{j=1}^{2m-1} A_j^i(\xi) D_\xi^j u^i = \mathcal{N}^i(u^i. g^i. \epsilon), \quad \text{in } \Sigma^i, \quad (4)$$

$$U^i(\xi^i, \tau) - U^{i+1}(\xi^i, \tau) = \delta^i(\tau), \quad \text{at } \Gamma^i. \quad (5)$$

$$u^i(\xi, 0) = u_0^i(\xi), \quad \text{in } \Omega^i. \quad (6)$$

Here $A_j^i(\xi) = D_j f(w^i(\xi, 0, \epsilon), D_\xi w^i(\xi, 0, \epsilon), \cdots)$ is the partial derivative of f with respect to the j-th variable. $u_0^i \in H^m(\Omega^i)$. $\mathcal{N}^i$ depends slowly on τ and $|\mathcal{N}^i|_{L^2(\Sigma^i)}$ $= O(|u^i|^2_{H^{2m,1}(\Sigma^i)} + |g^i|_{\Sigma^i} + |\epsilon\Delta\tau||u^i|_{H^{2m,1}(\Sigma^i)})$.

We look for a sequence of solutions $\{u^i\}_{-\infty}^{\infty}$ with $u^i \in H^{2m,1}(\Sigma^i)$. Let $H^k(I)$ be the usual Sobolev space and let $H_0^k(I)$ be the completion of C^∞ functions, which are zero in a neighborhood of $\tau = 0$, in the H^k norm. Define the product spaces

$$B^m(I) = \Pi_{k=0}^{2m-1} H^{1-\frac{2k+1}{4m}}(I), \quad B_0^m(I) = \Pi_{k=0}^{2m-1} H_0^{1-\frac{2k+1}{4m}}(I).$$

From the Trace Theorem, [14], the mapping

$$\xi \to U(\xi, \cdot), \quad \Omega^i \to B^m(I).$$

is continuous with

$$|U^i(\xi, \cdot)|_{B^m(I)} \le C|u^i|_{H^{2m,1}(\Sigma^i)}.$$

Therefore, $\delta^i \in B^m(I)$ in (3) and (5). Let $U_0^i = (u_0^i, \cdots, D_\xi^{m-1} u_0^i)^\tau$. Let π_j be the projection to the first j-tuples in the product space $\Pi_{j=1}^{2m} \mathbb{R}^n$, i.e., $\pi_j(u_1, \cdots, u_{2m}) = (u_1, \cdots, u_j)$. A compatibility condition is also assumed on the initial data and the jumps,

$$U_0^i(\xi^i) - U_0^{i+1}(\xi^i) = \pi_m \delta^i(0). \tag{7}$$

Notice that only the first m components of δ^i have well defined traces at $\tau = 0$.

We now discuss the method of solving (4)–(6) with the compatibility (7) in Sections **2**–**5**.

2. Consider

$$u_\tau + (-1)^m D_\xi^{2m} u = 0, \quad \xi \in \mathbb{R}, \ \tau \in \mathbb{R}^+,$$

with $u(\xi, 0) = 0$. Applying the Laplace transform, we have the so called dual system,

$$\hat{U}_\xi = J(s)\hat{U} = \begin{pmatrix} 0 & I & 0 & \cdots & 0 \\ 0 & 0 & I & \cdots & 0 \\ & & & \cdots & \\ (-1)^{m+1}s & 0 & 0 & \cdots & 0 \end{pmatrix} \hat{U}. \tag{8}$$

The matrix $J(s)$ has $2m$ eigenvalues $\lambda = [(-1)^{m+1}s]^{\frac{1}{2m}}$.

Consider a sector in $\mathbb{C}$,

$$S_\theta(M) = \{s : |s| \ge M, |arg(s)| \le \theta\}, \quad M > 0, \ \pi/2 < \theta < \pi.$$

When $s \in S_\theta(M)$, the eigenvalues λ are in $2m$ disjoint sectors of $\mathbb{C}$, with $|\text{Re}\lambda| \ge \cos(\frac{\pi-\theta}{2m}) \sqrt[2m]{|s|}$. There are m eigenvalues with positive real parts and m with negative real parts. Each eigenvalue has an n-dimensional eigenspace spanned by $(u, \lambda u, \cdots, \lambda^{2m-1} u)^\tau$, $u \in \mathbb{R}^n$.

Let $E^{m,\nu}(s)$ be the Banach space of points in $\mathbb{R}^{2mn}$ with an s-dependent norm,

$$|(u_0, u_1, \cdots, u_{2m-1})|_{E^{m,\nu}(s)} = \sum_{j=0}^{2m-1} (1 + |s|^{\nu + \frac{j}{2m}})|u_{2m-1-j}|_{\mathbb{R}^n}.$$

We actually will only use $\nu = 0$ or $\frac{1}{4m}$ in this paper. Let P_s and P_u be the projections in $\mathbb{R}^{2mn}$ to the stable and unstable spaces of $J(s)$, $s \in \mathcal{S}_\theta(M)$. The projections are of rank mn and can be constructed using eigenvalues and eigenvectors. Using the $E^{m,\nu}(s)$ norm, we can show that there exist K, α_1, $\alpha > 0$, such that

$$\begin{aligned}
|e^{J(s)\xi}P_s|_{E^{m,\nu}(s)} &\le Ke^{-\alpha_1 \sqrt[2m]{|s|}\xi} &\le Ke^{-\alpha(1 + \sqrt[2m]{|s|})\xi}, \quad \xi \ge 0, \\
|e^{J(s)\xi}P_u|_{E^{m,\nu}(s)} &\le Ke^{\alpha_1 \sqrt[2m]{|s|}\xi} &\le Ke^{\alpha(1 + \sqrt[2m]{|s|})\xi}, \quad \xi \le 0.
\end{aligned} \quad (9)$$

3. Consider

$$u_\tau + (-1)^m D_\xi^{2m}u - \sum_{j=1}^{2m-1} A_j^i(\xi)D_\xi^j u = 0, \quad \xi \in [\xi^{i-1}, \xi^i]. \quad (10)$$

with $u(\xi, 0) = 0$. Using the Laplace transform, we have a dual system,

$$\hat{U}_\xi = J(s)\hat{U} + (-1)^m \begin{pmatrix} 0 & 0 & \cdots & 0 \\ & & \cdots & \\ A_0^i(\xi) & A_1^i(\xi) & \cdots & A_{2m-1}^i(\xi) \end{pmatrix} \hat{U}. \quad (11)$$

Let $T^i(\xi, \zeta, s)$ be the solution matrix for system (11). System (11) is said to have an exponential dichotomy in $E^{m,\nu}(s)$ for $\xi \in [\xi^{i-1}, \xi^i]$, $s \in \mathcal{S}_\theta(M)$ if there exist projections $P_s^i(\xi, s) + P_u^i(\xi, s) = I$ in $E^{m,\nu}(s)$, continuous in ξ and analytic in s, and positive constants K, α, such that

$$\begin{aligned}
|T^i(\xi, \zeta, s)P_s^i(\zeta, s)|_{E^{m,\nu}(s)} &\le Ke^{-\alpha(1 + \sqrt[2m]{|s|})|\xi - \zeta|}, \quad \xi \ge \zeta, \\
|T^i(\xi, \zeta, s)P_u^i(\zeta, s)|_{E^{m,\nu}(s)} &\le Ke^{-\alpha(1 + \sqrt[2m]{|s|})|\xi - \zeta|}, \quad \xi \le \zeta.
\end{aligned}$$

Suppose that $\sup_{\xi \in [a,b], 0 \le k \le 2m-1} |A_k^i(\xi)| \le C$. From the Roughness of Exponential Dichotomy Theorem, [3], which is also valid in the Banach space $E^{m,\nu}(s)$, we find that there exists $M = M(C) > 0$, sufficiently large, such that (11) has an exponential dichotomy in $E^{m,\nu}(s)$ for $\xi \in [\xi^{i-1}, \xi^i]$ and $s \in \mathcal{S}_\theta(M)$. On the other hand, if M is fixed, then there exists $C = C(M) > 0$, sufficiently small, such that the same conclusion holds.

A function $f(s)$ is in the Hardy-Lebesgue class $\mathcal{H}(\gamma)$, $\gamma \in \mathbb{R}$, if
(i) $f(s)$ is analytic in $\text{Re}(s) > \gamma$;
(ii) $\{\sup_{\sigma > \gamma}(\int_{-\infty}^{\infty} |f(\sigma + i\omega)|^2 d\omega)\}^{1/2} < \infty$.
$\mathcal{H}(\gamma)$ is a Banach space with the norm defined by the left side of (ii). Based on the Paley-Wiener Theorem, [16], if $e^{-\gamma t}f(t) \in L^2(\mathbb{R}^+)$, then $\hat{f}(s) \in \mathcal{H}(\gamma)$, vice versa.

For $k \geq 0$ and $\gamma \in \mathbb{R}$, define a Banach space

$$\mathcal{H}^k(\gamma) = \{u(s) \mid u(s) \text{ and } (s - \gamma)^k u(s) \in \mathcal{H}(\gamma)\},$$
$$|u|_{\mathcal{H}^k(\gamma)} = |u|_{\mathcal{H}(\gamma)} + |(s - \gamma)^k u|_{\mathcal{H}(\gamma)}.$$

For any $\gamma \in \mathbb{R}$, $k \geq 0$, there exists $C = C(\gamma, k)$ such that

$$C^{-1}(1 + |s|^k) \leq 1 + |s - \gamma|^k \leq C(1 + |s|^k).$$

Therefore an equivalent norm for $\mathcal{H}^k(\gamma)$ is

$$|u|^2_{\mathcal{H}^k(\gamma)} = \sup_{\sigma > \gamma} \int_{-\infty}^{\infty} |u(\sigma + i\omega)|^2 (1 + |\sigma + i\omega|^{2k}) d\omega.$$

It can be shown that if $e^{-\gamma t} f(t) \in H_0^k(\mathbb{R}^+)$, then $\hat{f}(s) \in \mathcal{H}^k(\gamma)$.

We often use Banach spaces of functions with norms weighted by $e^{-\gamma\tau}$. For example $B_0^m(\mathbb{R}^+, \gamma) = \{\phi : e^{-\gamma\tau}\phi(\tau) \in B_0^m(\mathbb{R}+)\}$ with obvious norms. Other spaces like $L^2(\mathbb{R} \times \mathbb{R}^+, \gamma)$, etc. can be defined similarly.

Let $\mathcal{K}^m(\gamma) = \Pi_{k=0}^{2m-1} \mathcal{H}^{1-\frac{2k+1}{4m}}(\gamma)$. From the definition of $B_0^m(\mathbb{R}^+, \gamma)$, we see that

$$\phi \in B_0^m(\mathbb{R}^+, \gamma) \Leftrightarrow \hat{\phi} \in \mathcal{K}^m(\gamma).$$

Suppose now the system (11) has an exponential dichotomy in $E^{m, \frac{1}{4m}}(s)$ for $\xi \in [\xi^{i-1}\xi^i]$ and $\mathrm{Re}\, s > \gamma$. An equivalent norm for $\mathcal{K}^m(\gamma)$ is

$$|\phi|_{\mathcal{K}^m(\gamma)} \sim \sup_{\sigma > \gamma} \left[\int_{-\infty}^{\infty} |\phi(s)|_{E^{m, \frac{1}{4m}}(s)} d\omega \right]^{1/2}, \quad s = \sigma + i\omega.$$

Based on this, one can show that (11) also has an exponential dichotomy in $\mathcal{K}^m(\gamma)$ for $\xi \in [\xi^{i-1}, \xi^i]$. The definition for exponential dichotomies in a Banach space like $\mathcal{K}^m(\gamma)$ is standard, and can be found in [9]. Using the definitions of the function spaces and exponential dichotomies, it is not hard to show the following. (See [13] for the case $m = 1$.)

Lemma 1 *For any $\phi \in B_0^m(\mathbb{R}^+, \gamma)$, consider*

$$u_1 = \pi_1 \mathcal{L}^{-1}(T^i(\xi, \xi^{i-1}; s) P_s^i(\xi^{i-1}, s)\hat{\phi}(s)), \quad \xi \geq \xi^{i-1},$$
$$u_2 = \pi_1 \mathcal{L}^{-1}(T^i(\xi, \xi^i; s) P_u^i(\xi^i, s)\hat{\phi}(s)), \quad \xi \leq \xi^i.$$

If $\sup_{\xi \in [\xi^{i-1}, \xi^i], 0 \leq k \leq 2m-1} |A_k(\xi)| < \infty$, then $u_j \in H^{2m,1}([\xi^{i-1}, \xi^i] \times \mathbb{R}^+, \gamma)$, $j = 1, 2$ and is a solution to (10) with $u_j(\xi, 0) = 0$. Moreover

$$|u_j|_{H^{2m,1}(\gamma)} \leq C|\phi|_{B_0^m(\mathbb{R}^+, \gamma)}.$$

4. A sequence of functions $f^i \in F^i$, $i \in \mathbb{Z}$, where F^i is a Banach space, will be denoted by $\{f^i\}$. Define the norm $|\{f^i\}|_{F^i} = \sup_i \{|f^i|_{F^i}\}$.

Consider a linear system,

$$u_\tau^i + (-1)^m D_\xi^{2m} u^i - \sum_{j=1}^{2m-1} A_j^i(\xi) D_\xi^j u^i = h^i(\xi, \tau). \quad \text{in } \Sigma^i. \tag{12}$$

with jump conditions (5), initial conditions (6) and compatibilities (7). Assume that the coefficients $A_j^i(\xi)$ are extended to $\xi \in \mathbb{R}$ by constants outside Ω^i. After the extension, assume that the associated homogeneous dual system (11) has exponential dichotomies in $E^{m,\nu}(s)$, $\nu = 0, 1/4m$, for $\xi \in \mathbb{R}$ and $\mathrm{Re}(s) > \gamma$. Assume that $g^i \in L^2(\Sigma^i, \gamma)$, $\delta^i \in B^m(\mathbb{R}^+, \gamma)$ and $u_0^i \in H^m(\Omega^i)$. In this section $\Sigma^i = \Omega^i \times I$ with $I = \mathbb{R}^+$. Assume

$$|\{\delta^i\}|_{B^m(\mathbb{R}^+,\gamma)} + |\{h^i\}|_{L^2(\Sigma^i,\gamma)} + |\{u_0^i\}|_{H^m(\Omega^i)} < \infty.$$

We look for solutions $u^i \in H^{2m,1}(\Sigma^i)$, $i \in \mathbb{Z}$.

By a standard method, we can continuously extend h^i and u_0^i to $\xi \in \mathbb{R}$ so that

$$|h^i|_{L^2(\mathbb{R} \times \mathbb{R}^-)} \leq C|h^i|_{L^2(\Sigma^i)}, \quad |u_0^i|_{\in H^m(\mathbb{R})} \leq C|u_0^i|_{H^m(\Omega^i)}.$$

We first solve (12) in the domain $\mathbb{R} \times \mathbb{R}^+$, with an initial condition u_0^i but no jump conditions. Denote the solution by $\bar{u}^i$. From the existence of exponential dichotomy in $E^{m,0}(s)$. for $\mathrm{Re}(s) > \gamma$, we can prove that (12) defines a sectorial operator $\mathcal{A}^i$ in $L^2(\mathbb{R})$. Moreover,

$$|\lambda - \mathcal{A}^i|_{L^2(\mathbb{R})}^{-1} \leq \frac{C}{1 + |\lambda|}, \quad \mathrm{Re}\lambda > \gamma.$$

The proof of the case $m = 1$ can be found in [13]. The general case can be proved similarly. It is than easy to see that $\bar{u}^i$ can be solved by the analytic semigroup $e^{\mathcal{A}^i \tau}$ and the variation of constant formula. We have

$$|\bar{u}^i|_{H^{2m,1}(\mathbb{R} \times \mathbb{R}^+, \gamma)} \leq C(|h^i|_{L^2(\mathbb{R} \times \mathbb{R}^+, \gamma)} + |u_0^i|_{H^m}).$$

Let $\bar{U}^i = (\bar{u}^i, D_\xi \bar{u}^i, \cdots, D_\xi^{2m-1} \bar{u}^i)^\tau$, and $\bar{\delta}^i = \bar{U}^i(\xi^i, \cdot) - \bar{U}^{i+1}(\xi^i, \cdot)$. Then

$$|\bar{\delta}^i|_{B^m(\mathbb{R}^-,\gamma)} \leq C(|\{h^i\}|_{L^2(\Sigma^i,\gamma)} + |\{u_0^i\}|_{H^m(\Omega^i)}).$$

Let the solution to (12). (5)–(7) be $u^i = \bar{u}^i + v^i$. The function v^i satisfies (12) with $h^i = 0$ and $v^i(\xi, 0) = 0$. Let $\eta^i = \delta^i - \bar{\delta}^i$. Then $\eta^i \in B_0^m(\mathbb{R}^-, \gamma)$. Let $V^i = (v^i, D_\xi v^i, \cdots, D_\xi^{2m-1} v^i)^\tau$. Then the dual systems for V^i are

$$\hat{V}_\xi^i = J(s)\hat{V}^i + (-1)^m \begin{pmatrix} 0 & 0 & \cdots & 0 \\ & & \cdots & \\ A_0^i(\xi) & A_1^i(\xi) & \cdots & A_{2m-1}^i(\xi) \end{pmatrix} \hat{V}^i. \tag{13}$$

$$\hat{V}^i(\xi^i, \cdot) - \hat{V}^{i+1}(\xi^i, \cdot) = \hat{\eta}^i. \tag{14}$$

We want to solve (13) and (14) with $\hat{\eta}^i \in \mathcal{K}^m(\gamma)$.

$$|\hat{\eta}^i|_{\mathcal{K}^m(\gamma)} \leq C(|\delta^i|_{B^m(\mathbb{R}^+,\gamma)} + |\{h^i\}|_{L^2(\Sigma^i,\gamma)} + |\{u_0^i\}|_{H^m}).$$

For two subspaces $M \oplus N = \mathbb{R}^{2mn}$, denote by $P(M,N)$ the projection with the range and kernel being M and N respectively. Assume that at each ξ^i, we have

$$\mathcal{R}P_u^i(\xi^i,s) \oplus \mathcal{R}P_s^{i+1}(\xi^i,s) = \mathbb{R}^{2mn}.$$

Here $\mathcal{R}$ stands for the range of an operator, and P_u^i and P_s^{i+1} are projections associated to the exponential dichotomies in Ω^i and Ω^{i+1} respectively. Assume that the norms of the projections associated with the above splitting are uniformly bounded with respect to $i \in \mathbb{Z}$, $\mathrm{Re}\,s > \gamma$ in the $E^{m,\nu}(s)$ norm. Notice that the assumption is valid if we choose $\gamma > 0$ to be sufficiently large, due to the Roughness of Exponential Dichotomies again.

We now solve (13), (14) by an iteration method that is used to prove the *Temporal Shadowing Lemma*, [11]. As a first approximation, let

$$\phi_u^i(\xi^i,s) = P(\mathcal{R}P_u^i(\xi^i,s), \mathcal{R}P_s^{i+1}(\xi^i,s))\hat{\eta}^i(s),$$
$$\phi_s^i(\xi^{i-1},s) = -P(\mathcal{R}P_s^i(\xi^{i-1},s), \mathcal{R}P_u^{i-1}(\xi^{i-1},s))\hat{\eta}^{i-1}(s),$$
$$\phi^i(\xi,s) = T^i(\xi,\xi^{i-1},s)\phi_s^i(\xi^{i-1},s) + T^i(\xi,\xi^i,s)\phi_u^i(\xi,s).$$

From Lemma 1, $\pi_1 \mathcal{L}^{-1}(\phi^i(\xi,s)) \in H^{2m,1}(\Omega^i \times \mathbb{R}^+,\gamma)$ and is a solution for (12), with $h^i = 0$. However, at ξ^i, the jump is not exactly $\hat{\eta}^i$. Let $\phi^i(\xi^i,\cdot) - \phi^{i+1}(\xi^i,\cdot) = \hat{\eta}^i - \hat{\eta}_1^i$. Then

$$\hat{\eta}_1^i(s) = T^{i+1}(\xi^i,\xi^{i+1},s)\phi_u^{i+1}(\xi^i,s) - T^i(\xi^i,\xi^{i-1},s)\phi_s^i(\xi^{i-1},s).$$
$$|\{\hat{\eta}_1^i\}|_{\mathcal{K}^m(\gamma)} \leq Ce^{-\alpha d}|\{\hat{\eta}^i\}|_{\mathcal{K}^m(\gamma)}.$$

Here $d = \inf\{\xi^{i+1} - \xi^i\}$. The above process can be repeated with $\{\hat{\eta}^i\}$ replaced by $\{\hat{\eta}_1^i\}$, and the second approximation denoted by $\{\phi_1^i\}$. By iteration, we can have a sequence $\{\hat{\eta}_j^i\}$, $j \geq 1$ and a sequence of approximations $\{\phi_j^i\}$. Suppose now the constant $C_1 = Ce^{-\alpha d} < 1$, then the convergent series

$$\Phi^i = \phi^i + \sum_{j=1}^{\infty} \phi_j^i$$

is the desired solution to (13) and (14). $\pi_1 \mathcal{L}^{-1}(\Phi^i)$ is an exact solution for v^i. The uniqueness of $\{v^i\}$ can be proved by the exponential dichotomy argument and will be skipped here. Observe that by the Paley-Wiener Theorem,

$$\eta^i(\tau) = 0 \text{ for } \tau \leq \Delta\tau, i \in \mathbb{Z}, \quad \Rightarrow \quad v^i(\xi,\tau) = 0 \text{ for } \tau \leq \Delta\tau, i \in \mathbb{Z}. \quad (15)$$

This fact will be used in the next section.

Finally, the solution to the system (12), (5)-(7), $u^i = \bar{u}^i + v^i$, satisfies

$$|u^i|_{H^{2m,1}(\Sigma^i,\gamma)} \leq C(|\{\delta^i\}|_{B^m(\mathbb{R}^+,\gamma)} + |\{h^i\}|_{L^2(\Sigma^i,\gamma)} + |\{u_0^i\}|_{H^m(\Omega^i)}). \quad (16)$$

5. The nonlinear system (4)–(7) can be solved by using the result of §4 on the linear system and a contraction mapping in a suitable Banach space. The following *Local Spatial Shadowing Lemma* is the main result of Part I.

Theorem 2 *Assume that f is C^∞ with bounded derivatives with respect to all the variables, $\{w^i\}$ is a formal approximation with $|\{w^i\}|_{H^{2m,1}(\Sigma^i)} < \infty$. Let $I = [0, \Delta\tau]$. Assume that $\epsilon > 0$ is small so that a near identity change of coordinates can be made in $[0, \epsilon\Delta\tau]$ as in §1. Let $d = \inf_i\{\xi^{i+1} - \xi^i\} > 0$. Then there exist $\beta_0, \epsilon_0 > 0$ and a positive linear function $\mu^*(\beta), 0 < \beta \leq \beta_0$. If $0 < \epsilon < \epsilon_0$, and*

$$|\{u_0^i\}|_{H^m(\Omega^i)} + |\{g^i\}|_{L^2(\Sigma^i)} + |\{\delta^i\}|_{B^m(I)} \leq \mu^*(\beta).$$

then there is a unique solution $\{u^i\}$ to (4)–(7), with $|\{u^i\}|_{H^{2m,1}(\Sigma^i)} \leq \beta$. Moreover,

$$|\{u^i\}|_{H^{2m,1}(\Sigma^i)} \leq C(|\{u_0^i\}|_{H^m(\Omega^i)} + |\{g^i\}|_{L^2(\Sigma^i)} + |\{\delta^i\}|_{B^m(I)}).$$

Proof Let $h^i \in L^2(\Sigma^i)$. $i \in \mathbb{Z}$. Since $\gamma > 0$, it is easy to extend the domains of δ^i and h^i to $\tau \in \mathbb{R}^+$. so that

$$|\{h^i\}|_{L^2(\Omega^i \times \mathbb{R}^+, \gamma)} + |\{\delta^i\}|_{B^m(\mathbb{R}^+, \gamma)} \leq C(|\{h^i\}|_{L^2(\Sigma^i)} + |\{\delta^i\}|_{B^m(I)}).$$

Consider the associated linear system (12). From the assumptions, it is clear that $\sup_{\xi,i,k}|A_k^i(\xi)| < \infty$. Assume that the coefficients have been extended to $\xi \in \mathbb{R}$ by constants. then from §4, there exists $M_0 > 0$ such that if $M \geq M_0$ then (12) has an exponential dichotomy in $E^{m\nu}(s)$, $\nu = 0, \frac{1}{4m}$ for $\xi \in \mathbb{R}$ and $s \in \mathcal{S}_\theta(M)$. This also implies that (12) has an exponential dichotomy in $\mathcal{K}^m(\gamma)$ for $\gamma = M$, of which the exponential decay rate is $\tilde{\alpha} = \alpha(1 + {}^{2m}\!\sqrt{M})$. By choosing larger M, we have $C_1 = Ce^{-\tilde{\alpha}d} \leq 0.5$, where C_1 is as in §4. The result in §4 concerning the system (12), (5)–(7) is now valid. Let the unique solution be denoted by

$$\{u^i\} = \mathcal{F}_\gamma(\{u_0^i\}, \{\delta^i\}, \{h^i\}).$$

Restricting the solution to the finite interval $I = [0, \Delta\tau]$. in the unweighted norm, using (16), we have,

$$\begin{aligned}
|\{u^i\}|_{H^{2m,1}(\Sigma^i)} &\leq Ce^{\gamma\Delta\tau}|\{u^i\}|_{H^{2m,1}(\Omega^i \times \mathbb{R}^+, \gamma)} \\
&\leq Ce^{\gamma\Delta\tau}(|\{u_0^i\}|_{H^m(\Omega^i)} + |\{h^i\}|_{L^2(\Sigma^i)} + |\{\delta^i\}|_{B^m(I)}).
\end{aligned}$$

Let the solution in that finite time interval I be denoted by

$$\{u^i\} = \mathcal{F}_I(\{u_0^i\}, \{\delta^i\}, \{h^i\}).$$

Consider $Q(\beta) = \{\{u^i\} : u^i \in H^{2m,1}(\Sigma^i), |\{u^i\}|_{H^{2m,1}} \leq \beta\}$. Let $|\{u_0^i\}|_{H^m(\Omega^i}$ $+ |\{\delta^i\}|_{B^m(I)} + |\{g^i\}|_{L^2(\Sigma^i)} = \mu$. For $\{u^i\} \in Q(\beta)$, we have

$$\begin{aligned}
|\mathcal{N}^i(u^i, g^i, \epsilon)|_{L^2(\Sigma^i)} &\leq |g^i|_{L^2} + C(|u^i|^2 + \epsilon\Delta\tau|u^i|) \\
&\leq C(\beta^2 + \epsilon\Delta\tau\beta + \mu).
\end{aligned}$$

Consider the mapping

$$\{u_1^i\} = \mathcal{F}_I(\{u_0^i\}, \{\delta^i\}, \{\mathcal{N}^i(u^i, g^i, \epsilon)\}).$$

We have

$$|\{u_1^i\}|_{H^{2m,1}(\Sigma^i)} \leq C(\mu + \beta^2 + \epsilon \Delta \tau 3).$$

Let β be small such that $C\beta^2 < \frac{1}{3}\beta$. Let μ and ϵ_0 be small, depending on β, such that $C\mu < \frac{1}{3}\beta$ and $C\epsilon\Delta\tau < \frac{1}{3}$. Then $\mathcal{F}_I$ maps $Q(\beta)$ into itself. One can also verify that if β is small, then $\mathcal{F}_I$ is a contraction mapping. Therefore, there exists $\beta_0 > 0$ such that $\mathcal{F}_I : Q(\beta) \to Q(\beta)$ has a unique fixed point $\{u^i\}$.

Finally, the solution $\{u^i\}$ does not depend on the method of extending the domain of $\{\delta^i\}$, $\{g^i\}$ to $\tau \in \mathbb{R}^+$. This can be verified by using (15). $\qquad\square$

Remark. In many formal constructions, $d = \inf\{\xi^{i+1} - \xi^i\} \to \infty$ as $\epsilon \to 0$. Then the condition $C_1 = Ce^{-\bar{\alpha}d} \leq 0.5$ is valid if ϵ is small. We do not need to choose large $\gamma = M$ to make $\bar{\alpha}$ large.

1   Part II. Global Existence of Multiple Waves

6. The multiple wave solutions constructed in Part I exist only for a short time $t \in [0, \Delta t]$. $\Delta t = \epsilon\Delta\tau$. If $\{u^i\}$ is not too large, using the output of the previous interval as the input of the next time interval, the process can be repeated to obtain solutions in $[j\Delta t, (j+1)\Delta t]$, $j = 1, 2, \cdots$ recursively. It is shown, in [13], that if a formal approximation is defined for $t \in \mathbb{R}^+$s, under certain conditions, it is possible to obtain global solutions for $t \in \mathbb{R}^+$. In the second part of this paper, we summarize the results in [13].

Although the method should work for some higher order parabolic systems, such as the phase field equations, [7], to simplify the matter, we will consider a second order system,

$$\epsilon u_t = \epsilon^2 u_{xx} + f(u, x, \epsilon), \quad x \in \mathbb{R}, \, t \geq 0. \tag{17}$$

Assuming by the method of matched expansions , we have the formal series for the wave fronts,

$$\eta^\ell(t, \epsilon) = \sum_0^m \epsilon^j \eta_j^\ell(t), \quad \ell \in \mathbb{Z}.$$

and formal series solutions in the ℓ-th regular and singular layers,

$$u^{R\ell}(x, t, \epsilon) = \sum_0^m \epsilon^j u_j^{R\ell}(x, t),$$
$$u^{S\ell}(\xi, t, \epsilon) = \sum_0^m \epsilon^j u_j^{S\ell}(\xi, t).$$

Here "R" and "S" stand for regular and singular (internal) layers. and $\xi = (x - \eta^\ell(t,\epsilon))/\epsilon$.

For convenience. assume the the following **Periodicity Hypotheses**.

1. $f(u, x + x_p, \epsilon) = f(u, x, \epsilon)$;

2. $\eta^{\ell+\ell_p}(t, \epsilon) = \eta^\ell(t, \epsilon) + x_p$;

3. $u^{R(\ell+\ell_p)}(x, t, \epsilon) = u^{R\ell}(x - x_p, t, \epsilon)$;

4. $u^{S(\ell+\ell_p)}(\xi, t, \epsilon) = u^{S\ell}(\xi, t, \epsilon)$.

Here $x_p > 0$ and integer $\ell_p > 0$ are two constants. The periodicity hypotheses ensure that all the estimates obtained here are uniform with respect to layer index ℓ. They do not play any other rolls and are not necessary.

Let $0 < \beta < 1$ and let the width of the internal layers be $2\epsilon^{\beta-1}$. Define

$$y^{2\ell}(t) = \eta^\ell(t, \epsilon) + \epsilon^\beta,$$
$$y^{2\ell-1}(t) = \eta^\ell(t, \epsilon) - \epsilon^\beta.$$

The family of curves $\Gamma^i = \{(x, t) : x = y^i(t)\}$ divides the domain into regions Σ^i, $i \in \mathbb{Z}$, where Σ^i is between Γ^{i-1} and Γ^i. A formal approximation can be obtained from the matched expansions,

$$w^i(x, t. \epsilon) = \begin{cases} u^{R\ell}(x, t, \epsilon), & i = 2\ell - 1, \\ u^{S\ell}(\frac{x-\eta^\ell(t,\epsilon)}{\epsilon}, t, \epsilon), & i = 2\ell. \end{cases}$$

Here are the assumptions on w^i:

H 1 *There exist C. $\bar{\gamma} > 0$ such that for all small ϵ and i, $\ell \in \mathbb{Z}$,*

$$|w^i(x, t, \epsilon) - w^i(x, \infty, \epsilon)| \le Ce^{-\bar{\gamma}t}, \quad (x, t) \in \mathbb{R}^2.$$
$$|\eta^\ell(t, \epsilon) - \eta^\ell(\infty, \epsilon)| + |D_t\eta^\ell(t, \epsilon)| \le Ce^{-\bar{\gamma}t}, \quad t \in \mathbb{R}^-.$$

Here $w^i(x, \infty, \epsilon) = \lim_{t\to\infty} w^i(x, t, \epsilon)$, etc.

H 2 *There exists $\bar{\sigma} > 0$ such that in each regular layer Σ^i, $i = 2\ell - 1$,*

$$Re\sigma\{f_u(w^i(x, t, \epsilon), x, \epsilon)\} \le -\bar{\sigma}$$

uniformly for all $(x, t) \in \Sigma^i$, $i \in \mathbb{Z}$ and small $\epsilon > 0$.

H 3 *For an approximation $w^i(\xi, t, \epsilon)$ in an internal layer, as $\xi \to \pm\infty$ and $\epsilon \to 0$, both w^i and $\partial w^i/\partial \xi$ approach the corresponding values of w^{i+1} or w^{i-1} at common boundaries. More precisely, if $i = 2\ell$, then for any $\mu > 0$, there exist $N, \epsilon_0 > 0$ such that $\epsilon_0^{\beta-1} > N$, and for $0 < \epsilon < \epsilon_0$, $t \ge 0$,*

$$|W^i(\xi, t, \epsilon) - W^{i-1}(y^{i-1}(t), t, \epsilon)| \le \mu, \quad \text{for } -\epsilon^{\beta-1} \le \xi \le -N,$$
$$|W^i(\xi, t, \epsilon) - W^{i+1}(y^i(t), t, \epsilon)| \le \mu, \quad \text{for } \epsilon^{\beta-1} \ge \xi \ge N.$$

Here the function $W^i = (w^i, w^i_\xi)$ is expressed in the stretched variable $\xi = (x - \eta^\ell(t, \epsilon))/\epsilon$.

Let $\tilde{\xi} = \tilde{\xi}(\xi)$ be a function of ξ such that

$$\tilde{\xi} = \begin{cases} \xi, & \text{for } |\xi| \leq \epsilon^{\beta-1}. \\ -\epsilon^{\beta-1}, & \text{for } \xi < -\epsilon^{\beta-1}. \\ \epsilon^{\beta-1}, & \text{for } \xi > \epsilon^{\beta-1}. \end{cases}$$

For each $t \geq 0$, $i = 2\ell$, consider the operator $\mathcal{A}^i(t) : L^2(\mathbb{R}) \to L^2(\mathbb{R})$,

$$\mathcal{A}^i(t)u = u_{\xi\xi} + D_t\eta^\ell(t,\epsilon)u_\xi + f_u(w^i(\tilde{\xi},t,\epsilon), \eta^\ell(t.\epsilon) + \epsilon\tilde{\xi},\epsilon)u.$$

H 4 $\mathcal{A}^i(t)$. $i = 2\ell$, $t \leq 0$, *has a simple eigenvalue* $\lambda^i(\epsilon) = \epsilon\lambda_0^i(t) + O(\epsilon^2)$. *The rest of the spectrum is contained in* $\{Re\lambda \leq -\bar{\sigma}\}$, $\bar{\sigma}$ *as in* **H2**. *Moreover, for the limiting operator* $\mathcal{A}^i(\infty)$, *we have,*

$$\lambda_0^i(\infty) \leq \overline{\lambda}_0 < 0, \quad \textit{uniformly for all } i = 2\ell.$$

We look for exact solution u that is of the form $w^i + u^i$ in each Σ^i. The limit $u^i(x,\infty,\epsilon)$ describes the correction to $w^i(x,\infty,\epsilon)$ that yields a stationary solution, while $u^i(x.t.\epsilon) - u^i(x,\infty,\epsilon)$, together with $w^i(x,t,\epsilon) - w^i(x,\infty,\epsilon)$, describes how the solution approaches its limit as $t \to \infty$. Therefore, we define the following Banach spaces. For a constant $\gamma < 0$, let

$$X(\Omega \times \mathbb{R}^+,\gamma) = \{u : u = u_1 + u_2, \, u_1 \in L^2(\Omega). \, u_2 \in L^2(\Omega \times \mathbb{R}^+,\gamma)\}.$$
$$|u|_{X(\gamma)} = |u_1|_{L^2(\Omega)} + |u_2|_{L^2(\Omega \times \mathbb{R}^+,\gamma)}.$$
$$X^{2.1}(\Omega \times \mathbb{R}^+,\gamma) = \{u : u = u_1 + u_2, \, u_1 \in H^2(\Omega). \, u_2 \in H^{2,1}(\Omega \times \mathbb{R}^+,\gamma)\}.$$
$$|u|_{X^{2.1}(\gamma)} = |u_1|_{H^2(\Omega)} + |u_2|_{H^{2,1}(\Omega \times \mathbb{R}^+,\gamma)}.$$

It can be verified that for $u \in X(\Omega \times \mathbb{R}^+,\gamma)$ or $X^{2,1}(\Omega \times \mathbb{R}^+,\gamma)$, the decomposition $u = u_1 + u_2$ is unique.

For $\Sigma^i = \{(x,t) : y^{i-1}(t) < x < y^i(t), \, t \geq 0\}$, we say that $u \in L^2(\Sigma^i,\gamma)$, $H^{2,1}(\Sigma^i,\gamma)$, $X(\Sigma^i,\gamma)$ or $X^{2,1}(\Sigma^i,\gamma)$, etc., if u is the restriction of a function $\tilde{u} \in L^2(\mathbb{R} \times \mathbb{R}^+,\gamma)$. etc., to the domain Σ^i. The norms are defined by

$$|u|_{L^2(\Sigma^i,\gamma)} = \inf\{|\tilde{u}|_{L^2(\mathbb{R}\times\mathbb{R}^+,\gamma)}\}.$$
$$|u|_{X^{2.1}(\Sigma^i,\gamma)} = \inf\{|\tilde{u}|_{X^{2.1}(\mathbb{R}\times\mathbb{R}^+,\gamma)}\}.$$

Let

$$X^k(\gamma) = \{u \mid u = u_1 + u_2, \, u_1 \in \mathbb{R}^n, \, u_2 \in H^k(\gamma)\}, \, \gamma < 0.$$
$$X^{k_1 \times k_2}(\gamma) = X^{k_1}(\gamma) \times X^{k_2}(\gamma), \quad k_1 \geq 0. \, k_2 \geq 0.$$

Similar definitions can be given to spaces $L^2(\Omega \times I, \gamma)$, $H^{2,1}(\Omega \times I, \gamma)$, $H^{2,1}(\Sigma \cap I, \gamma)$, but not $X(\Omega \times I, \gamma)$ or $X^k(\gamma)$, if I is a finite time interval.

The main result of Part II is the following *Global Spatial Shadowing Lemma*.

Theorem 3 *Let* $\{\bar{w}^i\}$ *be a formal approximation of solutions for (17). Assume that the Hypotheses* **H1**–**H4** *are satisfied, and the constants* $\bar{\gamma}$ *and* $\bar{\sigma}$ *satisfy* $-\bar{\sigma} < -\bar{\gamma} < 0$. *Then there exist positive constants* j_0, J_2. ϵ_0 *and a negative constant* $\gamma = O(\epsilon)$

satisfying the following properties. Assume that $\{w^i\}$ is a formal approximation near $\{\bar{w}^i\}$, with

$$|w^i - \bar{w}^i|_{X^{2.1}(\Sigma^i,\gamma)} \leq C_1 \epsilon^{j_1}, \quad i \in \mathbb{Z}, \tag{18}$$

for some $C_1 > 0$ and $j_1 \geq 1$. Assume that for the approximation $\{u^i\}$, we have

$$|g^i|_{X(\Sigma^i,\gamma)} + |\delta^i|_{X^{0.75 \times 0.25}(\mathbb{R}^+,\gamma)} \leq C_2 \epsilon^{j_2}, \quad j_2 \geq j_0. \tag{19}$$

Then for $0 < \epsilon < \epsilon_0$, to any locally H^1 function u_0 with $|u_0 - w^i(0)|_{H^1(\Sigma^i \cap \{\tau=0\})} \leq C_2 \epsilon^{j_2}$, there exists a unique exact solution to (17) that satisfies $u_{exact}(x,0) = u_0(x)$, and

$$|u_{exact} - \bar{w}^i|_{X^{2.1}(\Sigma^i,\gamma))} = O(\epsilon^{j_3}), \quad i \in \mathbb{Z}.$$

Here $j_3 = \min\{j_1, j_2 - J_2\}$, $J_2 > 0$ is a constant that does not depends on ϵ. All the norms in this lemma are expressed by the stretched variables $\xi = x/\epsilon$, $\tau = t/\epsilon$.

Remark. (1) The constants j_0 and J_2 depend on $\{\bar{w}^i\}$. However the approximation may not satisfy (19). By adding higher order expansions, the new approximation $\{w^i\}$ will satisfy (18). Therefore, it has the same constants j_0 and J_2, which are stable with respect to perturbations. Moreover, if the order of expansion is sufficiently high, $\{w^i\}$ will also satisfy (19).

(2) The values of j_0 and J_2 are not important in applications. since $j_3 = j_1$ if j_2 is sufficiently large. Computing j_1 is relatively easy. One only need to compare $\{w^i\}$ with the next approximation.

(3) Hypotheses **H1**–**H4** are consequences of hypotheses used to construct formal expansions for system (17). See §11 for a brief discussion.

7. To prove the theorem, we use a partition of the time interval $t \in \mathbb{R}^+$,

$$\mathbb{R}^+ = \cup_{j=0}^{r-1}[j\Delta t, (j+1)\Delta t] \cup [t_f, \infty),$$

where $t_f = r\Delta t$. We assume that $\Delta t = \epsilon \Delta \tau$, where $\Delta \tau$ is independent of ϵ. t_f is large such that $e^{-\bar{\gamma} t} = O(\epsilon^2)$, where $\bar{\gamma}$ is as in **H1**. If ϵ is small the variations of boundary curves $x = y^{i-1}(t)$ and $y^i(t)$ are small such that a near identity change of coordinates can be made in each $\Sigma_j^i = \Sigma^i \cap \{t \in [j\Delta t, (j+1)\Delta t]\}$. $j \leq r-1$ or $\Sigma_r^i = \Sigma^i \cap \{t \in [t_f, \infty)\}$ to straighten Γ^{i-1} and Γ^i. In the sequel. we assume that $\Gamma^i = \{x = y_j^i\}$ in the jth time interval, where y_j^i is independent of t.

Define

$$\xi = [x - (y_j^{i-1} + y_j^i)/2]/\epsilon, \quad \text{in } \Sigma_j^i,$$
$$L_j^i(\epsilon) = (y_j^i - y_j^{i-1})/(2\epsilon),$$
$$\Omega_j^i = [-L_j^i(\epsilon), L_j^i(\epsilon)],$$
$$I^j = [0, \Delta\tau], 0 \leq j \leq r-1, \quad \text{and} \quad I^r = [0, \infty).$$

Let $\tau = (t - j\Delta t)/\epsilon$. Then $\Sigma_j^i = \Omega_j^i \times I^j$. In each Σ_j^i, $j \leq r-1$, we linearize around w^i at the fixed time $\tau = 0$, while in Σ_r^i, linearize around w^i at the fixed time $\tau = \infty$. Denoting the solutions in Σ_j^i by u_j^i. we have

$$u_{j\tau}^i = u_{j\xi\xi}^i + V_j^i(\xi)u_{j\xi}^i + A_j^i(\xi)u_j^i + \mathcal{N}_j^i(u_j^i.\xi.\tau,\epsilon), \quad \text{in } \Sigma_j^i, \tag{20}$$

$$U_j^i(L^i(\epsilon),\tau) - U_j^{i+1}(-L^{i+1}(\epsilon).\tau) = \delta_j^i(\tau). \quad \text{at } \Gamma^i, \tag{21}$$

$$u_{j+1}^i(\xi,0) = u_j^i(\xi_1,\Delta\tau), \; u_0^i(\xi,0) = u_0^i(\xi). \tag{22}$$

Here $\xi_1 = \Theta_j^i(\xi)$ is a near identity change of coordinates induced by the coordinates change that straightens the boundaries in each Σ_j^i. V_j^i is the constant wave speed in an internal layer, but is slowly ξ dependent in a regular layer due to the fact that the boundaries are not parallel there.

The idea of the proof is presented in the next three sections. In §8, we study the spectral properties of the linear systems and exponential dichotomies associated to the dual systems of the linear systems. In §9. we study the system in the last interval $[t_f,\infty)$. There we need to obtain asymptotic behavior of the solution $\{u_r^i\}$. Thus the space $X^{2,1}(\Sigma_r^i,\gamma)$ will be used. The result obtained there also serves as an upper bound of the accumulation error in the first r intervals $[j\Delta t,(j+1)\Delta t]$, $0 \leq j \leq r-1$. In §10, we study the system in finite time intervals. Since r is large, the estimates obtained in Part I is not accurate enough. More precise estimates will be derived which rely on the Hypotheses **H1–H4**, while the estimate in Part I does not.

8. Since nonlinear systems can be solved by contraction mappings, we study a linear system first. Drooping the subscript j. we consider.

$$u_\tau^i = u_{\xi\xi}^i + V^i(\xi)u_\xi^i + A^i(\xi)u^i + h^i(\xi,\tau), \tag{23}$$

$$U^i(L^i(\epsilon),\tau) - U^{i+1}(-L^{i+1}(\epsilon),\tau) = \delta^i(\tau), \tag{24}$$

$$u^i(\xi,0) = u_0^i(\xi). \tag{25}$$

The homogeneous dual system of (23) is

$$\hat{U}_\xi = \begin{pmatrix} 0 & I \\ sI - A^i(\xi) & -V^i(\xi) \end{pmatrix}\hat{U}. \tag{26}$$

Here $\hat{U} = (\hat{u},\hat{u}_\xi)^\tau$. It is crucial to study the exponential dichotomies for system (26).

We first discuss the system in regular layers. Freezing $\xi = \xi_0$ and consider (26) with constant coefficients. Using **H2**, we can verify that for $\text{Re}(s) \geq -\sigma_0$, $\sigma_0 = \bar{\sigma}/2$, the system is hyperbolic, having n eigenvalues with positive (negative) real parts.

Now recall that in regular layers, $A^i(\xi) = f_u(w^i((y^{i-1} + y^i)/2 + \epsilon\xi),(y^{i-1} + y^i)/2 + \epsilon\xi,\epsilon)$ depends slowly in ξ, i.e., $\frac{\partial A^i(\xi)}{\partial\xi} = O(\epsilon)$. Also, from the change of coordinates. cf. [13] for details, $V^i(\xi)$ also depends slowly on ϵ in regular layers. Then a theorem from [3] indicates that (26) has exponential dichotomy for $\xi \in \mathbb{R}$ and $\text{Re}(s) \geq -\sigma_0$.

In [13] it is shown that if (26) has an exponential dichotomy in R^{2n} for $\xi \in \mathbb{R}$ and for every $s \in \{\text{Re}(s) \geq -\sigma_0\}$, then it has an exponential dichotomy in the space $E^{1,\nu}(s), \nu = 0$ or 0.25.

Let the right hand side of (23) define a linear operator $\mathcal{A}^i(t)$. $t = j\Delta t$, $0 \leq j \leq r - 1$, or $t = \infty$. $j = r$. Using the fact that (26) has an exponential dichotomy in $E^{1,\nu}(s), \nu = 0$ for $\xi \in \mathbb{R}$ and $s \in \{\text{Re}(s) \geq -\sigma_0\}$, we can show that in regular layers, $\mathcal{A}^i(t)$ is an exponentially stable sectorial operator in $L^2(\mathbb{R})$. i.e.,

$$\text{Re}\{\sigma \mathcal{A}^i(t)\} \leq -\sigma_0,$$

for every $t \geq 0$.

We now discuss (26) in internal layers. Here the spectrum of $\mathcal{A}^i(t)$ is given by **H4**. Also $D_t \eta^\ell(t, \epsilon) = V_j^i$, $i = 2\ell$ if $t = j\Delta t$, $0 \leq j \leq r - 1$, and $V_r^i = 0$. From **H3**, if μ is small, then for $|\xi| \geq N$, the coefficients $A^i(\xi)$ and $V^i(\xi)$ are close to that of $A^i(\pm L^i(\epsilon))$ and $V^i(\pm L^i(\epsilon))$. But the system with constant coefficients $A^i(\pm L^i(\epsilon))$ and $V^i(\pm L^i(\epsilon))$ is hyperbolic. Therefore, from the Roughness of Exponential Dichotomies, which is also valid in spaces $E^{1,\nu}(s), \nu = 0, 0.25$. (26) has exponential dichotomies in $E^{1,\nu}(s), \nu = 0, 0.25$, for $\text{Re}(s) \geq -\sigma_0$ and $|\xi| \geq N$.

The exponential dichotomies in the region $\xi \leq -N$ and $\xi \geq N$ extend to $\xi \in \mathbb{R}^-$ and $\xi \in \mathbb{R}^+$ respectively. Let the projections be $P_s^i + P_u^i = I$. We can show that

$$\mathcal{R}P_u(0^-, s) \oplus \mathcal{R}P_s(0^+, s) = \mathbb{R}^{2n}, \quad \text{if } s \neq \lambda^i(\epsilon) \text{ and } \text{Re}(s) \geq -\sigma_0. \quad (27)$$

If (27) were not true. we could find a solution $\hat{U}(\xi, s)$ for (26) that decays exponentially as $\xi \to \pm\infty$. Then s would be an eigenvalue with $\hat{U}(\xi, s)$ as an eigen function. This contradicts to the fact $\lambda^i(\epsilon)$ is the only eigenvalue in the region $\text{Re}(s) \geq -\sigma_0$.

In [13], it is also shown that the projections defined by the splitting (27) is $O(1 + \frac{1}{|s - \lambda^i(\epsilon)|})$. Observe that **H4** also implies that for the linearization at $t = \infty$, (26) has an exponential dichotomy in $\xi \in \mathbb{R}$ in the region $\text{Re}(s) \geq \frac{\epsilon \bar{\lambda}_0}{2}$, where $\bar{\lambda}_0 < 0$ is as in **H4**.

9. We now study (23)–(25) in the time interval $[t_f, \infty)$. The procedure of solving them is similar to that used in Part I. The right hand side of (23) defines a sectorial operator $\mathcal{A}^i(\infty)$ in $L^2(\mathbb{R})$. Extend the domain of u_0^i and h^i to $\xi \in \mathbb{R}$. Let

$$\bar{u}^i = e^{\mathcal{A}^i(\infty)\tau} u_0^i + \int_0^\tau e^{\mathcal{A}^i(\infty)(\tau - \zeta)} h^i(\zeta) d\zeta.$$

As shown in §8,

$$\sigma \mathcal{A}^i(\infty) \subset \{\text{Re}(s) \leq -C\epsilon < 0\},$$

both in regular and internal layers. Because of the spectra of $\mathcal{A}^i(\infty)$, with some $\gamma = C\epsilon < 0$, we can show

$$|\bar{u}^i|_{X^{2.1}(\gamma)} \leq C(\epsilon^{-0.5}|u_0^i|_{H^1} + \epsilon^{-1.5}|h^i|_{X(\gamma)}). \quad (28)$$

LOCAL AND GLOBAL EXISTENCE OF MULTIPLE WAVES

The solution $\{u^i\}$ for (23)-(25) can be written as $u^i = \bar{u}^i + v^i$, with $\{v^i\}$ satisfying

$$v^i_\tau = v^i_{\xi\xi} + V^i(\xi)v^i_\xi + A^i(\xi)v^i, \tag{29}$$

$$V^i(L^i(\epsilon), \tau) - V^{i+1}(-L^{i+1}(\epsilon), \tau) = \eta^i(\tau), \tag{30}$$

$$v^i(\xi, 0) = 0. \tag{31}$$

Here $V = (v, v_\xi)^\tau$, η^i includes δ^i and corrections from the jumps of $\{\bar{u}^i\}$ at boundaries. We have

$$|\eta^i|_{X^{0.75 \times 0.25}(\gamma)} \leq C(|\delta^i|_{X^{0.75 \times 0.25}(\gamma)} + \epsilon^{-0.5}|\{u^i_0\}|_{H^1} + \epsilon^{-1.5}|\{h^i\}|_{L^2(\Sigma^i, \gamma)}).$$

Since for the dual system of (29)-(31), the exponential dichotomies exists in the region $\mathrm{Re}(s) \geq \gamma$, $\gamma < 0$, using the Laplace transform and the iterative scheme similar to that used in Part I, we can show that system (29)-(31) has a unique solution $\{v^i\}$, with

$$|v^i|_{X^{2,1}(\gamma)} \leq C|\{\eta^i\}|_{X^{0.75 \times 0.25}(\gamma)}.$$

Therefore, in the last interval, the solution $\{u^i\}$ satisfies

$$|u^i|_{X^{2,1}(\gamma)} \leq C(|\{\delta^i\}|_{X^{0.75 \times 0.25}(\gamma)} + \epsilon^{-0.5}|\{u^i_0\}|_{H^1} + \epsilon^{-1.5}|\{h^i\}|_{L^2(\Sigma^i, \gamma)}).$$

Using the above result, the nonlinear system can be solved by a contraction mapping in a ball of radius ϵ^{r_1}, $r_1 > 1.5$, in $X^{2,1}(\gamma)$. The following result is proved in [13].

Theorem 4 Let $|g^i|_{X(\gamma)} = o(\epsilon^{r_1+1.5})$, $|u^i_0|_{H^1} = o(\epsilon^{r_1+0.5})$, $|\delta^i|_{H^{0.75 \times 0.25}(\gamma)} = o(\epsilon^{r_1})$, where $r_1 > 0.5$. Let t_f be ϵ dependent such that $e^{-\bar{\gamma}t_f} \leq C_0\epsilon^2$, where $\bar{\gamma}$ is as in **H1** and C_0 is independent of ϵ. Then there exists $\epsilon_0 > 0$ such that for $0 < \epsilon < \epsilon_0$, the nonlinear system (20)- (22) has a unique solution $\{u^i\}$ satisfying the following estimate,

$$|\{u^i\}|_{X^{2,1}(\gamma)} \leq C\{|\{\delta^i\}|_{H^{0.75 \times 0.25}(\gamma)} + \epsilon^{-0.5}|\{u^i_0\}|_{H^1} + \epsilon^{-1.5}|\{g^i\}|_{X(\gamma)}\}. \tag{32}$$

10. We study system (20)-(22) in the finite intervals, $[j\Delta t, (j+1)\Delta t]$, $0 \leq j \leq r-1$. The existence of solutions in a finite interval has been discussed in Theorem 2. However, to guarantee the existence of solutions up to the last interval $[t_f, \infty)$, a stringent restriction on the accumulation errors in these finite intervals must be met. Since the critical eigenvalue $\lambda^i(\epsilon)$ may not be negative, u^i_j may grow as j increases. From Theorem 2, $\{u^i_j\}$ is determined by $\{u^i_j(0)\}$, $\{g^i_j\}$ and $\{\delta^i_j\}$. Among them only $\{u^i_j(0)\}$, carries the information from the previous j intervals. Therefore, it is crucial to control $|u^i_j(\Delta\tau)|$ in terms of $|\{u^i_j(0)\}|$.

Let $Q^{i,j}_0$ and $Q^{i,j}_s$ be the projections corresponding to the spectral set in $\lambda^i(\epsilon)$ and $\{\mathrm{Re}(s) \leq -\sigma_0\}$ in the ith internal layers and the jth time interval. Let

$Q_0^{i,j} u_j^i = \alpha_j^i \phi_j^i$ where ϕ_j^i is the eigenvalue corresponding to the eigenvalue $\lambda^i(\epsilon)$. Then

$$|\alpha_j^i| + |Q_s^{i,j} u_j^i|_{H^1},$$

is an equivalent norm for $|u_j^i|_{H^1}$. The following estimate has been proved in [13],

$$
\begin{aligned}
|\alpha_j^i(\Delta\tau)| + |Q_s^{i,j} u_j^i(\Delta\tau)|_{H^1} \le (1 + C\epsilon) \sup_i (|\alpha_j^i(0)| + |Q_s^{i,j} u_j^i(0)|_{H^1}) \\
+ C|\{\delta_j^i\}|_{H^{0.75 \times 0.25}(I^j)} + C|\{g_j^i\}|_{L^2}.
\end{aligned}
\tag{33}
$$

Since the spectral projections for the j and $(j+1)$th interval differ by $O(\epsilon)$, $|\alpha_{j+1}^i(0)| + |Q_s^{i,j+1} u_{j+1}^i(0)|_{H^1}$ is also bounded by the right hand side of (33). Therefore, one can show that

$$
\begin{aligned}
|\alpha_j^i(0)| + |Q_s^{i,j} u_j^i(0)|_{H^1} \le (1 + C\epsilon)^j \sup_i (|\alpha_0^i(0)| + |Q_s^{i,0} u_0^i|_{H^1}) \\
+ \frac{(1 + C\epsilon)^j}{\epsilon} \sup_{k<j} (|\{\delta_k^i\}|_{H^{0.75 \times 0.25}} + |\{g_k^i\}|_{L^2}).
\end{aligned}
$$

Using the fact $(1+C\epsilon)^{1/(C\epsilon)} < e$ and $j \le r \approx \log(\frac{1}{\epsilon})/\epsilon$, we see that $(1+C\epsilon)^j \le \epsilon^{-B}$ for some $B > 0$. Therefore, if all the terms in the right hand side are bounded by ϵ^M for some large $M > 0$, have the solution $\{u_j^i\}$ for all the finite intervals, and the initial condition for the infinite interval is small enough so that Theorem 4 can be used there. Here is the main theorem in this section.

Theorem 5 *There exists $\Delta\tau > 0$ such that if $r = [\log(\frac{1}{C_0\epsilon})/\epsilon\bar\gamma\Delta\tau] + 1)$, then we have the following concerning the solutions of (20)-(22) in I^j, $0 \le j \le r - 1$. There exist M_1 and $M_2 > 0$, such that if*

$$|\{u_0^i(0)\}|_{H^1} + |\{\delta_j^i\}|_{H^{0.75 \times 0.25}} + |\{g_j^i\}|_{L^2} = O(\epsilon^{M_1}),$$

uniformly for $0 \le j \le r-1$, then (20)-(22), $0 \le j \le r-1$ has a unique solution u_j^i for $0 \le \epsilon \le \epsilon_0$, where $\epsilon_0 > 0$ is a small constant. The solution in each I^j satisfies

$$
\begin{aligned}
|\alpha_j^i(0)| + |Q_s^{i,j} u_j^i(0)|_{H^1} \le C\epsilon^{-M_2}[\sup_i(|\alpha_0^i(0)| + |Q_s^{i,0} u_0^i|_{H^1}) \\
+ \sup_{k<j}(|\{\delta_k^i\}|_{H^{0.75 \times 0.25}} + |\{g_k^i\}|_{L^2}].
\end{aligned}
$$

In particular, if M_1 is sufficiently large, in the rth (infinite) interval $|\{u_r^i(0)\}|_{H^1} = o(\epsilon^{r_1+0.5})$, $r_1 > 1.5$. Therefore, Theorem 4 applies in the last infinite interval also.

To describe the idea of the proof of (33), it suffices to consider the linear system (23)-(25). For simplicity, we consider the 0th interval and drop the subscript j. The procedure of solving such system is similar to that used before. Define $\bar u^i$ as in Part I, and let $\bar u^i(\tau) = \bar\alpha^i(\tau)\phi^i + Q_s^{i,0} \bar u^i(\tau)$. In internal layers, $\lambda^i = O(\epsilon)$. Also in the stable space, the semigroup is exponentially stable. We can show,

$$
\begin{aligned}
|\bar\alpha^i(\Delta\tau)| &\le e^{(1+C\epsilon)\Delta\tau}(|\alpha^i(0)| + \sqrt{\Delta\tau}|h^i|_{L^2}), \\
|Q_s^{i,0} \bar u^{i'}(\Delta\tau)|_{H^1} &\le C(e^{-\sigma\Delta\tau}|Q_s^{i,0} u^i(0)|_{H^1} + |\alpha^i(0)|).
\end{aligned}
$$

for some $\sigma < 0$. Observe in the above, $\bar u^i(0) = u^i(0)$. Thus $\bar\alpha^i(0) = \alpha^i(0)$.

Let $\Delta\tau > 0$ be large so that $Ce^{-\sigma\Delta\tau} \leq 1/2$. We then find C_1 such that $e^{(1+C\epsilon)\Delta\tau} \leq 1 + C_1\epsilon$. We have

$$|\bar{\alpha}'(\Delta\tau)| + |Q_s^{i,0}\bar{u}^i(0)|_{H^1} \leq (1 + C_1\epsilon)|\alpha^i(0)| + 1/2|Q_s^{i,0}u^i(0)|_{H^1} + C|h^i|_{L^2}. \quad (34)$$

The above is also valid in regular layers with $\bar{\alpha}^i = 0$.

The solution for (23)–(25) again is written as $u^i = \bar{u}^i + v^i$, where $\{v^i\}$ satisfies (29)–(31). As in Part I, system (29)–(31) is solved by iterations, and $\{v^i\}$ is bounded by $\{\eta^i\}$ in suitable function spaces. An complication arises here since $\{\eta^i\}$ also include corrections of the jumps of $\{\bar{u}^i\}$ at common boundaries. The latter in term is bounded by $|\{\alpha^i(0)\}|$ and $|\{Q_s^{i,0}u^i(0)\}|$. of which the coefficients have to be carefully controlled.

It can be shown that $\{\eta^i\}$ depends weakly on $\alpha^i(0)$. since $|\alpha^i\phi^i| \leq Ce^{-\sigma|\xi|}$ for some $\sigma > 0$, and $L^i(\epsilon) \to \infty$ as $\epsilon \to 0$. However. it is a little tricky to control the dependence of $|\{\eta^i\}|$ on $|\{Q_s^{i,0}u^i(0)\}|$. The key here is to use the exponential decay of $e^{\mathcal{A}'\tau}$ in the stable subspace (which is the whole space in regular layers). To do so, weighted norms have been used, [13].

$$|U|_{H^{0.75\times 0.25}(I^j,\sigma)} = |e^{-\sigma\tau}U|_{H^{0.75\times 0.25}(I^j)}. \quad \sigma < 0,$$
$$|u|_{H^{2.1}(\Omega^i\times I^j,\sigma)} = |e^{-\sigma\tau}u|_{H^{2.1}(\Omega^i\times I^j)}. \quad \sigma < 0.$$

Observe here the interval $I^j = [0, \Delta\tau]$ is finite. It is shown that

$$|\eta^i|_{H^{0.75\times 0.25}(I^j,\sigma)} \leq Ce^{|\sigma|\Delta\tau}|\{\delta^i\}|_{H^{0.75\times 0.25}(I^j)}$$
$$+ \sup_i\{\epsilon|\alpha^i(0)| + C|Q_s^{i,0}u^i(0)| + Ce^{|\sigma|\Delta\tau}|h^i|_{L^2}\},$$

for some $\sigma < 0$, independent of ϵ. Also it is shown that the solution of the system (29)–(31) satisfies

$$|v^i|_{H^{2.1}(\Omega^i\times I^j,\sigma)} \leq C|\{\eta^i\}|_{H^{0.75\times 0.25}(I^j,\sigma)}.$$

Let $Q_0^{i,0}v^i = \bar{\alpha}^i\phi^i$.

$$|\bar{\alpha}^i(\Delta\tau)| + |Q_s^{i,0}v^i(\Delta\tau)|_{H^1} \leq C|v^i(\Delta\tau)|_{H^1}$$
$$\leq Ce^{\sigma\Delta\tau}|v^i|_{H^{2.1}(\Omega^i\times I^j,\sigma)}$$
$$\leq C|\{\delta^i\}|_{H^{0.75\times 0.25}(I^j)}$$
$$+ \sup_i\{C_1e^{\sigma\Delta\tau}|Q_s^{i,j}u^i(0)|_{H^1} + C\epsilon e^{\sigma\Delta\tau}|\alpha^i(0)| + C|h^i|_{L^2}\}.$$

Let $\Delta\tau$ be large so that $C_1e^{\sigma\Delta\tau}|Q_s^{i,j}u^i(0)|_{H^1} \leq 0.5$. Combine the above with (34), we have an estimate for the linear system,

$$|\alpha^i(\Delta\tau)| + |Q_s^{i,0}u^i(\Delta\tau)|_{H^1} \leq (1 + C\epsilon)\sup_i(|\alpha^i(0)| + |Q_s^{i,0}u^i(0)|_{H^1}) \qquad (35)$$
$$+ C|\{\delta^i\}|_{H^{0.75\times 0.25}(I^j)} + C|\{h^i\}|_{L^2}.$$

The estimate of the nonlinear system (33) can be obtained from (35) by replacing h^i with $\mathcal{N}^i$. and using the fact that if $|u^i|_{H^{2.1}} < \epsilon$.

$$|\mathcal{N}^i|_{L^2} \leq |g^i|_{L^2} + C\epsilon|u^i|_{H^{2.1}}.$$

11. Constructions of multiple waves in singularly perturbed equations have been discussed in many papers, see for example [5]. Expansions for system of reaction-diffusion equations to any order of ϵ can be found in [12]. The following hypotheses are used in that paper.

There is a partition
$$\mathbb{R} = \cup_{\ell=-\infty}^{\infty} [x^\ell, x^{\ell+1}]$$
that is periodic with respect to ℓ, compatible with the period of f. A C^∞ function $p^i(x)$ is defined on $[x^{i-1}, x^i]$ with $f_0(p^i(x), x) = 0$. Also assume that $p^{\ell+\ell_p}(x+x_p) = p^\ell(x)$, $x \in [x^{\ell-1}, x^\ell]$.

H* 1　$\mathrm{Re}\sigma\{f_{0u}(p^i(x), x)\} < 0$ for $x \in [x^{i-1}, x^i]$, $i \in \mathbb{Z}$.

Using a stretched variable $\xi = \dfrac{x - x^i}{\epsilon}$, we assume that the 0th order expansion of Eq. (17),
$$u_{\xi\xi} + f_0(u, x^i) = 0. \tag{36}$$
has a heteroclinic solution $q^i(\xi)$ connecting $p^i(x^i)$ to $p^{i+1}(x^i)$. Assume

H* 2　The linear homogeneous equation
$$\phi_{\xi\xi} + f_{0u}(q^i(\xi), x^i)\phi = 0,$$
has a unique bounded solution $q_\xi^i(\xi)$, up to constant multiples.

From **H*2**, we infer that the adjoint equation
$$\psi_{\xi\xi} + f_{0u}^\tau(q_\xi^i(\xi), x^i)\psi = 0$$
has a unique bounded solution $\psi_i(\xi)$ up to constant multiples.

The following assumption implies that the heteroclinic solution breaks as x moves away from x^i.

H* 3　$\int_{-\infty}^{\infty} \psi_i^\tau(\xi) f_{0x}(q^i(\xi), x^i) d\xi \neq 0$, $\quad i \in \mathbb{Z}$.

When x is in a neighborhood of x^i, we look for traveling wave solutions with wave speed $V^i(x)$. It is also of interest to find out conditions to ensure that the wave front moves towards x^i. Consider $\mathcal{A}^i : L^2(\mathbb{R}) \to L^2(\mathbb{R})$. defined as
$$\mathcal{A}^i u = u_{\xi\xi} + f_{0u}(q^i(\xi), x^i)u, \quad D(\mathcal{A}^i) = H^2(\mathbb{R}).$$

H* 4　$\lambda = 0$ is a simple eigenvalue for $\mathcal{A}^i$, $i \in \mathbb{Z}$. There exists $\sigma_0 > 0$ such that
$$\sigma(\mathcal{A}^i) \subset \{\lambda = 0\} \cup \{\mathrm{Re}\lambda \leq -\sigma_0\}.$$

Since $\lambda = 0$ is simple, we have
$$\int_{-\infty}^{\infty} \psi_i^\tau(\xi) q_\xi^i(\xi) d\xi \neq 0.$$

H* 5 $I = [\int_{-\infty}^{\infty} \psi_i^\tau(\xi) q_\xi^i(\xi) d\xi]^{-1} \int_{-\infty}^{\infty} \psi_i^\tau(\xi) f_{0x}(q^i(\xi), x^i) d\xi > 0$, $i \in \mathbb{Z}$.

Under Hypotheses **H*1**–**H*5**, it is shown, in [12], that formal expansions of wave front positions $\{\eta^i(t, \epsilon)\}$ and matched expansions of solutions $\{u^i(x, t, \epsilon)\}$ can be constructed. These hypotheses are general in the sense that **H*2** and **H*3** are generic assumptions. Hypothesis **H*1** is equivalent to the stability of $p^i(x)$ in a regular layer, as an equilibrium of an ODE, obtained from the 0th expansion of (17), i.e., by setting $\epsilon u_{xx} = 0$. Hypothesis **H*4** is equivalent to the stability of the traveling wave solution q^i in an internal layer, in the sense of Evans [4]. Hypothesis **H*5** implies that the wave front is moving towards x^i, if x is near x^i. In fact, if is shown in [12]

$$\frac{\partial V^i(x^i)}{\partial x} = -I \cdot (x - x^i).$$

Similar conditions like **H*5** was used by Fife [5]. Without such conditions, the wave front may move away from the stationary position and the formal solutions may not exist globally in time.

It is shown in [13] that Hypotheses **H*1**–**H*5** imply Hypotheses **H1**–**H4** of this paper. Only **H4** needs some explanation. When $\{w^i\}$ is a 0th order approximation, $\lambda = 0$ is always an eigenvalue. When higher order terms are added, the eigenvalue becomes $O(\epsilon)$. In [13], a precise formula expressing $\lambda^i(\epsilon)$ as a function of the wave speed and the wave front position was proved. In particular, when $t = \infty$, it yields

$$\frac{\partial \lambda^i}{\partial \epsilon} = \frac{\partial V^i}{\partial x}.$$

Here $V^i x)$ is the wave speed as a function of the wave position x. It is now clear that **H*5** implies **H4**.

Many authors have found that the signs of $\lambda^i(\epsilon)$ or the derivatives of the wave speed determine the stability of the multiple waves, [6]. The precise relation between the two quantities in general systems is interesting in its own right right.

In conclusion, the formal approximations obtained in [12] satisfies our hypotheses in Part II. Therefore, the *Global Spatial Shadowing Lemma* can be used to ensure the existence of a global solution near the formal approximation.

References

[1] N. Alikakos, P. Bates and G. Fusco, *Slow motion for the Cahn-Hilliard equation in one space dimension*, J. Differential Equations **90** (1991), 81–135.

[2] S.-N. Chow, X.-B. Lin and K. Palmer, *A shadowing lemma with applications to semilinear parabolic equations*, SIAM J. Math. Anal., **20**, (1989), 547–557.

[3] W. A. Coppel, *Dichotomies in stability theory*, A. Dold & B. Eckmann eds. Lecture Notes in Math. **629**, Spring-Verlag, New York, 1970.

[4] John W. Evans, *Nerve Axon Equations: I Linear Approximations*, Indiana Univ. Math. J., **21** (1972), 877–885

[5] P. C. Fife, *Pattern formation in reacting and diffusing systems.* J. Chem. Phys., **64** (1976), 554–564.

[6] G. Fusco and J. Hale. *Slow motion manifolds, dormant instability. and singular perturbations,* J. Dynamics and Differential Equations **1** (1989), 75–94.

[7] R. A. Gardner & C. K. R. T. Jones, *Traveling waves of a perturbed diffusion equation arising in a phase field model,* Indiana Univ. Math. J., **38** (1989), 1197–1222.

[8] J. Guckenheimer, J. Moser & S. Newhouse, Dynamical Systems, Birkhäuser, Boston, 1980.

[9] D. Henry, *Geometric theory of semilinear parabolic equations.* Lecture Notes in Math. vol **840**, Springer-Verlag. 1982.

[10] K. Kirchgässner, *Homoclinic bifurcation of perturbed reversible systems,* W. Knobloch and K. Smith, eds., Lecture Notes in Math., 1017 (1982). Springer-Verlag, New York, 328–363.

[11] X.-B. Lin, *Shadowing lemma and singularly perturbed boundary value problems,* SIAM J. Appl. Math., **49** (1989), 26–54.

[12] X.-B. Lin, *Asymptotic expansion for layer solutions of a singularly perturbed reaction-diffusion system,* preprint, 1994.

[13] X.-B. Lin, *Shadowing matching errors for wave-front like solutions.* preprint

[14] J. L. Lions and E. Magenes, *Non-homogeneous boundary value problems and applications I,* Springer-Verlag, New York, 1972.

[15] M. Renardy, *Bifurcation of singular solutions in reversible systems and applications to reaction-diffusion equations,* Advances in Mathematics **3**, (1982). 324–406.

[16] K. Yosida, *Functional Analysis,* Grundlehren der mathematischen Wissenschaften **123**, Springer, New York, 1980.

Progress in Nonlinear Differential Equations
and Their Applications, Vol. 19
© 1996 Birkhäuser Verlag Basel/Switzerland

Estimation of dimension and order of time series

Floris Takens[*]

1 Introduction

Since the beginning of the eighties there has been an expanding activity in analyzing time series in terms of (reconstructed) attractors and their dimensions. For
a surveys and further references, see [ABST,1993], [C,1991], and [GSS,1991]. The
notion of dimension, which was based on considerations concerning *deterministic*
time series, describes roughly how many parameters one needs to specify the possible states on the attractor of the underlying dynamical system. Another approach,
proposed by Cheng and Tong, is to analyze a time series primarily in terms of the
order, which can be heuristically described as the number of successive elements
of a time series which determine the state of the underlying system [CT,1992];
this notion was introduced in the context of non-deterministic systems. the *state*
should be interpreted as an (abstract) notion which summerizes the information
from the past, as far as it is of influence on the future; this notion of state makes
sense for both deterministic and stochastic systems. This same idea of order was
also studied by Savit and Green [SG,1991]. Their approach is closer to the above
mentioned considerations concerning deterministic systems. The notions of order
and dimension are different, but there are clear relations. e.g. the order should at
least be equal to the dimension. The main purpose of this paper is to give a survey
of these different approaches and to discuss in this context some new numerical
examples.

First we fix some notation and definitions. We consider *time series* $\mathcal{X} = \{X_n\}_{n=0}^{\infty}$, with time parametrized by the integers $n \in \mathbf{N}$, and taking values in some
finite dimensional vector space W, which, when not explicitly stated otherwise, is
assumed to be $\mathbf{R}$. Time series of finite length, i.e. with n running from 0 to some
N, will be considered as samples of (potentially) infinite time series. From a time
series one can obtain a new one by the process of *reconstruction*: for $\mathcal{X}$ as above,
the time series $\mathcal{R}_k(\mathcal{X})$, obtained by an order k, or a k-dimensional, reconstruction
of $\mathcal{X}$, has as its nth element $(\mathcal{R}_k(\mathcal{X}))_n = (X_n, \cdots, X_{n+k-1}) \in \mathbf{R}^k$ — these elements
are called k-dimensional *reconstruction vectors*. So each time series equals its order
one reconstruction.

[*]Department of Mathematics, University of Groningen, P.O.Box 800, 9700 AV Groningen,
The Netherlands

Although the present use of reconstructed time series, and the corresponding reconstruction measures. see below, was strongly motivated by results from the theory of deterministic dynamical systems, see [T,1981] and [PCFS,1980], they were already used in 1927 by Yule in studying the sunspot numbers [Y,1927].

Next we assume that all our time series are *stationary* in the sense that there are well defined measures, the so-called *reconstruction measures*. $\mu_k(\mathcal{X})$ on $\mathbf{R}^k$ corresponding to the density of the elements of $\mathcal{R}_k(\mathcal{X})$ in $\mathbf{R}^k$; a formal definition is given in section 2. We note that the measure $\mu_k(\mathcal{X})$ only describes the statistical properties of $\mathcal{R}_k(\mathcal{X})$. as opposed to the dynamical properties, in the sense that these measures remain the same if we apply a permutation $n \mapsto \sigma(n)$ to the indices of $\mathcal{R}_k(\mathcal{X})$, at least if the differences $n - \sigma(n)$ are uniformly bounded. The measure $\mu_{k+1}(\mathcal{X})$ contains both the statistical properties of $\mathcal{R}_k(\mathcal{X})$ and a certain amount of its dynamical properties: the statistical properties, because μ_k can be obtained from μ_{k+1} by taking the marginal probability distribution of the first k components of the vectors in $\mathbf{R}^{k+1}$. and the dynamical properties, because μ_{k+1} determines the probability distribution of elements of $\mathcal{R}_k(\mathcal{X})$ in terms of the element immediately preceding it. For the original time series $\mathcal{X}$, the reconstruction measure $\mu_{k+1}(\mathcal{X})$ determines the probability distribution of elements of $\mathcal{X}$ in terms of the last k elements immediately preceding it. The latter conditional probability measures are denoted by $\mu_{(x_1, \cdots, x_k)}$. where $x_1, \cdots, x_k$ stand for the k preceding elements of $\mathcal{X}$, as a function of which the probability distribution is given.

The dimensions. referred to in the beginning, are defined in terms of the reconstruction measures (one may think of the topological dimension of the support of these measures). In the next section we discuss this in more detail. The order of a time series if it exists can be defined in terms of the reconstruction measures. The formal definitions in the approaches of Cheng and Tong and of Savit and Green are different, but a heuristic description, which applies to both. is the following. The order is the lowest integer k such that, from the point of view of predicting future elements, on the basis of the reconstruction measures and a (finite) number of immediately preceding elements, nothing is gained by using more than the k immediately preceding elements.

2   Dimension of a probability measure and its estimation

We consider a probability measure μ on $\mathbf{R}^l$. If this measure is concentrated on some n-dimensional sub manifold, and if the measure has a continuous density with respect to the Lebesgue measure on that sub manifold, then the probability to find two randomly and independently chosen points within distance ε is proportional to ε^n, asymptotically for small ε. This suggests a definition of the dimension of a probability measure which we now describe. For the probability measure μ we first define the *correlation integrals* $P_\varepsilon(\mu)$ as the $\mu \times \mu$-measure of $\{(x, y) \in \mathbf{R}^{2l} | \|x - y\| < \varepsilon\}$ — so $P_\varepsilon(\mu)$ is the probability to find two points, μ-randomly and independently chosen, within distance ε. Then we define the dimension of μ as

$$D(\mu) = \lim_{\varepsilon \to 0} \frac{\ln(P_\varepsilon(\mu))}{\ln(\varepsilon)}.$$

If this limit does not converge, one may take the limsup or the liminf. In this paper we don't want to emphasize these convergence problems. The dimension, as defined here is called the *correlation dimension*.

Some remarks concerning this definition are in order:

1. For probability measures on $\mathbf{R}^l$ with continuous density the dimension is l, so for such measures this is not a useful invariant. Below we shall come back to this point.

2. The dimension, as defined above may take non-integer values.

3. There are many variations on this definition of dimension which are all based on a combination of a probability measure and a metric structure, i.e. using both probability and distance, see [HJKPS,1986].

4. This notion of dimension is often applied to reconstruction measures of time series as defined in the introduction. The motivation for this is that if a time series is produced by a deterministic system, then one can prove, under weak additional assumptions, that this dimension is the same for all reconstruction measures μ_k, for k sufficiently big, see [SYC.1991]. At the end of this section we discuss this in more detail, and in particular explain what we mean by 'a time series produced by a deterministic system'. For stochastic time series, i.e. due to systems with noise, this notion of dimension is not directly meaningful, since, in that case, the reconstruction measures usually have a continuous density and are not concentrated on a lower dimensional object.

Next we come to the *estimation* of this dimension. Since we are thinking of measures, like the reconstruction measures, which are defined as densities of (infinite) sequences of points, we describe the estimation problems also in terms of such (potentially) infinite sequences of points. So instead of a measure μ we start with a sequence $\{p_n\}_{n=0}^{\infty}$ of points in $\mathbf{R}^l$, which we assume to be bounded. The relation between the measure μ and the sequence $\{p_n\}_{n=0}^{\infty}$ is that for any continuous function f on $\mathbf{R}^l$, with bounded support, we have that

$$\int f d\mu = \lim_{n \to \infty} (n+1)^{-1} \sum_{i=0}^{n} f(p_i).$$

Since the dimension is defined in terms of correlation integrals, the straightforward thing to do is to first consider the estimation of these correlation integrals. These are estimated from finite sequences of points by just counting the fraction of pairs of different points which are within distance ε. If we use a sequence $\{p_i\}_{i=0}^{N}$ of a fixed length, then the estimates of $P_\varepsilon(\mu)$ will have greater relative variances for lower values of ε. Indeed, if all the distances between pairs of points p_i, p_j, $i < j$, can be considered as independent, then the standard deviation of the estimate of $P = P_\varepsilon(\mu)$ is

$$\sigma = \sqrt{\frac{2P(1-P)}{N(N-1)}}.$$

If we assume that $P = P_\varepsilon(\mu)$ tends to zero with ε, we see indeed that the relative standard deviation

$$\sigma/P = \sqrt{\frac{2(1-P)}{PN(1-N)}}$$

goes to ∞ for $\varepsilon \to 0$. So this direct approach of estimating the dimension fails when we try to take the limit $\varepsilon \to 0$.

As we mentioned, for the above reasoning it is essential that all the distances can be considered as independent. To justify this, we need first that the points p_n are uncorrelated and second that ε is sufficiently small, see [S.1992]. The independence of the points p_i is mainly a problem if we use oversampled time series from a process with continuous time — in this paper we shall not consider such examples. For a much more detailed discussion see [KS,1995]. In order to keep the arguments as simple as possible, we shall continue to use the above estimate for standard deviation of the error when estimating correlation integrals.

A different approach to estimating the dimension is the following. We first choose a cutoff distance ε_0 and then determine all the distances $r_{i,j}$ between pairs of different points p_i and p_j as far as they are smaller than ε_0. Then we *assume* that these distances are distributed proportionally to $r^{D-1}dr$ and estimate D by a maximum likelihood procedure [T,1985]. This leads to an estimate for D which equals

$$\hat{D} = -1/\mathcal{E}(ln(r_{i,j}/\varepsilon_0)),$$

where $\mathcal{E}$ stand for taking the average over all $0 \leq i < j \leq N$. for which $r_{i,j} \leq \varepsilon_0$. Assuming as before, when discussing the errors of estimates of the correlation integrals, that ε_0 is sufficiently small and that the points p_n are sufficiently uncorrelated, the standard deviation of the error in this last estimate, due to the finiteness of the time series, can be estimated as

$$\hat{\sigma} = \hat{D}/\sqrt{m},$$

where m is the number of distances $r_{i,j}$, $i < j$, which are smaller than ε_0.

Apart from the error due to the fact that the time series has finite length, there is the point that we still should take the limit $\varepsilon_0 \to 0$. Instead of trying to do this, we use this estimate as the basis of our definition of dimension *at the fixed length scale ε_0*. This avoids the problem that, for small values of ε_0, also this method gives big errors (this time because the number m of distances $r_{i,j}$ which are smaller than ε_0 decreases). In this way we obtain a definition of dimension which is still useful for reconstruction measures, which are, e.g. due to 'noise', not concentrated on a low dimensional object but only in a small neighbourhood of such an object — if this is an ε_1 neighbourhood, we should have $\varepsilon_1 << \varepsilon_0$. The formal definition of the dimension of μ at the length scale ε_0 is

$$D_{\varepsilon_0}(\mu) = -\frac{\int_{\|x-y\|<\varepsilon_0} 1 d(\mu \times \mu)}{\int_{\|x-y\|<\varepsilon_0} \ln(\|x-y\|/\varepsilon_0) d(\mu \times \mu)}.$$

We call this the ML (maximum likelihood) dimension of μ at the length scale ε_0. This is of course the quantity which is estimated by the above procedure.

Some remarks concerning this definition:

1. If the limit in the definition of $D(\mu)$ converges, then, for $\varepsilon_0 \to 0$, $D_{\varepsilon_0}(\mu)$ converges to $D(\mu)$.

2. For a probability measure μ on $\mathbf{R}^l$, we call $\bar{\mu}$ an ε-perturbation of μ if for each subset $A \subset \mathbf{R}^l$, with ε-neighbourhood A_ε. $\bar{\mu}(A_\varepsilon) \leq \mu(A)$. If $D_{\varepsilon'}(\mu)$ is smaller than the dimension l of the ambient space, uniformly for ε' in a δ-neighbourhood of ε_0, then also for a ε perturbation $\bar{\mu}$ of μ, $D_{\varepsilon_0}(\bar{\mu}) < l$, provided ε is sufficiently small (compared with ε_0 and δ). This means that, if these ML estimates of the dimensions at positive length scales are lower than the dimension of the ambient space, this has is persistent with respect to ε-perturbations, which correspond to the introduction of (small) noise.

As we mentioned before, these estimates of dimensions are applied to the reconstruction measures of time series. This was motivated by considerations concerning deterministic systems. Time series, generated by deterministic systems are defined in the following way. First, a (deterministic) dynamical system (with discrete time) consists of a state space Y and an evolution map (or generator) $\varphi : Y \to Y$, assigning to each state p the state $\varphi(p)$ which will occur one unit of time after the state p — we assume Y to be a finite dimensional vector space, φ to be differentiable and we assume positive orbits $\{\varphi^n(p)\}_{n\geq0}$ to be bounded. In order to describe the way in which such dynamical systems generate time series, we need to introduce a read-out map $y : Y \to \mathbf{R}$. assigning to each state the value which is recorded when the system is in that state: so the time series corresponding to the (positive) orbit $\{\varphi^n(p)\}_{n\geq0}$ is $\mathcal{X} = \{x_n = y(\varphi^n(p))\}_{n\geq0}$. We consider also the order k reconstructions of these time series. Instead of constructing these from $\mathcal{X}$, we can also obtain them by replacing the function y by $y_k = (y, y \circ \varphi, \cdots, y \circ \varphi^{k-1}) : Y \to \mathbf{R}^k$. Note that with these conventions, $y = y_1$. From the reconstruction theorem, see [T,1981] and [SYC,1991], we know that for generic φ and y, and for k sufficiently big, y_k defines an embedding of the state space Y in $\mathbf{R}^k$. This means that no two different points of the state space are mapped by y_k to the same point and that, if we restrict to a bounded part of Y, the ratio between distances in Y and the corresponding distances in $\mathbf{R}^k$ is uniformly bounded and bounded away from zero. This implies that, for a time series

$\mathcal{X}$ as above, the correlation dimensions of all the reconstruction measures μ_k, for k sufficiently big (in the sense of the reconstruction theorem), are the same. These dimensions may however depend on the initial point $p \in Y$ of the orbit used, since for different initial points one may be attracted to different attractors.

We noted earlier that for time series of stochastic systems, one expects the dimension of μ_k to be k. So we see how the estimation of the dimensions can be used to obtain an indication whether a time series is generated by a deterministic or by a stochastic system.

If we analyze the reconstruction measures in terms of the ML dimensions at a fixed length scale ε_0 we should find a similar difference between deterministic and stochastic time series, but, due to the finite (non-zero) length scale there are some differences:

1. Even for deterministic time series, the estimated dimension will in general increase with increasing order of the reconstruction. This is due to the fact that, for chaotic dynamical systems, the dependence of $\varphi^n(p)$ on p is extremely erratic even for values of n which are not very big. This means that if we look on a fixed length scale ε_0 we may not see the fine structure revealing the fact that all the reconstruction vectors are in a low dimensional set. We illustrate this in figure 1 where we plot the points $(\varphi^i(p), \varphi^{i+5}(p))$ for the case of the logistic map $\varphi(x) = 4x(1-x)$. From this diagram it is clear that if we analyze the reconstruction measures on a length scale of, say, $\varepsilon_0 = .2$, this configuration of points looks 2-dimensional. This means that if we consider reconstructions of higher order of a time series, generated by this logistic map (with read-out map $y = $ identity), we expect the estimated dimension (at the same length scale) to go up by at least one unit for every time we raise the order by five units. In our numerical simulations this effect can be seen in the analysis of the time series denoted by 2LOGIST.

2. When increasing the length scale ε_0 the estimated dimension will decrease, even in the case of a stochastic time series. For extreme values of ε_0 this follows from the fact that the values of $r_{i,j}/\varepsilon_0$ will all be close to zero.

3   Order estimation according to Cheng and Tong

The above discussion of dimension estimation was primarily motivated by time series from deterministic systems. The order estimation, which we briefly describe in this section, and which was proposed by Cheng and Tong, see [CT,1992] and [CT,1994], was primarily motivated by the consideration of time series produced by stochastic models of the form

$$X_n = f(X_{n-k}, \cdots, X_{n-1}) + \varepsilon_n,$$

where ε_n is a random element (independent for different values of n) with respect to a probability distribution with expectation value zero, which may depend on

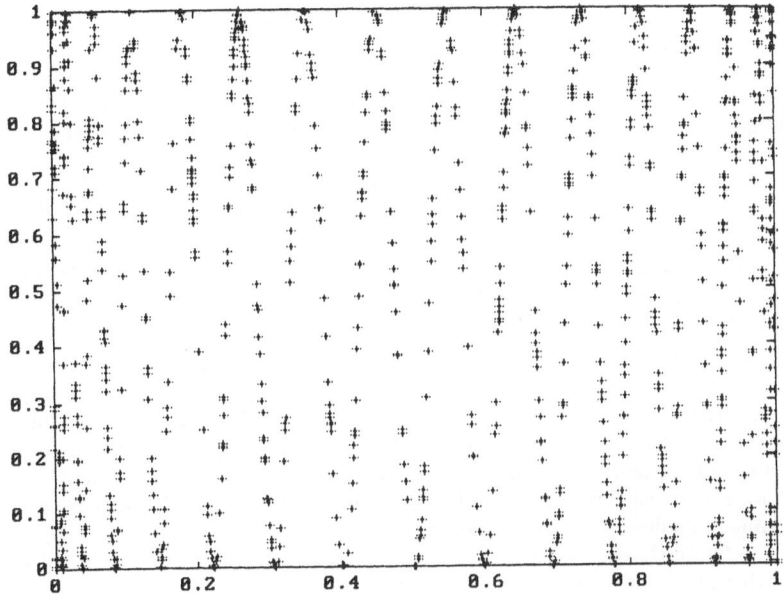

Figure 1: A plot of 1000 points $(\varphi^i(p), \varphi^{i+5}(p))$ for the logistic map

the values of $X_{n-k}, \cdots, X_{n-1}$, but not (explicitly) on n. For this class of models, and for the time series generated by these models, the order, as defined in the introduction, is k, at least if f depends in a non-trivial way on its first argument. Models of this form are called *nonlinear autoregressive models*.

The order estimation as we discuss it here is a procedure for estimating this order k from a time series of which the model is not known. The method is based on analyzing the predictability of the time series in terms of nonlinear regression.

This means that we first estimate in principle the functional relation f_l : $\mathbf{R}^l \to \mathbf{R}$ between l-dimensional reconstruction vectors and their next element in the time series, i.e. between $(\mathcal{R}_l(\mathcal{X}))_n = (X_n, \cdots, X_{n+l-1})$ and X_{n+l}, taking into account that the functional relation is not exact, but perturbed by noise. Without going into all details, the idea is that in order to estimate the value of f_l at the a vector Ξ. we consider the set of l-dimensional reconstruction vectors which are within some distance $h > 0$ from Ξ, where h will be referred to as the *bandwidth*. The value of $f_l(\Xi)$ is then estimated to be the average of the next elements of each of the selected (nearby) reconstruction vectors (in order to avoid discontinuities one considers all the reconstruction vectors, but with a weight function, depending on the distance from Ξ, putting the main emphasis on vectors within distance h). Note that the value of f_l in points, far away from the l-dimensional reconstruction vectors, say at a distance greater than h, is poorly estimated, but for the following of no importance. So this gives a procedure to predict a new element of the time series in terms of its l preceeding elements, which depends on the bandwidth h. We refer to this as nonlinear regression with bandwidth h.

Next we want to estimate the variance of the errors of these predictors. This is done by a so-called cross validation procedure: for each reconstruction vector $(\mathcal{R}_l(\mathcal{X}))_n$ we compare the value of the next element X_{n+l} of the time series with the value $\hat{f}_{l,n}((\mathcal{R}_l(\mathcal{X}))_n)$ and use this as an estimate of the prediction error at $(\mathcal{R}_l(\mathcal{X}))_n$. Here $\hat{f}_{l,n}$ is an estimate for f_l, essentially obtained as above, but without using the nth reconstruction vector $(\mathcal{R}_l(\mathcal{X}))_n$ and its successor (otherwise we would use a predictor based on foreknowledge). The average of all the squares of these estimated errors is then the estimate of the variance of the errors of the non-linear regression of order l and bandwidth h. Next we determine numerically the optimal value of h, being the value for which the estimate of the variance of the prediction error is minimal. This minimal value is then our estimate $\hat{\sigma}_l^2$ of the variance of the errors of nonlinear regression at order l.

This estimation is repeated for various values of l and the estimated order $\hat{k}$ is then the value for which $\hat{\sigma}_{\hat{k}}^2$ is minimal.

It was proved in [CT,1992], under some additional assumptions, that, for time series from nonlinear autoregressive models, the above procedure yields a consistent estimate for the order of a time series. Consistency means here that the probability that the estimate gives a wrong result approaches zero if the length of the time series, to which the procedure is applied, increases. In [CT,1994] it was proved that the data requirement for this order estimation is much less than for the usual dimension estimation: the required length of the time series is proportional to the square of the order, where for estimating the dimension one needs the length to be proportional to the exponential of the dimension [R,1990].

We conclude this section with some remarks on the relation between these nonlinear autoregressive models and the reconstruction measures defined before. If $\mathcal{X}$ is a time series generated by the autoregressive model

$$X_n = f(X_{n-k}, \cdots, X_{n-1}) + \varepsilon_n,$$

then $f(X_{n-k}, \cdots, X_{n-1})$ is just the expectation value of $\mu_{(X_{n-k}, \cdots, X_{n-1})}$, the conditional probability measure obtained from $\mu_{k+1}(\mathcal{X})$ as before, and the probability distribution, from which ε_n is taken, is, up to a translation in order to make the expectation value zero, equal to $\mu_{(X_{n-k}, \cdots, X_{n-1})}$. So, heuristically, one can say that the order of a time series is at most k if no new information, useful for making better prediction, can be obtained from the reconstruction measures $\mu_{k'}(\mathcal{X})$ for $k' \geq k+1$. Note that due to the reconstruction theorem, the order, in the present sense, of a time series is from a deterministic system is always finite.

4 Order estimation according to Savit and Green

In this section we give a description of the order estimation following the ideas of Sevit and Green, see [SG,1991]. It is based on the correlation integrals. In order not to overload the notation, we denote them here and in the next section by $P^k(\varepsilon)$ instead of $P_\varepsilon(\mu_k(\mathcal{X}))$; We define $P^0(\varepsilon)$ to be zero for all values of ε. Our presentation, in terms of reduction factors, differs somewhat from the original presentation: we found the presentation of the results of simulations in terms of the reduction factors convenient.

In the simplest case (order zero) this approach is the same as the B.D.S. test, see [BHL,1991], which can be summarized as follows. On the basis of the heuristic description of the notion of order at the end of the last section, one could say that a time series has order zero if it can be considered as a sequence of i.i.d. samples of a fixed distribution (i.i.d. meaning identically and independently distributed). Indeed in that case one obtains no new information, leading to better predictions, from the previous values of the time series — the only relevant information is in the order one reconstruction measure. Such a test for a time series to be i.i.d., in terms of correlation integrals, is based on the observation that if a time series is i.i.d., then for all ε and k, $P^k(\varepsilon) = (P^1(\varepsilon))^k$. Note that, for this relation to hold, we need to take as distance between reconstruction vectors the maximal difference of their components. From now on we assume that this maximum, or l_∞, norm is used whenever the correlation integrals of reconstruction measures are considered. The B.D.S. test consists of verifying whether these inequalities hold, within the limits of the expected estimation errors, for the estimated values of $P^k(\varepsilon)$.

This test is related with predictability in the following way. We define the *reduction factors* as $p_k(\varepsilon) = P^{k+1}(\varepsilon)/P^k(\varepsilon)$, for $k = 0, 1, \cdots$. This is a measure for the predictability of a time series *at length scale* ε in terms of the k preceding elements: if we want to predict a new element X_n we look for a (past) value of m such that the reconstruction vector $(X_{m-k}, \cdots, X_{m-1})$ is componentwise within distance ε from the reconstruction vector $X_{n-k}, \cdots, X_{n-1}$ — then the prediction that X_n will be within distance ε from X_m will be correct with a probability $p_k(\varepsilon)$. This follows directly from the definition of the correlation integrals.

If $P^k(\varepsilon) = (P^1(\varepsilon))^k$, then $p_k(\varepsilon)$ is independent of k which is consistent with the interpretation of order zero as meaning that prediction cannot be improved by using information about previous elements (other than in the form of information from the reconstruction measures).

Although we can use the reduction factors $p_k(\varepsilon)$ to compare the predictability, at a fixed length scale ε, in terms of the number of preceding terms used, it cannot be used easily to compare the predictability at different length scales: often for $\varepsilon < \varepsilon'$ one has that $p_k(\varepsilon) < p_k(\varepsilon')$, but this cannot be used as an indication that the predictions at the bigger length scale are better because it only means that the predictions at the bigger length scale have a greater probability of being correct, but, since the prediction interval is bigger, these predictions are weaker.

Here we point out that in [SG,1991] the time series are analyzed not in terms of the reduction factors, but in terms of the quantities

$$\delta_j(\varepsilon) = 1 - (P^j(\varepsilon))^2 / P^{j-1}(\varepsilon) P^{j+1}(\varepsilon)$$

and

$$S_j = P^1(\varepsilon)/(1 - \delta_1(\varepsilon))(1 - \delta_2(\varepsilon)) \cdots (1 - \delta_j(\varepsilon)).$$

We estimate $p_k(\varepsilon)$ by taking the quotient of the estimates for $P^{k+1}(\varepsilon)$ and $P^k(\varepsilon)$. In order to derive an estimate for the variance of the estimated reduction factor $p_k(\varepsilon)$ we proceed as follows. Suppose we find m different pairs of k-dimensional reconstruction vectors $((\mathcal{R}_k(\mathcal{X}))_i, (\mathcal{R}_k(\mathcal{X}))_j)$, with $i < j$, which are within distance ε. For each of these pairs we can test whether the $k+1$-dimensional reconstruction vectors, obtained by adding the next element in the time series to each of the reconstruction vectors, are still within distance ε. The probability that this is the case is by definition equal to $p_k(\varepsilon)$. If we take 1, resp. 0. as the outcome of the test if the new vectors are, resp. are not, within distance ε, the expectation value of the outcome is $p_k(\varepsilon)$ and its variance is $p_k(\varepsilon)(1 - p_k(\varepsilon))$. The estimation of $p_k(\varepsilon)$ is equivalent to performing this test for all the m pairs and taking the average. This leads to the expectation value $p_k(\varepsilon)$ and standard deviation $\sqrt{p_k(\varepsilon)(1 - p_k(\varepsilon))/m}$.

The above consideration suggests an alternative definition of order, this time depending on a length scale ε, as the smallest value k such that $p_{k'}(\varepsilon)$ is independent of k' as long as $k' \geq k$.

Some remarks concerning this order at length scale ε:

1. When estimating this order it is important to keep in mind that it can be underestimated due to lack of data: if the improvements of $p_k(\varepsilon)$ with increasing k are small, they may be invisible due to the errors when estimating $p_k(\varepsilon)$, especially for bigger values of k and smaller values of ε, where these errors are bigger. We shall not pursue this point here but, for our numerical simulations, just show the estimated values of $p_k(\varepsilon)$ as function of k.

2. We remind that the quantities $p_k(\varepsilon)$ as defined above are related with the correlation entropy: if the limit

$$\lim_{\varepsilon \to 0} \lim_{k \to \infty} p_k(\varepsilon)$$

exists, it equals the correlation entropy. Especially for time series generated by systems with noise, the last limit $\varepsilon \to 0$ usually diverges to ∞ while the first limit $k \to \infty$ usually exists for fixed values of ε.

3. In spatially extended systems it may be that the order is not defined (or infinite) but that the order at some positive length scale is well defined. Discussing the numerical simulations we shall come back to this.

5   Numerical simulations

In this section we discuss a number of numerical simulations related to the items discussed above, in particular the estimation of dimension and order in in the sense of Savit and Green. First we describe how the results are presented in the figures below, then we comment on the results for the different time series.

5.1   Organization of the figures

For each of the time series we show first a general survey containing all the computed correlation integrals, then we show the results of the estimations of dimensions an reduction factors at specified length scales. For each of the figures the underlying time series is specified by a name, indicating the dynamical system which generated the time series, and the length of the time series, e.g. HENON 1000 means that we used a time series obtained by iterating the Hénon map (with the standard parameter values $a = 1.4$ and $b = .3$) 1000 times.

Survey of correlation integrals   For each of the time series we first show this survey of all the estimated correlation integrals. Before computing the estimates of the correlation integrals, the time series were linearly scaled to make the difference between the maximum and minimum value equal to one. In all cases the scaling of the axis is the same. Horizontal is the ε axis which has a logarithmic scale running from $.9^{59} \sim .002$ to 1. The vertical axis shows the values of the estimated probabilities, i.e. the estimated values of the correlation integrals, in a logarithmic scale running from .00001 to 1. For each value of $k = 1, \cdots, 20$, the estimates of $P^k(\varepsilon)$ are plotted as a graph of a function of ε; in fact these estimates were calculated for $\varepsilon = .9^i$ with $i = 0. 1, \cdots, 59$. Since for each $k < k'$, we have $P^k(\varepsilon) > P^{k'}(\varepsilon)$, these graphs are ordered so that those with low reconstruction dimension are above those with high reconstruction dimension. The fact that the computations were carried out for ε equal to the successive powers of .9, explains the length scales used below: they are all powers of .9.

Estimated dimensions   In these figures we show the estimated dimensions (vertical axis) as function of the reconstruction dimension (horizontal axis). For each diagram, the value of ε is indicated in the title after 'at distance'. For each reconstruction dimension two values are shown: the estimated value plus and minus the estimated standard deviation. This estimate for the standard deviation was calculated essentially as indicated in section 2. In cases where there were no data, the values were omitted.

Estimated reduction factors   These figures show the estimated reduction factors $p_k(\varepsilon)$ (vertical axis) as function of the reconstruction dimension k (horizontal axis). The value of ε is given in the title after 'at distance'. Also here we show the estimated value plus and minus the estimated standard deviation, estimated as in section 4.

5.2   Remarks on the different time series

HENON 1000 Here we used a time series from the Hénon map, with the usual parameter values $a = 1.4$, $b = .3$, obtained by 1000 iterations. The first coordinate was used as read-out function. The figure surveying the correlation integrals is standard and already appeared in many publications. Note that for small values of the probabilities, i.e. near the bottom of the figure, the curves look shaky, due to the big relative estimation errors. The estimates of the dimensions were made at the length scale .1668 and are consistent with the values in the literature between 1.2 and 1.27, see [GAP,1983]. The reduction factors were estimated at three different length scales: .5905, .1668, and .08863. We expect this reduction factor to start low for reconstruction dimension 0 and then to increase till it reaches the final value — the reconstruction dimension where this final value is reached is then the order of the time series. This is indeed what we sees for the smaller length scales, but not for the biggest length scale where it decreases with time. This effect was also observed in [SG,1991] and attributed to the non-uniformness of expansions. We think that this effect requires further investigation. The same effect also appeared for the other time series which we investigated, usually for length scales bigger than the ones which we show in our figures.

Especially at the smallest length scale it is clear that the correct value of the order (two) can be deduced from the reduction factors.

HENON 100 These figures were obtained from the same Hénon map but now only using 100 iterations. This in order to show how the length of the time series affects the results.

2LOGIST 1000 For the next figures we used a time series obtained by adding two (independent) time series of 1000 iterations of the logistic map $x \mapsto 4x(1 - x)$ with different, and randomly chosen, initial points. From the results we see that the estimation of the dimension works only well at the length scale .08863. The fact that these dimension estimates have the strong tendency to increase with the reconstruction dimension is related with the rather high entropy if this map. One can show analytically that the order of this system is 2. This agrees with the estimates of the reduction factors, though at the length scales .0471 and .02503, one may doubt whether the order is 2 or 3. The fact that this way of estimating the order can also lead to an overestimation was observed in [SG,1991] and also attributed to the expansions not being uniform. It may be also due to the fact that two states, whose reconstruction vectors of dimension $k + 1$ are within distance ε are in general closer than two states whose k dimensional reconstruction vectors are within distance ε, even for systems with order $\leq k$. It may be interesting to observe the following. Since we know the entropy of this map to be $2 \ln 2 = \ln 4$, we know that the limiting value of the reduction factor (for sufficiently small ε and sufficiently big k) should be $1/4$. This is indeed what we find at the length scale .02503. On the much bigger length scale of .08863 however, the reduction factors give already a clear indication of the correct order.

12 CHENG TONG 1400 For the next sequence of figures, we used 1400 iterations of a system which was also used by Cheng and Tong to test their procedure to estimate the order of a time series. It is defined by the following equation:

$$X_n = .1X_{n-1} + (-.5 + .2\exp^{-.1X_{n-N}^2})X_{n-N} + 1.1\varepsilon_n,$$

with $N = 12$ and ε_n random elements from the uniform distribution in $[-.5, .5]$. Note that this equation is in the form of a nonlinear autoregressive model. Clearly the order of this system is $N = 12$. The dimension estimations don't give a clear hint to some specific dimension, only maybe an indication how the dimension estimates grow with the reconstruction dimension.

The estimates of the reduction factors reveal the following aspects. At the length scale .81, we see the same effect as for the time series from the Hénon map at a big length scale: the reduction factor decreases for increasing values of the reconstruction dimension (up to reconstruction dimension 11). Next, at the length scales .5905 and .4305 there is a clear indication that the order is 12. At the next length scale of .3138 no trace of the order 12 can be seen any more.

Comparing these results with the simulations of Cheng and Tong [CT,1994], we found that the method of estimating the order with reduction factors requires more data: with time series of this system with 700 (for which the Cheng Tong method is still reasonably reliable) or less iterations. the results were much less clear. Also we see that, at least in this case there is a rather narrow interval of length scales at which one can successfully detect the order.

SPACIAL HENON 2000 As final example we investigated a time series which was obtained by taking the sum of five independent time series of the Hénon map (with the usual parameters), but multiplied by different factors, namely 1, 1/4, 1/9, 1/16, and 1/25. This was done in order to simulate the effect of a spatially extended system with many (in principle infinitely many) independent sources where one observes a signal which is a sum of the contributions of these different sources, but with a weight depending on the distance. This time series has 2000 elements. The estimations of the dimensions do not seem to converge to a fixed value for increasing reconstruction dimension. It is however remarkable that the estimates of the dimensions at the length scale .3138 are much lower than those at the length scales .5905 and .1668 — I have no interpretation for this.

The estimated reduction factors seem to indicate an order around 10. This happens to agree well with the simple argument: Hénon has order 2, hence we expect an increase of the order by 2 units for each Hénon attractor which we add. This interpretation seems to me to be too simplistic. Also we see that these estimates of the reduction factors give important information about the predictability of this time series: they indicate clearly that the predictability is optimal at the smallest length scale .0471, in particular when basing the predictions on the 10 (or 8 to 11) most recent observations.

Acknowledgement. It was M. Casdagli, who made us aware of the work of Savit and Green.

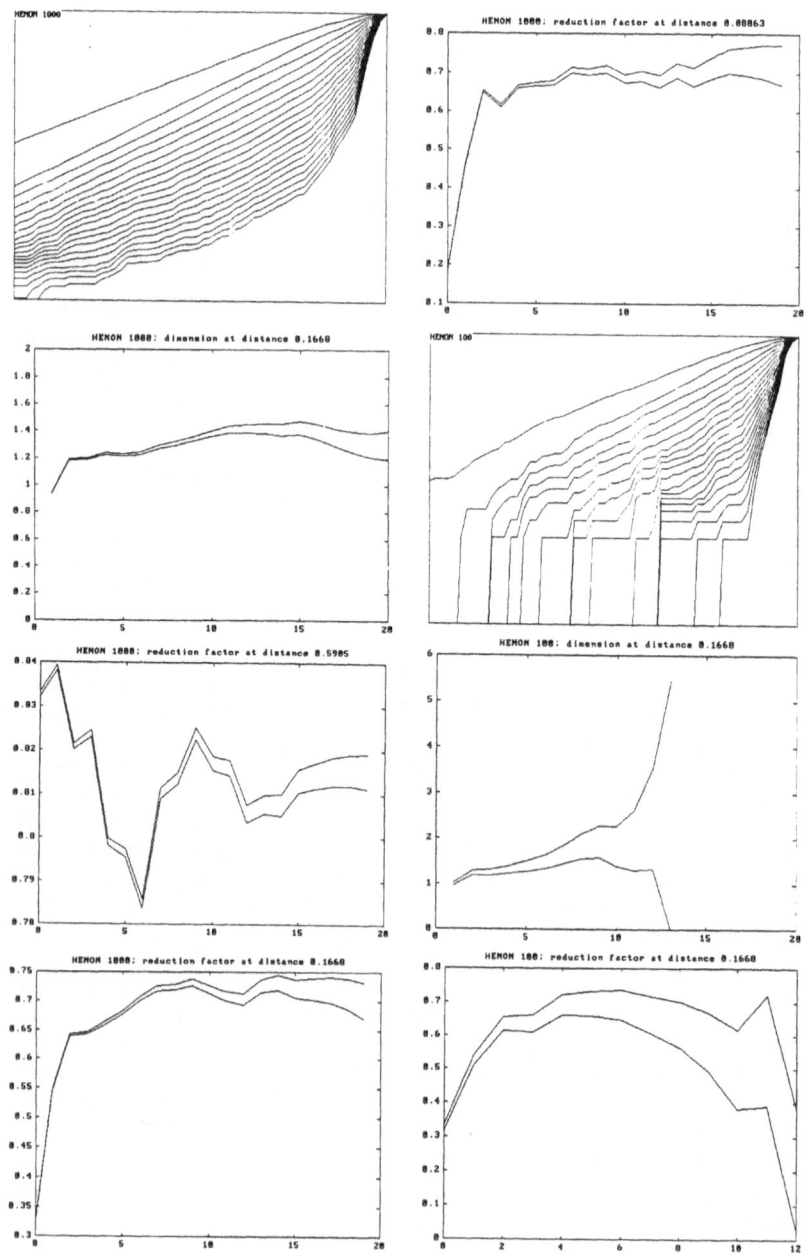

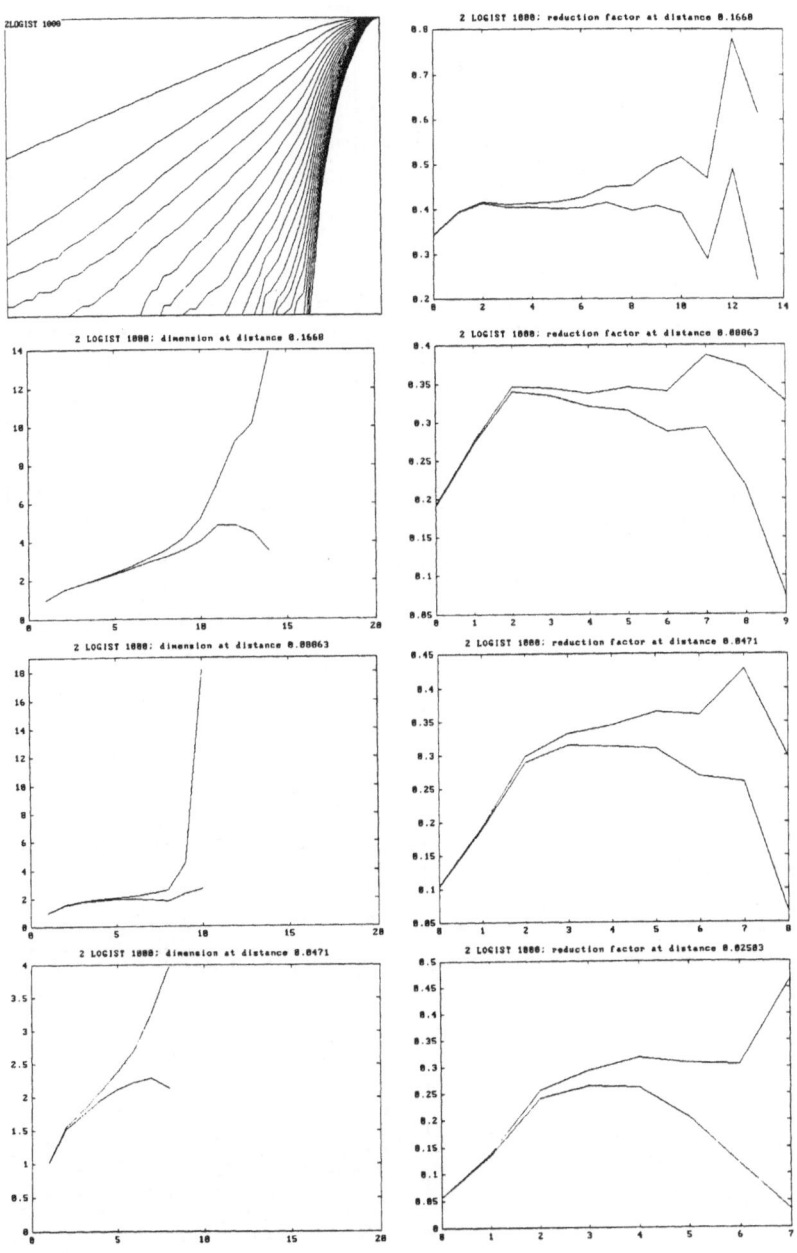

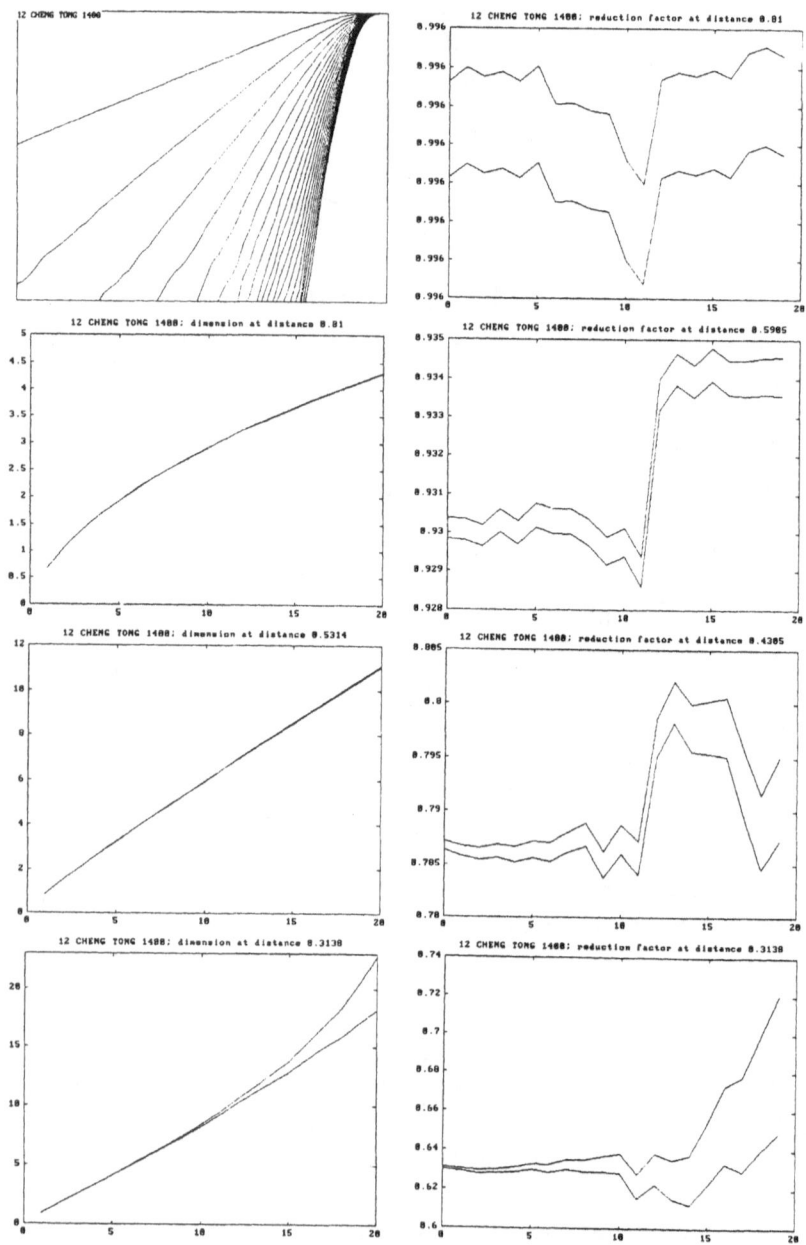

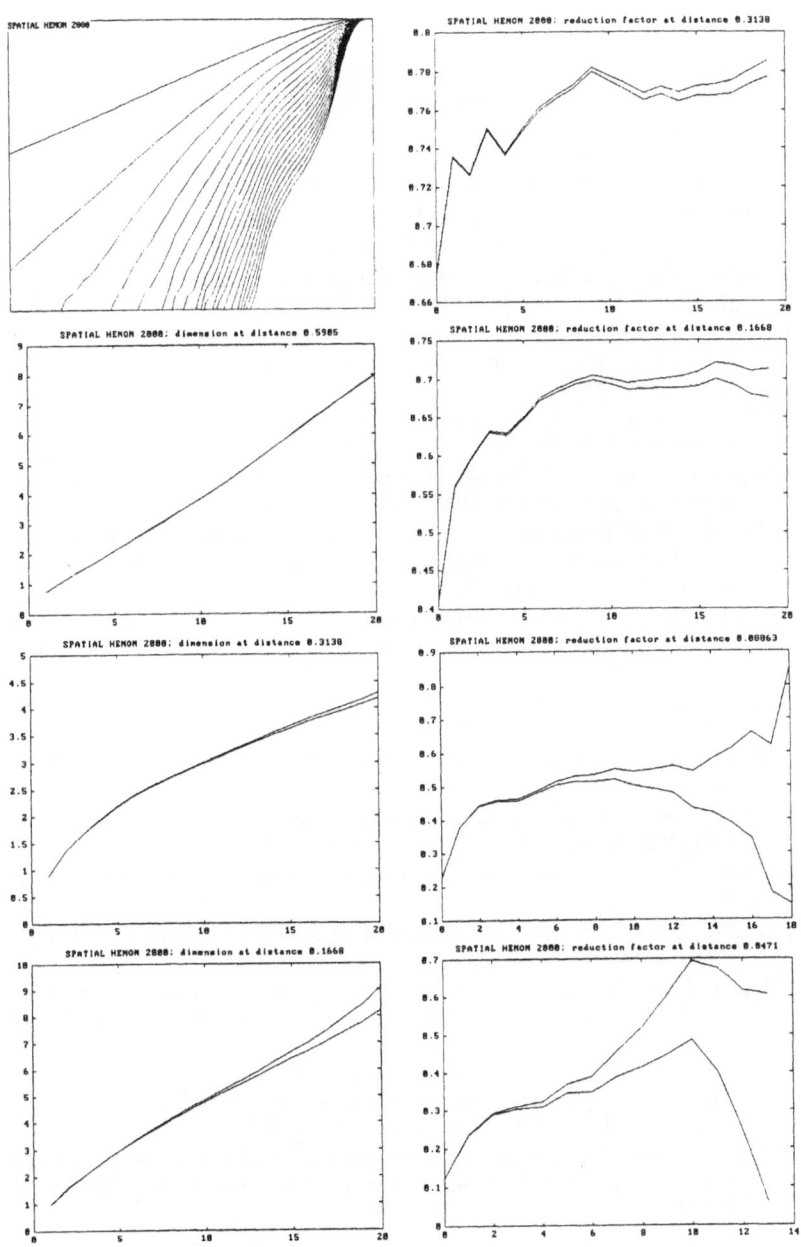

References

ABST,1993 H. D. I. Abarbanel, R. Brown, J. J. Sidorowich, L. S. Tsimring, The analysis of observed chaotic data in physical systems, Rev. of Modern Phys. **65** (4) (1993) 1331–1392.

BHL,1991 W. A. Brock, D. A. Hsieh, B. LeBaron, Nonlinear dynamics, chaos, and instability: statistical theory and economical evidence, MIT Press, 1991.

C,1991 C. D. Cutler, Some results on the behavior and estimation of the fractal dimensions of distributions on attractors, J. Statistical Physics **62** (3/4) (1991), 651–708.

CT,1992 B. Cheng, H. Tong, On consistent nonparametric order determination and chaos, J. R. Statist. Soc. B **54** (2) (1992) 427–449.

CT,1994 B. Cheng, H. Tong, Orthogonal projections, embedding dimensions and sample size in chaotic time series from a statistical perspective, Phil. Trans. R. Soc. Lond. A **348** (1994), 325–341.

GP,1983 P. Grassberger, I. Procaccia, Measuring the strangeness of strange attractors, Physica **9D** (1983), 189–208.

GSS,1991 P. Grassberger, T. Schreiber, C. Schaffrath, Nonlinear time sequence analysis, Bifurcation and Chaos **1** (3) (1991), 521–547.

HJKPS,1986 T. C. Halsey, M. H. Jensen, L. P. Kadanov, I. Procaccia, B. I. Shraiman, Fractal measures and their singularities: The characterization of strange sets, Phys. Rev. **A33** (2) (1986), 1141–1151.

KS,1995 G. Keller, R. Sporer, Remarks on the linear regression approach to dimension estimation, to appear in the proceedings of the KNAW workshop on dynamical systems, Amsterdam, 1994.

PCFS,1980 N. H. Packert, J. P. Crutchfield, J. D. Farmer, R. S. Shaw, Geometry from time series, Phys. Rev. Lett. **45** (1980), 712–716.

R,1990 D. Ruelle, Deterministic chaos: the science and the fiction, Proc. R. Soc. Lond. A, **427** (1990), 241.

S,1992 R. L. Smith, Estimating dimension in noisy chaotic time series, J. R. Statist. Soc. B **54** (2) (1992) 329–351.

SG,1991 R. Savit, M. Green, Time series and dependent variables, Physica D **50** (1991) 95–116.

SYC,1991 T. Sauer, J. A. Yorke, M. Casdagli, Embedology, J. Statistical Physics **65** (3/4) (1991), 579–616.

T,1981 F. Takens, Detecting strange attractors in turbulence, in Dynamical systems and turbulence, Warwick 1980, ed. D. A. Rand, L-S. Young, LNM **898**, Springer-Verlag, 1981.

T,1985 F. Takens, On the numerical determination of the dimension of an attractor, in: Dynamical systems and bifurcations, Groningen 1984, ed. B. L. J. Braaksma, H. W. Broer, F. Takens, LNM **1125**, Springer-Verlag, 1985.

Y,1927 G. U. Yule, On a method of investigating periodicities in disturbed series with special reference to Wolfer's sunspot numbers, Phil. Trans. R. Soc. **A227** (1927), 267–298.

Progress in Nonlinear Differential Equations
and Their Applications, Vol. 19
© 1996 Birkhäuser Verlag Basel/Switzerland

On the computation of normally
hyperbolic invariant manifolds

H.W. Broer   H.M. Osinga*   G. Vegter[†]

Abstract

We present a method for the numerical computation of normally hyperbolic invariant manifolds of discrete dynamical systems. The algorithm com-
putes the invariant manifolds $\mathbb{H}_\varepsilon$ for a C^1 family of diffeomorphisms f_ε,
parametrized by $\varepsilon \in \mathbb{R}$, provided the normally hyperbolic invariant manifold $\mathbb{H}_0$ of f_0 is given. The algorithm also computes the stable and unstable
manifolds of $\mathbb{H}_\varepsilon$. Examples illustrate the performance of the method.

1   Introduction

In this paper we continue the research started in Homburg, Osinga and Vegter [6],
by considering the problem of numerically computing normally hyperbolic invariant manifolds of C^1 diffeomorphisms, defined on a C^1 manifold M. More precisely, we consider perturbations of a C^1 diffeomorphism $f_0 : M \to M$, having a
1–normally hyperbolic invariant manifold H_0. For simplicity we assume that H_0
is compact. According to the *Invariant Manifold Theorem*, see [4], any C^1 diffeomorphism, that is near f_0 with respect to the C^1–norm, has an invariant manifold
near H_0. In particular, a one–parameter family f_ε of C^1 diffeomorphisms, depending continuously on the parameter $\varepsilon \in \mathbb{R}$, has a one–parameter family of normally
hyperbolic invariant C^1 manifolds H_ε near H_0, at least for small values of ε. Our
goal is to derive an algorithm that computes, for a given family f_ε, the invariant
manifold H_ε. As with many numerical methods we need a good initial guess of the
object we want to compute. In general this assumption is necessary to guarantee
uniqueness of the object we intend to compute. Therefore a representation of the
manifold H_0 is part of the input of the algorithm.

Hirsch, Pugh and Shub [4] is an extensive treatment of normally hyperbolic
invariant manifolds. For related work on invariant manifolds and hyperbolic dynamical systems we refer to Palis and Takens [7], Ruelle [8], and Shub [9].

---

*Supported by NWO grant 611-306-523

[†]Address of all authors: Department of Mathematics and Computing Science, University of
Groningen, P.O. Box 800, 9700 AV Groningen, The Netherlands. Email: broer@math.rug.nl,
osinga@cs.rug.nl, vegter@cs.rug.nl

Section 2.1 contains a brief review of the Invariant Manifold Theorem. We closely follow the approach of [4] in the development of our algorithm. However, computations involving geometric objects like manifolds, maps and bundles, require finite and unambiguous representations of these objects. To facilitate such representations we restrict ourselves to diffeomorphisms whose domain is an open subset of $\mathbb{R}^d$, see section 2.2. In this context we design an algorithm that has a straightforward implementation, and yet covers many interesting applications.

Section 3 describes a special version of the algorithm in the simple case of absence of normal expansion. It presents the *graph transform* as a key ingredient of the algorithm, see section 3.1. The graph transform, associated with the diffeomorphism f_ε, may be considered as a contraction, defined on the space of embeddings of H_0 in $\mathbb{R}^d$. (For brevity's sake we are cheating a little here, since the graph transform is actually defined on the space of sections of a certain normal bundle.) Its fixed point is an embedding, whose image $\mathbb{H}_\varepsilon$ is the invariant manifold of f_ε. The image of an embedding under the graph transform is defined, however, in terms of an implicit equation. To solve this equation efficiently we first derive a global version of Newton's method in section 3.2, that may be of some independent significance. This rather general method is applied to the computation of the normally hyperbolic invariant manifold $\mathbb{H}_\varepsilon$ of f_ε in section 3.3, that also contains rather precise estimates concerning the speed of convergence of the algorithm. The computation of the Df_ε-invariant splitting of the tangent bundle (see section 2.2 for a definition) along $\mathbb{H}_\varepsilon$ is described in section 3.4. We also indicate how our algorithm can be used to compute invariant manifolds in a continuation context, where the parameter ε ranges over an interval that is not necessarily small, see section 3.5. This setting arises frequently in applications.

We sketch the general case in section 4. Here we describe the computation of the normally hyperbolic invariant manifold, and its stable and unstable manifolds, when both normal expansion and normal contraction are present. Section 5 contains some numerical examples, illustrating the method first in the simple case of absence of normal expansion, see section 5.1, and subsequently in the general case, see section 5.2. Finally, we show, in section 5.3, how to apply the method to compute the invariant manifold of the Poincaré first–return map of a continuous system.

2   Normally hyperbolic submanifolds

2.1   The Invariant Manifold Theorem

First we present an overview of some basic definitions and results from [4]. Consider a C^1 diffeomorphism f_0 on a C^1 manifold M, having a 1–normally hyperbolic invariant manifold $H_0 \subset M$. Recall that H_0 is r–normally hyperbolic for f_0, $r \geq 1$, if there is a continuous Df_0–invariant splitting

$$T_{H_0}(M) = N^u(H_0) \oplus T(H_0) \oplus N^s(H_0).$$

and a Riemannian structure on the tangent bundle $T_{H_0}(M)$, such that, for $r \in H_0$ and $0 \leq k \leq r$:

$$
\begin{aligned}
\| Df_0 \,|\, N_r^s(H_0) \| \cdot \| (Df_0 \,|\, T_r(H_0))^{-1} \|^k &< 1, \\
\| (Df_0 \,|\, N_r^u(H_0))^{-1} \| \cdot \| Df_0 \,|\, T_r(H_0) \|^k &< 1.
\end{aligned}
\tag{1}
$$

Here the norms are associated with the Riemann structure on $T_{H_0}(M)$.

According to the *Invariant Manifold Theorem* a C^1 diffeomorphism f, that is C^1-near f_0, has a 1-normally hyperbolic invariant manifold H, that is C^1 and C^1-near H_0. In particular, there is a continuous Df-invariant splitting $T_H(M) = N^u(H) \oplus T(H) \oplus N^s(H)$, of the tangent bundle $T_H(M)$. Our primary goal is the computation of both H and the invariant splitting of $T_H(M)$. Furthermore the Invariant Manifold Theorem states that, for some neighborhood U of H, the sets

$$
W^s(H) = \bigcap_{n \geq 0} f^{-n}(U) \quad \text{and} \quad W^u(H) = \bigcap_{n \geq 0} f^n(U)
$$

are C^1 submanifolds of M, tangent to $T_r(H) \oplus N_r^s(H)$ and $T_r(H) \oplus N_r^u(H)$, at $r \in H$. These manifolds, called the *stable* and *unstable* manifold of H, can also be computed using the method developed in this paper, as we describe briefly in section 4.1.

2.2   Normally hyperbolic submanifolds of $\mathbb{R}^d$

Let $f_0 : U \subset \mathbb{R}^d \to \mathbb{R}^d$ be a diffeomorphism, defined on an open subset U of $\mathbb{R}^d$, having a 1-normally hyperbolic invariant manifold $H_0 \subset U$. (We usually write $f_0 : \mathbb{R}^d \to \mathbb{R}^d$, even though in general U is a proper subset of $\mathbb{R}^d$.) We assume throughout this paper that H_0 is compact. In this section we describe how to represent the geometric objects that show up in the computation of invariant manifolds, taking advantage of the fact that the ambient manifold is a euclidean space.

Representation of the invariant manifold

Let $\varphi_0 : H_0 \to \mathbb{R}^d$ be the canonical embedding of H_0. We distinguish between the *abstract manifold* H_0, and its *image* $\varphi_0(H_0)$, which is a submanifold of $\mathbb{R}^d$. To stress this distinction, we denote $\varphi_0(H_0)$ by $\mathbb{H}_0$. The tangent space $T_{\varphi_0(r)}(\varphi_0(H_0))$ is denoted by $\mathbb{T}_r(\mathbb{H}_0)$. Since f_0 leaves $\mathbb{H}_0$ invariant, there is a diffeomorphism $\sigma_0 : H_0 \to H_0$ such that $f_0(\varphi_0(r)) = \varphi_0(\sigma_0(r))$, for $r \in H_0$. Its inverse is denoted by ϱ_0. Note that σ_0 and ϱ_0 may be regarded as the restriction of f_0 and f_0^{-1} to H_0, respectively. See also figure 1. Although the distinction between the abstract manifold H_0 and its φ_0-image $\mathbb{H}_0$ in $\mathbb{R}^d$ involves rather extensive notation, our intention to develop algorithms that manipulate geometric objects like manifolds, maps, and bundles, requires that we are quite specific about the representation of these objects. So the 'user' of the algorithm may e.g. choose to represent points on H_0 by their coordinates in $\mathbb{R}^d$, in which case φ_0 is the inclusion map of H_0 in $\mathbb{R}^d$.

However, in some applications it may be more natural to represent the manifold H_0 by coordinates that are adapted to the dynamics of f_0 on H_0. like the case in which H_0 is a (higher–dimensional) torus, represented by angular coordinates.

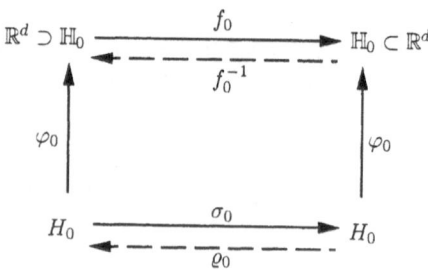

Figure 1: The abstract manifold H_0 and its embedding in $\mathbb{R}^d$.

Representation of normal bundles

The restricted context, in which the ambient manifold is $\mathbb{R}^d$, enables us to identify the stable and unstable normal bundles with certain subsets of $\mathbb{R}^d$. To see this, let the dimension of H_0 be denoted by c, and the dimension of the fibers of $N^s(H_0)$ and $N^u(H_0)$ by s and u, respectively. In particular, $c + s + u = d$. Note that, for $r \in H_0$, the space $T_r(H_0)$ corresponds to the affine subspace $\varphi_0(r) + \mathbb{T}_r(\mathbb{H}_0)$ of $\mathbb{R}^d$. Similarly $N_r^s(H_0)$ and $N_r^u(H_0)$ correspond to affine subspaces of $\mathbb{R}^d$, going through the point $\varphi_0(r) \in \mathbb{H}_0$. They are therefore of the form $\varphi_0(r) + \mathbb{N}_r^s(\mathbb{H}_0)$ and $\varphi_0(r) + \mathbb{N}_r^u(\mathbb{H}_0)$, respectively, where $\mathbb{N}_r^s(\mathbb{H}_0)$ and $\mathbb{N}_r^u(\mathbb{H}_0)$ are s–dimensional and u–dimensional subspaces of $\mathbb{R}^d$. We identify $T_{\varphi_0(r)}(\mathbb{R}^d)$ with $\mathbb{T}_r(\mathbb{H}_0) \oplus \mathbb{N}_r^s(\mathbb{H}_0) \oplus \mathbb{N}_r^u(\mathbb{H}_0)$. Finally

$$\Pi_r^c : \mathbb{R}^d \to \mathbb{T}_r(\mathbb{H}_0), \quad \Pi_r^s : \mathbb{R}^d \to \mathbb{N}_r^s(\mathbb{H}_0) \quad \text{and} \quad \Pi_r^u : \mathbb{R}^d \to \mathbb{N}_r^u(\mathbb{H}_0) \qquad (2)$$

are the canonical projections.

   The Riemannian metric on $\mathbb{R}^d$ induces an inner product on the spaces $\mathbb{T}_r(\mathbb{H}_0)$, $\mathbb{N}_r^s(\mathbb{H}_0)$ and $\mathbb{N}_r^u(\mathbb{H}_0)$. We represent this pointwise inner product by bases, consisting of vectors in $\mathbb{R}^d$, that are by definition orthonormal with respect to the Riemannian structure. More specifically, consider an orthonormal basis $v_1^s(r), \cdots, v_s^s(r)$ of $\mathbb{N}_r^s(\mathbb{H}_0)$, for $r \in H_0$. Note that in general the vector valued functions $v_i^s : H_0 \to \mathbb{R}^d$ are not globally continuous, since this amounts to triviality of the normal bundle $N^s(H_0)$. However, in this paper we make the following assumption:

Assumption 1 (Triviality of normal bundles)

There are C^0 functions $v_1^s, \cdots, v_s^s : H_0 \to \mathbb{R}^d$ such that $v_1^s(r), \cdots, v_s^s(r)$ form an orthonormal basis of $\mathbb{N}_r^s(\mathbb{H}_0)$, for $r \in H_0$. Similarly, there are C^0 functions $v_1^u, \cdots, v_u^u : H_0 \to \mathbb{R}^d$ such that $v_1^u(r), \cdots, v_u^u(r)$ form an orthonormal basis of $\mathbb{N}_r^u(\mathbb{H}_0)$, for $r \in H_0$.

The identification map $\iota_r^s : \mathbb{N}_r^s(\mathbb{H}_0) \to \mathbb{R}^s$ is defined by

$$\iota_r^s \left(\sum_{i=1}^{s} \eta_i^s v_i^s(r) \right) = (\eta_1^s, \cdots, \eta_s^s). \tag{3}$$

The identification map $\iota_r^u : \mathbb{N}_r^u(\mathbb{H}_0) \to \mathbb{R}^u$ is defined similarly.

Due to the triviality of the normal bundle the manifold H_0 has a neighborhood in $\mathbb{R}^d$ that is diffeomorphic to $H_0 \times \mathbb{R}^s \times \mathbb{R}^u$. More precisely, the map $\Phi : H_0 \times \mathbb{R}^s \times \mathbb{R}^u \to \mathbb{R}^d$, defined by

$$\Phi(r, \eta^s, \eta^u) = \varphi_0(r) + \sum_{i=1}^{s} \eta_i^s v_i^s(r) + \sum_{i=1}^{u} \eta_i^u v_i^u(r),$$

is a diffeomorphism from a neighborhood of $H_0 \times \{0\} \times \{0\}$ to a neighborhood of $\mathbb{H}_0$ in $\mathbb{R}^d$. Note that $\Phi(\{r\} \times \mathbb{R}^s \times \{0\}) = \mathbb{N}_r^s(\mathbb{H}_0)$, and $\Phi(\{r\} \times \{0\} \times \mathbb{R}^u) = \mathbb{N}_r^u(\mathbb{H}_0)$. The maps $\pi_c : \mathbb{R}^d \to H_0$, $\pi_s : \mathbb{R}^d \to \mathbb{R}^s$ and $\pi_u : \mathbb{R}^d \to \mathbb{R}^u$ are defined on a neighborhood of $\mathbb{H}_0$ by mapping inverse images under Φ onto H_0. $\mathbb{R}^s$ and $\mathbb{R}^u$, respectively, under the canonical projections. In this way we identify the stable normal bundle $N^s(H_0)$ with the space $H_0 \times \mathbb{R}^s$, and the unstable normal bundle $N^u(H_0)$ with $H_0 \times \mathbb{R}^u$. Therefore it is justifiable to refer to maps $\eta^s : H_0 \to \mathbb{R}^s$ and $\eta^u : H_0 \to \mathbb{R}^u$ as sections. With a pair of sections (η^s, η^u) we associate the embedding $\varphi : H_0 \to \mathbb{R}^d$, defined by $\varphi(r) = \Phi(r, \eta^s(r), \eta^u(r))$. In particular, the embedding φ_0 is associated with the 0-sections of the normal bundles. If f_0 is defined on a manifold other than $\mathbb{R}^d$, or if the normal bundles are not trivial, the methods of this paper still apply. However, the need for local coordinates introduces more complicated (multiple) representations of the geometric objects the algorithm manipulates, cf [8] for a proof of the Invariant Manifold Theorem along these lines. A different approach can be found in [4], where the exponential map, associated with the Riemannian metric, is used to identify a neighborhood of the 0-section in the normal bundle with a neighborhood of $\mathbb{H}_0$ in the ambient manifold. It seems hard to turn the latter method into an efficient algorithm.

Representation of derivatives

In computations it is important to have explicit representations for the *derivative* of e.g. f_0 in points of $\mathbb{H}_0$, cf (1). Since the linear spaces $\mathbb{N}_r^s(\mathbb{H}_0)$, $r \in H_0$, form a Df_0-invariant family, there are *globally defined* C^0 functions $\kappa_{ij}^s : H_0 \to \mathbb{R}^d$, $1 \leq i, j \leq s$, such that

$$Df_0(\varphi_0(r))(v_i^s(r)) = \sum_{j=1}^{s} \kappa_{ij}^s(r) v_j^s(\sigma_0(r)),$$

for $i = 1, \cdots, s$. Let $K_0^s(r)$ be the $s \times s$ matrix with entries $\kappa_{ij}^s(r)$. Similarly there are C^0 functions $\kappa_{ij}^u : H_0 \to \mathbb{R}^d$, $1 \leq i, j \leq u$, such that

$$Df_0(\varphi_0(r))(v_i^u(r)) = \sum_{j=1}^{s} \kappa_{ij}^u(r) v_j^u(\sigma_0(r)).$$

for $i = 1, \cdots, u$. Let $K_0^u(r)$ be the $u \times u$ matrix with entries $\kappa_{ij}^u(r)$.

Then *0–normal hyperbolicity* of $\mathbb{H}_0$ boils down to $\lambda^s := \sup_{r \in H_0} \| K_0^s(r) \| < 1$, and $\lambda^u := \sup_{r \in H_0} \| K_0^u(r)^{-1} \| < 1$. Here we take the matrix norm with respect to the standard inner product on $\mathbb{R}^s$ and $\mathbb{R}^u$, respectively. Although the matrices $K_0^s(r)$ and $K_0^u(r)$ do depend on the particular choice of the functions v_i^s, $1 \leq i \leq s$, and v_i^u, $1 \leq i \leq u$, their norms are independent of this choice.

To express *1–normal hyperbolicity*, let $v_1^c(r), \cdots, v_c^c(r)$ span the tangent space $\mathbb{T}_r(\mathbb{H}_0)$. (Note that in general v_i^c is not globally continuous, since this would amount to parallellizability of $\mathbb{H}_0$.) Since $\mathbb{H}_0$ is f_0–invariant, there are locally defined $\alpha_{ij}(r) \in \mathbb{R}$, $1 \leq i, j \leq c$, such that

$$Df_0(\varphi_0(r))(v_i^c(r)) = \sum_{j=1}^{c} \alpha_{ij}(r) v_j^c(\sigma_0(r)),$$

for $i = 1, \cdots, c$. Let $A_0(r)$ be the $c \times c$–matrix with entries $\alpha_{ij}(r)$. Then $\mu^s := \sup_{r \in H_0} \| K_0^s(r) \| \| A_0(r)^{-1} \| < 1$, and $\mu^u := \sup_{r \in H_0} \| K_0^u(r)^{-1} \| \| A_0(r) \| < 1$, since $\mathbb{H}_0$ is a 1–normally hyperbolic invariant manifold for f_0.

Perturbation context

We study diffeomorphisms on $\mathbb{R}^d$ that are C^1–near f_0. More specifically, we restrict to a *perturbation context* in which these diffeomorphisms occur in a C^1 family $f : \mathbb{R}^d \times \mathbb{R} \to \mathbb{R}^d$, such that $f_0(p) = f(p, 0)$, for $p \in \mathbb{R}^d$.

In this setting families of embeddings, sections of bundles, etc., are maps $g : X \times \mathbb{R} \to Y$, depending on $(x, \varepsilon) \in X$. Here $\varepsilon \in \mathbb{R}$ is considered as a parameter, ranging over some neighborhood of $0 \in \mathbb{R}$. Individual members of a family like g are denoted by subscripting the family name with the parameter name, e.g. $g_\varepsilon(x) = g(x, \varepsilon)$. This convention applies throughout the paper.

3 Special case: absence of normal expansion

In this section we develop an algorithm for the computation of the invariant manifold in the special case of absence of normal expansion, viz $N^u(H_0) = 0$. If no confusion is possible we drop the superscript s from our notation, by writing e.g. $K_0(r)$, $\kappa_{ij}(r)$, λ, ι_r instead of $K_0^s(r)$, $\kappa_{ij}^s(r)$, λ^s, ι_r^s, etc.

3.1   The graph transform

Our goal is to obtain the normally hyperbolic invariant manifold $\mathbb{H}_\varepsilon$ for f_ε by constructing an embedding $\varphi_\varepsilon : H_0 \to \mathbb{R}^d$ with $\mathbb{H}_\varepsilon = \varphi_\varepsilon H_0$. We follow [4], by considering special embeddings associated with sections $\eta_\varepsilon : H_0 \to \mathbb{R}^s$ according to $\varphi_\varepsilon(r) = \Phi(r, \eta_\varepsilon(r))$.

The graph of a section $\eta : H_0 \to \mathbb{R}^d$ is the subset $\mathrm{graph}(\eta)$ of $\mathbb{R}^d$, defined by $\mathrm{graph}(\eta) = \{\Phi(r, \eta(r)) \mid r \in H_0\}$. The *graph transform* Γ_{f_ε} is uniquely determined by the condition that it maps a section $h_\varepsilon : H_0 \to \mathbb{R}^s$ onto a section $\eta_\varepsilon : H_0 \to \mathbb{R}^s$, such that $f_\varepsilon(\mathrm{graph}(h_\varepsilon)) = \mathrm{graph}(\eta_\varepsilon)$. In other words, the point $\Phi(r, \eta(r, \varepsilon))$ is of the form $f(\Phi(\varrho(r, \varepsilon), h(\varrho(r, \varepsilon), \varepsilon)))$, where $\varrho(r, \varepsilon) \in H_0$. We define the graph transform Γ_f on families of sections, i.e. we take $\Gamma_f(\eta)(r, \varepsilon) = \Gamma_{f_\varepsilon}(\eta_\varepsilon)(r)$. Let $\Sigma(\varepsilon_0)$ be the space of continuous families of sections $h : H_0 \times [-\varepsilon_0, \varepsilon_0] \to \mathbb{R}^s$, with $h(r, 0) = 0$ for $r \in H_0$. Since $\mathbb{H}_0$ is f_0-invariant, the 0-section is a fixed point of Γ_{f_0}, and hence $\Sigma(\varepsilon_0)$ is invariant under Γ_f.

For $h \in \Sigma(\varepsilon_0)$ the family $\eta = \Gamma_f(h)$ is the second component of the solution $(\varrho(r, \varepsilon), \eta(r, \varepsilon))$ of the equation

$$F(\varrho, \eta, r, \varepsilon) = 0, \tag{4}$$

where $F : H_0 \times \mathbb{R}^s \times H_0 \times \mathbb{R} \to \mathbb{R}^d$ is defined by

$$F(\varrho, \eta, r, \varepsilon) = f(\Phi(\varrho, h(\varrho, \varepsilon)), \varepsilon) - \Phi(r, \eta), \tag{5}$$

see also figure 2. Note that F is defined on a neighborhood of $\{(\varrho_0(r), 0, r, 0) \mid r \in H_0\}$. Since $\pi_c\Phi(r, \eta) = r$, the solution of equation (4) can be obtained by first solving $\varrho(r, \varepsilon)$ from the equation

$$\sigma(\varrho, \varepsilon) = r, \tag{6}$$

where $\sigma : H_0 \times \mathbb{R} \to H_0$ is defined by

$$\sigma(\varrho, \varepsilon) = \pi_c f(\Phi(\varrho, h(\varrho, \varepsilon)), \varepsilon). \tag{7}$$

In other words: $\varrho_\varepsilon : H_0 \to H_0$ is the inverse of $\sigma_\varepsilon : H_0 \to H_0$. Then η is defined by

$$\eta(r, \varepsilon) = \pi_s f(\Phi(\varrho(r, \varepsilon), h(\varrho(r, \varepsilon), \varepsilon)), \varepsilon).$$

Using the fact that $h(\varrho, 0) = 0$ for $h \in \Sigma(\varepsilon_0)$, we see that

$$\sigma(\varrho, 0) = \pi_c f(\Phi(\varrho, 0), 0) = \pi_c f_0(\varphi_0(\varrho)) = \sigma_0(\varrho).$$

Consequently $\varrho(r, 0) = \varrho_0(r)$.

Equation (6) can be solved by introducing local coordinates on H_0 near $\varrho_0(r)$ and r, and by numerically constructing a solution, viz a local inverse to σ_ε, in terms of these local coordinates. However, we prefer to obtain η *globally*, exploiting the fact that we have identified the bundle $N^s(H_0)$ with a neighborhood of $\mathbb{H}_0$ in $\mathbb{R}^d$

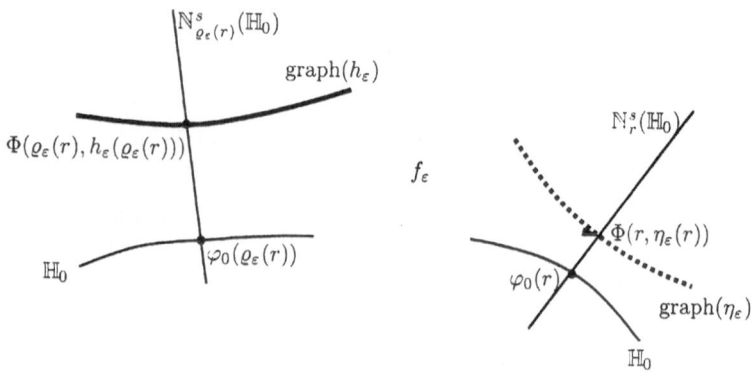

Figure 2: *The graph transform Γ_{f_ε} maps graph(h_ε) onto graph(η_ε). Its fixed point $\overline{h}_\varepsilon$ defines an embedding $\overline{\varphi}_\varepsilon : H_0 \to \mathbb{R}^d$ by $\overline{\varphi}_\varepsilon(r) = \Phi(r, \overline{h}_\varepsilon(r))$, whose image $\mathbb{H}_\varepsilon$ is the normally hyperbolic invariant manifold of f_ε.*

under the mapping Φ. To this end we transform equation (4) into an equation of the form

$$G(y, r, \varepsilon) = 0, \tag{8}$$

where $G : \mathbb{R}^d \times H_0 \times \mathbb{R} \to \mathbb{R}^d$ is a C^1-function defined on a neighborhood of $\varphi_0 H_0 \times H_0 \times \{0\}$, that satisfies, for $r \in H_0$:

$$G(\varphi_0(r), r, 0) = 0.$$

We construct G in section 3.3, but first we develop a global version of Newton's method for solving equations of the form (8).

3.2 A global version of Newton's method

In this section we develop a rather general method for solving equations of the form (8). This method, which may be considered as a global version of Newton's method for determining implicitly defined functions, may be of some independent significance. In this paper it provides a key subroutine for the algorithms that compute the normally hyperbolic submanifold and its stable and unstable manifolds.

First we consider in more detail the spaces of functions we are working with. In this general setting we consider a C^1-function $G : \mathbb{R}^d \times H_0 \times \mathbb{R} \to \mathbb{R}^d$, and a C^1-function $\overline{y}_0 : H_0 \to \mathbb{R}^d$ satisfying, for $r \in H_0$,

$$G(\overline{y}_0(r), r, 0) = 0, \quad D_y G(\overline{y}_0(r), r, 0) \text{ is invertible.} \tag{9}$$

(In our case $\overline{y}_0 = \varphi_0$.) Here $D_y G(\overline{y}_0(r), r, 0)$ is the restriction of $DG(\overline{y}_0(r), r, 0)$ to the space $T_{\overline{y}_0(r)}(\mathbb{R}^d) \times \{0\} \times \{0\}$; We denote this map by $L(r)$.

Note that the solution of equation (8) is a function $H_0 \times \mathbb{R} \to \mathbb{R}^d$, defined on a neighborhood of $H_0 \times \{0\}$, and near $\overline{y}_0$. The Newton operator $\mathcal{N}$ starts with such a function, and computes a better approximation to the solution of (8). More specifically, consider the Banach space $\mathcal{B}(\varepsilon_0)$ of continuous functions $y : H_0 \times [-\varepsilon_0, \varepsilon_0] \to \mathbb{R}^d$, endowed with the sup–norm. viz

$$\|y\| = \sup_{r \in H_0, |\varepsilon| \le \varepsilon_0} |y(r, \varepsilon)|.$$

Here $|y(r, \varepsilon)|$ is the length of $y(r, \varepsilon) \in \mathbb{R}^d$ with respect to the standard inner product on $\mathbb{R}^d$. We consider $\overline{y}_0$ as an element of this space by identifying it with the map $(r, \varepsilon) \mapsto \overline{y}_0(r)$. The Newton operator $\mathcal{N}$ is defined on $\mathcal{B}(\varepsilon_0)$ by

$$\mathcal{N}y(r, \varepsilon) = y(r, \varepsilon) - L(r)^{-1} \cdot G(y(r, \varepsilon), r, \varepsilon).$$

We first derive a precise expression for $\mathcal{N}y$, that is useful in the proof of later results.

Lemma 2 Let $G(y, r, \varepsilon) = G_0(y, r) + \varepsilon G_1(y, r) + O(\varepsilon^2)$, uniformly for (y, r) in some compact neighborhood of $\{(\overline{y}_0(r), r) \mid r \in H_0\}$. Then for $y \in \mathcal{B}(\varepsilon_0)$:

$$\mathcal{N}y(r, \varepsilon) = \overline{y}_0(r) - \varepsilon L(r)^{-1} \cdot G_1(\overline{y}_0(r), r) + O(\varepsilon^2 + \|y - \overline{y}_0\|^2).$$

Proof. Considering the Taylor series of $G(y, r, \varepsilon)$ at $(\overline{y}_0(r), r, 0)$ we see that

$$\begin{aligned} G(y, r, \varepsilon) &= G_0(\overline{y}_0(r), r) + D_y G_0(\overline{y}_0(r), r) \cdot (y - \overline{y}_0(r)) \\ &\quad + \varepsilon G_1(\overline{y}_0(r), r)) + O(\varepsilon^2 + \|y - \overline{y}_0\|^2). \end{aligned}$$

Since $G_0(\overline{y}_0(r), r) = 0$ and $D_y G_0(\overline{y}_0(r), r) = L(r)$, it follows from the definition of $\mathcal{N}$ that

$$\begin{aligned} \mathcal{N}y(r, \varepsilon) &= y(r, \varepsilon) - L(r)^{-1} \cdot L(r) \cdot (y(r, \varepsilon) - \overline{y}_0(r)) \\ &\quad - \varepsilon L(r)^{-1} \cdot G_1(\overline{y}_0(r), r) + O(\varepsilon^2 + \|y - \overline{y}_0\|^2) \\ &= \overline{y}_0(r) - \varepsilon L(r)^{-1} \cdot G_1(\overline{y}_0(r), r)) + O(\varepsilon^2 + \|y - \overline{y}_0\|^2). \end{aligned}$$

This completes the proof of the lemma. $\qquad\square$

The preceding lemma shows that $\mathcal{N}y(r, \varepsilon)$ is of the form $\overline{y}_0(r) + O(\varepsilon)$, provided $y(r, \varepsilon) = \overline{y}_0(r) + O(\varepsilon)$. To make this observation more precise we introduce the space

$$\mathcal{B}(\varepsilon_0, \beta) = \{y : H_0 \times [-\varepsilon_0, \varepsilon_0] \to \mathbb{R}^d \mid \sup_{r \in H_0, |\varepsilon| \le \varepsilon_0} |y(r, \varepsilon) - \overline{y}_0(r)| \le \beta \varepsilon_0\},$$

where ε_0 and β are positive constants; ε_0 is small, β is specified later. The space $\mathcal{B}(\varepsilon_0, \beta)$ is a closed subspace of $\mathcal{B}(\varepsilon_0)$, so in particular it is a complete metric space. Without proof we mention the following properties of the Newton operator, that are crucial in the derivation of our algorithm.

Theorem 3 *Let β be a constant such that $\beta > \sup_{r \in H_0} \| L(r)^{-1} \cdot G_1(\overline{y}_0(r), r) \|$.*
(i) For small values of ε_0 the space $\mathcal{B}(\varepsilon_0, \beta)$ is $\mathcal{N}$-invariant, i.e.:

$$\mathcal{N}(\mathcal{B}(\varepsilon_0, \beta)) \subset \mathcal{B}(\varepsilon_0, \beta).$$

(ii) For small values of ε_0 the Newton operator $\mathcal{N}$ is a contraction on $\mathcal{B}(\varepsilon_0, \beta)$ with contraction factor $O(\varepsilon_0)$. Its fixed point $\overline{y}$ satisfies

$$G(\overline{y}(r, \varepsilon), r, \varepsilon) = 0,$$

and is of the form $\overline{y}(r, \varepsilon) = \overline{y}_0(r) + \varepsilon \overline{y}_1(r) + O(\varepsilon^2)$, uniformly in $r \in H_0$, where

$$\overline{y}_1(r) = -L(r)^{-1} \cdot G_1(\overline{y}_0(r), r). \tag{10}$$

(iii) Let $\{y_n\} \subset \mathcal{B}(\varepsilon_0, \beta)$ be a sequence with $y_0 \in \mathcal{B}(\varepsilon_0, \beta)$, and $y_{n+1} = \mathcal{N} y_n$. Then, for all γ with $0 < \gamma < 1$, there is an $\varepsilon_0 > 0$ such that:

$$y_n(r, \varepsilon) = \overline{y}(r, \varepsilon) + O(\varepsilon^{\gamma n}),$$

uniformly for $|\varepsilon| \le \varepsilon_0$ and $r \in H_0$, as $n \to \infty$.

Theorem 3(ii) reveals that $y_1(r, \varepsilon) = \overline{y}_0(r) + \varepsilon \overline{y}_1(r)$ is a good initial guess for the solution of (8), and theorem 3(iii) guarantees that each application of the Newton operator $\mathcal{N}$ brings us closer to the fixed point roughly by a factor of $O(\varepsilon^\gamma)$. In the next section we apply these observations to the computation of the graph transform.

3.3   Computing the invariant manifold

In this section we apply the results of section 3.2 to compute the graph transform. To this end we first derive, in section 3.3.1, a more precise expression for equation (8), and apply our extension of Newton's method to solve it. It turns out that we can determine the image of the graph transform analytically up to terms of order ε^2, see section 3.3.2. This analysis enables us to iterate the graph transform starting from a good initial guess of the fixed point. A priori, the fixed point of the graph transform defines a C^0 invariant manifold $\mathbb{H}_\varepsilon$ of f_ε, for small values of ε. According to [4] it is even C^1. Although we can extend the analysis of this section to prove this stronger result as well, we abstain from doing so. since we are merely heading for an algorithm to compute the invariant manifold. In section 3.4 we present a method to compute the continuous Df_ε-invariant splitting of the tangent bundle of $\mathbb{H}_\varepsilon$.

We assume that (a representation of) the invariant splitting $\mathbb{T}_r(\mathbb{H}_0) \oplus \mathbb{N}_r^s(\mathbb{H}_0)$, the restrictions ϱ_0 and σ_0 of f_0^{-1} and f_0 to H_0, and the derivative $Df_0(r) = A_0(r) \oplus K_0(r)$ are given for all $r \in H_0$.

3.3.1 The Newton operator

First we transform equation (4) into an equation of the form (8). Ideally we like to find a function $G : \mathbb{R}^d \times H_0 \times \mathbb{R} \to \mathbb{R}^d$ such that $G_\varepsilon(y. r) = 0$ iff y is the point on graph(η_ε) above $r \in H_0$, where η_ε is the image of h_ε under the graph transform Γ_{f_ε}, see figure 2. In other words, $\eta_\varepsilon(r)$ is the second component of the solution $(\varrho_\varepsilon(r), \eta_\varepsilon(r))$ of equation (4). This could be achieved by designing a diffeomorphism $\psi_\varepsilon : H_0 \times \mathbb{R}^s \to \mathbb{R}^d$ such that $\psi_\varepsilon(\varrho_\varepsilon(r), \eta_\varepsilon(r)) = (r, \eta_\varepsilon(r))$, and by taking G such that $G_\varepsilon(\psi_\varepsilon(\xi, \eta), r) = F_\varepsilon(\xi, \eta, r)$. An obvious definition is $\psi_\varepsilon(\xi, \eta) = \Phi(\sigma_\varepsilon(\xi). \eta)$, with $\sigma_\varepsilon(\xi) = \sigma(\xi, \varepsilon)$ defined by (7). However, σ_ε is rather awkward to compute for $\varepsilon \neq 0$. In view of our assumption that (a representation of) σ_0 is given, we use ψ_0 instead of ψ_ε even for $\varepsilon \neq 0$, i.e. we consider the map $\Psi : H_0 \times \mathbb{R}^s \to \mathbb{R}^d$, defined by

$$\Psi(\xi, \eta) = \Phi(\sigma_0(\xi), \eta), \tag{11}$$

which is a diffeomorphism from a neighborhood of $H_0 \times \{0\}$ in $H_0 \times \mathbb{R}^s$ to a neighborhood of $\mathbb{H}_0$ in $\mathbb{R}^d$. Then define G by:

$$G(\Psi(\xi, \eta), r, \varepsilon) = F(\xi, \eta, r. \varepsilon). \tag{12}$$

Since $\varphi_0(r) = \Phi(r, 0) = \Psi(\varrho_0(r), 0)$, for $r \in H_0$, we see that $G(\varphi_0(r), r, 0) = F(\varrho_0(r), 0, r, 0) = 0$, so the first part of condition (9) is satisfied for $\overline{y}_0 = \varphi_0$. To check that the second part holds as well, we first derive an expression for $L(r) = D_y G(\varphi_0(r), r, 0) : \mathbb{R}^d \to \mathbb{R}^d$. It turns out that $L(r)$ has a very simple expression with respect to the splitting $\mathbb{T}_r(\mathbb{H}_0) \oplus \mathbb{N}_r^s(\mathbb{H}_0)$ on both its domain and its range. More precisely:

Lemma 4 For $r \in H_0$, the splitting $\mathbb{R}^d = \mathbb{T}_r(\mathbb{H}_0) \oplus \mathbb{N}_r^s(\mathbb{H}_0)$ is $L(r)$-invariant, and for $v_c \in \mathbb{T}_r(\mathbb{H}_0)$, $v_s \in \mathbb{N}_r^s(\mathbb{H}_0)$

$$L(r)(v_c \oplus v_s) = v_c \oplus (-v_s).$$

In particular $L(r)$ is invertible, and $L(r)^{-1} = L(r)$.

Proof. Recall that for $(\xi, \eta, r) \in H_0 \times \mathbb{R}^s \times H_0$:

$$F_0(\xi, \eta, r) = f_0(\Phi(\xi, 0)) - \Phi(r, \eta) = \Psi(\xi. 0) - \Psi(\varrho_0(r), \eta), \tag{13}$$

since $f_0(\Phi(\xi. 0)) = f_0(\varphi_0(\xi)) = \varphi_0(\sigma_0(\xi)) = \Psi(\xi, 0)$, and $\Phi(r, \eta) = \Phi(\sigma_0(\varrho_0(r)), \eta) = \Psi(\varrho_0(r), \eta)$. From (13) we derive

$$D_\xi F_0(\varrho_0(r), 0, r) = D_\xi \Psi(\varrho_0(r). 0),$$
$$D_\eta F_0(\varrho_0(r), 0, r) = -D_\eta \Psi(\varrho_0(r). 0).$$

Since

$$D_{(\xi, \eta)} F_0(\varrho_0(r), 0, r) = D_y G_0(\varphi_0(r), r) \cdot D\Psi(\varrho_0(r), 0) = L(r) \cdot D\Psi(\varrho_0(r). 0),$$

the proof is complete. $\qquad\square$

Lemma 4 yields the following straightforward method of computing $\mathcal{N}y$ for $y \in \mathcal{B}(\varepsilon_0, \beta)$.

Algorithm NEWTON
Input: $y : H_0 \times [-\varepsilon_0, \varepsilon_0] \to \mathbb{R}^d$.
Output: $\mathcal{N}y : H_0 \times [-\varepsilon_0, \varepsilon_0] \to \mathbb{R}^d$.
forall $r \in H_0, \varepsilon \in [-\varepsilon_0, \varepsilon_0]$ **do**
1      $x \leftarrow \pi_c(y(r, \varepsilon))$
            Comment: $x \in H_0$ and $y(r, \varepsilon) - \varphi_0(x) \in \mathbb{N}_x^s(\mathbb{H}_0)$
2      $\eta \leftarrow \iota_x(y(r, \varepsilon) - \varphi_0(x))$
            Comment: $y(r, \varepsilon) = \Phi(x, \eta)$
3      $Y \leftarrow F(\varrho_0(x), \eta, r, \varepsilon)$
            Comment: $Y = G(y(r, \varepsilon), r, \varepsilon)$
4      $Y^c \leftarrow \Pi_r^c(Y)$
        $Y^s \leftarrow \Pi_r^s(Y)$
5      $\mathcal{N}y(r, \varepsilon) \leftarrow y(r, \varepsilon) - Y^c + Y^s$
            Comment: $L(r)^{-1} \cdot G(y(r, \varepsilon), r, \varepsilon) = Y^c - Y^s$

A few further comments are in order. Execution of line 1 amounts to finding the point $\varphi_0(x) \in \mathbb{H}_0$ such that $y(r, \varepsilon) \in \varphi_0(x) + \mathbb{N}_x^s(\mathbb{H}_0)$. The maps ι_x, Π_x^c, Π_x^s and F have straightforward implementations, see their definitions (3), (2) and (5), respectively. Since also (a representation of) the map $\varrho_0 : H_0 \to H_0$ is given, lines 2, 3 and 4 can be implemented in a straightforward way. To justify the comment at line 3, observe that

$$
\begin{aligned}
G(y(r, \varepsilon), r, \varepsilon) &= G(\Phi(x, \eta), r, \varepsilon) \\
&= G(\Psi(\varrho_0(x), \eta), r, \varepsilon) \\
&= F(\varrho_0(x), \eta, r, \varepsilon) \\
&= Y.
\end{aligned}
$$

Finally the correctness of line 5 follows from lemma 4.

3.3.2   Using the graph transform to compute $\mathbb{H}_\varepsilon$

The map $\Psi : H_0 \times \mathbb{R}^s \to \mathbb{R}^d$, transforming F into G, also establishes a 1:1-correspondence between sections $\eta \in \Sigma(\varepsilon_0)$ and maps $y : H_0 \times \mathbb{R} \to \mathbb{R}^d$, defined by $y(r, \varepsilon) = \Phi(r, \eta(r, \varepsilon))$. To apply the Newton operator, we should restrict the domain of the graph transform to sections, corresponding to maps in the domain $\mathcal{B}(\varepsilon_0, \beta)$ of the Newton operator. Therefore we consider the subset $\Sigma(\varepsilon_0, \alpha)$ of $\Sigma(\varepsilon_0)$, defined by

$$
\Sigma(\varepsilon_0, \alpha) = \{ h \in \Sigma(\varepsilon_0) \mid \sup_{r \in H_0} |h(r, \varepsilon)| \leq \alpha\varepsilon \}.
$$

Since H_0 is compact, for $\beta > 0$ there is an $\alpha > 0$ such that a section in $\Sigma(\varepsilon_0, \alpha)$ corresponds to a map in $\mathcal{B}(\varepsilon_0, \beta)$. Hence, the image of a section $h \in \Sigma(\varepsilon_0, \alpha_0)$ under the graph transform Γ_f can be determined using algorithm NEWTON, designed in section 3.3.1. To obtain a good starting point for repeated application of the Newton operator, we first have to determine $G_1(\varphi_0(r), r) = \frac{\partial G}{\partial \varepsilon}(\varphi_0(r), r, 0)$, see theorem 3(ii), equation (10). To express G_1 in terms of the linear part of f and h, let

$$f(p, \varepsilon) = f_0(p) + \varepsilon f_1(p) + O(\varepsilon^2).$$

and

$$h(r, \varepsilon) = \varepsilon(h_1(r), \cdots, h_s(r)) + O(\varepsilon^2).$$

Lemma 5 *For* $r \in H_0$

$$G_1(\varphi_0(r), r) = \sum_{i,j=1}^{s} h_i(\varrho_0(r))\kappa_{ij}(\varrho_0(r))v_j^s(r) + f_1(\varphi_0(\varrho_0(r))).$$

Proof. Let $(y, r) \in \mathbb{R}^d \times H_0$, then $G_1(y, r) = \frac{\partial G}{\partial \varepsilon}(y, r, 0)$. Furthermore let $y = \Psi(\xi_0, \eta_0)$, for $(\xi_0, \eta_0) \in H_0 \times \mathbb{R}^s$, with Ψ as in (11), i.e.

$$y = \Phi(\sigma_0(\xi_0), \eta_0).$$

then $G(y, r, \varepsilon) = F(\xi_0, \eta_0, r, \varepsilon)$. Therefore

$$G(y, r, \varepsilon) = f(p(\varepsilon), \varepsilon) - \Phi(r, \eta_0).$$

where $p(\varepsilon) = \Phi(\xi_0, \varepsilon h(\xi_0, \varepsilon))$. In particular $p_0 := p(0) = \varphi_0(\xi_0)$. Hence

$$G_1(y, r) = Df_0(p_0) \cdot \dot{p}(0) + f_1(p_0).$$

with

$$\dot{p}(0) = \sum_{i=1}^{s} h_i(\xi_0)v_i^s(\xi_0).$$

Therefore

$$\begin{aligned}
G_1(y, r) &= Df_0(p_0) \cdot \left(\sum_{i=1}^{s} h_i(\xi_0)v_i^s(\xi_0)\right) + f_1(p_0) \\
&= \sum_{i,j=1}^{s} h_i(\xi_0)\kappa_{ij}(\xi_0)v_j^s(\sigma_0(\xi_0)) + f_1(p_0).
\end{aligned}$$

We obtain the desired expression by substituting $y = \varphi_0(r)$, in which case $\sigma_0(\xi_0) = r$ and hence $\xi_0 = \varrho_0(r)$. $\qquad\square$

For $p = \varphi_0(r)$, with $r \in H_0$, the curve $\varepsilon \mapsto f(\varphi_0(r), \varepsilon)$ passes through $f_0(\varphi_0(r)) = \varphi_0(\sigma_0(r))$. Therefore its tangent vector at this point, viz $f_1(\varphi_0(r))$, belongs to $T_{\varphi_0(\sigma_0(r))}(\mathbb{R}^d)$, which we identify with $\mathbb{T}_{\sigma_0(r)}(\mathbb{H}_0) \oplus \mathbb{N}_{\sigma_0(r)}^s(\mathbb{H}_0)$. Therefore there are unique C^0 functions $V^c, V^s \to \mathbb{R}^d$, with $V^c(r) \in \mathbb{T}_r(\mathbb{H}_0)$ and $V^s(r) \in \mathbb{N}_r^s(\mathbb{H}_0)$, such that $f_1(\varphi_0(r)) = V^c(\sigma_0(r)) + V^s(\sigma(r))$. Since ϱ_0 is the inverse of σ_0, we see that $f_1(\varphi_0(\varrho_0(r))) = V^c(r) + V^s(r)$, in other words:

$$V^c(r) = \Pi_r^c(f_1(\varphi_0(\varrho_0(r)))) \text{ and } V^s(r) = \Pi_r^s(f_1(\varphi_0(\varrho_0(r)))).$$

Corollary 6 *The fixed point $\overline{y}$ of $\mathcal{N}$ is of the form $\overline{y}(r, \varepsilon) = \varphi_0(r) + \varepsilon \overline{y}_1(r) + O(\varepsilon^2)$, where*

$$\overline{y}_1(r) = \sum_{i,j=1}^{s} h_i(\varrho_0(r)) \kappa_{ij}(\varrho_0(r)) v_j^s(r) - V^c(r) + V^s(r) \tag{14}$$

Algorithm Graph Transform
Input: $h \in \Sigma(\varepsilon_0, \alpha)$. $\delta > 0$ (maximal error)
Output: $\Gamma_f(h) \in \Sigma(\varepsilon_0, \alpha)$
forall $r \in H_0, \varepsilon \in [-\varepsilon_0, \varepsilon_0]$ **do**
1    $y(r, \varepsilon) \leftarrow \varphi_0(r) + \varepsilon \overline{y}_1(r)$
        Comment: cf corollary 6
2   **repeat**
3        $y_{\text{new}} \leftarrow \mathcal{N}y$
           Comment: Use algorithm Newton
4        $error \leftarrow \|y - y_{\text{new}}\|$
5        $y \leftarrow y_{\text{new}}$
6   **until** $error \leq \delta$
7    $\Gamma_f(h)(r, \varepsilon) \leftarrow \pi_s(y)$
        Comment: $\pi_s(y) = \pi_s(\Psi^{-1}(y))$

Remark 7 To compute $\overline{y}_1(r)$ in line 1 we use expression (14). In view of lemma 2 the variable y in algorithm Graph Transform satisfies the invariant $y = \varphi_0(r) + \varepsilon \overline{y}_1(r) + O(\varepsilon^2)$. In particular the output $\Gamma_f(h)$ is of the form

$$\Gamma_f(h)(r, \varepsilon) = \varepsilon(\overline{h}_1(r), \cdots, \overline{h}_s(r)) + O(\varepsilon^2).$$

where $(\overline{h}_1(r), \cdots, \overline{h}_s(r)) = \pi_s(\overline{y}_1(r))$, i.e.

$$\sum_{i=1}^{s} \overline{h}_i(r) v_i^s(r) = \sum_{i,j=1}^{s} h_i(\varrho_0(r)) \kappa_{ij}(\varrho_0(r)) v_j^s(r) + V^s(r). \tag{15}$$

This enables us to initialize $y(r, \varepsilon)$ properly upon repeated application of the graph transform Γ_f. In fact, we can even compute the fixed point of Γ_f up to terms of order ε^2 by repeated application of (15).

The crucial properties of the graph transform Γ_f are reflected by the following theorem.

Theorem 8 *For any constant $\overline{\lambda}$, such that $\lambda < \overline{\lambda} < 1$, there are values of α and ε_0 such that:*

(i) Γ_f *leaves* $\Sigma(\varepsilon_0, \alpha)$ *invariant, i.e.*

$$\Gamma_f(\Sigma(\varepsilon_0, \alpha)) \subset \Sigma(\varepsilon_0, \alpha).$$

(ii) Γ_f *is a contraction on* $\Sigma(\varepsilon_0, \alpha)$*, whose contraction factor does not exceed* $\overline{\lambda}$*.*

(iii) *The fixed point* $\overline{h}$ *of* Γ_f *defines a continuous family of* C^1 *embeddings* $\overline{\varphi}_\varepsilon : H_0 \times \to \mathbb{R}^d$ *by* $\overline{\varphi}_\varepsilon(r) = \Phi(r, \overline{h}_\varepsilon(r))$*, such that* $\mathbb{H}_\varepsilon := \overline{\varphi}_\varepsilon(H_0)$ *is the* 1*-normally hyperbolic invariant manifold of* f_ε*.*

We omit the proof from this version of the paper. Parts (i) and (ii) follow in a rather straightforward way from the general properties of the Newton operator, see theorem 3, and the expression for G_1, see lemma 5. Note that (i) and (ii) only guarantee that the fixed point $\overline{h}$ is a *continuous* section. Therefore, the map $\overline{\varphi}_\varepsilon$ is a C^0 embedding, and the set $\mathbb{H}_\varepsilon = \overline{\varphi}_\varepsilon(H_0)$ is a C^0 invariant manifold for f_ε. However, with a little more work we can even establish a similar result if we restrict the domain of the graph transform to Lipschitz-sections, see [4] or [9] for details. This stronger result is crucial for the proof of part (iii), viz that $\mathbb{H}_\varepsilon$ is a C^1 manifold, as we explain in the next section.

3.4   Computing the invariant splitting of the tangent bundle

In the previous subsection we derived an algorithm that computes the invariant manifold $\mathbb{H}_\varepsilon \subset \mathbb{R}^d$ of f_ε as the image of an embedding $\overline{\varphi}_\varepsilon : H_0 \to \mathbb{R}^d$. This algorithm computes a pair $(\overline{\varrho}, \overline{h})$, with $\overline{\varrho} : H_0 \times \mathbb{R} \to H_0$ and $\overline{h} : H_0 \times \mathbb{R} \to \mathbb{R}^s$, such that $\overline{\varphi}_\varepsilon(r) = \Phi(r, \overline{h}_\varepsilon(r))$, and

$$f_\varepsilon(\overline{\varphi}_\varepsilon(\overline{\varrho}_\varepsilon(r))) = \overline{\varphi}_\varepsilon(r).$$

The inverse of $\overline{\varrho}_\varepsilon$ is denoted by $\overline{\sigma}_\varepsilon(r)$. Therefore

$$f_\varepsilon(\overline{\varphi}_\varepsilon(r)) = \overline{\varphi}_\varepsilon(\overline{\sigma}_\varepsilon(r)).$$

Hence

$$\overline{\sigma}_\varepsilon(r) = \pi_c f_\varepsilon(\overline{\varphi}_\varepsilon(r)),$$

so $\overline{\sigma}_\varepsilon$ can easily be computed from $\overline{\varphi}_\varepsilon$.

Our goal in this section is to compute the Df_ε-invariant splitting $T(\mathbb{H}_\varepsilon) \oplus \mathbb{N}^s(\mathbb{H}_\varepsilon)$ of the tangent bundle $T_{\mathbb{H}_\varepsilon}(\mathbb{R}^d)$. To this end we write the map $Df_\varepsilon(\overline{\varphi}_\varepsilon(r))$ with respect to the splittings $\mathbb{T}_r(\mathbb{H}_0) \oplus \mathbb{N}_r^s(\mathbb{H}_0)$ and $\mathbb{T}_{\overline{\sigma}_\varepsilon(r)}(\mathbb{H}_0) \oplus \mathbb{N}_{\overline{\sigma}_\varepsilon(r)}^s(\mathbb{H}_0)$ as

$$\begin{pmatrix} A_\varepsilon(r) & B_\varepsilon(r) \\ C_\varepsilon(r) & K_\varepsilon(r) \end{pmatrix}.$$

where $A_\varepsilon(r) : \mathbb{T}_r(\mathbb{H}_0) \to \mathbb{T}_{\overline{\sigma}_\varepsilon(r)}(\mathbb{H}_0)$, $B_\varepsilon(r) : \mathbb{N}_r^s(\mathbb{H}_0) \to \mathbb{T}_{\overline{\sigma}_\varepsilon(r)}(\mathbb{H}_0)$, $C_\varepsilon(r) : \mathbb{T}_r(\mathbb{H}_0) \to \mathbb{N}_{\overline{\sigma}_\varepsilon(r)}^s(\mathbb{H}_0)$ and $K_\varepsilon(r) : \mathbb{N}_r^s(\mathbb{H}_0) \to \mathbb{N}_{\overline{\sigma}_\varepsilon(r)}^s(\mathbb{H}_0)$ are linear maps, depending continuously on (r, ε). Note in particular that $B_0(r) = 0$ and $C_0(r) = 0$, and $\| A_\varepsilon(r) \| = \| A_0(r) \| + O(\varepsilon)$, etc.

The algorithm that computes the Df_ε–invariant splitting of $T_{\overline{z}_\varepsilon}(\mathbb{R}^d)$ is again based on a graph transform. Consider, for $(r, \varepsilon) \in H_0 \times \mathbb{R}$. linear maps $\omega_\varepsilon(r)$: $\mathbb{N}_r^s(\mathbb{H}_0) \to \mathbb{T}_r(\mathbb{H}_0)$, depending continuously on (r, ε). The space of all such maps, defined for $(r, \varepsilon) \in H_0 \times [-\varepsilon_0, \varepsilon_0]$, is denoted by $\Omega_s(\varepsilon_0)$. It is a Banach space with norm defined by $\|\omega\| = \sup_{|\varepsilon| \le \varepsilon_0, r \in H_0} \|\omega_\varepsilon(r)\|$. Let $\Omega_s(\delta_0, \varepsilon_0)$ be the subspace consisting of those $\omega \in \Omega_s(\varepsilon_0)$ for which $\|\omega\| \le \delta_0$. Note that this is a closed subspace of $\Omega_s(\delta_0, \varepsilon_0)$. and therefore it is a complete metric space.

The graph of $\omega_\varepsilon(r)$ is the subspace graph$(\omega_\varepsilon(r))$ of $\mathbb{R}^d$, defined by

$$\mathrm{graph}(\omega_\varepsilon(r)) = \{\omega_\varepsilon(r) \cdot u \oplus u \mid u \in \mathbb{N}_r^s(\mathbb{H}_0)\}.$$

We define the operator T_s on $\Omega_s(\delta_0, \varepsilon_0)$ by the requirement that. for $\overline{\omega} = T_s(\omega)$,

$$\mathrm{graph}(\overline{\omega}_\varepsilon(r)) = Df_\varepsilon(\overline{\varphi}_\varepsilon(r))^{-1}\mathrm{graph}(\omega_\varepsilon(\overline{\sigma}_\varepsilon(r))). \tag{16}$$

Then (16) boils down to: for all $v \in \mathbb{N}_{\overline{\sigma}_\varepsilon(r)}^s(\mathbb{H}_0)$ there is a $\overline{v} \in \mathbb{N}_r^s(\mathbb{H}_0)$ such that

$$\left(\begin{array}{c} \overline{\omega} \cdot \overline{v} \\ \overline{v} \end{array} \right) = \left(\begin{array}{cc} A & B \\ C & K \end{array} \right)^{-1} \cdot \left(\begin{array}{c} \omega \cdot v \\ v \end{array} \right),$$

where $A = A_\varepsilon(r)$ (etc.). $\overline{\omega} = \overline{\omega}_\varepsilon(r)$ and $\omega = \omega_\varepsilon(\overline{\sigma}_\varepsilon(r))$. Eliminating v and $\overline{v}$ we see that

$$\overline{\omega} = (A - \omega \cdot C)^{-1} \cdot (-B + \omega \cdot K),$$

in other words

$$(T_s\omega)_\varepsilon(r) = (A_\varepsilon(r) - \omega_\varepsilon(\overline{\sigma}_\varepsilon(r)) \cdot C_\varepsilon(r))^{-1} \cdot (-B_\varepsilon(r) + \omega_\varepsilon(\overline{\sigma}_\varepsilon(r)) \cdot K_\varepsilon(r)). \tag{17}$$

Theorem 9 *Let $\overline{\lambda}$ and $\overline{\mu}$ be constants such that $\lambda < \overline{\lambda} < 1$ and $\mu < \overline{\mu} < 1$. Then, for δ_0 and ε_0 sufficiently small:*
(i) The space $\Omega_s(\delta_0, \varepsilon_0)$ is T_s–invariant, i.e.

$$T_s(\Omega_s(\delta_0, \varepsilon_0)) \subset \Omega_s(\delta_0, \varepsilon_0).$$

(ii) The operator T_s is a contraction, whose contraction factor does not exceed $\overline{\mu}$. Its fixed point $\overline{\omega}$ determines a Df_ε–invariant family $\{\mathbb{N}_r^s(\mathbb{H}_\varepsilon)\}_{r \in H_0}$ of subspaces of $\mathbb{R}^d$, defined by

$$\mathbb{N}_r^s(\mathbb{H}_\varepsilon) = \mathrm{graph}(\overline{\omega}_\varepsilon(r)).$$

(iii) For $r \in H_0$ and $v \in \mathbb{N}_r^s(\mathbb{H}_\varepsilon)$:

$$\|Df_\varepsilon(\overline{\varphi}_\varepsilon(r)) \cdot v\| \le \overline{\lambda}\|v\|.$$

Proof. The proof of the first part is rather straightforward, see the full version for details. So let $\omega_1, \omega_2 \in \Omega_s(\delta_0, \varepsilon_0)$. Writing again A instead of $A_\varepsilon(r)$. etc., we see that T_s is a contraction:

$$
\begin{aligned}
&\|T_s\omega_2 - T_s\omega_1\| \\
&\leq\ \|(A - \omega_2 \cdot C)^{-1} \cdot ((-B + \omega_2 \cdot K) - (-B + \omega_1 \cdot K))\| + \\
&\quad\ \|((A - \omega_2 \cdot C)^{-1} - (A - \omega_1 \cdot C)^{-1}) \cdot (-B + \omega_1 \cdot K)\| \\
&\leq\ \|(A - \omega_2 \cdot C)^{-1} \cdot (\omega_2 - \omega_1) \cdot K\| + \\
&\quad\ \|(A - \omega_2 \cdot C)^{-1} \cdot (\omega_1 - \omega_2) \cdot (A - \omega_1 \cdot C)^{-1}(-B + \omega_1 \cdot K)\| \\
&\leq\ (\|A^{-1}\| \|K\|(1 + O(\delta_0 + \varepsilon_0)) + \|A^{-1}\|^2 O(\delta_0 + \varepsilon_0))\|\omega_1 - \omega_2\| \\
&\leq\ (\mu + O(\delta_0 + \varepsilon_0))\|\omega_1 - \omega_2\| \\
&\leq\ \overline{\mu}\|\omega_1 - \omega_2\|,
\end{aligned}
$$

for δ_0 and ε_0 sufficiently small. (To derive the second inequality we use the identity $S_2^{-1} - S_1^{-1} = S_2^{-1} \cdot (S_1 - S_2) \cdot S_1^{-1}$.) Hence T_s is a contraction, whose contraction factor does not exceed $\overline{\mu}$.

(iii) Let $v \in \mathbb{N}_r^s(\mathbb{H}_\varepsilon)$, then $v = \overline{\omega}_\varepsilon(r) \cdot u \oplus u$, for some $u \in \mathbb{N}_r^s(\mathbb{H}_0)$. Since $\overline{\omega}$ is a fixed point of T_s it follows from (17) that $Df_\varepsilon(\overline{\varphi}_\varepsilon(r)) \cdot v = \overline{\omega}(\overline{\sigma}_\varepsilon(r)) \cdot w \oplus w$, where $w = (C_\varepsilon(r) \cdot \overline{\omega}_\varepsilon(r) + K_\varepsilon(r)) \cdot u$. Hence

$$
\begin{aligned}
\|Df_\varepsilon(\overline{\varphi}_\varepsilon(r)) \cdot v\| &\leq\ (\|K_\varepsilon(r)\| + O(\|\overline{\omega}_\varepsilon(r)\|))\|u\| \\
&\leq\ (\lambda + O(\delta_0 + \varepsilon_0))\|v\| \\
&\leq\ \overline{\lambda}\|v\|,
\end{aligned}
$$

for δ_0 and ε_0 sufficiently small. This completes the proof of the theorem. $\qquad\square$

To determine the tangent space of the invariant manifold $\mathbb{H}_\varepsilon$ of f_ε, we similarly introduce the space $\Omega_c(\varepsilon_0)$, consisting of families of linear maps $\omega_\varepsilon(r): \mathbb{T}_r(\mathbb{H}_0) \to \mathbb{N}_r^s(\mathbb{H}_0)$, depending continuously on $(r, \varepsilon) \in H_0 \times \mathbb{R}$. Its subspace $\Omega_c(\delta_0, \varepsilon_0)$ consists of those $\omega \in \Omega_c(\varepsilon_0)$ with $\|\omega\| \leq \delta_0$. The operator $T_c: \Omega_c(\varepsilon_0) \to \Omega_c(\varepsilon_0)$ is defined by the condition that $Df_\varepsilon(\overline{\varphi}_\varepsilon(\overline{\varrho}_\varepsilon(r)))$ maps the graph of $\omega_\varepsilon(\overline{\varrho}_\varepsilon(r))$ onto the graph of $\omega_\varepsilon(r)$. More precisely,

$$
\begin{pmatrix} \overline{v} \\ \overline{\omega} \cdot \overline{v} \end{pmatrix} = \begin{pmatrix} A & B \\ C & K \end{pmatrix} \cdot \begin{pmatrix} v \\ \omega \cdot v \end{pmatrix},
$$

where $A = A_\varepsilon(\overline{\varrho}_\varepsilon(r))$ (etc.), $\omega = \omega_\varepsilon(\overline{\varrho}_\varepsilon(r))$ and $\overline{\omega} = \overline{\omega}_\varepsilon(r)$. Elimination of v and $\overline{v}$ yields the following expression for T_c:

$$
(T_c\omega)_\varepsilon(r) = (C_\varepsilon(\overline{\varrho}_\varepsilon(r)) + K_\varepsilon(\overline{\varrho}_\varepsilon(r)) \cdot \omega_\varepsilon(\overline{\varrho}_\varepsilon(r)) \cdot (A_\varepsilon(\overline{\varrho}_\varepsilon(r)) + B_\varepsilon(\overline{\varrho}_\varepsilon(r)) \cdot \omega_\varepsilon(\overline{\varrho}_\varepsilon(r)))^{-1}.
$$

The following result is similar to theorem 9.

Theorem 10 Let $\overline{\lambda}$ and $\overline{\mu}$ be constants such that $\lambda < \overline{\lambda} < 1$ and $\mu < \overline{\mu} < 1$. Then, for δ_0 and ε_0 sufficiently small:
(i) The space $\Omega_c(\delta_0, \varepsilon_0)$ is T_c-invariant, i.e.

$$
T_c(\Omega_c(\delta_0, \varepsilon_0)) \subset \Omega_c(\delta_0, \varepsilon_0).
$$

(ii) The operator T_c is a contraction, whose contraction factor does not exceed $\overline{\mu}$. Its fixed point $\overline{\omega}$ defines the tangent bundle of $\mathbb{H}_\varepsilon$, i.e.

$$\mathbb{T}_r(\mathbb{H}_\varepsilon) = \{u \oplus \overline{\omega}(r) \cdot u \mid u \in \mathbb{T}_r(\mathbb{H}_0)\}.$$

(iii) For $r \in H_0$, let $\overline{A}_\varepsilon(r) \oplus \overline{K}_\varepsilon(r)$ be the expression for $Df_\varepsilon(\overline{\varphi}_\varepsilon(r))$ with respect to the invariant splitting $\mathbb{T}_r(\mathbb{H}_\varepsilon) \oplus \mathbb{N}_r^s(\mathbb{H}_\varepsilon)$ and $\mathbb{T}_{\overline{\sigma}_\varepsilon(r)}(\mathbb{H}_\varepsilon) \oplus \mathbb{N}_{\overline{\sigma}_\varepsilon(r)}^s(\mathbb{H}_\varepsilon)$ on domain and range. Then

$$\|(\overline{A}_\varepsilon(r))^{-1}\| \|\overline{K}_\varepsilon(r)\| \leq \overline{\mu}.$$

The proof of this result is similar to that of theorem 8. As we remarked earlier, it can proven that $\overline{\varphi}_\varepsilon$ is Lipschitz. Using the methods of [4], one can also prove that the space graph($\overline{\omega}_\varepsilon(r)$) is tangent to $\mathbb{H}_\varepsilon$ at $\overline{\varphi}_\varepsilon(r)$, for all $r \in H_0$. Therefore $\overline{\varphi}_\varepsilon$ is a C^1 embedding, whose image $\mathbb{H}_\varepsilon$ is therefore C^1 as well.

3.5 Continuation

In many examples one may want to compute a continuous family of invariant manifolds for a family f_ε of diffeomorphisms, where ε ranges over a parameter interval that is not necessarily small. To apply the algorithm to such *continuation problems* we increase the parameter in small steps (possibly adapting the step size near parameter values for which the normal hyperbolicity is weak), and adjust the invariant splitting after each increase of the parameter ε. In this setting the algorithm has to deliver output, that serves as input to the next step in the continuation process, viz the increase of the parameter ε. The input to the algorithm, that computes the invariant manifold, has been described at the beginning of section 3.3. In view of the condition that the output of the algorithm has to be of the same type as the input, we therefore require that for a certain value of ε the algorithm computes:

- An embedding $\overline{\varphi}_\varepsilon : H_0 \to \mathbb{R}^d$, whose image is the invariant manifold $\mathbb{H}_\varepsilon$ of f_ε. This embedding is computed by repeated application of algorithm GRAPH TRANSFORM, see section 3.3.2.

- A diffeomorphism $\overline{\varrho}_\varepsilon : H_0 \to H_0$, together with its inverse $\overline{\sigma}_\varepsilon$, such that $f_\varepsilon(\overline{\varphi}_\varepsilon(\overline{\varrho}_\varepsilon(r))) = \overline{\varphi}_\varepsilon(r)$ (as we have seen, $\overline{\varrho}_\varepsilon$ may be considered as the restriction of f_ε^{-1} to $\mathbb{H}_\varepsilon$). In fact, repeated application of algorithm GRAPH TRANSFORM not only yields the embedding $\overline{\varphi}_\varepsilon$, but also the map $\overline{\varrho}_\varepsilon$, see again section 3.3.2.

- The Df_ε–invariant splitting $\mathbb{T}(\mathbb{H}_\varepsilon) \oplus \mathbb{N}^s(\mathbb{H}_\varepsilon)$ of the tangent bundle $T_{\mathbb{H}_\varepsilon}(\mathbb{R}^d)$. In particular, we assume that $\mathbb{N}_r^s(\mathbb{H}_\varepsilon)$ is represented by the vectors $v_i^s(r, \varepsilon) \in \mathbb{R}^d$, $1 \leq i \leq s$, which define an orthonormal system with respect to the Riemannian metric on $\mathbb{R}^d$. The computation of this splitting is described in section 3.4.

Hence the algorithm can be applied without further adaptations to the computation of invariant manifolds in a continuation setting. We illustrate our method with several examples in section 5.

4 The general case

There are several ways to extend algorithm GRAPH TRANSFORM to compute the invariant manifold $\mathbb{H}_\varepsilon$ of f_ε in the case where both normal expansion and normal contraction are present. One extension computes $\mathbb{W}^s(\mathbb{H}_\varepsilon)$ and $\mathbb{W}^u(\mathbb{H}_\varepsilon)$, the stable and unstable manifolds of $\mathbb{H}_\varepsilon$, see section 2.1, and determines $\mathbb{H}_\varepsilon$ as the intersection of these manifolds. We describe this version in section 4.1. A drawback is the need for a separate algorithm to compute the intersection of submanifolds. Therefore we present, in section 4.2, a method that computes $\mathbb{H}_\varepsilon$ more directly. To the best of our knowledge this is the first algorithm that computes invariant manifolds for which the normal dynamics exhibits both contraction and expansion.

4.1 Computing $\mathbb{W}^s(\mathbb{H}_\varepsilon)$ and $\mathbb{W}^u(\mathbb{H}_\varepsilon)$

We construct the stable manifold $\mathbb{W}^s(\mathbb{H}_\varepsilon)$ as the graph of a function $y_\varepsilon : H_0 \times \mathbb{R}^s \to \mathbb{R}^u$ (Actually, the domain of y_ε is a neighborhood of $H_0 \times \{0\}$ in $H_0 \times \mathbb{R}^s$). This graph is defined by

$$\operatorname{graph}(y_\varepsilon) = \{\Phi(r, x, y_\varepsilon(r, x)) \mid (r, x) \in H_0 \times \mathbb{R}^s\}.$$

The graph transform Γ_{f_ε} is defined by the following geometric condition:

$$f_\varepsilon^{-1}(\operatorname{graph}(y_\varepsilon)) = \operatorname{graph}(\Gamma_{f_\varepsilon}(y_\varepsilon)). \tag{18}$$

Apart from technical details, we are now in the context of section 3.1. Therefore it is possible to translate condition (18) into an equation of the form (8). Solving this equation using algorithm NEWTON yields again a straightforward implementation of the graph transform, whose fixed point defines the invariant manifold $\mathbb{W}^s(\mathbb{H}_\varepsilon)$. Since the unstable manifold of $\mathbb{H}_\varepsilon$ is the stable manifold with respect to f_ε^{-1}, it can be computed similarly. For further details we refer to the full version of the paper.

4.2 Computing $\mathbb{H}_\varepsilon$

Let us assume that the stable manifold $\mathbb{W}^s(\mathbb{H}_\varepsilon)$ has been computed as the graph of a map $\overline{y}_\varepsilon : H_0 \times \mathbb{R}^s \to \mathbb{R}^u$. The manifold $\mathbb{H}_\varepsilon \subset \mathbb{W}^s(\mathbb{H}_\varepsilon)$ can then be determined as the graph of a map $\overline{h}_\varepsilon : H_0 \to \mathbb{R}^s$, i.e. as a set of the form

$$\operatorname{graph}(\overline{h}_\varepsilon) = \{\Phi(r, \overline{h}_\varepsilon(r), \overline{y}_\varepsilon(r, \overline{h}_\varepsilon(r))) \mid r \in H_0\}.$$

In fact, restricting to the stable manifold $\mathbb{W}^s(\mathbb{H}_\varepsilon)$ brings us back to the special case of absence of normal expansion. The map $\overline{y}_\varepsilon$ establishes a diffeomorphism between $H_0 \times \mathbb{R}^s$ and $\mathbb{W}^s(\mathbb{H}_\varepsilon)$. Proceeding as in section 3, we introduce the graph transform Γ_{f_ε} on the space of families of maps (sections) $H_0 \to \mathbb{R}^s$. More precisely, for a section $h_\varepsilon : H_0 \to \mathbb{R}^s$ the section $\overline{h}_\varepsilon = \Gamma_{f_\varepsilon}(h_\varepsilon)$ is defined by the condition

$$\mathrm{graph}(\overline{h}_\varepsilon) = f_\varepsilon(\mathrm{graph}(h_\varepsilon)).$$

The section $\overline{h}_\varepsilon$ is well-defined, since $\mathrm{graph}(\overline{y}_\varepsilon)$ is f_ε-invariant. The Df_ε-invariant splitting of $T_{\mathbb{H}_\varepsilon}(\mathbb{R}^d)$ can be computed as in the case of absence of normal expansion. In the full version of the paper we present more details, as well as some other methods for the computation of $\mathbb{H}_\varepsilon$, among others one that bypasses the computation of $\mathbb{W}^s(\mathbb{H}_\varepsilon)$.

5   Numerical examples

Finally we show the performance of our algorithm in some applications, all of which fall in the continuation context described in section 3.5. It turns out that in all examples the invariant manifolds are normally hyperbolic with respect to the Riemannian metric that coincides with the standard euclidean metric of the ambient space. Obviously, one can't assume this to be true in general.

5.1   The fattened Thom map

First we illustrate the algorithm in the simple case of absence of normal expansion, cf section 3. To this end consider the diffeomorphism f_ε, defined on $(\mathbb{R}/2\pi\mathbb{Z})^2 \times \mathbb{R}$ by

$$f_\varepsilon(x. y. z) = (2x + y + \varepsilon z, x + y + \varepsilon z, az + \varepsilon \sin x).$$

(We may consider f_ε as a diffeomorphism defined on $\mathbb{R}^3$, that is periodic in the first two coordinates.)

We fix the constant a, such that $0 < a < 1$ (more specifically, we take $a = 0.1$). Then the system f_0 has a normally hyperbolic invariant torus $\mathbb{H}_0 := (\mathbb{R}/2\pi\mathbb{Z})^2 \times \{0\}$. It may be considered as the image of $H_0 = (\mathbb{R}/2\pi\mathbb{Z})^2$ under the canonical embedding $\varphi_0(x,y) = (x,y,0)$. Note that the restriction of f_0 to this torus is the Thom automorphism $(x,y) \mapsto (2x + y, x + y)$. The tangent plane of $\mathbb{H}_0$ at $(x,y,0)$ is defined by $z = 0$, the space $\mathbb{N}^s_{(x,y,0)}(\mathbb{H}_0)$ is spanned by the unit vector in the z–direction. In figure 3 (left: top and bottom) the initial data, viz $\mathbb{H}_0$ and the splitting $\mathbb{T}(\mathbb{H}_0) \oplus \mathbb{N}^s(\mathbb{H}_0)$, is shown. The normally hyperbolic torus $\mathbb{H}_0$ is represented by a square mesh of 50×50, equidistant, points.

Numerically we detect that for $\varepsilon \approx 0.4699$ the normal behavior of f_ε ceases to dominate the tangential behavior, viz $\mu_s \approx 1$, cf section 2.1. (This observation is based on the computation of the eigenvalues of Df_ε at $(0.0.0) \in \mathbb{H}_\varepsilon$.)

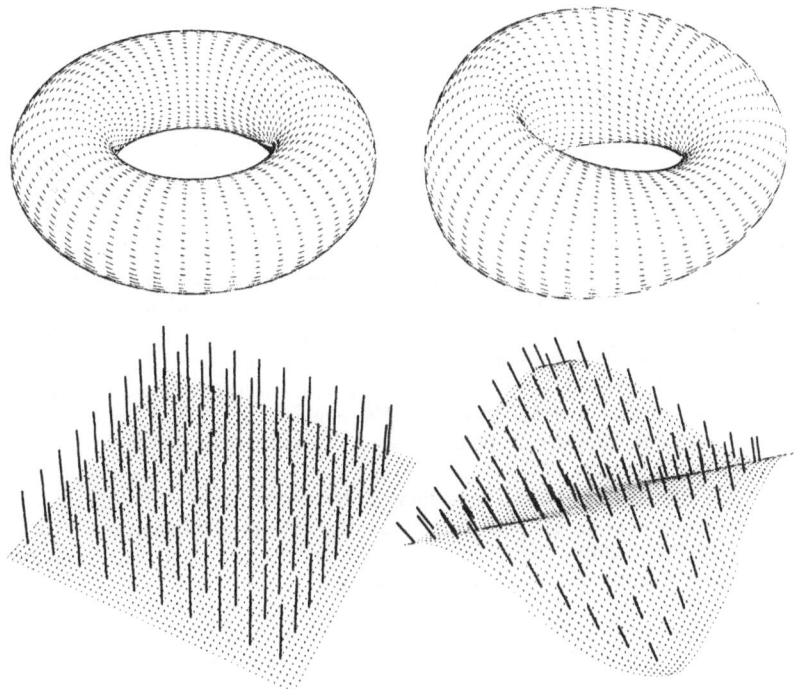

Figure 3: *The continuation of the Thom map in* $\mathbb{R}^3$: $a = 0.1$; ε *runs from 0 to* 0.4695.
Top: The torus for $\varepsilon = 0$ *(left) and* $\varepsilon = 0.4695$ *(right) embedded in* $\mathbb{R}^3$
Bottom: The data with the normal directions drawn at some of the mesh points for $\varepsilon = 0$ *(left) and* $\varepsilon = 0.4695$ *(right).*

Consequently, we can expect that the continuation cannot go past $\varepsilon = 0.4699$. Our algorithm computes a family $\mathbb{H}_\varepsilon$ of invariant tori. for ε ranging from 0 to 0.4695, with an estimated accuracy of order 10^{-4}. The initial increment of the continuation parameter ε is set to 0.02, but is adjusted (viz made smaller) as the normal hyperbolicity gets weaker. The square mesh. representing $\mathbb{H}_\varepsilon$, is fixed during the computations. Figure 3 (right: top and bottom) shows the last invariant torus we were able to compute. For this value of ε the contraction factor of one of the operators Γ_{f_ε}, see section 3.1 or $\mathcal{T}_s$, see section 3.4, is close to 1. In other words, the torus $\mathbb{H}_\varepsilon$ is about to lose its 1-normal hyperbolicity.

5.2 The fattened Arnol'd family

We now apply the general version of the algorithm, sketched in section 4. Consider the *fattened Arnol'd family* of diffeomorphisms on $(\mathbb{R}/2\pi\mathbb{Z}) \times \mathbb{R}^2$:

$$f_\varepsilon \left(\begin{array}{c} x \\ y \\ z \end{array} \right) = \left(\begin{array}{c} x + a + \varepsilon(y + z/2 + \sin x) \\ b(y + \sin x) \\ c(y + z + \sin x) \end{array} \right), \qquad (19)$$

where $x \in \mathbb{R}/2\pi\mathbb{Z}$. See also Broer, Simó and Tatjer [3] for a similar diffeomorphism on $\mathbb{S}^1 \times \mathbb{R}$. The constant a is defined modulo 2π, and the system f_0 has an invariant circle $\mathbb{H}_0$, on which the dynamics is (conjugate to) the rigid rotation $\sigma_0(x) = x + a$. Furthermore, for $0 < b < 1$ and $c > 1$, the invariant circle $\mathbb{H}_0$ is 1–normally hyperbolic with one–dimensional stable and unstable directions. (In our example we take $a = 0.1$, $b = 0.3$ and $c = 2.4$.) Consequently, the system has a 1–normally hyperbolic invariant circle $\mathbb{H}_\varepsilon$, for small ε. The dynamics of $f_\varepsilon \mid \mathbb{H}_\varepsilon$ is either periodic or quasi–periodic, the periodic behavior being characterized by the existence of so–called Arnol'd tongues, cf [1].

It is easy to represent the embedding φ_0 by determining a parametrization of the invariant circle $\mathbb{H}_0$, see [2]. Furthermore, the Df_0–invariant splitting $\mathbb{T}(\mathbb{H}_0) \oplus \mathbb{N}^s(\mathbb{H}_0) \oplus \mathbb{N}^u(\mathbb{H}_0)$ can also be determined explicitly, see figure 4 (left). The invariant circle $\mathbb{H}_0$ is represented by a mesh of 50 points.

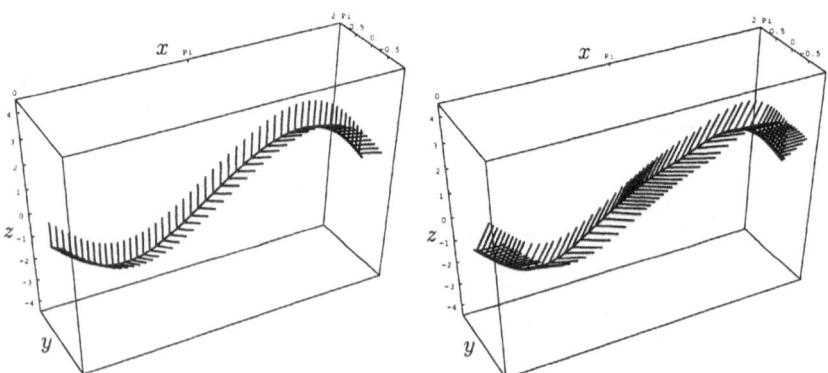

Figure 4: *The invariant circle of the fattened Arnol'd family:* $a = 0.1$, $b = 0.3$ *and* $c = 2.4$. *Left the initial data for* $\varepsilon = 0$ *and right the circle for* $\varepsilon = 0.7572$. *The normal directions are drawn for* 50 *mesh points (left) and every fifth of* 290 *mesh points (right).*

Two saddles appear on the invariant circle in a saddle–node bifurcation for $\varepsilon = 0.49$. Computation of the eigenvalues at these saddles reveals that for $\varepsilon \approx$

0.7761 the normal behavior of f_ε ceases to dominate the tangential behavior. So ε can't increase beyond 0.7761 during the continuation process. In fact, the algorithm computes a family $\mathbb{H}_\varepsilon$ of invariant circles, for ε ranging from 0 to 0.7572, with an estimated accuracy of order 10^{-4}. Again the initial increment of the continuation parameter ε is set to 0.02, and is adjusted (viz made smaller) as the normal hyperbolicity gets weaker. A picture of the invariant circle for $\varepsilon = 0.7572$ is shown in figure 4 (right). Notice the change in the normal directions near the inflection point, compared to the initial circle (left).

5.3 The forced Van der Pol oscillator

Finally we show how to apply the algorithm to compute the invariant manifold of the Poincaré first-return map of a continuous system. To this end consider the forced Van der Pol oscillator X_ε, a continuous system on the generalized phase space $\mathbb{R}^2 \times \mathbb{R}/2\pi\mathbb{Z}$:

$$\begin{cases} \dot{x} &= y \\ \dot{y} &= -x - a(x^2 - 1)y + \varepsilon\cos t \\ \dot{t} &= \omega. \end{cases} \tag{20}$$

Here a and ω are constants, with $a > 0$ and $0 < \omega < 2\pi$, and ε is the continuation parameter. We naturally get a diffeomorphism on the x, y-plane by considering the Poincaré map P_ε, the stroboscopic map of the 2π-periodic forcing term $\varepsilon\cos t$. For $\varepsilon = 0$ there is no forcing, so the the system decouples to the autonomous two-dimensional system, called the free Van der Pol oscillator:

$$\begin{cases} \dot{x} &= y \\ \dot{y} &= -x - a(x^2 - 1)y. \end{cases} \tag{21}$$

This planar autonomous system has a closed orbit, which is attracting for $a < 2$, see [5]. (In this example we take $a = 0.4$ and $\omega = 0.9$.) The closed orbit corresponds to an invariant circle of P_0, that is normally hyperbolic. Considered in the phase space $\mathbb{R}^2 \times \mathbb{R}/2\pi\mathbb{Z}$ of (20), this closed orbit yields an attracting invariant 2-torus. Due to normal hyperbolicity, the circle and, hence, the torus, is persistent for small values of ε. The invariant circle $\mathbb{H}_0$ of P_0 is a globally attracting limit cycle, and can therefore be computed by forward iteration of the planar system (21). A mesh of 50 points represents $\mathbb{H}_0$. The invariant splitting $\mathbb{T}_{\mathbb{H}_0}(\mathbb{R}^2) = \mathbb{T}(\mathbb{H}_0) \oplus \mathbb{N}^s(\mathbb{H}_0)$ is found by computing the eigenvectors of $D\phi_T(r)$, for $r \in \mathbb{H}_0$, where ϕ_T is the time T-map of the autonomous system (21), and T is the period of the limit cycle. The initial data is shown in figure 5 (left). Numerical computations show that for $\varepsilon \approx 0.3634$ a saddle on the circle and a source inside it disappear due to a *normal* saddle-node bifurcation, destroying the normally hyperbolic invariant circle. (The saddle is born earlier in the continuation process, due to a saddle-node bifurcation on the circle.) Hence, we expect the continuation process to break down for $\varepsilon \approx 0.3634$.

The algorithm computes the family $\mathbb{H}_\varepsilon$ of invariant circles for ε ranging from 0 to 0.3609, with an estimated accuracy of 10^{-4}. Figure 5 (right) shows the last

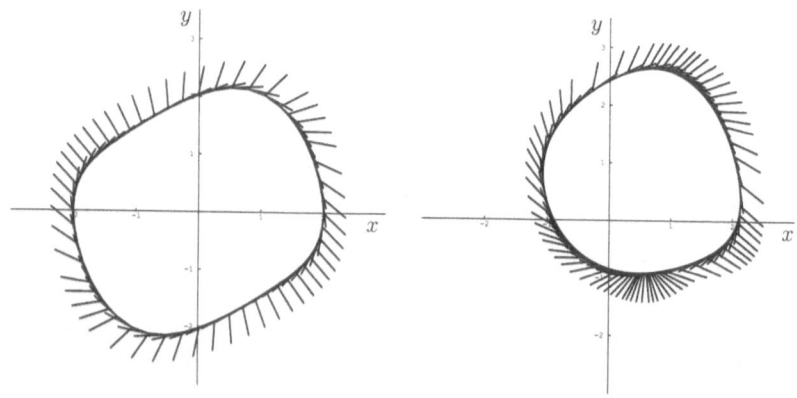

Figure 5: *The invariant circle of the forced Van der Pol oscillator. Left: the initial circle (ε = 0) with 50 mesh points. Right: the last circle (ε = 0.3609) has 356 mesh points; the tangent and normal directions are drawn for every fifth mesh point.*

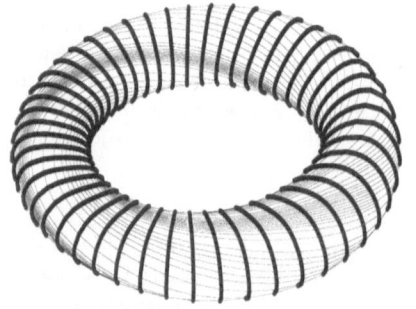

Figure 6: *The invariant torus of the forced Van der Pol oscillator.*

invariant circle we were able to compute. Again the algorithm refines the mesh and decreases the step size of the continuation parameter as the normal hyperbolicity gets weaker. Due to the automatic refinement of the mesh the final circle consists of 356 points. Figure 6 shows the result of saturating the final invariant circle, by computing the X_ε-orbit of every fifth mesh point. Representing each orbit by 50 points, corresponding to fixed length time intervals, we obtain 50 circles as an approximation of the X_ε-invariant torus in generalized phase space $\mathbb{R}^2 \times \mathbb{R}/2\pi\mathbb{Z}$.

References

[1] V.I. Arnol'd. Small denominators I. *Transl. Amer. Math. Soc., 2nd. series*, 46:213–284, 1965.

[2] H.W. Broer, H.M. Osinga, and G. Vegter. Computing normally hyperbolic invariant manifolds of discrete dynamical systems. In preparation.

[3] H.W. Broer, C. Simó, and J.C. Tatjer. Towards global models near homoclinic tangencies of dissipative diffeomorphisms. In preparation.

[4] M.W. Hirsch, C. Pugh, and M. Shub. *Invariant Manifolds*, volume 583 of *Lecture Notes in Mathematics*. Springer-Verlag, 1977.

[5] M.W. Hirsch and S. Smale. *Differential equations, dynamical systems, and linear algebra*. Academic Press, 1974.

[6] A.J. Homburg, H.M. Osinga, and G. Vegter. On the computation of invariant manifolds of fixed points. *Zeit. Angew. Math. Phys.*, 46:171–187, 1995.

[7] J. Palis and F. Takens. *Hyperbolicity & sensitive chaotic dynamics at homoclinic bifurcations*, volume 35 of *Cambridge studies in advanced mathematics*. Cambridge University Press, 1993.

[8] D. Ruelle. *Elements of differentiable dynamics and bifurcation theory*. Academic Press, 1989.

[9] M. Shub. *Global stability of dynamical systems*. Springer-Verlag, 1987.

Progress in Nonlinear Differential Equations
and Their Applications, Vol. 19
© 1996 Birkhäuser Verlag Basel/Switzerland

The Computation of Unstable Manifolds
Using Subdivision and Continuation

Michael Dellnitz*   Andreas Hohmann†

Abstract

Combining the subdivision algorithm for the computation of relative global
attractors as described in [1] with a box-oriented continuation technique,
we develop a new method for the computation of unstable manifolds. The
algorithm creates a covering of the unstable manifold, and it is in principle applicable to invariant manifolds of arbitrary dimension. As an example
we compute two-dimensional stable and unstable manifolds in the Lorenz
system.

1 Introduction

In [1] we have developed a method to approximate the global attractor of a dynamical system within a given region of phase space. This method is based on a
subdivision technique which allows to construct a covering of the (relative) global
attractor up to a prescribed accuracy. If there is a hyperbolic structure present
then the speed of convergence crucially depends on the contraction rate in the
stable direction (see [1], Proposition 4.10).

The fundamental observation for the method described in this paper is that
this subdivision algorithm — applied to a (small) neighborhood of a hyperbolic
periodic point — can be used to construct a covering of the corresponding local
unstable manifold up to a given accuracy. Such a covering is realized by the union
of boxes of a certain size. In our continuation algorithm the computation of this
covering is the *initialization step*. Then we proceed with the *continuation step*
and use a box-oriented continuation technique to extend the covering of the local

*Inst. für Angewandte Mathematik, Universität Hamburg, D-20146 Hamburg, Germany, Re-
search supported in part by the Deutsche Forschungsgemeinschaft, ONR Grant N00014-94-1-0317
and the Konrad-Zuse-Zentrum für Informationstechnik Berlin.

†Konrad-Zuse-Zentrum für Informationstechnik Berlin, D-10711 Berlin, Germany, Research
supported by the Deutsche Forschungsgemeinschaft.

unstable manifold successively to coverings of larger parts of the (global) unstable manifold.

In contrast to some of the existing methods for the computation of an unstable manifold (e.g. [8, 6]) our algorithm does, in principle, not depend on its dimension. In particular, we can approximate two-dimensional unstable manifolds — and we will illustrate this fact in Section 4 by numerical examples. However, there are limitations on the size of the dimension due to the fact that the storage of the covering may require too much memory. Consequently, for high dimensional invariant manifolds, just very rough approximations can be obtained.

The method presented here also has the topological advantage that we are working with *coverings* of invariant manifolds. Hence, in contrast to the situation in which one is approximating orbits or paths inside these manifolds — for a method based on the *graph transform* see [4] —, our approach guarantees that we cannot diverge away from them in the process of the continuation.

An outline of the paper is as follows. In Section 2 we introduce notation and review briefly the subdivision algorithm as developed in [1]. The continuation method as well as its implementation are described in Section 3. In particular we prove convergence of this method to a bounded part of the unstable manifold if the number of subdivision steps in the initialization is going to infinity (see Proposition 3.2). Finally, in Section 4, we consider as a numerical example the computation of two-dimensional invariant manifolds in the Lorenz system.

2 Brief Review of the Subdivision Algorithm

Preliminaries

We consider discrete dynamical systems,

$$x_{j+1} = f(x_j), \quad j = 0, 1, 2, \ldots,$$

where $f : \mathbb{R}^n \to \mathbb{R}^n$ is a diffeomorphism. Let p be a hyperbolic fixed point of f. The main purpose of this paper is to develop a continuation method for the computation of the *unstable manifold* $W^u(p)$. Recall that the hyperbolicity of p implies that this set is given by

$$W^u(p) = \{x \in \mathbb{R}^n : f^{-j}(x) \to p \text{ for } j \to \infty\}.$$

Locally, $W^u(p)$ is indeed a manifold inheriting the smoothness properties of f. The local part of $W^u(p)$ which contains the fixed point p, that is,

$$W^u_{\text{loc}}(p) = \{x \in \mathbb{R}^n : \|f^{-j}(x) - p\| < \epsilon \text{ for all } j \in \mathbb{N}\}.$$

is called the *local unstable manifold*. For details concerning the definitions and basic properties of invariant manifolds the reader is referred to [5, 2, 7].

Remark 2.1 To simplify the notation we restrict our considerations to the situation where the unstable manifold of a hyperbolic *fixed* point is of interest. However, it will be clear that the continuation method described in this article can more generally be applied to the computation of invariant manifolds of hyperbolic *periodic* points.

From [1] we recall the notion of a relative global attractor, which is the fundamental object of the subdivision algorithm described in the following subsection.

Definition 2.2 Let $Q \subset \mathbb{R}^n$ a bounded set. We define the *global attractor relative to Q* by

$$A_Q = \bigcap_{j \geq 0} f^j(Q). \tag{2.1}$$

Remark 2.3 Let Q be a neighborhood of a hyperbolic fixed point p. Then it is easily seen that the connected component of $W^u(p) \cap Q$ which contains p is part of A_Q. Moreover, provided Q is small enough, A_Q coincides with this part of the unstable manifold of p. We will use this fact in Section 3.

The Subdivision Algorithm

Our continuation method for the computation of unstable manifolds makes use of a subdivision algorithm which has been developed and investigated in [1]. For sake of completeness we are now going to summarize briefly this algorithm for the computation of relative global attractors, and we recall its convergence property. Concerning the details the reader is referred to [1].

The subdivision algorithm generates a sequence $\mathcal{B}_0, \mathcal{B}_1, \mathcal{B}_2, \ldots$ of finite collections of compact subsets of $\mathbb{R}^n$ with the property that for all integers k the set

$$Q_k = \bigcup_{B \in \mathcal{B}_k} B$$

is a covering of the relative global attractor under consideration. Moreover the sequence of coverings is constructed in such a way that the diameter

$$\text{diam}(\mathcal{B}_k) = \max_{B \in \mathcal{B}_k} \text{diam}(B)$$

converges to zero for $k \to \infty$.

Let us be more precise. Given an initial collection $\mathcal{B}_0$, one inductively obtains $\mathcal{B}_k$ from $\mathcal{B}_{k-1}$ for $k = 1, 2, \ldots$ in two steps.

1. *Subdivision:* Construct a new collection $\hat{\mathcal{B}}_k$ such that

$$\bigcup_{B \in \hat{\mathcal{B}}_k} B = \bigcup_{B \in \mathcal{B}_{k-1}} B \tag{2.2}$$

and

$$\text{diam}(\hat{\mathcal{B}}_k) \leq \theta \, \text{diam}(\mathcal{B}_{k-1}) \tag{2.3}$$

for some $0 < \theta < 1$.

2. *Selection:* Define the new collection $\mathcal{B}_k$ by

$$\mathcal{B}_k = \left\{ B \in \hat{\mathcal{B}}_k : f^{-1}(B) \cap \hat{B} \neq \emptyset \quad \text{for some} \quad \hat{B} \in \hat{\mathcal{B}}_k \right\}. \tag{2.4}$$

We now formulate the result which establishes the convergence of the algorithm to the relative global attractor (for a proof see [1]).

Proposition 2.4 *Let A_Q be a global attractor relative to the compact set Q, and let $\mathcal{B}_0$ be a finite collection of closed subsets with $Q_0 = \bigcup_{B \in \mathcal{B}_0} B = Q$. Then*

$$\lim_{k \to \infty} h(A_Q, Q_k) = 0,$$

where we denote by $h(B, C)$ the usual Hausdorff distance between two compact subsets $B, C \subset \mathbb{R}^n$.

3   The Continuation Method

Description of the Method

The continuation starts at a hyperbolic fixed point p with the unstable manifold $W^u(p)$. We fix once and for all a (large) compact set $Q \subset \mathbb{R}^n$ containing p, in which we want to approximate part of $W^u(p)$. To combine the subdivision process with a continuation method, we realize the subdivision using a family of partitions of Q. Here, we define a *partition* $\mathcal{P}$ of Q to be a finite family of subsets of Q such that

$$\bigcup_{B \in \mathcal{P}} B = Q \quad \text{and} \quad B \cap B' = \emptyset.$$

For a point $x \in Q$, let $\mathcal{P}(x) \in \mathcal{P}$ denote the element of $\mathcal{P}$ containing x. We consider a nested sequence $\mathcal{P}_\ell$, $\ell \in \mathbb{N}$, of successively finer partitions of Q, requiring that for all $B \in \mathcal{P}_\ell$ there exist $B_1, \ldots, B_m \in \mathcal{P}_{\ell+1}$ such that $B = \bigcup_i B_i$ and $\text{diam}(B_i) \leq \theta \, \text{diam}(B)$ for some $0 < \theta < 1$. A set $B \in \mathcal{P}_\ell$ is said to be of *level* ℓ. Given such a sequence $\{\mathcal{P}_\ell\}$, we obtain for any point $x \in Q$ a unique sequence $\{\mathcal{P}_\ell(x)\}$.

Now assume that $C = \mathcal{P}_\ell(p)$ is a neighborhood of the hyperbolic fixed point p such that the global attractor relative to C satisfies

$$A_C = W_{\text{loc}}^u(p) \cap C$$

(see Remark 2.3). Applying the subdivision algorithm with k subdivision steps to $\mathcal{B}_0 = \{C\}$, we obtain a covering $\mathcal{B}_k \subset \mathcal{P}_{\ell+k}$ of the local unstable manifold $W^u_{\mathrm{loc}}(p) \cap C$, that is,

$$A_C = W^u_{\mathrm{loc}}(p) \cap C \subset \bigcup_{B \in \mathcal{B}_k} B. \tag{3.1}$$

Note that, by Proposition 2.4, this covering converges to $W^u_{\mathrm{loc}}(p) \cap C$ for $k \to \infty$.

With these notions we are now in the position to describe our continuation algorithm as follows. For a fixed k we define a sequence $\mathcal{C}_0^{(k)}, \mathcal{C}_1^{(k)}, \dots$ of subsets $\mathcal{C}_j^{(k)} \subset \mathcal{P}_{\ell+k}$ by

1. *Initialization:*

$$\mathcal{C}_0^{(k)} = \mathcal{B}_k.$$

2. *Continuation:* For $j = 0, 1, 2, \dots$ define

$$\mathcal{C}_{j+1}^{(k)} = \left\{ B \in \mathcal{P}_{\ell+k} \; : \; B \cap f(B') \neq \emptyset \text{ for some } B' \in \mathcal{C}_j^{(k)} \right\}.$$

Remark 3.1 Observe that the unions

$$C_j^{(k)} = \bigcup_{B \in \mathcal{C}_j^{(k)}}$$

form nested sequences in k, i.e.,

$$C_j^{(0)} \supset C_j^{(1)} \supset \dots .$$

Intuitively it is clear that the algorithm, as constructed, generates an approximation of the unstable manifold $W^u(p)$. In particular, one expects that the bigger k and ℓ are chosen the better the approximation should be. However, since we restrict our attention to a compact subset $Q \subset \mathbb{R}^n$ it can just be guaranteed that the algorithm generates an approximation of a certain part of $W^u(p)$.

We now make the discussion in the previous paragraph precise, and we begin with the definition of the subset of $W^u(p)$ which is indeed approximated by the continuation method. We set $W_0 = W^u_{\mathrm{loc}}(p) \cap C$ and define inductively for $j = 0, 1, 2, \dots$

$$W_{j+1} = f(W_j) \cap Q.$$

With this notion we obtain the following convergence result.

Proposition 3.2 *The sets $C_j^{(k)}$ are coverings of W_j for all $j, k = 0, 1, \dots$. Moreover, for fixed j, $C_j^{(k)}$ converges to W_j in Hausdorff distance if the number k of subdivision steps in the initialization goes to infinity.*

Proof: The first statement follows directly from the fact that the set $\mathcal{B}_k$ obtained by the subdivision algorithm is always a covering of $W_0 = W^u_{\text{loc}}(p) \cap C$ (see (3.1)).

To prove the second statement, we first observe that by Proposition 2.4 $C_0^{(k)}$ converges to the relative global attractor $A_C = W^u_{\text{loc}}(p) \cap C = W_0$ for $k \to \infty$. Since j is fixed a continuity argument shows that the sets $C_j^{(k)}$ converge to W_j for $k \to \infty$, i.e.,

$$C_j^{(\infty)} = \bigcap_{k \geq 0} C_j^{(k)} = W_j.$$

$\square$

Remark 3.3   (a) It can in general not be guaranteed that the continuation method leads to an approximation of the entire set $W^u(p) \cap Q$. Rather it has to be expected that this is not the case. The reason is that the unstable manifold of the hyperbolic fixed point p may "leave" Q but may as well "wind back" into it. If this is the case then the continuation method, as described above, will not cover all of $W^u(p) \cap Q$.

(b) Observe that the convergence result in Proposition 3.2 does not require the existence of a hyperbolic structure along the unstable manifold. However, if we additionally assume its existence then we could establish results on the convergence behavior of the continuation method in a completely analogous way as in [1].

Implementation of the Algorithm

For the actual implementation of the continuation method, we have to choose the partitions $\mathcal{P}_\ell$. In the present code, we use generalized rectangles of the form

$$R(c, r) = \{y \in \mathbb{R}^n : c_i - r_i \leq y_i < c_i + r_i \text{ for } i = 1, \ldots, n\},$$

where $c, r \in \mathbb{R}^n$, $r_i > 0$ for $i = 1, \ldots, n$, are the center and the radius respectively. The first level is defined by $\mathcal{P}_0 = \{Q\}$. The next level $\mathcal{P}_\ell$, $\ell > 0$, is obtained by bisection with respect to the j-th coordinate, where j is varied cyclically. The subdivision of a generalized rectangle $R(c, r)$ leads to the two rectangles $R_-(c^-, \hat{r})$ and $R_+(c^+, \hat{r})$, where

$$\hat{r}_i = \begin{cases} r_i & \text{for } i \neq j \\ r_i/2 & \text{for } i = j \end{cases}, \quad c_i^{\pm} = \begin{cases} c_i & \text{for } i \neq j \\ c_i \pm r_i/2 & \text{for } i = j \end{cases}.$$

As described in [1], this permits a very efficient storage scheme: a collection $\mathcal{B} \in \bigcup_{\ell \in \mathbb{N}} \mathcal{P}_\ell$ is represented as a binary tree reflecting the subdivision structure.

The success of the continuation method depends on the implementation of the continuation step, where we have to check the intersection property $B \cap f(B') = \emptyset$

for two given boxes B, B'. As in the subdivision algorithm, we have to *discretize* this condition using a set of test points in each rectangle. The intersection property is then replaced by

$$f(x) \notin B \quad \text{for all test points } x \in B' . \tag{3.2}$$

With respect to the choice of test points, we obtained the best results using an equidistant grid of test points on all edges. This way, only a relatively small number of points is needed. For instance, in the three-dimensional examples in Section 4, we have used five points per edge leading to 60 test points in each generalized rectangle (not taking into account that an edge may be shared by several boxes).

4   Numerical Examples

As an example we compute approximations of two-dimensional invariant manifolds in the Lorenz system

$$\begin{aligned}
\dot{x} &= \sigma(y - x) \\
\dot{y} &= \rho x - y - xz \\
\dot{z} &= -\beta z + xy.
\end{aligned}$$

An easy computation shows that for positive β and for $\rho > 1$ the Lorenz system has three steady state solutions, namely

$$(0,0,0) \quad \text{and} \quad q_{\pm} = (\pm\sqrt{\beta(\rho-1)}, \pm\sqrt{\beta(\rho-1)}, \rho-1).$$

For our choices of parameter values these three steady states are hyperbolic. In the following subsections we present numerical results for approximations

— of a two-dimensional stable manifold of the origin, and

— of a two-dimensional unstable manifold of q_{+}.

Approximation of a Stable Manifold

In this computation we have chosen the "standard" set of parameter values, that is,

$$\sigma = 10, \quad \rho = 28 \quad \text{and} \quad \beta = 8/3.$$

With this choice a direct numerical simulation would lead to an approximation of the Lorenz attractor. (For illustrations as well as a discussion of topological properties of the Lorenz attractor the reader is referred to [3].)

Since we want to compute the two-dimensional stable manifold of the origin, we proceed backwards in time and apply the continuation method to the diffeomorphism given by the time-$(-T)$-map. Starting in a neighborhood of $(0, 0, 0)$ we approximate the stable manifold in $Q = [-25, 25]^3$. To demonstrate the continuation process, we begin with a rough approximation using the initial level $\ell = 9$ and $k = 3$ subdivision steps. In Figure 1 we display the coverings obtained by the algorithm after $j = 0, 1, 3, 5$ continuation steps. We remark that in this case the

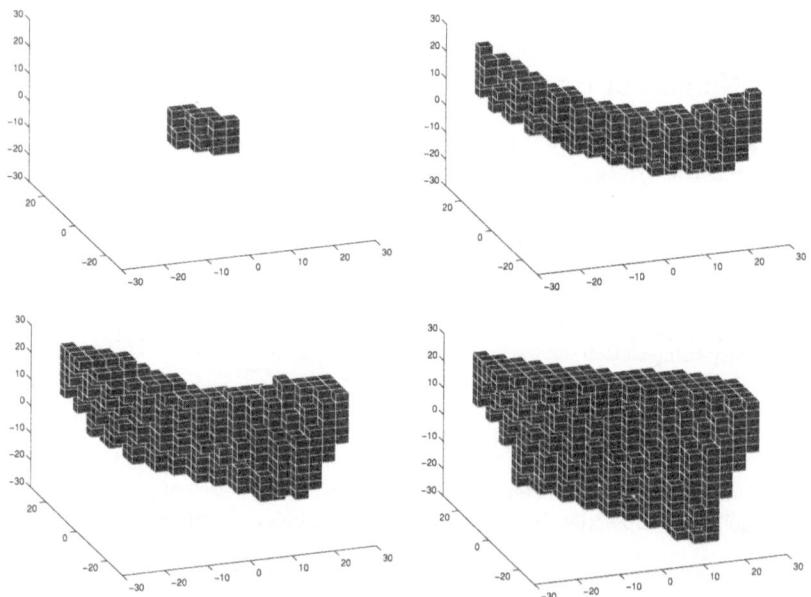

Figure 1: Continuation steps for the stable manifold of the origin in the Lorenz system for $j = 0, 1, 3, 5$.

stable eigenvalues are both real but the ratio of strong and weak contraction is relatively big. This is also reflected by the way the covering is growing (see Figure 1). However, the results illustrate that the algorithm can successfully be applied. A finer resolution ($\ell = 15, k = 3$) is shown in Figure 2. Here, we only plot the centers of the boxes.

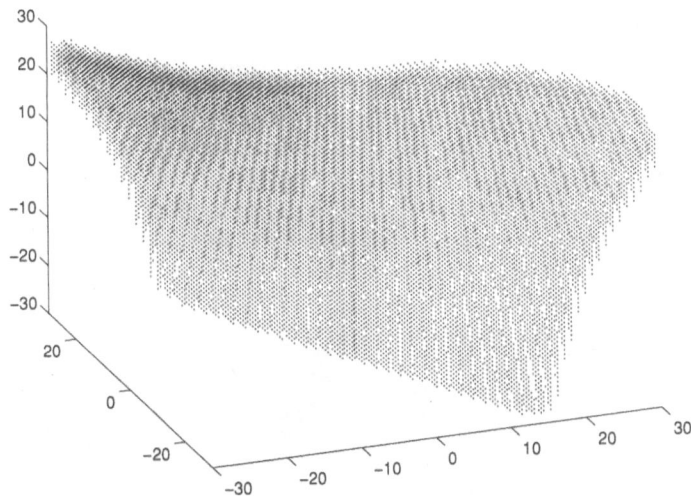

Figure 2: Two-dimensional stable manifold of the origin in the Lorenz system.

Approximation of an Unstable Manifold

In this computation we have chosen the following set of parameter values

$$\sigma = 10, \quad \rho = 28 \quad \text{and} \quad \beta = 0.4.$$

Direct simulation indicates that for this choice there is a globally stable periodic solution.

We apply the continuation method described in the previous sections to the diffeomorphism given by the time-T-map and start in a neighborhood of q_+. Relative to its two-dimensional unstable manifold q_+ is a spiral source.

We have chosen a discretization of level 21 and the resulting approximation of the unstable manifold is shown in Figure 3. The computations strongly suggest that the structure of this invariant manifold is very complicated. To give a better impression we show a two-dimensional cut through the manifold in Figure 4.

Acknowledgments. Both authors acknowledge the hospitality of the Department of Mathematics at the University of Houston where part of this work has been carried out.

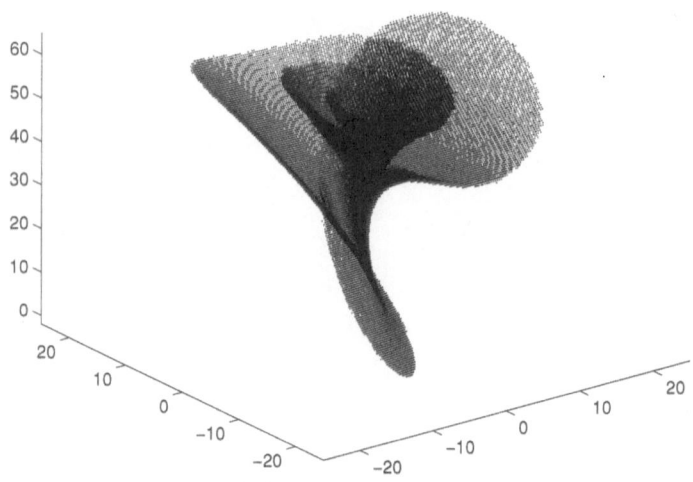

Figure 3: A two-dimensional unstable manifold in the Lorenz system.

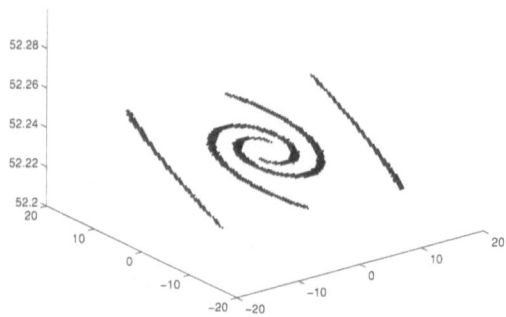

Figure 4: Cut through the unstable manifold.

References

[1] M. Dellnitz and A. Hohmann. A subdivision algorithm for the computation of unstable manifolds and global attractors, submitted to *Numerische Mathematik* (1995).

[2] J.-P. Eckmann and D. Ruelle. Ergodic theory of chaos and strange attractors. *Reviews of Modern Physics*, **57** (3) I (1985), 617–656.

[3] J. Guckenheimer and P. Holmes. *Nonlinear Oscillations, Dynamical Systems, and Bifurcations of Vector Fields*, (Springer, 1986).

[4] A.J. Homburg, H.M. Osinga and G. Vegter. On the computation of invariant manifolds of fixed points, *Z. angew. Math. Phys.* **46** (1995), 171–187.

[5] J. Palis and W. de Melo. *Geometric Theory of Dynamical Systems*, (Springer, 1982).

[6] T.S. Parker and L.O. Chua. *Practical Numerical Algorithms for Chaotic Systems*, Springer, New York, (1989).

[7] M. Shub. *Global Stability of Dynamical Systems*, (Springer, 1987).

[8] Z. You, E.J. Kostelich and J.A. Yorke. Calculating stable and unstable manifolds, *Int. J. Bif. Chaos* **1** (3) (1991), 605–623.

BIRKHÄUSER • MATHEMATICS

Progress in Nonlinear Differential Equations and Their Applications

Editor
Haim Brezis, Université P. et M. Curie 4, Place Jussieu, 75252 Paris Cedex 05, France and
Rutgers University New Brunswick, NJ 08903, U.S.A.

Progress in Nonlinear and Differential Equations and Their Applications is a book
series that lies at the interface of pure and applied mathematics. Many differential equations are motivated by problems arising in such diversified fields as Mechanics, Physics,
Differential Geometry, Engineering, Control Theory, Biology, and Economics. This series is
open to both the theoretical and applied aspects, hopefully stimulating a fruitful interaction between the two sides. It will publish monographs, polished notes arising from
lectures and seminars, graduate level texts, and proceedings of focused and refereed
conferences.

We encourage preparation of manuscripts in some form of TeX for delivery in camera-
ready copy, which leads to rapid publication, or in electronic form for interfacing with
laser printers or typesetters.

Proposals should be sent directly to the editor or to:
Birkhäuser Boston, 675 Massachusetts Avenue, Cambridge, MA 02139

**Please order through your
bookseller or write to:**
Birkhäuser Verlag AG
P.O. Box 133
CH-4010 Basel / Switzerland
FAX: ++41 / 61 / 271 76 66
e-mail: 00010.2310@compuserve.com

**For orders originating
in the USA or Canada:**
Birkhäuser
333 Meadowlands Parkway
Secaucus, NJ 07096-2491 / USA

Birkhäuser Verlag AG
Basel · Boston · Berlin

BIRKHÄUSER MATHEMATICS

Linear and Quasilinear Parabolic Problems

Volume I, Abstract Linear Theory Equation

Partial Differential Equations

H. Amann, Universität Zürich, Switzerland

MMA 89
Monographs in Mathematics

1995. 372 pages. Hardcover
ISBN 3-7643-5114-4

This treatise gives an exposition of the functional analytical approach to quasilinear parabolic evolution equations, developed to a large extent by the author during the last 10 years. This approach is based on the theory of linear nonautonomous parabolic evolution equations and on interpolation-extrapolation techniques. It is the only general method that applies to noncoercive quasilinear parabolic systems under nonlinear boundary conditions.

The present first volume is devoted to a detailed study of nonautonomous linear parabolic evolution equations in general Banach spaces. It contains a careful exposition of the constant domain case, leading to some improvements of the classical Sobolevskii-Tanabe results. It also includes recent results for equations possessing constant interpolation spaces. In addition, there are given systematic presentations of the theory of maximal regularity in spaces of continuous and Hölder continuous functions, and in Lebesgue spaces. It includes related recent theorems in the field of harmonic analysis in Banach spaces and on operators possessing bounded imaginary powers. Lastly, there is a complete presentation of the technique of interpolation-extrapolation spaces and of evolution equations in those spaces, containing many new results.

Please order through your bookseller or write to:
Birkhäuser Verlag AG
P.O. Box 133
CH-4010 Basel / Switzerland
FAX: ++41 / 61 / 271 76 66
e-mail:
100010.2310@compuserve.com

For orders originating in the USA or Canada:
Birkhäuser
333 Meadowlands Parkway
Secaucus, NJ 07094-2491
USA

Birkhäuser

Birkhäuser Verlag AG
Basel · Boston · Berlin

BIRKHÄUSER • MATHEMATICS

T.F. Nonnenmacher, Mathematische Physik, Universität Ulm, Germany /
G.A. Losa, Instituto Cantonale di Patologia, Locarno, Switzerland /
E.R. Weibel, Anatomisches Institut Universität Bern, Switzerland

Fractals in Biology and Medicine

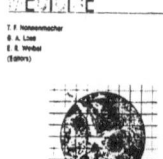

1994. 397 pages. Hardcover
ISBN 3-7643-2989-0

Fractals in Biology and Medicine explores the potential of fractal geometry for describing and understanding biological organisms, their development and growth as well as their structural design and functional properties. It extends these notions to assess changes associated with disease in the hope to contribute to the understanding of pathogenetic processes in medicine.

The book is the first comprehensive presentation of the importance of the new concept of fractal geometry for biological and medical sciences. It collates in a logical sequence extended papers based on invited lectures and free communications presented at a symposium in Ascona, Switzerland, attended by leading scientists in this field, among them the originator of fractal geometry, Benoît Mandelbrot.

Fractals in Biology and Medicine begins by asking how the theoretical construct of fractal geometry can be applied to biomedical sciences and then addresses the role of fractals in the design and morphogenesis of biological organisms as well as in molecular and cell biology. The consideration of fractal structure in understanding metabolic functions and pathological changes is a particularly promising avenue for future research.

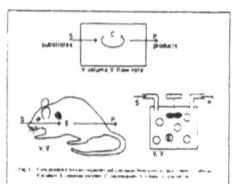

Please order through your bookseller or write to:
Birkhäuser Verlag AG
P.O. Box 133
CH-4010 Basel / Switzerland
FAX: ++41 / 61 / 271 76 66
e-mail: 100010.2310@compuserve.com

For orders originating in the USA or Canada:
Birkhäuser
333 Meadowlands Parkway
Secaucus, NJ 07096-2491 / USA

Birkhäuser

Birkhäuser Verlag AG
Basel · Boston · Berlin

BAT
Birkhäuser Advanced Texts
Basler Lehrbücher

H. Hofer / E. Zehnder, ETH Mathematik, Zürich, Switzerland

Symplectic Invariants
and Hamiltonian Dynamics

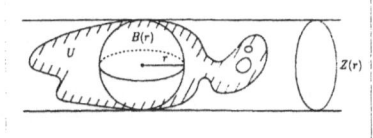

1994. 342 pages. Hardcover
ISBN 3-7643-5066-0

"Symplectic Topology has become a fascinating subject of research over the past fifteen years... This book is written by two experienced researchers, will certainly fill in a gap in the theory of symplectic topology. The authors have taken part in the development of such a theory by themselves or by their collaboration with other outstanding people in the area... All the chapters have a nice introduction with the historic developement of the subject and with a perfect description of the state of the art."

ZENTRALBLATT MATHEMATIK, 1995

Please order through your
bookseller or write to:
Birkhäuser Verlag AG
P.O. Box 133
CH-4010 Basel / Switzerland
FAX: ++41 / 61 / 271 76 66
e-mail: 100010.2310@compuserve.com

For orders originating
in the USA or Canada:
Birkhäuser
333 Meadowlands Parkway
Secaucus, NJ 07096-2491 / USA

Birkhäuser

Birkhäuser Verlag AG
Basel · Boston · Berlin